Grundlehren der mathematischen Wissenschaften 332

A Series of Comprehensive Studies in Mathematics

T0178571

Masaki Kashiwara
Pierre Schapira

Categories and Sheaves

 Springer

Masaki Kashiwara

Research Institute for Mathematical Sciences
Kyoto University
Kitashirakawa-Oiwake-cho
606-8502 Kyoto
Japan
E-mail: masaki@kurims.kyoto-u.ac.jp

Pierre Schapira

Institut de Mathématiques
Université Pierre et Marie Curie
4, place Jussieu
75252 Paris Cedex 05,
France
E-mail: schapira@math.jussieu.fr

Mathematics Subject Classification (2000): 18A, 18E, 18F10, 18F20, 18G

ISSN 0072-7830

ISBN:13 978-3-642-06620-7   e-ISBN:13 978-3-540-27950-1

Springer is a part of Springer Science+Business Media
springeronline.com
© Springer-Verlag Berlin Heidelberg 2006
Softcover reprint of the hardcover 1st edition 2006

Cover design: *design & production* GmbH, Heidelberg

Preface

The language of Mathematics has changed drastically since the middle of the twentieth century, in particular after Grothendieck's ideas spread from algebraic geometry to many other subjects. As an enrichment for the notions of sets and functions, categories and sheaves are new tools which appear almost everywhere nowadays, sometimes simply in the role of a useful language, but often as the natural approach to a deeper understanding of mathematics.

Category theory, initiated by Eilenberg and Mac Lane in the forties (see [19, 20]), may be seen as part of a wider movement transcending mathematics, of which structuralism in various areas of knowledge is perhaps another facet. Before the advent of categories, people were used to working with a given set endowed with a given structure (a topological space for example) and to studying its properties. The categorical point of view is essentially different. The stress is placed not upon the objects, but on the relations (the morphisms) between objects within the category. The language is natural and allows one to unify various branches of mathematics and to make unexpected links between seemingly different subjects.

Category theory is elementary in the sense that there are few prerequisites to its study, though it may appear forbiddingly abstract to many people. Indeed, the usual course of mathematical education is not conducive to such a conceptual way of thinking. Most mathematicians are used to manipulating spaces and functions, computing integrals and so on, fewer understand the importance of the difference between an equality and an isomorphism or appreciate the beauty and efficiency of diagrams.

Another fundamental idea is that of a sheaf. Sheaves provide a tool for passing from local to global situations and a good deal of mathematics (and physics) revolves around such questions. Sheaves allow us to study objects that exist locally but not globally, such as the holomorphic functions on the Riemann sphere or the orientation on a Möbius strip, and the cohomology of sheaves measures in some sense the obstruction to passing from local to global.

Jean Leray invented sheaves on a topological space in the forties (see [46] and Houzel's historical notes in [38]). Their importance, however, became more evident through the Cartan Seminar and the work of Serre. Subsequently, Serre's work [62] on the local triviality of algebraic fiber bundles led Grothendieck to the realization that the usual notion of a topological space was not appropriate for algebraic geometry (there being an insufficiency of open subsets), and introduced sites, that is, categories endowed with "Grothendieck topologies" and extended sheaf theory to sites.

The development of homological algebra is closely linked to that of category and sheaf theory. Homological algebra is a vast generalization of linear algebra and a key tool in all parts of mathematics dealing with linear phenomena, for example, representations, abelian sheaves, and so forth. Two milestones are the introduction of spectral sequences by Leray (loc. cit.) and the introduction of derived categories by Grothendieck in the sixties.

In this book, we present categories, homological algebra and sheaves in a systematic and exhaustive manner starting from scratch and continuing with full proofs to an exposition of the most recent results in the literature, and sometimes beyond. We also present the main features and key results of related topics that would deserve a whole book for themselves (e.g., tensor categories, triangulated categories, stacks).

Acknowledgments

All along the preparation of the manuscript, we benefited from numerous constructive comments, remarks and suggestions from Andrea D'Agnolo, Pierre Deligne, Stéphane Guillermou, Bernhard Keller, Amnon Neeman, Pietro Polesello, Raphael Rouquier, Valerio Toledano and Ingo Waschkies. David Coyle has assisted us in questions of English idiom and Liliane Beaulieu in historical questions. It is a pleasure to thank all of them here.

Kyoto, Japan   *Masaki Kashiwara*
Paris, France   *Pierre Schapira*
May 2005

Contents

Introduction

The aim of this book is to describe the topics outlined in the preface, categories, homological algebra and sheaves. We also present the main features and key results in related topics which await a similar full-scale treatment such as, for example, tensor categories, triangulated categories, stacks.

The general theory of categories and functors, with emphasis on inductive and projective limits, tensor categories, representable functors, ind-objects and localization is dealt with in Chaps. 1–7.

Homological algebra, including additive, abelian, triangulated and derived categories, is treated in Chaps. 8–15. Chapter 9 provides the tools (using transfinite induction) which will be used later for presenting unbounded derived categories.

Sheaf theory is treated in Chaps. 16–19 in the general framework of Grothendieck topologies. In particular, the results of Chap. 14 are applied to the study of the derived category of the category of sheaves on a ringed site. We also sketch an approach to the more sophisticated subject of stacks (roughly speaking, sheaves with values in the 2-category of categories) and introduce the important notion of twisted sheaves.

Of necessity we have excluded many exciting developments and applications such as n-categories, operads, A_∞-categories, model categories, among others. Without doubt these new areas will soon be intensively treated in the literature, and it is our hope that the present work will provide a basis for their understanding.

We now proceed to a more detailed outline of the contents of the book.

Chapter 1. We begin by defining the basic notions of categories and functors, illustrated with many classical examples. There are some set–theoretical dangers and to avoid contradictions, we work in a given universe. Universes are presented axiomatically, referring to [64] for a more detailed treatment. Among other concepts introduced in this chapter are morphisms of functors, equivalences of categories, representable functors, adjoint functors and so on. We introduce in particular the category $\mathrm{Fct}(I, \mathcal{C})$ of functors from a small category I to a category $\mathcal{C}$ in a universe $\mathcal{U}$, and look briefly at the 2-category $\mathcal{U}$-**Cat** of all $\mathcal{U}$-categories.

Here, the key result is the Yoneda lemma showing that a category $\mathcal{C}$ may be embedded in the category $\mathcal{C}^\wedge$ of all contravariant functors from $\mathcal{C}$ to **Set**, the category of sets. This allows us in a sense to reduce category theory to set theory and leads naturally to the notion of a representable functor. The category $\mathcal{C}^\wedge$ enjoys most of the properties of the category **Set**, and it is often extremely convenient, if not necessary, to replace $\mathcal{C}$ by $\mathcal{C}^\wedge$, just as in analysis, we are lead to replace functions by generalized functions.

Chapters 2 and 3. Inductive and projective limits are the most important concepts dealt with in this book. They can be seen as the essential tool of category theory, corresponding approximately to the notions of union and intersection in set theory. Since students often find them difficult to master, we provide many detailed examples. The category **Set** is not equivalent to its opposite category, and projective and inductive limits in **Set** behave very differently. Note that inductive and projective limits in a category are both defined as representable functors of *projective* limits in the category **Set**.

Having reached this point we need to construct the *Kan extension* of functors. Consider three categories $J, I, \mathcal{C}$ and a functor $\varphi\colon J \to I$. The functor φ defines by composition a functor φ_* from $\mathrm{Fct}(I, \mathcal{C})$ to $\mathrm{Fct}(J, \mathcal{C})$, and we can construct a right or left adjoint for this functor by using projective or inductive limits. These constructions will systematically be used in our presentation of sheaf theory and correspond to the operations of direct or inverse images of sheaves.

Next, we cover two essential tools for the study of limits in detail: cofinal functors (roughly analogous to the notion of extracted sequences in analysis) and filtrant[1] categories (which generalizes the notion of a directed set). As we shall see in this book, filtrant categories are of fundamental importance.

We define right exact functors (and similarly by reversing the arrows, left exact functors). Given that finite inductive limits exist, a functor is right exact if and only if it commutes with such limits.

Special attention is given to the category **Set** and to the study of filtrant inductive limits in **Set**. We prove in particular that inductive limits in **Set** indexed by a small category I commute with finite projective limits if and only if I is filtrant.

Chapter 4. Tensor categories axiomatize the properties of tensor products of vector spaces. Nowadays, tensor categories appear in many areas, mathematical physics, knot theory, computer science among others. They acquired popular attention when it was found that quantum groups produce rich examples of non-commutative tensor categories. Tensor categories and their applications in themselves merit an extended treatment, but we content ourselves

---

[1] Some authors use the terms "filtered" or "filtering". We have chosen to keep the French word.

here with a rapid treatment referring the reader to [15, 40] and [59] from the vast literature on this subject.

Chapter 5. We give various criteria for a functor with values in **Set** to be representable and, as a by-product, obtain criteria under which a functor will have an adjoint. This necessitates the introduction of two important notions: strict morphisms and systems of generators (and in particular, a generator) in a category $\mathcal{C}$. References are made to [64].

Chapter 6. The Yoneda functor, which sends a category $\mathcal{C}$ to $\mathcal{C}^\wedge$, enjoys many pleasing properties, such as that of being fully faithful and commuting with projective limits, but it is not right exact.

The category $\mathrm{Ind}(\mathcal{C})$ of ind-objects of $\mathcal{C}$ is the subcategory of $\mathcal{C}^\wedge$ consisting of small and filtrant inductive limits of objects in $\mathcal{C}$. This category has many remarkable properties: it contains $\mathcal{C}$ as a full subcategory, admits small filtrant inductive limits, and the functor from $\mathcal{C}$ to $\mathrm{Ind}(\mathcal{C})$ induced by the Yoneda functor is now right exact. On the other hand, we shall show in Chap. 15 that in the abelian case, $\mathrm{Ind}(\mathcal{C})$ does not in general have enough injective objects when we remain in a given universe.

This theory, introduced in [64] (see also [3] for complementary material) was not commonly used until recently, even by algebraic geometers, but matters are rapidly changing and ind-objects are increasingly playing an important role.

Chapter 7. The process of localization appears everywhere and in many forms in mathematics. Although natural, the construction is not easy in a categorical setting. As usual, it is easier to embed than to form quotient.

If a category $\mathcal{C}$ is localized with respect to a family of morphisms $\mathcal{S}$, the morphisms of $\mathcal{S}$ become isomorphisms in the localized category $\mathcal{C}_\mathcal{S}$ and if $F\colon \mathcal{C} \to \mathcal{A}$ is a functor which sends the morphisms in $\mathcal{S}$ to isomorphisms in $\mathcal{A}$, then F will factor uniquely through the natural functor $Q\colon \mathcal{C} \to \mathcal{C}_\mathcal{S}$. This is the aim of localization. We construct the localization of $\mathcal{C}$ when $\mathcal{S}$ satisfies suitable conditions, namely, when $\mathcal{S}$ is a (right or left) multiplicative system.

Interesting features appear when we try to localize a functor F that is defined on $\mathcal{C}$ with values in some category $\mathcal{A}$, and does not map the arrows in $\mathcal{S}$ to isomorphisms in $\mathcal{A}$. Even in this case, we can define the right or left localization of the functor F under suitable conditions. We interpret the right localization functor as a left adjoint to the composition with the functor Q, and this adjoint exists if $\mathcal{A}$ admits inductive limits. It is then a natural idea to replace the category $\mathcal{A}$ with that of ind-objects of $\mathcal{A}$, and check whether the localization of F at $X \in \mathcal{C}$ is representable in $\mathcal{A}$. This is the approach taken by Deligne [17] which we follow here.

Localization is an essential step in constructing derived categories. A classical reference for localization is [24].

Chapter 8. The standard example of abelian categories is the category $\mathrm{Mod}(R)$ of modules over a ring R. Additive categories present a much weaker structure which appears for example when considering special classes of modules (e.g. the category of projective modules over the ring R is additive but not abelian).

The concept of abelian categories emerged in the early 1950s (see [13]). They inherit all the main properties of the category $\mathrm{Mod}(R)$ and form a natural framework for the development of homological algebra, as is shown in the subsequent chapters. Of particular importance are the Grothendieck categories, that is, abelian categories which admit (exact) small filtrant inductive limits and a generator. We prove in particular the Gabriel-Popescu theorem (see [54]) which asserts that a Grothendieck category may be embedded into the category of modules over the ring of endomorphisms of a generator.

We also study the abelian category $\mathrm{Ind}(\mathcal{C})$ of ind-objects of an abelian category $\mathcal{C}$ and show in particular that the category $\mathrm{Ind}(\mathcal{C})$ is abelian and that the natural functor $\mathcal{C} \to \mathrm{Ind}(\mathcal{C})$ is exact. Finally we prove that under suitable hypotheses, the Kan extension of a right (or left) exact functor defined on an additive subcategory of an abelian category is also exact. Classical references are the book [14] by Cartan-Eilenberg, and Grothendieck's paper [28] which stresses the role of abelian categories, derived functors and injective objects.

An important source of historical information on this period is given in [16] by two of the main contributors.

Chapter 9. In this chapter we extend many results on filtrant inductive limits to the case of π-filtrant inductive limits, for an infinite cardinal π. An object X is π-accessible if $\mathrm{Hom}_{\mathcal{C}}(X, \bullet)$ commutes with π-filtrant inductive limits. We specify conditions which ensure that the category $\mathcal{C}_{\pi}$ of π-accessible objects is small and that the category of its ind-objects is equivalent $\mathcal{C}$. These techniques are used to prove that, under suitable hypotheses, given a family $\mathcal{F}$ of morphisms in a category $\mathcal{C}$, there are *enough $\mathcal{F}$-injective* objects.

Some arguments developed here were initiated in Grothendieck's paper [28] and play an essential role in the theory of model categories (see [56] and [32]). They are used in Chap. 14 in proving that the derived category of a Grothendieck category admits enough homotopically injective objects.

Here, we give two important applications. The first one is the fact that a Grothendieck category possesses enough injective objects. The second one is the Freyd-Mitchell theorem which asserts that any small abelian category may be embedded in the category of modules over a suitable ring. References are made to [64]. Accessible objects are also discussed in [1, 23] and [49].

Chapter 10. Triangulated categories first appeared implicitly in papers on stable homotopy theory after the work of Puppe [55], until Verdier axiomatized the properties of these categories (we refer to the preface by L. Illusie of [69] for more historical comments). Triangulated categories are now very popular and are part of the basic language in various branches of mathematics, especially algebraic geometry (see e.g. [57, 70]), algebraic topology and representation theory (see e.g. [35]). They appeared in analysis in the early 1970s under the influence of Mikio Sato (see [58]) and more recently in symplectic geometry after Kontsevich expressed mirror symmetry (see [43]) using this language.

A category endowed with an automorphism T is called here a category with translation. In such a category, a triangle is a sequence of morphisms $X \to Y \to Z \to T(X)$. A triangulated category is an additive category with translation endowed with a family of so-called *distinguished triangles* satisfying certain axioms. Although the first example of a triangulated category only appears in the next chapter, it seems worthwhile to develop this very elegant and easy formalism here for its own sake.

In this chapter, we study the localization of triangulated categories and the construction of cohomological functors in some detail. We also give a short proof of the Brown representability theorem [11], in the form due to Neeman [53], which asserts that, under suitable hypotheses, a contravariant cohomological functor defined on a triangulated category which sends small direct sums to products is representable.

We do not treat t-structures here, referring to the original paper [4] (see [38] for an expository treatment).

Chapter 11. It is perhaps the main idea of homological algebra to replace an object in a category $\mathcal{C}$ by a complex of objects of $\mathcal{C}$, the components of which have "good properties". For example, when considering the tensor product and its derived functors, we replace a module by a complex of projective (or flat) modules and, when considering the global-section functor and its derived functors, we replace a sheaf by a complex of flabby sheaves.

It is therefore natural to study the category $C(\mathcal{C})$ of complexes of objects of an additive category $\mathcal{C}$. This category inherits an automorphism, the *shift functor*, called the "suspension" by algebraic topologists. Other basic constructions borrowed from algebraic topology are that of the *mapping cone* of a morphism and that of *homotopy* of complexes. In fact, in order to be able to work, i.e., to form commutative diagrams, we have to make morphisms in $C(\mathcal{C})$ which are homotopic to zero, actually isomorphic to zero. This defines the homotopy category $K(\mathcal{C})$ and the main result (stated in the slightly more general framework of additive categories with translation) is that $K(\mathcal{C})$ is triangulated.

Many complexes, such as Čech complexes in sheaf theory (see Chap. 18 below), are obtained naturally by simplicial construction. Here, we construct complexes associated with simplicial objects and give a criterion for these complexes to be homotopic to zero.

When considering bifunctors on additive categories, we are rapidly lead to consider the category $C(C(\mathcal{C}))$ of complexes of complexes (i.e., double complexes), and so on. We explain here how a diagonal procedure allows us, under suitable hypotheses, to reduce a double complex to a simple one. Delicate questions of signs arise and necessitate careful treatment.

Chapter 12. When $\mathcal{C}$ is abelian, we can define the j-th cohomology object $H^j(X)$ of a complex X. The main result is that the functor H^j is cohomological, that is, sends distinguished triangles in K($\mathcal{C}$) to long exact sequences in $\mathcal{C}$.

When a functor F with values in $\mathcal{C}$ is defined on the category of finite sets, it is possible to attach to F a complex in $\mathcal{C}$, generalizing the classical notion of Koszul complexes. We provide the tools needed to calculate the cohomology of such complexes and treat some examples such as distributive families of subobjects.

We also study the cohomology of a double complex, replacing the Leray's traditional spectral sequences by an intensive use of the truncation functors. We find this approach much easier and perfectly adequate in practice.

Chapter 13. Constructing the derived category of an abelian category is easy with the tools now at hand. It is nothing more than the localization of the homotopy category K($\mathcal{C}$) with respect to exact complexes.

Here we give the main constructions and results concerning derived categories and functors, including some new results.

Despite their popularity, derived categories are sometimes supposed difficult. A possible reason for this reputation is that to date there has been no systematic, pedagogical treatment of the theory. The classical texts on derived categories are the famous Hartshorne Notes [31], or Verdier's résumé of his thesis [68] (of which the complete manuscript has been published recently [69]). Apart from these, there are a few others which may be found in particular in the books [25, 38] and [71]. Recall that the original idea of derived categories goes back to Grothendieck.

Chapter 14. Using the results of Chap. 9, we study the (unbounded) derived category D($\mathcal{C}$) of a Grothendieck category $\mathcal{C}$. First, we show that any complex in a Grothendieck category is quasi-isomorphic to a *homotopically injective complex* and we deduce the existence of right derived functors in D($\mathcal{C}$). We then prove that the Brown representability theorem holds in D($\mathcal{C}$) and discuss the existence of left derived functors, as well as the composition of (right or left) derived functors and derived adjunction formulas.

Spaltenstein [65] was the first to consider unbounded complexes and the corresponding derived functors. The (difficult) result which asserts that the Brown representability theorem holds in the derived category of a Grothendieck category seems to be due to independently to [2] and [21] (see also [6, 42, 53] and [44]). Note that most of the ideas presented here come from topology, in which context the names of Adams, Bousfield, Kan, Thomason among others should be mentioned.

Chapter 15. We study here the derived category of the category Ind($\mathcal{C}$) of ind-objects of an abelian category $\mathcal{C}$. Things are not easy since in the simple case where $\mathcal{C}$ is the category of vector spaces over a field k, the category Ind($\mathcal{C}$) does not have enough injective objects. In order to overcome this difficulty, we introduce the notion of quasi-injective objects. We show that under suitable hypotheses, there are enough such objects and that they allow us to derive functors. We also study some links between the derived category of $\mathrm{Ind}(\mathcal{C})$ and that of ind-objects of the derived category of $\mathcal{C}$. Note that the category of ind-objects of a triangulated category does not seem to be triangulated.

Most of the results in this chapter are new and we hope that they may be useful. They are so when applied to the construction of ind-sheaves, for which we refer to [39].

Chapter 16. The notion of sheaves relies on that of coverings and a Grothendieck topology on a category is defined by axiomatizing the notion of coverings.

In this chapter we give the axioms for Grothendieck topologies using sieves and then introduce the notions of local epimorphisms and local isomorphisms. We give several examples and study the properties of the family of local isomorphisms in detail, showing in particular that this family is stable under inductive limits. The classical reference is [64].

Chapter 17. A site X is a category $\mathcal{C}_X$ endowed with a Grothendieck topology. A presheaf F on X with values in a category $\mathcal{A}$ is a contravariant functor on $\mathcal{C}_X$ with values in $\mathcal{A}$, and a presheaf F is a sheaf if, for any local isomorphism $A \to U$, $F(U) \to F(A)$ is an isomorphism. When $\mathcal{C}_X$ is the category of open subsets of a topological space X, we recover a familiar notion.

Here, we construct the sheaf F^a associated with a presheaf F with values in a category $\mathcal{A}$ satisfying suitable properties. We also study restriction and extension of sheaves, direct and inverse images, and internal $\mathcal{H}om$. However, we do not enter the theory of Topos, referring to [64] (see also [48] for further exciting developments).

Chapter 18. When $\mathcal{O}_X$ is a sheaf of rings on a site X, we define the category $\mathrm{Mod}(\mathcal{O}_X)$ of sheaves of $\mathcal{O}_X$-modules. This is a Grothendieck category to which we may apply the tools obtained in Chap. 14.

In this Chapter, we construct the unbounded derived functors $R\mathcal{H}om_{\mathcal{O}_X}$ of internal hom, $\overset{L}{\otimes}_{\mathcal{O}_X}$ of tensor product, Rf_* of direct image and Lf^* of inverse image (these two last functors being associated with a morphism f of ringed sites) and we study their relations. Such constructions are well-known in the case of bounded derived categories, but the unbounded case, initiated by Spaltenstein [65], is more delicate.

We do not treat proper direct images and duality for sheaves. Indeed, there is no such theory for sheaves on abstract sites, where the construction in the algebraic case for which we refer to [17], differs from that in the topological case for which we refer to [38].

Chapter 19. The notion of constant functions is not local and it is more natural (and useful) to consider locally constant functions. The presheaf of such functions is in fact a sheaf, called a constant sheaf. There are however sheaves which are locally, but not globally, isomorphic to this constant sheaf, and this leads us to the fundamental notion of locally constant sheaves, or local systems. The orientation sheaf on a real manifold is a good example of such a sheaf. We consider similarly categories which are locally equivalent to the category of sheaves, which leads us to the notions of stacks and twisted sheaves.

A stack on a site X is, roughly speaking, a sheaf of categories, or, more precisely, a sheaf with values in the 2-category of all $\mathcal{U}$-categories of a given universe $\mathcal{U}$. Indeed, it would be possible to consider higher objects (n-stacks), but we do not pursue this matter here. This new field of mathematics was first explored in the sixties by Grothendieck and Giraud (see [26]) and after having been long considered highly esoteric, it is now the object of intense activity from algebraic geometry to theoretical physics. Note that 2-categories were first introduced by Bénabou (see [5]), a student of an independent-minded category theorist, Charles Ehresmann.

This last chapter should be understood as a short presentation of possible directions in which the theory may develop.

1

The Language of Categories

A set E is a collection of elements, and given two elements x and y in E there are no relations between x and y. The notion of a category is more sophisticated. A category $\mathcal{C}$ possesses objects similarly as a set possesses elements, but now for each pair of objects X and Y in $\mathcal{C}$, one is given a set $\mathrm{Hom}_{\mathcal{C}}(X, Y)$ called the morphisms from X to Y, representing possible relations between X and Y.

Once we have the notion of a category, it is natural to ask what are the morphisms from a category to another, and this lead to the notion of functors. We can also define the morphisms of functors, and as a byproduct, the notion of an equivalence of categories. At this stage, it would be tempting to define the notion of a 2-category, but this is out of the scope of this book.

The cornerstone of Category Theory is the Yoneda lemma. It asserts that a category $\mathcal{C}$ may be embedded in the category $\mathcal{C}^{\wedge}$ of all contravariant functors from this category to the category **Set** of sets, the morphisms in **Set** being the usual maps. This allows us, in some sense, to reduce Category Theory to Set Theory. The Yoneda lemma naturally leads to the notion of representable functor, and in particular to that of adjoint functor.

To a category $\mathcal{C}$, we can associate its opposite category $\mathcal{C}^{\mathrm{op}}$ obtained by reversing the arrows, and in this theory most of the constructions have their counterparts, monomorphism and epimorphism, right adjoint and left adjoint, etc. Of course, when a statement may be deduced from another one by reversing the arrows, we shall simply give one of the two statements. But the category **Set** is not equivalent to its opposite category, and **Set** plays a very special role in the whole theory. For example, inductive and projective limits in categories are constructed by using projective limits in **Set**.

A first example of a category would be the category **Set** mentioned above. But at this stage one encounters a serious difficulty, namely that of manipulating "all" sets. Moreover, we constantly use the category of all functors from a given category to **Set**. In this book, to avoid contradictions, we work in a given universe. Here, we shall begin by briefly recalling the axioms of universes, referring to [64] for more details.

1.1 Preliminaries: Sets and Universes

The aim of this section is to fix some notations and to recall the axioms of universes. We do not intend neither to enter Set Theory, nor to say more about universes than what we need. For this last subject, references are made to [64].

For a set u, we denote as usual by $\mathcal{P}(u)$ the set of subsets of u: $\mathcal{P}(u) = \{x; x \subset u\}$. For $x_1, \ldots x_n$, we denote as usual by $\{x_1, \ldots, x_n\}$ the set whose elements are $x_1, \ldots x_n$.

Definition 1.1.1. *A* universe $\mathcal{U}$ *is a set satisfying the following properties:*

(i) $\emptyset \in \mathcal{U}$,
(ii) $u \in \mathcal{U}$ *implies* $u \subset \mathcal{U}$, *(equivalently, $x \in \mathcal{U}$ and $y \in x$ implies $y \in \mathcal{U}$, or else $\mathcal{U} \subset \mathcal{P}(\mathcal{U})$),*
(iii) $u \in \mathcal{U}$ *implies* $\{u\} \in \mathcal{U}$,
(iv) $u \in \mathcal{U}$ *implies* $\mathcal{P}(u) \in \mathcal{U}$,
(v) *if* $I \in \mathcal{U}$ *and* $u_i \in \mathcal{U}$ *for all* $i \in I$, *then* $\bigcup_{i \in I} u_i \in \mathcal{U}$,
(vi) $\mathbb{N} \in \mathcal{U}$.

As a consequence we have

(vii) $u \in \mathcal{U}$ implies $\bigcup_{x \in u} x \in \mathcal{U}$,
(viii) $u, v \in \mathcal{U}$ implies $u \times v \in \mathcal{U}$,
(ix) $u \subset v \in \mathcal{U}$ implies $u \in \mathcal{U}$,
(x) if $I \in \mathcal{U}$ and $u_i \in \mathcal{U}$ for all $i \in I$, then $\prod_{i \in I} u_i \in \mathcal{U}$.

Following Grothendieck, we shall add an axiom to the Zermelo-Fraenkel theory, asking that for any set X there exists a universe $\mathcal{U}$ such that $X \in \mathcal{U}$. For more explanations, refer to [64].

Definition 1.1.2. *Let $\mathcal{U}$ be a universe.*

(i) *A set is called a $\mathcal{U}$-set if it belongs to $\mathcal{U}$.*
(ii) *A set is called $\mathcal{U}$-small if it is isomorphic to a set belonging to $\mathcal{U}$.*

Definition 1.1.3.   (i) *An* order *on a set I is a relation $\leq$ which is:*
    (a) *reflexive, that is, $i \leq i$ for all $i \in I$,*
    (b) *transitive, that is, $i \leq j$, $j \leq k \Rightarrow i \leq k$,*
    (c) *anti-symmetric, that is, $i \leq j$, $j \leq i \Rightarrow i = j$.*
(ii) *An order is* directed *(we shall also say "filtrant") if I is non empty and if for any $i, j \in I$, there exists $k \in I$ such that $i \leq k$ and $j \leq k$.*
(iii) *An order is* total *(some authors say "linear") if for any $i, j \in I$, one has $i \leq j$ or $j \leq i$.*
(iv) *An ordered set I is* inductively ordered *if any totally ordered subset J of I has an upper bound (i.e., there exists $a \in I$ such that $j \leq a$ for all $j \in J$).*
(v) *If $\leq$ is an order on I, $<$ is the relation given by $x < y$ if and only if $x \leq y$ and $x \neq y$. We also write $x \geq y$ if $y \leq x$ and $x > y$ if $y < x$.*

Recall that Zorn's lemma asserts that any inductively ordered set admits a maximal element.

Notations 1.1.4. (i) We denote by {pt} a set with one element, and this single element is often denoted by pt. We denote by ∅ the set with no element.
(ii) In all this book, a ring means an associative ring with unit, and the action of a ring on a module is unital. If there is no risk of confusion, we simply denote by 0 the module with a single element. A field is a non-zero commutative ring in which every non-zero element is invertible.
(iii) We shall often denote by k a commutative ring. A k-algebra is a ring R endowed with a morphism of rings $\varphi \colon k \to R$ such that the image of k is contained in the center of R. We denote by $k^\times$ the group of invertible elements of k.
(iv) As usual, we denote by $\mathbb{Z}$ the ring of integers and by $\mathbb{Q}$ (resp. $\mathbb{R}$, resp. $\mathbb{C}$) the field of rational numbers (resp. real numbers, resp. complex numbers). We denote by $\mathbb{N}$ the set of non-negative integers, that is, $\mathbb{N} = \{n \in \mathbb{Z} \,;\, n \geq 0\}$.
(v) We denote by $k[x_1, \ldots, x_n]$ the ring of polynomials in the variables $x_1, \ldots, x_n$ over a commutative ring k.
(vi) We denote by δ_{ij} the Kronecker symbol, $\delta_{ij} = 1$ if $i = j$ and $\delta_{ij} = 0$ otherwise.

1.2 Categories and Functors

Definition 1.2.1. *A* category $\mathcal{C}$ *consists of* :

(i) *a set* $\mathrm{Ob}(\mathcal{C})$,
(ii) *for any* $X, Y \in \mathrm{Ob}(\mathcal{C})$, *a set* $\mathrm{Hom}_{\mathcal{C}}(X, Y)$,
(iii) *for any* $X, Y, Z \in \mathrm{Ob}(\mathcal{C})$, *a map:*

$$\mathrm{Hom}_{\mathcal{C}}(X, Y) \times \mathrm{Hom}_{\mathcal{C}}(Y, Z) \to \mathrm{Hom}_{\mathcal{C}}(X, Z)$$

called the composition and denoted by $(f, g) \mapsto g \circ f$,

these data satisfying:

(a) $\circ$ *is associative, i.e., for* $f \in \mathrm{Hom}_{\mathcal{C}}(X, Y)$, $g \in \mathrm{Hom}_{\mathcal{C}}(Y, Z)$ *and* $h \in \mathrm{Hom}_{\mathcal{C}}(Z, W)$, *we have* $(h \circ g) \circ f = h \circ (g \circ f)$,
(b) *for each* $X \in \mathrm{Ob}(\mathcal{C})$, *there exists* $\mathrm{id}_X \in \mathrm{Hom}\,(X, X)$ *such that* $f \circ \mathrm{id}_X = f$ *for all* $f \in \mathrm{Hom}_{\mathcal{C}}(X, Y)$ *and* $\mathrm{id}_X \circ g = g$ *for all* $g \in \mathrm{Hom}_{\mathcal{C}}(Y, X)$.

An element of $\mathrm{Ob}(\mathcal{C})$ is called an *object* of $\mathcal{C}$ and for $X, Y \in \mathrm{Ob}(\mathcal{C})$, an element of $\mathrm{Hom}_{\mathcal{C}}(X, Y)$ is called a *morphism* (from X to Y) in $\mathcal{C}$. The morphism id_X is called the *identity morphism* (or the identity, for short) of X. Note that there is a unique $\mathrm{id}_X \in \mathrm{Hom}_{\mathcal{C}}(X, X)$ satisfying the condition in (b).

A category $\mathcal{C}$ is called a $\mathcal{U}$-category if $\mathrm{Hom}_{\mathcal{C}}(X, Y)$ is $\mathcal{U}$-small for any $X, Y \in \mathrm{Ob}(\mathcal{C})$.

A $\mathcal{U}$-small category is a $\mathcal{U}$-category $\mathcal{C}$ such that $\mathrm{Ob}(\mathcal{C})$ is $\mathcal{U}$-small.

Notation 1.2.2. We often write $X \in \mathcal{C}$ instead of $X \in \mathrm{Ob}(\mathcal{C})$, and $f \colon X \to Y$ or else $f \colon Y \leftarrow X$ instead of $f \in \mathrm{Hom}_{\mathcal{C}}(X, Y)$. We say that X is the *source* and Y the *target* of f. We sometimes call f an *arrow* instead of "a morphism".

We introduce the *opposite category* $\mathcal{C}^{\mathrm{op}}$ by setting:

$$\mathrm{Ob}(\mathcal{C}^{\mathrm{op}}) = \mathrm{Ob}(\mathcal{C}), \quad \mathrm{Hom}_{\mathcal{C}^{\mathrm{op}}}(X, Y) = \mathrm{Hom}_{\mathcal{C}}(Y, X),$$

and defining the new composition $g \overset{\mathrm{op}}{\circ} f$ of $f \in \mathrm{Hom}_{\mathcal{C}^{\mathrm{op}}}(X, Y)$ and $g \in \mathrm{Hom}_{\mathcal{C}^{\mathrm{op}}}(Y, Z)$ by $g \overset{\mathrm{op}}{\circ} f = f \circ g$. For an object X or a morphism f in $\mathcal{C}$, we shall sometimes denote by X^{op} or f^{op} its image in $\mathcal{C}^{\mathrm{op}}$. In the sequel, we shall simply write $\circ$ instead of $\overset{\mathrm{op}}{\circ}$.

A morphism $f \colon X \to Y$ is an *isomorphism* if there exists $g \colon X \leftarrow Y$ such that $f \circ g = \mathrm{id}_Y$, $g \circ f = \mathrm{id}_X$. Such a g, which is unique, is called the *inverse* of f and is denoted by f^{-1}. If $f \colon X \to Y$ is an isomorphism, we write $f \colon X \overset{\sim}{\to} Y$. If there is an isomorphism $X \overset{\sim}{\to} Y$, we say that X and Y are isomorphic and we write $X \simeq Y$.

An *endomorphism* is a morphism with same source and target, that is, a morphism $f \colon X \to X$.

An *automorphism* is an endomorphism which is an isomorphism.

Two morphisms f and g are *parallel* if they have same source and same target, visualized by $f, g \colon X \rightrightarrows Y$.

A morphism $f \colon X \to Y$ is a *monomorphism* if for any pair of parallel morphisms $g_1, g_2 \colon Z \rightrightarrows X$, $f \circ g_1 = f \circ g_2$ implies $g_1 = g_2$.

A morphism $f \colon X \to Y$ is an *epimorphism* if $f^{\mathrm{op}} \colon Y^{\mathrm{op}} \to X^{\mathrm{op}}$ is a monomorphism in $\mathcal{C}^{\mathrm{op}}$. Hence, f is an epimorphism if and only if for any pair of parallel morphisms $g_1, g_2 \colon Y \rightrightarrows Z$, $g_1 \circ f = g_2 \circ f$ implies $g_1 = g_2$.

Note that f is a monomorphism if and only if the map $f \circ \colon \mathrm{Hom}_{\mathcal{C}}(Z, X) \to \mathrm{Hom}_{\mathcal{C}}(Z, Y)$ is injective for any object Z, and f is an epimorphism if and only if the map $\circ f \colon \mathrm{Hom}_{\mathcal{C}}(Y, Z) \to \mathrm{Hom}_{\mathcal{C}}(X, Z)$ is injective for any object Z.

Also note that if $X \overset{f}{\to} Y \overset{g}{\to} Z$ are morphisms and if f and g are monomorphisms (resp. epimorphisms, resp. isomorphisms), then $g \circ f$ is a monomorphism (resp. epimorphism, resp. isomorphism).

We sometimes write $f \colon X \rightarrowtail Y$ or else $f \colon X \hookrightarrow Y$ to denote a monomorphism and $f \colon X \twoheadrightarrow Y$ to denote an epimorphism.

For two morphisms $f \colon X \to Y$ and $g \colon Y \to X$ satisfying $f \circ g = \mathrm{id}_Y$, f is called a *left inverse* of g and g is called a *right inverse* of f. We also say that g is a *section* of f or f is a *cosection* of g. In such a situation, f is an epimorphism and g a monomorphism.

A category $\mathcal{C}'$ is a *subcategory* of $\mathcal{C}$, denoted by $\mathcal{C}' \subset \mathcal{C}$, if: $\mathrm{Ob}(\mathcal{C}') \subset \mathrm{Ob}(\mathcal{C})$, $\mathrm{Hom}_{\mathcal{C}'}(X, Y) \subset \mathrm{Hom}_{\mathcal{C}}(X, Y)$ for any $X, Y \in \mathcal{C}'$, the composition in $\mathcal{C}'$ is induced by the composition in $\mathcal{C}$ and the identity morphisms in $\mathcal{C}'$ are identity morphisms in $\mathcal{C}$. A subcategory $\mathcal{C}'$ of $\mathcal{C}$ is *full* if $\mathrm{Hom}_{\mathcal{C}'}(X, Y) = \mathrm{Hom}_{\mathcal{C}}(X, Y)$ for all $X, Y \in \mathcal{C}'$. A full subcategory $\mathcal{C}'$ of $\mathcal{C}$ is *saturated* if $X \in \mathcal{C}$ belongs to $\mathcal{C}'$ whenever X is isomorphic to an object of $\mathcal{C}'$.

A category is *discrete* if all the morphisms are the identity morphisms.

A category $\mathcal{C}$ is *non empty* if $\mathrm{Ob}(\mathcal{C})$ is non empty.

A category $\mathcal{C}$ is a *groupoid* if all morphisms are isomorphisms.

A category $\mathcal{C}$ is *finite* if the set of all morphisms in $\mathcal{C}$ (hence, in particular, the set of objects) is a finite set.

A category $\mathcal{C}$ is *connected* if it is non empty and for any pair of objects $X, Y \in \mathcal{C}$, there is a finite sequence of objects $(X_0, \ldots, X_n)$, $X_0 = X$, $X_n = Y$, such that at least one of the sets $\mathrm{Hom}_{\mathcal{C}}(X_j, X_{j+1})$ or $\mathrm{Hom}_{\mathcal{C}}(X_{j+1}, X_j)$ is non empty for any $j \in \mathbb{N}$ with $0 \le j \le n - 1$.

Remark that a *monoid M* (i.e., a set endowed with an internal product with associative and unital law) is nothing but a category with only one object. (To M, associate the category $\mathcal{M}$ with the single object a and morphisms $\mathrm{Hom}_{\mathcal{M}}(a, a) = M$.) Similarly, a group G defines a groupoid, namely the category $\mathcal{G}$ with a single object a and morphisms $\mathrm{Hom}_{\mathcal{G}}(a, a) = G$.

A *diagram* in a category $\mathcal{C}$ is a family of symbols representing objects of $\mathcal{C}$ and a family of arrows between these symbols representing morphisms of these objects. One defines in an obvious way the notion of a *commutative diagram*. For example, consider the diagrams

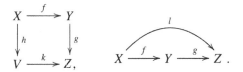

Then the first diagram is commutative if and only if $g \circ f = k \circ h$ and the second diagram is commutative if and only if $g \circ f = l$.

Notation 1.2.3. We shall also encounter diagrams such as:

$$(1.2.1) \qquad\qquad Z \underset{g_2}{\overset{g_1}{\rightrightarrows}} X \xrightarrow{\ f\ } Y \ .$$

We shall say that the two compositions coincide if $f \circ g_1 = f \circ g_2$.

We shall also encounter diagrams of categories. (See Remark 1.3.6 below.)

Examples 1.2.4. (i) **Set** is the category of $\mathcal{U}$-sets and maps, $\mathbf{Set}^f$ the full subcategory consisting of finite $\mathcal{U}$-sets. If we need to emphasize the universe $\mathcal{U}$, we write $\mathcal{U}$-**Set** instead of **Set**. Note that the category of all sets is not a category since the collection of all sets is not a set. This is one of the reasons why we have to introduce a universe $\mathcal{U}$.

(ii) The category **Rel** of binary relations is defined by: $\mathrm{Ob}(\mathbf{Rel}) = \mathrm{Ob}(\mathbf{Set})$ and $\mathrm{Hom}_{\mathbf{Rel}}(X, Y) = \mathcal{P}(X \times Y)$, the set of subsets of $X \times Y$. The composition law is defined as follows. If $f : X \to Y$ and $g : Y \to Z$, $g \circ f$ is the set

$\{(x, z) \in X \times Z;$ there exists y such that $(x, y) \in f, (y, z) \in g\}$.

Of course, id_X is the diagonal set of $X \times X$.

Notice that **Set** is a subcategory of **Rel**, but is not a full subcategory.

(iii) **pSet** is the category of pointed $\mathcal{U}$-sets. An object of **pSet** is a pair (X, x) with a $\mathcal{U}$-set X and $x \in X$. A morphism $f\colon (X, x) \to (Y, y)$ is a map $f\colon X \to Y$ such that $f(x) = y$.

(iv) Let R be a ring (with $R \in \mathcal{U}$). The category of left R-modules belonging to $\mathcal{U}$ and R-linear maps is denoted $\mathrm{Mod}(R)$. Hence, by definition, $\mathrm{Hom}_{\mathrm{Mod}(R)}(\cdot, \cdot) = \mathrm{Hom}_R(\cdot, \cdot)$. Recall that right R-modules are left R^{op}-modules, where R^{op} denotes the ring R with the opposite multiplicative structure. Note that $\mathrm{Mod}(\mathbb{Z})$ is the category of abelian groups.

We denote by $\mathrm{End}_R(M)$ the ring of R-linear endomorphisms of an R-module M and by $\mathrm{Aut}_R(M)$ the group of R-linear automorphisms of M.

We denote by $\mathrm{Mod}^f(R)$ the full subcategory of $\mathrm{Mod}(R)$ consisting of *finitely generated* R-modules. (Recall that M is finitely generated if there exists a surjective R-linear map $u\colon R^{\oplus n} \twoheadrightarrow M$ for some integer $n \geq 0$.) One also says *of finite type* instead of "finitely generated".

We denote by $\mathrm{Mod}^{\mathrm{fp}}(R)$ the full subcategory of $\mathrm{Mod}^f(R)$ consisting of R-modules of *finite presentation*. (Recall that M is of finite presentation if it is of finite type and moreover the kernel of the linear map u above is of finite type.)

(v) Let $(I, \leq)$ be an ordered set. We associate to it a category $\mathcal{I}$ as follows.

$$\mathrm{Ob}(\mathcal{I}) = I$$
$$\mathrm{Hom}_{\mathcal{I}}(i, j) = \begin{cases} \{\mathrm{pt}\} & \text{if } i \leq j , \\ \emptyset & \text{otherwise} . \end{cases}$$

In other words, the set of morphisms from i to j has a single element if $i \leq j$, and is empty otherwise. Note that $\mathcal{I}^{\mathrm{op}}$ is the category associated to $(I, \leq^{\mathrm{op}})$, where $x \leq^{\mathrm{op}} y$ if and only if $y \leq x$. In the sequel, we shall often simply write I instead of $\mathcal{I}$. (See Exercise 1.3 for a converse construction.)

(vi) We denote by **Top** the category of topological spaces belonging to $\mathcal{U}$ and continuous maps.

The set of all morphisms of a category $\mathcal{C}$ may be endowed with a structure of a category.

Definition 1.2.5. *Let $\mathcal{C}$ be a category. We denote by $\mathrm{Mor}(\mathcal{C})$ the category whose objects are the morphisms in $\mathcal{C}$ and whose morphisms are described as follows. Let $f\colon X \to Y$ and $g\colon X' \to Y'$ belong to $\mathrm{Mor}(\mathcal{C})$. Then $\mathrm{Hom}_{\mathrm{Mor}(\mathcal{C})}(f, g) = \{u\colon X \to X', v\colon Y \to Y'; g \circ u = v \circ f\}$. The composition and the identity in $\mathrm{Mor}(\mathcal{C})$ are the obvious ones.*

A morphism $f \to g$ in $\mathrm{Mor}(\mathcal{C})$ is visualized by the commutative diagram:

$$\begin{array}{ccc} X & \xrightarrow{\ f\ } & Y \\ {\scriptstyle u}\downarrow & & \downarrow{\scriptstyle v} \\ X' & \xrightarrow{\ g\ } & Y'. \end{array}$$

Definition 1.2.6. (i) *An object $P \in \mathcal{C}$ is called* initial *if for all $X \in \mathcal{C}$, $\mathrm{Hom}_{\mathcal{C}}(P, X) \simeq \{\mathrm{pt}\}$. We often denote by $\emptyset_{\mathcal{C}}$ an initial object in $\mathcal{C}$. (Note that if P_1 and P_2 are initial, then there is a unique isomorphism $P_1 \simeq P_2$.)*
 (ii) *We say that P is* terminal *in $\mathcal{C}$ if P is initial in $\mathcal{C}^{\mathrm{op}}$, i.e., for all $X \in \mathcal{C}$, $\mathrm{Hom}_{\mathcal{C}}(X, P) \simeq \{\mathrm{pt}\}$. We often denote by $\mathrm{pt}_{\mathcal{C}}$ a terminal object in $\mathcal{C}$.*
 (iii) *We say that P is a* zero *object if it is both initial and terminal (see Exercise 1.1). Such a P is often denoted by 0. If $\mathcal{C}$ has a zero object 0, for any objects $X, Y \in \mathcal{C}$, the morphism obtained as the composition $X \to 0 \to Y$ is still denoted by $0\colon X \to Y$. (Note that the composition of $0\colon X \to Y$ and any morphism $f\colon Y \to Z$ is $0\colon X \to Z$.)*

Examples 1.2.7. (i) In the category **Set**, $\emptyset$ is initial and $\{\mathrm{pt}\}$ is terminal.
(ii) In the category **pSet**, the object $(\{\mathrm{pt}\}, \mathrm{pt})$ is a zero object.
(iii) The zero module 0 is a zero object in $\mathrm{Mod}(R)$.
(iv) The category associated with the ordered set $(\mathbb{Z}, \leq)$ has neither initial nor terminal object.

Notations 1.2.8. (i) We shall denote by **Pt** a category with a single object and a single morphism (the identity of this object).
(ii) We shall simply denote by $\emptyset$ the empty category with no objects (and hence, no morphisms).
(iii) We shall often represent by the diagram $\bullet \to \bullet$ the category which consists of two objects, say, a and b, and one morphism $a \to b$ other than id_a and id_b. We denote this category by **Arr**.
(iv) We represent by $\bullet \rightrightarrows \bullet$ the category with two objects, say $\{a, b\}$, and two parallel morphisms $a \rightrightarrows b$ other than $\mathrm{id}_a, \mathrm{id}_b$.
(v) We shall denote by **Pr** a category with a single object c and one morphism $p\colon c \to c$ other than id_c, satisfying $p^2 = p$.

Example 1.2.9. Let R be a ring. Let $N \in \mathrm{Mod}(R^{\mathrm{op}})$ and $M \in \mathrm{Mod}(R)$. Define a category $\mathcal{C}$ as follows. The objects of $\mathcal{C}$ are the pairs (f, L) where $L \in \mathrm{Mod}(\mathbb{Z})$ and f is a bilinear map from $N \times M$ to L (i.e., it is $\mathbb{Z}$-bilinear and satisfies $f(na, m) = f(n, am)$ for all $a \in R$). A morphism from $f\colon N \times M \to L$ to $g\colon N \times M \to K$ is a linear map $h\colon L \to K$ such that $h \circ f = g$. Since any bilinear map $f\colon N \times M \to L$ (i.e., any object of $\mathcal{C}$) factorizes uniquely through $u\colon N \times M \to N \otimes_R M$, the object $(u, N \otimes_R M)$ is initial in $\mathcal{C}$.

Definition 1.2.10. (i) *Let C and C' be two categories. A functor $F: C \to C'$ consists of a map $F: \mathrm{Ob}(C) \to \mathrm{Ob}(C')$ and of maps $F: \mathrm{Hom}_C(X, Y) \to \mathrm{Hom}_{C'}(F(X), F(Y))$ for all $X, Y \in C$, such that*

$$F(\mathrm{id}_X) = \mathrm{id}_{F(X)} \ \text{ for all } X \in C,$$
$$F(g \circ f) = F(g) \circ F(f) \text{ for all } f: X \to Y, \ g: Y \to Z.$$

A contravariant functor from C to C' is a functor from C^{op} to C'. In other words, it satisfies $F(g \circ f) = F(f) \circ F(g)$.

(ii) *For categories C, C', C'' and functors $F: C \to C'$, $G: C' \to C''$ their composition $G \circ F: C \to C''$ is the functor defined by $(G \circ F)(X) = G(F(X))$ for all $X \in C$ and $(G \circ F)(f) = G(F(f))$ for all morphism f in C.*

If one wishes to put the emphasis on the fact that a functor is not contravariant, one says it is *covariant*.

It is convenient to introduce the contravariant functor

(1.2.2) $\mathrm{op}: C \to C^{\mathrm{op}}$

defined by the identity of C.

Note that a functor $F: C \to C'$ naturally induces a functor

(1.2.3) $F^{\mathrm{op}}: C^{\mathrm{op}} \to C'^{\mathrm{op}}$.

Definition 1.2.11. *Let $F: C \to C'$ be a functor.*

(i) *We say that F is* faithful (*resp.* full, fully faithful) *if*

$$\mathrm{Hom}_C(X, Y) \to \mathrm{Hom}_{C'}(F(X), F(Y))$$

is injective (resp. surjective, bijective) for any X, Y in C.

(ii) *We say that F is* essentially surjective *if for each $Y \in C'$ there exist $X \in C$ and an isomorphism $F(X) \xrightarrow{\sim} Y$.*

(iii) *We say that F is* conservative *if a morphism f in C is an isomorphism as soon as $F(f)$ is an isomorphism in C'.*

Note that properties (i)–(iii) are closed by composition of functors. In other words, if $C \xrightarrow{F} C' \xrightarrow{G} C''$ are functors and if F and G satisfy the property (i) (resp. (ii), resp. (iii)), then so does $G \circ F$.

Proposition 1.2.12. *Let $F: C \to C'$ be a faithful functor and let $f: X \to Y$ be a morphism in C. If $F(f)$ is an epimorphism (resp. a monomorphism), then f is an epimorphism (resp. a monomorphism).*

Proof. Assume that $F(f)$ is an epimorphism and consider a pair of parallel arrows $g, h: Y \rightrightarrows Z$ such that $g \circ f = h \circ f$. Then $F(g) \circ F(f) = F(h) \circ F(f)$. If $F(f)$ is an epimorphism, we deduce $F(g) = F(h)$ and if F is faithful, this implies $g = h$.

The case of a monomorphism is treated similarly.   q.e.d.

Definition 1.2.13. *Consider a family* $\{\mathcal{C}_i\}_{i \in I}$ *of categories indexed by a set* I.

(i) *We define the* product category $\prod_{i \in I} \mathcal{C}_i$ *by setting:*

$$\mathrm{Ob}(\prod_{i \in I} \mathcal{C}_i) = \prod_{i \in I} \mathrm{Ob}(\mathcal{C}_i) \ ,$$

$$\mathrm{Hom}_{\prod_{i \in I} \mathcal{C}_i}(\{X_i\}_i, \{Y_i\}_i) = \prod_{i \in I} \mathrm{Hom}_{\mathcal{C}_i}(X_i, Y_i) \ .$$

(ii) *We define the* disjoint union category $\bigsqcup_{i \in I} \mathcal{C}_i$ *by setting:*

$$\mathrm{Ob}(\bigsqcup_{i \in I} \mathcal{C}_i) = \{(X, i); i \in I, X \in \mathrm{Ob}(\mathcal{C}_i)\} \ ,$$

$$\mathrm{Hom}_{\bigsqcup_{i \in I} \mathcal{C}_i}((X, i), (Y, j)) = \begin{cases} \mathrm{Hom}_{\mathcal{C}_i}(X, Y) & \text{if } i = j \ , \\ \varnothing & \text{if } i \neq j \ . \end{cases}$$

As usual, if I has two elements, say $I = \{1, 2\}$, we denote the product by $\mathcal{C}_1 \times \mathcal{C}_2$ and the disjoint union by $\mathcal{C}_1 \sqcup \mathcal{C}_2$.

If $\{F_i \colon \mathcal{C}_i \to \mathcal{C}'_i\}_{i \in I}$ is a family of functors, we define naturally the functor $\prod_{i \in I} F_i$ from $\prod_{i \in I} \mathcal{C}_i$ to $\prod_{i \in I} \mathcal{C}'_i$ and the functor $\bigsqcup_{i \in I} F_i$ from $\bigsqcup_{i \in I} \mathcal{C}_i$ to $\bigsqcup_{i \in I} \mathcal{C}'_i$.

A functor $F \colon \mathcal{C} \times \mathcal{C}' \to \mathcal{C}''$ is called a *bifunctor*. This is equivalent to saying that for $X \in \mathcal{C}$ and $X' \in \mathcal{C}'$, $F(X, \cdot) \colon \mathcal{C}' \to \mathcal{C}''$ and $F(\cdot, X') \colon \mathcal{C} \to \mathcal{C}''$ are functors, and moreover for any morphisms $f \colon X \to Y$ in $\mathcal{C}$, $g \colon X' \to Y'$ in $\mathcal{C}'$, the diagram below commutes:

$$
\begin{array}{ccc}
F(X, X') & \xrightarrow{\ F(X,g)\ } & F(X, Y') \\
{\scriptstyle F(f,X')}\big\downarrow & & \big\downarrow{\scriptstyle F(f,Y')} \\
F(Y, X') & \xrightarrow{\ F(Y,g)\ } & F(Y, Y').
\end{array}
$$

Indeed, $(f, g) = (\mathrm{id}_Y, g) \circ (f, \mathrm{id}_{X'}) = (f, \mathrm{id}_{Y'}) \circ (\mathrm{id}_X, g)$.

Examples 1.2.14. (i) If $\mathcal{C}$ is a $\mathcal{U}$-category, $\mathrm{Hom}_{\mathcal{C}}(\cdot, \cdot) \colon \mathcal{C}^{\mathrm{op}} \times \mathcal{C} \to \mathbf{Set}$ is a bifunctor.

(ii) Let R be a k-algebra. We have the two bifunctors:

$$\cdot \otimes_R \cdot \ : \ \mathrm{Mod}(R^{\mathrm{op}}) \times \mathrm{Mod}(R) \to \mathrm{Mod}(k) \ ,$$

$$\mathrm{Hom}_R(\cdot, \cdot) \ : \ \mathrm{Mod}(R)^{\mathrm{op}} \times \mathrm{Mod}(R) \to \mathrm{Mod}(k) \ .$$

(iii) The forgetful functor *for*: $\mathbf{Top} \to \mathbf{Set}$ which associates its underlying set to a topological space is faithful but not fully faithful.

Notations 1.2.15. (i) Let I and $\mathcal{C}$ be two categories, and let $X \in \mathcal{C}$. We denote by Δ_X^I, or simply Δ_X if there is no risk of confusion, the *constant functor* from I to $\mathcal{C}$ given by $I \ni i \mapsto X$ and $\mathrm{Mor}(I) \ni (i \to j) \mapsto \mathrm{id}_X$.

(ii) Let $\mathcal{C}$ be a category, $\mathcal{C}'$ a subcategory. The natural functor $\mathcal{C}' \to \mathcal{C}$ is often called the *embedding* functor.

We end this section with a few definitions.

Definition 1.2.16. *Let $F: \mathcal{C} \to \mathcal{C}'$ be a functor and let $A \in \mathcal{C}'$.*

(i) *The category $\mathcal{C}_A$ is given by*

$$\mathrm{Ob}(\mathcal{C}_A) = \{(X, s); X \in \mathcal{C}, \ s: F(X) \to A\},$$
$$\mathrm{Hom}_{\mathcal{C}_A}((X, s), (Y, t)) = \{f \in \mathrm{Hom}_{\mathcal{C}}(X, Y); s = t \circ F(f)\}.$$

(ii) *The category $\mathcal{C}^A$ is given by*

$$\mathrm{Ob}(\mathcal{C}^A) = \{(X, s); X \in \mathcal{C}, s: A \to F(X)\},$$
$$\mathrm{Hom}_{\mathcal{C}^A}((X, s), (Y, t)) = \{f \in \mathrm{Hom}_{\mathcal{C}}(X, Y); t = F(f) \circ s\}.$$

We define the faithful functors

(1.2.4) $\qquad\qquad$ $j_A: \mathcal{C}_A \to \mathcal{C}$ $\quad$ *by setting* $j_A(X, s) = X$,

(1.2.5) $\qquad\qquad$ $j^A: \mathcal{C}^A \to \mathcal{C}$ $\quad$ *by setting* $j^A(X, s) = X$.

For an object (X, s) in $\mathcal{C}_A$ (resp. in $\mathcal{C}^A$), we sometimes write $(F(X) \to A)$ (resp. $(A \to F(X))$) or simply X.

The categories $\mathcal{C}_A$ and $\mathcal{C}^A$ depend on the functor F, but we do not mention F in the notation. Definition 1.2.16 will be generalized in Definition 3.4.1.

Definition 1.2.17. *For a category $\mathcal{C}$, denote by $\sim$ the equivalence relation on $\mathrm{Ob}(\mathcal{C})$ generated by the relation $X \sim Y$ if $\mathrm{Hom}_{\mathcal{C}}(X, Y) \neq \emptyset$. We denote by $\pi_0(\mathcal{C})$ the set of equivalence classes of $\mathrm{Ob}(\mathcal{C})$.*

Regarding $\pi_0(\mathcal{C})$ as a discrete category, there is a natural functor $\mathcal{C} \to \pi_0(\mathcal{C})$. Then, for $a \in \pi_0(\mathcal{C})$, $\mathcal{C}^a$ and $\mathcal{C}_a$ are equivalent, they are connected, and the set of their objects is the set of objects in the equivalence class a. In particular, $\mathcal{C}$ is connected if and only if $\pi_0(\mathcal{C})$ consists of a single element.

Two monomorphisms $f: Y \rightarrowtail X$ and $g: Z \rightarrowtail X$ with the same target are isomorphic if there exists an isomorphism $h: Y \to Z$ such that $f = g \circ h$. In other words, $f: Y \to X$ and $g: Z \to X$ are isomorphic in $\mathcal{C}_X$. Note that such an h is unique. Similarly, two epimorphisms $X \twoheadrightarrow Y$ and $X \twoheadrightarrow Z$ are isomorphic if they are isomorphic in $\mathcal{C}^X$.

Definition 1.2.18. *Let $\mathcal{C}$ be a category and let $X \in \mathcal{C}$.*

(i) *An isomorphism class of a monomorphism with target X is called a* sub-object *of X.*

(ii) *An isomorphism class of an epimorphism with source X is called a* quotient *of X.*

Note that the set of subobjects of X is an ordered set by the relation $(f: Y \rightarrowtail X) \leq (f': Y' \rightarrowtail X)$ if there exists a morphism $h: Y \to Y'$ such that $f = f' \circ h$. (If such an h exists, then it is unique.)

1.3 Morphisms of Functors

Definition 1.3.1. *Let $\mathcal{C}$ and $\mathcal{C}'$ be two categories and let F_1 and F_2 be two functors from $\mathcal{C}$ to $\mathcal{C}'$. A* morphism of functors $\theta \colon F_1 \to F_2$ *consists of a morphism $\theta_X \colon F_1(X) \to F_2(X)$ (also denoted by $\theta(X)$) for all $X \in \mathcal{C}$ such that for all $f \colon X \to Y$, the diagram below commutes:*

(1.3.1)
$$
\begin{array}{ccc}
F_1(X) & \xrightarrow{\;\theta_X\;} & F_2(X) \\
{\scriptstyle F_1(f)}\downarrow & & \downarrow{\scriptstyle F_2(f)} \\
F_1(Y) & \xrightarrow{\;\theta_Y\;} & F_2(Y).
\end{array}
$$

Example 1.3.2. Assume that k is a field and denote by $*$ the duality functor from $\mathrm{Mod}(k)^{\mathrm{op}}$ to $\mathrm{Mod}(k)$, which associates $V^* = \mathrm{Hom}_k(V, k)$ to a vector space V. Then $\mathrm{id} \to {}^{**}$ is a morphism of functors from $\mathrm{Mod}(k)$ to itself.

If $\theta \colon F_1 \to F_2$ and $\lambda \colon F_2 \to F_3$ are morphisms of functors, we define naturally the morphism of functors $\lambda \circ \theta \colon F_1 \to F_3$ by $(\lambda \circ \theta)_X = \lambda_X \circ \theta_X$.

Notations 1.3.3. (i) Let $\mathcal{C}$ and $\mathcal{C}'$ be two categories. We shall denote by $\mathrm{Fct}(\mathcal{C}, \mathcal{C}')$ the category of functors from $\mathcal{C}$ to $\mathcal{C}'$. Hence, if F_1 and F_2 are two functors from $\mathcal{C}$ to $\mathcal{C}'$, $\mathrm{Hom}_{\mathrm{Fct}(\mathcal{C},\mathcal{C}')}(F_1, F_2)$ denotes the set of morphisms from F_1 to F_2. If $\mathcal{C}$ is small and $\mathcal{C}'$ is a $\mathcal{U}$-category, then $\mathrm{Fct}(\mathcal{C}, \mathcal{C}')$ is a $\mathcal{U}$-category. (ii) We also use the short notation $\mathcal{C}^I$ instead of $\mathrm{Fct}(I, \mathcal{C})$.

Note that if $\mathcal{C}, \mathcal{C}', \mathcal{C}''$ are three categories, the composition of functors defines a bifunctor

(1.3.2)
$$
\mathrm{Fct}(\mathcal{C}, \mathcal{C}') \times \mathrm{Fct}(\mathcal{C}', \mathcal{C}'') \to \mathrm{Fct}(\mathcal{C}, \mathcal{C}'') \, .
$$

A morphism of functors is visualized by a diagram:

(1.3.3)
$$
\mathcal{C} \underset{F_2}{\overset{F_1}{\rightrightarrows}} {\Downarrow\theta}\; \mathcal{C}'.
$$

Remark 1.3.4. Morphisms of functors may be composed "horizontally" and "vertically". More precisely:
(i) Consider three categories $\mathcal{C}, \mathcal{C}', \mathcal{C}''$ and functors $F_1, F_2 \colon \mathcal{C} \to \mathcal{C}'$ and $G_1, G_2 \colon \mathcal{C}' \to \mathcal{C}''$. If $\theta \colon F_1 \to F_2$ and $\lambda \colon G_1 \to G_2$ are morphisms of functors, the morphism of functors $\lambda \circ \theta \colon G_1 \circ F_1 \to G_2 \circ F_2$ is naturally defined. It is visualized by the diagram

$$
\mathcal{C} \underset{F_2}{\overset{F_1}{\rightrightarrows}}{\Downarrow\theta}\; \mathcal{C}' \underset{G_2}{\overset{G_1}{\rightrightarrows}}{\Downarrow\lambda}\; \mathcal{C}'' \quad\rightsquigarrow\quad \mathcal{C} \underset{G_2\circ F_2}{\overset{G_1\circ F_1}{\rightrightarrows}}{\Downarrow\lambda\circ\theta}\; \mathcal{C}'' \, .
$$

If $\lambda = \mathrm{id}_{G_1}$, we write $G_1 \circ \theta$ instead of $\mathrm{id}_{G_1} \circ \theta \colon G_1 \circ F_1 \to G_1 \circ F_2$ and if $\theta = \mathrm{id}_{F_1}$, we write $\lambda \circ F_1$ instead of $\lambda \circ \mathrm{id}_{F_1} \colon G_1 \circ F_1 \to G_2 \circ F_1$.

(ii) Consider three functors $F_1, F_2, F_3 \colon \mathcal{C} \to \mathcal{C}'$, and morphisms of functors $\theta \colon F_1 \to F_2$ and $\lambda \colon F_2 \to F_3$. The morphism of functors $\lambda \circ \theta \colon F_1 \to F_3$ is naturally defined. It is visualized by the diagram

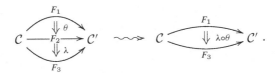

Remark 1.3.5. Consider the category $\mathcal{U}\text{-}\mathbf{Cat}$ whose objects are the small $\mathcal{U}$-categories and the morphisms are the functors of such categories, that is,

$$\mathrm{Hom}_{\mathcal{U}-\mathbf{Cat}}(\mathcal{C}, \mathcal{C}') = \mathrm{Fct}(\mathcal{C}, \mathcal{C}') \,.$$

Since $\mathrm{Hom}_{\mathcal{U}-\mathbf{Cat}}(\mathcal{C}, \mathcal{C}')$ is not only a set, but is in fact a category, $\mathcal{U}\text{-}\mathbf{Cat}$ is not only a category, it has a structure of a so-called 2-category. We shall not develop the theory of 2-categories in this book.

Remark 1.3.6. We shall sometimes use diagrams where symbols represent categories and arrows represent functors. In such a case we shall abusively say that the diagram commutes if it commutes up to isomorphisms of functors, or better, we shall say that the diagram quasi-commutes or is *quasi-commutative*.

Notation 1.3.7. Let $\mathcal{C}$ be a category. We denote by $\mathrm{id}_{\mathcal{C}} \colon \mathcal{C} \to \mathcal{C}$ the identity functor of $\mathcal{C}$. We denote by $\mathrm{End}\,(\mathrm{id}_{\mathcal{C}})$ the set of endomorphisms of the identity functor $\mathrm{id}_{\mathcal{C}} \colon \mathcal{C} \to \mathcal{C}$, that is,

$$\mathrm{End}\,(\mathrm{id}_{\mathcal{C}}) = \mathrm{Hom}\,_{\mathrm{Fct}(\mathcal{C},\mathcal{C})}(\mathrm{id}_{\mathcal{C}}, \mathrm{id}_{\mathcal{C}}) \,.$$

We denote by $\mathrm{Aut}(\mathrm{id}_{\mathcal{C}})$ the subset of $\mathrm{End}\,(\mathrm{id}_{\mathcal{C}})$ consisting of isomorphisms from $\mathrm{id}_{\mathcal{C}}$ to $\mathrm{id}_{\mathcal{C}}$.

Clearly, $\mathrm{End}\,(\mathrm{id}_{\mathcal{C}})$ is a monoid and $\mathrm{Aut}(\mathrm{id}_{\mathcal{C}})$ is a group.

Lemma 1.3.8. *The composition law on* $\mathrm{End}\,(\mathrm{id}_{\mathcal{C}})$ *is commutative.*

Proof. Let θ and λ belong to $\mathrm{End}\,(\mathrm{id}_{\mathcal{C}})$. Let $X \in \mathcal{C}$ and consider the morphism $\lambda_X \colon X \to X$. The desired assertion follows from the commutativity of the diagram (1.3.1) with $F_1 = F_2 = \mathrm{id}_{\mathcal{C}}$, $Y = X$ and $f = \lambda_X$, because $F_1(f) = F_2(f) = \lambda_X$.   q.e.d.

Consider three categories $\mathcal{I}, \mathcal{C}, \mathcal{C}'$ and a functor

$$\varphi \colon \mathcal{C} \to \mathcal{C}' \,.$$

Then φ defines a functor

(1.3.4) $\qquad\qquad \varphi \circ \colon \mathrm{Fct}(\mathcal{I}, \mathcal{C}) \to \mathrm{Fct}(\mathcal{I}, \mathcal{C}'), \quad F \mapsto \varphi \circ F$.

We shall use the lemma below, whose proof is obvious and left to the reader.

Lemma 1.3.9. *If φ is faithful (resp. fully faithful), then so is the functor $\varphi \circ$ in (1.3.4).*

We have the notion of an isomorphism of categories. A functor $F \colon \mathcal{C} \to \mathcal{C}'$ is an isomorphism of categories if there exists $G \colon \mathcal{C}' \to \mathcal{C}$ such that $G \circ F(X) = X$ and $F \circ G(Y) = Y$ for all $X \in \mathcal{C}$, all $Y \in \mathcal{C}'$, and similarly for the morphisms. In practice, such a situation almost never appears and there is an important weaker notion that we introduce now.

Definition 1.3.10. *A functor $F \colon \mathcal{C} \to \mathcal{C}'$ is an equivalence of categories if there exist $G \colon \mathcal{C}' \to \mathcal{C}$ and isomorphisms of functors $\alpha \colon G \circ F \xrightarrow{\sim} \mathrm{id}_{\mathcal{C}}$, $\beta \colon F \circ G \xrightarrow{\sim} \mathrm{id}_{\mathcal{C}'}$. In such a situation, we write $F \colon \mathcal{C} \xrightarrow{\sim} \mathcal{C}'$, we say that F and G are quasi-inverse to each other and we say that G is a quasi-inverse to F. (See Exercise 1.16.)*

Lemma 1.3.11. *Consider a functor $F \colon \mathcal{C} \to \mathcal{C}'$ and a full subcategory $\mathcal{C}_0'$ of $\mathcal{C}'$ such that for each $X \in \mathcal{C}$, there exist $Y \in \mathcal{C}_0'$ and an isomorphism $F(X) \simeq Y$. Denote by ι' the embedding $\mathcal{C}_0' \to \mathcal{C}'$. Then there exist a functor $F_0 \colon \mathcal{C} \to \mathcal{C}_0'$ and an isomorphism of functors $\theta_0 \colon F \xrightarrow{\sim} \iota' \circ F_0$. Moreover, F_0 is unique up to unique isomorphism. More precisely, given another isomorphism $\theta_1 \colon F \xrightarrow{\sim} \iota' \circ F_1$, there exists a unique isomorphism of functors $\theta \colon F_1 \xrightarrow{\sim} F_0$ such that $\theta_0 = (\iota' \circ \theta) \circ \theta_1$.*

Proof. Using Zorn's Lemma, for each $X \in \mathcal{C}$, choose $Y \in \mathcal{C}_0'$ and an isomorphism $\varphi_X \colon Y \xrightarrow{\sim} F(X)$, and set $F_0(X) = Y$. If $f \colon X \to X'$ is a morphism in $\mathcal{C}$, define $F_0(f) \colon F_0(X) \to F_0(X')$ as the composition $F_0(X) \xrightarrow[\varphi_X]{\sim} F(X) \xrightarrow{F(f)} F(X') \xleftarrow[\varphi_{X'}]{\sim} F_0(X')$. The fact that F_0 commutes with the composition of morphisms is visualized by

$$
\begin{array}{ccccc}
F(X) & \xrightarrow{\ F(f)\ } & F(X') & \xrightarrow{\ F(g)\ } & F(X'') \\[2pt]
\varphi_X \Big\uparrow \wr & & \varphi_{X'} \Big\uparrow \wr & & \varphi_{X''} \Big\uparrow \wr \\[2pt]
Y = F_0(X) & \xrightarrow{\ F_0(f)\ } & Y' = F_0(X') & \xrightarrow{\ F_0(g)\ } & Y'' = F_0(X'') .
\end{array}
$$

The other assertions are obvious. $\qquad$ q.e.d.

Lemma 1.3.12. *Let $\mathcal{C}$ be a category. There exists a full subcategory $\mathcal{C}_0$ such that the embedding functor $\iota \colon \mathcal{C}_0 \to \mathcal{C}$ is an equivalence of categories and $\mathcal{C}_0$ has the property that any two isomorphic objects in $\mathcal{C}_0$ are equal.*

Proof. In the set $\mathrm{Ob}(\mathcal{C})$, consider the equivalence relation $X \sim Y$ if and only if there exists an isomorphism $X \simeq Y$. By Zorn's lemma, we may pick an object in each of the equivalence classes of $\sim$. The full subcategory $\mathcal{C}_0$ of $\mathcal{C}$ consisting of such objects has the required properties. Indeed, denote by $\iota \colon \mathcal{C}_0 \to \mathcal{C}$ the embedding functor. Applying Lemma 1.3.11 to $\mathrm{id}_{\mathcal{C}} \colon \mathcal{C} \to \mathcal{C}$, there exists a functor $F_0 \colon \mathcal{C} \to \mathcal{C}_0$ such that $\iota \circ F_0$ is isomorphic to $\mathrm{id}_{\mathcal{C}}$. Since

$$\iota \circ (F_0 \circ \iota) = (\iota \circ F_0) \circ \iota \simeq \mathrm{id}_{\mathcal{C}} \circ \iota \simeq \iota \simeq \iota \circ \mathrm{id}_{\mathcal{C}_0}$$

and ι is fully faithful, $F_0 \circ \iota$ is isomorphic to $\mathrm{id}_{\mathcal{C}_0}$.   q.e.d.

Proposition 1.3.13. *A functor $F \colon \mathcal{C} \to \mathcal{C}'$ is an equivalence of categories if and only if F is fully faithful and essentially surjective.*

Proof. The necessity of the condition is clear. Let us prove the converse statement. By Lemma 1.3.12, there exists a full subcategory $\mathcal{C}_0$ of $\mathcal{C}$ such that $\iota \colon \mathcal{C}_0 \to \mathcal{C}$ is an equivalence and if two objects of $\mathcal{C}_0$ are isomorphic, then they are equal. Let κ be a quasi-inverse of ι. We proceed similarly with $\mathcal{C}'$, and construct $\mathcal{C}'_0, \iota'$ and κ'. Then the composition of functors

$$\kappa' \circ F \circ \iota \colon \mathcal{C}_0 \to \mathcal{C}'_0$$

is an isomorphism. Denote by K its inverse and set $G = \iota \circ K \circ \kappa'$. Clearly, G is a quasi-inverse to F.   q.e.d.

Corollary 1.3.14. *Let $F \colon \mathcal{C} \to \mathcal{C}'$ be a fully faithful functor. Then there exist a full subcategory $\mathcal{C}'_0$ of $\mathcal{C}'$ and an equivalence of categories $F' \colon \mathcal{C} \xrightarrow{\sim} \mathcal{C}'_0$ such that F is isomorphic to $\iota' \circ F'$, where $\iota' \colon \mathcal{C}'_0 \to \mathcal{C}'$ is the embedding functor.*

Proof. Define $\mathcal{C}'_0$ as the full subcategory of $\mathcal{C}'$ whose objects are the image by F of the objects of $\mathcal{C}$ and apply Proposition 1.3.13.   q.e.d.

Examples 1.3.15. (i) Let k be a field and let $\mathcal{C}$ denote the category defined by $\mathrm{Ob}(\mathcal{C}) = \mathbb{N}$ and $\mathrm{Hom}_{\mathcal{C}}(n, m) = M_{m,n}(k)$, the space of matrices of type (m, n) with entries in k. The composition of morphisms in $\mathcal{C}$ is given by the composition of matrices. Define the functor $F \colon \mathcal{C} \to \mathrm{Mod}^{\mathrm{f}}(k)$ as follows. Set $F(n) = k^n$, and if A is a matrix of type (m, n), let $F(A)$ be the linear map from k^n to k^m associated with A. Then F is an equivalence of categories.

(ii) Let $\mathcal{C}$ and $\mathcal{C}'$ be two categories. There is an isomorphism of categories:

$$(1.3.5) \qquad \mathrm{Fct}(\mathcal{C}, \mathcal{C}')^{\mathrm{op}} \simeq \mathrm{Fct}(\mathcal{C}^{\mathrm{op}}, \mathcal{C}'^{\mathrm{op}}), \quad F \mapsto \mathrm{op} \circ F \circ \mathrm{op}\,.$$

(iii) Consider a family $\{\mathcal{C}_i\}_{i \in I}$ of categories indexed by a small set I. If I is the empty set, then $\prod_{i \in I} \mathcal{C}_i$ is equivalent to the category $\mathbf{Pt}$ and $\bigsqcup_{i \in I} \mathcal{C}_i$ is equivalent to the empty category.

Definition 1.3.16. *A category is* essentially $\mathcal{U}$-small *if it is equivalent to a $\mathcal{U}$-small category.*

Remark that $\mathcal{C}$ is essentially $\mathcal{U}$-small if and only if $\mathcal{C}$ is a $\mathcal{U}$-category and there exists a $\mathcal{U}$-small subset S of $\mathrm{Ob}(\mathcal{C})$ such that any object of $\mathcal{C}$ is isomorphic to an object in S.

One shall be aware that if $F\colon \mathcal{C} \to \mathcal{C}'$ is faithful, it may not exist a sub-category $\mathcal{C}'_0$ of $\mathcal{C}'$ and an equivalence $F'\colon \mathcal{C} \xrightarrow{\sim} \mathcal{C}'_0$ such that F is isomorphic to $\iota' \circ F'$, where $\iota'\colon \mathcal{C}'_0 \to \mathcal{C}'$ is the embedding functor (see Exercise 1.18). That is the reason why we introduce Definition 1.3.17 below.

Definition 1.3.17. (i) *Let* $F\colon \mathcal{C} \to \mathcal{C}'$ *be a functor. We say that* F *is* half-full *if for any pair of objects* $X, Y \in \mathcal{C}$ *such that* $F(X)$ *and* $F(Y)$ *are isomorphic in* $\mathcal{C}'$, *there exists an isomorphism* $X \simeq Y$ *in* $\mathcal{C}$. *(We do not ask the isomorphism in* $\mathcal{C}'$ *to be the image by* F *of the isomorphism in* $\mathcal{C}$.)
(ii) *We say that a subcategory* $\mathcal{C}_0$ *of* $\mathcal{C}$ *is* half-full *if the embedding functor is half-full.*

Proposition 1.3.18. *Let* $F\colon \mathcal{C} \to \mathcal{C}'$ *be a faithful and half-full functor. Then there exists a subcategory* $\mathcal{C}'_0$ *of* $\mathcal{C}'$ *such that* $F(\mathrm{Ob}(\mathcal{C})) \subset \mathrm{Ob}(\mathcal{C}'_0)$, $F(\mathrm{Mor}(\mathcal{C})) \subset \mathrm{Mor}(\mathcal{C}'_0)$ *and* F *induces an equivalence of categories* $\mathcal{C} \simeq \mathcal{C}'_0$. *Moreover, the embedding functor* $\mathcal{C}'_0 \to \mathcal{C}'$ *is faithful and half-full.*

Proof. Let us define the category $\mathcal{C}'_0$ as follows:

$$\mathrm{Ob}(\mathcal{C}'_0) = \{F(X); X \in \mathrm{Ob}(\mathcal{C})\},$$
$$\mathrm{Hom}_{\mathcal{C}'_0}(F(X), F(Y)) = F(\mathrm{Hom}_{\mathcal{C}}(X, Y)) \subset \mathrm{Hom}_{\mathcal{C}'}(F(X), F(Y)).$$

It is immediately checked that the definition of $\mathrm{Hom}_{\mathcal{C}'_0}(F(X), F(Y))$ does not depend on the choice of X, Y, thanks to the hypothesis that F is half-full, and hence the family of morphisms in $\mathcal{C}'_0$ is closed by composition. By its construction, the functor $F\colon \mathcal{C} \to \mathcal{C}'_0$ is fully faithful and essentially surjective. It is thus an equivalence.   q.e.d.

1.4 The Yoneda Lemma

Convention 1.4.1. We start with a given universe $\mathcal{U}$, and do not mention it when unnecessary. In this book, *a category means a* $\mathcal{U}$-*category, small means* $\mathcal{U}$-*small, and* **Set** *denotes the category of* $\mathcal{U}$-*sets*, unless otherwise mentioned. However, some constructions force us to deal with a category which is not necessarily a $\mathcal{U}$-category. We call such a category a *big* category. If this has no implications for our purpose, we do not always mention it. Note that any category is $\mathcal{V}$-small for some universe $\mathcal{V}$.

Definition 1.4.2. *Let* $\mathcal{C}$ *be a* $\mathcal{U}$-*category. We define the big categories*

$$\mathcal{C}^\wedge_\mathcal{U} : \text{ the category of functors from } \mathcal{C}^{\mathrm{op}} \text{ to } \mathcal{U}\text{-}\mathbf{Set},$$
$$\mathcal{C}^\vee_\mathcal{U} : \text{ the category of functors from } \mathcal{C}^{\mathrm{op}} \text{ to } (\mathcal{U}\text{-}\mathbf{Set})^{\mathrm{op}},$$

and the functors

$$h_{\mathcal{C}} : \mathcal{C} \to \mathcal{C}_{\mathcal{U}}^{\wedge}, \quad X \mapsto \mathrm{Hom}_{\mathcal{C}}(\cdot, X),$$
$$k_{\mathcal{C}} : \mathcal{C} \to \mathcal{C}_{\mathcal{U}}^{\vee}, \quad X \mapsto \mathrm{Hom}_{\mathcal{C}}(X, \cdot).$$

Since $\mathrm{Hom}_{\mathcal{C}}(X, Y) \in \mathcal{U}$ for all $X, Y \in \mathcal{C}$, the functors $h_{\mathcal{C}}$ and $k_{\mathcal{C}}$ are well-defined. They are often called the "Yoneda functors". Hence

$$\mathcal{C}_{\mathcal{U}}^{\wedge} = \mathrm{Fct}(\mathcal{C}^{\mathrm{op}}, \mathcal{U}\text{-}\mathbf{Set}),$$
$$\mathcal{C}_{\mathcal{U}}^{\vee} = \mathrm{Fct}(\mathcal{C}^{\mathrm{op}}, \mathcal{U}\text{-}\mathbf{Set}^{\mathrm{op}}) \simeq \mathrm{Fct}(\mathcal{C}, \mathcal{U}\text{-}\mathbf{Set})^{\mathrm{op}}.$$

Note that $\mathcal{C}_{\mathcal{U}}^{\wedge}$ and $\mathcal{C}_{\mathcal{U}}^{\vee}$ are not $\mathcal{U}$-categories in general. If $\mathcal{C}$ is $\mathcal{U}$-small, then $\mathcal{C}_{\mathcal{U}}^{\wedge}$ and $\mathcal{C}_{\mathcal{U}}^{\vee}$ are $\mathcal{U}$-categories.

In the sequel, we shall write $\mathcal{C}^{\wedge}$ and $\mathcal{C}^{\vee}$ for short. By (1.3.5) there is a natural isomorphism

$$(1.4.1) \qquad \mathcal{C}^{\vee} \simeq \mathcal{C}^{\mathrm{op}\wedge\mathrm{op}}$$

and $\mathcal{C}^{\vee}$ is the opposite big category to the category of functors from $\mathcal{C}$ to $\mathbf{Set}$. Hence, for $X \in \mathcal{C}$, $k_{\mathcal{C}}(X) = (h_{\mathcal{C}^{\mathrm{op}}}(X^{\mathrm{op}}))^{\mathrm{op}}$.

The next result, although it is elementary, is crucial for the understanding of the rest of the book. In the sequel, we write $\mathbf{Set}$ for $\mathcal{U}$-$\mathbf{Set}$.

Proposition 1.4.3. [The Yoneda lemma]

(i) *For $A \in \mathcal{C}^{\wedge}$ and $X \in \mathcal{C}$, $\mathrm{Hom}_{\mathcal{C}^{\wedge}}(h_{\mathcal{C}}(X), A) \simeq A(X)$.*
(ii) *For $B \in \mathcal{C}^{\vee}$ and $X \in \mathcal{C}$, $\mathrm{Hom}_{\mathcal{C}^{\vee}}(B, k_{\mathcal{C}}(X)) \simeq B(X)$.*

Moreover, these isomorphisms are functorial with respect to X, A, B, that is, they define isomorphisms of functors from $\mathcal{C}^{\mathrm{op}} \times \mathcal{C}^{\wedge}$ to $\mathbf{Set}$ or from $\mathcal{C}^{\vee\mathrm{op}} \times \mathcal{C}$ to $\mathbf{Set}$.

Proof. By (1.4.1) is enough to prove one of the two statements. Let us prove (i).

The map $\varphi \colon \mathrm{Hom}_{\mathcal{C}^{\wedge}}(h_{\mathcal{C}}(X), A) \to A(X)$ is constructed by the chain of maps: $\mathrm{Hom}_{\mathcal{C}^{\wedge}}(h_{\mathcal{C}}(X), A) \to \mathrm{Hom}_{\mathbf{Set}}(\mathrm{Hom}_{\mathcal{C}}(X, X), A(X)) \to A(X)$, where the last map is associated with id_X.

To construct $\psi : A(X) \to \mathrm{Hom}_{\mathcal{C}^{\wedge}}(h_{\mathcal{C}}(X), A)$, it is enough to associate with $s \in A(X)$ and $Y \in \mathcal{C}$ a map $\psi(s)_Y \colon \mathrm{Hom}_{\mathcal{C}}(Y, X) \to A(Y)$. It is defined by the chain of maps $\mathrm{Hom}_{\mathcal{C}}(Y, X) \to \mathrm{Hom}_{\mathbf{Set}}(A(X), A(Y)) \to A(Y)$ where the last map is associated with $s \in A(X)$. Clearly, $\psi(s)$ satisfies (1.3.1).

It is easily checked that φ and ψ are inverse to each other.   q.e.d.

The next results will be of constant use.

Corollary 1.4.4. *The two functors $h_{\mathcal{C}}$ and $k_{\mathcal{C}}$ are fully faithful.*

Proof. For X and Y in $\mathcal{C}$, we have $\mathrm{Hom}_{\mathcal{C}^{\wedge}}(h_{\mathcal{C}}(X), h_{\mathcal{C}}(Y)) \simeq h_{\mathcal{C}}(Y)(X) = \mathrm{Hom}_{\mathcal{C}}(X, Y)$.   q.e.d.

Hence, it is possible to regard C as a full subcategory of either $C^\wedge$ or $C^\vee$.

Notation 1.4.5. By identifying $X \in C$ with $h_C(X) \in C^\wedge$, it is natural to set

(1.4.2) $$X(Y) = \mathrm{Hom}_C(Y, X).$$

Similarly, for A and B in $C^\wedge$, we shall sometimes write $A(B)$ instead of $\mathrm{Hom}_{C^\wedge}(B, A)$.

Corollary 1.4.6. *Let $F: C \to C'$ be a functor of U-categories and assume that C is U-small. For $A \in C'^\wedge$, the category C_A associated with $C \to C' \to C'^\wedge$ (see Definition 1.2.16) is U-small.*

Similarly, for $B \in C'^\vee$, the category C^B associated with $C \to C' \to C'^\vee$ is U-small.

Proof. By the Yoneda lemma, for a given $X \in C$, the family of morphisms $h_C \circ F(X) \to A$ is the set $A(F(X))$. Hence, C_A is the category of pairs (X, s) of $X \in C$ and $s \in A(F(X))$. If C is small, then the set $\bigsqcup_{X \in C} A(F(X))$ is small. The case of C^B is similar.   q.e.d.

Corollary 1.4.7. *Let C be a category, $f: X \to Y$ a morphism in C. Assume that for each $W \in C$, the morphism $\mathrm{Hom}_C(W, X) \xrightarrow{f \circ} \mathrm{Hom}_C(W, Y)$ (resp. $\mathrm{Hom}_C(Y, W) \xrightarrow{\circ f} \mathrm{Hom}_C(X, W)$) is an isomorphism. Then f is an isomorphism.*

Proof. By hypothesis, $h_C(f): h_C(X) \to h_C(Y)$ (resp. $k_C(f): k_C(Y) \to k_C(X)$) is an isomorphism. Hence, the result follows from the Yoneda lemma (Corollary 1.4.4).   q.e.d.

Definition 1.4.8. *A functor F from C^{op} to **Set** (resp. C to **Set**) is representable if there is an isomorphism $h_C(X) \xrightarrow{\sim} F$ (resp. $F \xrightarrow{\sim} k_C(X)$) for some $X \in C$. Such an object X is called a representative of F.*

It follows from Corollary 1.4.4 that the isomorphism $F \simeq h_C(X)$ (resp. $F \simeq k_C(X)$) determines X up to unique isomorphism.

Assume that $F \in C^\wedge$ is represented by $X_0 \in C$. Then $\mathrm{Hom}_{C^\wedge}(h_C(X_0), F) \simeq F(X_0)$ gives an element $s_0 \in F(X_0)$. Moreover, for any $Y \in C$ and $t \in F(Y)$, there exists a unique morphism $f: X_0 \to Y$ such that $t = F(f)(s_0)$. Conversely, for $X_0 \in C$ and $s_0 \in F(X_0)$, (X_0, s_0) defines a morphism $h_C(X_0) \to F$. If it is an isomorphism, that is, if the map $\mathrm{Hom}_C(Y, X_0) \to F(Y)$ given by $f \mapsto F(f)(s_0)$ is bijective for all $Y \in C$, then F is representable by X_0.

Corollary 1.4.9. *Let $F: C \to C'^\wedge$ be a functor. If $F(X)$ is isomorphic to an object of C' for any $X \in C$, then there exists a unique (up to unique isomorphism) functor $F_0: C \to C'$ such that $F \simeq h_{C'} \circ F_0$.*

Proof. This follows from Corollary 1.4.4 and Lemma 1.3.11.   q.e.d.

Proposition 1.4.10. *Let* $F \in \mathcal{C}^\wedge$. *Then* F *is representable if and only if* $\mathcal{C}_F$ *has a terminal object.*

Proof. Let $(X, s) \in \mathcal{C}_F$, that is, $X \in \mathcal{C}$ and $s \in F(X)$. For any $(Y, t) \in \mathcal{C}_F$,

$$\mathrm{Hom}_{\mathcal{C}_F}((Y, t), (X, s)) \simeq \{u \in \mathrm{Hom}_\mathcal{C}(Y, X); F(u)(s) = t\}\,.$$

Hence, (X, s) is a terminal object of $\mathcal{C}_F$ if and only if $\mathrm{Hom}_{\mathcal{C}_F}((Y, t), (X, s)) \simeq \{\mathrm{pt}\}$ for any $Y \in \mathcal{C}$ and $t \in F(Y)$, and this condition is equivalent to saying that the map $\mathrm{Hom}_\mathcal{C}(Y, X) \to F(X)$ given by $u \mapsto F(u)(s)$ is bijective for any $Y \in \mathcal{C}$. $\hspace{4cm}$ q.e.d.

Representable functors is a categorical language to deal with universal problems. Let us illustrate this by an example.

Example 1.4.11. Consider the situation of Example 1.2.9. Denote by $B(N \times M, L)$ the set of bilinear maps from $N \times M$ to L. Then the functor $F \colon L \mapsto B(N \times M, L)$ is representable by $N \otimes_R M$, since $F(L) = B(N \times M, L) \simeq \mathrm{Hom}_\mathbb{Z}(N \otimes_R M, L)$.

If a functor $F \colon \mathcal{C} \to \mathbf{Set}$ takes its values in a category defined by some algebraic structure (we do not intend to give a precise meaning to such a sentence) and if this functor is representable by some object X, then X will be endowed with morphisms which will mimic this algebraic structure. For example if F takes its values in the category **Group** of groups, then X will be endowed with a structure of a "group-object". This notion will be discussed in Sect. 8.1.

We shall see in Chap. 2 that the notion of representable functor allows us to define projective and inductive limits in categories.

We conclude this section with a technical result which shall be useful in various parts of this book.

Lemma 1.4.12. *Let* $\mathcal{C}$ *be a category and let* $A \in \mathcal{C}^\wedge$. *There is a natural equivalence of big categories* $(\mathcal{C}_A)^\wedge \simeq (\mathcal{C}^\wedge)_A$ *such that the diagram of big categories and functors below quasi-commutes:*

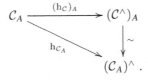

Proof. (i) We construct a functor $\lambda \colon (\mathcal{C}^\wedge)_A \to (\mathcal{C}_A)^\wedge$ as follows. Let $G \in \mathcal{C}^\wedge$ and $t \colon G \to A$. For $(X \overset{s}{\to} A) \in \mathcal{C}_A$, we set

$$\left(\lambda(G \overset{t}{\to} A)\right)(X \overset{s}{\to} A) = \mathrm{Hom}_{(\mathcal{C}^\wedge)_A}((X, s), (G, t))$$
$$= \{u \in G(X); t_X(u) = s \in A(X)\}\,.$$

(ii) We construct a functor $\mu\colon (\mathcal{C}_A)^\wedge \to (\mathcal{C}^\wedge)_A$ as follows. Let $F \in (\mathcal{C}_A)^\wedge$ and $X \in \mathcal{C}$. Set

$$\mu(F)(X) = \{(x,s); s \in A(X), x \in F(X \xrightarrow{s} A)\}$$

and define $(\mu(F) \to A) \in (\mathcal{C}^\wedge)_A$ by

$$\mu(F)(X) \ni (x,s) \mapsto s \in A(X) \text{ for } X \in \mathcal{C} .$$

(iii) It is easily checked that the functors λ and μ are quasi-inverse to each other.   q.e.d.

Remark 1.4.13. One shall be aware that the category $\mathcal{C}^\wedge$ associated with the $\mathcal{U}$-category $\mathcal{C}$ depends on the universe $\mathcal{U}$. Let $\mathcal{V}$ be another universe with $\mathcal{U} \subset \mathcal{V}$. Since the functor from $\mathcal{U}$-**Set** to $\mathcal{V}$-**Set** is fully faithful, it follows from Lemma 1.3.9 that the functor

(1.4.3) $\iota_{\mathcal{V},\mathcal{U}}\colon \mathcal{C}_{\mathcal{U}}^\wedge \to \mathcal{C}_{\mathcal{V}}^\wedge$

is fully faithful.

Hence $F \in \mathcal{C}_{\mathcal{U}}^\wedge$ is representable if and only if $\iota_{\mathcal{V},\mathcal{U}}(F)$ is representable.

1.5 Adjoint Functors

Consider a functor $F\colon \mathcal{C} \to \mathcal{C}'$. It defines a functor

(1.5.1) $F_*\colon \mathcal{C}'^\wedge \to \mathcal{C}^\wedge$,

$$F_*(B)(X) = B(F(X)) \text{ for } B \in \mathcal{C}'^\wedge, X \in \mathcal{C} .$$

If there is no risk of confusion, we still denote by $F_*\colon \mathcal{C}' \to \mathcal{C}^\wedge$ the restriction of F_* to $\mathcal{C}'$, that is, we write F_* instead of $F_* \circ h_{\mathcal{C}'}$. Hence,

$$F_*(Y)(X) = h_{\mathcal{C}'}(Y)(F(X)) = \mathrm{Hom}_{\mathcal{C}'}(F(X), Y) .$$

In other words, F_* is the functor

$$F_*\colon \mathcal{C}' \to \mathcal{C}^\wedge, \quad Y \mapsto \mathrm{Hom}_{\mathcal{C}'}(F(\,\cdot\,), Y) .$$

Applying Corollary 1.4.9, we obtain:

Theorem 1.5.1. *Assume that the functor $F_*(Y)$ is representable for each $Y \in \mathcal{C}'$. Then there exists a functor $G\colon \mathcal{C}' \to \mathcal{C}$ such that $F_* \simeq h_{\mathcal{C}} \circ G$, and the functor G is unique up to unique isomorphism.*

The uniqueness of G means the following. Consider two isomorphisms of functors $\theta_0\colon F_* \xrightarrow{\sim} h_{\mathcal{C}} \circ G_0$ and $\theta_1\colon F_* \xrightarrow{\sim} h_{\mathcal{C}} \circ G_1$. Then there exists a unique isomorphism of functors $\theta\colon G_0 \to G_1$ such that $\theta_1 = (h_{\mathcal{C}} \circ \theta) \circ \theta_0$.

Proof. Applying Lemma 1.3.11 to the functor $F_* \colon \mathcal{C}' \to \mathcal{C}^\wedge$ and the full subcategory $\mathcal{C}$ of $\mathcal{C}^\wedge$, we get a functor $G \colon \mathcal{C}' \to \mathcal{C}$ such that $F_* \xrightarrow{\sim} h_{\mathcal{C}} \circ G$, and this functor G is unique up to unique isomorphism, again by this lemma.   q.e.d.

In the situation of Theorem 1.5.1, we get:

$$(1.5.2) \qquad \mathrm{Hom}_{\mathcal{C}}(X, G(Y)) \simeq F_*(Y)(X) \simeq \mathrm{Hom}_{\mathcal{C}'}(F(X), Y) \,.$$

Consider the functor

$$G_* \colon \mathcal{C} \to \mathcal{C}'^\vee, \quad X \mapsto \mathrm{Hom}_{\mathcal{C}}(X, G(\bullet)) \,.$$

Then for each $X \in \mathcal{C}$, $G_*(X)$ is representable by $F(X)$.

For the reader's convenience, we change our notations, replacing F with L and G with R.

Definition 1.5.2. *Let $L \colon \mathcal{C} \to \mathcal{C}'$ and $R \colon \mathcal{C}' \to \mathcal{C}$ be two functors. The pair (L, R) is a pair of* adjoint *functors, or L is a* left adjoint *functor to R, or R is a* right adjoint *functor to L, if there exists an isomorphism of bifunctors from $\mathcal{C}^{\mathrm{op}} \times \mathcal{C}'$ to* **Set***:*

$$(1.5.3) \qquad \mathrm{Hom}_{\mathcal{C}'}(L(\bullet), \bullet) \simeq \mathrm{Hom}_{\mathcal{C}}(\bullet, R(\bullet)) \,.$$

We call the isomorphism in (1.5.3) the *adjunction isomorphism* .

With the language of adjoint functors, we can reformulate Theorem 1.5.1 as follows.

Theorem 1.5.3. *Let $L \colon \mathcal{C} \to \mathcal{C}'$ and $R \colon \mathcal{C}' \to \mathcal{C}$ be two functors. If L (resp. R) admits a right (resp. left) adjoint functor, this adjoint functor is unique up to unique isomorphism. Moreover, a functor L admits a right adjoint if and only if the functor $\mathrm{Hom}_{\mathcal{C}'}(L(\bullet), Y)$ is representable for any $Y \in \mathcal{C}'$.*

Let $X \in \mathcal{C}$. Applying the isomorphism (1.5.3) with X and $L(X)$, we find the isomorphism $\mathrm{Hom}_{\mathcal{C}'}(L(X), L(X)) \simeq \mathrm{Hom}_{\mathcal{C}}(X, R \circ L(X))$ and the identity of $L(X)$ defines a morphism $X \to R \circ L(X)$. Similarly, we construct $L \circ R(Y) \to Y$ and these morphisms are functorial with respect to X and Y. Hence, we have constructed morphisms of functors

$$(1.5.4) \qquad\qquad\qquad \varepsilon \colon \mathrm{id}_{\mathcal{C}} \to R \circ L \,,$$
$$(1.5.5) \qquad\qquad\qquad \eta \colon L \circ R \to \mathrm{id}_{\mathcal{C}'} \,.$$

By this construction, we have commutative diagrams for $Y, Y' \in \mathcal{C}'$ and $X, X' \in \mathcal{C}$

$$(1.5.6) \qquad \mathrm{Hom}_{\mathcal{C}'}(Y, Y') \xrightarrow{\ R\ } \mathrm{Hom}_{\mathcal{C}}(R(Y), R(Y'))$$

$$\eta_Y \searrow \qquad \sim \downarrow ad$$

$$\mathrm{Hom}_{\mathcal{C}'}(LR(Y), Y'),$$

(1.5.7)

$$\mathrm{Hom}_{\mathcal{C}}(X, X') \xrightarrow{\quad L \quad} \mathrm{Hom}_{\mathcal{C}'}(L(X), L(X'))$$

with a diagonal arrow labeled $\varepsilon_{X'}$ from $\mathrm{Hom}_{\mathcal{C}}(X, X')$ to $\mathrm{Hom}_{\mathcal{C}}(X, RL(X'))$, and a vertical arrow labeled $\sim$ with ad down to

$$\mathrm{Hom}_{\mathcal{C}}(X, RL(X')) \ .$$

It is easily checked that

(1.5.8) $(\eta \circ L) \circ (L \circ \varepsilon): L \to L \circ R \circ L \to L$ is id_L ,

(1.5.9) $(R \circ \eta) \circ (\varepsilon \circ R): R \to R \circ L \circ R \to R$ is id_R .

Proposition 1.5.4. *Let $L: \mathcal{C} \to \mathcal{C}'$ and $R: \mathcal{C}' \to \mathcal{C}$ be two functors and let ε and η be two morphisms of functors as in (1.5.4) and (1.5.5) satisfying (1.5.8) and (1.5.9). Then (L, R) is a pair of adjoint functors.*

Proof. We leave to the reader to check that the two composite morphisms

$$\mathrm{Hom}_{\mathcal{C}'}(L(X), Y) \xrightarrow{R} \mathrm{Hom}_{\mathcal{C}}(R \circ L(X), R(Y)) \xrightarrow{\varepsilon_X} \mathrm{Hom}_{\mathcal{C}}(X, R(Y))$$

and

$$\mathrm{Hom}_{\mathcal{C}}(X, R(Y)) \xrightarrow{L} \mathrm{Hom}_{\mathcal{C}'}(L(X), L \circ R(Y)) \xrightarrow{\eta_Y} \mathrm{Hom}_{\mathcal{C}'}(L(X), Y)$$

are inverse to each other.   q.e.d.

In the situation of Proposition 1.5.4, we say that $\langle L, R, \eta, \varepsilon \rangle$ is *an adjunction* and that ε and η are the *adjunction morphisms*.

Proposition 1.5.5. *Let $\mathcal{C}, \mathcal{C}'$ and $\mathcal{C}''$ be categories and let $\mathcal{C} \underset{R}{\overset{L}{\rightleftarrows}} \mathcal{C}' \underset{R'}{\overset{L'}{\rightleftarrows}} \mathcal{C}''$ be functors. If (L, R) and (L', R') are pairs of adjoint functors, then $(L' \circ L, R \circ R')$ is a pair of adjoint functors.*

Proof. For $X \in \mathcal{C}$ and $Y \in \mathcal{C}''$, we have functorial isomorphisms:

$$\mathrm{Hom}_{\mathcal{C}''}(L'L(X), Y) \simeq \mathrm{Hom}_{\mathcal{C}'}(L(X), R'(Y))$$
$$\simeq \mathrm{Hom}_{\mathcal{C}}(X, RR'(Y)) \ .$$

q.e.d.

Proposition 1.5.6. *Let $\langle L, R, \eta, \varepsilon \rangle$ be an adjunction.*

(i) *The functor R is fully faithful if and only if the morphism $\eta: L \circ R \to \mathrm{id}_{\mathcal{C}'}$ is an isomorphism.*

(ii) *The functor L is fully faithful if and only if the morphism $\varepsilon: \mathrm{id}_{\mathcal{C}} \to R \circ L$ is an isomorphism.*

(iii) *The conditions below are equivalent*

    (a) *L is an equivalence of categories,*

(b) R *is an equivalence of categories,*
(c) L *and* R *are fully faithful.*
In such a case, L *and* R *are quasi-inverse one to each other, and* (1.5.4), (1.5.5) *are isomorphisms.*

Proof. (i) Let $Y, Y' \in C$ and consider the diagram (1.5.6). We find that the map $\mathrm{Hom}_{C'}(Y, Y') \to \mathrm{Hom}_{C}(R(Y), R(Y'))$ is bijective if and only if the map $\mathrm{Hom}_{C'}(Y, Y') \to \mathrm{Hom}_{C'}(L \circ R(Y), Y')$ is bijective. Therefore R is fully faithful if and only if $L \circ R(Y) \to Y$ is an isomorphism for all Y, and this proves (i). (ii) is dual, and (iii) follows immediately from (i) and (ii).   q.e.d.

Remark 1.5.7. If $F : C \to C'$ is an equivalence of categories and if G is a quasi-inverse to F, then G is both a right and a left adjoint to F.

Examples 1.5.8. (i) For $X, Y, Z \in \mathbf{Set}$, there is a natural isomorphism

$$\mathrm{Hom}_{\mathbf{Set}}(X \times Y, Z) \simeq \mathrm{Hom}_{\mathbf{Set}}(X, \mathrm{Hom}_{\mathbf{Set}}(Y, Z))$$

and this isomorphism is functorial with respect to X, Y, Z. Hence the functors $\cdot \times Y$ and $\mathrm{Hom}_{\mathbf{Set}}(Y, \cdot)$ are adjoint.
(ii) Let R be a k-algebra (see Notation 1.1.4). Let $K \in \mathrm{Mod}(k)$ and $M, N \in \mathrm{Mod}(R)$. The formula:

$$\mathrm{Hom}_R(N \otimes_k K, M) \simeq \mathrm{Hom}_R(N, \mathrm{Hom}_k(K, M))$$

tells us that the functors $\cdot \otimes_k K$ and $\mathrm{Hom}_k(K, \cdot)$ from $\mathrm{Mod}(R)$ to $\mathrm{Mod}(R)$ are adjoint.
In the preceding situation, denote by $for \colon \mathrm{Mod}(R) \to \mathrm{Mod}(k)$ the forgetful functor which associates the underlying k-module to an R-module M. Applying the above formula with $N = R$, we get

$$\mathrm{Hom}_R(R \otimes_k K, M) \simeq \mathrm{Hom}_k(K, for(M)) \ .$$

Hence, the functor $R \otimes_k \cdot$ (extension of scalars) is a left adjoint to for.
    Similarly, the functor $\mathrm{Hom}_k(R, \cdot) \colon \mathrm{Mod}(k) \to \mathrm{Mod}(R)$ is a right adjoint to for.

Exercises

Exercise 1.1. Let C be a category which has an initial object $\emptyset_C$ and a terminal object pt_C. Prove that if $\mathrm{Hom}_C(\mathrm{pt}_C, \emptyset_C)$ is not empty, then $\mathrm{pt}_C \simeq \emptyset_C$.

Exercise 1.2. Prove that the categories $\mathbf{Set}$ and $\mathbf{Set}^{\mathrm{op}}$ are not equivalent. (Hint: any morphism $X \to \emptyset$ is an isomorphism in $\mathbf{Set}$.)

Exercise 1.3. Let C be a category such that for any $X, Y \in \mathrm{Ob}(C)$, the set $\mathrm{Hom}_C(X, Y)$ has at most one element. Prove that C is equivalent to the category associated with an ordered set.

Exercise 1.4. (i) Prove that a morphism f in the category **Set** is a monomorphism (resp. an epimorphism) if and only if it is injective (resp. surjective).
(ii) Prove that the morphism $\mathbb{Z} \to \mathbb{Q}$ is a monomorphism and an epimorphism in the category **Ring** of rings belonging to $\mathcal{U}$ and morphisms of rings.
(iii) Prove that $\mathbb{Z}$ is an initial object and $\{0\}$ is a terminal object in the category **Ring**.

Exercise 1.5. (i) Let $\mathcal{C}$ be a non-empty category such that for any $X, Y \in \mathcal{C}$, X and Y are isomorphic. Let us choose $X \in \mathcal{C}$ and set $M = \mathrm{Hom}_{\mathcal{C}}(X, X)$. Prove that $\mathcal{C}$ is equivalent to the category associated with the monoid M.
(ii) Let $\mathcal{C}$ be a connected groupoid. Prove that $\mathcal{C}$ is equivalent to the category associated with a group.

Exercise 1.6. Let $\mathcal{C}$ be a category and let $X \in \mathcal{C}$. Prove that the full subcategory of $\mathcal{C}_X$ consisting of monomorphisms is equivalent to the category associated with the ordered set of subobjects of X.

Exercise 1.7. Let $\mathcal{C}$ be a category and let $f \colon X \to Y$ and $g \colon Y \to Z$ be morphisms in $\mathcal{C}$. Assume that $g \circ f$ is an isomorphism and g is a monomorphism. Prove that f and g are isomorphisms.

Exercise 1.8. Let $\mathcal{C}$ be a category with a zero object denoted by 0 and let $X \in \mathcal{C}$. Prove that if $\mathrm{id}_X = 0$ (*i.e.*, id_X is the composition $X \to 0 \to X$) then $X \simeq 0$.

Exercise 1.9. Let $F \colon \mathcal{C} \to \mathcal{C}'$ be an equivalence of categories and let G be a quasi-inverse. Let $H \colon \mathcal{C} \to$ **Set** be a representable functor, X a representative. Prove that $H \circ G$ is representable by $F(X)$.

Exercise 1.10. Let $F \colon \mathcal{C} \to \mathcal{C}'$ be a functor. Prove that F has a right adjoint if and only if the category $\mathcal{C}_Y$ has a terminal object for any $Y \in \mathcal{C}'$.

Exercise 1.11. Prove that the category $\mathcal{C}$ is equivalent to the opposite category $\mathcal{C}^{\mathrm{op}}$ in the following cases:

(a) $\mathcal{C}$ is the category of finite abelian groups,
(b) $\mathcal{C}$ is the category **Rel** of relations (see Example 1.2.4 (ii)).

Exercise 1.12. (i) Let $\mathcal{C} = \emptyset$ be the empty category. Prove that $\mathcal{C}^\wedge = \mathbf{Pt}$ (see Notation 1.2.8).
(ii) Let $\mathcal{C} = \mathbf{Pt}$. Prove that $\mathcal{C}^\wedge \simeq$ **Set**.

Exercise 1.13. Let $\mathcal{C}$ be a category.
(i) Prove that the terminal object $\mathrm{pt}_{\mathcal{C}^\wedge}$ of $\mathcal{C}^\wedge$ is the constant functor with values $\{\mathrm{pt}\} \in$ **Set** and that the initial object $\emptyset_{\mathcal{C}^\wedge}$ of $\mathcal{C}^\wedge$ is the constant functor with values $\emptyset \in$ **Set**.
(ii) Prove that $Z \in \mathcal{C}$ is a terminal object of $\mathcal{C}$ if and only if $h_{\mathcal{C}}(Z)$ is a terminal object of $\mathcal{C}^\wedge$.

Exercise 1.14. Let $F: \mathcal{C} \to \mathcal{C}'$ be a functor, and assume that F admits a right adjoint R and a left adjoint L. Prove that R is fully faithful if and only if L is fully faithful.
(Hint: use Proposition 1.5.6 with the morphisms of functors $\varepsilon: \mathrm{id} \to FL$, $\varepsilon': \mathrm{id} \to RF$, $\eta: LF \to \mathrm{id}$ and $\eta': FR \to \mathrm{id}$. Then consider the commutative diagram below.)

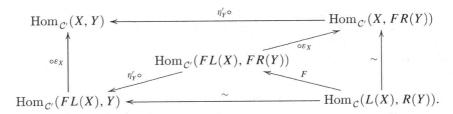

Exercise 1.15. Let $F: \mathcal{C} \to \mathcal{C}'$ be a fully faithful functor, let $G: \mathcal{C}' \to \mathcal{C}$ be a functor and let $\varepsilon: \mathrm{id}_{\mathcal{C}'} \to F \circ G$ be a morphism of functors. Assume that $\varepsilon \circ F: F \to F \circ G \circ F$ and $G \circ \varepsilon: G \to G \circ F \circ G$ are isomorphisms. Prove that G is left adjoint to F.

Exercise 1.16. Assume that $F: \mathcal{C} \to \mathcal{C}'$ and $G: \mathcal{C}' \to \mathcal{C}$ are equivalences of categories quasi-inverse to each other. Prove that there are isomorphisms of functors $\alpha: G \circ F \xrightarrow{\sim} \mathrm{id}_{\mathcal{C}}$ and $\beta: F \circ G \xrightarrow{\sim} \mathrm{id}_{\mathcal{C}'}$ such that $F \circ \alpha = \beta \circ F$ and $\alpha \circ G = G \circ \beta$, that is, $F(\alpha_X) = \beta_{F(X)}$ in $\mathrm{Hom}_{\mathcal{C}'}(F \circ G \circ F(X), F(X))$, and $\alpha_{G(Y)} = G(\beta_Y)$ in $\mathrm{Hom}_{\mathcal{C}}(G \circ F \circ G(Y), G(Y))$.

Exercise 1.17. Let $\mathcal{C}$ be a category and let S be a set. Consider the constant functor $\Delta_S: \mathcal{C}^{\mathrm{op}} \to \mathbf{Set}$ with values S (see Notations 1.2.15). Prove that if Δ_S is representable by $Z \in \mathcal{C}$, then $S \simeq \{\mathrm{pt}\}$ and Z is a terminal object in $\mathcal{C}$.

Exercise 1.18. Let $\mathcal{C}$ be a category and S a non empty set. Define the category $\widetilde{S}$ by setting $\mathrm{Ob}(\widetilde{S}) = S$ and $\mathrm{Hom}_{\widetilde{S}}(a, b) = \{\mathrm{pt}\}$ for any $a, b \in S$.
(i) Prove that the functor $\theta: \mathcal{C} \times \widetilde{S} \to \mathcal{C}$, $(X, a) \mapsto X$ is an equivalence.
(ii) Let $\mathbf{Arr}$ be the category $\bullet \to \bullet$ (see Notations 1.2.8 (iii)). Let $\varphi: \mathbf{Arr} \to \mathbf{Pr}$ be the natural functor. Prove that φ is faithful but there exists no subcategory of $\mathbf{Pr}$ equivalent to $\mathbf{Arr}$.
(iii) Let $F: \mathcal{C}' \to \mathcal{C}$ be a faithful functor. Prove that there exist a non empty set S, a subcategory $\mathcal{C}_0$ of $\mathcal{C} \times \widetilde{S}$ and an equivalence $\lambda: \mathcal{C}' \xrightarrow{\sim} \mathcal{C}_0$ such that F is isomorphic to the composition $\mathcal{C}' \xrightarrow{\lambda} \mathcal{C}_0 \to \mathcal{C} \times \widetilde{S} \xrightarrow{\theta} \mathcal{C}$.

Exercise 1.19. Let $\mathcal{C}$, $\mathcal{C}'$ be categories and $L_\nu: \mathcal{C} \to \mathcal{C}'$, $R_\nu: \mathcal{C}' \to \mathcal{C}$ be functors such that (L_ν, R_ν) is a pair of adjoint functors ($\nu = 1, 2$). Let $\varepsilon_\nu: \mathrm{id}_{\mathcal{C}} \to R_\nu \circ L_\nu$ and $\eta_\nu: L_\nu \circ R_\nu \to \mathrm{id}_{\mathcal{C}'}$ be the adjunction morphisms. Prove that the two maps λ, μ:

$$\mathrm{Hom}_{\mathrm{Fct}(\mathcal{C},\mathcal{C}')}(L_1, L_2) \underset{\mu}{\overset{\lambda}{\rightleftarrows}} \mathrm{Hom}_{\mathrm{Fct}(\mathcal{C}',\mathcal{C})}(R_2, R_1)$$

given by

$$\lambda(\varphi) \colon R_2 \xrightarrow{\;\varepsilon_1 \circ R_2\;} R_1 \circ L_1 \circ R_2 \xrightarrow{\;R_1 \circ \varphi \circ R_2\;} R_1 \circ L_2 \circ R_2 \xrightarrow{\;R_1 \circ \eta_2\;} R_1$$

$$\text{for } \varphi \in \operatorname{Hom}_{\operatorname{Fct}(\mathcal{C},\mathcal{C}')}(L_1, L_2),$$

$$\mu(\psi) \colon L_1 \xrightarrow{\;L_1 \circ \varepsilon_2\;} L_1 \circ R_2 \circ L_2 \xrightarrow{\;L_1 \circ \psi \circ L_2\;} L_1 \circ R_1 \circ L_2 \xrightarrow{\;\varepsilon_1 \circ L_2\;} L_2$$

$$\text{for } \psi \in \operatorname{Hom}_{\operatorname{Fct}(\mathcal{C}',\mathcal{C})}(R_2, R_1).$$

are inverse to each other.

Exercise 1.20. Consider three categories $J, I, \mathcal{C}$ and a functor $\varphi \colon J \to I$. Assume that φ is essentially surjective. Prove that the functor $\circ \varphi \colon \operatorname{Fct}(I, \mathcal{C}) \to \operatorname{Fct}(J, \mathcal{C})$ is faithful and conservative. (See Lemma 7.1.3 for refinements of this result.)

Exercise 1.21. The simplicial category $\mathbf{\Delta}$ is defined as follows. The objects of $\mathbf{\Delta}$ are the finite totally ordered sets and the morphisms are the order-preserving maps. Let $\widetilde{\mathbf{\Delta}}$ be the subcategory of $\mathbf{\Delta}$ consisting of non-empty sets and

$$\operatorname{Hom}_{\widetilde{\mathbf{\Delta}}}(\sigma, \tau) =$$
$$\left\{ u \in \operatorname{Hom}_{\mathbf{\Delta}}(\sigma, \tau) \, ; \, \begin{matrix} u \text{ sends the smallest (resp. the largest)} \\ \text{element of } \sigma \text{ to the smallest (resp. the} \\ \text{largest) element of } \tau \end{matrix} \right\}.$$

For integers n, m denote by $[n, m]$ the totally ordered set $\{k \in \mathbb{Z}; \, n \le k \le m\}$.
(i) Prove that the natural functor $\mathbf{\Delta} \to \mathbf{Set}^f$ is half-full and faithful.
(ii) Prove that the full subcategory of $\mathbf{\Delta}$ consisting of objects $\{[0, n]\}_{n \ge -1}$ is equivalent to $\mathbf{\Delta}$.
(iii) Prove that $\mathbf{\Delta}$, as well as $\widetilde{\mathbf{\Delta}}$, admit an initial object and a terminal object.
(iv) For $\sigma \in \mathbf{\Delta}$, let us endow $\mathcal{S}(\sigma) := \operatorname{Hom}_{\mathbf{\Delta}}(\sigma, [0, 1])$ with a structure of an ordered set by setting for $\xi, \eta \in \mathcal{S}(\sigma)$, $\xi \le \eta$ if $\xi(i) \le \eta(i)$ for all $i \in \sigma$. Prove that $\mathcal{S}(\sigma)$ is a totally ordered set.
(v) Prove that the functor $\varphi \colon \mathbf{\Delta} \to \widetilde{\mathbf{\Delta}}^{\operatorname{op}}$ given by $\sigma \mapsto \operatorname{Hom}_{\mathbf{\Delta}}(\sigma, [0, 1])$ and the functor $\psi \colon \widetilde{\mathbf{\Delta}}^{\operatorname{op}} \to \mathbf{\Delta}$ given by $\tau \mapsto \operatorname{Hom}_{\widetilde{\mathbf{\Delta}}}(\tau, [0, 1])$ are quasi-inverse to each other and give an equivalence $\mathbf{\Delta} \simeq \widetilde{\mathbf{\Delta}}^{\operatorname{op}}$.
(vi) Denote by $\mathbf{\Delta}_{inj}$ (resp. $\widetilde{\mathbf{\Delta}}_{sur}$) the subcategory of $\mathbf{\Delta}$ (resp. of $\widetilde{\mathbf{\Delta}}$) such that $\operatorname{Ob}(\mathbf{\Delta}_{inj}) = \operatorname{Ob}(\mathbf{\Delta})$, (resp. $\operatorname{Ob}(\widetilde{\mathbf{\Delta}}_{sur}) = \operatorname{Ob}(\widetilde{\mathbf{\Delta}})$) the morphisms being the injective (resp. surjective) order-preserving maps. Prove that $\mathbf{\Delta}_{inj}$ and $(\widetilde{\mathbf{\Delta}}_{sur})^{\operatorname{op}}$ are equivalent.
(vii) Denote by $\iota \colon \widetilde{\mathbf{\Delta}} \to \mathbf{\Delta}$ the canonical functor and by $\kappa \colon \mathbf{\Delta} \to \widetilde{\mathbf{\Delta}}$ the functor $\tau \mapsto \{0\} \sqcup \tau \sqcup \{\infty\}$ (with 0 the smallest element in $\{0\} \sqcup \tau \sqcup \{\infty\}$ and ∞ the largest). Prove that (κ, ι) is a pair of adjoint functors and the diagram below quasi-commutes:

$$\begin{array}{ccc}
\boldsymbol{\Delta} & \xrightarrow{\ \kappa\ } & \widetilde{\boldsymbol{\Delta}} \\
{\scriptstyle\sim}\downarrow{\scriptstyle\varphi} & & {\scriptstyle\sim}\downarrow{\scriptstyle\psi^{\mathrm{op}}} \\
\widetilde{\boldsymbol{\Delta}}^{\,\mathrm{op}} & \xrightarrow{\ \iota^{\mathrm{op}}\ } & \boldsymbol{\Delta}^{\mathrm{op}}.
\end{array}$$

(Remark: the simplicial category will be used in §11.4.)

2

Limits

Inductive and projective limits are at the heart of category theory. They are an essential tool, if not the only one, to construct new objects and new functors. Inductive and projective limits in categories are constructed by using *projective* limits in **Set**. In fact, if $\beta\colon J^{\mathrm{op}} \to \mathcal{C}$ is a functor, its projective limit is a representative of the functor which associates the projective limit of $\mathrm{Hom}_{\mathcal{C}}(Z, \beta)$ to Z, and if $\alpha\colon J \to \mathcal{C}$ is a functor, its inductive limit is a representative of the functor which associates the projective limit of $\mathrm{Hom}_{\mathcal{C}}(\alpha, Z)$ to Z.

In this chapter we construct these limits and describe with some details particular cases, such as products, kernels, fiber products, etc. as well as the dual notions (coproducts, etc.).

Given a functor $\varphi\colon J \to I$ and a category $\mathcal{C}$, the composition by φ defines a functor $\varphi_*\colon \mathrm{Fct}(I, \mathcal{C}) \to \mathrm{Fct}(J, \mathcal{C})$. Projective and inductive limits are the tools to construct a right or left adjoint to the functor φ_*. This procedure is known as the "Kan extension" of functors. When applying this construction to the Yoneda functor, we get an equivalence of categories between functors defined on $\mathcal{C}$ and functors defined on $\mathcal{C}^\wedge$ and commuting with small inductive limits.

We pay special attention to inductive limits in the category **Set**, but the reader will have to wait until Chap. 3 to encounter filtrant inductive limits, these limits being often much easier to manipulate.

It is well-known, already to the students, that the limit of a convergent sequence of real numbers remains unchanged when the sequence is replaced by a subsequence. There is a similar phenomena in Category Theory which leads to the notion of cofinal functor. A functor of small categories $\varphi\colon J \to I$ is cofinal if, for any functor $\alpha\colon I \to \mathcal{C}$, the limits of α and $\alpha \circ \varphi$ are isomorphic. We prove here that φ is cofinal if and only if, for any $i \in I$, the category J^i, whose objects are the pairs (j, u) of $j \in J$ and $u\colon i \to \varphi(j)$, is connected.

We also introduce ind-limits and pro-limits, that is, inductive and projective limits in the categories $\mathcal{C}^\wedge$ and $\mathcal{C}^\vee$, respectively.

2.1 Limits

Recall Convention 1.4.1.

In this section, I, J, K etc. will denote small categories. Let $\mathcal{C}$ be a category. A functor $\alpha\colon I \to \mathcal{C}$ (resp. $\beta\colon I^{\mathrm{op}} \to \mathcal{C}$) is sometimes called an *inductive system* (resp. a *projective system*) in $\mathcal{C}$ indexed by I.

Assume first that $\mathcal{C}$ is the category **Set** and let us consider projective systems. In other words, β is an object of $I^\wedge$. Denote by $\mathrm{pt}_{I^\wedge}$ the constant functor from I^{op} to **Set**, defined by $\mathrm{pt}_{I^\wedge}(i) = \{\mathrm{pt}\}$ for all $i \in I$. Note that $\mathrm{pt}_{I^\wedge}$ is a terminal object of $I^\wedge$. We define a set, called the projective limit of β, by

$$(2.1.1) \qquad \varprojlim \beta = \mathrm{Hom}_{I^\wedge}(\mathrm{pt}_{I^\wedge}, \beta) \, .$$

The family of morphisms:

$$\mathrm{Hom}_{I^\wedge}(\mathrm{pt}_{I^\wedge}, \beta) \to \mathrm{Hom}_{\mathbf{Set}}(\mathrm{pt}_{I^\wedge}(i), \beta(i)) \simeq \beta(i), \quad i \in I \, ,$$

defines the map $\varprojlim \beta \to \prod_i \beta(i)$, and it is immediately checked that:

$$(2.1.2) \quad \varprojlim \beta \simeq \left\{ \{x_i\}_i \in \prod_i \beta(i) \, ; \, \beta(s)(x_j) = x_i \text{ for all } s \in \mathrm{Hom}_I(i, j) \right\} \, .$$

Since I and $\beta(i)$ are small, $\varprojlim \beta$ is a small set. The next result is obvious.

Lemma 2.1.1. *Let $\beta\colon I^{\mathrm{op}} \to \mathbf{Set}$ be a functor and let $X \in \mathbf{Set}$. There is a natural isomorphism*

$$\mathrm{Hom}_{\mathbf{Set}}(X, \varprojlim \beta) \xrightarrow{\sim} \varprojlim \mathrm{Hom}_{\mathbf{Set}}(X, \beta) \, ,$$

where $\mathrm{Hom}_{\mathbf{Set}}(X, \beta)$ denotes the functor $I^{\mathrm{op}} \to \mathbf{Set}$, $i \mapsto \mathrm{Hom}_{\mathbf{Set}}(X, \beta(i))$.

Let $\varphi\colon J \to I$ and $\beta\colon I^{\mathrm{op}} \to \mathbf{Set}$ be functors. Denote by $\varphi^{\mathrm{op}}\colon J^{\mathrm{op}} \to I^{\mathrm{op}}$ the associated functor. Using (2.1.1), we get a natural morphism:

$$(2.1.3) \qquad \varprojlim \beta \to \varprojlim (\beta \circ \varphi^{\mathrm{op}}) \, .$$

Now let α (resp. β) be a functor from I (resp. I^{op}) to a category $\mathcal{C}$. For $X \in \mathcal{C}$, $\mathrm{Hom}_{\mathcal{C}}(\alpha, X)$ and $\mathrm{Hom}_{\mathcal{C}}(X, \beta)$ are functors from I^{op} to **Set**. We can then define inductive and projective limits as functors from $\mathcal{C}$ or $\mathcal{C}^{\mathrm{op}}$ to **Set** as follows.

Recall that $\mathcal{C}^\wedge$ and $\mathcal{C}^\vee$ are given in Definition 1.4.2.

Definition 2.1.2.   (i) *We define $\varinjlim \alpha \in \mathcal{C}^\vee$ and $\varprojlim \beta \in \mathcal{C}^\wedge$ by the formulas*

$$(2.1.4) \quad \varinjlim \alpha\colon X \mapsto \varprojlim \mathrm{Hom}_{\mathcal{C}}(\alpha, X) = \varprojlim(\mathrm{h}_{\mathcal{C}}(X) \circ \alpha) \in \mathbf{Set} \, ,$$

$$(2.1.5) \quad \varprojlim \beta\colon X \mapsto \varprojlim \mathrm{Hom}_{\mathcal{C}}(X, \beta) = \varprojlim(\mathrm{k}_{\mathcal{C}}(X) \circ \beta) \in \mathbf{Set} \, .$$

(ii) *If these functors are representable, we keep the same notations to denote one of their representatives in C, and we call these representatives the inductive or projective limit, respectively.*

(iii) *If for every functor α from I (resp. I^{op}) to $\mathcal{C}$, $\varinjlim \alpha$ (resp. $\varprojlim \alpha$) is representable, we say that $\mathcal{C}$ admits inductive (resp. projective) limits indexed by I.*

(iv) *We say that a category $\mathcal{C}$ admits finite (resp. small) projective limits if it admits projective limits indexed by finite (resp. small) categories, and similarly, replacing "projective limits" with "inductive limits".*

When $\mathcal{C} = \mathbf{Set}$, this definition of $\varprojlim \beta$ coincides with the former one, in view of Lemma 2.1.1.

Remark 2.1.3. The definitions of $\mathcal{C}^{\wedge}$ and $\mathcal{C}^{\vee}$ depend on the choice of the universe $\mathcal{U}$. However, given a functor $\alpha \colon I \to \mathcal{C}$, the fact that $\varinjlim \alpha$ is representable as well as its representative does not depend on the choice of the universe $\mathcal{U}$ such that I is $\mathcal{U}$-small and $\mathcal{C}$ is a $\mathcal{U}$-category, and similarly for projective limits.

Notations 2.1.4. (i) We shall sometimes use a more intuitive notation, writing $\varinjlim_{i\in I} \alpha(i)$ or $\varinjlim_{i} \alpha(i)$ instead of $\varinjlim \alpha$. We may also write $\varprojlim_{i\in I} \beta(i)$ or $\varprojlim_{i\in I^{\mathrm{op}}} \beta(i)$ or $\varprojlim_{i} \beta(i)$ instead of $\varprojlim \beta$.

(ii) Notice that in the literature, lim is sometimes used for the projective limit, and colim for the inductive limit, and one writes lim β and colim α instead of $\varprojlim \beta$ and $\varinjlim \alpha$.

Remark 2.1.5. Let I be a small set and $\alpha \colon I \to \mathcal{C}$ a functor. It defines a functor $\alpha^{\mathrm{op}} \colon I^{\mathrm{op}} \to \mathcal{C}^{\mathrm{op}}$ and there is a natural isomorphism

$$(\varinjlim \alpha)^{\mathrm{op}} \simeq \varprojlim \alpha^{\mathrm{op}} .$$

Hence, results on projective limits may be deduced from results on inductive limits, and conversely.

Moreover, a functor $\alpha \colon I \to \mathcal{C}$ defines a functor $\beta \colon (I^{\mathrm{op}})^{\mathrm{op}} \to \mathcal{C}$ and an inductive system indexed by I is the same as a projective system indexed by I^{op}. However one shall be aware that the inductive limit of α has no relation in general with the projective limit of β. (See the examples below, in particular when I is discrete.)

By Definition 2.1.2, if $\varinjlim \alpha$ or $\varprojlim \beta$ are representable, we get:

$$(2.1.6) \qquad \mathrm{Hom}_{\mathcal{C}}(\varinjlim \alpha, X) \simeq \varprojlim \mathrm{Hom}_{\mathcal{C}}(\alpha, X) ,$$

$$(2.1.7) \qquad \mathrm{Hom}_{\mathcal{C}}(X, \varprojlim \beta) \simeq \varprojlim \mathrm{Hom}_{\mathcal{C}}(X, \beta) .$$

Note that the right-hand sides are the projective limits in **Set**.

Assume that $\varinjlim \alpha$ is representable by $Y \in C$. We get:

$$\varprojlim_i \mathrm{Hom}_C(\alpha(i), Y) \simeq \mathrm{Hom}_C(Y, Y)$$

and the identity of Y defines a family of morphisms

$$\rho_i : \alpha(i) \to Y = \varinjlim \alpha \quad \text{with } \rho_j \circ \alpha(s) = \rho_i \text{ for all } s : i \to j .$$

Consider a family of morphisms $f_i : \alpha(i) \to X$ in C satisfying the natural compatibility conditions, visualized by the commutative diagram, with $s : i \to j$

$$\begin{array}{ccc}
\alpha(i) & \xrightarrow{\;f_i\;} & X \\
{\scriptstyle \alpha(s)}\big\downarrow & \nearrow{\scriptstyle f_j} & \\
\alpha(j) & &
\end{array}$$

This family of morphisms is nothing but an element of $\varprojlim\limits_i \mathrm{Hom}\,(\alpha(i), X)$, hence by (2.1.6) it gives an element of $\mathrm{Hom}\,(Y, X)$. Therefore there exists a unique morphism $g : Y \to X$ such that $f_i = g \circ \rho_i$.

Similarly, if $\varprojlim \beta$ is representable, we obtain a family of morphisms $\rho_i : \varprojlim \beta \to \beta(i)$ and any family of morphisms from X to the $\beta(i)$'s satisfying the natural compatibility conditions will factorize uniquely through $\varprojlim \beta$. This is visualized by the commutative diagrams:

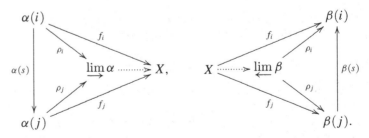

If $\theta : \alpha \to \alpha'$ is a morphism of functors, it induces a morphism $\varinjlim \alpha \to \varinjlim \alpha'$ in $C^\vee$.

It follows from (2.1.3) that if $\varphi : J \to I$, $\alpha : I \to C$ and $\beta : I^{\mathrm{op}} \to C$ are functors, we have natural morphisms:

(2.1.8) $$\varinjlim (\alpha \circ \varphi) \to \varinjlim \alpha ,$$

(2.1.9) $$\varprojlim (\beta \circ \varphi^{\mathrm{op}}) \leftarrow \varprojlim \beta .$$

Proposition 2.1.6. *Let I be a category and assume that C admits inductive limits (resp. projective limits) indexed by I. Then for any category J, the* *big category $\mathcal{C}^J$ admits inductive limits (resp. projective limits) indexed by I. Moreover, for $j \in J$, denote by $\rho_j : \mathcal{C}^J \to \mathcal{C}$ the functor which associates $\gamma(j)$ to a functor $\gamma : J \to \mathcal{C}$. Then, if $\alpha : I \to \mathcal{C}^J$ (resp. $\beta : I^{\mathrm{op}} \to \mathcal{C}^J$) is a functor, its inductive (resp. projective) limit is given by*

$$(\varinjlim \alpha)(j) = \varinjlim (\rho_j \circ \alpha) \ \text{for any } j \in J$$
$$(resp. \ (\varprojlim \beta)(j) = \varprojlim (\rho_j \circ \beta) \ \text{for any } j \in J) \ .$$

In other words:

$$(\varinjlim \alpha)(j) = \varinjlim (\alpha(j)) \ ,$$
$$(resp. \ (\varprojlim \beta)(j) = \varprojlim (\beta(j))) \ .$$

The proof is obvious.

For a small category I and a functor $\alpha : I \to \mathcal{C}$, $\varinjlim (\mathrm{k}_{\mathcal{C}} \circ \alpha) \in \mathcal{C}^{\vee}$ exists and coincides with $\varinjlim \alpha \in \mathcal{C}^{\vee}$ given in Definition 2.1.2. Then $\varinjlim \alpha$ exists if and only if $\varinjlim (\mathrm{k}_{\mathcal{C}} \circ \alpha)$ is representable, and in this case, $\varinjlim \alpha$ is its representative. There is a similar remark for $\varprojlim$, replacing $\mathcal{C}^{\vee}$ with $\mathcal{C}^{\wedge}$.

We shall consider inductive or projective limits associated with bifunctors.

Proposition 2.1.7. *Let I and J be two small categories and assume that $\mathcal{C}$ admits inductive limits indexed by I and J. Consider a bifunctor $\alpha : I \times J \to \mathcal{C}$ and let $\alpha_J : I \to \mathcal{C}^J$ and $\alpha_I : J \to \mathcal{C}^I$ be the functors induced by α. Then $\varinjlim \alpha$ exists and we have the isomorphisms*

$$\varinjlim \alpha \simeq \varinjlim(\varinjlim \alpha_J) \simeq \varinjlim(\varinjlim \alpha_I) \ .$$

Similarly, if $\beta : I^{\mathrm{op}} \times J^{\mathrm{op}} \to \mathcal{C}$ is a bifunctor, then β defines functors $\beta_J : I^{\mathrm{op}} \to \mathcal{C}^{J^{\mathrm{op}}}$ and $\beta_I : J^{\mathrm{op}} \to \mathcal{C}^{I^{\mathrm{op}}}$ and we have the isomorphisms

$$\varprojlim \beta \simeq \varprojlim(\varprojlim \beta_J) \simeq \varprojlim(\varprojlim \beta_I) \ .$$

In other words:

$$\varinjlim_{i,j} \alpha(i, j) \simeq \varinjlim_{j}(\varinjlim_{i} \alpha(i, j)) \simeq \varinjlim_{i}(\varinjlim_{j} \alpha(i, j)) \ ,$$
$$\varprojlim_{i,j} \beta(i, j) \simeq \varprojlim_{j}(\varprojlim_{i} \beta(i, j)) \simeq \varprojlim_{i}(\varprojlim_{j} \beta(i, j)) \ .$$

The proof is obvious.

Definition 2.1.8. *Let $F : \mathcal{C} \to \mathcal{C}'$ be a functor and I a category.*

(i) *Assume that $\mathcal{C}$ admits inductive limits indexed by I. We say that F commutes with such limits if for any $\alpha : I \to \mathcal{C}$, $\varinjlim (F \circ \alpha)$ exits in $\mathcal{C}'$ and is represented by $F(\varinjlim \alpha)$.*

(ii) *Similarly if C admits projective limits indexed by I, we say that F commutes with such limits if for any $\beta\colon I^{\mathrm{op}} \to C$, $\varprojlim(F\circ\beta)$ exists and is represented by $F(\varprojlim\beta)$.*

Note that if C admits inductive (resp. projective) limits indexed by I, there is a natural morphism $\varinjlim(F\circ\alpha) \to F(\varinjlim\alpha)$ in $C'^\vee$ (resp. $F(\varprojlim\beta) \to \varprojlim(F\circ\beta)$ in $C'^\wedge$). Then (i) (resp. (ii)) means that this morphism is an isomorphism for any functor α (resp. β).

Example 2.1.9. Let k be a field, $C = C' = \mathrm{Mod}(k)$, and let $X \in C$. Then the functor $\mathrm{Hom}_k(X,\cdot)$ commutes with small inductive limit if X is finite-dimensional, and it does not if X is infinite-dimensional. Of course, it always commutes with small projective limits.

If C admits projective limits indexed by a category I, the Yoneda functor $h_C\colon C \to C^\wedge$ commutes with such projective limits by the definition, but one shall be aware that even if C admits inductive limits, the functors h_C does not commute with inductive limits in general (see Exercises 2.19 and 3.7).

Proposition 2.1.10. *Let $F\colon C \to C'$ be a functor. Assume that:*

(i) *F admits a left adjoint $G\colon C' \to C$,*
(ii) *C admits projective limits indexed by a small category I.*

Then F commutes with projective limits indexed by I, that is, the natural morphism $F(\varprojlim\beta) \to \varprojlim F(\beta)$ is an isomorphism for any $\beta\colon I^{\mathrm{op}} \to C$.

Proof. For any $Y \in C'$, there is the chain of isomorphisms

$$
\begin{aligned}
\mathrm{Hom}_{C'}(Y, F(\varprojlim\beta)) &\simeq \mathrm{Hom}_C(G(Y), \varprojlim\beta) \\
&\simeq \varprojlim \mathrm{Hom}_C(G(Y), \beta) \\
&\simeq \varprojlim \mathrm{Hom}_{C'}(Y, F(\beta)) \\
&\simeq \mathrm{Hom}_{C'^\wedge}(Y, \varprojlim F(\beta)) \,.
\end{aligned}
$$

Then the result follows by the Yoneda lemma.   q.e.d.

Of course there is a similar result for inductive limits. If C admits inductive limits indexed by I and F admits a right adjoint, then F commutes with such limits.

The next results will be useful.

Lemma 2.1.11. *Let C be a category and let $\alpha\colon C \to C$ be the identity functor. If $\varinjlim\alpha$ is representable by an object S of C, then S is a terminal object of C.*

Proof. For $X \in C$ denote by a_X the natural morphism $X \to \varinjlim\alpha \simeq S$. The family of morphisms a_X satisfies:

(a) for every $f: X \to Y$, $a_Y \circ f = a_X$,
(b) if a pair of parallel arrows $u, u': S \rightrightarrows Z$ satisfy $u \circ a_X = u' \circ a_X$ for all $X \in \mathcal{C}$, then $u = u'$.

First, we shall show that $a_S = \mathrm{id}_S$. Applying (a) to $f = a_X$, we get $a_S \circ a_X = a_X$. Hence $a_S \circ a_X = \mathrm{id}_S \circ a_X$, and this implies $a_S = \mathrm{id}_S$ by (b).

We can now complete the proof. Let $f: X \to S$. By (i), $f = a_S \circ f = a_X$. Hence, $\mathrm{Hom}_{\mathcal{C}}(X, S) \simeq \{a_X\}$.   q.e.d.

Recall (see Notations 1.2.15) that $\Delta_X: I \to \mathcal{C}$ is the constant functor with values $X \in \mathcal{C}$.

Lemma 2.1.12. *Let I and $\mathcal{C}$ be two categories and assume that I is connected. Let $X \in \mathcal{C}$. Then $X \xrightarrow{\sim} \varprojlim \Delta_X$ and $\varinjlim \Delta_X \xrightarrow{\sim} X$.*

Proof. (i) Assume first that $\mathcal{C} = \mathbf{Set}$. By (2.1.2), $\varprojlim \Delta_X$ is the subset of X^I consisting of the $\{x_i\}_{i \in I}$ (with $x_i \in X$) such that $x_i = x_{i'}$ if there exists an arrow $i \to i'$. Then the x_i's are equal to one another since $\mathcal{C}$ is connected, and we obtain $X \xrightarrow{\sim} \varprojlim \Delta_X$.

(ii) By (i), we have the isomorphisms for $Y \in \mathcal{C}$

$$\mathrm{Hom}_{\mathcal{C}^\vee}(\varinjlim \Delta_X, Y) \simeq \varprojlim \mathrm{Hom}_{\mathcal{C}}(\Delta_X, Y) \simeq \varprojlim \Delta_{\mathrm{Hom}_{\mathcal{C}}(X,Y)} \simeq \mathrm{Hom}_{\mathcal{C}}(X, Y) ,$$
$$\mathrm{Hom}_{\mathcal{C}^\wedge}(Y, \varprojlim \Delta_X) \simeq \varprojlim \mathrm{Hom}_{\mathcal{C}}(Y, \Delta_X) \simeq \varprojlim \Delta_{\mathrm{Hom}_{\mathcal{C}}(Y,X)} \simeq \mathrm{Hom}_{\mathcal{C}}(Y, X) .$$

Hence, the results follow from the Yoneda lemma.   q.e.d.

(See Corollary 2.4.5 for a converse statement.)

Let $A \in \mathcal{C}^\wedge$, and let $\mathcal{C}_A$ denote the category associated with the Yoneda functor $h_{\mathcal{C}}: \mathcal{C} \to \mathcal{C}^\wedge$ (see Definition 1.2.16). Hence, $\mathcal{C}_A$ is the category of pairs (X, u) of $X \in \mathcal{C}$ and $u \in A(X)$.

Lemma 2.1.13. *Let I be a category and assume that $\mathcal{C}$ admits inductive limits indexed by I.*

(i) *If $A: \mathcal{C}^{\mathrm{op}} \to \mathbf{Set}$ commutes with projective limits indexed by I (i.e., $A(\varinjlim_{i \in I} X_i) \simeq \varprojlim_{i \in I} A(X_i)$ for any inductive system $\{X_i\}_{i \in I}$ in $\mathcal{C}$), then $\mathcal{C}_A$ admits inductive limits indexed by I and $j_A: \mathcal{C}_A \to \mathcal{C}$ commutes with such limits.*
(ii) *If a functor $F: \mathcal{C} \to \mathcal{C}'$ commutes with inductive limits indexed by I, then for any $Y \in \mathcal{C}'$, $\mathcal{C}_Y$ admits inductive limits indexed by I and $\mathcal{C}_Y \to \mathcal{C}$ commutes with such limits.*

Proof. (i) Let $\{(X_i, u_i)\}_{i \in I}$ be an inductive system in $\mathcal{C}_A$ indexed by I. Then $u := \{u_i\}_i \in \varprojlim_i A(X_i) \simeq A(\varinjlim_i X_i)$ gives an object $(\varinjlim_i X_i, u)$ of $\mathcal{C}_A$. It is easily checked that it is an inductive limit of $\{(X_i, u_i)\}_{i \in I}$.

(ii) Let A be the functor from $\mathcal{C}^{op}$ to **Set** given by $A(X) = \mathrm{Hom}_{\mathcal{C}'}(F(X), Y)$. Then A commutes with projective limits indexed by I and $\mathcal{C}_A$ is equivalent to $\mathcal{C}_Y$.   q.e.d.

Definition 2.1.14. *Let us denote by* $\mathrm{Mor}_0(\mathcal{C})$ *the category whose objects are the morphisms in $\mathcal{C}$ and whose morphisms are described as follows. Let* $f: X \to Y$ *and* $g: X' \to Y'$ *belong to* $\mathrm{Mor}(\mathcal{C})$. *Then* $\mathrm{Hom}_{\mathrm{Mor}_0(\mathcal{C})}(f, g) = \{u: X \to X', v: Y' \to Y; f = v \circ g \circ u\}$. *The composition and the identity in* $\mathrm{Mor}_0(\mathcal{C})$ *are the obvious ones.*

A morphism $f \to g$ in $\mathrm{Mor}_0(\mathcal{C})$ is visualized by the commutative diagram:

$$
\begin{array}{ccc}
X & \xrightarrow{\;f\;} & Y \\
{\scriptstyle u}\big\downarrow & & \big\uparrow{\scriptstyle v} \\
X' & \xrightarrow[\;g\;]{} & Y'.
\end{array}
$$

Lemma 2.1.15. *Let I and $\mathcal{C}$ be two categories and let $\alpha, \beta \in \mathrm{Fct}(I, \mathcal{C})$. Then* $(i \to j) \mapsto \mathrm{Hom}_{\mathcal{C}}(\alpha(i), \beta(j))$ *is a functor from* $\mathrm{Mor}_0(I)^{op}$ *to* **Set**, *and there is a natural isomorphism*

$$(2.1.10) \qquad \mathrm{Hom}_{\mathrm{Fct}(I,\mathcal{C})}(\alpha, \beta) \xrightarrow{\sim} \varprojlim_{(i \to j)\in\mathrm{Mor}_0(I)} \mathrm{Hom}_{\mathcal{C}}(\alpha(i), \beta(j)).$$

Proof. The first statement, as well as the construction of the map (2.1.10) is clear. This map is obviously injective. Let us show that it is surjective. Let

$$\varphi := \{\varphi(i \to j)\}_{(i\to j)\in\mathrm{Mor}_0(I)} \in \varprojlim_{(i \to j)\in\mathrm{Mor}_0(I)} \mathrm{Hom}_{\mathcal{C}}(\alpha(i), \beta(j)).$$

Then $\varphi(i \xrightarrow{\mathrm{id}_i} i)$ defines the morphism $\theta_i: \alpha(i) \to \beta(i)$. Let us show that $\theta := \{\theta_i\}_{i\in I}$ is a morphism of functors from α to β.

Let $f: i \to j$ be a morphism in I. To f we associate the two morphisms in $\mathrm{Mor}_0(I)$:

$$
\begin{array}{ccc}
i & \xrightarrow{\;f\;} & j \\
{\scriptstyle \mathrm{id}_i}\big\downarrow & & \big\uparrow{\scriptstyle f} \\
i & \xrightarrow[\;\mathrm{id}_i\;]{} & i,
\end{array}
\qquad
\begin{array}{ccc}
i & \xrightarrow{\;f\;} & j \\
{\scriptstyle f}\big\downarrow & & \big\uparrow{\scriptstyle \mathrm{id}_j} \\
j & \xrightarrow[\;\mathrm{id}_j\;]{} & j.
\end{array}
$$

In the diagram below

$$
\begin{array}{ccc}
\alpha(i) & \xrightarrow{\;\theta_i\;} & \beta(i) \\
\big\downarrow & \searrow{\scriptstyle \varphi(i\to j)} & \big\downarrow \\
\alpha(j) & \xrightarrow[\;\theta_j\;]{} & \beta(j)
\end{array}
$$

the two triangles commute, and hence the square commutes. Hence, $\theta \in$ Hom $_{\mathrm{Fct}(I,\mathcal{C})}(\alpha, \beta)$ and φ is the image of θ.   q.e.d.

2.2 Examples

Empty Limits

If I is the empty category and $\alpha\colon I \to \mathcal{C}$ is a functor, then $\varinjlim \alpha$ is representable if and only if $\mathcal{C}$ has an initial object $\emptyset_{\mathcal{C}}$, and in this case $\varinjlim \alpha \simeq \emptyset_{\mathcal{C}}$. Similarly, $\varprojlim \alpha$ is representable if and only if $\mathcal{C}$ has a terminal object $\mathrm{pt}_{\mathcal{C}}$, and in this case $\varprojlim \alpha \simeq \mathrm{pt}_{\mathcal{C}}$.

Terminal Object

If I has a terminal object, say pt_I, and $\alpha\colon I \to \mathcal{C}$ (resp. $\beta\colon I^{\mathrm{op}} \to \mathcal{C}$) is a functor, then

$$\varinjlim \alpha \simeq \alpha(\mathrm{pt}_I)\,,$$

$$(\text{resp.} \quad \varprojlim \beta \simeq \beta(\mathrm{pt}_I).)$$

Sums and Products

Consider a family $\{X_i\}_{i\in I}$ of objects in $\mathcal{C}$ indexed by a set I. We may regard I as a discrete category and associate to this family the functor $\alpha\colon I \to \mathcal{C}$ given by $\alpha(i) = X_i$.

Definition 2.2.1. *Consider a family $\{X_i\}_{i\in I}$ of objects in $\mathcal{C}$ identified with a functor $\alpha\colon I \to \mathcal{C}$.*

  (i) *The* coproduct *of the X_i's, denoted by $\coprod_i X_i$, is given by $\coprod_i X_i := \varinjlim \alpha$.*
  (ii) *The* product *of the X_i's, denoted by $\prod_i X_i$, is given by $\prod_i X_i := \varprojlim \alpha$.*

Hence we have isomorphisms, functorial with respect to $Y \in \mathcal{C}$:

(2.2.1) $$\mathrm{Hom}_{\mathcal{C}}(\coprod_i X_i, Y) \simeq \prod_i \mathrm{Hom}_{\mathcal{C}}(X_i, Y)\,,$$

(2.2.2) $$\mathrm{Hom}_{\mathcal{C}}(Y, \prod_i X_i) \simeq \prod_i \mathrm{Hom}_{\mathcal{C}}(Y, X_i)\,.$$

The natural morphism $X_j \to \coprod_i X_i$ is called the *j-th coprojection*. Similarly, the natural morphism $\prod_i X_i \to X_j$ is called the *j-th projection*.

When $X_i = X$ for all $i \in I$, we simply denote the coproduct by $X^{\coprod I}$ and we denote the product by $X^{\prod I}$. We also write $X^{(I)}$ and X^I instead of $X^{\coprod I}$ and $X^{\prod I}$, respectively.

If $X^{\coprod I}$ exists, we have

(2.2.3) $\mathrm{Hom}_{\mathbf{Set}}(I, \mathrm{Hom}_{\mathcal{C}}(X, Y)) \simeq \mathrm{Hom}_{\mathcal{C}}(X^{\coprod I}, Y)$.

If $X^{\prod I}$ exists, we have

(2.2.4) $\mathrm{Hom}_{\mathbf{Set}}(I, \mathrm{Hom}_{\mathcal{C}}(Y, X)) \simeq \mathrm{Hom}_{\mathcal{C}}(Y, X^{\prod I})$.

If $I = \{0, 1\}$, the coproduct and product (if they exist) are denoted by $X_0 \coprod X_1$ and $X_0 \prod X_1$, respectively. Moreover, one usually writes $X_0 \sqcup X_1$ and $X_0 \times X_1$ instead of $X_0 \coprod X_1$ and $X_0 \prod X_1$, respectively.

The coproduct and product of two objects are visualized by the commutative diagrams:

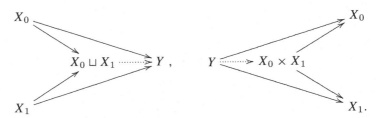

In other words, any pair of morphisms from (resp. to) X_0 and X_1 to (resp. from) Y factors uniquely through $X_0 \sqcup X_1$ (resp. $X_0 \times X_1$). If $\mathcal{C}$ is the category **Set**, $X_0 \sqcup X_1$ is the disjoint union and $X_0 \times X_1$ is the product of the two sets X_0 and X_1.

Cokernels and Kernels

Consider the category I with two objects and two parallel morphisms other than identities (see Notations 1.2.8 (iv)), visualized by

(2.2.5) $\bullet \rightrightarrows \bullet$

A functor $\alpha \colon I \to \mathcal{C}$ is nothing but two parallel arrows in $\mathcal{C}$:

(2.2.6) $f, g : X_0 \rightrightarrows X_1.$

In the sequel we shall identify such a functor with a diagram (2.2.6).

Definition 2.2.2. *Consider two parallel arrows $f, g \colon X_0 \rightrightarrows X_1$ in $\mathcal{C}$ identified with a functor $\alpha \colon I \to \mathcal{C}$.*

(i) *The* cokernel (*also called the* co-equalizer) *of the pair (f, g), denoted by* $\mathrm{Coker}(f, g)$, *is given by* $\mathrm{Coker}(f, g) := \varinjlim \alpha$.

(ii) *The* kernel (*also called the* equalizer) *of the pair (f, g), denoted by* $\mathrm{Ker}(f, g)$, *is given by* $\mathrm{Ker}(f, g) := \varprojlim \alpha$.

(iii) *A sequence $X_0 \rightrightarrows X_1 \to Z$ (resp. $Z \to X_0 \rightrightarrows X_1$) is exact if Z is isomorphic to the cokernel (resp. kernel) of $X_0 \rightrightarrows X_1$.*

Hence we have the isomorphisms, functorial with respect to $Y \in \mathcal{C}$:

(2.2.7) $\operatorname{Hom}_{\mathcal{C}}(\operatorname{Coker}(f, g), Y) \simeq \{u \in \operatorname{Hom}_{\mathcal{C}}(X_1, Y); u \circ f = u \circ g\}$,

(2.2.8) $\operatorname{Hom}_{\mathcal{C}}(Y, \operatorname{Ker}(f, g)) \simeq \{u \in \operatorname{Hom}_{\mathcal{C}}(Y, X_0); f \circ u = g \circ u\}$.

The cokernel L is visualized by the commutative diagram:

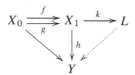

which means that any morphism $h \colon X_1 \to Y$ such that $h \circ f = h \circ g$ factors uniquely through k. Note that

(2.2.9) k is an epimorphism .

Dually, the kernel K is visualized by the commutative diagram:

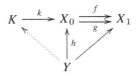

and

(2.2.10) k is a monomorphism.

Proposition 2.2.3. *Let $F \colon \mathcal{C} \to \mathcal{C}'$ be a functor.*

(i) *Assume that F is conservative and assume one of the hypotheses* (a) *or* (b) *below:*
(a) $\mathcal{C}$ *admits kernels and F commutes with kernels,*
(b) $\mathcal{C}$ *admits cokernels and F commutes with cokernels.*
Then F is faithful.
(ii) *Assume that F is faithful and assume that any morphism in $\mathcal{C}$ which is both a monomorphism and an epimorphism is an isomorphism. Then F is conservative.*

Proof. (i) Assume (a). Let $f, g \colon X \rightrightarrows Y$ be a pair of parallel arrows such that $F(f) = F(g)$. Let $N := \operatorname{Ker}(f, g)$. Denote by $u \colon N \to X$ the natural morphism. Then $F(N) \simeq \operatorname{Ker}(F(f), F(g))$. Hence $F(u)$ is an isomorphism. Since F is conservative, we get $N \xrightarrow{\sim} X$ and this implies $f = g$. Hence, F is faithful.
Assuming (b) instead of (a), the proof is the same by reversing the arrows.

(ii) Let $f \colon X \to Y$ be a morphism such that $F(f)$ is an isomorphism. Then f is both a monomorphism and an epimorphism by Proposition 1.2.12. It follows from the hypothesis that f is an isomorphism.   q.e.d.

Fiber Products and Coproducts

Consider the category I with three objects and two morphisms other than the identity morphisms visualized by the diagram

Let α be a functor from I to $\mathcal{C}$. Hence α is characterized by a diagram:

$$Y_0 \xleftarrow{\ f_0\ } X \xrightarrow{\ f_1\ } Y_1.$$

The inductive limit of α, if it exists, is called the *fiber coproduct* of Y_0 and Y_1 over X and denoted by $Y_0 \sqcup_X Y_1$.

Hence, for any $Z \in \mathcal{C}$, $\mathrm{Hom}_{\mathcal{C}}(Y_0 \sqcup_X Y_1, Z) \simeq \{(u_0, u_1); u_0 \in \mathrm{Hom}_{\mathcal{C}}(Y_0, Z),$ $u_1 \in \mathrm{Hom}_{\mathcal{C}}(Y_1, Z), u_0 \circ f_0 = u_1 \circ f_1\}$.

The fiber coproduct is visualized by the commutative diagram:

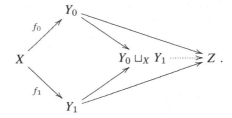

This means that if two morphisms from Y_0 and Y_1 to Z coincide after composition with f_0 and f_1 respectively, then they factorize uniquely through $Y_0 \sqcup_X Y_1$. We shall sometimes call the morphism $Y_i \to Y_0 \sqcup_X Y_1$ the *i-th coprojection*.

The fiber products over Y is defined by reversing the arrows. If β is a functor from I^{op} to $\mathcal{C}$, it is characterized by a diagram:

$$X_0 \xrightarrow{\ g_0\ } Y \xleftarrow{\ g_1\ } X_1.$$

The projective limit of β, if it exists, is called the *fiber product* of X_0 and X_1 over Y and denoted by $X_0 \times_Y X_1$. It is visualized by the commutative diagram:

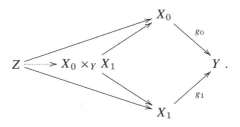

We shall sometimes call the morphism $X_0 \times_Y X_1 \to X_i$ the *i-th projection*.

Clearly, we have:

Proposition 2.2.4. (i) *Consider morphisms* $f_0 \colon X \to Y_0$ *and* $f_1 \colon X \to Y_1$.
If $Y_0 \sqcup Y_1$ *and* $Y_0 \sqcup_X Y_1$ *exist in* $\mathcal{C}$, *then the sequence* $X \rightrightarrows Y_0 \sqcup Y_1 \to Y_0 \sqcup_X Y_1$
is exact (*see* Definition 2.2.2).

(ii) *Consider morphisms* $g_0 \colon X_0 \to Y$ *and* $g_1 \colon X_1 \to Y$. *If* $Y_0 \times Y_1$ *and*
$Y_0 \times_X Y_1$ *exist in* $\mathcal{C}$, *then the sequence* $X_0 \times_Y X_1 \to X_0 \times X_1 \rightrightarrows Y$ *is*
exact.

Remark 2.2.5. The fiber coproduct (resp. fiber product) may also be formulated using the usual coproduct (resp. product).

(i) Let $Y \in \mathcal{C}$ and recall that j_Y denotes the forgetful functor $\mathcal{C}_Y \to \mathcal{C}$. Assume
that $\mathcal{C}_Y$ admits products indexed by a category I.

Consider a family $\{X_i \xrightarrow{f_i} Y\}_{i \in I}$ of objects of $\mathcal{C}_Y$. The fiber product over Y
of the X_i's, denoted by $\prod_{Y,i} X_i$, is given by

$$\prod_{Y,i} X_i := j_Y \big(\prod_i (X_i \xrightarrow{f_i} Y) \big)$$

where the product on the right hand side is the product in $\mathcal{C}_Y$. Clearly one
recovers $X_0 \times_Y X_1$ when $I = \{0, 1\}$.

The natural morphism $\prod_{Y,i} X_i \to X_j$ is again called the *j-th projection*.

(ii) One defines similarly the fiber coproduct over Y of a family $\{Y \xrightarrow{f_i} X_i\}_{i \in I}$
of objects of $\mathcal{C}^Y$, and one denotes it by $\bigsqcup_{Y,i} X_i$. The natural morphism $X_j \to$
$\bigsqcup_{Y,i} X_i$ is again called the *j-th coprojection*.

Recall that if a category $\mathcal{C}$ admits inductive limits indexed by a category I
and $Z \in \mathcal{C}$, then $\mathcal{C}_Z$ admits inductive limits indexed by I. (See Lemma 2.1.13.)

Definition 2.2.6. *Let* $\mathcal{C}$ *be a category which admits fiber products and induc-*
tive limits indexed by a category I.

(i) *We say that inductive limits in* $\mathcal{C}$ *indexed by* I *are* stable by base change
if for any morphism $Y \to Z$ *in* $\mathcal{C}$, *the base change functor* $\mathcal{C}_Z \to \mathcal{C}_Y$ *given*
by $\mathcal{C}_Z \ni (X \to Z) \mapsto (X \times_Z Y \to Y) \in \mathcal{C}_Y$ *commutes with inductive limits*
indexed by I.
This is equivalent to saying that for any inductive system $\{X_i\}_{i \in I}$ *in* $\mathcal{C}$
and any pair of morphisms $Y \to Z$ *and* $\varinjlim_{i \in I} X_i \to Z$ *in* $\mathcal{C}$, *we have the*
isomorphism

(2.2.11) $$\varinjlim_{i \in I}(X_i \times_Z Y) \xrightarrow{\sim} (\varinjlim_{i \in I} X_i) \times_Z Y \,.$$

(ii) *If* $\mathcal{C}$ *admits small inductive limits and* (2.2.11) *holds for any small cate-*
gory I, *we say that small inductive limits in* $\mathcal{C}$ *are* stable by base change.

The category **Set** admits small inductive limits and such limits are stable
by base change (see Exercise 2.7), but one shall be aware that in the category $\text{Mod}(\mathbb{Z})$, even finite inductive limits are not stable by base change. (See
Exercise 2.26.)

Definition 2.2.7. *Let us consider a commutative diagram in* $\mathcal{C}$*:*

(2.2.12)
$$
\begin{array}{ccc}
Y & \longrightarrow & X_0 \\
\downarrow & & \downarrow \\
X_1 & \longrightarrow & Z.
\end{array}
$$

(i) *The square* (2.2.12) *is* co-Cartesian *if* $X_0 \sqcup_Y X_1 \xrightarrow{\sim} Z$.

(ii) *The square* (2.2.12) *is* Cartesian *if* $Y \xrightarrow{\sim} X_0 \times_Z X_1$.

Assume that $\mathcal{C}$ admits finite coproducts. Then (2.2.12) is co-Cartesian if and only if the sequence below is exact (see Definition 2.2.2):

$$ Y \rightrightarrows X_0 \sqcup X_1 \to Z \, . $$

Assume that $\mathcal{C}$ admits finite products. Then (2.2.12) is Cartesian if and only if the sequence below is exact:

$$ Y \to X_0 \times X_1 \rightrightarrows Z \, . $$

Notations 2.2.8. Let $f \colon X \to Y$ be a morphism in a category $\mathcal{C}$.
(i) Assume that $\mathcal{C}$ admits fiber coproducts and denote by $i_1, i_2 \colon Y \rightrightarrows Y \sqcup_X Y$ the coprojections. We denote by $\sigma_Y \colon Y \sqcup_X Y \to Y$ (or simply σ) the natural morphism associated with $\mathrm{id}_Y \colon Y \to Y$, that is, $\sigma_Y \circ i_1 = \sigma_Y \circ i_2 = \mathrm{id}_Y$. We call σ_Y the *codiagonal morphism.*
(ii) Assume that $\mathcal{C}$ admits fiber products and denote by $p_1, p_2 \colon X \times_Y X \rightrightarrows X$ the projections. We denote by $\delta_X \colon X \to X \times_Y X$ (or simply δ) the natural morphism associated with $\mathrm{id}_X \colon X \to X$, that is, $p_1 \circ \delta_X = p_2 \circ \delta_X = \mathrm{id}_X$. We call δ_X the *diagonal morphism.*

Consider a category $\mathcal{C}$ which admits finite products and let $X \in \mathcal{C}$. We construct a functor

(2.2.13)
$$ X^{\Pi} \colon (\mathbf{Set}^f)^{\mathrm{op}} \to \mathcal{C} $$

as follows. For $I \in \mathbf{Set}^f$, we set

$$ X^{\Pi}(I) := X^{\Pi I} \ (\text{in particular, } X^{\Pi}(\emptyset) = \mathrm{pt}_{\mathcal{C}}) \, , $$

and for $(f \colon J \to I) \in \mathrm{Mor}(\mathcal{C})$,

$$ X^{\Pi}(f) \colon X^{\Pi I} \to X^{\Pi J} $$

is the morphism whose composition with the j-th projection $X^{\Pi J} \to X$ is the $f(j)$-th projection $X^{\Pi I} \to X$. Equivalently, for any $Z \in \mathcal{C}$, we have a map

$$\operatorname{Hom}_{\mathcal{C}}(Z, X^{\prod I}) \;\simeq\; \operatorname{Hom}_{\mathbf{Set}}(I, \operatorname{Hom}_{\mathcal{C}}(Z, X))$$
$$\xrightarrow{\;\circ f\;} \operatorname{Hom}_{\mathbf{Set}}(J, \operatorname{Hom}_{\mathcal{C}}(Z, X))$$
$$\simeq \operatorname{Hom}_{\mathcal{C}}(Z, X^{\prod J})\,,$$

which induces a morphism $X^{\prod I} \to X^{\prod J}$ by the Yoneda Lemma.

When $\mathcal{C}$ admits coproducts, we construct similarly a functor

(2.2.14) $$X^{\coprod}\colon \mathbf{Set}^f \to \mathcal{C}.$$

Thanks to Remark 2.2.5, these constructions extend to fiber coproducts and fiber products. If $\mathcal{C}$ admits fiber products and $u\colon X \to Y$ is a morphism in $\mathcal{C}$, we get a functor

(2.2.15) $$X^{\prod_Y}\colon (\mathbf{Set}^f)^{\mathrm{op}} \to \mathcal{C}$$
$$\mathbf{Set}^f \ni I \mapsto X^{\prod_Y I}\,,$$

and similarly with fiber coproducts.

Limits as Kernels and Products

We have seen that coproducts and cokernels (resp. products and kernels) are particular cases of inductive (resp. projective) limits. We shall show that, conversely, it is possible to construct inductive (resp. projective) limits using coproducts and cokernels (resp. products and kernels), when such objects exist.

Recall that $\mathrm{Mor}(I)$ denote the category of morphisms in I. There are two natural maps (source and target) from $\mathrm{Ob}(\mathrm{Mor}(I))$ to $\mathrm{Ob}(I)$:

$$\sigma \;:\; \mathrm{Ob}(\mathrm{Mor}(I)) \to \mathrm{Ob}(I), \quad (s\colon i \to j) \mapsto i\,,$$
$$\tau \;:\; \mathrm{Ob}(\mathrm{Mor}(I)) \to \mathrm{Ob}(I), \quad (s\colon i \to j) \mapsto j\,.$$

For a functor $\alpha\colon I \to \mathcal{C}$ and a morphism $s\colon i \to j$ in I, we get two morphisms in $\mathcal{C}^\vee$:

$$\alpha(i) \underset{\alpha(s)}{\overset{\mathrm{id}_{\alpha(i)}}{\rightrightarrows}} \alpha(i) \sqcup \alpha(j)$$

from which we deduce two morphisms in $\mathcal{C}^\vee$: $\alpha(\sigma(s)) \rightrightarrows \coprod_{i \in I} \alpha(i)$. These morphisms define the two morphisms in $\mathcal{C}^\vee$:

(2.2.16) $$\coprod_{s \in \mathrm{Mor}(I)} \alpha(\sigma(s)) \underset{b}{\overset{a}{\rightrightarrows}} \coprod_{i \in I} \alpha(i).$$

Similarly, if $\beta\colon I^{\mathrm{op}} \to \mathcal{C}$ is a functor and $s\colon i \to j$, we get two morphisms in $\mathcal{C}^\wedge$:

$$\beta(i) \times \beta(j) \underset{\beta(s)}{\overset{\mathrm{id}_{\beta(i)}}{\rightrightarrows}} \beta(i)$$

from which we deduce two morphisms in $\mathcal{C}^\wedge$: $\prod_{i \in I} \beta(i) \rightrightarrows \beta(\sigma(s))$. These morphisms define the two morphisms in $\mathcal{C}^\wedge$:

$$(2.2.17) \qquad \prod_{i \in I} \beta(i) \underset{b}{\overset{a}{\rightrightarrows}} \prod_{s \in \mathrm{Mor}(I)} \beta(\sigma(s)).$$

Proposition 2.2.9. (i) $\varinjlim \alpha$ is the cokernel of (a, b) in (2.2.16),
  (ii) $\varprojlim \beta$ is the kernel of (a, b) in (2.2.17).

Proof. Replacing $\mathcal{C}$ with $\mathcal{C}^{\mathrm{op}}$, it is enough to prove (ii).
  When $\mathcal{C} = \mathbf{Set}$, (ii) is nothing but the definition of projective limits in **Set**. Therefore, for $Z \in \mathcal{C}$, the projective limit $\varprojlim \mathrm{Hom}_\mathcal{C}(Z, \beta)$ in **Set** is the kernel of

$$\prod_{i \in I} \mathrm{Hom}_\mathcal{C}(Z, \beta(i)) \underset{b}{\overset{a}{\rightrightarrows}} \prod_{s \in \mathrm{Mor}(I)} \mathrm{Hom}_\mathcal{C}(Z, \beta(\sigma(s))).$$

The result follows by the Yoneda lemma.   q.e.d.

Corollary 2.2.10. *A category $\mathcal{C}$ admits small projective limits if and only if it satisfies:*

  (i) *$\mathcal{C}$ admits small products,*
  (ii) *for any pair of parallel arrows $f, g \colon X \rightrightarrows Y$ in $\mathcal{C}$, its kernel exists in $\mathcal{C}$.*

Corollary 2.2.11. *A category $\mathcal{C}$ admits finite projective limits if and only if it satisfies:*

  (i) *$\mathcal{C}$ admits a terminal object,*
  (ii) *for any $X, Y \in \mathrm{Ob}(\mathcal{C})$, their product $X \times Y$ exists in $\mathcal{C}$,*
  (iii) *for any pair of parallel arrows $f, g \colon X \rightrightarrows Y$ in $\mathcal{C}$, its kernel exists in $\mathcal{C}$.*

There is a similar result for finite inductive limits, replacing a terminal object by an initial object, products by coproducts and kernels by cokernels. (See also Exercise 2.6.)

2.3 Kan Extension of Functors

Definition 2.3.1. *Consider three categories $J, I, \mathcal{C}$ and a functor $\varphi \colon J \to I$.*

  (i) *The functor $\varphi_* \in \mathrm{Fct}\big(\mathrm{Fct}(I, \mathcal{C}), \mathrm{Fct}(J, \mathcal{C})\big)$ is defined by*

$$\varphi_* \alpha = \alpha \circ \varphi \text{ for } \alpha \in \mathrm{Fct}(I, \mathcal{C}) \, .$$

  (ii) *If the functor φ_* admits a left adjoint, we denote it by $\varphi^\dagger$. In such a case we have $\varphi^\dagger \in \mathrm{Fct}\big(\mathrm{Fct}(J, \mathcal{C}), \mathrm{Fct}(I, \mathcal{C})\big)$, and for $\alpha \in \mathrm{Fct}(I, \mathcal{C})$, $\beta \in \mathrm{Fct}(J, \mathcal{C})$ there is an isomorphism*

$$(2.3.1) \qquad \mathrm{Hom}_{\mathrm{Fct}(I, \mathcal{C})}(\varphi^\dagger \beta, \alpha) \simeq \mathrm{Hom}_{\mathrm{Fct}(J, \mathcal{C})}(\beta, \varphi_* \alpha) \, .$$

(iii) *If the functor φ_* admits a right adjoint, we denote it by $\varphi^\ddagger$. In such a case we have $\varphi^\ddagger \in \mathrm{Fct}\big(\mathrm{Fct}(J,\mathcal{C}),\mathrm{Fct}(I,\mathcal{C})\big)$, and for $\alpha \in \mathrm{Fct}(I,\mathcal{C})$, $\beta \in \mathrm{Fct}(J,\mathcal{C})$ there is an isomorphism*

(2.3.2) $$\mathrm{Hom}_{\,\mathrm{Fct}(I,\mathcal{C})}(\alpha, \varphi^\ddagger\beta) \simeq \mathrm{Hom}_{\,\mathrm{Fct}(J,\mathcal{C})}(\varphi_*\alpha, \beta) \,.$$

These functors of big categories are visualized by the diagram

$$\mathrm{Fct}(I,\mathcal{C}) \xrightleftharpoons[\;\varphi^\dagger\;]{\overset{\varphi^\ddagger}{\underset{\varphi_*}{\longleftarrow}}} \mathrm{Fct}(J,\mathcal{C}).$$

We have the adjunction morphisms

(2.3.3) $$\mathrm{id} \to \varphi_* \circ \varphi^\dagger \,,$$
(2.3.4) $$\varphi_* \circ \varphi^\ddagger \to \mathrm{id} \,.$$

For $\beta \in \mathrm{Fct}(J,\mathcal{C})$, the functors $\varphi^\dagger\beta$ and $\varphi^\ddagger\beta$ are visualized by the diagram:

The functors $\varphi^\dagger$ and $\varphi^\ddagger$ may be deduced one from the other by using the equivalence $\mathrm{Fct}(I,\mathcal{C})^{\mathrm{op}} \simeq \mathrm{Fct}(I^{\mathrm{op}},\mathcal{C}^{\mathrm{op}})$. Namely, we have the quasi-commutative diagram (assuming that $\varphi^\dagger$ exists):

(2.3.5)
$$
\begin{array}{ccc}
\mathrm{Fct}(J,\mathcal{C})^{\mathrm{op}} & \xrightarrow{\;(\varphi^\dagger)^{\mathrm{op}}\;} & \mathrm{Fct}(I,\mathcal{C})^{\mathrm{op}} \\
\Big\downarrow{\scriptstyle\sim} & & \Big\downarrow{\scriptstyle\sim} \\
\mathrm{Fct}(J^{\mathrm{op}},\mathcal{C}^{\mathrm{op}}) & \xrightarrow{\;(\varphi^{\mathrm{op}})^{\ddagger}\;} & \mathrm{Fct}(I^{\mathrm{op}},\mathcal{C}^{\mathrm{op}}).
\end{array}
$$

Definition 2.3.1 may be generalized as follows.

Definition 2.3.2. *Let $\beta \in \mathrm{Fct}(J,\mathcal{C})$.*

(a) *If the functor*

$$\mathrm{Fct}(I,\mathcal{C}) \ni \alpha \mapsto \mathrm{Hom}_{\,\mathrm{Fct}(J,\mathcal{C})}(\beta, \varphi_*\alpha) \in \mathbf{Set}$$

is representable, we denote by $\varphi^\dagger\beta \in \mathrm{Fct}(I,\mathcal{C})$ its representative, and we say that $\varphi^\dagger\beta$ exists.

(b) *Similarly, if the functor*

$$\mathrm{Fct}(I,\mathcal{C}) \ni \alpha \mapsto \mathrm{Hom}_{\,\mathrm{Fct}(J,\mathcal{C})}(\varphi_*\alpha, \beta) \in \mathbf{Set}$$

is representable, we denote by $\varphi^\ddagger\beta \in \mathrm{Fct}(I,\mathcal{C})$ its representative, and we say that $\varphi^\ddagger\beta$ exists.

Here **Set** should be understood as $\mathcal{V}$-**Set** for a sufficiently large universe $\mathcal{V}$.

If $\varphi^\dagger\beta$ (resp. $\varphi^\ddagger\beta$) exists, the isomorphism (2.3.1) (resp. (2.3.2)) holds for any $\alpha \in \mathrm{Mor}(I, \mathcal{C})$. It is obvious that if $\varphi^\dagger\beta$ (resp. $\varphi^\ddagger\beta$) exists for all $\beta \in \mathrm{Fct}(J, \mathcal{C})$, then the functor $\varphi^\dagger$ (resp. $\varphi^\ddagger$) exists.

Theorem 2.3.3. *Let $\varphi \colon J \to I$ be a functor and $\beta \in \mathrm{Fct}(J, \mathcal{C})$.*

(i) *Assume that* $\varinjlim_{(\varphi(j)\to i)\in J_i} \beta(j)$ *exists in $\mathcal{C}$ for any $i \in I$. Then $\varphi^\dagger\beta$ exists and we have*

$$(2.3.6) \qquad \varphi^\dagger\beta(i) \simeq \varinjlim_{(\varphi(j)\to i)\in J_i} \beta(j) \quad for\ i \in I \ .$$

In particular, if $\mathcal{C}$ admits small inductive limits and J is small, then $\varphi^\dagger$ exists. If moreover φ is fully faithful, then $\varphi^\dagger$ is fully faithful and there is an isomorphism $\mathrm{id}_{\mathrm{Fct}(J,\mathcal{C})} \xrightarrow{\sim} \varphi_*\varphi^\dagger$.

(ii) *Assume that* $\varprojlim_{(i\to\varphi(j))\in J^i} \beta(j)$ *exists for any $i \in I$. Then $\varphi^\ddagger\beta \in \mathrm{Fct}(I, \mathcal{C})$ exists and we have*

$$(2.3.7) \qquad \varphi^\ddagger\beta(i) \simeq \varprojlim_{(i\to\varphi(j))\in J^i} \beta(j) \quad for\ i \in I \ .$$

In particular, if $\mathcal{C}$ admits small projective limits and J is small, then $\varphi^\ddagger$ exists. If moreover φ is fully faithful, then $\varphi^\ddagger$ is fully faithful and there is an isomorphism $\varphi_*\varphi^\ddagger \xrightarrow{\sim} \mathrm{id}_{\mathrm{Fct}(J,\mathcal{C})}$.

Proof. (i) (a) Let us define $\varphi^\dagger\beta(i)$ by (2.3.6). For a morphism $u \colon i \to i'$, the morphism $\varphi^\dagger\beta(u)\colon \varphi^\dagger\beta(i) \to \varphi^\dagger\beta(i')$ is given as follows. Let $j \in J$ together with a morphism $\varphi(j) \to i$. It defines $\varphi(j) \to i \xrightarrow{u} i'$, hence a morphism

$$\beta(j) \to \varinjlim_{(\varphi(j')\to i')\in J_{i'}} \beta(j') = \varphi^\dagger\beta(i') \ .$$

Passing to the inductive limit with respect to $(\varphi(j) \to i) \in J_i$, we get the morphism $\varphi^\dagger\beta(u)\colon \varphi^\dagger\beta(i) \to \varphi^\dagger\beta(i')$. Thus $\varphi^\dagger\beta$ is a functor.

(i) (b) We shall show that (2.3.1) holds for the functor $\varphi^\dagger\beta$ defined by (2.3.6). It would be possible to use Lemma 2.1.15 but we prefer to give a direct proof.

First, we construct a map

$$\Phi \colon \mathrm{Hom}_{\mathrm{Fct}(J,\mathcal{C})}(\beta, \varphi_*\alpha) \to \mathrm{Hom}_{\mathrm{Fct}(I,\mathcal{C})}(\varphi^\dagger\beta, \alpha) \ .$$

An element $u \in \mathrm{Hom}_{\mathrm{Fct}(J,\mathcal{C})}(\beta, \varphi_*\alpha)$ gives a morphism $\beta(j) \to \alpha(\varphi(j)) \to \alpha(i)$ for any $i \in I$ and $\varphi(j) \to i \in J_i$. Hence we obtain a morphism

$$\varphi^\dagger\beta(i) = \varinjlim_{\varphi(j)\to i} \beta(j) \to \alpha(i) \ .$$

Clearly, the family of morphisms $\varphi^\dagger \beta(i) \to \alpha(i)$ so constructed is functorial in $i \in I$, hence defines $\Phi(u) \in \mathrm{Hom}_{\mathrm{Fct}(I,\mathcal{C})}(\varphi^\dagger \beta, \alpha)$.

Next, we construct a map

$$\Psi : \mathrm{Hom}_{\mathrm{Fct}(I,\mathcal{C})}(\varphi^\dagger \beta, \alpha) \to \mathrm{Hom}_{\mathrm{Fct}(J,\mathcal{C})}(\beta, \varphi_* \alpha) \ .$$

An element $v \in \mathrm{Hom}_{\mathrm{Fct}(I,\mathcal{C})}(\varphi^\dagger \beta, \alpha)$ defines a morphism for $j \in J$:

$$\beta(j) \to \varinjlim_{\varphi(j') \to \varphi(j)} \beta(j') \simeq \varphi^\dagger \beta(\varphi(j)) \xrightarrow{v} \alpha(\varphi(j)) \ .$$

Clearly, the family of morphisms $\beta(j) \to \varphi_* \alpha(j)$ so constructed is functorial in $j \in J$, hence defines $\Psi(v) \in \mathrm{Hom}_{\mathrm{Fct}(J,\mathcal{C})}(\beta, \varphi_* \alpha)$.

It is left to the reader to check that the maps Φ and Ψ are inverse to each other.

(i) (c) Assume that φ is fully faithful, $\mathcal{C}$ admits small inductive limits and J is small. Let $\beta \in \mathrm{Fct}(J, \mathcal{A})$ and $j \in J$. Since $J_j \to J_{\varphi(j)}$ is an equivalence of categories, we have

$$(\varphi_* \varphi^\dagger \beta)(j) \simeq (\varphi^\dagger \beta)(\varphi(j)) \simeq \varinjlim_{\varphi(j') \to \varphi(j)} \beta(j')$$

$$\simeq \varinjlim_{j' \to j} \beta(j') \simeq \beta(j) \ .$$

We deduce that $\varphi^\dagger$ is fully faithful by Proposition 1.5.6.

(ii) is equivalent to (i) by (2.3.5).   q.e.d.

Let $\alpha \colon J \to \mathcal{C}$ and $\beta \colon J^{\mathrm{op}} \to \mathcal{C}$ be functors. The morphisms (2.1.8) or (2.1.9) give morphisms

(2.3.8) $$\varinjlim \varphi_* \varphi^\dagger \alpha \to \varinjlim \varphi^\dagger \alpha \ ,$$

(2.3.9) $$\varprojlim (\varphi^{\mathrm{op}})^\ddagger \beta \to \varprojlim (\varphi^{\mathrm{op}})_* (\varphi^{\mathrm{op}})^\ddagger \beta \ .$$

Together with (2.3.3) and (2.3.4) we obtain the morphisms

(2.3.10) $$\varinjlim \alpha \to \varinjlim \varphi^\dagger \alpha \ ,$$

(2.3.11) $$\varprojlim (\varphi^{\mathrm{op}})^\ddagger \beta \to \varprojlim \beta \ .$$

Corollary 2.3.4. *Let $\varphi \colon J \to I$ be a functor of small categories.*

   (i) *Assume that $\mathcal{C}$ admits small inductive limits and let $\alpha \colon J \to \mathcal{C}$ be a functor. Then (2.3.10) is an isomorphism.*

  (ii) *Assume that $\mathcal{C}$ admits small projective limits and let $\beta \colon J^{\mathrm{op}} \to \mathcal{C}$ be a functor. Then (2.3.11) is an isomorphism.*

More intuitively, isomorphisms (2.3.10) and (2.3.11) may be written as

$$\varinjlim_{j \in J} \alpha(j) \xrightarrow{\sim} \varinjlim_{i \in I} (\varinjlim_{\varphi(j) \to i} \alpha(j)) \, ,$$

$$\varprojlim_{j \in J} \beta(j) \xleftarrow{\sim} \varprojlim_{i \in I} (\varprojlim_{\varphi(j) \to i} \beta(j)) \, .$$

Proof. For $X \in \mathcal{C}$, consider the constant functor $\Delta_X^I \colon I \to \mathcal{C}$ (see Notation 1.2.15). We have $\varphi_* \Delta_X^I \simeq \Delta_X^J$. Using the result of Exercise 2.8 we get the chain of isomorphisms

$$\mathrm{Hom}_{\mathcal{C}}(\varinjlim \alpha, X) \simeq \mathrm{Hom}_{\mathcal{C}^J}(\alpha, \Delta_X^J) \simeq \mathrm{Hom}_{\mathcal{C}^J}(\alpha, \varphi_* \Delta_X^I)$$

$$\simeq \mathrm{Hom}_{\mathcal{C}^I}(\varphi^\dagger \alpha, \Delta_X^I) \simeq \mathrm{Hom}_{\mathcal{C}}(\varinjlim \varphi^\dagger \alpha, X) \, .$$

q.e.d.

2.4 Inductive Limits in the Category Set

We have already noticed that the category **Set** admits small projective limits. Recall that $\bigsqcup$ denotes the disjoint union of sets.

Proposition 2.4.1. *The category* **Set** *admits small inductive limits. More precisely, if I is a small category and $\alpha \colon I \to$ **Set** *is a functor, then*

$$\varinjlim \alpha \simeq (\bigsqcup_{i \in I} \alpha(i)) / \sim \, ,$$

where $\sim$ is the equivalence relation generated by $\alpha(i) \ni x \sim y \in \alpha(j)$ if there exists $s \colon i \to j$ with $\alpha(s)(x) = y$.

Proof. Let $S \in$ **Set**. By the definition of the projective limit in **Set**, we get:

$$\varprojlim \mathrm{Hom}(\alpha, S) \simeq \big\{ \{p(i)\}_{i \in I} \, ; \, p(i) \in \mathrm{Hom}_{\mathbf{Set}}(\alpha(i), S), \ p(i) = p(j) \circ \alpha(s)$$
$$\text{for any } s \colon i \to j \big\}$$

$$\simeq \big\{ p \in \mathrm{Hom}_{\mathbf{Set}}(\bigsqcup_{i \in I} \alpha(i), S) \, ; \, p(x) = p(y) \text{ if } x \sim y \big\} \, .$$

The result follows.   q.e.d.

Notation 2.4.2. In the category **Set**, the notation $\bigsqcup$ is preferred to $\coprod$.

Let $\mathcal{C}$ be a category. Applying Proposition 2.1.6 we get:

Corollary 2.4.3. *The big category $\mathcal{C}^\wedge$ admits small inductive and small projective limits. If I is a small category and $\alpha \colon I \to \mathcal{C}^\wedge$ is a functor, we have the isomorphism for $X \in \mathcal{C}$*

$$(\varinjlim_i \alpha(i))(X) \simeq \varinjlim_i (\alpha(i)(X)) \ .$$

Similarly, if $\beta\colon I^{\mathrm{op}} \to \mathcal{C}^{\wedge}$ is a functor, we have the isomorphism

$$(\varprojlim_i \beta(i))(X) \simeq \varprojlim_i (\beta(i)(X)) \ .$$

There is a similar result for $\mathcal{C}^{\vee}$.

Recall that the terminal (resp. initial) object $\mathrm{pt}_{\mathcal{C}^{\wedge}}$ (resp. $\emptyset_{\mathcal{C}^{\wedge}}$) of $\mathcal{C}^{\wedge}$ is given by $\mathrm{pt}_{\mathcal{C}^{\wedge}}(X) = \{\mathrm{pt}\}$ (resp. $\emptyset_{\mathcal{C}^{\wedge}}(X) = \emptyset$).

Corollary 2.4.4.   (i) *The coproduct in* **Set** *is the disjoint union.*
  (ii) *The cokernel of $f, g\colon X \rightrightarrows Y$ in* **Set** *is the quotient set $Y/\sim$, where $\sim$ is the equivalence relation generated by $y \sim y'$ if there exists $x \in X$ such that $f(x) = y$ and $g(x) = y'$.*
 (iii) *Let I be a small category, let $S \in$ **Set**, and consider the constant functor $\Delta_S\colon I \to$ **Set** with values S. Then $\varinjlim \Delta_S \simeq S^{\sqcup \pi_0(I)}$ (see Definition 1.2.17). In particular, if $S = \{\mathrm{pt}\}$, then $\varinjlim \Delta_S \simeq \pi_0(I)$.*
 (iv) *Let I be a small category and consider a functor $\alpha\colon I \to$ **Set**. Set $I(\alpha):= I^{\{\mathrm{pt}\}}$, that is,*

$$\mathrm{Ob}(I(\alpha)) = \{(i, x); i \in I, x \in \alpha(i)\} \ ,$$
$$\mathrm{Hom}_{I(\alpha)}((i, x), (j, y)) = \{s \in \mathrm{Hom}_I(i, j); \alpha(s)(x) = y\} \ .$$

Then $\varinjlim \alpha \simeq \pi_0(I(\alpha))$.

Proof. (i) and (ii) are particular cases of Proposition 2.4.1.
(iii) Consider $\pi_0(I)$ as a discrete category. Then the functor Δ_S decomposes as

$$I \xrightarrow{\theta} \pi_0(I) \xrightarrow{\widetilde{\Delta}_S} \textbf{Set} \ ,$$

where $\widetilde{\Delta}_S$ is the constant functor with values S. Since I_a is connected for $a \in \pi_0(I)$, Lemma 2.1.12 implies that $\theta^{\dagger}\Delta_S \simeq \widetilde{\Delta}_S$. Applying Corollary 2.3.4, we get

$$\varinjlim \Delta_S \simeq \varinjlim \widetilde{\Delta}_S \simeq S^{\sqcup \pi_0(I)} \ .$$

(iv) By its definition,

$$\pi_0(I(\alpha)) = \bigsqcup_{(i,x) \in I(\alpha)} \{(i, x); i \in I, x \in \alpha(i)\}/ \sim$$

where $\sim$ is the equivalence relation generated by $(i, x) \sim (j, y)$ if there exists $s\colon i \to j$ with $\alpha(s)(x) = y$. This set is isomorphic to the set given in Proposition 2.4.1. q.e.d.

Corollary 2.4.5. *Let I be a small category and let $\Delta_{\{pt\}} \colon I \to \mathbf{Set}$ denote the constant functor with values $\{pt\}$. Then I is connected if and only if $\varinjlim \Delta_{\{pt\}} \simeq \{pt\}$.*

Proof. Apply Corollary 2.4.4 (iii).   q.e.d.

Corollary 2.4.6. *Let $F \colon C \to C'$ and $G \colon C \to C''$ be two functors, let $A \in C'$ and let $B \in C''$. We have the isomorphism*

$$(2.4.1) \quad \varinjlim_{(G(X) \to B) \in C_B} \mathrm{Hom}_{C'}(A, F(X)) \simeq \varinjlim_{(A \to F(X)) \in C^A} \mathrm{Hom}_{C''}(G(X), B) \,.$$

Proof. Consider the two functors $\varphi \colon C_B \to \mathbf{Set}$ and $\psi \colon (C^A)^{\mathrm{op}} \to \mathbf{Set}$ given by $\varphi(G(X) \to B) = \mathrm{Hom}_{C'}(A, F(X))$ and $\psi(A \to F(X)) = \mathrm{Hom}_{C''}(G(X), B)$. Define the category J as follows.

$$\mathrm{Ob}(J) = \big\{ (X, s, t) \, ; \, X \in C, \, s \colon A \to F(X), \, t \colon G(X) \to B \big\} \,,$$

$$\mathrm{Hom}_J((X, s, t), (X', s', t'))$$
$$= \big\{ f \colon X \to X' \, ; \, \text{the diagrams below commute} \big\}$$

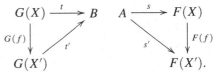

Using the notations and the result of Corollary 2.4.4 (iv), we have $J \simeq (C_B)(\varphi)$, $J^{\mathrm{op}} \simeq ((C^A)^{\mathrm{op}})(\psi)$, and $\varinjlim \varphi$ and $\varinjlim \psi$ are respectively isomorphic to $\pi_0(J)$ and $\pi_0(J^{\mathrm{op}}) \simeq \pi_0(J)$.   q.e.d.

The next result will be used in the sequel.

Lemma 2.4.7. *Let I be a small category and let $i_0 \in I$. Let $\alpha \colon I \to \mathbf{Set}$ be the functor $i \mapsto \mathrm{Hom}_I(i_0, i)$. Then $\varinjlim \alpha \simeq \{pt\}$.*

Proof. It is enough to show that the composition

$$\{pt\} \to \mathrm{Hom}_I(i_0, i_0) \to \varinjlim \alpha \,,$$
$$\mathrm{pt} \mapsto \mathrm{id}_{i_0}$$

is a surjective map. For $i \in I$ and $u \in \alpha(i) = \mathrm{Hom}_I(i_0, i)$, we have $u = \alpha(u) \circ \mathrm{id}_{i_0}$. Consider the maps:

$$\mathrm{Hom}_I(i_0, i_0) \xrightarrow{\alpha(u)} \mathrm{Hom}_I(i_0, i) \to \varinjlim \alpha \,.$$

The image of u in $\varinjlim \alpha$ is the image of id_{i_0}.   q.e.d.

2.5 Cofinal Functors

Definition 2.5.1.   (i) *A functor $\varphi: J \to I$ is* cofinal *if the category J^i is connected for any $i \in I$.*
  (ii) *A functor $\varphi: J \to I$ is* co-cofinal *if $\varphi^{op}: J^{op} \to I^{op}$ is cofinal, that is, if the category J_i is connected for any $i \in I$.*

We shall also say that J is cofinal to I by φ, or that J is cofinal to I.

Proposition 2.5.2. *Let $\varphi: J \to I$ be a functor of small categories. The conditions below are equivalent.*

  (i) *φ is cofinal,*
 (ii) *for any functor $\beta: I^{op} \to \mathbf{Set}$, the natural map $\varprojlim \beta \to \varprojlim (\beta \circ \varphi^{op})$ is bijective,*
(iii) *for any category $\mathcal{C}$ and any functor $\beta: I^{op} \to \mathcal{C}$, the natural morphism $\varprojlim \beta \to \varprojlim(\beta \circ \varphi^{op})$ is an isomorphism in $\mathcal{C}^\wedge$,*

 (iv) *for any functor $\alpha: I \to \mathbf{Set}$, the natural map $\varinjlim (\alpha \circ \varphi) \to \varinjlim \alpha$ is bijective,*
  (v) *for any category $\mathcal{C}$ and any functor $\alpha: I \to \mathcal{C}$, the natural morphism $\varinjlim(\alpha \circ \varphi) \to \varinjlim \alpha$ is an isomorphism in $\mathcal{C}^\vee$,*

 (vi) *for any $i \in I$, $\varinjlim\limits_{j \in J} \mathrm{Hom}\,_I(i, \varphi(j)) \simeq \{\mathrm{pt}\}$.*

Proof. (i) $\Rightarrow$ (v). Let us show that the natural morphism

$$\lambda: \varinjlim (\alpha \circ \varphi) \to \varinjlim \alpha$$

is an isomorphism. For $i_0 \in I$, let $j^{i_0}: J^{i_0} \to J$ be the forgetful functor. For $j \in J^{i_0}$, the morphism $i_0 \to \varphi(j)$ induces a morphism $\alpha(i_0) \to \alpha(\varphi(j))$. Hence, identifying $\alpha(i_0)$ with the constant functor $\Delta_{\alpha(i_0)}: J^{i_0} \to \mathcal{C}$, we obtain a chain of morphisms

$$\varinjlim_{j \in J^{i_0}} \alpha(i_0) \to \varinjlim_{j \in J^{i_0}} \alpha(\varphi(j)) \simeq \varinjlim \alpha \circ \varphi \circ j^{i_0} \to \varinjlim \alpha \circ \varphi \,.$$

Since J^{i_0} is connected, $\varinjlim\limits_{j \in J^{i_0}} \alpha(i_0) \simeq \alpha(i_0)$ by Lemma 2.1.12. Hence, we obtain a morphism $\alpha(i_0) \to \varinjlim \alpha \circ \varphi$. Taking the inductive limit with respect to $i_0 \in I$, we get a morphism $\mu: \varinjlim \alpha \to \varinjlim \alpha \circ \varphi$. Hence, for any $i \in I$, the composition $\alpha(i) \to \varinjlim \alpha \xrightarrow{\mu} \varinjlim \alpha \circ \varphi$ is given by $\alpha(i) \xrightarrow{\alpha(u)} \alpha(\varphi(j)) \to \varinjlim \alpha \circ \varphi$ by taking $j \in J$ and a morphism $u: i \to \varphi(j)$. It is easily checked that λ and μ are inverse to each other.

(ii) $\Rightarrow$ (iii). Let $X \in \mathcal{C}$. By the hypothesis, there is an isomorphism

$$\varprojlim \mathrm{Hom}_{\mathcal{C}}(X, \beta) \xrightarrow{\sim} \varprojlim \mathrm{Hom}_{\mathcal{C}}(X, \beta \circ \varphi) \ .$$

To conclude, apply Corollary 1.4.7.

(iii) $\Rightarrow$ (ii), (iii) $\Leftrightarrow$ (v) and (v) $\Rightarrow$ (iv) are obvious.

(iv) $\Rightarrow$ (vi) follows from Lemma 2.4.7.

(vi) $\Rightarrow$ (i). Let $i \in I$. Let $\beta : J \to \mathbf{Set}$ be the constant functor with values $\{\mathrm{pt}\}$. Then $J_{\{\mathrm{pt}\}} \simeq J$ (here, $J_{\{\mathrm{pt}\}}$ is associated to β), and we have

$$\{\mathrm{pt}\} \simeq \varinjlim_{j \in J} \mathrm{Hom}_I(i, \varphi(j))$$

$$\simeq \varinjlim_{j \in J^i} \mathrm{Hom}_{\mathbf{Set}}(\beta(j), \{\mathrm{pt}\}) \simeq \varinjlim_{j \in J^i} \beta(j) \simeq \pi_0(J^i) \ .$$

Here the first isomorphism follows from the hypothesis, the second from Corollary 2.4.6 and the last from Corollary 2.4.4 (iii).   q.e.d.

Corollary 2.5.3. *Let $\varphi : J \to I$ be a cofinal functor of small categories. Then I is connected if and only if J is connected.*

Proof. Denote by $\Delta^I_{\{\mathrm{pt}\}}$ the constant functor $I \to \mathbf{Set}$ with values $\{\mathrm{pt}\}$, and similarly with J instead of I. Then $\Delta^J_{\{\mathrm{pt}\}} \simeq \Delta^I_{\{\mathrm{pt}\}} \circ \varphi$. Since $\varinjlim \Delta^J_{\{\mathrm{pt}\}} \simeq \varinjlim \Delta^I_{\{\mathrm{pt}\}}$, the result follows from Corollary 2.4.5.   q.e.d.

Proposition 2.5.4. *Let $\psi : K \to J$ and $\varphi : J \to I$ be two functors.*

(i) *If φ and ψ are cofinal, then so is $\varphi \circ \psi$.*
(ii) *If $\varphi \circ \psi$ and ψ are cofinal, then so is φ.*
(iii) *If φ is fully faithful and $\varphi \circ \psi$ is cofinal, then φ and ψ are cofinal.*

Proof. By taking a larger universe, we may assume that I, J, K are small. Consider a functor $\alpha : I \to \mathbf{Set}$. We get functors

$$K \xrightarrow{\psi} J \xrightarrow{\varphi} I \xrightarrow{\alpha} \mathbf{Set}$$

and maps

$$\varinjlim \alpha \circ \varphi \circ \psi \xrightarrow{a_\psi(\alpha \circ \varphi)} \varinjlim (\alpha \circ \varphi) \xrightarrow{a_\varphi(\alpha)} \varinjlim \alpha \ .$$

(i) Clearly, if $a_\varphi(\alpha)$ and $a_\psi(\alpha \circ \varphi)$ are bijective for all α, then $a_{\varphi \circ \psi}(\alpha) = a_\varphi(\alpha) \circ a_\psi(\alpha \circ \varphi)$ is bijective for all α.

(ii) Assume that $a_{\varphi \circ \psi}(\alpha)$ and $a_\psi(\alpha \circ \varphi)$ are bijective for all α. Then $a_\varphi(\alpha)$ is bijective for all α.

(iii) For $j \in J$, $K^j \simeq K^{\varphi(j)}$ and this category is connected. Hence, ψ is cofinal. Then φ is cofinal by (ii).   q.e.d.

Definition 2.5.5. (i) *A category I is* cofinally small *if there exist a small category J and a cofinal functor $\varphi\colon J \to I$.*
(ii) *A category I is* co-cofinally small *if I^{op} is cofinally small.*

Corollary 2.5.6. *Assume that I is cofinally small. Then there exists a small full subcategory J of I cofinal to I.*

Proof. Let $\theta\colon K \to I$ be a cofinal functor with K small and let J denote the full subcategory of I whose objects are the images of $\mathrm{Ob}(K)$ by θ. Then J is small. Denote by $\psi\colon K \to J$ the functor induced by θ. Then Proposition 2.5.4 (iii) implies that the embedding functor $J \to I$ is cofinal.   q.e.d.

Note that if $\mathcal{C}$ is a category which admits small inductive limits and I is cofinally small, then $\mathcal{C}$ admits inductive limits indexed by I, and similarly for projective limits.

2.6 Ind-lim and Pro-lim

Let $\mathcal{C}$ be a category. Recall that the Yoneda lemma implies that the functor $h_{\mathcal{C}}\colon \mathcal{C} \to \mathcal{C}^{\wedge}$ is fully faithful, which allows us to identify $\mathcal{C}$ with a full subcategory of $\mathcal{C}^{\wedge}$. Hence, when there is no risk of confusion, we shall not write the functor $h_{\mathcal{C}}$.

Recall that in Notation 1.4.5 we have set $A(X) = \mathrm{Hom}_{\mathcal{C}^{\wedge}}(X, A)$ for $A \in \mathcal{C}^{\wedge}$ and $X \in \mathcal{C}$, and more generally, $A(B) = \mathrm{Hom}_{\mathcal{C}^{\wedge}}(B, A)$ for $A, B \in \mathcal{C}^{\wedge}$. In particular, we identify an element $s \in A(X)$ with a morphism $s\colon X \to A$.

We have already noticed in Corollary 2.4.3 that the big category $\mathcal{C}^{\wedge}$ admits small projective and inductive limits. Whenever $\mathcal{C}$ admits small projective limits, the functor $h_{\mathcal{C}}$ commutes with such limits, but even when $\mathcal{C}$ admits small inductive limits, the functor $h_{\mathcal{C}}$ does not commute with $\varinjlim$.

In order to avoid any confusion, we introduce the following notations.

Notations 2.6.1. (i) We denote by "$\varinjlim$" and "$\coprod$" the inductive limit and the coproduct in $\mathcal{C}^{\wedge}$, respectively.
(ii) We sometimes write X "$\coprod$" Y instead of X "$\coprod$" Y.
(iii) If I is small and $\alpha\colon I \to \mathcal{C}^{\wedge}$ is a functor, we sometimes write "$\varinjlim_{i \in I}$" $\alpha(i)$ or "$\varinjlim_{i}$" $\alpha(i)$ instead of "$\varinjlim$" α. Recall that $(\,"\varinjlim_{i}" \, \alpha(i))(X) \simeq \varinjlim_{i} \big((\alpha(i))(X)\big)$ for any $X \in \mathcal{C}$.
(iv) If I is small and $\alpha\colon I \to \mathcal{C}$ is a functor, we set for short "$\varinjlim$" $\alpha = $ "$\varinjlim$"$(h_{\mathcal{C}} \circ \alpha)$.

(v) We call "$\varinjlim$" α the ind-lim of α.

Notations 2.6.2. (i) Similarly, we denote by "$\varprojlim$" and "$\prod$" the projective limit and the product in $\mathcal{C}^\vee$.

(ii) If I is small and $\beta\colon I^{\mathrm{op}} \to \mathcal{C}^\vee$ is a functor, we sometimes write "$\varprojlim_{i \in I}$" $\beta(i)$, "$\varprojlim_{i}$" $\beta(i)$ or "$\varprojlim_{i \in I^{\mathrm{op}}}$" $\beta(i)$ instead of "$\varprojlim$" β.

(iii) If I is small and $\beta\colon I^{\mathrm{op}} \to \mathcal{C}$ is a functor, we set for short "$\varprojlim$" $\beta =$ "$\varprojlim$"$(k_{\mathcal{C}} \circ \beta)$.

(iv) We call "$\varprojlim$" β the pro-lim of β.

With these notations, if I is small and $\alpha\colon I \to \mathcal{C}^\wedge$ and $\beta\colon I^{\mathrm{op}} \to \mathcal{C}^\vee$ are functors, we have for $X \in \mathcal{C}$

$$(2.6.1) \qquad \operatorname{Hom}_{\mathcal{C}^\wedge}(X, \text{``}\varinjlim\text{''}\,\alpha) = \varinjlim \operatorname{Hom}_{\mathcal{C}^\wedge}(X, \alpha) \,,$$

$$(2.6.2) \qquad \operatorname{Hom}_{\mathcal{C}^\vee}(\text{``}\varprojlim\text{''}\,\beta, X) = \varinjlim \operatorname{Hom}_{\mathcal{C}^\vee}(\beta, X) \,.$$

One shall be aware that isomorphism (2.6.1) (resp. (2.6.2)) is no more true for $X \in \mathcal{C}^\wedge$ (resp. $X \in \mathcal{C}^\vee$) in general, even if α (resp. β) takes its values in $\mathcal{C}$.

For $A \in \mathcal{C}^\wedge$ and $B \in \mathcal{C}^\vee$, we have

$$\operatorname{Hom}_{\mathcal{C}^\wedge}(\text{``}\varinjlim\text{''}\,\alpha, A) \simeq \varprojlim \operatorname{Hom}_{\mathcal{C}^\wedge}(\alpha, A) \,,$$

$$\operatorname{Hom}_{\mathcal{C}^\vee}(B, \text{``}\varprojlim\text{''}\,\beta) \simeq \varprojlim \operatorname{Hom}_{\mathcal{C}^\vee}(B, \beta) \,.$$

Notice that the inductive limit of $\alpha\colon I \to \mathcal{C}$ is an object of $\mathcal{C}^\vee$ while the ind-lim of α is an object of $\mathcal{C}^\wedge$, and the projective limit of $\beta\colon I^{\mathrm{op}} \to \mathcal{C}$ is an object of $\mathcal{C}^\wedge$ while the pro-lim of β is an object of $\mathcal{C}^\vee$.

Let β be a contravariant functor from I to $\mathcal{C}$, that is, a functor $I^{\mathrm{op}} \to \mathcal{C}$. Then we get a functor $\beta^{\mathrm{op}}\colon I \to \mathcal{C}^{\mathrm{op}}$, and we have:

$$(2.6.3) \qquad \text{``}\varprojlim\text{''}\,\beta \simeq \xi(\text{``}\varinjlim\text{''}\,(\beta^{\mathrm{op}})) \,,$$

where ξ is the contravariant functor $(\mathcal{C}^{\mathrm{op}})^\wedge \to \mathcal{C}^\vee$.

From now on, we shall concentrate our study on "$\varinjlim$", the results on "$\varprojlim$" being deduced using (2.6.3).

Assume that $\mathcal{C}$ admits small inductive limits. Then, for a functor $\alpha\colon I \to \mathcal{C}$, the natural map $\varinjlim \operatorname{Hom}_{\mathcal{C}}(X, \alpha) \to \operatorname{Hom}_{\mathcal{C}}(X, \varinjlim \alpha)$ defines the morphism in $\mathcal{C}^\wedge$:

$$\text{``}\varinjlim\text{''}\,\alpha \to h_{\mathcal{C}}(\varinjlim \alpha) \,.$$

If $\alpha\colon I \to \mathcal{C}$ and $\beta\colon J \to \mathcal{C}$ are functors defined on small categories, there are isomorphisms:

$$\operatorname{Hom}_{\mathcal{C}^\wedge}(\text{``}\varinjlim_{i}\text{''}\,\alpha(i), \text{``}\varinjlim_{j}\text{''}\,\beta(j)) \simeq \varprojlim_{i} \operatorname{Hom}_{\mathcal{C}^\wedge}(\alpha(i), \text{``}\varinjlim_{j}\text{''}\,\beta(j))$$

$$(2.6.4) \qquad\qquad\qquad\qquad\qquad \simeq \varprojlim_{i} \varinjlim_{j} \operatorname{Hom}_{\mathcal{C}}(\alpha(i), \beta(j)) \,.$$

Proposition 2.6.3. *Let $\mathcal{C}$ be a category.*

(i) *Let $A \in \mathcal{C}^\wedge$. Then $\underset{(V \to A) \in \mathcal{C}_A}{\text{"$\varinjlim$"}} V$ exists in $\mathcal{C}^\wedge$ and $\underset{(V \to A) \in \mathcal{C}_A}{\text{"$\varinjlim$"}} V \simeq A$.*

(ii) *Let I be a small category and $\alpha \colon I \to \mathcal{C}$ a functor. Set $A = \text{"$\varinjlim$"}\, \alpha$. Then the functor $\widetilde{\alpha} \colon I \to \mathcal{C}_A$ associated with α is cofinal.*

Using the functor $j_A \colon \mathcal{C}_A \to \mathcal{C}$ given in Definition 1.2.16, (i) is translated as: $A \simeq \text{"$\varinjlim$"}\, j_A$. Note that $\mathcal{C}_A$ is not essentially small in general.

Proof. (i) follows from the fact that, for any $B \in \mathcal{C}^\wedge$, the map

$$\text{Hom}_{\mathcal{C}^\wedge}(A, B) \to \underset{(V \to A) \in \mathcal{C}_A}{\varprojlim} \text{Hom}_{\mathcal{C}^\wedge}(V, B)$$

$$\simeq \underset{(V \to A) \in \mathcal{C}_A}{\varprojlim} B(V)$$

is bijective by the definition of a morphism of functors.

(ii) The functor $h_{\mathcal{C}} \colon \mathcal{C} \to \mathcal{C}^\wedge$ induces a functor $(h_{\mathcal{C}})_A \colon \mathcal{C}_A \to (\mathcal{C}^\wedge)_A$. By Lemma 1.4.12, there exists an equivalence $\lambda \colon (\mathcal{C}^\wedge)_A \simeq (\mathcal{C}_A)^\wedge$ such that $h_{\mathcal{C}_A} \simeq \lambda \circ (h_{\mathcal{C}})_A$, visualized by the diagram

$$
\begin{array}{ccccccc}
I & \xrightarrow{\ \widetilde{\alpha}\ } & \mathcal{C}_A & \xrightarrow{\ (h_{\mathcal{C}})_A\ } & (\mathcal{C}^\wedge)_A & \longrightarrow & \mathcal{C}^\wedge \\
 & & & \searrow{\scriptstyle h_{\mathcal{C}_A}} & \ \downarrow{\scriptstyle \sim}\lambda & & \\
 & & & & (\mathcal{C}_A)^\wedge. & &
\end{array}
$$

By Lemma 2.1.13, the functor $(\mathcal{C}^\wedge)_A \to \mathcal{C}^\wedge$ commutes with small inductive limits. Since $\text{"$\varinjlim$"}\,(h_{\mathcal{C}} \circ \alpha) \simeq A$, it follows that $\lambda^{-1}\big(\text{"$\varinjlim$"}\,(h_{\mathcal{C}_A} \circ \widetilde{\alpha})\big)$ is isomorphic to $(A \xrightarrow{\text{id}} A)$, the terminal object of $(\mathcal{C}^\wedge)_A$. Hence, $\text{"$\varinjlim$"}\,(h_{\mathcal{C}_A} \circ \widetilde{\alpha}) \simeq \text{pt}_{(\mathcal{C}_A)^\wedge}$, i.e.,

$$\underset{i \in I}{\varinjlim}\, \text{Hom}_{\mathcal{C}_A}\big((X, s), \widetilde{\alpha}(i)\big) \simeq \{\text{pt}\}$$

for any $(X, s) \in \mathcal{C}_A$. This implies that $\widetilde{\alpha} \colon I \to \mathcal{C}_A$ is cofinal by Proposition 2.5.2 (i)$\Leftrightarrow$(vi).   q.e.d.

Let us compare the inductive limits "$\varinjlim$" in $\mathcal{C}^\wedge$ and $\varinjlim$ in $\mathcal{C}$.

Proposition 2.6.4. *Let I be a small category and $\alpha \colon I \to \mathcal{C}$ a functor. Assume that $\text{"$\varinjlim$"}\,(h_{\mathcal{C}} \circ \alpha) \in \mathcal{C}^\wedge$ is isomorphic to an object $X \in \mathcal{C}$. Then for any functor $F \colon \mathcal{C} \to \mathcal{C}'$, $\varinjlim (F \circ \alpha) \simeq F(X)$.*

Proof. It is enough to prove the isomorphism

$$\varprojlim \text{Hom}_{\mathcal{C}'}(F \circ \alpha, Y) \simeq \text{Hom}_{\mathcal{C}'}(F(X), Y)\,,$$

functorially in $Y \in \mathcal{C}'$. Let us define $\mathcal{H}om\,(F, Y) \in \mathcal{C}^\wedge$ by

(2.6.5) $\mathcal{H}om\,(F, Y)(Z) = \mathrm{Hom}_{\mathcal{C}'}(F(Z), Y)$ for $Z \in \mathcal{C}$.

Then

$$\varprojlim \mathrm{Hom}_{\mathcal{C}'}(F \circ \alpha, Y) \simeq \varprojlim \mathrm{Hom}_{\mathcal{C}^\wedge}(\alpha, \mathcal{H}om\,(F, Y))$$
$$\simeq \mathrm{Hom}_{\mathcal{C}^\wedge}(\text{``}\varinjlim\text{''}\,\alpha, \mathcal{H}om\,(F, Y))$$
$$\simeq \mathrm{Hom}_{\mathcal{C}^\wedge}(X, \mathcal{H}om\,(F, Y))$$
$$\simeq \mathrm{Hom}_{\mathcal{C}'}(F(X), Y)\ .$$

q.e.d.

This shows that "$\varinjlim$" $\alpha \simeq X$ implies $\varinjlim \alpha \simeq X$, but the first assertion is much stronger (see Exercise 2.25 and also Proposition 6.2.1).

Remark 2.6.5. Let $\mathcal{U} \subset \mathcal{V}$ be two universes and $\mathcal{C}$ a $\mathcal{U}$-category. With the notations of Remark 1.4.13, we have a fully faithful functor $\iota_{\mathcal{V},\mathcal{U}} \colon \mathcal{C}_{\mathcal{U}}^\wedge \to \mathcal{C}_{\mathcal{V}}^\wedge$. This functor commutes with inductive and projective limits indexed by $\mathcal{U}$-small categories.

2.7 Yoneda Extension of Functors

In this section, we apply Theorem 2.3.3 to the particular case where $\varphi \colon J \to I$ is the Yoneda functor $h_\mathcal{C} \colon \mathcal{C} \to \mathcal{C}^\wedge$. Hence, we assume

(2.7.1)   the category $\mathcal{C}$ is small .

Proposition 2.7.1. *Let $F \colon \mathcal{C} \to \mathcal{A}$ be a functor, assume (2.7.1) and assume that $\mathcal{A}$ admits small inductive limits. Then the functor $h_\mathcal{C}^\dagger\, F \colon \mathcal{C}^\wedge \to \mathcal{A}$ exists, commutes with small inductive limits and satisfies $h_\mathcal{C}^\dagger\, F \circ h_\mathcal{C} \simeq F$.*
Conversely if a functor $\widetilde{F} \colon \mathcal{C}^\wedge \to \mathcal{A}$ satisfies the following two conditions:

(a) $\widetilde{F} \circ h_\mathcal{C} \simeq F$,
(b) $\widetilde{F}$ *commutes with small inductive limits with values in $\mathcal{C}$ (i.e., for any functor $\alpha \colon I \to \mathcal{C}$ with I small, $\widetilde{F}(\text{``}\varinjlim\text{''}\,\alpha) \simeq \varinjlim (F \circ \alpha))$,*

then $\widetilde{F} \simeq h_\mathcal{C}^\dagger\, F$.

Proof. We set $\widetilde{F} = h_\mathcal{C}^\dagger\, F$. By Theorem 2.3.3, this functor exists and we have

(2.7.2) $\widetilde{F}(A) = \varinjlim_{(U \to A) \in \mathcal{C}_A} F(U)$ for $A \in \mathcal{C}^\wedge$.

Since $h_\mathcal{C}$ is fully faithful, the same theorem implies $\widetilde{F} \circ h_\mathcal{C} \simeq F$.

For $M \in \mathcal{A}$, recall (see (2.6.5)) that $\mathcal{H}om\,(F, M) \in \mathcal{C}^\wedge$ is given by the formula

$$\mathrm{Hom}_{\mathcal{C}^\wedge}(U, \mathcal{H}om\,(F, M)) = \mathrm{Hom}_{\mathcal{A}}(F(U), M) \text{ for } U \in \mathcal{C}\,.$$

For $A \in \mathcal{C}^\wedge$, we get

$$\mathrm{Hom}_{\mathcal{C}^\wedge}(A, \mathcal{H}om\,(F, M)) \simeq \mathrm{Hom}_{\mathcal{C}^\wedge}(\underset{U \to A}{\text{``}\varinjlim\text{''}}\, U, \mathcal{H}om\,(F, M))$$

$$\simeq \varprojlim_{U \to A} \mathrm{Hom}_{\mathcal{C}^\wedge}(U, \mathcal{H}om\,(F, M)) \simeq \varprojlim_{U \to A} \mathrm{Hom}_{\mathcal{A}}(F(U), M)$$

$$\simeq \mathrm{Hom}_{\mathcal{A}}(\varinjlim_{U \to A} F(U), M) \simeq \mathrm{Hom}_{\mathcal{A}}(\widetilde{F}(A), M)\,.$$

Let I be a small category and $\alpha\colon I \to \mathcal{C}^\wedge$ a functor. Set $A = \text{``}\varinjlim\text{''}\,\alpha$. We get

$$\mathrm{Hom}_{\mathcal{A}}(\widetilde{F}(A), M) \simeq \mathrm{Hom}_{\mathcal{C}^\wedge}(A, \mathcal{H}om\,(F, M)) \simeq \varprojlim \mathrm{Hom}_{\mathcal{C}^\wedge}(\alpha, \mathcal{H}om\,(F, M))$$

$$\simeq \varprojlim \mathrm{Hom}_{\mathcal{A}}(\widetilde{F}(\alpha), M) \simeq \mathrm{Hom}_{\mathcal{A}}(\varinjlim \widetilde{F}(\alpha), M)\,.$$

Therefore the natural morphism $\varinjlim \widetilde{F}(\alpha) \to \widetilde{F}(A)$ is an isomorphism by Corollary 1.4.7.

The uniqueness is obvious since (a) and (b) imply (2.7.2) by Proposition 2.6.3.   q.e.d.

Notation 2.7.2. If $F\colon \mathcal{C} \to \mathcal{C}'$ is a functor of small categories, we shall denote by

$$\widehat{F}\colon \mathcal{C}^\wedge \to (\mathcal{C}')^\wedge$$

the functor $\mathrm{h}_{\mathcal{C}}^\dagger\,(\mathrm{h}_{\mathcal{C}'} \circ F)$ associated with $\mathrm{h}_{\mathcal{C}'} \circ F\colon \mathcal{C} \to (\mathcal{C}')^\wedge$. Hence, for $A \in \mathcal{C}^\wedge$ and $V \in \mathcal{C}'$,

$$\left(\widehat{F}(A)\right)(V) \simeq \varinjlim_{(U \to A) \in \mathcal{C}_A} \mathrm{Hom}_{\mathcal{C}'}(V, F(U))\,.$$

By Proposition 2.7.1, $\widehat{F}$ commutes with small inductive limits.

Notation 2.7.3. We denote by $\mathrm{Fct}^{il}(\mathcal{C}^\wedge, \mathcal{A})$ the full big subcategory of the big category $\mathrm{Fct}(\mathcal{C}^\wedge, \mathcal{A})$ consisting of functors which commute with small inductive limits.

Corollary 2.7.4. *Assume* (2.7.1) *and assume that* $\mathcal{A}$ *admits small inductive limits. Then* $\mathrm{h}_{\mathcal{C}*}\colon \mathrm{Fct}(\mathcal{C}^\wedge, \mathcal{A}) \to \mathrm{Fct}(\mathcal{C}, \mathcal{A})$ *induces an equivalence of categories*

$$(2.7.3) \qquad\qquad \mathrm{h}_{\mathcal{C}*}\colon \mathrm{Fct}^{il}(\mathcal{C}^\wedge, \mathcal{A}) \xrightarrow{\sim} \mathrm{Fct}(\mathcal{C}, \mathcal{A})\,,$$

and a quasi-inverse is given by $\mathrm{h}_{\mathcal{C}}^\dagger$.

Proof. By Proposition 2.7.1, the functor $h_{\mathcal{C}}^{\dagger}$ takes its values in $\mathrm{Fct}^{il}(\mathcal{C}^{\wedge}, \mathcal{A})$ and any $G \in \mathrm{Fct}^{il}(\mathcal{C}^{\wedge}, \mathcal{A})$ is isomorphic to $h_{\mathcal{C}}^{\dagger}(h_{\mathcal{C}*}G)$. Hence, $h_{\mathcal{C}}^{\dagger}$ is essentially surjective. Since $h_{\mathcal{C}}^{\dagger}$ is fully faithful by Theorem 2.3.3, the result follows. q.e.d.

Let $F \colon \mathcal{C} \to \mathcal{C}'$ be a functor of small categories. We have defined $\widehat{F} \colon \mathcal{C}^{\wedge} \to \mathcal{C}'^{\wedge}$ in Notation 2.7.2 and we have defined $(F^{\mathrm{op}})^{\dagger} \colon \mathcal{C}^{\wedge} \to \mathcal{C}'^{\wedge}$ in Definition 2.3.1 with $\mathcal{A} = \mathbf{Set}$.

Proposition 2.7.5. *Let $F \colon \mathcal{C} \to \mathcal{C}'$ be a functor of small categories. There is an isomorphism $\widehat{F} \simeq (F^{\mathrm{op}})^{\dagger}$ in $\mathrm{Fct}(\mathcal{C}^{\wedge}, (\mathcal{C}')^{\wedge})$.*

Proof. Let $A \in (\mathcal{C}')^{\wedge}$ and $V \in \mathcal{C}'$. Applying Corollary 2.4.6, we obtain

$$
\begin{aligned}
(\widehat{F}(A))(V) &\simeq \varinjlim_{(U \to A) \in \mathcal{C}_A} \mathrm{Hom}_{\mathcal{C}'}(V, F(U)) \\
&\simeq \varinjlim_{(V \to F(U)) \in \mathcal{C}^V} \mathrm{Hom}_{\mathcal{C}^{\wedge}}(U, A) \simeq ((F^{\mathrm{op}})^{\dagger}(A))(V) \, .
\end{aligned}
$$

q.e.d.

Exercises

Exercise 2.1. Let $F \colon \mathcal{C} \to \mathcal{C}'$ be an equivalence of categories and let G be a quasi-inverse. Assume that $\mathcal{C}$ admits inductive limits indexed by a category I. Prove that $\mathcal{C}'$ has the same property and that if $\alpha \colon I \to \mathcal{C}$ is a functor, then $\varinjlim(F \circ \alpha) \simeq F(\varinjlim \alpha)$.

Exercise 2.2. Let $f \colon X \twoheadrightarrow Y$ be an epimorphism in a category $\mathcal{C}$ and let $s_1, s_2 \colon Y \rightrightarrows Z$ be a pair of parallel arrows. Prove that the natural morphism $\mathrm{Coker}(s_1 \circ f, s_2 \circ f) \to \mathrm{Coker}(s_1, s_2)$ is an isomorphism in $\mathcal{C}$ if these cokernels exist.

Exercise 2.3. Let $f \colon X \to Y$ be a morphism in $\mathbf{Set}$, and set $Z = Y \sqcup_X Y$. Prove that

$$
Z \simeq f(X) \sqcup (Y \setminus f(X)) \sqcup (Y \setminus f(X)) \, .
$$

Exercise 2.4. Let $\mathcal{C}$ be a category which admits fiber products and let $f \colon X \to Y$ be a morphism in $\mathcal{C}$. Denote by p_1, p_2 the projections $X \times_Y X \rightrightarrows X$ and by δ the diagonal morphism $X \to X \times_Y X$ (see Notations 2.2.8).
(i) Prove that δ is a monomorphism and p_1, p_2 are epimorphisms.
(ii) Prove the equivalences

$$
\begin{aligned}
f \text{ is a monomorphism} &\Longleftrightarrow \delta \text{ is an isomorphism} \\
&\Longleftrightarrow \delta \text{ is an epimorphism} \Longleftrightarrow p_1 = p_2 \, .
\end{aligned}
$$

(iii) Dually, assume that $\mathcal{C}$ admits fiber coproducts and denote by i_1, i_2 the coprojections $Y \rightrightarrows Y \sqcup_X Y$ and by σ the codiagonal morphism $Y \sqcup_X Y \to Y$. Prove the equivalences

$$f \text{ is an epimorphism} \Longleftrightarrow \sigma \text{ is an isomorphism}$$
$$\Longleftrightarrow \sigma \text{ is a monomorphism} \Longleftrightarrow i_1 = i_2 \ .$$

Exercise 2.5. Let $\mathcal{C}$ be a category and consider a pair of parallel arrows $f, g \colon X \rightrightarrows Y$.
(i) Assume that $\mathcal{C}$ admits finite inductive limits. Prove that

$$X \underset{X \sqcup X}{\sqcup} Y \xrightarrow{\sim} \mathrm{Coker}(f, g) \ .$$

Here $X \sqcup X \to X$ is the codiagonal morphism and $X \sqcup X \to Y$ is the morphism associated to f, g.
(ii) Dually, assume that $\mathcal{C}$ admits finite projective limits. Prove that

$$\mathrm{Ker}(f, g) \xrightarrow{\sim} X \underset{Y \times Y}{\times} Y \ .$$

Here $Y \to Y \times Y$ is the diagonal morphism and $X \to Y \times Y$ is the morphism associated to f, g.

Exercise 2.6. Let $\mathcal{C}$ be a category, and consider the following conditions.

  (i) $\mathcal{C}$ admits small projective limits,
  (ii) $\mathcal{C}$ admits finite projective limits,
 (iii) $\mathcal{C}$ admits small products,
 (iv) $\mathcal{C}$ admits finite products,
  (v) $\mathcal{C}$ has a terminal object,
 (vi) for every X, Y in $\mathcal{C}$, $X \times Y$ exists in $\mathcal{C}$,
(vii) for every pair of parallel arrows $f, g \colon X \rightrightarrows Y$ in $\mathcal{C}$, $\mathrm{Ker}(f, g)$ exists in $\mathcal{C}$,
(viii) for every pair of morphisms $X \to Z$ and $Y \to Z$ in $\mathcal{C}$, $X \times_Z Y$ exists in $\mathcal{C}$.

Prove the following implications:
(i) $\Leftrightarrow$ (iii) + (vii) $\Leftrightarrow$ (iii) + (viii),
(ii) $\Leftrightarrow$ (iv) + (vii) $\Leftrightarrow$ (iv) + (viii) $\Leftrightarrow$ (v) + (viii),
(iv) $\Leftrightarrow$ (v) + (vi).

Exercise 2.7. Let $Z \in \mathbf{Set}$.
(i) Prove that the category $\mathbf{Set}_Z$ admits products (denoted here by $X \times_Z Y$) and that the functor $\bullet \times_Z Y \colon \mathbf{Set}_Z \to \mathbf{Set}_Z$ is left adjoint to the functor $\mathcal{H}om_Z(Y, \bullet)$ given by $\mathcal{H}om_Z(Y, X) = \bigsqcup_{z \in Z} \mathrm{Hom}_{\mathbf{Set}}(Y_z, X_z)$, where X_z is the fiber of $X \to Z$ over $z \in Z$.
(ii) Deduce that small inductive limits in $\mathbf{Set}$ are stable by base change (see Definition 2.2.6).

Exercise 2.8. Let I and $\mathcal{C}$ be two categories and denote by Δ the functor from $\mathcal{C}$ to $\mathcal{C}^I$ which associates to $X \in \mathcal{C}$ the constant functor Δ_X (see Notations 1.2.15). Assume that $\mathcal{C}$ admits inductive limits indexed by I.
(i) Prove that $\varinjlim \colon \mathcal{C}^I \to \mathcal{C}$ is a functor.

(ii) Prove that $(\varinjlim, \Delta)$ is a pair of adjoint functors, i.e.,

$$\mathrm{Hom}_{\mathcal{C}}(\varinjlim \alpha, Y) \simeq \mathrm{Hom}_{\mathcal{C}^I}(\alpha, \Delta_Y) \text{ for } \alpha \colon I \to \mathcal{C} \text{ and } Y \in \mathcal{C} .$$

(iii) Replacing I with the opposite category, deduce the formula (assuming projective limits exist):

$$\mathrm{Hom}_{\mathcal{C}}(X, \varprojlim \beta) \simeq \mathrm{Hom}_{\mathcal{C}^{I^{\mathrm{op}}}}(\Delta_X, \beta) .$$

Exercise 2.9. Let $\mathcal{C}$ be a category, X an object of $\mathcal{C}$, and let $q \colon X \to X$ be a projector i.e., a morphism satisfying $q^2 = q$. Prove that the conditions below are equivalent.

(i) q factorizes as $q = g \circ f$ with an epimorphism $f \colon X \twoheadrightarrow Y$ and a monomorphism $g \colon Y \rightarrowtail X$.
(ii) There exist $Y \in \mathcal{C}$ and morphisms $f \colon X \to Y$, $g \colon Y \to X$ such that $g \circ f = q$ and $f \circ g = \mathrm{id}_Y$.
(iii) Let $\mathbb{Z}$ be endowed with its natural order, and let $\alpha \colon \mathbb{Z} \to \mathcal{C}$ be the functor $\alpha(n) = X, \alpha(n \to m) = q$ for $m > n$. Then $\varinjlim \alpha$ exists in $\mathcal{C}$.

(iv) Let α be as in (iii). Then $\varprojlim \alpha$ exists in $\mathcal{C}$. (Here, we identify α with a functor $(\mathbb{Z}^{\mathrm{op}})^{\mathrm{op}} \to \mathcal{C}$.)
(v) Let $\mathbf{Pr}$ be the category defined in Notations 1.2.8. Let $\beta \colon \mathbf{Pr} \to \mathcal{C}$ be the functor $\beta(c) = X, \beta(p) = q$. Then $\varinjlim \beta$ exists in $\mathcal{C}$.

(vi) Let β be as in (v). Then $\varprojlim \beta$ exists in $\mathcal{C}$.

(A category in which any projector $q \colon X \to X$ satisfies the equivalent conditions above is said to be idempotent complete, or else, is called a Karoubi category.)

Exercise 2.10. Let $\varphi \colon J \to I$ be a functor and assume that φ admits a right adjoint ψ. Prove that $\varphi^{\dagger} \simeq \psi_*$ and $\psi^{\ddagger} \simeq \varphi_*$.

Exercise 2.11. Let $\mathcal{C}$ be a category. Prove that $\mathcal{C}_X$ admits finite projective limits for any $X \in \mathcal{C}$ if and only if $\mathcal{C}$ admits fiber products.

Exercise 2.12. Let $\mathcal{C}$ be a category, let $X \in \mathcal{C}$ and denote as usual by $\mathrm{j}_X \colon \mathcal{C}_X \to \mathcal{C}$ the canonical functor.
(i) Prove that if $\mathcal{C}$ admits inductive limits indexed by a small category I, then so does $\mathcal{C}_X$ and j_X commutes with such limits. (See Lemma 2.1.13.)
(ii) Prove that if $\mathcal{C}$ admits projective limits indexed by a small connected category I, then so does $\mathcal{C}_X$ and j_X commutes with such limits.

(iii) Assume now that $\mathcal{C}$ admits finite (resp. small) projective limits. Prove that $\mathcal{C}_X$ admits finite (resp. small) projective limits, and prove that if j_X commutes with such limits then X is a terminal object.

Exercise 2.13. Let $\alpha\colon I \to \mathcal{C}$ be a functor, let $\varphi_1, \varphi_2\colon J \rightrightarrows I$ be two functors and $\theta\colon \varphi_1 \to \varphi_2$ a morphism of functors. Assume that $\varinjlim \alpha$, $\varinjlim (\alpha \circ \varphi_1)$ and $\varinjlim (\alpha \circ \varphi_2)$ exist. Prove that the diagram below commutes.

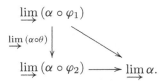

Exercise 2.14. Let I be a category and let $\mathcal{C}$ be a category which admits inductive limits indexed by I. Let $\alpha\colon I \to \mathcal{C}$ and $\varphi\colon I \to I$ be two functors and let $\theta\colon \mathrm{id}_I \to \varphi$ be a morphism of functors. Assume that $\theta_{\varphi(i)} = \varphi(\theta_i)$ as elements of $\mathrm{Hom}_I(\varphi(i), \varphi(\varphi(i)))$ for every $i \in I$.

Let us denote by $\eta\colon \varinjlim (\alpha \circ \varphi) \to \varinjlim \alpha$ the natural morphism (see (2.1.8)) and by $\xi\colon \varinjlim \alpha \to \varinjlim (\alpha \circ \varphi)$ the morphism induced by $\alpha \circ \theta\colon \alpha \to \alpha \circ \varphi$. Prove that η and ξ are isomorphisms inverse to each other.

Exercise 2.15. Let $\mu\colon J \to I$ and $\lambda\colon I \to J$ be two functors, and assume that μ is right adjoint to λ. Denote by $\theta\colon \mathrm{id}_I \to \mu \circ \lambda$ the canonical morphism of functors.

Let $\mathcal{C}$ be a category which admits inductive limits indexed by I and let $\alpha\colon I \to \mathcal{C}$ be a functor. Consider the sequence of morphisms in $\mathcal{C}$:

$$\varinjlim \alpha \xrightarrow{u} \varinjlim (\alpha \circ \mu \circ \lambda) \xrightarrow{v} \varinjlim (\alpha \circ \mu) \xrightarrow{w} \varinjlim \alpha \,,$$

where the morphism u is induced by θ and v, w are the canonical ones (see (2.1.8)). Prove that:
(i) μ is cofinal,
(ii) the composition $w \circ v \circ u$ is the identity, and $v \circ u$ and w are isomorphisms,
(iii) all morphisms u, v, w are isomorphisms if $\lambda(\theta_i)\colon \lambda(i) \to \lambda \circ \mu \circ \lambda(i)$ is an isomorphism for all $i \in I$.
(Hint: use the result of Exercise 2.14.)

Exercise 2.16. Let $F\colon \mathcal{C} \to \mathcal{C}'$ be a functor. Assume that $\mathcal{C}$ has a terminal objects $\mathrm{pt}_{\mathcal{C}}$. Prove that $F(\mathrm{pt}_{\mathcal{C}})$ is a terminal object of $\mathcal{C}'$ if and only if F is cofinal.

Exercise 2.17. Let I and J be small categories and let $\alpha\colon J \to I$ be a functor. Prove that α is cofinal if and only if the object "$\varinjlim$"α of $I^\wedge$ is a terminal object.

Exercise 2.18. Let $\mathcal{C}$ be a category, $Z \in \mathcal{C}$, $\Delta_Z \colon \mathbf{Pt} \to \mathcal{C}$ the unique functor with value Z.
(i) Prove that for $X \in \mathcal{C}$, the category $\mathbf{Pt}^X$ is equivalent to the discrete category associated with the set $\mathrm{Hom}_{\mathcal{C}}(X, Z)$.
(ii) Prove that Δ_Z is cofinal if and only if Z is a terminal object.

Exercise 2.19. Let $\mathcal{C}$ be a category and let $X, Y \in \mathcal{C}$. Prove that $X \text{ "}\bigsqcup\text{" } Y \in \mathcal{C}^\wedge$ is never isomorphic to an object of $\mathcal{C}$. Here, "$\bigsqcup$" denotes the coproduct in $\mathcal{C}^\wedge$.

Exercise 2.20. Let $\mathcal{C}$ be a category and denote by D the set $\mathrm{Mor}(\mathcal{C}) \sqcup (\mathrm{Ob}(\mathcal{C}) \times \{0,1\})$. The set D is endowed with the order $\leq$ given by $f \leq (X, 0)$, $f \leq (Y, 1)$ for any $f \colon X \to Y$ (together with the trivial relation $x \leq x$ for any $x \in D$). Denote by $\mathcal{D}$ the category associated with the ordered set $(D, \leq)$. Let $\varphi \colon \mathcal{D} \to \mathcal{C}$ be a functor given as follows: $\varphi\big((X, n)\big) = X$ for $n = 0, 1$, $\varphi(f) = X$ for $f \colon X \to Y$, $\varphi(f) \to \varphi\big((X, 0)\big)$ is id_X and $\varphi(f) \to \varphi\big((Y, 1)\big)$ is f. Prove that φ is well-defined and it is a cofinal functor.

Exercise 2.21. Let $\mathcal{C}$ be a category admitting small inductive limits and let I be a small category. Let $\varphi \colon I \to \mathcal{C}$ be a functor.
(i) Define $\psi \colon \mathrm{Ob}(I) \to \mathrm{Ob}(\mathcal{C})$ by setting $\psi(i) = \bigsqcup_{(i' \to i) \in I_i} \varphi(i')$. Extend ψ to a functor from I to $\mathcal{C}$.
(ii) Prove that $\varinjlim \psi \simeq \bigsqcup_{i \in I} \varphi(i)$. (Hint: letting I_d be the discrete category associated with $\mathrm{Ob}(I)$, apply Corollary 2.3.4 to the natural functor $\theta \colon I_d \to I$.)

Exercise 2.22. Consider a Cartesian square

$$\begin{array}{ccc} X' & \xrightarrow{f'} & Y' \\ {\scriptstyle g'}\downarrow & & \downarrow{\scriptstyle g} \\ X & \xrightarrow{f} & Y \, . \end{array}$$

Prove that if f is a monomorphism, then f' is a monomorphism.

Exercise 2.23. Let $\mathcal{C}$ be a category and let $\varphi \colon F \to G$ be a morphism in $\mathcal{C}^\wedge$. Prove that
(i) φ is a monomorphism if and only if $\varphi(X) \colon F(X) \to G(X)$ is injective for any $X \in \mathcal{C}$,
(ii) φ is an epimorphism if and only if $\varphi(X) \colon F(X) \to G(X)$ is surjective for any $X \in \mathcal{C}$.
(iii) Deduce that a morphism $u \colon A \to U$ in $\mathcal{C}^\wedge$ with $U \in \mathcal{C}$ is an epimorphism in $\mathcal{C}^\wedge$ if and only if u admits a section.
(iv) Assume that $\mathcal{C}$ is small and denote as usual by $\mathrm{pt}_{\mathcal{C}^\wedge}$ a terminal object of $\mathcal{C}^\wedge$. Prove that "$\bigsqcup_{U \in \mathcal{C}}$" $U \to \mathrm{pt}_{\mathcal{C}^\wedge}$ is an epimorphism in $\mathcal{C}^\wedge$.

(Hint: use the isomorphisms

$$(F \times_G F)(X) \simeq F(X) \times_{G(X)} F(X), \ (G \sqcup_F G)(X) \simeq G(X) \sqcup_{F(X)} G(X)$$

and Exercise 2.4.)

Exercise 2.24. Let $\mathcal{C}$ be a small category, and $u : A \twoheadrightarrow B$ an epimorphism in $\mathcal{C}^\wedge$. Prove that $A \times_B A \rightrightarrows A \to B$ is exact in $\mathcal{C}^\wedge$, that is, the sequence $S(B) \to S(A) \rightrightarrows S(A \times_B A)$ is exact in **Set** for any object S of $\mathcal{C}^\wedge$.

Exercise 2.25. Let $X \overset{f}{\underset{g}{\rightrightarrows}} Y \overset{h}{\longrightarrow} Z$ be a diagram in a category $\mathcal{C}$ such that the two compositions coincide. Prove that the conditions (i) and (ii) below are equivalent:

(i) the sequence $X \rightrightarrows Y \to Z$ is exact in $\mathcal{C}^\wedge$,
(ii) there exists $s : Z \to Y$ which satisfies the two following conditions:
    (a) $h \circ s = \mathrm{id}_Z$,
    (b) there exist an integer $n \geq 0$ and $u_0, \ldots, u_n$ in $\mathrm{Hom}_{\mathcal{C}}(Y, X)$ such that $f \circ u_0 = \mathrm{id}_Y$, $f \circ u_k = g \circ u_{k-1}$ ($1 \leq k \leq n$), $g \circ u_n = s \circ h$.

(Hint: use the exact sequence $\mathrm{Hom}_{\mathcal{C}}(Z, X) \rightrightarrows \mathrm{Hom}_{\mathcal{C}}(Z, Y) \to \mathrm{Hom}_{\mathcal{C}}(Z, Z)$ in **Set**.)

Exercise 2.26. Let $\mathcal{C}$ be a category which admits finite inductive limits and finite projective limits. Assume that finite inductive limits are stable by base change. Let $\emptyset_{\mathcal{C}}$ be an initial object in $\mathcal{C}$.
(i) Prove that $\emptyset_{\mathcal{C}} \times X \simeq \emptyset_{\mathcal{C}}$ for any $X \in \mathcal{C}$. (Hint: consider the empty inductive limit.)
(ii) Prove that any morphism $X \to \emptyset_{\mathcal{C}}$ is an isomorphism. (Hint: consider $\emptyset_{\mathcal{C}} \leftarrow X \overset{\mathrm{id}_X}{\longrightarrow} X$ and apply (i) to show that id_X factorizes through $\emptyset_{\mathcal{C}}$.)

Exercise 2.27. Consider two commutative diagrams in a category $\mathcal{C}$:

$$
\begin{array}{ccc}
X & \overset{f}{\longrightarrow} & Y \\
{\scriptstyle g}\downarrow & & \downarrow \\
Z & \longrightarrow & U,
\end{array}
\qquad
\begin{array}{ccc}
X & \overset{f}{\longrightarrow} & Y \\
{\scriptstyle g}\downarrow & & \downarrow \\
Z & \longrightarrow & V .
\end{array}
$$

Assume that the left square is Cartesian and the right square is co-Cartesian. Prove that the right square is Cartesian.

Exercise 2.28. Let $\mathcal{C}$ be a category admitting fiber products and let $f : X \to Z$, $g : Y \to Z$ and $u : Z \to Z'$ be morphisms in $\mathcal{C}$. Denote by $v : X \times_Z Y \to X \times_{Z'} Y$ the induced morphism. Prove that if u is a monomorphism, then v is an isomorphism.

3

Filtrant Limits

The notion of filtrant categories, which generalizes that of directed ordered set, plays an essential role in Category Theory and will be used all along this book. We prove here that a small category I is filtrant if and only if inductive limits defined on I with values in **Set** commute with finite projective limits.

We introduce also the IPC-property on a category $\mathcal{C}$, a property which asserts, in some sense, that filtrant inductive limits commute with small products. This property is satisfied by **Set**, as well as by $\mathcal{C}^\wedge$ for any small category $\mathcal{C}$.

We introduce the notion of (right or left) exact (resp. small) functor. For example, a functor $F \colon \mathcal{C} \to \mathcal{C}'$ will be called right exact if, for any $Y \in \mathcal{C}'$, the category $\mathcal{C}_Y$ (whose objects are the pairs (X, u) of $X \in \mathcal{C}$ and $u \colon F(X) \to Y$) is filtrant. When $\mathcal{C}$ admits finite inductive limits, we recover the classical definition: F is right exact if and only if it commutes with finite inductive limits.

In this chapter, we study the links between various properties of categories and functors, such as being cofinal, being filtrant, being exact, etc.

We also introduce the category $M[I \xrightarrow{\varphi} K \xleftarrow{\psi} J]$ associated with two functors $\varphi \colon I \to K$ and $\psi \colon J \to K$ and study its properties with some details.

The notion of a filtrant category will be generalized in Chap. 9 in which we will study π-filtrant categories, π being an infinite cardinal.

3.1 Filtrant Inductive Limits in the Category Set

If *for* denotes the forgetful functor from the category $\mathrm{Mod}(\mathbb{Z})$ to the category **Set**, which associates to a $\mathbb{Z}$-module M the underlying set M, then *for* commutes with $\varprojlim$ but not with $\varinjlim$. Indeed, if M_0 and M_1 are two modules, their coproduct in the category of modules is their direct sum, not their disjoint union. The reason is that the functor $\varinjlim \colon \mathrm{Fct}(I, \mathbf{Set}) \to \mathbf{Set}$ does not commute with finite projective limits for small categories I in general. Indeed, if it commuted, then for any inductive system $\{M_i\}_{i \in I}$ in $\mathrm{Mod}(\mathbb{Z})$, the addition maps would give $(\varinjlim for(M_i)) \times (\varinjlim for(M_i)) \simeq \varinjlim (for(M_i) \times for(M_i)) \to \varinjlim for(M_i)$, and $\varinjlim for(M_i)$ would have a structure of a $\mathbb{Z}$-module.

We shall introduce a property on I such that inductive limits indexed by I commute with finite projective limits.

Definition 3.1.1. *A category I is* filtrant *if it satisfies the conditions* (i)–(iii) *below.*

  (i) *I is non empty,*
  (ii) *for any i and j in I, there exist $k \in I$ and morphisms $i \to k$, $j \to k$,*
  (iii) *for any parallel morphisms $f, g \colon i \rightrightarrows j$, there exists a morphism $h \colon j \to k$ such that $h \circ f = h \circ g$.*

A category I is cofiltrant *if I^{op} is filtrant.*

The conditions (ii)–(iii) above are visualized by the diagrams:

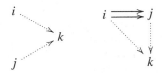

Note that an ordered set $(I, \leq)$ is directed if the associated category I is filtrant.

Lemma 3.1.2. *A category I is filtrant if and only if, for any finite category J and any functor $\varphi \colon J \to I$, there exists $i \in I$ such that $\varprojlim_{j \in J} \mathrm{Hom}_I(\varphi(j), i) \neq \emptyset$.*

Proof. (i) Assume that I is filtrant and let J and φ be as in the statement. Since J is finite, there exist $i_0 \in I$ and morphisms $s(j) \colon \varphi(j) \to i_0$ for all $j \in J$. Moreover, there exist $k(j) \in I$ and a morphism $\lambda(j) \colon i_0 \to k(j)$ such that the composition

$$\varphi(j) \xrightarrow{\varphi(t)} \varphi(j') \xrightarrow{s(j')} i_0 \xrightarrow{\lambda(j)} k(j)$$

does not depend on $t \colon j \to j'$. Now, there exist $i_1 \in I$ and morphisms $\xi(j) \colon k(j) \to i_1$. Finally, take a morphism $i_1 \to i_2$ such that the composition $i_0 \to k(j) \to i_1 \to i_2$ does not depend on j. The family of morphisms $u_j \colon \varphi(j) \to i_0 \to k(j) \to i_1 \to i_2$ defines an element of $\varprojlim_{j \in J} \mathrm{Hom}_I(\varphi(j), i_2)$.

(ii) Conversely, let us check the conditions (i)–(iii) of Definition 3.1.1. By taking for J the empty category we obtain (i). By taking for J the category $\mathbf{Pt} \sqcup \mathbf{Pt}$ (the category with two objects and no morphisms other than the identities) we obtain (ii). By taking for J the category $\bullet \rightrightarrows \bullet$ (see Notation 1.2.8 (iv)) we obtain (iii).   q.e.d.

Proposition 3.1.3. *Let* $\alpha: I \to$ **Set** *be a functor with* I *small and filtrant. Define the relation* $\sim$ *on* $\coprod_i \alpha(i)$ *as follows:* $\alpha(i) \ni x \sim y \in \alpha(j)$ *if there exist* $s: i \to k$ *and* $t: j \to k$ *such that* $\alpha(s)(x) = \alpha(t)(y)$. *Then*

(i) *the relation* $\sim$ *is an equivalence relation,*
(ii) $\varinjlim \alpha \simeq \coprod_i \alpha(i)/\sim$.

Proof. (i) Assuming that $x_j \in \alpha(i_j)$ $(j = 1, 2, 3)$ satisfy $x_1 \sim x_2$ and $x_2 \sim x_3$, let us show that $x_1 \sim x_3$. There exist morphisms visualized by the solid diagram:

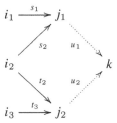

such that $\alpha(s_1)x_1 = \alpha(s_2)x_2$, $\alpha(t_2)x_2 = \alpha(t_3)x_3$. By Lemma 3.1.2, we can complete the solid diagram to a commutative diagram with the dotted arrows. Then $\alpha(u_1 \circ s_1)x_1 = \alpha(u_1 \circ s_2)x_2 = \alpha(u_2 \circ t_2)x_2 = \alpha(u_2 \circ t_3)x_3$. Hence $x_1 \sim x_3$.

(ii) follows from Proposition 2.4.1.   q.e.d.

Corollary 3.1.4. *Let* $\alpha: I \to$ **Set** *be a functor with* I *small and filtrant.*

(i) *Let* S *be a finite subset in* $\varinjlim \alpha$. *Then there exists* $i \in I$ *such that* S *is contained in the image of* $\alpha(i)$ *by the natural map* $\alpha(i) \to \varinjlim \alpha$.
(ii) *Let* $i \in I$ *and let* x *and* y *be elements of* $\alpha(i)$ *with the same image in* $\varinjlim \alpha$. *Then there exists* $s: i \to j$ *such that* $\alpha(s)(x) = \alpha(s)(y)$ *in* $\alpha(j)$.

The proof is left as an exercise.

Notice that the result of Corollary 3.1.4 does not hold in general if I is not filtrant.

Corollary 3.1.5. *Let* R *be a ring and denote by* for *the forgetful functor* $\mathrm{Mod}(R) \to$ **Set**. *Then the functor* for *commutes with small filtrant inductive limits. In other words, if* I *is small and filtrant and* $\alpha: I \to \mathrm{Mod}(R)$ *is a functor, then*

$$for(\varinjlim_i \alpha(i)) = \varinjlim_i (for(\alpha(i))).$$

The proof is left as an exercise.

Inductive limits with values in **Set** indexed by small filtrant categories commute with finite projective limits. More precisely:

Theorem 3.1.6. *Let I be a small category. The two conditions below are equivalent.*

(a) *I is filtrant,*
(b) *for any finite category J and any functor $\alpha\colon I \times J^{\mathrm{op}} \to \mathbf{Set}$, the natural morphism*

(3.1.1) $$\varinjlim_i \varprojlim_j \alpha(i, j) \to \varprojlim_j \varinjlim_i \alpha(i, j)$$

is an isomorphism. In other words, the functor $\varinjlim\colon \mathrm{Fct}(I, \mathbf{Set}) \to \mathbf{Set}$ commutes with finite projective limits .

Proof. (a) $\Rightarrow$ (b). Assume that I is filtrant. It is enough to prove that $\varinjlim$ commutes with kernels and with finite products.

(i) $\varinjlim$ commutes with kernels. Let $\alpha, \beta\colon I \to \mathbf{Set}$ be two functors and let $f, g\colon \alpha \rightrightarrows \beta$ be two morphisms of functors. Define γ as the kernel of (f, g), that is, we have exact sequences

$$\gamma(i) \to \alpha(i) \rightrightarrows \beta(i) .$$

Let Z denote the kernel of $\varinjlim_i \alpha(i) \rightrightarrows \varinjlim_i \beta(i)$. We have to prove that the natural map $\lambda\colon \varinjlim_i \gamma(i) \to Z$ is bijective.

(i)(1) The map λ is surjective. Indeed for $x \in Z$, represent x by some $x_i \in \alpha(i)$. Then $f_i(x_i)$ and $g_i(x_i)$ in $\beta(i)$ having the same image in $\varinjlim \beta$, there exists $s\colon i \to j$ such that $\beta(s)f_i(x_i) = \beta(s)g_i(x_i)$. Set $x_j = \alpha(s)x_i$. Then $f_j(x_j) = g_j(x_j)$, which means that $x_j \in \gamma(j)$. Clearly, $\lambda(x_j) = x$.

(i)(2) The map λ is injective. Indeed, let $x, y \in \varinjlim \gamma$ with $\lambda(x) = \lambda(y)$. We may represent x and y by elements x_i and y_i of $\gamma(i)$ for some $i \in I$. Since x_i and y_i have the same image in $\varinjlim \alpha$, there exists $i \to j$ such that they have the same image in $\alpha(j)$. Therefore their images in $\gamma(j)$ will be the same.

(ii) $\varinjlim$ commutes with finite products. The proof is similar to the preceding one and left to the reader.

(b) $\Rightarrow$ (a). In order to prove that I is filtrant, we shall apply Lemma 3.1.2. Consider a finite category J and a functor $\varphi\colon J \to I$. Let us show that there exists $i \in I$ such that $\varprojlim_{j\in J} \mathrm{Hom}_I(\varphi(j), i) \neq \emptyset$. By the assumption, we have a bijection

(3.1.2) $$\varinjlim_{i\in I} \varprojlim_{j\in J} \mathrm{Hom}_I(\alpha(j), i) \xrightarrow{\sim} \varprojlim_{j\in J} \varinjlim_{i\in I} \mathrm{Hom}_I(\alpha(j), i) .$$

By Lemma 2.4.7, $\varinjlim_{i\in I} \mathrm{Hom}_I(\alpha(j), i) \simeq \{\mathrm{pt}\}$, which implies that the right-hand side of (3.1.2) is isomorphic to $\{\mathrm{pt}\}$. Hence, there exists $i \in I$ such that $\varprojlim_{j\in J} \mathrm{Hom}_I(\alpha(j), i) \neq \emptyset$.   q.e.d.

Applying this result together with Corollary 3.1.5, we obtain:

Corollary 3.1.7. *Let R be a ring and let I be a small filtrant category. Then the functor $\varinjlim : \mathrm{Mod}(R)^I \to \mathrm{Mod}(R)$ commutes with finite projective limits.*

Proposition 3.1.8. *Let $\psi : K \to I$ and $\varphi : J \to I$ be functors. Assume that ψ is cofinal.*

(i) *If $J_{\psi(k)} \to I_{\psi(k)}$ is cofinal for every $k \in K$, then φ is cofinal.*
(ii) *If K is filtrant and $J_{\psi(k)}$ is filtrant for every $k \in K$, then J is filtrant.*

Proof. By replacing the universe $\mathcal{U}$ with a bigger one, we may assume that I, J and K are small categories.
(i) For any functor $\alpha : I \to \mathbf{Set}$, there is a chain of isomorphisms

$$\varinjlim \alpha \circ \varphi \simeq \varinjlim_{i \in I} \varinjlim_{j \in J_i} \alpha \circ \varphi(j) \simeq \varinjlim_{k \in K} \varinjlim_{j \in J_{\psi(k)}} \alpha \circ \varphi(j)$$

$$\simeq \varinjlim_{k \in K} \varinjlim_{i \in I_{\psi(k)}} \alpha(i) \simeq \varinjlim \alpha \circ \psi \simeq \varinjlim \alpha .$$

Here, the first and fourth isomorphisms follow from Corollary 2.3.4, the second and fifth isomorphisms follow from the fact that ψ is cofinal, and the third isomorphism follows from the fact that $J_{\psi(k)} \to I_{\psi(k)}$ is cofinal.
(ii) For any functor $\alpha : J \to \mathbf{Set}$, we have by Corollary 2.3.4

$$\varinjlim_{j \in J} \alpha(j) \simeq \varinjlim_{i \in I} \varinjlim_{j \in J_i} \alpha(j)$$

$$\simeq \varinjlim_{k \in K} \varinjlim_{j \in J_{\psi(k)}} \alpha(j) .$$

Since $\varinjlim_{j \in J_{\psi(k)}}$ and $\varinjlim_{k \in K}$ commute with finite projective limits, the functor $\varinjlim :$
$\mathrm{Fct}(J, \mathbf{Set}) \to \mathbf{Set}$ commutes with finite projective limits. The result then follows from Theorem 3.1.6.   q.e.d.

The IPC Property

Theorem 3.1.6 does not hold anymore when removing the hypothesis that J is finite. However, when J is small and discrete there is a useful result which is satisfied by many categories and that we describe now.

We consider a category $\mathcal{A}$ and we make the hypothesis:

(3.1.3) $\mathcal{A}$ admits small products and small filtrant inductive limits .

Let $\{I_s\}_{s \in S}$ be a family of small and filtrant categories indexed by a small set S. Consider the product category

(3.1.4)
$$K = \prod_{s \in S} I_s \,.$$

It is easily checked that K is filtrant (see Proposition 3.2.1 below).
  For $s \in S$, denote by π_s the projection functor $\pi_s \colon K \to I_s$.
  Consider a family of functors

(3.1.5)
$$\alpha = \{\alpha_s\}_{s \in S} \text{ with } \alpha_s \colon I_s \to \mathcal{A} \,.$$

Define the functor

(3.1.6)
$$\varphi \colon K \to \mathcal{A} \,,$$
$$\varphi = \prod_s \alpha_s \circ \pi_s,$$

that is, for $k = \{\pi_s(k)\}_s \in K$

$$\varphi(k) = \prod_{s \in S} \alpha_s(\pi_s(k)) \,.$$

The object $\prod_{s \in S} \varinjlim_{i \in I_s} \alpha_s(i)$ is well defined in $\mathcal{A}$, and the family of morphisms
$\alpha_s(\pi_s(k)) \to \varinjlim \alpha_s$ defines the morphism $\prod_{s \in S} \alpha_s(\pi_s(k)) \to \prod_{s \in S} \varinjlim \alpha_s$, hence
the morphism

(3.1.7)
$$\varinjlim \varphi \to \prod_{s \in S} \varinjlim \alpha_s,$$

or equivalently,

(3.1.8)
$$\varinjlim_{k \in K} \left(\prod_{s \in S} \alpha_s(\pi_s(k))\right) \to \prod_{s \in S} \left(\varinjlim_{i_s \in I_s} \alpha_s(i_s)\right) \,.$$

The morphism of functors (3.1.8) is visualized by the diagram (see (1.3.3)):

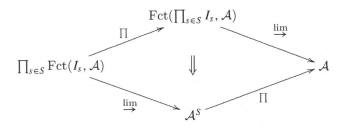

Example 3.1.9. Assume that $I_s = I$ for all $s \in S$. Then $K = I^S$ and α is a
functor $I \times S \to \mathcal{A}$. Morphism (3.1.8) may be written as:

(3.1.9)
$$\varinjlim_{k \in K} \prod_{s \in S} \alpha(k(s), s) \to \prod_{s \in S} \varinjlim_{i \in I} \alpha(i, s) \,.$$

Definition 3.1.10. *Let $\mathcal{A}$ be a category satisfying* (3.1.3). *We say that $\mathcal{A}$ satisfies the* IPC-*property* (*inductive-limit-product commutation property*) *if the morphism* (3.1.7) *is an isomorphism for any family $\{I_s\}_{s \in S}$ of small and filtrant categories indexed by a small set S and any family of functors $\alpha_s \colon I_s \to \mathcal{A}$ indexed by $s \in S$.*

Proposition 3.1.11. (i) *If categories $\mathcal{A}_i$ $(i \in I)$ satisfy the* IPC-*property, then so does the product category $\prod_{i \in I} \mathcal{A}_i$.*
(ii) *The category* **Set** *satisfies the* IPC-*property.*
(iii) *Let $\mathcal{A}$ be a category satisfying* (3.1.3). *Assume that there exist a set I and a functor $\lambda \colon \mathcal{A} \to \mathbf{Set}^I$ such that*

(3.1.10)
$$\begin{cases} \lambda \text{ commutes with small products,} \\ \lambda \text{ commutes with small filtrant inductive limits,} \\ \lambda \text{ is conservative (see Definition 1.2.11).} \end{cases}$$

Then $\mathcal{A}$ satisfies the IPC-*property.*

Proof. (i) is obvious.
(ii) Consider a family of functors $\{\alpha_s\}_{s \in S}$ with $\alpha_s \colon I_s \to \mathbf{Set}$. We keep the notations (3.1.4)–(3.1.6) and we set $A_s = \varinjlim \alpha_s$. Let

$$u \colon \varinjlim \varphi \to \prod_{s \in S} A_s$$

denote the natural map.
(a) u is surjective. Indeed, let $x := \{x_s\}_{s \in S} \in \prod_s A_s$. For each $s \in S$, there exist $i_s \in I_s$ and $x_{i_s} \in \alpha_s(i_s)$ whose image in A_s is x_s. Set $k = \{i_s\}_{s \in S}$ and $y_k := \{x_{i_s}\}_{s \in S} \in \varphi(k)$. Denote by y the image of y_k in $\varinjlim \varphi$. Then $u(y) = x$.
(b) u is injective. Let $y, y' \in \varinjlim \varphi$ with $u(y) = u(y')$. Since K is filtrant, there exist $k = \{i_s\}_{s \in S} \in K$ and $y_k := \{x_{i_s}\}_{s \in S} \in \varphi(k)$, $y'_k := \{x'_{i_s}\}_{s \in S} \in \varphi(k)$ such that y is the image of y_k and y' is the image of y'_k. For each $s \in S$, x_{i_s} and x'_{i_s} have the same image in A_s. Since I_s is filtrant, there exists $i_s \to i'_s$ such that x_{i_s} and x'_{i_s} have the same image in $\alpha_s(i'_s)$. Set $k' = \{i'_s\}_{s \in S}$. Then y_k and y'_k have the same image in $\varphi(k')$. Hence $y = y'$.

(iii) Let $\alpha_s \colon I_s \to \mathcal{A}$ be a functor $(s \in S)$. We consider the functor $\lambda \circ \alpha_s \colon I_s \to \mathbf{Set}^I$. Using the hypothesis that λ commutes with small products and small filtrant inductive limits, we get the isomorphism

$$\lambda(\varinjlim_{k \in K}(\prod_{s \in S} \alpha_s(\pi_s(k)))) \xrightarrow{\sim} \lambda(\prod_{s \in S}(\varinjlim \alpha_s)) \,,$$

because $\mathbf{Set}^I$ satisfies the IPC-property. The result follows since λ is conservative. q.e.d.

Corollary 3.1.12. *Let $\mathcal{C}$ be a small category. Then $\mathcal{C}^{\wedge}$ satisfies the* IPC-*property.*

Corollary 3.1.13. *Let R be a ring. Then $\mathrm{Mod}(R)$ satisfies the* IPC-*property.*

3.2 Filtrant Categories

We have introduced filtrant categories in Definition 3.1.1. We shall now study their properties.

Proposition 3.2.1.   (i) *If a category has a terminal object, then it is filtrant.*
(ii) *If a category admits finite inductive limits, then it is filtrant.*
(iii) *A product of filtrant categories is filtrant.*
(iv) *If a category is filtrant, then it is connected.*

The proof is obvious.
    Proposition 2.5.2 may be formulated slightly differently when J is filtrant.

Proposition 3.2.2. *Assume that J is filtrant and let $\varphi \colon J \to I$ be a functor. Then the conditions below are equivalent:*

(i) *φ is cofinal,*
(ii) *J^i is filtrant for every $i \in I$,*
(iii) *the following two conditions hold:*
    (a) *for each $i \in I$ there exist $j \in J$ and a morphism $s \colon i \to \varphi(j)$ (i.e., J^i is non empty),*
    (b) *for any $i \in I$, any $j \in J$, and any pair of parallel morphisms $s, s' \colon i \rightrightarrows \varphi(j)$ in I, there exists a morphism $t \colon j \to k$ in J such that $\varphi(t) \circ s = \varphi(t) \circ s'$.*

Moreover, if these equivalent conditions are satisfied, then I is filtrant.

Proof. (iii) $\Rightarrow$ (ii). Let $i \in I$. Let us check conditions (i)–(iii) of Definition 3.1.1 for J^i. First, J^i is non empty by (a). Then, consider morphisms $s \colon i \to \varphi(j)$ and $s' \colon i \to \varphi(j')$. Since J is filtrant, there exist $t \colon j \to k$ and $t' \colon j' \to k$. Applying the hypothesis (b) above to the morphisms $\varphi(t) \circ s$ and $\varphi(t') \circ s'$, we may assume that $\varphi(t) \circ s = \varphi(t') \circ s'$. Hence, t, t' induce morphisms in J^i. Finally, let $i \to \varphi(j_1)$ and $i \to \varphi(j_2)$ be two objects of J^i and let $\xi, \eta \colon j_1 \rightrightarrows j_2$ be two parallel arrows in J^i. There exists a morphism $j_2 \to j_3$ in J such that the two compositions $j_1 \rightrightarrows j_2 \to j_3$ coincide. Hence, the composition $i \to \varphi(j_2) \to \varphi(j_3)$ defines an object of J^i, and the two compositions

$$(i \to \varphi(j_1)) \rightrightarrows (i \to \varphi(j_2)) \to (i \to \varphi(j_3))$$

coincide.
(ii) $\Rightarrow$ (i). If J^i is filtrant, then it is connected.
(i) $\Rightarrow$ (iii). By Definition 2.5.1, J^i is connected and in particular non empty. Hence, (a) is satisfied.
    Let us prove (b). For $i \in I$, let $\alpha \colon J \to \mathbf{Set}$ be the functor $j \mapsto \mathrm{Hom}_I(i, \varphi(j))$. Then Proposition 2.5.2 implies that $\varinjlim \alpha \simeq \{\mathrm{pt}\}$. Consider a pair of parallel morphisms $s, s' \colon i \rightrightarrows \varphi(j)$. Hence, $s, s' \in \alpha(j) = \mathrm{Hom}_I(i, \varphi(j))$. Applying Proposition 3.1.3, there exists a morphism $t \colon j \to k$ in J such that s and s' have the same image in $\mathrm{Hom}_I(i, \varphi(k))$.

The last assertion easily follows from (iii).   q.e.d.

Corollary 3.2.3. *Let I be a filtrant category.*

(i) *For any $i \in I$, I^i is filtrant and the functor $\mathrm{j}^i \colon I^i \to I$ is cofinal.*
(ii) *The diagonal functor $I \to I \times I$ is cofinal.*

Proof. (i) Applying Proposition 3.2.2 (ii) to id: $I \to I$, I^i is filtrant. To prove that $\varphi := \mathrm{j}^i \colon I^i \to I$ is cofinal, let us check the conditions in Proposition 3.2.2 (iii).
(a) For any $i_1 \in I$, there exist $k \in I$ and arrows $i \to k$, $i_1 \to k$. Then we have a morphism $i_1 \to \varphi(i \to k)$.
(b) Let $i_1 \in I$, $u \colon i \to i_2$ and let $s, s' \colon i_1 \rightrightarrows i_2 = \varphi((i \xrightarrow{u} i_2))$. There exists a morphism $t \colon i_2 \to i_3$ such that the two compositions $i_1 \rightrightarrows i_2 \to i_3$ coincide. The morphism $\widetilde{t} \colon (i \xrightarrow{u} i_2) \to (i \xrightarrow{t \circ u} i_3)$ in I^i induced by t satisfies $\varphi(\widetilde{t}) \circ s = \varphi(\widetilde{t}) \circ s'$.

(ii) For $(i_1, i_2) \in I \times I$, $I^{(i_1, i_2)} \simeq (I^{i_1})^{i_2}$. By (i), I^{i_1} as well as $(I^{i_1})^{i_2}$ are filtrant. This implies that the functor $I \to I \times I$ is cofinal by Proposition 3.2.2. q.e.d.

Proposition 3.2.4. *Let $\varphi \colon J \to I$ be a functor. Assume that I is filtrant, φ is fully faithful, and for any $i \in I$ there exists a morphism $i \to \varphi(j)$ with $j \in J$. Then J is filtrant and φ is cofinal.*

Proof. (a) J is filtrant. Clearly, condition (i) of Definition 3.1.1 is satisfied. Let us check condition (ii), the proof of (iii) being similar to this case. Let $j_1, j_2 \in J$. Since I is filtrant, there exist $i \in I$ and morphisms $\varphi(j_1) \to i$ and $\varphi(j_2) \to i$. By the assumption there exist $j_3 \in J$ and a morphism $i \to \varphi(j_3)$. Since the functor $\varphi \colon J \to I$ is fully faithful, the composition $\varphi(j_a) \to i \to \varphi(j_3)$, $(a = 1, 2)$ is the image by φ of a morphism $j_a \to j_3$ $(a = 1, 2)$.

(b) φ is cofinal. Condition (iii) (a) of Proposition 3.2.2 is satisfied by the hypothesis. Condition (iii) (b) is proved as in (a) above. q.e.d.

The next technical results will be useful in the sequel.

Proposition 3.2.5. *Let $I \xrightarrow{\varphi} J \xrightarrow{\psi} K$ be cofinal functors with I, J, K filtrant categories. Then*

(i) *for any $j \in J$, $I^j \to I$ is cofinal,*
(ii) *for any $i \in I$, $I^i \to I^{\varphi(i)}$ is cofinal,*
(iii) *for any $k \in K$, $I^k \to J^k$ is cofinal,*
(iv) *for any morphism $u \colon k \to \psi(j)$ in K, the induced functor $u^* \colon I^j \to I^k$ is cofinal.*

Proof. (i) It is enough to show that for any $i \in I$, $(I^j)^i$ is filtrant. By Corollary 3.2.3, the composition $I^i \to I \to J$ is cofinal, and thus $(I^i)^j \simeq (I^j)^i$ is filtrant.
(iii) For any object $a = (k \to \psi(j)) \in J^k$, we have $(I^k)^a \simeq I^j$ and this category is filtrant (since φ is cofinal). Hence, by Proposition 3.2.2, $I^k \to J^k$ is cofinal.

(iv) By (iii), $I^k \to J^k$ is cofinal. Regarding u as an object of J^k, $(I^k)^u \simeq I^j$ and $(I^k)^u \to I^k$ is cofinal by (i).
(ii) is a particular case of (iv).   q.e.d.

Proposition 3.2.6. *A filtrant category I is cofinally small if and only if there exists a small subset S of $\mathrm{Ob}(I)$ such that for any $i \in I$ there exists a morphism $i \to j$ with $j \in S$.*

Proof. (i) let $\varphi: J \to I$ be a cofinal functor with J small, and let S be the image of $\mathrm{Ob}(J)$ by the functor φ. Then S is small, and the condition in the statement is satisfied since φ is cofinal.
(ii) Conversely, consider S as a full subcategory of I. Then $S \to I$ is cofinal by Proposition 3.2.4.   q.e.d.

Remark 3.2.7. By the result of Exercise 2.20, for any category $\mathcal{C}$, there exists an ordered set $(D, \leq)$ such that, denoting by $\mathcal{D}$ the associated category, there exists a cofinal functor $\varphi: \mathcal{D} \to \mathcal{C}$. There is also a result of Deligne (see [30]) which asserts that if I is small and filtrant, then there exists a small ordered filtrant set J cofinal to I.

Lemma 3.2.8. *Let I be a small ordered set, $\alpha: I \to \mathcal{C}$ a functor. Let $\mathcal{J}$ denote the set of finite subsets of I, ordered by inclusion. To each $J \in \mathcal{J}$, associate the restriction $\alpha_J: J \to \mathcal{C}$ of α to J. Then $\mathcal{J}$ is small and filtrant and moreover*

$$(3.2.1) \qquad \varinjlim \alpha \simeq \varinjlim_{J \in \mathcal{J}} (\varinjlim \alpha_J) .$$

Proof. (i) Clearly, $\mathcal{J}$ is small and filtrant.
(ii) Let us prove the isomorphism (3.2.1). Let K be the ordered subset of the ordered set $I \times \mathcal{J}$ consisting of pairs (i, J) with $i \in J$. The projection $I \times \mathcal{J} \to I$ defines a functor $\varphi: K \to I$.
(a) φ is cofinal. Indeed, for any $i_0 \in I$, $K^{i_0} \simeq \{(i, J) \in K; i_0 \leq i\}$. For any $(i, J) \in K^{i_0}$, we have $(i, J) \leq (i, J \cup \{i_0\})$ and $(i_0, \{i_0\}) \leq (i, J \cup \{i_0\})$. Hence K^{i_0} is connected.
(b) Applying Proposition 2.5.2 and Corollary 2.3.4, we obtain

$$\varinjlim \alpha \simeq \varinjlim \alpha \circ \varphi \simeq \varinjlim_{J \in \mathcal{J}} \varinjlim_{k \in K_J} \alpha \circ \varphi(k) .$$

(c) Let $\xi_J: J \to K_J$ be the functor $J \ni j \mapsto (j, J)$. The functor ξ_J is cofinal. Indeed, for $k = (j_1, J_1) \in K_J$, we have $j_1 \in J_1 \subset J$ and hence $J^k \simeq \{j \in J; j_1 \leq j\}$ is connected since j_1 is the smallest element.
(d) We deduce the isomorphisms

$$\varinjlim_{k \in K_J} \alpha \circ \varphi(k) \simeq \alpha \circ \varphi \circ \xi_J \simeq \varinjlim \alpha_J .$$

q.e.d.

By Lemma 3.2.8, inductive limits indexed by small ordered sets can be decomposed into filtrant inductive limits and finite inductive limits. Using Exercise 2.20 or Proposition 2.2.9, the same result holds for any small inductive limit.

Hence, many properties on small inductive limits decompose into properties on small filtrant inductive limits and properties on finite inductive limits. In particular:

Lemma 3.2.9. *If C admits small filtrant inductive limits and finite inductive limits (resp. finite coproducts), then C admits small inductive limits (resp. small coproducts). Moreover, if a functor $F : C \to C'$ commutes with small filtrant inductive limits and finite inductive limits (resp. finite coproducts), then F commutes with small inductive limits (resp. small coproducts).*

Recall that if C admits finite inductive limits and small coproducts, then C admits small inductive limits and if a functor $F : C \to C'$ commutes with finite inductive limits and small coproducts, then F commutes with small inductive limits. This follows from Proposition 2.2.9.

Notation 3.2.10. (i) We shall sometimes use the sketchy terminology "a filtrant inductive system". It means a functor $\alpha : I \to C$ where the category I is filtrant. We use similar formulations such as "a small filtrant inductive system", etc.
(ii) We shall also use the formulation "a filtrant projective system". Our convention is

(3.2.2) a filtrant projective system is a functor $\beta : J^{\mathrm{op}} \to C$ with J filtrant .

3.3 Exact Functors

Let $F : C \to C'$ be a functor. Recall that for $U \in C'$, C_U denotes the category whose objects are the pairs (X, u) of $X \in C$ and $u : F(X) \to U$, and C^U denotes the category whose objects are the pairs (X, v) of $X \in C$ and $v : U \to F(X)$. The natural functors $\mathrm{j}_U : C_U \to C$ and $\mathrm{j}^U : C^U \to C$ are faithful (see Definition 1.2.16).

Definition 3.3.1. *Let $F : C \to C'$ be a functor.*

(i) *We say that F is* right exact *if the category C_U is filtrant for any $U \in C'$.*
(ii) *We say that F is* left exact *if $F^{\mathrm{op}} : C^{\mathrm{op}} \to C'^{\mathrm{op}}$ is right exact or equivalently the category C^U is cofiltrant for any $U \in C'$.*
(iii) *We say that F is* exact *if it is both right and left exact.*

Proposition 3.3.2. *Let $F : C \to C'$ be a left exact functor, let J be a finite category and let $\beta : J^{\mathrm{op}} \to C$ be a functor. Assume that $\varprojlim \beta$ exists in C. Then $\varprojlim(F \circ \beta)$ exists in C' and is isomorphic to $F(\varprojlim \beta)$. In particular, left exact functors commute with finite projective limits if C admits such limits.*

There is a similar statement for right exact functors and inductive limits.

Proof. Using Corollary 2.4.6, we get the chain of isomorphisms for $X \in \mathcal{C}$ and $U \in \mathcal{C}'$

(3.3.1)
$$\operatorname{Hom}_{\mathcal{C}'}(U, F(X)) \simeq \varinjlim_{Z \to X} \operatorname{Hom}_{\mathcal{C}'}(U, F(Z))$$
$$\simeq \varinjlim_{(U \to F(Z)) \in \mathcal{C}^U} \operatorname{Hom}_{\mathcal{C}}(Z, X) .$$

Hence we have for any $U \in \mathcal{C}'$

$$\varprojlim_{j} \operatorname{Hom}_{\mathcal{C}'}(U, F(\beta(j))) \simeq \varprojlim_{j} \varinjlim_{Z \in \mathcal{C}^U} \operatorname{Hom}_{\mathcal{C}}(Z, \beta(j))$$
$$\simeq \varinjlim_{Z \in \mathcal{C}^U} \varprojlim_{j} \operatorname{Hom}_{\mathcal{C}}(Z, \beta(j))$$
$$\simeq \varinjlim_{Z \in \mathcal{C}^U} \operatorname{Hom}_{\mathcal{C}}(Z, \varprojlim_{j} \beta(j))$$
$$\simeq \operatorname{Hom}_{\mathcal{C}'}(U, F(\varprojlim_{j} \beta(j))) ,$$

where the second isomorphism follows from Theorem 3.1.6 because Z ranges over the filtrant category $(\mathcal{C}^U)^{\mathrm{op}}$. Hence, $F(\varprojlim \beta)$ represents the projective limit of $F(\beta(j))$.   q.e.d.

The next result is a partial converse to Proposition 3.3.2.

Proposition 3.3.3. *Let $F : \mathcal{C} \to \mathcal{C}'$ be a functor and assume that $\mathcal{C}$ admits finite projective limits. Then F is left exact if and only if it commutes with such limits.*

There is a similar statement for right exact functors and inductive limits.

Proof. (i) Assume that F is left exact. Then it commutes with finite projective limits by Proposition 3.3.2.
(ii) Assume that F commutes with finite projective limits. Then $\mathcal{C}^U$ admits finite projective limits by Lemma 2.1.13, hence is cofiltrant by Proposition 3.2.1 (ii).   q.e.d.

Corollary 3.3.4. *Assume that $\mathcal{C}$ admits finite projective limits. Then $F : \mathcal{C} \to \mathcal{C}'$ is left exact if and only if it satisfies:*

   (i) *F sends a terminal object of $\mathcal{C}$ to a terminal object of $\mathcal{C}'$,*
  (ii) *for any $X, Y \in \mathcal{C}$, $F(X) \times F(Y)$ exists in $\mathcal{C}'$ and $F(X \times Y) \xrightarrow{\sim} F(X) \times F(Y)$,*
 (iii) *F commutes with kernels, i.e., for any parallel arrows $f, g : X \rightrightarrows Y$ in $\mathcal{C}$, $F(\operatorname{Ker}(f, g))$ is a kernel of the parallel arrows $(F(f), F(g))$.*

Moreover, assuming (i), condition (ii) + (iii) is equivalent to

 (iv) *F commutes with fiber products, i.e., $F(X \times_Z Y) \xrightarrow{\sim} F(X) \times_{F(Z)} F(Y)$ for any pair of morphisms $X \to Z$ and $Y \to Z$ in $\mathcal{C}$.*

Proof. The result follows immediately from Propositions 3.3.3 and 2.2.9 and Exercise 2.6.   q.e.d.

Example 3.3.5. Let R be a ring. The forgetful functor $for: \mathrm{Mod}(R) \to \mathbf{Set}$ is left exact, but *for* is not right exact since it does not respect initial objects.

Proposition 3.3.6. *Let* $F: \mathcal{C} \to \mathcal{C}'$ *be a functor. If* F *admits a right (resp. left) adjoint, then* F *is right (resp. left) exact.*

Proof. Denote by G the right adjoint to F. Let $V \in \mathcal{C}'$. Then for any $U \in \mathcal{C}$, there is an isomorphism $\mathrm{Hom}_{\mathcal{C}'}(F(U), V) \simeq \mathrm{Hom}_{\mathcal{C}}(U, G(V))$. Hence the category $\mathcal{C}_V$ of arrows $F(U) \to V$ is equivalent to the category $\mathcal{C}_{G(V)}$ of arrows $U \to G(V)$. This last category having a terminal object, namely $\mathrm{id}_{G(V)}$, it is filtrant.   q.e.d.

Proposition 3.3.7. (i) *Let* $\mathcal{C}$ *be a category which admits finite inductive limits and finite projective limits. Then the functor* $\mathrm{Hom}_{\mathcal{C}}: \mathcal{C}^{\mathrm{op}} \times \mathcal{C} \to \mathbf{Set}$ *is left exact in each argument.*
(ii) *Let* $\mathcal{C}$ *be a category admitting inductive limits indexed by a category* I. *Then the functor* $\varinjlim: \mathrm{Fct}(I, \mathcal{C}) \to \mathcal{C}$ *is right exact. Similarly, if* $\mathcal{C}$ *admits projective limits indexed by a category* J, *the functor* $\varprojlim: \mathrm{Fct}(J^{\mathrm{op}}, \mathcal{C}) \to \mathcal{C}$ *is left exact.*
(iii) *A small product of left (resp. right) exact functors is left (resp. right) exact. More precisely, if* $F_i: \mathcal{C}_i \to \mathcal{C}'_i$ *is a family of left (resp. right) exact functors indexed by a small set* I, *then the functor* $\prod_i F_i: \prod_i \mathcal{C}_i \to \prod_i \mathcal{C}'_i$ *is left (resp. right) exact.*
(iv) *Let* I *be a filtrant category. The functor* $\varinjlim: \mathrm{Fct}(I, \mathbf{Set}) \to \mathbf{Set}$ *as well as the functor* $\varinjlim: \mathrm{Fct}(I, \mathrm{Mod}(k)) \to \mathrm{Mod}(k)$ *are exact.*
(v) *Let* I *be a small set. Then the functor* $\prod: \mathrm{Mod}(k)^I \to \mathrm{Mod}(k)$ *is exact.*

Proof. (i) follows immediately from (2.1.6) and (2.1.7).
(ii) The functor $\varinjlim$ admits a right adjoint (see Exercise 2.8).
(iii) follows from Proposition 3.2.1 (iii).
(iv) follows from Proposition 3.1.6 and Corollary 3.1.7.
(v) is well-known and obvious.   q.e.d.

Definition 3.3.8. *Let* $\mathcal{C}$ *be a category and* I *a small category. Assume that* $\mathcal{C}$ *admits inductive limits indexed by* I. *If the functor* $\varinjlim: \mathcal{C}^I \to \mathcal{C}$ *is exact, we say that inductive limits indexed by* I *are exact in* $\mathcal{C}$. *If inductive limits indexed by any small filtrant category are exact in* $\mathcal{C}$, *we say that small filtrant inductive limits are exact in* $\mathcal{C}$.

Lemma 3.3.9. *Let* $\mathcal{C}$ *be a category which admits finite projective limits and inductive limits indexed by a connected category* I. *Assume that inductive limits indexed by* I *are exact. Then inductive limits indexed by* I *are stable by base changes (see Definition 2.2.6).*

Proof. Consider an inductive system $\{X_i\}_{i \in I}$ and a pair of morphisms $Y \to Z$ and $\varinjlim_{i \in I} X_i \to Z$ in $\mathcal{C}$. Let $\{Y_i\}_{i \in I}$ and $\{Z_i\}_{i \in I}$ denote the constant inductive systems with $Y_i = Y$ and $Z_i = Z$ for all $i \in I$ (and the identity morphisms associated with the morphisms in I). We have the isomorphisms

$$
\begin{aligned}
\varinjlim_{i \in I}(X_i \times_Z Y) &\simeq \varinjlim_{i \in I}(X_i \times_{Z_i} Y_i) \\
&\simeq \varinjlim_{i \in I} X_i \times_{\varinjlim_{i \in I} Z_i} \varinjlim_{i \in I} Y_i \\
&\simeq (\varinjlim_{i \in I} X_i) \times_Z Y .
\end{aligned}
$$

Here, the second isomorphism follows from the hypothesis that $\varinjlim$ is exact and the third isomorphism from the hypothesis that I is connected together with Lemma 2.1.12.   q.e.d.

We shall prove in Corollary 3.4.6 that if a functor $F \colon \mathcal{C} \to \mathcal{C}'$ is right exact, then the associated functor $\mathrm{Mor}(\mathcal{C}) \to \mathrm{Mor}(\mathcal{C}')$ is again right exact and we shall prove in Corollary 3.3.19 that if a functor $F \colon \mathcal{C} \to \mathcal{C}'$ is left exact, it extends to an exact functor $\mathcal{C}^\wedge \to (\mathcal{C}')^\wedge$.

Lemma 3.3.10. *Let $\varphi \colon J \to I$ be a left exact functor. Then φ is cofinal. In particular, if φ admits a left adjoint, then it is cofinal.*

Proof. A cofiltrant category is connected. The second assertion then follows from Proposition 3.3.6.   q.e.d.

Proposition 3.3.11. *Let $\varphi \colon J \to I$ be a functor. Assume that I is filtrant and φ is right exact. Then J is filtrant.*

Proof. This follows from Proposition 3.1.8 (ii).   q.e.d.

Proposition 3.3.12. *Let $F \colon \mathcal{C} \to \mathcal{C}'$ and $G \colon \mathcal{C}' \to \mathcal{C}''$ be two functors. If F and G are right exact, then $G \circ F$ is right exact.*

There is a similar result for left exact functors.

Proof. Since G is right exact, $\mathcal{C}'_Z$ is filtrant for any $Z \in \mathcal{C}''$. The functor $\mathcal{C}_Z \to \mathcal{C}'_Z$ is again right exact. Indeed, for any $Y \in \mathcal{C}'_Z$, $(\mathcal{C}_Z)_Y \simeq \mathcal{C}_Y$ is filtrant because F is right exact. Hence, Proposition 3.3.11 implies that $\mathcal{C}_Z$ is filtrant.   q.e.d.

Recall that for a category $\mathcal{C}$ and for $A \in \mathcal{C}^\wedge$, $\mathcal{C}_A$ is the category of pairs (X, u) of $X \in \mathcal{C}$ and $u \in A(X)$.

Proposition 3.3.13. *Assume that a category $\mathcal{C}$ admits finite inductive limits. Let $A \in \mathcal{C}^\wedge$. Then $A \colon \mathcal{C}^{\mathrm{op}} \to \mathbf{Set}$ is left exact if and only if the category $\mathcal{C}_A$ is filtrant.*

Proof. The proof is similar to that of Proposition 3.3.2.

(i) Assume that $\mathcal{C}_A$ is filtrant. By Proposition 2.6.3,

$$A(X) \simeq \varinjlim_{(Y \to A) \in \mathcal{C}_A} \mathrm{Hom}_{\mathcal{C}}(X, Y) \text{ for } X \in \mathcal{C} .$$

Since the functor from $\mathcal{C}^{\mathrm{op}}$ to **Set** given by $X \mapsto \mathrm{Hom}_{\mathcal{C}}(X, Y)$ commutes with finite projective limits, and small filtrant inductive limits commute with finite projective limits in **Set** (Proposition 3.1.6), A commutes with finite projective limits.

(ii) Conversely, assume that A is left exact and let us prove that $\mathcal{C}_A$ is filtrant. Since A commutes with finite projective limits, $\mathcal{C}_A$ admits finite inductive limits by Lemma 2.1.13, hence is filtrant by Proposition 3.2.1 (ii).   q.e.d.

Small Functors

Definition 3.3.14. *Let $F: \mathcal{C} \to \mathcal{C}'$ be a functor.*

(i) *We say that F is* right small *if for any $U \in \mathcal{C}'$, the category $\mathcal{C}_U$ is cofinally small.*

(ii) *We say that F is* left small *if $F^{\mathrm{op}}: \mathcal{C}^{\mathrm{op}} \to \mathcal{C}'^{\mathrm{op}}$ is right small or equivalently, if the category $\mathcal{C}^U$ is co-cofinally small.*

Note that if a category $\mathcal{C}$ is essentially small, then any functor $F: \mathcal{C} \to \mathcal{C}'$ is right small and left small.

Proposition 3.3.15. *Let $F: \mathcal{C} \to \mathcal{C}'$ be a right small functor and assume that $\mathcal{C}'$ is cofinally small. Then $\mathcal{C}$ is cofinally small.*

Proof. By Corollary 2.5.6, $\mathcal{C}'$ contains a small full subcategory $\mathcal{S}$ cofinal to $\mathcal{C}'$. For any $S \in \mathcal{S}$, $\mathcal{C}_S$ is cofinally small by the assumption, and this implies that there exists a small full subcategory $\mathcal{A}(S)$ of $\mathcal{C}_S$ cofinal to $\mathcal{C}_S$. Let $j_S: \mathcal{C}_S \to \mathcal{C}$ be the forgetful functor. Denote by $\mathcal{A}$ the full subcategory of $\mathcal{C}$ such that

$$\mathrm{Ob}(\mathcal{A}) = \bigcup_{S \in \mathcal{S}} j_S(\mathrm{Ob}(\mathcal{A}(S))) .$$

Then $\mathcal{A}$ is small. For $S \in \mathcal{S}$, we have functors $\mathcal{A}(S) \to \mathcal{A}_S \hookrightarrow \mathcal{C}_S$, and it follows from Proposition 2.5.4 (iii) that $\mathcal{A}_S \hookrightarrow \mathcal{C}_S$ is cofinal. Hence, Proposition 3.1.8 (i) implies that the functor $\mathcal{A} \to \mathcal{C}$ is cofinal.   q.e.d.

Proposition 3.3.16. *Let $F: \mathcal{C} \to \mathcal{C}'$ and $G: \mathcal{C}' \to \mathcal{C}''$ be two functors. If F and G are right small, then $G \circ F$ is right small.*

There is a similar result for left small functors.

Proof. (i) For any $W \in \mathcal{C}''$, $\mathcal{C}_W \to \mathcal{C}'_W$ is right small. Indeed, for any $(G(V) \to W) \in \mathcal{C}'_W$, $(\mathcal{C}_W)_{(G(V) \to W)} \simeq \mathcal{C}_V$ is cofinally small since F is right small.

(ii) Since G is right small, $\mathcal{C}'_W$ is cofinally small, and this implies that $\mathcal{C}_W$ is itself cofinally small by (i) and Proposition 3.3.15.   q.e.d.

Proposition 3.3.17. *Let $F: \mathcal{C} \to \mathcal{C}'$ be a functor. If F admits a right (resp. left) adjoint, then F is right (resp. left) small.*

The proof goes as for Proposition 3.3.6

Kan Extension of Functors, Revisited

We shall reformulate Theorem 2.3.3 using the notion of small functors and we shall discuss the right exactness of the functors we have constructed. For sake of brevity, we only treat the functor $\varphi^\dagger$. By reversing the arrows, (i.e., by using diagram 2.3.5) there is a similar result for the functor $\varphi^\ddagger$.

Theorem 3.3.18. *Let $\varphi: J \to I$ be a functor and let $\mathcal{C}$ be a category.*

(a) *Assume*

$$(3.3.2) \qquad \begin{cases} \varphi \text{ is right small,} \\ \mathcal{C} \text{ admits small inductive limits,} \end{cases}$$

or

$$(3.3.3) \qquad \begin{cases} \varphi \text{ is right exact and right small,} \\ \mathcal{C} \text{ admits small filtrant inductive limits.} \end{cases}$$

Then a left adjoint $\varphi^\dagger$ to the functor φ_* exists and (2.3.6) holds.
(b) *Assume (3.3.3) and also*

$$(3.3.4) \qquad \text{small filtrant inductive limits are exact in } \mathcal{C},$$

$$(3.3.5) \qquad \mathcal{C} \text{ admits finite projective limits .}$$

Then the functor $\varphi^\dagger$ is exact.

Proof. (a) By Theorem 2.3.3, it is enough to show that

$$\varinjlim_{(\varphi(j) \to i) \in J_i} \beta(j)$$

exists for $i \in I$ and $\beta \in \mathrm{Fct}(J, \mathcal{C})$. This follows by the assumption.

(b) Since $\varphi^\dagger$ admits a right adjoint, it is right exact. By hypothesis (3.3.5), the big category $\mathrm{Fct}(J, \mathcal{C})$ admits finite projective limits, and it is enough to check that the functor $\varphi^\dagger$ commutes with such limits.

Consider a finite projective system $\{\beta_k\}_{k \in K}$ in $\mathrm{Fct}(J, \mathcal{C})$. Let $i \in I$. There is a chain of isomorphisms:

$$\begin{aligned}
\left(\varphi^\dagger(\varprojlim_k \beta_k)\right)(i) &\simeq \varinjlim_{(\varphi(j) \to i) \in J_i} \varprojlim_k (\beta_k(j)) \\
&\simeq \varprojlim_k \varinjlim_{(\varphi(j) \to i) \in J_i} (\beta_k(j)) \\
&\simeq \varprojlim_k \left((\varphi^\dagger \beta_k)(i)\right),
\end{aligned}$$

where the second isomorphism follows from hypotheses (3.3.4) and (3.3.3). q.e.d.

Corollary 3.3.19. *Let $F: J \to I$ be a functor of small categories and assume that F is left exact. Then $\widehat{F}: J^{\wedge} \to I^{\wedge}$ (see § 2.7) is exact.*

Proof. Apply Propositions 2.7.5 and 3.3.18.   q.e.d.

3.4 Categories Associated with Two Functors

It is convenient to generalize Definition 1.2.16. Consider functors

$$I \xrightarrow{\varphi} K \xleftarrow{\psi} J .$$

Definition 3.4.1. *The category $M[I \xrightarrow{\varphi} K \xleftarrow{\psi} J]$ is given by*

$$\mathrm{Ob}(M[I \xrightarrow{\varphi} K \xleftarrow{\psi} J]) = \left\{(i, j, u)\,;\, i \in I,\, j \in J,\, u \in \mathrm{Hom}_K(\varphi(i), \psi(j))\right\}$$

$$\mathrm{Hom}_{M[I \xrightarrow{\varphi} K \xleftarrow{\psi} J]}((i, j, u), (i', j', u'))$$

$$= \Big\{(v_1, v_2) \in \mathrm{Hom}_I(i, i') \times \mathrm{Hom}_J(j, j')\,;\, the\ diagram$$

$$\begin{array}{ccc} \varphi(i) & \xrightarrow{\ \ u\ \ } & \psi(j) \\ \Big\downarrow{\varphi(v_1)} & & \Big\downarrow{\psi(v_2)} \\ \varphi(i') & \xrightarrow{\ \ u'\ \ } & \psi(j') \end{array} \ commutes\Big\} .$$

If there is no risk of confusion, we shall write $M[I \to K \leftarrow J]$ instead of $M[I \xrightarrow{\varphi} K \xleftarrow{\psi} J]$.

Let $F: \mathcal{C} \to \mathcal{C}'$ be a functor and let $A \in \mathcal{C}'$. Recall that **Pt** denote the category with a single object and a single morphism and denote by $\Delta_A: \mathbf{Pt} \to \mathcal{C}'$ the unique functor with values A. Then

$$\mathcal{C}_A \simeq M[\mathcal{C} \xrightarrow{F} \mathcal{C}' \xleftarrow{\Delta_A} \mathbf{Pt}] ,$$
$$\mathcal{C}^A \simeq M[\mathbf{Pt} \xrightarrow{\Delta_A} \mathcal{C}' \xleftarrow{F} \mathcal{C}] .$$

Suppose that we have a diagram of functors

(3.4.1)

$$\begin{array}{ccccc} I_1 & \xrightarrow{\varphi_1} & K_1 & \xleftarrow{\psi_1} & J_1 \\ \Big\downarrow{F} & & \Big\downarrow{H} & & \Big\downarrow{G} \\ I_2 & \xrightarrow{\varphi_2} & K_2 & \xleftarrow{\psi_2} & J_2 \end{array}$$

and that this diagram commutes up to isomorphisms of functors, that is, this diagram is quasi-commutative (see Remark 1.3.6). It allows us to define naturally a functor

$$(3.4.2) \qquad \theta : M[I_1 \to K_1 \leftarrow J_1] \to M[I_2 \to K_2 \leftarrow J_2] \ .$$

Proposition 3.4.2. *Consider the quasi-commutative diagram of categories* (3.4.1) *and the functor* θ *in* (3.4.2).

(i) *If* F *and* G *are faithful, then* θ *is faithful.*
(ii) *If* F *and* G *are fully faithful and* H *is faithful, then* θ *is fully faithful.*
(iii) *If* F *and* G *are equivalences of categories and* H *is fully faithful, then* θ *is an equivalence of categories.*

The proof is left as an exercise.

Proposition 3.4.3. *Let* I, J, K *be three categories and let* $\varphi : I \to K$ *and* $\psi : J \to K$ *be two functors.*

(i) *For any category* $\mathcal{C}$ *and any functor* $\alpha : M[I \to K \leftarrow J] \to \mathcal{C}$, *we have*
$$\varinjlim \alpha \simeq \varinjlim_{j \in J} \varinjlim_{i \in I_{\psi(j)}} \alpha((i, j, \varphi(i) \to \psi(j))).$$
(ii) *If* ψ *is cofinal, then* $M[I \to K \leftarrow J] \to I$ *is cofinal.*
(iii) *If* I *is connected and* ψ *is cofinal, then* $M[I \to K \leftarrow J]$ *is connected.*

Proof. (i) Set $M := M[I \to K \leftarrow J]$. Then, for every $j_0 \in J$, the canonical functor $\xi : I_{\psi(j_0)} \to M_{j_0}$ admits a left adjoint η given by

$$M_{j_0} \ni (i, j, \varphi(i) \xrightarrow{t} \psi(j), j \xrightarrow{s} j_0) \mapsto (i, \varphi(i) \xrightarrow{t} \psi(j) \xrightarrow{\psi(s)} \psi(j_0)) \in I_{\psi(j_0)} \ .$$

Hence, ξ is cofinal by Lemma 3.3.10. It remains to apply Corollary 2.3.4.

(ii) For any functor $\alpha : I \to \mathbf{Set}$, denote by β the composition of functors $M[I \to K \leftarrow J] \to I \to \mathbf{Set}$. By Corollary 2.3.4, we have $\varinjlim \alpha \simeq \varinjlim_{k \in K} \varinjlim_{i \in I_k} \alpha(i)$.

Since $\psi : J \to K$ is cofinal, we obtain

$$\varinjlim \alpha \simeq \varinjlim_{j \in J} \varinjlim_{i \in I_{\psi(j)}} \alpha(i) \simeq \varinjlim_{j \in J} \varinjlim_{i \in I_{\psi(j)}} \beta((i, j, \varphi(i) \to \psi(j))) \simeq \varinjlim \beta \ ,$$

where the last isomorphism follows from (i).

(iii) follows from (ii) and Corollary 2.5.3.   q.e.d.

Proposition 3.4.4. *Consider the quasi-commutative diagram of categories* (3.4.1) *and the functor* θ *in* (3.4.2). *Assume*

(i) *the category* J_ν *is filtrant and the functor* ψ_ν *is cofinal for* $\nu = 1, 2$,
(ii) *the functors* F *and* G *are cofinal.*

Then the functor θ *is cofinal.*

Note that the hypotheses imply that K_ν is filtrant for $\nu = 1, 2$ by Proposition 3.2.2 and H is cofinal by Proposition 2.5.4.

Proof. We shall write M_ν ($\nu = 1, 2$) instead of $M[I_\nu \to K_\nu \leftarrow J_\nu]$ for short. Let $a = (i_2, j_2, u_2) \in M_2$. We shall check that M_1^a is connected. Let

$$\varphi' : (I_1)^{i_2} \to (K_1)^{\varphi_2(i_2)}$$

denote the canonical functor. The morphism $u_2 \colon \varphi_2(i_2) \to \psi_2(j_2)$ defines the functor

$$\psi' : (J_1)^{j_2} \to (K_1)^{\varphi_2(i_2)}$$

by associating to an object $(j_2 \to G(j_1)) \in (J_1)^{j_2}$ the object $(\varphi_2(i_2) \to \psi_2(j_2) \to \psi_2(G(j_1)) \simeq H(\psi_1(j_1)))$ of $(K_1)^{\varphi_2(i_2)}$. The equivalence

$$(M_1)^a \simeq M[(I_1)^{i_2} \xrightarrow{\varphi'} (K_1)^{\varphi_2(i_2)} \xleftarrow{\psi'} (J_1)^{j_2}]$$

is easily checked. The category $(I_1)^{i_2}$ is connected. By Proposition 3.4.3, it is enough to show that ψ' is cofinal. The functor ψ' decomposes as

$$(J_1)^{j_2} \to (J_1)^{\psi_2(j_2)} \to (K_1)^{\varphi_2(i_2)} \; ,$$

and these arrows are cofinal by Proposition 3.2.5.   q.e.d.

Proposition 3.4.5. *Let I, J, K be three categories and let $\varphi \colon I \to K$ and $\psi \colon J \to K$ be two functors. Assume that I, J are filtrant and ψ is cofinal. Then*

(i) *the category $M[I \xrightarrow{\varphi} K \xleftarrow{\psi} J]$ is filtrant,*

(ii) *the canonical projection functors from $M[I \xrightarrow{\varphi} K \xleftarrow{\psi} J]$ to I, J and $I \times J$ are cofinal,*

(iii) *if I and J are cofinally small, then $M[I \xrightarrow{\varphi} K \xleftarrow{\psi} J]$ is cofinally small.*

Proof. (i) By Proposition 3.3.11, it is enough to show that the functor $M := M[I \to K \leftarrow J] \to I$ is right exact. For every $i \in I$, M_i is equivalent to $M[I_i \to K \leftarrow J]$. On the other hand, since $i \in I_i$ is a terminal object of I_i, the functor $\xi \colon \mathbf{Pt} \to I_i$, pt $\mapsto i$, is cofinal. Applying Proposition 3.4.4, we get that the functor $\theta \colon M[\mathbf{Pt} \to K \leftarrow J] \to M[I_i \to K \leftarrow J]$ is cofinal. Since $M[\mathbf{Pt} \to K \leftarrow J] \simeq J^{\varphi(i)}$, this category is filtrant by Proposition 3.2.2, and this statement also implies that $M_i \simeq M[I_i \to K \leftarrow J]$ is filtrant.

(ii) There are natural equivalences of categories

$$I \simeq M[I \to \mathbf{Pt} \leftarrow \mathbf{Pt}], \quad I \times J \simeq M[I \to \mathbf{Pt} \leftarrow J], \quad J \simeq M[\mathbf{Pt} \to \mathbf{Pt} \leftarrow J] \; .$$

Hence, the result follows from Proposition 3.4.4.

(iii) By the hypothesis, there exist small filtrant categories I', J' and cofinal functors $I' \to I$ and $J' \to J$. Then $M[I' \to K \leftarrow J'] \to M[I \to K \leftarrow J]$ is cofinal by Proposition 3.4.4. Since $M[I' \to K \leftarrow J']$ is small, the result follows.   q.e.d.

Let $F: \mathcal{C} \to \mathcal{C}'$ be a functor. We denote by

(3.4.3) $$\mathrm{Mor}(F): \mathrm{Mor}(\mathcal{C}) \to \mathrm{Mor}(\mathcal{C}')$$

the functor naturally associated with F.

Corollary 3.4.6. *Let $F: \mathcal{C} \to \mathcal{C}'$ be a right exact functor. Then*

(i) *the functor $\mathrm{Mor}(F)$ in (3.4.3) is right exact,*
(ii) *for any morphism $f: Y \to Y'$ in $\mathcal{C}'$, the canonical projection functors from $\mathrm{Mor}(\mathcal{C})_f$ to $\mathcal{C}_Y$, $\mathcal{C}_{Y'}$ and $\mathcal{C}_Y \times \mathcal{C}_{Y'}$ are cofinal,*
(iii) *if moreover $\mathcal{C}_Y$ and $\mathcal{C}_{Y'}$ are cofinally small, then $\mathrm{Mor}(\mathcal{C})_f$ is cofinally small. In particular, if F is right small then $\mathrm{Mor}(F)$ is right small.*

Proof. It is enough to remark that $\mathrm{Mor}(\mathcal{C})_f \simeq M[\mathcal{C}_Y \to \mathcal{C}_{Y'} \xleftarrow{\mathrm{id}_{\mathcal{C}_{Y'}}} \mathcal{C}_{Y'}]$ and to apply the preceding results.   q.e.d.

Exercises

Exercise 3.1. Let $\mathcal{C}$ be the category with two objects $\{a, b\}$ and whose morphisms other than identities are a morphism $f: a \to b$, a morphism $g: b \to a$ and a morphism $p: b \to b$, these morphisms satisfying $f \circ g = p$, $g \circ f = \mathrm{id}_a$, $p \circ p = p$. Prove that $\mathcal{C}$ admits filtrant inductive and filtrant projective limits.

Exercise 3.2. Let $\mathcal{C}$ be a category.
(i) Prove that small filtrant inductive limits commute with finite projective limits in $\mathcal{C}^\wedge$ (i.e., Proposition 3.1.6 holds with $\mathcal{C}^\wedge$ instead of **Set**).
(ii) Prove that small inductive limits are stable by base change in $\mathcal{C}^\wedge$ (see Definition 2.2.6).
(Hint: use Exercise 2.7.)

Exercise 3.3. Let **Pt** and **Pr** be the categories introduced in Notations 1.2.8. Let $\varphi: \mathbf{Pt} \to \mathbf{Pr}$ be the unique functor from **Pt** to **Pr**.
(i) Prove that **Pt** and **Pr** are filtrant.
(ii) Prove that φ satisfies condition (a) in Proposition 3.2.2 (iii), but that φ is not cofinal.
(iii) Prove that $\mathbf{Pt}^c \simeq \mathbf{Pt} \sqcup \mathbf{Pt}$ (a set with two elements regarded as a discrete category).

Exercise 3.4. Let $F: \mathcal{C} \to \mathcal{C}'$ be a functor.
(i) Assume that F is left (resp. right) exact and let $f: X \to Y$ be a monomorphism (resp. an epimorphism) in $\mathcal{C}$. Prove that $F(f)$ is a monomorphism (resp. an epimorphism).
(ii) Deduce that if $\mathcal{C}$ and $\mathcal{C}'$ are small categories and $u: A \to B$ is an epimorphism in $\mathcal{C}^\wedge$, then $\widehat{F}(u): \widehat{F}(A) \to \widehat{F}(B)$ is an epimorphism (the functor $\widehat{F}: \mathcal{C}^\wedge \to (\mathcal{C}')^\wedge$ is defined in Notation 2.7.2). (Hint: use Proposition 2.7.1.)

Exercise 3.5. Let $F \colon \mathcal{C} \to \mathcal{C}'$ be a functor of small categories. Prove that F is left exact if and only if $\widehat{F} \colon \mathcal{C}^\wedge \to \mathcal{C}'^\wedge$ is exact.

Exercise 3.6. Assume that $\mathcal{C}$ is idempotent complete (see Exercise 2.9). Prove that the Yoneda functor $h_\mathcal{C} \colon \mathcal{C} \to \mathcal{C}^\wedge$ is left exact if and only if $\mathcal{C}$ admits finite projective limits.

Exercise 3.7. Let $\mathcal{C}$ be a category admitting an initial object. Denote by $\emptyset_\mathcal{C}$ and $\emptyset_{\mathcal{C}^\wedge}$ the initial object of $\mathcal{C}$ and $\mathcal{C}^\wedge$, respectively.
(i) Show that $\emptyset_{\mathcal{C}^\wedge}(X) = \emptyset$ for any $X \in \mathcal{C}$ and deduce that $h_\mathcal{C}(\emptyset_\mathcal{C})$ and $\emptyset_{\mathcal{C}^\wedge}$ are not isomorphic.
(ii) Prove that the Yoneda functor $h_\mathcal{C} \colon \mathcal{C} \to \mathcal{C}^\wedge$ is not right exact for any category $\mathcal{C}$.

Exercise 3.8. Let I be a filtrant category such that $\operatorname{Mor}(I)$ is countable. Prove that there exists a cofinal functor $\mathbb{N} \to I$. Here, $\mathbb{N}$ is regarded as the category associated with its natural order.

Exercise 3.9. Let $\mathcal{C}$ be a finite filtrant category. Prove that there exists a cofinal functor $\mathbf{Pr} \to \mathcal{C}$. (See Notations 1.2.8 (v).)

4

Tensor Categories

This chapter is devoted to tensor categories which axiomatize the properties of tensor products of vector spaces. Its importance became more evident when quantum groups produced rich examples of non commutative tensor categories and this notion is now used in many areas, mathematical physics, knot theory, computer sciences, etc. Tensor categories and their applications deserve at least a whole book, and we shall be extremely superficial and sketchy here. Among the vast literature on this subject, let us only quote [15, 40].

We begin this chapter by introducing projectors in categories. Then we define and study tensor categories, dual pairs, braidings and the Yang-Baxter equations. We also introduce the notions of a ring in a tensor category and a module over this ring in a category on which the tensor category operates. As a particular case we treat monads, and finally we prove the Bar-Beck theorem.

Most of the notions introduced in this Chapter (with the exception of §4.1) are not necessary for the understanding of the rest of the book, and this chapter may be skipped.

4.1 Projectors

The notion of a projector in linear algebra has its counterpart in Category Theory.

Definition 4.1.1. *Let C be a category. A* projector *(P, ε) on C is the data of a functor $P \colon C \to C$ and a morphism $\varepsilon \colon \mathrm{id}_C \to P$ such that the two morphisms of functors $\varepsilon \circ P$, $P \circ \varepsilon \colon P \rightrightarrows P^2$ are isomorphisms. Here, $P^2 := P \circ P$.*

Lemma 4.1.2. *If (P, ε) is a projector, then $\varepsilon \circ P = P \circ \varepsilon$.*

Proof. For any $X \in C$, we have a commutative diagram with solid arrows:

$$X \xrightarrow{\varepsilon_X} P(X)$$

(4.1.1)

$$\varepsilon_X \downarrow \quad u \quad \downarrow P(\varepsilon_X)$$

$$P(X) \xrightarrow[\varepsilon_{P(X)}]{\sim} P^2(X).$$

Since $\varepsilon_{P(X)}$ is an isomorphism, we can find a morphism $u\colon P(X) \to P(X)$ such that $\varepsilon_{P(X)} \circ u = P(\varepsilon_X)$. Then $u \circ \varepsilon_X = \varepsilon_X$ and the commutative diagram

$$P(X) \xrightarrow{P(\varepsilon_X)} P^2(X)$$

$$P(\varepsilon_X) \downarrow \quad \nearrow P(u)$$

$$P^2(X)$$

implies that $P(u) = \mathrm{id}_{P^2(X)}$. Since $\varepsilon_{P(X)}$ is an isomorphism, we conclude that $u = \mathrm{id}_{P(X)}$ by the commutative diagram

$$P(X) \xrightarrow{u} P(X)$$

$$\varepsilon_{P(X)} \downarrow \qquad \downarrow \varepsilon_{P(X)}$$

$$P^2(X) \xrightarrow{P(u)} P^2(X).$$

q.e.d.

Proposition 4.1.3. *Let (P, ε) be a projector on $\mathcal{C}$.*

(i) *For any $X, Y \in \mathcal{C}$, the map*

$$\mathrm{Hom}_{\mathcal{C}}(P(X), P(Y)) \xrightarrow{\circ \varepsilon_X} \mathrm{Hom}_{\mathcal{C}}(X, P(Y))$$

is bijective.

(ii) *The following three conditions on $X \in \mathcal{C}$ are equivalent:*
   (a) *$\varepsilon_X\colon X \to P(X)$ is an isomorphism,*
   (b) *$\mathrm{Hom}_{\mathcal{C}}(P(Y), X) \xrightarrow{\circ \varepsilon_Y} \mathrm{Hom}_{\mathcal{C}}(Y, X)$ is bijective for any $Y \in \mathcal{C}$,*
   (c) *the map in (b) is surjective for $Y = X$.*

(iii) *Let $\mathcal{C}_0$ be the full subcategory of $\mathcal{C}$ consisting of objects $X \in \mathcal{C}$ satisfying the equivalent conditions in (ii). Then $P(X) \in \mathcal{C}_0$ for any $X \in \mathcal{C}$ and P induces a functor $\mathcal{C} \to \mathcal{C}_0$ which is left adjoint to the inclusion functor $\iota\colon \mathcal{C}_0 \to \mathcal{C}$.*

Proof. (i) The composition

$$\theta\colon \mathrm{Hom}_{\mathcal{C}}(X, P(Y)) \to \mathrm{Hom}_{\mathcal{C}}(P(X), P^2(Y)) \xleftarrow{\sim} \mathrm{Hom}_{\mathcal{C}}(P(X), P(Y)),$$

where the second map is given by $\varepsilon_{P(Y)}$, is an inverse of the map $\circ\varepsilon_X$. Indeed, $\theta \circ (\cdot \circ \varepsilon_X)$ and $(\cdot \circ \varepsilon_X) \circ \theta$ are the identities, as seen by the commutative diagrams below.

$$P(X) \xrightarrow{u} P(Y) \qquad X \xrightarrow{v} P(Y)$$

(diagram, left)

$$\mathrm{id}_{P(X)} \nearrow \quad \varepsilon_{P(X)} \downarrow \quad \sim \downarrow \varepsilon_{P(Y)}$$

$$P(X) \xrightarrow[P(\varepsilon_X)]{} P^2(X) \xrightarrow[P(u)]{} P^2(Y),$$

(diagram, right)

$$\varepsilon_X \downarrow \quad \theta(v) \nearrow \quad \sim \downarrow \varepsilon_{P(Y)}$$

$$P(X) \xrightarrow[P(v)]{} P^2(Y).$$

(ii) (a) $\Rightarrow$ (b) follows from (i).

(b) $\Rightarrow$ (c) is obvious.

(c) $\Rightarrow$ (a). There exists a morphism $u \colon P(X) \to X$ such that $u \circ \varepsilon_X = \mathrm{id}_X$. Since $(\varepsilon_X \circ u) \circ \varepsilon_X = \varepsilon_X \circ \mathrm{id}_X = \mathrm{id}_{P(X)} \circ \varepsilon_X$, we have $\varepsilon_X \circ u = \mathrm{id}_{P(X)}$ by (i) with $Y = X$. Hence, ε_X is an isomorphism.

(iii) Since $\varepsilon_{P(X)}$ is an isomorphism, $P(X) \in \mathcal{C}_0$ for any $X \in \mathcal{C}$ and P induces a functor $\mathcal{C} \to \mathcal{C}_0$. This functor is a left adjoint to $\iota \colon \mathcal{C}_0 \to \mathcal{C}$ by (i).   q.e.d.

Proposition 4.1.4. *Let $R \colon \mathcal{C}' \to \mathcal{C}$ be a fully faithful functor and assume that R admits a left adjoint $L \colon \mathcal{C} \to \mathcal{C}'$. Let $\varepsilon \colon \mathrm{id}_\mathcal{C} \to R \circ L$ and $\eta \colon L \circ R \to \mathrm{id}_{\mathcal{C}'}$ be the adjunction morphisms. Set $P = R \circ L \colon \mathcal{C} \to \mathcal{C}$. Then*

(i) *(P, ε) is a projector,*

(ii) *for any $X \in \mathcal{C}$, the following conditions are equivalent:*
   (a) *$\varepsilon_X \colon X \to RL(X)$ is an isomorphism,*
   (b) *$\mathrm{Hom}_\mathcal{C}(RL(Y), X) \xrightarrow{\circ \varepsilon_Y} \mathrm{Hom}_\mathcal{C}(Y, X)$ is bijective for any $Y \in \mathcal{C}$.*

(iii) *Let $\mathcal{C}_0$ be the full subcategory of $\mathcal{C}$ consisting of objects X satisfying the equivalent conditions in (ii). Then $\mathcal{C}'$ is equivalent to $\mathcal{C}_0$.*

Proof. Since R is fully faithful, η is an isomorphism.

(i) The two compositions

$$P \overset{\varepsilon \circ P}{\underset{P \circ \varepsilon}{\rightrightarrows}} P^2 \xrightarrow{R \eta L} P$$

are equal to id_P. Since $R \circ \eta \circ L \colon RLRL \to RL$ is an isomorphism, it follows that $P \circ \varepsilon$ and $\varepsilon \circ P$ are isomorphisms.

(ii) follows from Proposition 4.1.3.

(iii) For $X \in \mathcal{C}'$, the morphism $R(\eta_X) \colon PR(X) = RLR(X) \to R(X)$ is an isomorphism. Since the composition

$$R(X) \xrightarrow{\varepsilon_{R(X)}} PR(X) \xrightarrow{R(\eta_X)} R(X)$$

is $\mathrm{id}_{R(X)}$, $\varepsilon_{R(X)}$ is an isomorphism. Hence, R sends $\mathcal{C}'$ to $\mathcal{C}_0$. This functor is fully faithful, and it is essentially surjective since $Y \simeq RL(Y)$ for any $Y \in \mathcal{C}_0$.
    q.e.d.

4.2 Tensor Categories

Definition 4.2.1. *A tensor category is the data of a category $\mathcal{T}$, a bifunctor $\cdot \otimes \cdot : \mathcal{T} \times \mathcal{T} \to \mathcal{T}$ and an isomorphism of functors $a \in \mathrm{Mor}(\mathrm{Fct}(\mathcal{T} \times \mathcal{T} \times \mathcal{T}, \mathcal{T}))$,*

$$a(X, Y, Z) : (X \otimes Y) \otimes Z \xrightarrow{\sim} X \otimes (Y \otimes Z)$$

such that the diagram below is commutative for any $X, Y, Z, W \in \mathcal{T}$:

(4.2.1)

$$
\begin{array}{ccc}
((X \otimes Y) \otimes Z) \otimes W & \xrightarrow{\ a(X \otimes Y, Z, W)\ } & (X \otimes Y) \otimes (Z \otimes W) \\
{\scriptstyle a(X,Y,Z) \otimes W} \downarrow & & \downarrow {\scriptstyle a(X,Y,Z \otimes W)} \\
(X \otimes (Y \otimes Z)) \otimes W & & \\
{\scriptstyle a(X,Y \otimes Z, W)} \downarrow & & \\
X \otimes ((Y \otimes Z) \otimes W) & \xrightarrow{\ X \otimes a(Y,Z,W)\ } & X \otimes (Y \otimes (Z \otimes W)).
\end{array}
$$

Examples 4.2.2. The following $(\mathcal{T}, \otimes, a)$ (with a the obvious one) are tensor categories.

(i) k is a commutative ring, $\mathcal{T} = \mathrm{Mod}(k)$ and $\otimes = \otimes_k$.

(ii) M is a monoid, $\mathcal{T}$ is the discrete category with $\mathrm{Ob}(\mathcal{T}) = M$, $a \otimes b = ab$ for $a, b \in M$.

(iii) A is a k-algebra, $\mathcal{T} = \mathrm{Mod}(A \otimes_k A^{\mathrm{op}})$ and $\otimes = \otimes_A$.

(iv) $\mathcal{C}$ is a category, $\mathcal{T} = \mathrm{Fct}(\mathcal{C}, \mathcal{C})$ and $\otimes = \circ$.

(v) $\mathcal{T}$ is a category which admits finite products and $\otimes = \times$.

(vi) $\mathcal{T}$ is a category which admits finite coproducts and $\otimes = \sqcup$.

(vii) G is a group, k is a field, $\mathcal{T}$ is the category of G-modules over k, that is, the category whose objects are the pairs (V, φ), $V \in \mathrm{Mod}(k)$, $\varphi : G \to \mathrm{Aut}_k(V)$ is a morphism of groups, and the morphisms are the natural ones. For $V, W \in \mathcal{T}$, $V \otimes W$ is the tensor product in $\mathrm{Mod}(k)$ endowed with the diagonal action of G given by $g(v \otimes w) = gv \otimes gw$.

(viii) I is a category, $\mathcal{T} = \mathcal{S}(I)$ is the category defined as follows. The objects of $\mathcal{S}(I)$ are the finite sequences of objects of I of length ≥ 1. For $X = (x_1, \ldots, x_n)$ and $Y = (y_1, \ldots, y_p)$ in $\mathcal{S}(I)$,

$$\mathrm{Hom}_{\mathcal{S}(I)}(X, Y) = \begin{cases} \prod_{i=1}^n \mathrm{Hom}_I(x_i, y_i) & \text{if } n = p, \\ \emptyset & \text{otherwise}. \end{cases}$$

Hence, $\mathcal{S}(I) \simeq \bigsqcup_{n \geq 1} I^n$.

For two objects $X = (x_1, \ldots, x_n)$ and $Y = (y_1, \ldots, y_p)$ of $\mathcal{S}(I)$, define $X \otimes Y$ as the sequence $(x_1, \ldots, x_n, y_1, \ldots, y_p)$.

(ix) k is a commutative ring and, with the notations of Chap. 11, $\mathcal{T} = \mathrm{C}^{\mathrm{b}}(\mathrm{Mod}(k))$ is the category of bounded complexes of k-modules and $X \otimes Y$ is the simple complex associated with the double complex $X \otimes_k Y$.

Let $(\mathcal{T}, \otimes, a)$ be a tensor category. Then $\mathcal{T}^{\mathrm{op}}$ has a structure of a tensor category in an obvious way. Another tensor category structure on $\mathcal{T}$ is obtained as follows. For $X, Y \in \mathcal{T}$, define

$$X \overset{r}{\otimes} Y := Y \otimes X .$$

For $X, Y, Z \in \mathcal{T}$, define

$$a^r(X, Y, Z) : (X \overset{r}{\otimes} Y) \overset{r}{\otimes} Z \overset{\sim}{\to} X \overset{r}{\otimes} (Y \overset{r}{\otimes} Z)$$

by

$$(X \overset{r}{\otimes} Y) \overset{r}{\otimes} Z = Z \otimes (Y \otimes X) \xrightarrow{a(Z,Y,X)^{-1}} (Z \otimes Y) \otimes X = X \overset{r}{\otimes} (Y \overset{r}{\otimes} Z) .$$

Then $(\mathcal{T}, \overset{r}{\otimes}, a^r)$ is a tensor category. We call it the *reversed* tensor category of $(\mathcal{T}, \otimes, a)$.

Tensor Functors

Definition 4.2.3. *Let $\mathcal{T}$ and $\mathcal{T}'$ be two tensor categories. A functor of tensor categories (or, a* tensor functor*) is a pair (F, ξ_F) where $F \colon \mathcal{T} \to \mathcal{T}'$ is a functor and ξ_F is an isomorphism of bifunctors*

$$\xi_F \colon F(\cdot \otimes \cdot) \overset{\sim}{\to} F(\cdot) \otimes F(\cdot)$$

such that the diagram below commutes for all $X, Y, Z \in \mathcal{T}$:

(4.2.2)

$$\begin{array}{ccc}
F((X \otimes Y) \otimes Z) & \xrightarrow{\;\;F(a(X,Y,Z))\;\;} & F(X \otimes (Y \otimes Z)) \\
{\scriptstyle \xi_F(X \otimes Y, Z)} \downarrow & & \downarrow {\scriptstyle \xi_F(X, Y \otimes Z)} \\
F(X \otimes Y) \otimes F(Z) & & F(X) \otimes F(Y \otimes Z) \\
{\scriptstyle \xi_F(X,Y) \otimes F(Z)} \downarrow & & \downarrow {\scriptstyle F(X) \otimes \xi_F(Y,Z)} \\
(F(X) \otimes F(Y)) \otimes F(Z) & \xrightarrow{\;\;a(F(X),F(Y),F(Z))\;\;} & F(X) \otimes (F(Y) \otimes F(Z)).
\end{array}$$

In practice, we omit to write ξ_F.

For two tensor functors $F, G \colon \mathcal{T} \to \mathcal{T}'$, a morphism of tensor functors $\theta \colon F \to G$ is a morphism of functors such that the diagram below commutes for all $X, Y \in \mathcal{T}$:

$$\begin{array}{ccc}
F(X \otimes Y) & \xrightarrow{\;\;\xi_F(X,Y)\;\;} & F(X) \otimes F(Y) \\
{\scriptstyle \theta_{X \otimes Y}} \downarrow & & \downarrow {\scriptstyle \theta_X \otimes \theta_Y} \\
G(X \otimes Y) & \xrightarrow{\;\;\xi_G(X,Y)\;\;} & G(X) \otimes G(Y) .
\end{array}$$

Recall that to a category I we have associated a tensor category $\mathcal{S}(I)$ in Example 4.2.2 (viii). Let us denote by $\iota \colon I \to \mathcal{S}(I)$ the canonical functor.

Lemma 4.2.4. *let $\mathcal{T}$ be a tensor category, let I be a category and let $\varphi\colon I \to \mathcal{T}$ be a functor. There exists a functor of tensor categories $\Phi\colon \mathcal{S}(I) \to \mathcal{T}$ such that $\Phi \circ \iota \simeq \varphi$. Moreover, Φ is unique up to unique isomorphism.*

Proof. We define by induction on n

$$\Phi\big((i_1,\ldots,i_n)\big) = \Phi\big((i_1,\ldots,i_{n-1})\big) \otimes \varphi(i_n) \ .$$

We define the isomorphism

$$\xi_\Phi\colon \Phi\big((i_1,\ldots,i_n)\otimes(j_1,\ldots,j_m)\big) \xrightarrow{\sim} \Phi\big((i_1,\ldots,i_n)\big) \otimes \Phi\big((j_1,\ldots,j_m)\big)$$

by the induction on m as follows:

$$
\begin{aligned}
&\Phi\big((i_1,\ldots,i_n)\otimes(j_1,\ldots,j_m)\big) \\
&\simeq \Phi\big((i_1,\ldots,i_n,j_1,\ldots,j_m)\big) \\
&\simeq \Phi\big((i_1,\ldots,i_n,j_1,\ldots,j_{m-1})\big) \otimes \varphi(j_m) \\
&\simeq \Phi\big((i_1,\ldots,i_n)\otimes(j_1,\ldots,j_{m-1})\big) \otimes \varphi(j_m) \\
&\simeq \Big(\Phi\big((i_1,\ldots,i_n)\big)\otimes\Phi\big((j_1,\ldots,j_{m-1})\big)\Big) \otimes \varphi(j_m) \\
&\simeq \Phi\big((i_1,\ldots,i_n)\big)\otimes\Big(\Phi\big((j_1,\ldots,j_{m-1})\big)\otimes\varphi(j_m)\Big) \\
&\simeq \Phi\big((i_1,\ldots,i_n)\big)\otimes\Phi\big((j_1,\ldots,j_m)\big) \ .
\end{aligned}
$$

It is left to the reader to check that this defines a functor of tensor categories.
q.e.d.

Hence, in a tensor category $\mathcal{T}$, it is possible to define the tensor product $X_1 \otimes \cdots \otimes X_n$ for $X_1,\ldots,X_n \in \mathcal{T}$ by the formula

$$X_1 \otimes \cdots \otimes X_n = (\cdots((X_1 \otimes X_2) \otimes X_3)\otimes\cdots)\otimes X_n$$

and this does not depend on the order of the parentheses, up to a unique isomorphism.

In the sequel, we shall often omit the parentheses.

Unit Object

Definition 4.2.5. *A* unit object *of a tensor category $\mathcal{T}$ is an object $\mathbf{1}$ of $\mathcal{T}$ endowed with an isomorphism $\varrho\colon \mathbf{1}\otimes\mathbf{1}\xrightarrow{\sim}\mathbf{1}$ such that the functors from $\mathcal{T}$ to $\mathcal{T}$ given by $X \mapsto X\otimes\mathbf{1}$ and $X \mapsto \mathbf{1}\otimes X$ are fully faithful.*

Lemma 4.2.6. *Let $(\mathbf{1},\varrho)$ be a unit object of $\mathcal{T}$. Then there exist unique functorial isomorphisms $\alpha(X)\colon X\otimes\mathbf{1}\xrightarrow{\sim}X$ and $\beta(X)\colon \mathbf{1}\otimes X\xrightarrow{\sim}X$ satisfying the following properties*

(a) $\alpha(\mathbf{1}) = \beta(\mathbf{1}) = \varrho$,

(b) *the two morphisms* $X \otimes Y \otimes \mathbf{1} \underset{X \otimes \alpha(Y)}{\overset{\alpha(X \otimes Y)}{\rightrightarrows}} X \otimes Y$ *coincide,*

(c) *the two morphisms* $\mathbf{1} \otimes X \otimes Y \underset{\beta(X) \otimes Y}{\overset{\beta(X \otimes Y)}{\rightrightarrows}} X \otimes Y$ *coincide,*

(d) *the two morphisms* $X \otimes \mathbf{1} \otimes Y \underset{X \otimes \beta(Y)}{\overset{\alpha(X) \otimes Y}{\rightrightarrows}} X \otimes Y$ *coincide,*

(e) *the diagram*

$$
\begin{array}{ccc}
\mathbf{1} \otimes X \otimes \mathbf{1} & \xrightarrow{\mathbf{1} \otimes \alpha(X)} & \mathbf{1} \otimes X \\
{\scriptstyle \beta(X) \otimes \mathbf{1}} \downarrow & & \downarrow {\scriptstyle \beta(X)} \\
X \otimes \mathbf{1} & \xrightarrow{\alpha(X)} & X
\end{array}
$$
commutes .

Proof. If such α and β exist, then (a) and (d) imply $\alpha(X) \otimes \mathbf{1} = X \otimes \beta(\mathbf{1}) = X \otimes \varrho$, $\alpha(X)$ is uniquely determined because $X \mapsto X \otimes \mathbf{1}$ is fully faithful, and similarly with β.

Proof of the existence of α, β. Since $X \mapsto X \otimes \mathbf{1}$ is fully faithful, there exists a unique morphism $\alpha(X)\colon X \otimes \mathbf{1} \to X$ such that $\alpha(X) \otimes \mathbf{1}\colon X \otimes \mathbf{1} \otimes \mathbf{1} \to X \otimes \mathbf{1}$ coincides with $X \otimes \varrho$. Since $X \otimes \varrho$ is an isomorphism, $\alpha(X)$ is an isomorphism. The morphism β is constructed similarly by $\mathbf{1} \otimes \beta(X) = \varrho \otimes X$.

Proof of (b)–(c). The morphism $X \otimes Y \otimes \varrho \colon X \otimes Y \otimes \mathbf{1} \otimes \mathbf{1} \to X \otimes Y \otimes \mathbf{1}$ coincides with $\alpha(X \otimes Y) \otimes \mathbf{1}$ and also with $X \otimes \alpha(Y) \otimes \mathbf{1}$. Hence, $\alpha(X \otimes Y) = X \otimes \alpha(Y)$. The proof of (c) is similar.

Proof of (e). By the functoriality of α, the diagram in (e) commutes when replacing $\mathbf{1} \otimes \alpha(X)$ in the top row with $\alpha(\mathbf{1} \otimes X)$. Since $\alpha(\mathbf{1} \otimes X) = \mathbf{1} \otimes \alpha(X)$ by (b), we conclude.

Proof of (d). Consider the diagram

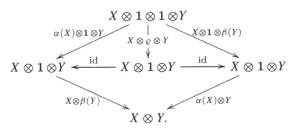

Since the upper two triangles commute as well as the big square, we obtain $X \otimes \beta(Y) = \alpha(X) \otimes Y$.

Proof of (a). By (d), one has $\alpha(\mathbf{1}) \otimes \mathbf{1} = \mathbf{1} \otimes \beta(\mathbf{1})$. On the other hand, $\alpha(\mathbf{1}) \otimes \mathbf{1} = \mathbf{1} \otimes \varrho$ by the construction of α. Hence, $\mathbf{1} \otimes \beta(\mathbf{1}) = \mathbf{1} \otimes \varrho$. This implies that $\beta(\mathbf{1}) = \varrho$. The proof for α is similar.   q.e.d.

Remark 4.2.7. If $(\mathbf{1}, \varrho)$ and $(\mathbf{1}', \varrho')$ are unit objects, then there exists a unique isomorphism $\iota\colon \mathbf{1} \to \mathbf{1}'$ compatible with ϱ and ϱ', that is, the diagram

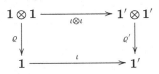

commutes. Indeed, $\mathbf{1} \xleftarrow{\sim} \mathbf{1} \otimes \mathbf{1}' \xrightarrow{\sim} \mathbf{1}'$ gives ι which satisfies the desired properties.

Remark that all tensor categories in Examples 4.2.2 except (viii) admit a unit object.

Definition 4.2.8. *Let $\mathcal{T}$ be a tensor category with a unit object $(\mathbf{1}, \varrho)$. A tensor functor $F\colon \mathcal{T} \to \mathcal{T}'$ is called* unital *if $(F(\mathbf{1}), F(\varrho))$ is a unit object of $\mathcal{T}'$.*

More precisely, $F(\mathbf{1}) \otimes F(\mathbf{1}) \xrightarrow{\sim} F(\mathbf{1})$ is given as the composition $F(\mathbf{1}) \otimes F(\mathbf{1}) \xleftarrow[\xi_F(\mathbf{1},\mathbf{1})]{\sim} F(\mathbf{1} \otimes \mathbf{1}) \xrightarrow[F(\varrho)]{\sim} F(\mathbf{1})$.

Definition 4.2.9. *Let $\mathcal{T}$ be a tensor category. An action of $\mathcal{T}$ on a category $\mathcal{C}$ is a tensor functor $F\colon \mathcal{T} \to \mathrm{Fct}(\mathcal{C}, \mathcal{C})$. If $\mathcal{T}$ has a unit object and $\mathcal{T} \to \mathrm{Fct}(\mathcal{C}, \mathcal{C})$ is unital, the action is called unital.*

For $X \in \mathcal{T}$ and $W \in \mathcal{C}$, set $X \otimes W := F(X)(W)$. To give isomorphisms $\xi_F(X, Y)\colon F(X \otimes Y) \xrightarrow{\sim} F(X) \circ F(Y)$ is thus equivalent to give isomorphisms $(X \otimes Y) \otimes W \xrightarrow{\sim} X \otimes (Y \otimes W)$. Hence, to give an action of $\mathcal{T}$ on $\mathcal{C}$ is equivalent to giving a bifunctor $\otimes\colon \mathcal{T} \times \mathcal{C} \to \mathcal{C}$ and isomorphisms $a(X, Y, W)\colon (X \otimes Y) \otimes W \simeq X \otimes (Y \otimes W)$ functorial in $X, Y \in \mathcal{T}$ and $W \in \mathcal{C}$ such that the diagram (4.2.1) commutes for $X, Y, Z \in \mathcal{T}$ and $W \in \mathcal{C}$. In this language, the action is unital if there exists an isomorphism $\eta(X)\colon \mathbf{1} \otimes X \xrightarrow{\sim} X$ functorially in $X \in \mathcal{C}$ such that the diagram

$$
\begin{array}{ccc}
\mathbf{1} \otimes \mathbf{1} \otimes X & \xrightarrow{\varrho \otimes X} & \mathbf{1} \otimes X \\
{\scriptstyle \mathbf{1} \otimes \eta(X)}\big\downarrow & & \big\downarrow {\scriptstyle \eta(X)} \\
\mathbf{1} \otimes X & \xrightarrow{\eta(X)} & X
\end{array}
$$

commutes. (See Exercise 4.8.)

Examples 4.2.10. (i) For a category $\mathcal{C}$, the tensor category $\mathrm{Fct}(\mathcal{C}, \mathcal{C})$ acts on $\mathcal{C}$.
(ii) If $\mathcal{T}$ is a tensor category, then $\mathcal{T}$ acts on itself.

Dual Pairs

We shall now introduce the notion of a dual pair and the reader will notice some similarities with that of adjoint functors (see Sect. 4.3).

Definition 4.2.11. *Let T be a tensor category with a unit object 1. Let $X, Y \in T$ be two objects and $\varepsilon \colon 1 \to Y \otimes X$ and $\eta \colon X \otimes Y \to 1$ two morphisms. We say that (X, Y) is a* dual pair *or that X is a* left dual *to Y or Y is a* right dual *to X if the conditions* (a) *and* (b) *below are satisfied:*

(a) *the composition $X \simeq X \otimes 1 \xrightarrow{X \otimes \varepsilon} X \otimes Y \otimes X \xrightarrow{\eta \otimes X} 1 \otimes X \simeq X$ is the identity of X,*

(b) *the composition $Y \simeq 1 \otimes Y \xrightarrow{\varepsilon \otimes Y} Y \otimes X \otimes Y \xrightarrow{Y \otimes \eta} Y \otimes 1 \simeq Y$ is the identity of Y.*

Lemma 4.2.12. *If (X, Y) is a dual pair, then for any $Z, W \in T$, there is an isomorphisms $\mathrm{Hom}_T(Z, W \otimes X) \simeq \mathrm{Hom}_T(Z \otimes Y, W)$ and $\mathrm{Hom}_T(X \otimes Z, W) \simeq \mathrm{Hom}_T(Z, Y \otimes W)$.*

Proof. We shall only prove the first isomorphism.
First, we construct a map $A \colon \mathrm{Hom}_T(Z, W \otimes X) \to \mathrm{Hom}_T(Z \otimes Y, W)$ as follows. Let $u \in \mathrm{Hom}_T(Z, W \otimes X)$. Then $A(u)$ is the composition $Z \otimes Y \xrightarrow{u \otimes Y} W \otimes X \otimes Y \xrightarrow{W \otimes \eta} W \otimes 1 \simeq W$.
Next, we construct a map $B \colon \mathrm{Hom}_T(Z \otimes Y, W) \to \mathrm{Hom}_T(Z, W \otimes X)$ as follows. Let $v \in \mathrm{Hom}_T(Z \otimes Y, W)$. Then $B(v)$ is the composition $Z \xrightarrow{\sim} Z \otimes 1 \xrightarrow{Z \otimes \varepsilon} Z \otimes Y \otimes X \xrightarrow{v \otimes X} W \otimes X$.
It is easily checked that A and B are inverse to each other.   q.e.d.

Remark 4.2.13. (i) Y is a representative of the functor $Z \mapsto \mathrm{Hom}_T(X \otimes Z, 1)$ as well as a representative of the functor $W \mapsto \mathrm{Hom}_T(1, W \otimes X)$.
(ii) $(\cdot \otimes Y, \cdot \otimes X)$ is a pair of adjoint functors, as well as $(X \otimes \cdot, Y \otimes \cdot)$.

Braiding

Definition 4.2.14. *A* braiding, *also called an* R-matrix, *is an isomorphism $X \otimes Y \xrightarrow{\sim} Y \otimes X$ functorially in $X, Y \in T$, such that the diagrams*

(4.2.3)
$$X \otimes Y \otimes Z \xrightarrow{R(X,Y) \otimes Z} Y \otimes X \otimes Z$$

with $R(X, Y \otimes Z)$ going diagonally to $Y \otimes Z \otimes X$, and $Y \otimes R(X, Z)$ going down from $Y \otimes X \otimes Z$ to $Y \otimes Z \otimes X$.

and

$$(4.2.4)$$

$$X \otimes Y \otimes Z \xrightarrow{\ X \otimes R(Y,Z)\ } X \otimes Z \otimes Y$$

$$R(X \otimes Y, Z) \searrow \qquad \downarrow R(X,Z) \otimes Y$$

$$Z \otimes X \otimes Y$$

commute for all $X, Y, Z \in \mathcal{T}$.

Consider the diagram

$$(4.2.5)$$

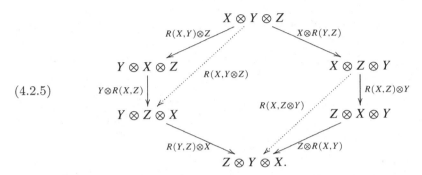

Lemma 4.2.15. *If R is a braiding, then the solid diagram* (4.2.5) *commutes.*

The commutativity of this diagram may be translated by the so-called "Yang-Baxter equation"

$$(4.2.6) \quad \begin{aligned} &(R(Y, Z) \otimes X) \circ (Y \otimes R(X, Z)) \circ (R(X, Y) \otimes Z) \\ &= (Z \otimes R(X, Y)) \circ (R(X, Z) \otimes Y) \circ (X \otimes R(Y, Z)) \,. \end{aligned}$$

Proof. Consider the diagram (4.2.5) with the dotted arrows. The triangles $(X \otimes Y \otimes Z,\ Y \otimes X \otimes Z,\ Y \otimes Z \otimes X)$ and $(X \otimes Z \otimes Y,\ Z \otimes X \otimes Y,\ Z \otimes Y \otimes X)$ commute by the definition of a braiding. The square $(X \otimes Y \otimes Z,\ X \otimes Z \otimes Y,\ Y \otimes X \otimes Z,\ Z \otimes Y \otimes X)$ commutes by the functoriality of R.   q.e.d.

Note that if R is a braiding, then

$$R(Y, X)^{-1} \colon X \otimes Y \xrightarrow{\sim} Y \otimes X$$

is also a braiding. We denote it by R^{-1}.

Definition 4.2.16. *A tensor category with a braiding R is called a commutative tensor category if $R = R^{-1}$, i.e., the composition $X \otimes Y \xrightarrow{R(X,Y)} Y \otimes X \xrightarrow{R(Y,X)} X \otimes Y$ is equal to* $\mathrm{id}_{X \otimes Y}$.

Remark 4.2.17. Commutative tensor categories are called "tensor categories" by some authors and tensor categories are then called *monoidal categories*.

4.3 Rings, Modules and Monads

By mimicking the definition of a monoid in the tensor category **Set** (by Example 4.2.2 (v)), or of a ring in the tensor category $\mathrm{Mod}(\mathbb{Z})$ (see Example 4.2.2 (i)), we introduce the following notion.

Definition 4.3.1. *Let $\mathcal{T}$ be a tensor category with a unit* **1**. *A ring in $\mathcal{T}$ is a triplet $(A, \mu_A, \varepsilon_A)$ of an object $A \in \mathcal{T}$ and two morphisms $\mu_A \colon A \otimes A \to A$ and $\varepsilon_A \colon \mathbf{1} \to A$ such that the diagrams below commute:*

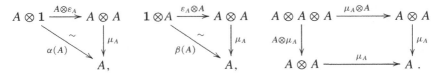

Note that ε_A is a unit and μ_A is a composition in the case of rings in $\mathrm{Mod}(k)$.

Remark 4.3.2. Some authors call $(A, \mu_A, \varepsilon_A)$ a monoid.

Definition 4.3.3. *Let $\mathcal{T}$ be a tensor category with a unit* **1** *acting unitally on a category $\mathcal{C}$ (see Definition 4.2.9). Let $(A, \mu_A, \varepsilon_A)$ be a ring in $\mathcal{T}$.*

(i) *An A-module in $\mathcal{C}$ is a pair (M, μ_M) of an object $M \in \mathcal{C}$ and a morphism $\mu_M \colon A \otimes M \to M$ such that the diagrams below in $\mathcal{C}$ commute:*

$$
\begin{array}{ccc}
\mathbf{1} \otimes M \xrightarrow{\ \varepsilon_A \otimes M\ } A \otimes M & \qquad & A \otimes A \otimes M \xrightarrow{\ \mu_A \otimes M\ } A \otimes M \\
\searrow \quad \downarrow{\mu_M} & & {\scriptstyle A \otimes \mu_M}\downarrow \qquad \downarrow{\mu_M} \\
M, & & A \otimes M \xrightarrow{\ \mu_M\ } M.
\end{array}
$$

(ii) *For two A-modules (M, μ_M) and (N, μ_N), a morphism $u \colon (M, \mu_M) \to (N, \mu_N)$ is a morphism $u \colon M \to N$ making the diagram below commutative:*

$$
\begin{array}{ccc}
A \otimes M & \xrightarrow{\ A \otimes u\ } & A \otimes N \\
\downarrow{\mu_M} & & \downarrow{\mu_N} \\
M & \xrightarrow{\ u\ } & N .
\end{array}
$$

Clearly, the family of A-modules in $\mathcal{C}$ forms a category $\mathrm{Mod}(A, \mathcal{C})$ and the forgetful functor $for \colon \mathrm{Mod}(A, \mathcal{C}) \to \mathcal{C}$ is faithful.

Lemma 4.3.4. *Let $\mathcal{T}$ and $\mathcal{C}$ be as in Definition 4.3.3, let $(A, \mu_A, \varepsilon_A)$ be a ring in $\mathcal{T}$ and let (M, μ_M) be an A-module in $\mathcal{C}$. Then the diagram below is exact in $\mathcal{C}^\wedge$:*

$$
A \otimes A \otimes M \underset{A \otimes \mu_M}{\overset{\mu_A \otimes M}{\rightrightarrows}} A \otimes M \xrightarrow{\ \mu_M\ } M .
$$

Proof. The morphisms $s\colon M \simeq 1 \otimes M \xrightarrow{\varepsilon_A \otimes M} A \otimes M$ and $u\colon A \otimes M \simeq 1 \otimes A \otimes M \xrightarrow{\varepsilon_A \otimes A \otimes M} A \otimes A \otimes M$ satisfy

$$\mu_M \circ s = \mathrm{id}_M, \quad (A \otimes \mu_M) \circ u = s \circ \mu_M, \quad (\mu_A \otimes M) \circ u = \mathrm{id}_{A \otimes M}\ .$$

Hence, it is enough to apply the result of Exercise 2.25.   q.e.d.

Recall that, for a category $\mathcal{C}$, the tensor category $\mathrm{Fct}(\mathcal{C}, \mathcal{C})$ acts on $\mathcal{C}$.

Definition 4.3.5. *Let $\mathcal{C}$ be a category. A ring in the tensor category $\mathrm{Fct}(\mathcal{C}, \mathcal{C})$ is called a* monad *in $\mathcal{C}$.*

The following lemma gives examples of monads and A-modules.

Lemma 4.3.6. *Let $\mathcal{C} \underset{R}{\overset{L}{\rightleftarrows}} \mathcal{C}'$ be functors such that (L, R) is a pair of adjoint functors. Let $\varepsilon\colon \mathrm{id}_{\mathcal{C}} \to R \circ L$ and $\eta\colon L \circ R \to \mathrm{id}_{\mathcal{C}'}$ be the adjunction morphisms.*

(a) *Set $A := R \circ L$, $\varepsilon_A := \varepsilon$ and $\mu_A := R \circ \eta \circ L$. (Hence, $\mu_A\colon A \circ A = R \circ L \circ R \circ L \to R \circ L = A$.) Then $(A, \mu_A, \varepsilon_A)$ is a monad in $\mathcal{C}$.*

(b) *Let $Y \in \mathcal{C}'$. Set $X = R(Y) \in \mathcal{C}$ and $\mu_X = R(\eta_Y)\colon A(X) = R \circ L \circ R(Y) \xrightarrow{R(\eta(Y))} R(Y) = X$. Then (X, μ_X) is an A-module and the correspondence $Y \mapsto (X, \mu_X)$ defines a functor $\Phi\colon \mathcal{C}' \to \mathrm{Mod}(A, \mathcal{C})$.*

Proof. Leaving the rest of the proof to the reader, we shall only prove the associativity of μ_A, that is, the commutativity of the diagram

$$
\begin{array}{ccc}
A \circ A \circ A(X) & \xrightarrow{\ \mu_A(A(X))\ } & A \circ A(X) \\
{\scriptstyle A(\mu_A(X))} \big\downarrow & & \big\downarrow {\scriptstyle \mu_A(X)} \\
A \circ A(X) & \xrightarrow{\ \mu_A(X)\ } & A(X).
\end{array}
$$

We have $A(\mu_A(X)) = R \circ L \circ R(\eta(L(X)))$, $\mu_A(A(X)) = R(\eta(L \circ R \circ L(X)))$ and $\mu_A(X) = R(\eta(L(X)))$. Setting $B := L \circ R$ and $Y := L(X)$, the above diagram is the image by R of the diagram below

$$
\begin{array}{ccc}
B \circ B(Y) & \xrightarrow{\ \eta(B(Y))\ } & B(Y) \\
{\scriptstyle B(\eta(Y))} \big\downarrow & & \big\downarrow {\scriptstyle \eta(Y)} \\
B(Y) & \xrightarrow{\ \eta(Y)\ } & Y.
\end{array}
$$

The commutativity of this diagram follows from the fact that $\eta\colon B \to \mathrm{id}_{\mathcal{C}'}$ is a morphism of functors.   q.e.d.

Lemma 4.3.7. *Let $(A, \mu_A, \varepsilon_A)$ be a monad in $\mathcal{C}$.*

(a) *For any $X \in \mathcal{C}$, $(A(X), \mu_A(X))$ is an A-module.*
(b) *The functor $\mathcal{C} \to \mathrm{Mod}(A, \mathcal{C})$ given by $X \mapsto (A(X), \mu_A(X))$ is a left adjoint of the forgetful functor for$: \mathrm{Mod}(A, \mathcal{C}) \to \mathcal{C}$.*

Proof. (i) is left to the reader.
(ii) We define maps

$$\mathrm{Hom}_{\mathrm{Mod}(A,\mathcal{C})}((A(Y), \mu_A(Y)), (X, \mu_X)) \underset{\beta}{\overset{\alpha}{\rightleftarrows}} \mathrm{Hom}_{\mathcal{C}}(Y, X)$$

as follows. To $v\colon (A(Y), \mu_A(Y)) \to (X, \mu_X)$ we associate $\alpha(v)$, the composition $Y \xrightarrow{\varepsilon_A(Y)} A(Y) \xrightarrow{v} X$.

To $u\colon Y \to X$, we associate $\beta(u)$, the composition $A(Y) \xrightarrow{A(u)} A(X) \xrightarrow{\mu_X} X$. It is easily checked that α and β are well defined and inverse to each other.
q.e.d.

The next theorem is due to Barr and Beck.

Theorem 4.3.8. *Let $\mathcal{C} \underset{R}{\overset{L}{\rightleftarrows}} \mathcal{C}'$ be functors such that (L, R) is a pair of adjoint functors. Let $(A = R \circ L, \varepsilon_A, \mu_A)$ and $\Phi\colon \mathcal{C}' \to \mathrm{Mod}(A, \mathcal{C})$ be as in Lemma 4.3.6. Then the following conditions are equivalent.*

(i) *Φ is an equivalence of categories,*
(ii) *the following two conditions hold:*
  (a) *R is conservative,*
  (b) *for any pair of parallel arrows $f, g\colon X \rightrightarrows Y$ in $\mathcal{C}'$, if $\mathrm{Coker}(R(f), R(g))$ exists in $\mathcal{C}$ and $R(X) \underset{R(g)}{\overset{R(f)}{\rightrightarrows}} R(Y) \longrightarrow \mathrm{Coker}(R(f), R(g))$ is exact in $\mathcal{C}^\wedge$ (see Exercise 2.25), then $\mathrm{Coker}(f, g)$ exists and $\mathrm{Coker}(R(f), R(g)) \xrightarrow{\sim} R(\mathrm{Coker}(f, g))$.*

In particular, if $\mathcal{C}'$ admits finite inductive limits and R is conservative and exact, then $\Phi\colon \mathcal{C}' \to \mathrm{Mod}(A, \mathcal{C})$ is an equivalence of categories.

Proof. (i) $\Rightarrow$ (ii). We may assume that A is a monad in $\mathcal{C}$ and R is the forgetful functor $\mathcal{C}' = \mathrm{Mod}(A, \mathcal{C}) \to \mathcal{C}$. Hence, L is the functor $X \mapsto (A(X), \mu_A(X))$ by Lemma 4.3.7. Then (a) is obvious. Let us show (b). Let $f, g\colon (X, \mu_X) \rightrightarrows (Y, \mu_Y)$ be a pair of parallel arrows and assume that $X \rightrightarrows Y \to Z$ is exact in $\mathcal{C}^\wedge$. Then $A(X) \rightrightarrows A(Y) \to A(Z)$ as well as $A^2(X) \rightrightarrows A^2(Y) \to A^2(Z)$ are exact by Proposition 2.6.4. By the commutativity of the solid diagram with exact rows

$$A(X) \rightrightarrows A(Y) \longrightarrow A(Z)$$

$$\downarrow{\mu_X} \qquad \downarrow{\mu_Y} \qquad \vdots{w}$$

$$X \rightrightarrows Y \longrightarrow Z,$$

we find the morphism $w\colon A(Z) \to Z$. It is easily checked that (Z, w) is an A-module and $(Z, w) \simeq \mathrm{Coker}(f, g)$ in $\mathrm{Mod}(A, \mathcal{C})$.

(ii) $\Rightarrow$ (i). Let us construct a quasi-inverse $\Psi\colon \mathrm{Mod}(A, \mathcal{C}) \to \mathcal{C}'$ of Φ. Let $(X, \mu_X) \in \mathrm{Mod}(A, \mathcal{C})$. Applying L to $\mu_X\colon A(X) \to X$, we obtain

$$(4.3.1) \qquad L \circ R \circ L(X) \underset{\eta(L(X))}{\overset{L(\mu_X)}{\rightrightarrows}} L(X).$$

Applying R to this diagram we get

$$R \circ L \circ R \circ L(X) \underset{R(\eta(L(X)))}{\overset{R \circ L(\mu_X)}{\rightrightarrows}} R \circ L(X)$$

which is equal to the diagram $A \circ A(X) \underset{\mu_A(X)}{\overset{A(\mu_X)}{\rightrightarrows}} A(X)$.
The sequence

$$(4.3.2) \qquad A \circ A(X) \underset{\mu_A(X)}{\overset{A(\mu_X)}{\rightrightarrows}} A(X) \overset{\mu_X}{\longrightarrow} X$$

is exact in $\mathcal{C}^{\wedge}$ by Lemma 4.3.4. Therefore, (b) implies that (4.3.1) has a cokernel

$$(4.3.3) \qquad L \circ R \circ L(X) \underset{\eta(L(X))}{\overset{L(\mu_X)}{\rightrightarrows}} L(X) \overset{\varphi}{\longrightarrow} Y,$$

and there exists a commutative diagram

$$A(X) = R \circ L(X) \overset{\mu_X}{\longrightarrow} X$$

$$R(\varphi) \downarrow \qquad \overset{\sim}{\underset{\psi}{\swarrow}}$$

$$R(Y).$$

We set $\Phi((X, \mu_X)) = Y$. Since the following diagram commutes

$$X \underset{\varepsilon_A(X)=\varepsilon(X)}{\overset{\mathrm{id}_X}{\rightrightarrows}} A(X) \underset{\mu_X}{\longrightarrow} X$$

$$R(\varphi) \downarrow \qquad \underset{\psi}{\swarrow}$$

$$R(Y),$$

φ and ψ correspond by the adjunction isomorphism $\text{Hom}_{\mathcal{C}'}(L(X), Y) \simeq \text{Hom}_{\mathcal{C}}(X, R(Y))$. This implies that the diagram

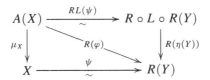

commutes. Hence, $\Phi\Psi((X, \mu_X)) \simeq (X, \mu_X)$.

Conversely, for $Y \in \mathcal{C}'$, let us set $(X, \mu_X) = \Phi(Y) = (R(Y), R(\eta(Y))) \in \text{Mod}(A, \mathcal{C})$. Then the two compositions coincide:

(4.3.4) $$L \circ R \circ L \circ R(Y) \underset{\eta(L \circ R(Y))}{\overset{L \circ R(\eta(Y))}{\rightrightarrows}} L \circ R(Y) \xrightarrow{\eta(Y)} Y.$$

Applying R to this diagram, we find the sequence $A \circ A(X) \rightrightarrows A(X) \to X$ which is exact in $\mathcal{C}^\wedge$ by Lemma 4.3.4. Hence, (b) implies that

$$R(Y) = X \simeq R\big(\text{Coker}(L \circ R \circ L \circ R(Y) \rightrightarrows L \circ R(Y))\big).$$

Then (a) implies that $Y \simeq \text{Coker}(L \circ R \circ L \circ R(Y) \rightrightarrows L \circ R(Y))$. Hence, $\Psi(\Phi(Y)) \simeq Y$.   q.e.d.

Exercises

Exercise 4.1. Let **Pr** be the category given in Notations 1.2.8 (v). Let $F: \mathbf{Pr} \to \mathbf{Pr}$ be the functor given by $F(u) = \text{id}_c$ for any $u \in \text{Mor}(\mathbf{Pr})$. Let $\varepsilon: \text{id}_{\mathbf{Pr}} \to F$ be the morphism of functors given by $\varepsilon_c = p$.
(i) Prove that F and ε are well-defined.
(ii) Prove that $F \circ \varepsilon: F \to F^2$ is an isomorphism but $\varepsilon \circ F: F \to F^2$ is not an isomorphism.

Exercise 4.2. Let $\mathcal{T}$ be a tensor category with a unit object $\mathbf{1}$. Let $X \in \mathcal{T}$ and $\alpha: \mathbf{1} \to X$. Prove that if the compositions $X \simeq \mathbf{1} \otimes X \xrightarrow{\alpha \otimes X} X \otimes X$ and $X \simeq X \otimes \mathbf{1} \xrightarrow{X \otimes \alpha} X \otimes X$ are isomorphisms, then they are equal and the inverse morphism $\mu: X \otimes X \to X$ gives a ring structure on X.

Exercise 4.3. Prove that if a tensor category has a unit object, then this object is unique up to unique isomorphism. More precisely, prove the statement in Remark 4.2.7. Also prove that if $(\mathbf{1}, \varrho)$ is a unit object, then $\varrho \otimes \mathbf{1} = \mathbf{1} \otimes \varrho$.

Exercise 4.4. Let $\mathcal{T}$ be a tensor category with a unit $\mathbf{1}$ and a braiding R.
(i) Prove that the diagram below commutes:

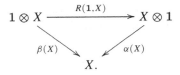

(ii) Prove that $R(\mathbf{1}, \mathbf{1}) = \mathrm{id}_{\mathbf{1} \otimes \mathbf{1}}$.

Exercise 4.5. Let k be a field and recall that $k^\times$ denotes the group of its invertible elements. Let L be an additive group and denote by $\mathcal{C}$ the category whose objects are the families

$$\mathrm{Ob}(\mathcal{C}) = \{X = \{X_l\}_{l \in L}; X_l \in \mathrm{Mod}(k), \ X_l = 0 \text{ for all but finitely many } l\},$$

the morphisms in $\mathcal{C}$ being the natural ones. For $X = \{X_l\}_{l \in L}$ and $Y = \{Y_l\}_{l \in L}$, define $X \otimes Y$ by $(X \otimes Y)_l = \oplus_{l = l' + l''} X_{l'} \otimes Y_{l''}$.

(i) Let $c: L \times L \times L \to k^\times$ be a function. For $X, Y, Z \in \mathcal{C}$, let

$$a_c(X, Y, Z): (X \otimes Y) \otimes Z \to X \otimes (Y \otimes Z)$$

be the isomorphism induced by

$$(X_{l_1} \otimes Y_{l_2}) \otimes Z_{l_3} \xrightarrow{c(l_1, l_2, l_3)} X_{l_1} \otimes (Y_{l_2} \otimes Z_{l_3}).$$

Prove that $(\mathcal{C}, \otimes, a_c)$ is a tensor category if and only if c satisfies the cocycle condition:

$$(4.3.5) \quad c(l_1 + l_2, l_3, l_4)c(l_1, l_2, l_3 + l_4) = c(l_1, l_2, l_3)c(l_1, l_2 + l_3, l_4)c(l_2, l_3, l_4).$$

If c satisfies the cocycle condition (4.3.5), we shall denote by $\otimes_c$ the tensor product in the tensor category $(\mathcal{C}, \otimes, a_c)$.

(ii) Let b and c be two functions from $L \times L \times L$ to $k^\times$ both satisfying (4.3.5). Let $\varphi: L \times L \to k^\times$ be a function and for $X, Y \in \mathcal{C}$, let $\xi(X, Y): X \otimes Y \to X \otimes Y$ be the isomorphism in $\mathcal{C}$ given by

$$X_l \otimes Y_{l'} \xrightarrow{\varphi(l, l')} X_l \otimes Y_{l'}.$$

Prove that $(\mathrm{id}_{\mathcal{C}}, \xi)$ is a tensor functor from $(\mathcal{C}, \otimes_b, a_b)$ to $(\mathcal{C}, \otimes_c, a_c)$ if and only if

$$(4.3.6) \quad c(l_1, l_2, l_3) = \frac{\varphi(l_2, l_3)\varphi(l_1, l_2 + l_3)}{\varphi(l_1, l_2)\varphi(l_1 + l_2, l_3)} b(l_1, l_2, l_3).$$

(iii) Assume that c satisfies the cocycle condition (4.3.5) and let $\rho: L \times L \to k^\times$ be a function. Let

$$R(X, Y): X \otimes_c Y \to Y \otimes_c X$$

be the isomorphism induced by

$$X_l \otimes Y_{l'} \xrightarrow{\rho(l, l')} Y_{l'} \otimes X_l.$$

(a) Prove that R satisfies the Yang-Baxter equation (4.2.6) if

$$c(l_1, l_2, l_3)c(l_2, l_3, l_1)c(l_3, l_1, l_2) = c(l_1, l_3, l_2)c(l_3, l_2, l_1)c(l_2, l_1, l_3) .$$

(b) Prove that R is a braiding if and only if

(4.3.7) $\qquad \dfrac{c(l_1, l_2, l_3)c(l_2, l_3, l_1)}{c(l_2, l_1, l_3)} = \dfrac{\rho(l_1, l_2)\rho(l_1, l_3)}{\rho(l_1, l_2 + l_3)} = \dfrac{\rho(l_2 + l_3, l_1)}{\rho(l_2, l_1)\rho(l_3, l_1)} .$

(iv) Let $\psi : L \to k$ be a function. Define $\theta : \mathrm{id}_{\mathcal{C}} \to \mathrm{id}_{\mathcal{C}}$ by setting $\theta_X|_{X_l} = \psi(l)\,\mathrm{id}_{X_l}$. Prove that θ is a morphism of tensor functors if and only if

$$\psi(l_1 + l_2) = \psi(l_1)\psi(l_2) .$$

(v) Let $L = \mathbb{Z}/2\mathbb{Z}$.

(a) Prove that the function c given by

(4.3.8) $\qquad c(l_1, l_2, l_3) = \begin{cases} -1 & \text{if } l_1 = l_2 = l_3 = 1 \bmod 2 , \\ 1 & \text{otherwise} \end{cases}$

satisfies the cocycle condition (4.3.5).

(b) Assume that there exists an element $i \in k^{\times}$ such that $i^2 = -1$ and let c be as in (4.3.8). Prove that the solutions of (4.3.7) are given by

$$\rho(l, l') = \begin{cases} \pm i & \text{if } l = l' = 1 \bmod 2 , \\ 1 & \text{otherwise.} \end{cases}$$

(vi) Let $L = \mathbb{Z}/2\mathbb{Z}$. Prove that two tensor categories $(\mathcal{C}, \otimes_c, a_c)$ and $(\mathcal{C}, \otimes_b, a_b)$ with c as in (4.3.8) and $b(l_1, l_2, l_3) = 1$, are not equivalent when k is a field of characteristic different from 2.

(vii) Let $L = \mathbb{Z}/2\mathbb{Z}$, and b as in (vi). Let R be the braiding given by $\rho(l, l') = -1$ or 1 according that $l = l' = 1 \bmod 2$ or not. Prove that $(\mathcal{C}, \otimes_b, a_b)$ is a commutative tensor category. (The objects of $\mathcal{C}$ are called *super* vector spaces.)

Exercise 4.6. Let $\mathcal{T}$ be a tensor category with a unit object $\mathbf{1}$. Prove that if $\theta : \mathrm{id}_{\mathcal{T}} \to \mathrm{id}_{\mathcal{T}}$ is an isomorphism of tensor functors, then $\theta_{\mathbf{1}} = \mathrm{id}_{\mathbf{1}}$.

Exercise 4.7. Let $\mathcal{T}$ be a tensor category with a unit object. Prove that if (X, Y) and (X, Y') are dual pairs, then Y and Y' are isomorphic.

Exercise 4.8. Let $\mathcal{T}$ be a tensor category with a unit object $\mathbf{1}$ and acting on a category $\mathcal{C}$. Prove that this action is unital if and only if the functor $\mathcal{C} \ni X \mapsto \mathbf{1} \otimes X \in \mathcal{C}$ is fully faithful.

Exercise 4.9. Let $\mathbf{\Delta}$ be the category of finite totally ordered sets and order-preserving maps (see Definition 11.4.1 and Exercise 1.21).

(i) For $\sigma, \tau \in \Delta$, define $\sigma \otimes \tau$ as the set $\sigma \sqcup \tau$ endowed with the total order such that $i < j$ for any i in the image of σ and j in the image of τ and $\sigma \to \sigma \sqcup \tau$ and $\tau \to \sigma \sqcup \tau$ are order-preserving. Prove that Δ is a tensor category with a unit object.

(ii) Let $R(\sigma, \tau) \colon \sigma \otimes \tau \to \tau \otimes \sigma$ denote the unique isomorphism of these two objects in Δ. Prove that R defines a commutative tensor category structure on Δ.

(iii) Let $\mathcal{T}$ be a tensor category with a unit object. Prove that the category of rings in $\mathcal{T}$ is equivalent to the category of unital tensor functors from Δ to $\mathcal{T}$.

Exercise 4.10. Let G be a group and let us denote by $\mathcal{G}$ the associated discrete category. A structure of a tensor category on $\mathcal{G}$ is defined by setting $g_1 \otimes g_2 = g_1 g_2$ $(g_1, g_2 \in G)$. Let $\mathcal{C}$ be a category. An action of G on $\mathcal{C}$ is a unital action $\psi \colon \mathcal{G} \to \mathrm{Fct}(\mathcal{C}, \mathcal{C})$ of the tensor category $\mathcal{G}$ on $\mathcal{C}$.

(i) Let $T \colon \mathcal{C} \to \mathcal{C}$ be an auto-equivalence. Show that there exists an action ψ of $\mathbb{Z}$ on $\mathcal{C}$ such that $\psi(1) = T$.

(ii) Let T_1 and T_2 be two auto-equivalences of $\mathcal{S}$ and let $\varphi_{12} \colon T_1 \circ T_2 \xrightarrow{\sim} T_2 \circ T_1$ be an isomorphism of functors. Show that there exists an action ψ of $\mathbb{Z}^2$ on $\mathcal{C}$ such that $\psi((1, 0)) = T_1$ and $\psi((0, 1)) = T_2$.

(iii) More generally, let $T_1, \ldots, T_n$ be n auto-equivalences of $\mathcal{C}$ for a non-negative integer n, and let $\varphi_{ij} \colon T_i \circ T_j \xrightarrow{\sim} T_j \circ T_i$ be isomorphisms of functors for $1 \le i < j \le n$. Assume that for any $1 \le i < j < k \le n$, the diagram below commutes

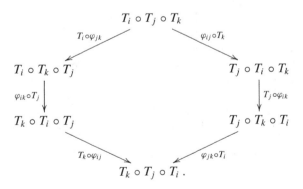

Denote by $u_1, \ldots, u_n$ the canonical basis of $\mathbb{Z}^n$. Prove that there exists an action ψ of $\mathbb{Z}^n$ on $\mathcal{C}$ such that $\psi(u_i) = T_i$ and the composition $T_i \circ T_j \simeq \psi(u_i \otimes u_j) = \psi(u_j \otimes u_i) \xrightarrow{\sim} T_j \circ T_i$ coincides with φ_{ij}.

Exercise 4.11. Let $\mathcal{T}$ be a tensor category with a unit object $(\mathbf{1}, \varrho)$. Let $a \in \mathrm{End}_{\mathcal{T}}(\mathbf{1})$.

(i) Prove that the diagram

$$
\begin{array}{ccc}
1 \otimes 1 & \xrightarrow{\;1 \otimes a\;} & 1 \otimes 1 \\
\varrho \downarrow & & \downarrow \varrho \\
1 & \xrightarrow{\;a\;} & 1
\end{array}
$$

commutes and that $1 \otimes a = a \otimes 1$.

(ii) Prove that $\mathrm{End}_T(1)$ is commutative.

(iii) Define

$$
R : \mathrm{End}_T(1) \to \mathrm{End}_{\mathrm{Fct}(T,T)}(\cdot \otimes 1) \xleftarrow{\sim} \mathrm{End}_{\mathrm{Fct}(T,T)}(\mathrm{id}_T) ,
$$
$$
L : \mathrm{End}_T(1) \to \mathrm{End}_{\mathrm{Fct}(T,T)}(1 \otimes \cdot) \xleftarrow{\sim} \mathrm{End}_{\mathrm{Fct}(T,T)}(\mathrm{id}_T) ,
$$

where $R(a)_X \otimes 1 = X \otimes a$ and $1 \otimes L(a)_X = a \otimes X$. Prove that if T has a braiding, then $R = L$.

Exercise 4.12. Let T be a tensor category with a unit object $(1, \varrho)$. Let $X, Y \in T$ and assume that $X \otimes Y \simeq 1$ and $Y \otimes X \simeq 1$. Prove that there exist isomorphisms $\xi : X \otimes Y \xrightarrow{\sim} 1$ and $\eta : Y \otimes X \xrightarrow{\sim} 1$ such that the diagrams below commute.

$$
\begin{array}{ccc}
X \otimes Y \otimes X & \xrightarrow{\;\xi \otimes X\;} & 1 \otimes X \\
X \otimes \eta \downarrow & & \downarrow \\
X \otimes 1 & \longrightarrow & X,
\end{array}
\qquad
\begin{array}{ccc}
Y \otimes X \otimes Y & \xrightarrow{\;Y \otimes \xi\;} & Y \otimes 1 \\
\eta \otimes Y \downarrow & & \downarrow \\
1 \otimes Y & \longrightarrow & Y.
\end{array}
$$

Exercise 4.13. Let T be a tensor category with a unit object $(1, \varrho)$. Assume to be given $X \in T$, a positive integer n and an isomorphism $\lambda : X^{\otimes n} \xrightarrow{\sim} 1$. Consider the diagram

(4.3.9)
$$
\begin{array}{ccc}
X^{\otimes(n+1)} & \xrightarrow{\;X \otimes \lambda\;} & X \otimes 1 \\
\lambda \otimes X \downarrow & & \downarrow \\
1 \otimes X & \longrightarrow & X.
\end{array}
$$

(i) Assume that (4.3.9) commutes. Prove that there exists a unital functor $\varphi : \mathbb{Z}/n\mathbb{Z} \to T$ such that $\varphi(1) = X$. Here, the group $\mathbb{Z}/n\mathbb{Z}$ is regarded as a tensor category as in Exercise 4.10.

(ii) Prove that if T has a braiding, the fact that the diagram (4.3.9) commutes does not depend on the choice of the isomorphism $\lambda : X^{\otimes n} \xrightarrow{\sim} 1$. (Hint: use Exercise 4.11 (iii).)

(iii) Give an example of a braided tensor category T and (X, λ) such that (4.3.9) does not commute. (Hint: use Exercise 4.5 (v).)

5

Generators and Representability

The aim of this chapter is to give various criteria for a functor with values in **Set** to be representable, and as a by-product, criteria for a functor to have an adjoint.

For that purpose, we need to introduce two important notions. The first one is that of a strict morphism for a category $\mathcal{C}$ which admits finite inductive and finite projective limits. In such a category, there are natural definitions of the coimage and of the image of a morphism, and the morphism is strict if the coimage is isomorphic to the image. A crucial fact for our purpose here is that if $\mathcal{C}$ admits a generator (see below), then the family of strict quotients of any object is a small set.

The second important notion is that of a system of generators (and in particular, a generator) in a category $\mathcal{C}$. If $\mathcal{C}$ admits small inductive limits and G is a generator, then any object $X \in \mathcal{C}$ is a quotient of a small coproduct of copies of G, similarly as any module over a ring A is a quotient of $A^{\oplus I}$ for a small set I.

With these tools in hands, it is then possible to state various theorems of representability. For example, we prove that if $\mathcal{C}$ admits small inductive limits, finite projective limits, a generator and small filtrant inductive limits are stable by base change, then any contravariant functor from $\mathcal{C}$ to **Set** is representable as soon as it sends small inductive limits to projective limits (Theorem 5.3.9).

Many of these results are classical and we refer to [64].

5.1 Strict Morphisms

Definition 5.1.1. *Let $\mathcal{C}$ be a category which admits finite inductive and finite projective limits and let $f : X \to Y$ be a morphism in $\mathcal{C}$.*

(i) *The* coimage *of f, denoted by $\operatorname{Coim} f$, is given by*

$$\operatorname{Coim} f = \operatorname{Coker}(X \times_Y X \rightrightarrows X) \,.$$

(ii) *The* image *of* f*, denoted by* $\operatorname{Im} f$*, is given by*

$$\operatorname{Im} f = \operatorname{Ker}(Y \rightrightarrows Y \sqcup_X Y) .$$

Note that the natural morphism $X \to \operatorname{Coim} f$ is an epimorphism and the natural morphism $\operatorname{Im} f \to Y$ is a monomorphism.

Proposition 5.1.2. *Let* $\mathcal{C}$ *be a category which admits finite inductive and finite projective limits and let* $f \colon X \to Y$ *be a morphism in* $\mathcal{C}$*.*

(i) *There is an isomorphism* $X \underset{X \times_Y X}{\sqcup} X \xrightarrow{\sim} \operatorname{Coim} f$*.*

(ii) *There is an isomorphism* $\operatorname{Im} f \xrightarrow{\sim} Y \underset{Y \sqcup_X Y}{\times} Y$*.*

(iii) *There is a unique morphism*

$$(5.1.1) \qquad\qquad u \colon \operatorname{Coim} f \to \operatorname{Im} f$$

such that the composition $X \to \operatorname{Coim} f \xrightarrow{u} \operatorname{Im} f \to Y$ *is* f*.*

(iv) *The following three conditions are equivalent:*
    (a) f *is an epimorphism,*
    (b) $\operatorname{Im} f \to Y$ *is an isomorphism,*
    (c) $\operatorname{Im} f \to Y$ *is an epimorphism.*

Proof. (ii) Set $Z = Y \sqcup_X Y$. We shall prove the isomorphism $\operatorname{Ker}(i_1, i_2 \colon Y \rightrightarrows Z) \simeq Y \times_Z Y$. For any $U \in \mathcal{C}$, we have

$$\operatorname{Hom}_{\mathcal{C}}(U, Y \times_Z Y) = \bigl\{(y_1, y_2) ; \, y_1, y_2 \in Y(U), i_1(y_1) = i_2(y_2)\bigr\} .$$

The codiagonal morphism $\sigma \colon Z \to Y$ satisfies $\sigma \circ i_1 = \sigma \circ i_2 = \operatorname{id}_Y$. Hence, $i_1(y_1) = i_2(y_2)$ implies $y_1 = \sigma \circ i_1(y_1) = \sigma \circ i_2(y_2) = y_2$. Therefore we obtain

$$\begin{aligned}\operatorname{Hom}_{\mathcal{C}}(U, Y \times_Z Y) &\simeq \bigl\{y \in Y(U) ; \, i_1(y) = i_2(y)\bigr\} \\ &\simeq \operatorname{Hom}_{\mathcal{C}}(U, \operatorname{Ker}(i_1, i_2 \colon Y \rightrightarrows Z)) .\end{aligned}$$

(i) follows from (ii) by reversing the arrows.
(iii) Consider the diagram

$$X \times_Y X \underset{p_2}{\overset{p_1}{\rightrightarrows}} X \xrightarrow{f} Y \underset{i_2}{\overset{i_1}{\rightrightarrows}} Y \sqcup_X Y .$$

with $s \colon X \to \operatorname{Coim} f$, $\tilde f$, and $u \colon \operatorname{Coim} f \dashrightarrow \operatorname{Im} f$.

Since $f \circ p_1 = f \circ p_2$, f factors uniquely as $X \xrightarrow{s} \operatorname{Coim} f \xrightarrow{\tilde f} Y$. Since $i_1 \circ f = i_1 \circ \tilde f \circ s$ and $i_2 \circ f = i_2 \circ \tilde f \circ s$ are equal and s is an epimorphism, we obtain $i_1 \circ \tilde f = i_2 \circ \tilde f$. Hence $\tilde f$ factors through $\operatorname{Im} f$.

The uniqueness follows from the fact that $X \to \operatorname{Coim} f$ is an epimorphism and $\operatorname{Im} f \to Y$ is a monomorphism.

(iv) Assume that f is an epimorphism. By the construction, the two morphisms $i_1, i_2 \colon Y \to Y \sqcup_X Y$ satisfy $i_1 \circ f = i_2 \circ f$. Since f is an epimorphism, it follows that $i_1 = i_2$. Therefore, $\operatorname{Ker}(i_1, i_2) \simeq Y$.

Conversely, assume that $w \colon \operatorname{Im} f \to Y$ is an epimorphism. Since $i_1 \circ w = i_2 \circ w$, we have $i_1 = i_2$. Consider two morphisms $g_1, g_2 \colon Y \rightrightarrows Z$ such that $g_1 \circ f = g_2 \circ f$. These two morphisms define $g \colon Y \sqcup_X Y \to Z$ and $g_1 = i_1 \circ g = i_2 \circ g = g_2$.

q.e.d.

Examples 5.1.3. (i) Let $\mathcal{C} = \mathbf{Set}$. In this case, the morphism (5.1.1) is an isomorphism, and $\operatorname{Im} f \simeq f(X)$, the set-theoretical image of f.

(ii) Let $\mathcal{C}$ denote the category of topological spaces and let $f \colon X \to Y$ be a continuous map. Then, $\operatorname{Coim} f$ is the space $f(X)$ endowed with the quotient topology of X and $\operatorname{Im} f$ is the space $f(X)$ endowed with topology induced by Y. Hence, (5.1.1) is not an isomorphism in general.

Definition 5.1.4. *Let $\mathcal{C}$ be a category which admits finite inductive limits and finite projective limits. A morphism f is* strict *if* $\operatorname{Coim} f \to \operatorname{Im} f$ *is an isomorphism.*

Proposition 5.1.5. *Let $\mathcal{C}$ be a category which admits finite inductive limits and finite projective limits and let $f \colon X \to Y$ be a morphism in $\mathcal{C}$.*

(i) *The following five conditions are equivalent*
  (a) *f is a strict epimorphism,*
  (b) $\operatorname{Coim} f \xrightarrow{\sim} Y$,
  (c) *the sequence $X \times_Y X \rightrightarrows X \to Y$ is exact,*
  (d) *there exists a pair of parallel arrows $g, h \colon Z \rightrightarrows X$ such that $f \circ g = f \circ h$ and $\operatorname{Coker}(g, h) \to Y$ is an isomorphism,*
  (e) *for any $Z \in \mathcal{C}$, $\operatorname{Hom}_{\mathcal{C}}(Y, Z)$ is isomorphic to the set of morphisms $u \colon X \to Z$ satisfying $u \circ v_1 = u \circ v_2$ for any pair of parallel morphisms $v_1, v_2 \colon W \rightrightarrows X$ such that $f \circ v_1 = f \circ v_2$.*

(ii) *If f is both a strict epimorphism and a monomorphism, then f is an isomorphism.*

(iii) *The morphism $X \to \operatorname{Coim} f$ is a strict epimorphism.*

Proof. (i) (a) $\Rightarrow$ (b) since $\operatorname{Im} f \xrightarrow{\sim} Y$ by Proposition 5.1.2 (iv).

(i) (b) $\Rightarrow$ (a) is obvious.

(i) (b) $\Leftrightarrow$ (c) is obvious.

(i) (d) $\Rightarrow$ (b). Assume that the sequence $Z \rightrightarrows X \xrightarrow{f} Y$ is exact. Consider the solid diagram

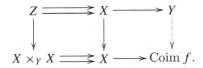

We get a morphism $Y \to \mathrm{Coim}\, f$ which is inverse to the natural morphism $\mathrm{Coim}\, f \to Y$.

(i) (c) $\Rightarrow$ (d) is obvious.

(i) (c) $\Leftrightarrow$ (e). The condition on u in (e) is equivalent to saying that the two compositions $X \times_Y X \rightrightarrows X \xrightarrow{u} Z$ coincide.

(ii) The morphism f decomposes as $X \to \mathrm{Coim}\, f \to Y$. The first arrow is an isomorphism by Proposition 5.1.2 (iv) (with the arrows reversed) and the second arrow is an isomorphism by (i).

(iii) follows from (i) (d) by the definition of $\mathrm{Coim}\, f$.   q.e.d.

Remark that in Proposition 5.1.5, it is not necessary to assume that $\mathcal{C}$ admits finite inductive and projective limits to formulate condition (i) (e).

Definition 5.1.6. *Let $\mathcal{C}$ be a category. A morphism $f \colon X \to Y$ is a* strict epimorphism *if condition* (i) (e) *in Proposition 5.1.5 is satisfied.*

Note that condition (i) (e) in Proposition 5.1.5 is equivalent to saying that the map
$$\mathrm{Hom}_{\mathcal{C}}(Y, Z) \to \mathrm{Hom}_{\mathcal{C}^\wedge}(\mathrm{Im}\, \mathrm{h}_{\mathcal{C}}(f), \mathrm{h}_{\mathcal{C}}(Z))$$
is an isomorphism for any $Z \in \mathcal{C}$.

The notion of a strict monomorphism is defined similarly.

Proposition 5.1.7. *Let $\mathcal{C}$ be a category which admits finite inductive limits and finite projective limits. Assume that any epimorphism in $\mathcal{C}$ is strict. Let $f \colon X \to Y$ be a morphism in $\mathcal{C}$.*

(i) *The morphism $\mathrm{Coim}\, f \to Y$ is a monomorphism.*

(ii) *If f decomposes as $X \xrightarrow{u} I \xrightarrow{v} Y$ with an epimorphism u and a monomorphism v, then I is isomorphic to $\mathrm{Coim}\, f$.*

Proof. (i) Set $I = \mathrm{Coim}\, f$ and let $X \xrightarrow{u} I \xrightarrow{v} Y$ be the canonical morphisms. Let w denote the composition $X \to I \to \mathrm{Coim}\, v$. Since w is a strict epimorphism, $\mathrm{Coim}\, w$ is isomorphic to $\mathrm{Coim}\, v$. For a pair of parallel arrows $\varphi, \psi \colon W \rightrightarrows X$, the condition $u \circ \varphi = u \circ \psi$ is equivalent to the condition $f \circ \varphi = f \circ \psi$. Indeed, if $f \circ \varphi = f \circ \psi$, then (φ, ψ) gives a morphism $W \to X \times_Y X$, and the two compositions $W \to X \times_Y X \rightrightarrows X \to I$ are equal and coincide with $u \circ \varphi$ and $u \circ \psi$.

Hence, these two conditions are also equivalent to $w \circ \varphi = w \circ \psi$. This implies $X \times_{\mathrm{Coim}\, v} X \simeq X \times_Y X$, and hence

$$I \simeq \mathrm{Coker}(X \times_Y X \rightrightarrows X) \simeq \mathrm{Coker}(X \times_{\mathrm{Coim}\, v} X \rightrightarrows X)$$
$$\simeq \mathrm{Coim}\, w \simeq \mathrm{Coim}\, v \,.$$

Then Proposition 5.1.2 (iv) (with the arrows reversed) implies that v is a monomorphism.

(ii) Since v is a monomorphism, the canonical morphism $X \times_I X \to X \times_Y X$ is an isomorphism. Hence,

$$\mathrm{Coim}\, f \simeq \mathrm{Coker}(X \times_Y X \rightrightarrows X) \simeq \mathrm{Coker}(X \times_I X \rightrightarrows X)$$
$$\simeq \mathrm{Coim}(X \to I) \simeq I \,,$$

where the last isomorphism follows from the fact that u is a strict epimorphism.   q.e.d.

Similarly as in Definition 1.2.18, we set:

Definition 5.1.8. *Let $\mathcal{C}$ be a category and let $X \in \mathcal{C}$.*

(i) *An isomorphism class of a strict epimorphism with source X is called a* strict quotient *of X.*

(ii) *An isomorphism class of a strict monomorphism with target X is called a* strict subobject *of X.*

5.2 Generators and Representability

Recall that, unless otherwise specified, a category means a $\mathcal{U}$-category. In particular, we denote by **Set** the category of $\mathcal{U}$-sets.

Definition 5.2.1. *Let $\mathcal{C}$ be a category.*

(i) *A* system of generators *in $\mathcal{C}$ is a family of objects $\{G_i\}_{i \in I}$ of $\mathcal{C}$ such that I is small and the functor $\mathcal{C} \to \mathbf{Set}$ given by $X \mapsto \prod_{i \in I} \mathrm{Hom}_{\mathcal{C}}(G_i, X)$ is conservative, that is, a morphism $f : X \to Y$ is an isomorphism as soon as $\mathrm{Hom}_{\mathcal{C}}(G_i, X) \to \mathrm{Hom}_{\mathcal{C}}(G_i, Y)$ is an isomorphism for all $i \in I$.*
If the family $\{G_i\}_{i \in I}$ consists of a single object G, G is called a generator.

(ii) *A* system of cogenerators *(resp. a* cogenerator*) in $\mathcal{C}$ is a system of generators (resp. is a generator) in $\mathcal{C}^{\mathrm{op}}$.*

Note that if $\mathcal{C}$ admits small coproducts and a system of generators $\{G_i\}_{i \in I}$, then it admits a generator, namely $\coprod_i G_i$.

Examples 5.2.2. (i) The object $\{\mathrm{pt}\}$ is a generator in **Set**, and a set consisting of two elements is a cogenerator in **Set**.

(ii) Let A be a ring. Then A is a generator in $\mathrm{Mod}(A)$.

(iii) Let $\mathcal{C}$ be a small category. Then $\mathrm{Ob}(\mathcal{C})$ is a system of generators in $\mathcal{C}^{\wedge}$, by Corollary 1.4.7.

We shall concentrate our study on categories having a generator. By reversing the arrows, the reader will deduce the corresponding results for categories having a cogenerator.

For $G \in \mathcal{C}$, we shall denote by φ_G the functor

$$\varphi_G := \operatorname{Hom}_{\mathcal{C}}(G, \cdot) \colon \mathcal{C} \to \mathbf{Set} \ .$$

Note that for $X \in \mathcal{C}$, the identity element of

$$\operatorname{Hom}_{\mathbf{Set}}(\operatorname{Hom}_{\mathcal{C}}(G, X), \operatorname{Hom}_{\mathcal{C}}(G, X)) \simeq \operatorname{Hom}_{\mathcal{C}^{\vee}}(G^{\coprod \operatorname{Hom}(G,X)}, X)$$

defines a canonical morphism in $\mathcal{C}^{\vee}$

(5.2.1) $$G^{\coprod \operatorname{Hom}(G,X)} \to X \ .$$

Proposition 5.2.3. *Assume that $\mathcal{C}$ admits finite projective limits, small coproducts and a generator G. Then:*

  (i) *the functor $\varphi_G = \operatorname{Hom}_{\mathcal{C}}(G, \cdot)$ is faithful,*
 (ii) *a morphism $f \colon X \to Y$ in $\mathcal{C}$ is a monomorphism if and only if $\varphi_G(f) \colon \operatorname{Hom}_{\mathcal{C}}(G, X) \to \operatorname{Hom}_{\mathcal{C}}(G, Y)$ is injective,*
(iii) *a morphism $f \colon X \to Y$ in $\mathcal{C}$ is an epimorphism if $\varphi_G(f) \colon \operatorname{Hom}_{\mathcal{C}}(G, X) \to \operatorname{Hom}_{\mathcal{C}}(G, Y)$ is surjective,*
 (iv) *for any $X \in \mathcal{C}$ the canonical morphism $G^{\coprod \operatorname{Hom}(G,X)} \to X$ defined in (5.2.1) is an epimorphism in $\mathcal{C}$,*
  (v) *for any $X \in \mathcal{C}$, the family of subobjects (see Definition 1.2.18) of X is a small set.*

Proof. (i) follows from Proposition 2.2.3 and the fact that $\operatorname{Hom}_{\mathcal{C}}(G, \cdot)$ is left exact.

(ii)–(iii) follow from (i) and Proposition 1.2.12.

(iv) By (iii) it is enough to check that $\operatorname{Hom}_{\mathcal{C}}(G, G^{\coprod \operatorname{Hom}(G,X)}) \to \operatorname{Hom}_{\mathcal{C}}(G, X)$ is an epimorphism, which is obvious.

(v) We have a map from the family of subobjects of X to the set of subsets of $\varphi_G(X)$. Since $\varphi_G(X)$ is a small set, it is enough to show that this map is injective. For two subobjects $Y_1 \hookrightarrow X$ and $Y_2 \hookrightarrow X$, $Y_1 \times_X Y_2$ is a subobject of X. Assuming that $\operatorname{Im}(\varphi_G(Y_1) \to \varphi_G(X)) = \operatorname{Im}(\varphi_G(Y_2) \to \varphi_G(X))$, we find

$$\varphi_G(Y_1 \times_X Y_2) \simeq \varphi_G(Y_1) \times_{\varphi_G(X)} \varphi_G(Y_2) \simeq \varphi_G(Y_1) \simeq \varphi_G(Y_2) \ .$$

Hence, $Y_1 \times_X Y_2 \xrightarrow{\sim} Y_i$ for $i = 1, 2$. Therefore, Y_1 and Y_2 are isomorphic. q.e.d.

Proposition 5.2.4. *Let $\mathcal{C}$ be a category which admits finite projective limits and small coproducts, and assume that any morphism which is both an epimorphism and a monomorphism is an isomorphism. For an object G of $\mathcal{C}$, the following conditions are equivalent.*

  (i) *G is a generator,*

(ii) φ_G *is faithful,*
(iii) *for any $X \in \mathcal{C}$, there exist a small set I and an epimorphism $G^{\sqcup I} \to X$.*

Proof. We know by Proposition 5.2.3 that (i) $\Rightarrow$ (ii) & (iii).
(ii) $\Rightarrow$ (i). Let $f: X \to Y$ and assume that $\varphi_G(f)$ is an isomorphism. By Proposition 1.2.12, f is a monomorphism and an epimorphism. We conclude that f is an isomorphism by the third hypothesis.
(iii) $\Rightarrow$ (ii). Let $f, g: X \rightrightarrows Y$ and assume that $\varphi_G(f) = \varphi_G(g)$. For any small set I and any morphism $u: G^{\sqcup I} \to X$, the two compositions $G^{\sqcup I} \to X \rightrightarrows Y$ are equal. If u is an epimorphism, this implies $f = g$.   q.e.d.

Theorem 5.2.5. *Let $\mathcal{C}$ be a category which admits small inductive limits and let $F: \mathcal{C}^{\mathrm{op}} \to \mathbf{Set}$ be a functor. Then F is representable if and only if the two conditions below are satisfied:*

(a) *F commutes with small projective limits (i.e., F sends inductive limits in $\mathcal{C}$ to projective limits in $\mathbf{Set}$),*
(b) *the category $\mathcal{C}_F$ is cofinally small. (The category $\mathcal{C}_F$ is associated with $F \in \mathcal{C}^{\wedge}$ and $h_{\mathcal{C}}: \mathcal{C} \to \mathcal{C}^{\wedge}$ as in Definition 1.2.16. In particular, its objects are the pairs (X, u) of $X \in \mathcal{C}$ and $u \in F(X)$.)*

Proof. (i) Condition (a) is obviously necessary. Moreover, if F is representable, let us say by $Y \in \mathcal{C}$, then the category $\mathcal{C}_F \simeq \mathcal{C}_Y$ admits a terminal object, namely (Y, id_Y).
(ii) Conversely, assume that F satisfies (a) and (b).
By hypothesis (a) and Lemma 2.1.13, $\mathcal{C}_F$ admits small inductive limits.
By hypothesis (b), $\mathcal{C}_F$ is cofinally small. Hence the inductive limit of the identity functor is well-defined in $\mathcal{C}_F$. Denote this object of $\mathcal{C}_F$ by X_0:

$$X_0 = \varinjlim_{X \in \mathcal{C}_F} X \ .$$

Since X_0 is a terminal object of $\mathcal{C}_F$ by Lemma 2.1.11, X_0 is a representative of F by Lemma 1.4.10.   q.e.d.

We shall give a condition in order that the condition (b) of Theorem 5.2.5 is satisfied.

Theorem 5.2.6. *Let $\mathcal{C}$ be a category satisfying:*

(i) *$\mathcal{C}$ admits a generator G,*
(ii) *$\mathcal{C}$ admits small inductive limits,*
(iii) *for any $X \in \mathcal{C}$ the family of quotients of X is a small set.*

Then any functor $F: \mathcal{C}^{\mathrm{op}} \to \mathbf{Set}$ which commutes with small projective limits is representable.

Remark 5.2.7. The hypotheses (iii) is not assumed in [64], but the authors could not follow the argument of loc. cit.

Proof. By Theorem 5.2.5, it is enough to check that the category $\mathcal{C}_F$ is cofinally small. Note that F being left exact, this category is filtrant by Proposition 3.3.13.

Set $Z_0 = G \sqcup F(G)$. By the assumption on F, we have

$$F(Z_0) \simeq F(G)^{F(G)} \simeq \mathrm{Hom}_{\mathbf{Set}}(F(G), F(G)) \,.$$

Denote by $u_0 \in F(Z_0)$ the image of $\mathrm{id}_{F(G)}$. Hence, (Z_0, u_0) belongs to $\mathcal{C}_F$. Let $(X, u) \in \mathcal{C}_F$ and set $X_1 = G \sqcup \mathrm{Hom}(G,X)$. Then the natural morphism $X_1 \to X$ is an epimorphism by Proposition 5.2.3 (iv).

Consider the maps $\mathrm{Hom}_{\mathcal{C}}(G, X) \to \mathrm{Hom}_{\mathbf{Set}}(F(X), F(G)) \to F(G)$ where the second one is associated with $u \in F(X)$. They define the morphism $X_1 = G \sqcup \mathrm{Hom}(G,X) \to Z_0 = G \sqcup F(G)$ and the commutative diagram in $\mathcal{C}^\wedge$

Define X' as $X \sqcup_{X_1} Z_0$ and consider the diagram below in which the square is co-Cartesian:

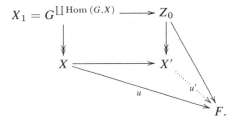

Since F commutes with projective limits, the dotted arrow may be completed. Since $X_1 \to X$ is an epimorphism, $Z_0 \to X'$ is an epimorphism by Exercise 2.22. Hence, for any $(X, u) \in \mathcal{C}_F$ we have found a morphism $(X, u) \to (X', u')$ in $\mathcal{C}_F$ such that there exists an epimorphism $Z_0 \twoheadrightarrow X'$. By hypothesis (iii) and Proposition 3.2.6, $\mathcal{C}_F$ is cofinally small.   q.e.d.

Proposition 5.2.8. *Let $\mathcal{C}$ be a category which admits small inductive limits. Assume that any functor $F: \mathcal{C}^{\mathrm{op}} \to \mathbf{Set}$ is representable if it commutes with small projective limits. Then:*

(i) *$\mathcal{C}$ admits small projective limits,*
(ii) *a functor $F: \mathcal{C} \to \mathcal{C}'$ admits a right adjoint if and only if it commutes with small inductive limits.*

Proof. (i) Let $\beta: I^{\mathrm{op}} \to \mathcal{C}$ be a projective system indexed by a small category I. Consider the object $F \in \mathcal{C}^\wedge$ given by

$$F(X) = \varprojlim_i \operatorname{Hom}_{\mathcal{C}}(X, \beta(i)) \ .$$

This functor from $\mathcal{C}^{\mathrm{op}}$ to **Set** commutes with small projective limits in $\mathcal{C}^{\mathrm{op}}$, and hence it is representable.
(ii) For any $Y \in \mathcal{C}'$, the functor $X \mapsto \operatorname{Hom}_{\mathcal{C}'}(F(X), Y)$ commutes with small projective limits, and hence it is representable.   q.e.d.

Proposition 5.2.9. *Assume that $\mathcal{C}$ admits finite inductive limits, finite projective limits, and a generator. Then the family of strict quotients of an object $X \in \mathcal{C}$ is a small set.*

Proof. Recall that $f \colon X \to Y$ is a strict epimorphism if and only if the sequence $X \times_Y X \rightrightarrows X \to Y$ is exact. Hence, we may identify the family of strict quotients of X with a family of subobjects of $X \times X$, and this is a small set by Proposition 5.2.3 (v).

q.e.d.

Corollary 5.2.10. *Assume that the category $\mathcal{C}$ admits small inductive limits, finite projective limits and a generator. Assume moreover that any epimorphism in $\mathcal{C}$ is strict. Then a functor $F \colon \mathcal{C}^{\mathrm{op}} \to$ **Set** is representable if and only if it commutes with small projective limits.*

Examples 5.2.11. The hypotheses of Corollary 5.2.10 are satisfied by the category **Set** as well as by the category $\mathrm{Mod}(R)$ of modules over a ring R.

5.3 Strictly Generating Subcategories

In Sect. 5.2 we obtained representability results in a category $\mathcal{C}$ when assuming either that the family of quotients of any object is small or that any epimorphism is strict. In this section, we shall get rid of this kind of hypotheses.

Let $\mathcal{C}$ be a category and $\mathcal{F}$ a small full subcategory of $\mathcal{C}$. Then we have the natural functor

(5.3.1) $\varphi \colon \mathcal{C} \to \mathcal{F}^{\wedge},$

which associates with $X \in \mathcal{C}$ the functor $\mathcal{F} \ni Y \mapsto \operatorname{Hom}_{\mathcal{C}}(Y, X)$. By the Yoneda Lemma, we have

$$\operatorname{Hom}_{\mathcal{F}^{\wedge}}(\varphi(X), \varphi(Y)) \simeq \operatorname{Hom}_{\mathcal{C}}(X, Y)$$

for $X \in \mathcal{F}$ and $Y \in \mathcal{C}$.

By the definition, φ is conservative if and only if $\mathrm{Ob}(\mathcal{F})$ is a system of generators. If moreover $\mathcal{C}$ admits finite projective limits, then φ is faithful by Proposition 2.2.3.

Definition 5.3.1. *Let C be a category and $\mathcal{F}$ an essentially small full sub-category of C. We say that $\mathcal{F}$ is* strictly generating *in C if the functor φ in (5.3.1) is fully faithful.*

Note that if $\mathcal{F}$ is a strictly generating full subcategory, then $\mathrm{Ob}(\mathcal{F})$ is a system of generators.

Lemma 5.3.2. *Let C be a category, and let $\mathcal{F}$ and $\mathcal{G}$ be small full subcategories of C. Assume that $\mathcal{F} \subset \mathcal{G}$ and $\mathcal{F}$ is strictly generating. Then $\mathcal{G}$ is also strictly generating.*

Proof. Let $\varphi_{\mathcal{F}} \colon C \to \mathcal{F}^{\wedge}$ and $\varphi_{\mathcal{G}} \colon C \to \mathcal{G}^{\wedge}$ be the natural functors. Then $\varphi_{\mathcal{F}}$ is fully faithful and it decomposes as

$$C \xrightarrow{\ \varphi_{\mathcal{G}}\ } \mathcal{G}^{\wedge} \xrightarrow{\ \iota\ } \mathcal{F}^{\wedge}.$$

Hence $\varphi_{\mathcal{G}}$ is faithful. Let us show that the map

$$\mathrm{Hom}_{C}(X, Y) \to \mathrm{Hom}_{\mathcal{G}^{\wedge}}(\varphi_{\mathcal{G}}(X), \varphi_{\mathcal{G}}(Y))$$

is surjective for any $X, Y \in C$. Let $\xi \in \mathrm{Hom}_{\mathcal{G}^{\wedge}}(\varphi_{\mathcal{G}}(X), \varphi_{\mathcal{G}}(Y))$. Since $\varphi_{\mathcal{F}}$ is fully faithful, there exists $f \in \mathrm{Hom}_{C}(X, Y)$ such that

$$(5.3.2) \qquad \iota(\xi) = \varphi_{\mathcal{F}}(f) \quad \text{as elements of } \mathrm{Hom}_{\mathcal{F}^{\wedge}}(\varphi_{\mathcal{F}}(X), \varphi_{\mathcal{F}}(Y)).$$

Let us show that $\xi = \varphi_{\mathcal{G}}(f)$. It is enough to show that, for any $Z \in \mathcal{G}$, the map induced by ξ

$$\xi_{Z} \colon \mathrm{Hom}_{C}(Z, X) \longrightarrow \mathrm{Hom}_{C}(Z, Y)$$

coincides with the map $v \mapsto f \circ v$.

Let $v \in \mathrm{Hom}_{C}(Z, X)$. Then for any $S \in \mathcal{F}$ and $s \colon S \to Z$:

$$\xi_{Z}(v) \circ s = \xi_{S}(v \circ s) = \iota(\xi)_{S}(v \circ s) = f \circ v \circ s ,$$

where the last equality follows from (5.3.2). Hence $\varphi_{\mathcal{F}}(\xi_{Z}(v)) = \varphi_{\mathcal{F}}(f \circ v)$ as elements of $\mathrm{Hom}_{\mathcal{F}^{\wedge}}(\varphi_{\mathcal{F}}(Z), \varphi_{\mathcal{F}}(Y))$, and the faithfulness of $\varphi_{\mathcal{F}}$ implies $\xi_{Z}(v) = f \circ v$.   q.e.d.

Lemma 5.3.3. *Let C be a category which admits small inductive limits and let $\mathcal{F}$ be a small full subcategory of C. Then the functor $\varphi \colon C \to \mathcal{F}^{\wedge}$ admits a left adjoint $\psi \colon \mathcal{F}^{\wedge} \to C$ and for $F \in \mathcal{F}^{\wedge}$, we have*

$$\psi(F) \simeq \varinjlim_{(Y \to F) \in \mathcal{F}_{F}} Y .$$

Proof. For $X \in \mathcal{C}$ and $F \in \mathcal{F}^\wedge$, we have the chain of isomorphisms

$$\mathrm{Hom}_{\mathcal{C}}(\varinjlim_{(Y \to F) \in \mathcal{F}_F} Y, X) \simeq \varprojlim_{(Y \to F) \in \mathcal{F}_F} \mathrm{Hom}_{\mathcal{C}}(Y, X)$$

$$\simeq \varprojlim_{(Y \to F) \in \mathcal{F}_F} \mathrm{Hom}_{\mathcal{F}^\wedge}(\varphi(Y), \varphi(X))$$

$$\simeq \mathrm{Hom}_{\mathcal{F}^\wedge}(\varinjlim_{(Y \to F) \in \mathcal{F}_F} \varphi(Y), \varphi(X)),$$

and $\varinjlim_{(Y \to F) \in \mathcal{F}_F} \varphi(Y) \simeq F$ by Proposition 2.6.3.   q.e.d.

Proposition 5.3.4. *Let $\mathcal{C}$ be a category which admits small inductive limits and let $\mathcal{F}$ be a small strictly generating full subcategory of $\mathcal{C}$. Let $\mathcal{E}$ denote the full subcategory of $\mathcal{F}^\wedge$ consisting of objects $F \in \mathcal{F}^\wedge$ such that the functor $\mathcal{C} \ni X \mapsto \mathrm{Hom}_{\mathcal{F}^\wedge}(\varphi(X), F) \in \mathbf{Set}$ commutes with small projective limits. Then $\mathcal{C}$ is equivalent to $\mathcal{E}$ by φ.*

Proof. It is obvious that φ sends $\mathcal{C}$ to $\mathcal{E}$. Hence, it is enough to show that any $F \in \mathcal{E}$ is isomorphic to the image of an object of $\mathcal{C}$ by φ. Let ψ denote the left adjoint to φ constructed in Lemma 5.3.3. By Proposition 4.1.4, it is enough to prove the isomorphism

$$\mathrm{Hom}_{\mathcal{F}^\wedge}(\varphi\psi(G), F) \xrightarrow{\sim} \mathrm{Hom}_{\mathcal{F}^\wedge}(G, F)$$

for any $G \in \mathcal{F}^\wedge$ and $F \in \mathcal{E}$. We have the chain of isomorphisms

$$\mathrm{Hom}_{\mathcal{F}^\wedge}(\varphi\psi(G), F) \simeq \mathrm{Hom}_{\mathcal{F}^\wedge}(\varphi(\varinjlim_{(X \to G) \in \mathcal{F}_G} X), F)$$

$$\simeq \varprojlim_{(X \to G) \in \mathcal{F}_G} \mathrm{Hom}_{\mathcal{F}^\wedge}(\varphi(X), F)$$

$$\simeq \mathrm{Hom}_{\mathcal{F}^\wedge}(\varinjlim_{(X \to G) \in \mathcal{F}_G} \varphi(X), F)$$

$$\simeq \mathrm{Hom}_{\mathcal{F}^\wedge}(G, F),$$

where the second isomorphism follows from the hypothesis $F \in \mathcal{E}$ and the last isomorphism follows from Proposition 2.6.3 (i).   q.e.d.

Proposition 5.3.5. *Let $\mathcal{C}$ be a category which admits small inductive limits and assume that there exists a small strictly generating full subcategory of $\mathcal{C}$. Let $F \colon \mathcal{C}^{\mathrm{op}} \to \mathbf{Set}$ be a functor. If F commutes with small projective limits, then F is representable.*

Proof. Let $\mathcal{F}$ be a small strictly generating full subcategory of $\mathcal{C}$. Let $\widetilde{F} \in \mathcal{F}^\wedge$ be the restriction of F to $\mathcal{F}$. For $X \in \mathcal{C}$, we have

$$\operatorname{Hom}_{\mathcal{F}^\wedge}(\varphi(X), \widetilde{F}) \simeq \operatorname{Hom}_{\mathcal{F}^\wedge}(\varinjlim_{(Y \to X) \in \mathcal{F}_X} \varphi(Y), \widetilde{F})$$

$$\simeq \varprojlim_{(Y \to X) \in \mathcal{F}_X} \operatorname{Hom}_{\mathcal{F}^\wedge}(\varphi(Y), \widetilde{F})$$

$$\simeq \varprojlim_{(Y \to X) \in \mathcal{F}_X} F(Y) \simeq F(\varinjlim_{(Y \to X) \in \mathcal{F}_X} Y).$$

Since φ is fully faithful, we have $\varinjlim_{(Y \to X) \in \mathcal{F}_X} Y \simeq \psi\varphi(X) \simeq X$. Hence, we obtain

(5.3.3) $$F(X) \xrightarrow{\sim} \operatorname{Hom}_{\mathcal{F}^\wedge}(\varphi(X), \widetilde{F}) \quad \text{for any } X \in \mathcal{C}.$$

It follows that the functor $\mathcal{C} \ni X \mapsto \operatorname{Hom}_{\mathcal{F}^\wedge}(\varphi(X), \widetilde{F})$ sends small inductive limits to projective limits, and by Proposition 5.3.4 there exists $X_0 \in \mathcal{C}$ such that $\widetilde{F} \simeq \varphi(X_0)$. Then (5.3.3) implies that

$$F(X) \simeq \operatorname{Hom}_{\mathcal{F}^\wedge}(\varphi(X), \widetilde{F})$$
$$\simeq \operatorname{Hom}_{\mathcal{F}^\wedge}(\varphi(X), \varphi(X_0)) \simeq \operatorname{Hom}_{\mathcal{C}}(X, X_0)$$

for any $X \in \mathcal{C}$.   q.e.d.

We shall give several criteria for a small full subcategory $\mathcal{F}$ to be strictly generating.

Theorem 5.3.6. *Let $\mathcal{C}$ be a category satisfying the conditions* (i)–(iii) *below:*

(i) $\mathcal{C}$ *admits small inductive limits and finite projective limits,*

(ii) *small filtrant inductive limits are stable by base change* (*see* Definition 2.2.6),

(iii) *any epimorphism is strict.*

Let $\mathcal{F}$ be an essentially small full subcategory of $\mathcal{C}$ such that

(a) $\operatorname{Ob}(\mathcal{F})$ *is a system of generators,*

(b) $\mathcal{F}$ *is closed by finite coproducts in $\mathcal{C}$.*

Then $\mathcal{F}$ is strictly generating.

Proof. We may assume from the beginning that $\mathcal{F}$ is small.

(i) As already mentioned, the functor φ in (5.3.1) is conservative and faithful.

(ii) By Proposition 1.2.12, a morphism f in $\mathcal{C}$ is an epimorphism as soon as $\varphi(f)$ is an epimorphism.

(iii) Let us fix $X \in \mathcal{C}$. For a small filtrant inductive system $\{Y_i\}_{i \in I}$ in $\mathcal{C}_X$, we have

(5.3.4) $$\varinjlim_i \operatorname{Coim}(Y_i \to X) \xrightarrow{\sim} \operatorname{Coim}(\varinjlim_i Y_i \to X).$$

Indeed, setting $Y_\infty = \varinjlim_i Y_i$, we have

$$\varinjlim_i (Y_i \times_X Y_i) \simeq \varinjlim_{i_1, i_2} (Y_{i_1} \times_X Y_{i_2}) \simeq \varinjlim_{i_1} \varinjlim_{i_2} (Y_{i_1} \times_X Y_{i_2})$$
$$\simeq \varinjlim_{i_1} (Y_{i_1} \times_X Y_\infty) \simeq Y_\infty \times_X Y_\infty .$$

Here the first isomorphism follows from Corollary 3.2.3 (ii), and the last two isomorphisms follow from hypothesis (ii). Hence we obtain

$$\mathrm{Coim}(Y_\infty \to X) \simeq \mathrm{Coker}(Y_\infty \times_X Y_\infty \rightrightarrows Y_\infty)$$
$$\simeq \mathrm{Coker}\big(\varinjlim_i (Y_i \times_X Y_i) \rightrightarrows \varinjlim_i Y_i\big)$$
$$\simeq \varinjlim_i \mathrm{Coker}(Y_i \times_X Y_i \rightrightarrows Y_i)$$
$$\simeq \varinjlim_i \mathrm{Coim}(Y_i \to X) .$$

(iv) For $Z \in \mathcal{F}_X$, set

$$\eta(Z) = \mathrm{Coim}(Z \to X) := \mathrm{Coker}(Z \times_X Z \rightrightarrows Z) .$$

Then η defines a functor $\mathcal{F}_X \to \mathcal{C}_X$. For any $Y \in \mathcal{C}$, we have

$$\mathrm{Hom}_{\mathcal{C}}(\eta(Z), Y) \simeq \mathrm{Ker}\big(\mathrm{Hom}_{\mathcal{C}}(Z, Y) \rightrightarrows \mathrm{Hom}_{\mathcal{C}}(Z \times_X Z, Y)\big) .$$

We have $\mathrm{Hom}_{\mathcal{C}}(Z, Y) \simeq \mathrm{Hom}_{\mathcal{F}^\wedge}(\varphi(Z), \varphi(Y))$ by the Yoneda Lemma. On the other hand, the map $\mathrm{Hom}_{\mathcal{C}}(Z \times_X Z, Y) \to \mathrm{Hom}_{\mathcal{F}^\wedge}(\varphi(Z \times_X Z), \varphi(Y)) \simeq \mathrm{Hom}_{\mathcal{F}^\wedge}(\varphi(Z) \times_{\varphi(X)} \varphi(Z), \varphi(Y))$ is injective since φ is faithful. Hence we obtain

$\mathrm{Hom}_{\mathcal{C}}(\eta(Z), Y)$
$$\simeq \mathrm{Ker}\big(\mathrm{Hom}_{\mathcal{F}^\wedge}(\varphi(Z), \varphi(Y)) \rightrightarrows \mathrm{Hom}_{\mathcal{F}^\wedge}(\varphi(Z) \times_{\varphi(X)} \varphi(Z), \varphi(Y)))$$
$$\simeq \mathrm{Hom}_{\mathcal{F}^\wedge}\big(\mathrm{Coker}(\varphi(Z) \times_{\varphi(X)} \varphi(Z) \rightrightarrows \varphi(Z)), \varphi(Y)\big)$$
$$\simeq \mathrm{Hom}_{\mathcal{F}^\wedge}\big(\mathrm{Im}(\varphi(Z) \to \varphi(X)), \varphi(Y)\big) .$$

(v) Let us denote by I the set of finite subsets of $\mathrm{Ob}(\mathcal{F}_X)$, ordered by inclusion. Regarding I as a category, it is small and filtrant. For $A \in I$, $\xi(A) := \sqcup_{Z \in A} Z$ belongs to $\mathcal{F}_X$ by (b), and ξ defines a functor $I \to \mathcal{F}_X$. Then

(5.3.5) $\varinjlim_{A \in I} \varphi(\xi(A)) \to \varphi(X)$ is an epimorphism .

Indeed, for any $S \in \mathcal{F}$ and $u \in \varphi(X)(S) = \mathrm{Hom}_{\mathcal{C}}(S, X)$, u is in the image of $\varphi(\xi(A))(S)$ with $A = \{(S, u)\}$.

(vi) Since $\varinjlim_{A \in I} \varphi(\xi(A)) \to \varphi(X)$ factors through $\varphi(\varinjlim_{A \in I} \xi(A))$, the morphism $\varphi(\varinjlim_{A \in I} \xi(A)) \to \varphi(X)$ is an epimorphism, and (ii) implies that $\varinjlim_{A \in I} \xi(A) \to X$

is an epimorphism, hence a strict epimorphism by the hypothesis. Proposition 5.1.5 (i) implies $\mathrm{Coim}(\varinjlim_{A \in I} \xi(A) \to X) \simeq X$. By (iii), we have

$$
\varinjlim_{A \in I} \eta(\xi(A)) = \varinjlim_{A \in I} \mathrm{Coim}(\xi(A) \to X)
$$
$$
\simeq \mathrm{Coim}(\varinjlim_{A \in I} \xi(A) \to X) \simeq X \ .
$$

(vii) For any $Y \in \mathcal{C}$, we obtain the chain of isomorphisms

$$
\mathrm{Hom}_{\mathcal{C}}(X, Y) \simeq \mathrm{Hom}_{\mathcal{C}}(\varinjlim_{A \in I} \eta(\xi(A)), Y)
$$
$$
\simeq \varprojlim_{A \in I} \mathrm{Hom}_{\mathcal{C}}(\eta(\xi(A)), Y)
$$
$$
\simeq \varprojlim_{A \in I} \mathrm{Hom}_{\mathcal{F}^\wedge}(\mathrm{Im}(\varphi(\xi(A)) \to \varphi(X)), \varphi(Y))
$$
$$
\simeq \mathrm{Hom}_{\mathcal{F}^\wedge}\big(\varinjlim_{A \in I}(\mathrm{Im}(\varphi(\xi(A)) \to \varphi(X))), \varphi(Y)\big)
$$
$$
\simeq \mathrm{Hom}_{\mathcal{F}^\wedge}(\varphi(X), \varphi(Y)) \ ,
$$

where the last isomorphism follows from (5.3.5).   q.e.d.

Remark 5.3.7. See Exercises 5.5–5.8 which show that it is not possible to drop conditions (ii), (iii) or (b) in Theorem 5.3.6.

Theorem 5.3.8. *Let $\mathcal{C}$ be a category and consider the conditions below:*

(i) *$\mathcal{C}$ admits small inductive limits and finite projective limits,*
(ii) *small inductive limits in $\mathcal{C}$ are stable by base change,*
(ii)' *small filtrant inductive limits in $\mathcal{C}$ are stable by base change.*

Let us consider the conditions on an essentially small full subcategory $\mathcal{F}$ of $\mathcal{C}$:

(a) *$\mathrm{Ob}(\mathcal{F})$ is a system of generators,*
(b) *the inclusion functor $\mathcal{F} \hookrightarrow \mathcal{C}$ is right exact.*

Assume either (i), (ii) and (a) or (i), (ii)', (a) and (b). Then $\mathcal{F}$ is strictly generating.

Proof. We already know that $\varphi \colon \mathcal{C} \to \mathcal{F}^\wedge$ is conservative and faithful.
Assuming (i), let $\psi \colon \mathcal{F}^\wedge \to \mathcal{C}$ be the functor

$$
\mathcal{F}^\wedge \ni F \mapsto \varinjlim_{(X \to F) \in \mathcal{F}_F} X \in \mathcal{C} \ .
$$

Then ψ is left adjoint to φ by Lemma 5.3.3. By Proposition 1.5.6 (i), it is enough to show that $\psi \circ \varphi \to \mathrm{id}_{\mathcal{C}}$ is an isomorphism.

(A) First, we assume (i), (ii) and (a).

(A1) We begin by proving that

$$(5.3.6) \begin{cases} \text{for any } X \in \mathcal{C} \text{ and any small inductive system } \{X_i\}_{i \in I} \text{ in } \mathcal{F}_X, \text{ if} \\ \varinjlim_i \varphi(X_i) \to \varphi(X) \text{ is an isomorphism, then } \varinjlim_i X_i \to X \text{ is an iso-} \\ \text{morphism.} \end{cases}$$

Set $X_0 = \varinjlim_i X_i \in \mathcal{C}$ and let $u \colon X_0 \to X$ be the canonical morphism. Since the composition $\varinjlim_i \varphi(X_i) \to \varphi(X_0) \to \varphi(X)$ is an isomorphism, $\varphi(u) \colon \varphi(X_0) \to \varphi(X)$ is an epimorphism. Since φ is conservative by (a), it remains to show that $\varphi(u)$ is a monomorphism.

For $i_1, i_2 \in I$, the two compositions $X_{i_1} \times_X X_{i_2} \to X_{i_\nu} \to X_0$ ($\nu = 1, 2$) give two morphisms $\xi_1, \xi_2 \colon X_{i_1} \times_X X_{i_2} \rightrightarrows X_0$. Then we have a diagram

$$\varphi(X_{i_1} \times_X X_{i_2}) \rightrightarrows \varinjlim_i \varphi(X_i) \xrightarrow{\hspace{1cm}} \varphi(X_0) \xrightarrow{\hspace{1cm}} \varphi(X).$$

Hence, the two arrows $\varphi(X_{i_1} \times_X X_{i_2}) \rightrightarrows \varinjlim_i \varphi(X_i)$ coincide, which implies $\varphi(\xi_1) = \varphi(\xi_2)$. Thus we obtain $\xi_1 = \xi_2$. It means that

$$X_{i_1} \times_{X_0} X_{i_2} \to X_{i_1} \times_X X_{i_2}$$

is an isomorphism for any $i_1, i_2 \in I$.

On the other hand, the condition (ii) implies that

$$(5.3.7) \qquad \begin{aligned} \varinjlim_{i_1, i_2} (X_{i_1} \times_{X_0} X_{i_2}) &\simeq \varinjlim_{i_1} (X_{i_1} \times_{X_0} \varinjlim_{i_2} X_{i_2}) \\ &\simeq (\varinjlim_{i_1} X_{i_1}) \times_{X_0} (\varinjlim_{i_2} X_{i_2}), \end{aligned}$$

and similarly,

$$(5.3.8) \qquad \varinjlim_{i_1, i_2} (X_{i_1} \times_X X_{i_2}) \simeq (\varinjlim_{i_1} X_{i_1}) \times_X (\varinjlim_{i_2} X_{i_2}).$$

Hence, we obtain the isomorphisms

$$\varinjlim_{i_1, i_2} (X_{i_1} \times_{X_0} X_{i_2}) \simeq X_0,$$

$$\varinjlim_{i_1, i_2} (X_{i_1} \times_X X_{i_2}) \simeq X_0 \times_X X_0.$$

Hence, $X_0 \to X_0 \times_X X_0$ is an isomorphism, and this means that $X_0 \to X$ is a monomorphism by Exercise 2.4.

We have proved that $\varphi(X_0) \to \varphi(X)$ is a monomorphism and this completes the proof of (5.3.6).

(A2) Finally we shall show that $\psi \circ \varphi \to \mathrm{id}_{\mathcal{C}}$ is an isomorphism. For any $X \in \mathcal{C}$, we have $\varinjlim_{(Y \to X) \in \mathcal{F}_X} \varphi(Y) \xrightarrow{\sim} \varphi(X)$ by Proposition 2.6.3 (i), and (5.3.6) implies that $\psi\varphi(X) \simeq \varinjlim_{(Y \to X) \in \mathcal{F}_X} Y \simeq X$.

(B) Now, we assume (i), (ii)', (a) and (b). The proof is similar to the former case (A). For $X \in \mathcal{C}$, $\mathcal{F}_X$ is filtrant by (b). Hence, in step (A2), we only need (5.3.6) when I is filtrant. On the other hand, (5.3.6) in the filtrant case follows from (ii)' by the same argument as in (A1). Note that, in case (A), the condition (ii) is used only in proving (5.3.7) and (5.3.8).   q.e.d.

Theorem 5.3.9. *Let $\mathcal{C}$ be a category satisfying:*

(i) *$\mathcal{C}$ admits small inductive limits and finite projective limits,*
(ii) *small filtrant inductive limits in $\mathcal{C}$ are stable by base change,*
(iii) *$\mathcal{C}$ admits a generator.*

Then any functor $F \colon \mathcal{C}^{\mathrm{op}} \to \mathbf{Set}$ which commutes with small projective limits is representable.

Proof. Let $\emptyset_{\mathcal{C}}$ be an initial object of $\mathcal{C}$ and let G be a generator of $\mathcal{C}$. We construct by induction an increasing sequence $\{\mathcal{F}_n\}_{n \geq 0}$ of small full subcategories as follows.

$$\mathrm{Ob}(\mathcal{F}_0) = \{\emptyset_{\mathcal{C}}, G\}$$
$$\mathrm{Ob}(\mathcal{F}_n) = \mathrm{Ob}(\mathcal{F}_{n-1}) \bigsqcup \{Y_1 \sqcup_X Y_2 \, ; \, X \to Y_1 \text{ and } X \to Y_2 \text{ are morphisms}$$
$$\text{in } \mathcal{F}_{n-1}\} \quad \text{for } n > 0.$$

Let $\mathcal{F}$ be the full subcategory of $\mathcal{C}$ with $\mathrm{Ob}(\mathcal{F}) = \bigcup_n \mathrm{Ob}(\mathcal{F}_n)$. Then $\mathcal{F}$ is a small category, $\mathrm{Ob}(\mathcal{F})$ is a system of generators, and $\mathcal{F}$ is closed by finite inductive limits. Hence, Proposition 3.3.3 implies that $\mathcal{F} \to \mathcal{C}$ is right exact, and $\mathcal{F}$ is strictly generating by Theorem 5.3.8. It remains to apply Corollary 5.3.5.
          q.e.d.

Note that if small filtrant inductive limits in $\mathcal{C}$ are exact, then such limits are stable by base change by Lemma 3.3.9.

Exercises

Exercise 5.1. Let $\mathcal{C}$ be one of the categories $\mathcal{C} = \mathbf{Set}$, $\mathcal{C} = \mathrm{Mod}(R)$ for a ring R, or $\mathcal{C} = \mathcal{D}^{\wedge}$ for a small category $\mathcal{D}$. Prove that any morphism in $\mathcal{C}$ is strict. Also prove that, when $\mathcal{C} = \mathcal{D}^{\wedge}$ and f is a morphism in $\mathcal{C}$, $\mathrm{Im}\, f$ is the functor $\mathcal{D} \ni Z \mapsto \mathrm{Im}(f(Z))$.

Exercise 5.2. Assume that a category $\mathcal{C}$ admits finite projective limits and finite inductive limits. Let $f \colon X \to Y$ be a morphism in $\mathcal{C}$. Prove the isomorphism $\mathrm{Hom}_{\mathcal{C}}(\mathrm{Coim}(f), Z) \simeq \mathrm{Hom}_{\mathcal{C}^{\wedge}}(\mathrm{Im}(\mathrm{h}_{\mathcal{C}}(f)), \mathrm{h}_{\mathcal{C}}(Z))$ for any $Z \in \mathcal{C}$.

Exercise 5.3. Let $\mathcal{C}$ be a category which admits finite inductive limits and finite projective limits. Consider the following conditions on $\mathcal{C}$:

(a) any morphism is strict,
(b) any epimorphism is strict,
(c) for any morphism $f \colon X \to Y$, $\operatorname{Coim} f \to Y$ is a monomorphism,
(d) any morphism which is both an epimorphism and a monomorphism is an isomorphism,
(e) for any strict epimorphisms $f \colon X \to Y$ and $g \colon Y \to Z$, their composition $g \circ f$ is a strict epimorphism.

Prove that (a) $\Rightarrow$ (b) $\Leftrightarrow$ (c) + (d) and that (c) $\Leftrightarrow$ (e).
(Hint: (e) $\Rightarrow$ (c). Adapt the proof of Proposition 5.1.7.
(c) $\Rightarrow$ (e). Consider $W = \operatorname{Coim}(g \circ f)$. Using the fact that $W \to Z$ is a monomorphism, deduce that $Y \times_W Y \to Y \times_Z Y$ is an isomorphism.)

Exercise 5.4. Let $\mathcal{C}$ be a category which admits finite inductive limits and finite projective limits. Let $f \colon X \to Y$ be the composition $X \xrightarrow{g} Z \xrightarrow{h} Y$ where g is a strict epimorphism. Prove that h factors uniquely through $\operatorname{Coim} f \to Y$ such that the composition $X \to Z \to \operatorname{Coim} f$ coincides with the canonical morphism.

Exercise 5.5. Let k be a field and set $\mathcal{F} := \operatorname{Mod}^{\mathrm{f}}(k)$, the full subcategory of $\operatorname{Mod}(k)$ consisting of finite-dimensional vector spaces. For $V \in \operatorname{Mod}(k)$, set $V^* = \operatorname{Hom}_k(V, k)$.
(i) Prove that the functor $V \mapsto V^*$ induces an equivalence of categories $\mathcal{F} \simeq \mathcal{F}^{\mathrm{op}}$.
(ii) Let $V \in \operatorname{Mod}(k)$. Prove the isomorphism $\varprojlim_{(V \to W) \in \mathcal{F}^V} W \simeq V^{**}$.
(iii) Prove that $\mathcal{F}$ is a strictly generating full subcategory of $\operatorname{Mod}(k)$.
(iv) Prove that $\operatorname{Mod}(k)^{\mathrm{op}}$ and $\mathcal{F}^{\mathrm{op}}$ satisfy all hypotheses of Theorem 5.3.6 except condition (ii).
(v) Prove that the functor $\varphi \colon \operatorname{Mod}(k)^{\mathrm{op}} \to (\mathcal{F}^{\mathrm{op}})^{\wedge}$ defined in (5.3.1) decomposes as $\operatorname{Mod}(k)^{\mathrm{op}} \xrightarrow{*} \operatorname{Mod}(k) \to \mathcal{F}^{\wedge} \xrightarrow{\sim} (\mathcal{F}^{\mathrm{op}})^{\wedge}$.
(vi) Prove that the functor $\varphi \colon \operatorname{Mod}(k)^{\mathrm{op}} \to (\mathcal{F}^{\mathrm{op}})^{\wedge}$ is not fully faithful.

Exercise 5.6. Let k be a field and denote by $\mathcal{F}$ the full subcategory of $\operatorname{Mod}(k)$ consisting of the single object $\{k\}$. Prove that $\operatorname{Mod}(k) \to \mathcal{F}^{\wedge}$ is not fully faithful.

Exercise 5.7. Let A be a ring and denote by $\mathcal{F}$ the full subcategory of $\operatorname{Mod}(A)$ consisting of the two objects $\{A, A^{\oplus 2}\}$. Prove that $\operatorname{Mod}(A) \to \mathcal{F}^{\wedge}$ is fully faithful.

Exercise 5.8. Let k be a field, let $A = k[x, y]$ and let $\mathcal{C} = \operatorname{Mod}(A)$. Let $\mathfrak{a}$ denote the ideal $\mathfrak{a} = Ax + Ay$. (See also Exercises 8.27–8.29.) Let $\mathcal{C}_0$ be the full subcategory of $\mathcal{C}$ consisting of objects X such that there exists an epimorphism

$\mathfrak{a}^{\oplus I} \twoheadrightarrow X$ for some small set I. Let $\mathcal{F}$ be the full subcategory of $\mathcal{C}_0$ consisting of the objects $\left\{ \mathfrak{a}^{\oplus n} \, ; \, n \geq 0 \right\}$. Let $\mathcal{G}$ be the full subcategory of $\mathcal{C}$ consisting of the objects $\left\{ A^{\oplus n} \, ; \, n \geq 0 \right\}$.

(i) Prove that $\mathcal{F}$ and $\mathcal{G}$ are equivalent.

(ii) Prove that the functor $\varphi \colon \mathcal{C}_0 \to \mathcal{F}^\wedge$ given by

$$\mathcal{C}_0 \ni X \mapsto (\mathcal{F} \ni Y \mapsto \mathrm{Hom}_{\mathcal{C}}(Y, X))$$

decomposes as $\mathcal{C}_0 \xrightarrow{\xi} \mathrm{Mod}(A) \xrightarrow{\eta} \mathcal{F}^\wedge$ where $\xi(X) = \mathrm{Hom}_A(\mathfrak{a}, X)$ and $\eta(M)(Y) = \mathrm{Hom}_A(Y, \mathfrak{a}) \otimes_A M$ for $Y \in \mathcal{F}$. (In other words, $\eta(M) \in \mathcal{F}^\wedge$ is the functor $\mathcal{F} \ni \mathfrak{a}^{\oplus n} \mapsto M^{\oplus n}$.)

(iii) Prove that η is fully faithful. (Hint: use (i) and Theorem 5.3.6.)

(iv) Prove that φ is not fully faithful.

(v) Prove that $(\mathcal{C}_0, \mathcal{F})$ satisfies all the conditions in Theorem 5.3.6 except condition (iii).

(vi) Prove that any functor $F \colon \mathcal{C}_0^{\mathrm{op}} \to \mathbf{Set}$ commuting with small projective limits is representable. (Hint: use Theorem 5.2.6 or Theorem 5.3.9.)

Exercise 5.9. Let $\mathcal{C}$ be a category with a generator and satisfying the conditions (i) and (ii) in Theorem 5.3.8. Prove that for any $X, Y \in \mathcal{C}$, there exists an object $\mathcal{H}om\,(X, Y)$ in $\mathcal{C}$ which represents the functor $\mathcal{C} \ni Z \mapsto \mathrm{Hom}_{\mathcal{C}}(Z \times X, Y)$.

Exercise 5.10. (i) Let $\mathbf{Arr}$ be the category given in Notations 1.2.8 (iii), with two objects a and b and one morphism from a to b. Prove that $\mathbf{Arr}$ satisfies the conditions (i) and (ii) in Theorem 5.3.8, and b is a generator.

(ii) Conversely, let $\mathcal{C}$ be a category which satisfies the conditions (i) and (ii) in Theorem 5.3.8. Moreover assume that there exists a generator G such that $\mathrm{End}_{\mathcal{C}}(G) = \{\mathrm{id}_G\}$. Prove that $\mathcal{C}$ is equivalent to either $\mathbf{Set}$, or $\mathbf{Arr}$ or $\mathbf{Pt}$. (Hint: apply Theorem 5.3.8.)

Exercise 5.11. Prove that a functor $F \colon \mathbf{Set} \to \mathbf{Set}$ is representable if F commutes with small projective limits.

6

Indization of Categories

In this chapter we develop the theory of ind-objects. The basic reference is [64] where most, if not all, the results which appear here were already obtained (see also [3]). Apart from loc. cit., and despite its importance, it seems difficult to find in the literature a concise exposition of this subject. This chapter is an attempt in this direction.

6.1 Indization of Categories and Functors

Recall that a universe $\mathcal{U}$ is given. When we consider a category, it means a $\mathcal{U}$-category and **Set** is the category of $\mathcal{U}$-sets (see Convention 1.4.1). As far as this has no implications, we will skip this point.

Recall that for a category $\mathcal{C}$, inductive limits in $\mathcal{C}^\wedge := \mathrm{Fct}(\mathcal{C}^{\mathrm{op}}, \mathbf{Set})$ are denoted by "$\varinjlim$".

Definition 6.1.1. (i) *Let $\mathcal{C}$ be a $\mathcal{U}$-category. An* ind-object *in $\mathcal{C}$ is an object $A \in \mathcal{C}^\wedge$ which is isomorphic to "$\varinjlim$" α for some functor $\alpha \colon I \to \mathcal{C}$ with I filtrant and $\mathcal{U}$-small.*

(ii) *We denote by $\mathrm{Ind}^{\mathcal{U}}(\mathcal{C})$ (or simply $\mathrm{Ind}(\mathcal{C})$ if there is no risk of confusion) the full big subcategory of $\mathcal{C}^\wedge$ consisting of ind-objects, and call it the indization of $\mathcal{C}$. We denote by $\iota_{\mathcal{C}} \colon \mathcal{C} \to \mathrm{Ind}(\mathcal{C})$ the natural functor (induced by $\mathrm{h}_{\mathcal{C}}$).*

(iii) *Similarly, a* pro-object *in $\mathcal{C}$ is an object $B \in \mathcal{C}^\vee$ which is isomorphic to "$\varprojlim$" β for some functor $\beta \colon I^{\mathrm{op}} \to \mathcal{C}$ with I filtrant and small.*

(iv) *We denote by $\mathrm{Pro}^{\mathcal{U}}(\mathcal{C})$ (or simply $\mathrm{Pro}(\mathcal{C})$) the full big subcategory of $\mathcal{C}^\vee$ consisting of pro-objects.*

Lemma 6.1.2. *The categories $\mathrm{Ind}(\mathcal{C})$ and $\mathrm{Pro}(\mathcal{C})$ are $\mathcal{U}$-categories.*

Proof. It is enough to treat $\mathrm{Ind}(\mathcal{C})$. Let $A, B \in \mathrm{Ind}(\mathcal{C})$. We may assume that $A \simeq \underset{i \in I}{\text{"}\varinjlim\text{"}}\, \alpha(i)$ and $B \simeq \underset{j \in J}{\text{"}\varinjlim\text{"}}\, \beta(j)$ for small and filtrant categories I and J. In this case $\mathrm{Hom}_{\mathcal{C}}(A, B)$ is isomorphic to a small set by (2.6.4).   q.e.d.

We may replace "filtrant and small" by "filtrant and cofinally small" in the above definition.

There is an equivalence

$$\mathrm{Pro}(\mathcal{C}) \simeq (\mathrm{Ind}(\mathcal{C}^{\mathrm{op}}))^{\mathrm{op}} \ .$$

Hence, we may restrict our study to ind-objects.

Example 6.1.3. Let k be a field and let V denote an infinite-dimensional k-vector space. Consider the contravariant functor on $\mathrm{Mod}(k)$, $W \mapsto V \otimes \mathrm{Hom}_k(W, k)$. It defines an ind-object of $\mathrm{Mod}(k)$ which is not in $\mathrm{Mod}(k)$. Notice that this functor is isomorphic to the functor $V \mapsto \underset{V' \subset V}{\text{"}\varinjlim\text{"}}\, V'$ where V' ranges over the filtrant set of finite-dimensional vector subspaces of V.

Notation 6.1.4. We shall often denote by the capital letters A, B, C, etc. objects of $\mathcal{C}^{\wedge}$ and as usual by X, Y, Z objects of $\mathcal{C}$.

Recall that for $A \in \mathcal{C}^{\wedge}$, we introduced the category $\mathcal{C}_A$ and the forgetful functor $j_A \colon \mathcal{C}_A \to \mathcal{C}$, and proved the isomorphism $A \simeq \text{"}\varinjlim\text{"}\, j_A$ (see Proposition 2.6.3).

Proposition 6.1.5. *Let $A \in \mathcal{C}^{\wedge}$. Then $A \in \mathrm{Ind}(\mathcal{C})$ if and only if $\mathcal{C}_A$ is filtrant and cofinally small.*

Proof. This follows immediately from Proposition 2.6.3 and Proposition 3.2.2.   q.e.d.

Applying Definitions 3.3.1 and 3.3.14, we get:

Corollary 6.1.6. *The functor $\iota_{\mathcal{C}} \colon \mathcal{C} \to \mathrm{Ind}(\mathcal{C})$ is right exact and right small.*

Proposition 6.1.7. *Assume that a category $\mathcal{C}$ admits finite inductive limits. Then $\mathrm{Ind}(\mathcal{C})$ is the full subcategory of $\mathcal{C}^{\wedge}$ consisting of functors $A \colon \mathcal{C}^{\mathrm{op}} \to \mathbf{Set}$ such that A is left exact and $\mathcal{C}_A$ is cofinally small.*

Proof. Apply Propositions 3.3.13 and 6.1.5.   q.e.d.

Theorem 6.1.8. *Let $\mathcal{C}$ be a category. The category $\mathrm{Ind}(\mathcal{C})$ admits small filtrant inductive limits and the natural functor $\mathrm{Ind}(\mathcal{C}) \to \mathcal{C}^{\wedge}$ commutes with such limits.*

Similarly $\mathrm{Pro}(\mathcal{C})$ admits small filtrant projective limits and the natural functor $\mathrm{Pro}(\mathcal{C}) \to \mathcal{C}^{\vee}$ commutes with such limits.

Proof. Let $\alpha\colon I \to \mathrm{Ind}(\mathcal{C})$ be a functor with I small and filtrant and let $A = \text{``}\varinjlim\text{''}\,\alpha \in \mathcal{C}^{\wedge}$. It is enough to show that A belongs to $\mathrm{Ind}(\mathcal{C})$. We shall use Proposition 6.1.5.

(i) $\mathcal{C}_A$ is filtrant. By Lemma 3.1.2, it is enough to show that for any finite category J and any functor $\beta\colon J \to \mathcal{C}_A$, there exists $Z \in \mathcal{C}_A$ such that $\varprojlim \mathrm{Hom}_{\mathcal{C}_A}(\beta, Z) \neq \emptyset$. For any $X \in \mathcal{C}_A$, we have

$$\mathrm{Hom}_{(\mathcal{C}^{\wedge})_A}(X, A) \simeq \varinjlim_{i\in I} \mathrm{Hom}_{(\mathcal{C}^{\wedge})_A}(X, \alpha(i))$$

$$\simeq \varinjlim_{i\in I}\ \varinjlim_{Y\in\mathcal{C}_{\alpha(i)}}\ \mathrm{Hom}_{\mathcal{C}_A}(X, Y)\,.$$

Since I and $\mathcal{C}_{\alpha(i)}$ are filtrant, $\varinjlim\limits_{i\in I}$ and $\varinjlim\limits_{Y\in\mathcal{C}_{\alpha(i)}}$ commute with finite projective limits by Theorem 3.1.6. Hence, we obtain

$$\{\mathrm{pt}\} \simeq \varprojlim_{j\in J} \mathrm{Hom}_{(\mathcal{C}^{\wedge})_A}(\beta(j), A)$$

$$\simeq \varinjlim_{i\in I}\ \varinjlim_{Y\in\mathcal{C}_{\alpha(i)}}\ \varprojlim_{j\in J} \mathrm{Hom}_{\mathcal{C}_A}(\beta(j), Y)\,.$$

Hence, there exist $i \in I$ and $Y \in \mathcal{C}_{\alpha(i)}$ such that $\varprojlim \mathrm{Hom}_{\mathcal{C}_A}(\beta, Y) \neq \emptyset$.

(ii) $\mathcal{C}_A$ is cofinally small. By Proposition 3.2.6, for any $i \in I$, there exists a small subset S_i of $\mathrm{Ob}(\mathcal{C}_{\alpha(i)})$ such that for any $X \in \mathcal{C}_{\alpha(i)}$ there exists a morphism $X \to Y$ with $Y \in S_i$. Let $\varphi_i\colon \mathcal{C}_{\alpha(i)} \to \mathcal{C}_A$ be the canonical functor. Then $S = \bigcup_{i\in I} \varphi_i(S_i)$ is a small subset of $\mathrm{Ob}(\mathcal{C}_A)$ and for any $X \in \mathcal{C}_A$ there exists a morphism $X \to Y$ with $Y \in S$.   q.e.d.

Proposition 6.1.9. *Let $F\colon \mathcal{C} \to \mathcal{C}'$ be a functor. There exists a unique functor $IF : \mathrm{Ind}(\mathcal{C}) \to \mathrm{Ind}(\mathcal{C}')$ such that:*

(i) *the restriction of IF to $\mathcal{C}$ is F,*
(ii) *IF commutes with small filtrant inductive limits, that is, if $\alpha\colon I \to \mathrm{Ind}(\mathcal{C})$ is a functor with I small and filtrant, then we have*

$$IF(\text{``}\varinjlim\text{''}\,\alpha) \xrightarrow{\ \sim\ } \text{``}\varinjlim\text{''}(IF \circ \alpha)\,.$$

The proof goes as the one of Proposition 2.7.1 and we do not repeat it. The functor IF is given by

$$IF(A) = \text{``}\varinjlim_{(U\to A)\in\mathcal{C}_A}\text{''}\ F(U) \quad \text{for } A \in \mathrm{Ind}(\mathcal{C})\,.$$

Proposition 6.1.9 (i) may be visualized by the commutative diagram below:

$$
\begin{array}{ccc}
\mathcal{C} & \xrightarrow{\ F\ } & \mathcal{C}' \\
\downarrow & & \downarrow \\
\mathrm{Ind}(\mathcal{C}) & \xrightarrow{\ IF\ } & \mathrm{Ind}(\mathcal{C}')\,.
\end{array}
$$

Recall that if $A \simeq \text{``}\varinjlim_i\text{''}\, \alpha(i)$, $B \simeq \text{``}\varinjlim_j\text{''}\, \beta(j)$, then (see (2.6.4))

$$\text{Hom}_{\text{Ind}(\mathcal{C})}(A, B) \simeq \varprojlim_i \varinjlim_j \text{Hom}_{\mathcal{C}}(\alpha(i), \beta(j)) .$$

The map $IF\colon \text{Hom}_{\text{Ind}(\mathcal{C})}(A, B) \to \text{Hom}_{\text{Ind}(\mathcal{C}')}(IF(A), IF(B))$ is given by

(6.1.1) $\varprojlim_i \varinjlim_j \text{Hom}_{\mathcal{C}}(\alpha(i), \beta(j)) \to \varprojlim_i \varinjlim_j \text{Hom}_{\mathcal{C}'}(F(\alpha(i)), F(\beta(j))) .$

Remark that if $\mathcal{C}$ is small, the diagram below commutes.

$$
\begin{array}{ccc}
\text{Ind}(\mathcal{C}) & \xrightarrow{\ IF\ } & \text{Ind}(\mathcal{C}') \\
\downarrow & & \downarrow \\
\mathcal{C}^{\wedge} & \xrightarrow{\ \widehat{F}\ } & \mathcal{C}'^{\wedge} .
\end{array}
$$

(The functor $\widehat{F}$ is defined in Proposition 2.7.1 and Notation 2.7.2.)

Proposition 6.1.10. *Let $F\colon \mathcal{C} \to \mathcal{C}'$. If F is faithful (resp. fully faithful), so is IF.*

Proof. This follows from (6.1.1).   q.e.d.

Proposition 6.1.11. *Let $F\colon \mathcal{C} \to \mathcal{C}'$ and $G\colon \mathcal{C}' \to \mathcal{C}''$ be two functors. Then $I(G \circ F) \simeq IG \circ IF$.*

Proof. The proof is obvious.   q.e.d.

Let $\mathcal{C}$ and $\mathcal{C}'$ be two categories. By Proposition 6.1.9, the projection functors $\mathcal{C} \times \mathcal{C}' \to \mathcal{C}$ and $\mathcal{C} \times \mathcal{C}' \to \mathcal{C}'$ define the functor

(6.1.2) $\theta\colon \text{Ind}(\mathcal{C} \times \mathcal{C}') \to \text{Ind}(\mathcal{C}) \times \text{Ind}(\mathcal{C}')$

Proposition 6.1.12. *The functor θ in (6.1.2) is an equivalence.*

Proof. A quasi-inverse to θ is constructed as follows. To $A \in \text{Ind}(\mathcal{C})$ and $A' \in \text{Ind}(\mathcal{C}')$, associate $\varinjlim_{((X \to A),(X' \to A')) \in \mathcal{C}_A \times \mathcal{C}_{A'}} (X, X')$. Since $\mathcal{C}_A \times \mathcal{C}_{A'}$ is cofinally small and filtrant, it belongs to $\text{Ind}(\mathcal{C} \times \mathcal{C}')$.   q.e.d.

Proposition 6.1.13. *Let $\alpha\colon I \to \mathcal{C}$ and $\beta\colon J \to \mathcal{C}$ be functors with I and J small and filtrant. Let $f\colon \text{``}\varinjlim\text{''}\, \alpha \to \text{``}\varinjlim\text{''}\, \beta$ be a morphism in $\text{Ind}(\mathcal{C})$. Then there exist a small and filtrant category K, cofinal functors $p_I\colon K \to I$, $p_J\colon K \to J$ and a morphism of functors $\varphi\colon \alpha \circ p_I \to \beta \circ p_J$ making the diagram below commutative*

(6.1.3)

$$
\begin{array}{ccc}
\text{``}\varinjlim\text{''}(\alpha \circ p_I) & \xrightarrow{\text{``}\varinjlim\text{''}\varphi} & \text{``}\varinjlim\text{''}(\beta \circ p_J) \\
\Big\downarrow{\sim} & & \Big\downarrow{\sim} \\
\text{``}\varinjlim\text{''}\alpha & \xrightarrow{\quad f \quad} & \text{``}\varinjlim\text{''}\beta.
\end{array}
$$

Proof. Set $A = \text{``}\varinjlim\text{''}\alpha$, $B = \text{``}\varinjlim\text{''}\beta$, and denote by $\tilde{\alpha}\colon I \to \mathcal{C}_A$, $\tilde{\beta}\colon J \to \mathcal{C}_B$ and $\tilde{f}\colon \mathcal{C}_A \to \mathcal{C}_B$ the functors induced by α, β and f. Consider the category $K := M[I \xrightarrow{\tilde{f}\circ\tilde{\alpha}} \mathcal{C}_B \xleftarrow{\tilde{\beta}} J]$ (see Definition 3.4.1).

The functor $\tilde{\beta}$ is cofinal by Proposition 2.6.3 (ii), and the categories I and J are small and filtrant by the hypotheses. Proposition 3.4.5 then implies that the category K is filtrant, cofinally small and the projection functors p_I and p_J from K to I and J are cofinal.

We may identify K with the category whose objects are the triplets (i, j, g) of $i \in I$, $j \in J$ and $g\colon \alpha(i) \to \beta(j)$ such that the diagram below commutes

$$
\begin{array}{ccc}
\alpha(i) & \xrightarrow{\quad g \quad} & \beta(j) \\
\Big\downarrow & & \Big\downarrow \\
\text{``}\varinjlim\text{''}\alpha & \xrightarrow{\quad f \quad} & \text{``}\varinjlim\text{''}\beta,
\end{array}
$$

and the morphisms are the natural ones. Then g defines a morphism of functors $\varphi\colon \alpha \circ p_I \to \beta \circ p_J$ such that the diagram (6.1.3) commutes.   q.e.d.

Corollary 6.1.14. *Let $f\colon A \to B$ be a morphism in $\mathrm{Ind}(\mathcal{C})$. Then there exist a small and filtrant category I and a morphism $\varphi\colon \alpha \to \beta$ of functors from I to $\mathcal{C}$ such that $A \simeq \text{``}\varinjlim\text{''}\alpha$, $B \simeq \text{``}\varinjlim\text{''}\beta$ and $f = \text{``}\varinjlim\text{''}\varphi$.*

We shall extend this result to the case of a pair of parallel arrows. A more general statement for finite diagrams will be given in Sect. 6.4.

Corollary 6.1.15. *Let $f, g\colon A \rightrightarrows B$ be two morphisms in $\mathrm{Ind}(\mathcal{C})$. Then there exist a small and filtrant category I and morphisms $\varphi, \psi\colon \alpha \rightrightarrows \beta$ of functors from $I \to \mathcal{C}$ such that $A \simeq \text{``}\varinjlim\text{''}\alpha$, $B \simeq \text{``}\varinjlim\text{''}\beta$, $f = \text{``}\varinjlim\text{''}\varphi$ and $g = \text{``}\varinjlim\text{''}\psi$.*

Proof. Let I and J be small filtrant categories and let $\alpha\colon I \to \mathcal{C}$ and $\beta\colon J \to \mathcal{C}$ be two functors such that $A \simeq \text{``}\varinjlim\text{''}\alpha$ and $B \simeq \text{``}\varinjlim\text{''}\beta$. Denote by $\tilde{\alpha}\colon I \to \mathcal{C}\times\mathcal{C}$ the functor $i \mapsto \alpha(i)\times\alpha(i)$, and similarly with $\tilde{\beta}$. Then $(A, A) \simeq \text{``}\varinjlim\text{''}\tilde{\alpha}$ and $(B, B) \simeq \text{``}\varinjlim\text{''}\tilde{\beta}$.

By Proposition 6.1.12, the morphism $(f, g)\colon A \times A \to B \times B$ in $\mathrm{Ind}(\mathcal{C}) \times \mathrm{Ind}(\mathcal{C})$ defines a morphism in $\mathrm{Ind}(\mathcal{C} \times \mathcal{C})$. We still denote this morphism by (f, g) and apply Proposition 6.1.13. We find a small and filtrant category K, functors $p_I \colon K \to I$, $p_J \colon K \to J$ and a morphism of functors (φ, ψ) from $\tilde{\alpha} \circ p_I$ to $\tilde{\beta} \circ p_J$ such that $(f, g) = \text{"}\varinjlim\text{"}\,(\varphi, \psi)$. It follows that $f = \text{"}\varinjlim\text{"}\,\varphi$ and $g = \text{"}\varinjlim\text{"}\,\psi$.

q.e.d.

Proposition 6.1.16. (i) *Assume that for any pair of parallel arrows in $\mathcal{C}$, its kernel in $\mathcal{C}^{\wedge}$ belongs to $\mathrm{Ind}(\mathcal{C})$. Then, for any pair of parallel arrows in $\mathrm{Ind}(\mathcal{C})$, its kernel in $\mathcal{C}^{\wedge}$ is its kernel in $\mathrm{Ind}(\mathcal{C})$.*

(ii) *Let J be a small set and assume that the product in $\mathcal{C}^{\wedge}$ of any family indexed by J of objects of $\mathcal{C}$ belongs to $\mathrm{Ind}(\mathcal{C})$. Then, for any family indexed by J of objects of $\mathrm{Ind}(\mathcal{C})$, its product in $\mathcal{C}^{\wedge}$ is its product in $\mathrm{Ind}(\mathcal{C})$.*

Proof. (i) Let $f, g \colon A \rightrightarrows B$ be a pair of parallel arrows in $\mathrm{Ind}(\mathcal{C})$. With the notations of Corollary 6.1.14, we may assume that $A = \text{"}\varinjlim\text{"}\,\alpha$, $B = \text{"}\varinjlim\text{"}\,\beta$ and there exist morphisms of functors $\varphi, \psi \colon \alpha \rightrightarrows \beta$ such that $f = \text{"}\varinjlim\text{"}\,\varphi$ and $g = \text{"}\varinjlim\text{"}\,\psi$. Let γ denote the kernel of (φ, ψ). Then $\text{"}\varinjlim\text{"}\,\gamma$ is a kernel of (f, g) in $\mathcal{C}^{\wedge}$ and belongs to $\mathrm{Ind}(\mathcal{C})$.

(ii) Let $A_j \in \mathrm{Ind}(\mathcal{C})$, $j \in J$. For each $j \in J$, there exist a small and filtrant category I_j and a functor $\alpha_j \colon I_j \to \mathcal{C}$ such that $A_j \simeq \text{"}\varinjlim\text{"}\,\alpha_j$. Define the small filtrant category $K = \prod_{j \in J} I_j$ and denote by $\pi_j \colon K \to I_j$ the natural functor.

Using Corollary 3.1.12 we get the isomorphisms in $\mathcal{C}^{\wedge}$

$$\prod_{j \in J} A_j \simeq \prod_{j \in J} \text{"}\varinjlim_{i \in I_j}\text{"}\,\alpha_j(i) \simeq \text{"}\varinjlim_{k \in K}\text{"}\,\prod_{j \in J} \alpha_j(\pi_j(k))\,.$$

q.e.d.

Corollary 6.1.17. (i) *Assume that the category $\mathcal{C}$ admits finite projective limits. Then the category $\mathrm{Ind}(\mathcal{C})$ admits finite projective limits. Moreover, the natural functors $\mathcal{C} \to \mathrm{Ind}(\mathcal{C})$ and $\mathrm{Ind}(\mathcal{C}) \to \mathcal{C}^{\wedge}$ are left exact.*

(ii) *Assume that the category $\mathcal{C}$ admits small projective limits. Then the category $\mathrm{Ind}(\mathcal{C})$ admits small projective limits and the natural functors $\mathcal{C} \to \mathrm{Ind}(\mathcal{C})$ and $\mathrm{Ind}(\mathcal{C}) \to \mathcal{C}^{\wedge}$ commute with small projective limits.*

Proposition 6.1.18. (i) *Assume that the category $\mathcal{C}$ admits cokernels, that is, the cokernel of any pair of parallel arrows exists in $\mathcal{C}$. Then $\mathrm{Ind}(\mathcal{C})$ admits cokernels.*

(ii) *Assume that $\mathcal{C}$ admits finite coproducts. Then $\mathrm{Ind}(\mathcal{C})$ admits small coproducts.*

(iii) *Assume that the category $\mathcal{C}$ admits finite inductive limits. Then $\mathrm{Ind}(\mathcal{C})$ admits small inductive limits.*

Proof. (i) Let $f, g \colon A \rightrightarrows B$ be arrows in $\mathrm{Ind}(\mathcal{C})$. With the notations of Corollary 6.1.14, we may assume that $A = \text{"}\varinjlim\text{"}\,\alpha$, $B = \text{"}\varinjlim\text{"}\,\beta$ and there exist morphisms of functors $\varphi, \psi \colon \alpha \rightrightarrows \beta$ such that $f = \text{"}\varinjlim\text{"}\,\varphi$ and $g = \text{"}\varinjlim\text{"}\,\psi$. Let λ_i denote the cokernel of $(\alpha(i), \beta(i))$ and let $L \in \mathrm{Ind}(\mathcal{C})$.

Then $\mathrm{Hom}_{\mathcal{C}^\wedge}(\lambda(i), L)$ is the kernel of $\mathrm{Hom}_{\mathcal{C}^\wedge}(\beta(i), L) \rightrightarrows \mathrm{Hom}_{\mathcal{C}^\wedge}(\alpha(i), L)$. Applying the left exact functor $\varprojlim$, we conclude that "$\varinjlim$" λ is a cokernel of ("$\varinjlim$" φ, "$\varinjlim$" ψ).

(ii) The proof that $\mathrm{Ind}(\mathcal{C})$ admits finite coproducts is similar to the proof in (i). The general case follows by Lemma 3.2.9.

(iii) follows from (i), (ii) and the same lemma.   q.e.d.

Recall that if $\mathcal{C}$ admits cokernels (resp. finite coproducts, resp. finite inductive limits), then the functor $\iota_{\mathcal{C}} \colon \mathcal{C} \to \mathrm{Ind}(\mathcal{C})$ commutes with such limits by Corollary 6.1.6 and Proposition 3.3.2.

Proposition 6.1.19. *Assume that $\mathcal{C}$ admits finite inductive limits and finite projective limits. Then small filtrant inductive limits are exact in $\mathrm{Ind}(\mathcal{C})$.*

Proof. It is enough to check that small filtrant inductive limits commute with finite projective limits in $\mathrm{Ind}(\mathcal{C})$. Since the embedding $\mathrm{Ind}(\mathcal{C}) \to \mathcal{C}^\wedge$ commutes with small filtrant inductive limits and with finite projective limits, this follows from the fact that small filtrant inductive limits are exact in $\mathcal{C}^\wedge$ (see Exercise 3.2).   q.e.d.

Remark 6.1.20. (i) The natural functor $\mathrm{Ind}(\mathcal{C}) \to \mathcal{C}^\wedge$ commutes with filtrant inductive limits (Theorem 6.1.8), but it does not commute with inductive limits in general. Indeed, it does not commute with finite coproducts (see Exercise 6.3). Hence, when writing "$\varinjlim$" for an inductive system indexed by a non filtrant category I, the limit should be understood in $\mathcal{C}^\wedge$.
(ii) If $\mathcal{C}$ admits finite inductive limits, then $\mathrm{Ind}(\mathcal{C})$ admits small inductive limits and $\iota_{\mathcal{C}} \colon \mathcal{C} \to \mathrm{Ind}(\mathcal{C})$ commutes with finite inductive limits (Corollary 6.1.6 and Proposition 6.1.18) but if $\mathcal{C}$ admits small filtrant inductive limits, $\iota_{\mathcal{C}}$ does not commute with such limits in general. We may summarize these properties by the table below. Here, "o" means that the functors commute, and "×" they do not.

	$\mathcal{C} \to \mathrm{Ind}(\mathcal{C})$	$\mathrm{Ind}(\mathcal{C}) \to \mathcal{C}^\wedge$
finite inductive limits	o	×
finite coproducts	o	×
small filtrant inductive limits	×	o
small coproducts	×	×
small inductive limits	×	×

Since the definition of $\mathrm{Ind}(\mathcal{C})$ makes use of the notion of being small, it depends on the choice of the universe. However, the result below tells us that when replacing a universe $\mathcal{U}$ with a bigger one $\mathcal{V}$, the category of ind-objects of $\mathcal{C}$ in $\mathcal{U}$ is a full subcategory of that of ind-objects of $\mathcal{C}$ in $\mathcal{V}$.

More precisely, consider two universes $\mathcal{U}$ and $\mathcal{V}$ with $\mathcal{U} \subset \mathcal{V}$, and let $\mathcal{C}$ denote a $\mathcal{U}$-category.

Proposition 6.1.21. *The natural functor* $\mathrm{Ind}^{\mathcal{U}}(\mathcal{C}) \to \mathrm{Ind}^{\mathcal{V}}(\mathcal{C})$ *is fully faithful. If $\mathcal{C}$ admits finite inductive limits, then this functor commutes with $\mathcal{U}$-small inductive limits. If $\mathcal{C}$ admits finite (resp. $\mathcal{U}$-small) projective limits, then this functor commutes with such projective limits.*

Proof. The first statement follows from isomorphisms (2.6.4). The functor $\mathrm{Ind}^{\mathcal{U}}(\mathcal{C}) \to \mathrm{Ind}^{\mathcal{V}}(\mathcal{C})$ commutes with finite inductive limits as seen in the proof of Proposition 6.1.18. Since it commutes with $\mathcal{U}$-small filtrant inductive limits, it commutes with $\mathcal{U}$-small inductive limits. Recall that the natural functor $\mathcal{C}_{\mathcal{U}}^{\wedge} \to \mathcal{C}_{\mathcal{V}}^{\wedge}$ commutes with $\mathcal{U}$-small projective limits (see Remark 2.6.5). Then the functor $\mathrm{Ind}^{\mathcal{U}}(\mathcal{C}) \to \mathrm{Ind}^{\mathcal{V}}(\mathcal{C})$ commutes with finite (resp. $\mathcal{U}$-small projective) limits by Proposition 6.1.16 if $\mathcal{C}$ admits such limits.   q.e.d.

6.2 Representable Ind-limits

Let $\alpha \colon I \to \mathcal{C}$ be a functor with I small and filtrant. We shall study under which conditions the functor "$\varinjlim$" is representable in $\mathcal{C}$.

For each $i \in I$, let us denote by $\rho_i \colon \alpha(i) \to$ "$\varinjlim$" α the natural functor. It satisfies

$$(6.2.1) \qquad \rho_j \circ \alpha(s) = \rho_i \quad \text{for any } s \colon i \to j \,.$$

Proposition 6.2.1. *Let $\alpha \colon I \to \mathcal{C}$ be a functor with I small and filtrant and let $Z \in \mathcal{C}$. The conditions below are equivalent:*

(i) "$\varinjlim$" α *is representable by Z,*

(ii) *there exist an $i_0 \in I$ and a morphism $\tau_0 \colon Z \to \alpha(i_0)$ satisfying the property: for any morphism $s \colon i_0 \to i$, there exist a morphism $g \colon \alpha(i) \to Z$ and a morphism $t \colon i \to j$ satisfying*
    (a) $g \circ \alpha(s) \circ \tau_0 = \mathrm{id}_Z$,
    (b) $\alpha(t) \circ \alpha(s) \circ \tau_0 \circ g = \alpha(t)$.

Proof. (i) $\Rightarrow$ (ii)   Let $\varphi \colon Z \xrightarrow{\sim}$ "$\varinjlim$" α be an isomorphism. Since we have $\mathrm{Hom}_{\mathrm{Ind}(\mathcal{C})}(Z, Z) \simeq \varinjlim_i \mathrm{Hom}_{\mathcal{C}}(Z, \alpha(i))$, there exist $i_0 \in I$ and $\tau_0 \colon Z \to \alpha(i_0)$ such that $\varphi = \rho_{i_0} \circ \tau_0$. For any $i \in I$, the chain of morphisms $\alpha(i) \to$ "$\varinjlim$" $\alpha \xleftarrow{\sim} Z$ defines a morphism $g_i \colon \alpha(i) \to Z$ with $\varphi \circ g_i = \rho_i$. Hence, for any $s \colon i_0 \to i$, we have

$$\varphi \circ g_i \circ \alpha(s) \circ \tau_0 = \rho_i \circ \alpha(s) \circ \tau_0 = \rho_{i_0} \circ \tau_0 = \varphi \,.$$

This shows (ii)-(a). Since I is filtrant and

$$\rho_i \circ \mathrm{id}_{\alpha(i)} = \rho_i = \varphi \circ g_i = \rho_i \circ \alpha(s) \circ \tau_0 \circ g_i ,$$

there exists $t : i \to j$ satisfying $\alpha(t) \circ \mathrm{id}_{\alpha(i)} = \alpha(t) \circ (\alpha(s) \circ \tau_0 \circ g_i)$. This is visualized by the diagram

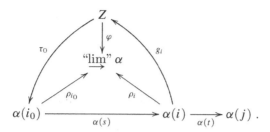

(ii) $\Rightarrow$ (i)   The morphism $\tau_0 : Z \to \alpha(i_0)$ defines the morphism

$$\varphi = \rho_{i_0} \circ \tau_0 : Z \to \text{``$\varinjlim$''} \, \alpha .$$

To prove that φ is an isomorphism, it is enough to check that φ induces an isomorphism

$$\varphi_X : \mathrm{Hom}_{\mathcal{C}}(X, Z) \xrightarrow{\sim} \varinjlim_i \mathrm{Hom}_{\mathcal{C}}(X, \alpha(i)) \quad \text{for any } X \in \mathcal{C} .$$

Injectivity of φ_X. Let $u, v \in \mathrm{Hom}_{\mathcal{C}}(X, Z)$ with $\varphi_X(u) = \varphi_X(v)$. There exists $s : i_0 \to i$ such that $\alpha(s) \circ \tau_0 \circ u = \alpha(s) \circ \tau_0 \circ v$. Then for $g \in \mathrm{Hom}_{\mathcal{C}}(\alpha(i), Z)$ as in (ii),

$$u = g \circ \alpha(s) \circ \tau_0 \circ u = g \circ \alpha(s) \circ \tau_0 \circ v = v .$$

Surjectivity of φ_X. Let $w \in \mathrm{Hom}_{\mathcal{C}}(X, \alpha(i))$ and let $s : i_0 \to i$. Take $g : \alpha(i) \to Z$ and $t : i \to j$ as in (ii). Then

$$\alpha(t) \circ w = \alpha(t) \circ \alpha(s) \circ \tau_0 \circ g \circ w .$$

The image of w in $\varinjlim_j \mathrm{Hom}_{\mathcal{C}}(X, \alpha(j))$ is $\varphi_X(g \circ w)$.   q.e.d.

6.3 Indization of Categories Admitting Inductive Limits

In this section we shall study $\mathrm{Ind}(\mathcal{C})$ in the case where $\mathcal{C}$ admits small filtrant inductive limits. Recall that $\iota_{\mathcal{C}} : \mathcal{C} \to \mathrm{Ind}(\mathcal{C})$ denotes the natural embedding functor.

Proposition 6.3.1. *Assume that $\mathcal{C}$ admits small filtrant inductive limits.*

(i) *The functor $\iota_{\mathcal{C}} : \mathcal{C} \to \mathrm{Ind}(\mathcal{C})$ admits a left adjoint $\sigma_{\mathcal{C}} : \mathrm{Ind}(\mathcal{C}) \to \mathcal{C}$, and if $A \simeq \text{``$\varinjlim$''} \, \alpha$, then $\sigma_{\mathcal{C}}(A) \simeq \varinjlim \alpha$.*

(ii) *We have $\sigma_{\mathcal{C}} \circ \iota_{\mathcal{C}} \simeq \mathrm{id}_{\mathcal{C}}$.*

Proof. (i) Let $A \in \mathrm{Ind}(\mathcal{C})$ and let us show that the functor

$$\mathcal{C} \ni X \mapsto \mathrm{Hom}_{\mathrm{Ind}(\mathcal{C})}(A, \iota_{\mathcal{C}}(X))$$

is representable. Let $\alpha \colon I \to \mathcal{C}$ be a functor with I small and filtrant such that $A \simeq \text{"}\varinjlim\text{"} \alpha$. Then

$$\mathrm{Hom}_{\mathrm{Ind}(\mathcal{C})}(\text{"}\varinjlim\text{"} \alpha, \iota_{\mathcal{C}}(X)) \simeq \varprojlim_{i} \mathrm{Hom}_{\mathcal{C}}(\alpha(i), X)$$
$$\simeq \mathrm{Hom}_{\mathcal{C}}(\varinjlim \alpha, X) \,.$$

(ii) is obvious.   q.e.d.

Corollary 6.3.2. *Assume that $\mathcal{C}$ admits small filtrant inductive limits. Then for any functor $F \colon \mathcal{J} \to \mathcal{C}$ there exists a unique (up to unique isomorphism) functor $JF \colon \mathrm{Ind}(\mathcal{J}) \to \mathcal{C}$ such that JF commutes with small filtrant inductive limits and the composition $\mathcal{J} \to \mathrm{Ind}(\mathcal{J}) \to \mathcal{C}$ is isomorphic to F.*

Indeed, JF is given by the composition $\mathrm{Ind}(\mathcal{J}) \xrightarrow{IF} \mathrm{Ind}(\mathcal{C}) \xrightarrow{\sigma_{\mathcal{C}}} \mathcal{C}$.
    The next definition will be generalized in Definition 9.2.7.

Definition 6.3.3. *Assume that $\mathcal{C}$ admits small filtrant inductive limits. We say that an object X of $\mathcal{C}$ is of finite presentation if for any $\alpha \colon I \to \mathcal{C}$ with I small and filtrant, the natural morphism $\varinjlim \mathrm{Hom}_{\mathcal{C}}(X, \alpha) \to \mathrm{Hom}_{\mathcal{C}}(X, \varinjlim \alpha)$*

is an isomorphism, that is, if $\mathrm{Hom}_{\mathrm{Ind}(\mathcal{C})}(X, A) \to \mathrm{Hom}_{\mathcal{C}}(X, \sigma_{\mathcal{C}}(A))$ is an isomorphism for any $A \in \mathrm{Ind}(\mathcal{C})$.

Some authors use the term "compact" instead of "of finite presentation".
    Note that any object of a category $\mathcal{C}$ is of finite presentation in $\mathrm{Ind}(\mathcal{C})$.

Proposition 6.3.4. *Let $F \colon \mathcal{J} \to \mathcal{C}$ be a functor and assume:*

(i) *$\mathcal{C}$ admits small filtrant inductive limits,*
(ii) *F is fully faithful,*
(iii) *for any $Y \in \mathcal{J}$, $F(Y)$ is of finite presentation.*

Then $JF \colon \mathrm{Ind}(\mathcal{J}) \to \mathcal{C}$ is fully faithful.

Proof. Let $\alpha \colon I \to \mathcal{J}$ and $\beta \colon J \to \mathcal{J}$ be two functors with I and J both small and filtrant. Using the hypothesis that $F(\beta(j))$ is of finite presentation for any $j \in J$, we get the chain of isomorphisms

$$\mathrm{Hom}_{\mathrm{Ind}(\mathcal{J})}(\text{"}\varinjlim_{j}\text{"} \beta(j), \text{"}\varinjlim_{i}\text{"} \alpha(i)) \simeq \varprojlim_{j} \varinjlim_{i} \mathrm{Hom}_{\mathcal{J}}(\beta(j), \alpha(i))$$
$$\simeq \varprojlim_{j} \varinjlim_{i} \mathrm{Hom}_{\mathcal{C}}(F(\beta(j)), F(\alpha(i)))$$
$$\simeq \varprojlim_{j} \mathrm{Hom}_{\mathcal{C}}(F(\beta(j)), \varinjlim_{i} F(\alpha(i)))$$
$$\simeq \mathrm{Hom}_{\mathcal{C}}(\varinjlim_{j} F(\beta(j)), \varinjlim_{i} F(\alpha(i)))$$
$$\simeq \mathrm{Hom}_{\mathcal{C}}(JF(\text{"}\varinjlim_{j}\text{"} \beta(j)), JF(\text{"}\varinjlim_{i}\text{"} \alpha(i))) \,.$$
                                                                                          q.e.d.

Let $\mathcal{C}$ be a category which admits small filtrant inductive limits. We denote by $\mathcal{C}^{\mathrm{fp}}$ the full subcategory of $\mathcal{C}$ consisting of objects of finite presentation and by $\rho: \mathcal{C}^{\mathrm{fp}} \to \mathcal{C}$ the natural functor. The functor ρ induces a fully faithful functor $I\rho: \mathrm{Ind}(\mathcal{C}^{\mathrm{fp}}) \to \mathrm{Ind}(\mathcal{C})$ and we have the diagram of functors

(6.3.1)

$$
\begin{array}{ccc}
\mathcal{C}^{\mathrm{fp}} & \xrightarrow{\ \rho\ } & \mathcal{C} \\
{\scriptstyle \iota_{\mathcal{C}}}\downarrow & {\scriptstyle J\rho}\nearrow \ {\scriptstyle \iota_{\mathcal{C}}}\downarrow\ \uparrow{\scriptstyle \sigma_{\mathcal{C}}} & \\
\mathrm{Ind}(\mathcal{C}^{\mathrm{fp}}) & \xrightarrow[\ I\rho\ ]{} & \mathrm{Ind}(\mathcal{C}).
\end{array}
$$

Note that the functors $J\rho$ and $I\rho$ are fully faithful. Also note that the diagram (6.3.1) is not commutative in general. More precisely:

(6.3.2) $$\iota_{\mathcal{C}} \circ J\rho \neq I\rho$$

in general (see Exercise 6.6).

Corollary 6.3.5. *Let $\mathcal{C}$ be a category admitting small filtrant inductive limits and assume that any object of $\mathcal{C}$ is a small filtrant inductive limit of objects of finite presentation. Then the functor $J\rho: \mathrm{Ind}(\mathcal{C}^{\mathrm{fp}}) \to \mathcal{C}$ is an equivalence of categories.*

Indeed, the functor $J\rho$ is fully faithful by Proposition 6.3.4 and is essentially surjective by the hypothesis.

A related result to Corollary 6.3.5 will be given in Proposition 9.2.19 below in the framework of π-accessible objects.

Examples 6.3.6. (i) There are equivalences $\mathbf{Set}^f \simeq (\mathbf{Set})^{\mathrm{fp}}$ and $\mathrm{Ind}(\mathbf{Set}^f) \simeq \mathbf{Set}$.
(ii) There are equivalences $\mathrm{Mod}^{\mathrm{fp}}(R) \simeq (\mathrm{Mod}(R))^{\mathrm{fp}}$ and $\mathrm{Ind}(\mathrm{Mod}^{\mathrm{fp}}(R)) \simeq \mathrm{Mod}(R)$ for any ring R (see Exercise 6.8).

Corollary 6.3.7. *In the situation of Corollary 6.3.5, the functor $\sigma_{\mathcal{C}}$ admits a left adjoint $\kappa_{\mathcal{C}}: \mathcal{C} \to \mathrm{Ind}(\mathcal{C})$. Moreover:*

(i) *If $\xi: I \to \mathcal{C}^{\mathrm{fp}}$ is a functor with I small and filtrant and $X \simeq \varinjlim \xi$ in $\mathcal{C}$, then $\kappa_{\mathcal{C}}(X) \simeq \text{"}\varinjlim\text{"}\, \rho \circ \xi$,*

(ii) *we have $\sigma_{\mathcal{C}} \circ \kappa_{\mathcal{C}} \simeq \mathrm{id}$,*
(iii) *$\kappa_{\mathcal{C}}$ is fully faithful.*

If there is no risk of confusion, we shall write κ instead of $\kappa_{\mathcal{C}}$.

Proof. (i) Denote by κ' a quasi-inverse of $J\rho$ and set $\kappa = I\rho \circ \kappa'$.
Let $X \in \mathcal{C}$ and let us show that the functor

$$\mathrm{Ind}(\mathcal{C}) \ni A \mapsto \mathrm{Hom}_{\mathcal{C}}(X, \sigma_{\mathcal{C}}(A))$$

is representable by $\kappa(X)$. In the sequel we shall not write $I\rho$ for short.

There exists $\xi \colon J \to \mathcal{C}^{\mathrm{fp}}$ with J small and filtrant such that $X \simeq \varinjlim \xi$. Then $\kappa(X) \simeq \text{``}\varinjlim\text{''}\, \xi$. We get the chain of isomorphisms

$$\mathrm{Hom}_{\mathrm{Ind}(\mathcal{C})}(\kappa(X), A) \simeq \mathrm{Hom}_{\mathrm{Ind}(\mathcal{C})}(\text{``}\varinjlim_{j}\text{''}\, \xi(j), A)$$

$$\simeq \varprojlim_{j} \mathrm{Hom}_{\mathrm{Ind}(\mathcal{C})}(\xi(j), A)$$

$$\simeq \varprojlim_{j} \mathrm{Hom}_{\mathcal{C}}(\xi(j), \sigma_{\mathcal{C}}(A))$$

$$\simeq \mathrm{Hom}_{\mathcal{C}}(\varinjlim_{j} \xi(j), \sigma_{\mathcal{C}}(A))$$

$$\simeq \mathrm{Hom}_{\mathcal{C}}(X, \sigma_{\mathcal{C}}(A)) \, .$$

The other assertions are obvious.   q.e.d.

6.4 Finite Diagrams in Ind($\mathcal{C}$)

Let K be a small category. The canonical functor $\mathcal{C} \to \mathrm{Ind}(\mathcal{C})$ defines the functor

(6.4.1) $\Phi_0 \colon \mathrm{Fct}(K, \mathcal{C}) \to \mathrm{Fct}(K, \mathrm{Ind}(\mathcal{C}))$.

Since $\mathrm{Fct}(K, \mathrm{Ind}(\mathcal{C}))$ admits small filtrant inductive limits, we may apply Corollary 6.3.2, and extend the functor Φ_0 to a functor

(6.4.2) $\Phi \colon \mathrm{Ind}(\mathrm{Fct}(K, \mathcal{C})) \to \mathrm{Fct}(K, \mathrm{Ind}(\mathcal{C}))$

which commutes with small filtrant inductive limits.

Proposition 6.4.1. *Assume that K is a finite category. Then the functor Φ in (6.4.2) is fully faithful.*

Proof. We shall apply Proposition 6.3.4 to Φ_0. Clearly, the functor Φ_0 is fully faithful and $\mathrm{Fct}(K, \mathrm{Ind}(\mathcal{C}))$ admits small filtrant inductive limits. Hence, it remains to check that given a small and filtrant category I, a functor $\alpha \colon I \to \mathrm{Fct}(K, \mathrm{Ind}(\mathcal{C}))$ and an object $\psi \in \mathrm{Fct}(K, \mathcal{C})$, the map

(6.4.3) $\varinjlim_{i} \mathrm{Hom}_{\mathrm{Fct}(K, \mathrm{Ind}(\mathcal{C}))}(\psi, \alpha(i)) \to \mathrm{Hom}_{\mathrm{Fct}(K, \mathrm{Ind}(\mathcal{C}))}(\psi, \varinjlim_{i} \alpha(i))$

is bijective. This follows from Lemma 2.1.15 and the chain of isomorphisms

$$\varinjlim_{i} \mathrm{Hom}_{\mathrm{Fct}(K,\mathrm{Ind}(\mathcal{C}))}(\psi, \alpha(i)) \simeq \varinjlim_{i} \varprojlim_{(a \to b) \in \mathrm{Mor}_0(K)} \mathrm{Hom}_{\mathrm{Ind}(\mathcal{C})}(\psi(a), \alpha(i)(b))$$

$$\simeq \varprojlim_{(a \to b) \in \mathrm{Mor}_0(K)} \varinjlim_{i} \mathrm{Hom}_{\mathrm{Ind}(\mathcal{C})}(\psi(a), \alpha(i)(b))$$

$$\simeq \varprojlim_{(a \to b) \in \mathrm{Mor}_0(K)} \mathrm{Hom}_{\mathrm{Ind}(\mathcal{C})}(\psi(a), \varinjlim_{i} \alpha(i)(b))$$

$$\simeq \mathrm{Hom}_{\mathrm{Fct}(K,\mathrm{Ind}(\mathcal{C}))}(\psi, \varinjlim_{i} \alpha(i)) \ .$$

Here, we have used the fact that in the category **Set**, small filtrant inductive limits commute with finite projective limits (Theorem 3.1.6).   q.e.d.

We shall give a condition in order that the functor Φ in (6.4.2) is an equivalence. We need some preparation.

Consider the category $M[\mathcal{C}_1 \xrightarrow{F} \mathcal{C}_0 \xleftarrow{G} \mathcal{C}_2]$ associated with functors $\mathcal{C}_1 \xrightarrow{F} \mathcal{C}_0 \xleftarrow{G} \mathcal{C}_2$ (see Definition 3.4.1). We set for short:

$$M_0 = M[\mathcal{C}_1 \to \mathcal{C}_0 \leftarrow \mathcal{C}_2] \ ,$$
$$M_1 = M[\mathrm{Ind}(\mathcal{C}_1) \to \mathrm{Ind}(\mathcal{C}_0) \leftarrow \mathrm{Ind}(\mathcal{C}_2)] \ .$$

Then M_1 admits small filtrant inductive limits, and by Proposition 3.4.2 there is a canonical fully faithful functor $M_0 \to M_1$ which thus extend to a functor

(6.4.4) $$\Psi : \mathrm{Ind}(M_0) \to M_1$$

commuting with small filtrant inductive limits.

Proposition 6.4.2. *The functor Ψ in (6.4.4) is an equivalence of categories.*

Proof. (i) Ψ is fully faithful. Since Ψ commutes with small filtrant inductive limits, it is enough to show that for $X \in M_0$ and a small filtrant inductive system $\{Y_i\}_{i \in I}$ in M_0, we have

(6.4.5) $$\varinjlim_{i} \mathrm{Hom}_{M_0}(X, Y_i) \xrightarrow{\sim} \mathrm{Hom}_{M_1}(X, \varinjlim_{i} \Psi(Y_i)) \ .$$

Let us write $X = (X_1, X_2, u)$ with $X_\nu \in \mathcal{C}_\nu$ $(\nu = 1, 2)$, $u \colon F(X_1) \to G(X_2)$, and let $Y_i = (Y_1^i, Y_2^i, v_i)$ with $Y_\nu^i \in \mathcal{C}_\nu$, $v_i \colon F(Y_1^i) \to G(Y_2^i)$.

Define the morphisms

$$\alpha_i : \mathrm{Hom}_{\mathcal{C}_1}(X_1, Y_1^i) \to \mathrm{Hom}_{\mathcal{C}_0}(F(X_1), G(Y_2^i))$$

$$f \mapsto (F(X_1) \xrightarrow{F(f)} F(Y_1^i) \xrightarrow{v_i} G(Y_2^i)) \ ,$$

$$\beta_i : \mathrm{Hom}_{\mathcal{C}_2}(X_2, Y_2^i) \to \mathrm{Hom}_{\mathcal{C}_0}(F(X_1), G(Y_2^i))$$

$$g \mapsto (F(X_1) \xrightarrow{u} G(X_2) \xrightarrow{G(g)} G(Y_2^i)) \ .$$

Then

$$\mathrm{Hom}_{M_0}(X, Y_i) = \mathrm{Hom}_{\mathcal{C}_1}(X_1, Y_1^i) \times_{\mathrm{Hom}_{\mathcal{C}_0}(F(X_1), G(Y_2^i))} \mathrm{Hom}_{\mathcal{C}_2}(X_2, Y_2^i) \,.$$

Since filtrant inductive limits commute with fiber products, we have

$$\mathrm{Hom}_{M_1}(X, \text{“}\varinjlim_i\text{”} \, Y_i)$$

$$\simeq \mathrm{Hom}_{\mathrm{Ind}(\mathcal{C}_1)}(X_1, \text{“}\varinjlim_i\text{”} \, Y_1^i)$$

$$\times_{\mathrm{Hom}_{\mathrm{Ind}(\mathcal{C}_0)}(F(X_1), \text{“}\varinjlim_i\text{”} \, G(Y_2^i))} \mathrm{Hom}_{\mathrm{Ind}(\mathcal{C}_2)}(X_2, \text{“}\varinjlim_i\text{”} \, Y_2^i)$$

$$\simeq \varinjlim_i \left(\mathrm{Hom}_{\mathcal{C}_1}(X_1, Y_1^i) \times_{\mathrm{Hom}_{\mathcal{C}_0}(F(X_1), G(Y_2^i))} \mathrm{Hom}_{\mathcal{C}_2}(X_2, Y_2^i) \right)$$

$$\simeq \varinjlim_i \mathrm{Hom}_{M_0}(X, Y_i) \,.$$

(ii) Ψ is essentially surjective. Let $(X_1, X_2, u) \in M_1$ with $X_1 = \text{“}\varinjlim_{i \in I}\text{”} X_1^i$, $X_2 = \text{“}\varinjlim_{j \in J}\text{”} X_2^j$, and $u \colon \text{“}\varinjlim_i\text{”} F(X_1^i) \to \text{“}\varinjlim_j\text{”} G(X_2^j)$. By Proposition 6.1.13 there exist a filtrant category K, cofinal functors $p_I \colon K \to I$ and $p_J \colon K \to J$ and a morphism of functors $v = \{v_k\}_{k \in K}$, $v_k \colon F(X_1^{p_I(k)}) \to G(X_2^{p_J(k)})$ such that $\text{“}\varinjlim_k\text{”} v_k = u$. Define $Z^k = (X_1^{p_I(k)}, X_2^{p_J(k)}, v_k)$. Then $Z^k \in M_0$ and $\Psi(\text{“}\varinjlim_k\text{”} Z^k) \simeq (X_1, X_2, u)$.   q.e.d.

Theorem 6.4.3. *Let K be a finite category such that $\mathrm{Hom}_K(a, a) = \{\mathrm{id}_a\}$ for any $a \in K$. Then the natural functor Φ in (6.4.2) is an equivalence.*

Proof. We may assume from the beginning that if two objects in K are isomorphic, then they are identical. Then $\mathrm{Ob}(K)$ has a structure of an ordered set as follows: $a \leq b$ if and only if $\mathrm{Hom}_K(a, b) \neq \emptyset$.

Indeed, if $a \leq b$ and $b \leq a$, then there are morphisms $u \colon a \to b$ and $v \colon b \to a$. Since $v \circ u = \mathrm{id}_a$ and $u \circ v = \mathrm{id}_b$, a and b are isomorphic, hence $a = b$.

We shall prove the result by induction on the cardinal of $\mathrm{Ob}(K)$. If this number is zero, the result is obvious. Otherwise, take a maximal element a of $\mathrm{Ob}(K)$. Then $\mathrm{Hom}_K(a, b) = \emptyset$ for any $b \neq a$. Denote by L the full subcategory of K such that $\mathrm{Ob}(L) = \mathrm{Ob}(K) \setminus \{a\}$ and denote by L_a the category of arrows $b \to a$, with $b \in L$. There is a natural functor $F \colon \mathrm{Fct}(L, \mathcal{C}) \to \mathrm{Fct}(L_a, \mathcal{C})$ associated with $L_a \to L$ and a natural functor $G \colon \mathcal{C} \simeq \mathrm{Fct}(\mathbf{Pt}, \mathcal{C}) \to \mathrm{Fct}(L_a, \mathcal{C})$ associated with the constant functor $L_a \to \mathbf{Pt}$.

There is an equivalence

$$(6.4.6) \qquad \mathrm{Fct}(K, \mathcal{C}) \simeq M[\mathrm{Fct}(L, \mathcal{C}) \xrightarrow{F} \mathrm{Fct}(L_a, \mathcal{C}) \xleftarrow{G} \mathcal{C}] \,.$$

Replacing $\mathcal{C}$ with $\mathrm{Ind}(\mathcal{C})$ and applying Proposition 6.4.2 we get the equivalences

(6.4.7) $\mathrm{Fct}(K, \mathrm{Ind}(\mathcal{C})) \simeq$

$$M[\mathrm{Fct}(L, \mathrm{Ind}(\mathcal{C})) \xrightarrow{IF} \mathrm{Fct}(L_a, \mathrm{Ind}(\mathcal{C})) \xleftarrow{IG} \mathrm{Ind}(\mathcal{C})] ,$$

(6.4.8) $\mathrm{Ind}(\mathrm{Fct}(K, \mathcal{C})) \simeq$

$$M[\mathrm{Ind}(\mathrm{Fct}(L, \mathcal{C})) \xrightarrow{IF} \mathrm{Ind}(\mathrm{Fct}(L_a, \mathcal{C})) \xleftarrow{IG} \mathrm{Ind}(\mathcal{C})] .$$

Consider the diagram

$$
\begin{array}{ccccc}
\mathrm{Ind}(\mathrm{Fct}(L, \mathcal{C})) & \longrightarrow & \mathrm{Ind}(\mathrm{Fct}(L_a, \mathcal{C})) & \longleftarrow & \mathrm{Ind}(\mathcal{C}) \\
\theta_1 \downarrow & & \theta_0 \downarrow & & \mathrm{id}_{\mathrm{Ind}(\mathcal{C})} \downarrow \\
\mathrm{Fct}(L, \mathrm{Ind}(\mathcal{C})) & \longrightarrow & \mathrm{Fct}(L_a, \mathrm{Ind}(\mathcal{C})) & \longleftarrow & \mathrm{Ind}(\mathcal{C}) .
\end{array}
$$

By the induction hypothesis θ_1 is an equivalence, and by Proposition 6.4.1, θ_0 is fully faithful. It follows that

$$\theta : M[\mathrm{Ind}(\mathrm{Fct}(L, \mathcal{C})) \to \mathrm{Ind}(\mathrm{Fct}(L_a, \mathcal{C})) \leftarrow \mathrm{Ind}(\mathcal{C})]$$
$$\longrightarrow M[\mathrm{Fct}(L, \mathrm{Ind}(\mathcal{C})) \to \mathrm{Fct}(L_a, \mathrm{Ind}(\mathcal{C})) \leftarrow \mathrm{Ind}(\mathcal{C})]$$

is an equivalence of categories by Proposition 3.4.2. The left hand side is equivalent to $\mathrm{Ind}(\mathrm{Fct}(K, \mathcal{C}))$ by (6.4.8), and the right hand side is equivalent to $\mathrm{Fct}(K, \mathrm{Ind}(\mathcal{C}))$ by (6.4.7).   q.e.d.

Corollary 6.4.4. *For any category* $\mathcal{C}$, *the natural functor* $\mathrm{Ind}(\mathrm{Mor}(\mathcal{C})) \to \mathrm{Mor}(\mathrm{Ind}(\mathcal{C}))$ *is an equivalence.*

Proof. Apply Theorem 6.4.3 by taking as K the category $\bullet \to \bullet$.   q.e.d.

Exercises

Exercise 6.1. (i) Let $\mathcal{C}$ be a small category and let $A \in \mathrm{Ind}(\mathcal{C})$. Prove that the two conditions below are equivalent.

(a) The functor $\mathrm{Hom}_{\mathrm{Ind}(\mathcal{C})}(A, \cdot)$ from $\mathrm{Ind}(\mathcal{C})$ to **Set** commutes with small filtrant inductive limits, i.e., A is of finite presentation in $\mathrm{Ind}(\mathcal{C})$.

(b) There exist $X \in \mathcal{C}$ and morphisms $A \xrightarrow{i} X \xrightarrow{p} A$ such that $p \circ i = \mathrm{id}_A$.

(ii) Prove that any $A \in \mathcal{C}^\wedge$ which satisfies (b) belongs to $\mathrm{Ind}(\mathcal{C})$.
(iii) Prove that $\mathcal{C} \to (\mathrm{Ind}(\mathcal{C}))^{\mathrm{fp}}$ is an equivalence if and only if $\mathcal{C}$ is idempotent complete (see Exercise 2.9).

Exercise 6.2. Prove that if X is an initial (resp. terminal) object in $\mathcal{C}$, then $\iota_\mathcal{C}(X)$ is an initial (resp. terminal) object in $\mathrm{Ind}(\mathcal{C})$.

Exercise 6.3. Let $\mathcal{C}$ be a small category and denote by $\emptyset_{\mathcal{C}^\wedge}$ and $\mathrm{pt}_{\mathcal{C}^\wedge}$ the initial and terminal objects of $\mathcal{C}^\wedge$, respectively.
(i) Prove that $\emptyset_{\mathcal{C}^\wedge} \notin \mathrm{Ind}(\mathcal{C})$. (Hint: see Exercise 3.7.)
(ii) Prove that $\mathrm{pt}_{\mathcal{C}^\wedge} \in \mathrm{Ind}(\mathcal{C})$ if and only if $\mathcal{C}$ is filtrant and cofinally small.

Exercise 6.4. Let $\mathcal{C}$ be a category which admits finite inductive limits and denote by $\alpha : \mathrm{Ind}(\mathcal{C}) \to \mathcal{C}^\wedge$ the natural functor. Prove that the functor α does not commute with finite inductive limits (see Exercise 6.3).

Exercise 6.5. Prove that $\mathrm{Pro}(\mathbf{Set}^f)$ is equivalent to the category of Hausdorff totally disconnected compact spaces. (Recall that on such spaces, any point has an open and closed neighborhood system.)

Exercise 6.6. Let k be a field, $\mathcal{C} = \mathrm{Mod}(k)$. Let $V = k^{\oplus \mathbb{Z}}$ and $V_n = k^{\oplus I_n}$ where $I_n = \{i \in \mathbb{Z} \, ; \, |i| \leq n\}$.
(i) Construct the natural morphism "$\varinjlim_n$" $V_n \to V$.

(ii) Show that this morphism is a monomorphism and not an epimorphism.

Exercise 6.7. Let $\mathcal{C}$ be a category which admits small filtrant inductive limits. Let us say that an object X of $\mathcal{C}$ is of finite type if for any functor $\alpha : I \to \mathcal{C}$ with I small and filtrant, the natural map $\varinjlim \mathrm{Hom}_{\mathcal{C}}(X, \alpha) \to \mathrm{Hom}_{\mathcal{C}}(X, \varinjlim \alpha)$ is injective. Prove that this definition coincides with the usual one when $\mathcal{C} = \mathrm{Mod}(R)$ for a ring R (see Examples 1.2.4 (iv)).

Exercise 6.8. Let R be a ring.
(i) Prove that $M \in \mathrm{Mod}(R)$ is of finite presentation in the sense of Definition 6.3.3 if and only if it is of finite presentation in the classical sense (see Examples 1.2.4 (iv)), that is, if there exists an exact sequence $R^{\oplus n_1} \to R^{\oplus n_0} \to M \to 0$.
(ii) Prove that any R-module M is a small filtrant inductive limit of modules of finite presentation. (Hint: consider the full subcategory of $(\mathrm{Mod}(A))_M$ consisting of modules of finite presentation and prove it is essentially small and filtrant.)
(iii) Deduce that the functor $J\rho$ defined in Diagram (6.3.1) induces an equivalence $J\rho \colon \mathrm{Ind}(\mathrm{Mod}^{\mathrm{fp}}(R)) \xrightarrow{\sim} \mathrm{Mod}(R)$.

Exercise 6.9. Let $\mathcal{C}$ be a small category, $F : \mathcal{C} \to \mathcal{C}'$ a functor and denote by $F_* : \mathcal{C}' \to \mathcal{C}^\wedge$ the functor given by $F_*(Y)(U) = \mathrm{Hom}_{\mathcal{C}'}(F(U), Y)$ for $Y \in \mathcal{C}'$, $U \in \mathcal{C}$. Prove that the functor F is right exact if and only if F_* sends $\mathcal{C}'$ to $\mathrm{Ind}(\mathcal{C})$.

Exercise 6.10. Let $\mathcal{C}$ be a category and consider the functor

$$\Phi : \mathrm{Ind}(\mathcal{C}) \to \mathcal{C}^\vee \quad \text{given by} \quad A \mapsto \varinjlim_{(X \to A) \in \mathcal{C}_A} k_{\mathcal{C}}(X) \, .$$

(i) Prove that Φ commutes with small filtrant inductive limits and prove that the composition $\mathcal{C} \xrightarrow{\iota_{\mathcal{C}}} \mathrm{Ind}(\mathcal{C}) \xrightarrow{\Phi} \mathcal{C}^\vee$ is isomorphic to the Yoneda functor $k_{\mathcal{C}}$.
(ii) Assume that $\mathcal{C}$ admits filtrant inductive limits. Prove that the functor Φ factorizes as $\mathrm{Ind}(\mathcal{C}) \xrightarrow{\sigma_{\mathcal{C}}} \mathcal{C} \xrightarrow{k_{\mathcal{C}}} \mathcal{C}^\vee$, where $\sigma_{\mathcal{C}}$ is defined in the course of Proposition 6.3.1.

Exercise 6.11. Let $\mathcal{J}$ be a full subcategory of a category $\mathcal{C}$ and let $A \in \mathrm{Ind}(\mathcal{C})$. Prove that A is isomorphic to the image of an object of $\mathrm{Ind}(\mathcal{J})$ if and only if any morphism $X \to A$ in $\mathrm{Ind}(\mathcal{C})$ with $X \in \mathcal{C}$ factors through an object of $\mathcal{J}$.

Exercise 6.12. Let G be a group and let $\mathcal{G}$ be the category with one object denoted by c and morphisms $\mathrm{Hom}_{\mathcal{G}}(c, c) = G$. A G-set is a set S with an action of G. If S and S' are G-sets, a G-equivariant map $f : S \to S'$ is a map satisfying $f(gs) = gf(s)$ for all $s \in S$ and all $g \in G$. We denote by G-**Set** the category of G-sets and G-equivariant maps.
(i) Prove that $\mathcal{G}^{\mathrm{op}}$ is equivalent to $\mathcal{G}$.
(ii) Prove that $\mathcal{G}^\wedge$ is equivalent to G-**Set** and that the object c of $\mathcal{G}$ corresponds to the G-set G endowed with the left action of G.
(iii) For a G-set X, prove that $\mathcal{G}_X$ is equivalent to the category $\mathcal{C}$ given by $\mathrm{Ob}(\mathcal{C}) = X$ and $\mathrm{Hom}_{\mathcal{C}}(x, y) = \{g \in G ; y = gx\}$ for $x, y \in X$.
(iv) Prove that $\mathcal{G} \xrightarrow{\sim} \mathrm{Ind}(\mathcal{G})$.

7

Localization

Consider a category $\mathcal{C}$ and a family $\mathcal{S}$ of morphisms in $\mathcal{C}$. The aim of localization is to find a new category $\mathcal{C}_\mathcal{S}$ and a functor $Q: \mathcal{C} \to \mathcal{C}_\mathcal{S}$ which sends the morphisms belonging to $\mathcal{S}$ to isomorphisms in $\mathcal{C}_\mathcal{S}$, $(\mathcal{C}_\mathcal{S}, Q)$ being "universal" for such a property.

In this chapter, we shall construct the localization of a category when $\mathcal{S}$ satisfies suitable conditions. A classical reference is [24].

We discuss with some details the localization of functors. When considering a functor F from $\mathcal{C}$ to a category $\mathcal{A}$ which does not necessarily send the morphisms in $\mathcal{S}$ to isomorphisms in $\mathcal{A}$, it is possible to define the right (resp. the left) localization of F, a functor $R_\mathcal{S}F$ (resp. $L_\mathcal{S}F$) from $\mathcal{C}_\mathcal{S}$ to $\mathcal{A}$. Such a right localization always exists if $\mathcal{A}$ admits filtrant inductive limits.

We also discuss an important situation where a functor is localizable. This is when there exists a full subcategory $\mathcal{I}$ of $\mathcal{C}$ whose localization is equivalent to that of $\mathcal{C}$ and such that F sends the morphisms of $\mathcal{S}$ belonging to $\mathcal{I}$ to isomorphisms. This is the case that we shall encounter when deriving functors in derived categories in Chap. 13.

We do not treat in this book the theory of model categories of Quillen which would allow us to consider the quotient of categories in a more general framework (cf. [32, 56]).

7.1 Localization of Categories

Let $\mathcal{C}$ be a category and let $\mathcal{S}$ be a family of morphisms in $\mathcal{C}$.

Definition 7.1.1. *A localization of $\mathcal{C}$ by $\mathcal{S}$ is the data of a big category $\mathcal{C}_\mathcal{S}$ and a functor $Q: \mathcal{C} \to \mathcal{C}_\mathcal{S}$ satisfying:*

(a) *for all $s \in \mathcal{S}$, $Q(s)$ is an isomorphism,*
(b) *for any big category $\mathcal{A}$ and any functor $F: \mathcal{C} \to \mathcal{A}$ such that $F(s)$ is an isomorphism for all $s \in \mathcal{S}$, there exist a functor $F_\mathcal{S}: \mathcal{C}_\mathcal{S} \to \mathcal{A}$ and an isomorphism $F \simeq F_\mathcal{S} \circ Q$ visualized by the diagram*

$$\begin{array}{ccc} \mathcal{C} & \xrightarrow{\;F\;} & \mathcal{A}, \\ {\scriptstyle Q}\downarrow & \nearrow & \\ \mathcal{C}_\mathcal{S} & {\scriptstyle F_S} & \end{array}$$

(c) *if G_1 and G_2 are two objects of* $\mathrm{Fct}(\mathcal{C}_\mathcal{S}, \mathcal{A})$, *then the natural map*

(7.1.1) $\mathrm{Hom}_{\mathrm{Fct}(\mathcal{C}_\mathcal{S},\mathcal{A})}(G_1, G_2) \to \mathrm{Hom}_{\mathrm{Fct}(\mathcal{C},\mathcal{A})}(G_1 \circ Q, G_2 \circ Q)$

is bijective.

Note that (c) means that the functor $\circ Q \colon \mathrm{Fct}(\mathcal{C}_\mathcal{S}, \mathcal{A}) \to \mathrm{Fct}(\mathcal{C}, \mathcal{A})$ is fully faithful. This implies that F_S in (b) is unique up to unique isomorphism.

Proposition 7.1.2.   (i) *If $\mathcal{C}_\mathcal{S}$ exists, it is unique up to equivalence of categories.*
(ii) *If $\mathcal{C}_\mathcal{S}$ exists, then, denoting by $\mathcal{S}^{\mathrm{op}}$ the image of $\mathcal{S}$ in $\mathcal{C}^{\mathrm{op}}$ by the functor* op, $(\mathcal{C}^{\mathrm{op}})_{\mathcal{S}^{\mathrm{op}}}$ *exists and there is an equivalence of categories:*

$$(\mathcal{C}_\mathcal{S})^{\mathrm{op}} \simeq (\mathcal{C}^{\mathrm{op}})_{\mathcal{S}^{\mathrm{op}}}.$$

The proof is obvious.

Lemma 7.1.3. *Consider three categories $\mathcal{C}$, $\mathcal{C}'$, $\mathcal{A}$ and two functors Q, G :*

$$\mathcal{C} \xrightarrow{\;Q\;} \mathcal{C}' \xrightarrow{\;G\;} \mathcal{A}.$$

Assume the following condition. For any $X \in \mathcal{C}'$, there exist $Y \in \mathcal{C}$ and a morphism $s \colon X \to Q(Y)$ which satisfy the following two properties (a) *and* (b):

(a) $G(s)$ *is an isomorphism,*
(b) *for any $Y' \in \mathcal{C}$ and any morphism $t \colon X \to Q(Y')$, there exist $Y'' \in \mathcal{C}$ and morphisms $s' \colon Y' \to Y''$ and $t' \colon Y \to Y''$ in $\mathcal{C}$ such that $G(Q(s'))$ is an isomorphism and the diagram below commutes*

$$\begin{array}{ccc} X & \xrightarrow{\;s\;} & Q(Y) \\ {\scriptstyle t}\downarrow & & \downarrow{\scriptstyle Q(t')} \\ Q(Y') & \xrightarrow{\;Q(s')\;} & Q(Y''). \end{array}$$

Then $Q^\ddagger Q_* G$ exists and is isomorphic to G (see Definition 2.3.2), that is, the natural map $\mathrm{Hom}_{\mathrm{Fct}(\mathcal{C}',\mathcal{A})}(F, G) \to \mathrm{Hom}_{\mathrm{Fct}(\mathcal{C},\mathcal{A})}(F \circ Q, G \circ Q)$ is bijective for any functor $F \colon \mathcal{C}' \to \mathcal{A}$.

Remark 7.1.4. Since the conclusion of the lemma still holds when replacing the categories with the opposite categories, the similar result holds when reversing the arrows.

Proof. (i) The map is injective. Let θ_1 and θ_2 be two morphisms from F to G and assume $\theta_1(Q(Y)) = \theta_2(Q(Y))$ for all $Y \in \mathcal{C}$. For $X \in \mathcal{C}'$, choose a morphism $s \colon X \to Q(Y)$ such that $G(s)$ is an isomorphism.

Consider the commutative diagram where $i = 1, 2$:

$$
\begin{array}{ccc}
F(X) & \xrightarrow{\theta_i(X)} & G(X) \\
{\scriptstyle F(s)}\big\downarrow & & \big\downarrow{\scriptstyle G(s)} \\
F(Q(Y)) & \xrightarrow{\theta_i(Q(Y))} & G(Q(Y)) \,.
\end{array}
$$

Since $G(s)$ is an isomorphism, we find $\theta_1(X) = \theta_2(X)$.

(ii) The map is surjective. Let $\theta \colon F \circ Q \to G \circ Q$ be a morphism of functors. For each $X \in \mathcal{C}'$, choose a morphism $s \colon X \to Q(Y)$ satisfying the conditions (a) and (b). Then define $\tilde{\theta}(X) \colon F(X) \to G(X)$ as $\tilde{\theta}(X) = (G(s))^{-1} \circ \theta(Y) \circ F(s)$. Let us prove that this construction is functorial, and in particular, does not depend on the choice of the morphism $s \colon X \to Q(Y)$. (Take $f = \mathrm{id}_X$ in the proof below.)

Let $f \colon X_1 \to X_2$ be a morphism in $\mathcal{C}'$. For any choice of morphisms $s_1 \colon X_1 \to Q(Y_1)$ and $s_2 \colon X_2 \to Q(Y_2)$ satisfying the conditions (a) and (b), apply the condition (b) to $s_1 \colon X_1 \to Q(Y_1)$ and $s_2 \circ f \colon X_1 \to Q(Y_2)$. Then, there are morphisms $Y_1 \xrightarrow{t_1} Y_3$ and $Y_2 \xrightarrow{t_2} Y_3$ such that $G(Q(t_2))$ is an isomorphism and $Q(t_1) \circ s_1 = Q(t_2) \circ s_2 \circ f$. We get the diagram

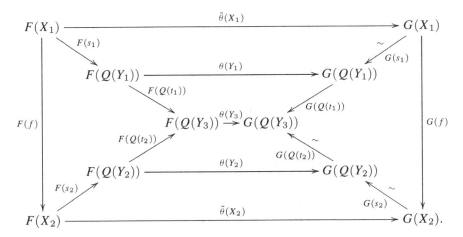

Since all the internal diagrams commute, the square with vertices $F(X_1)$, $G(X_1)$, $F(X_2)$, $G(X_2)$ commutes.   q.e.d.

Definition 7.1.5. *The family* $\mathcal{S} \subset \mathrm{Mor}(\mathcal{C})$ *is a right multiplicative system if it satisfies the axioms* S1–S4 *below.*

S1 *Any isomorphism in* $\mathcal{C}$ *belongs to* $\mathcal{S}$.

S2 *If two morphisms $f\colon X \to Y$ and $g\colon Y \to Z$ belong to $\mathcal{S}$, then $g \circ f$ belongs to $\mathcal{S}$.*

S3 *Given two morphisms $f\colon X \to Y$ and $s\colon X \to X'$ with $s \in \mathcal{S}$, there exist $t\colon Y \to Y'$ and $g\colon X' \to Y'$ with $t \in \mathcal{S}$ and $g \circ s = t \circ f$. This is visualized by the diagram:*

$$
\begin{array}{ccc}
X & \xrightarrow{\ f\ } & Y \\
{\scriptstyle s}\downarrow & & \downarrow{\scriptstyle t} \\
X' & \dashrightarrow[g]{} & Y' \,.
\end{array}
$$

S4 *Let $f, g\colon X \rightrightarrows Y$ be two parallel morphisms. If there exists $s\colon W \to X$ in $\mathcal{S}$ such that $f \circ s = g \circ s$, then there exists $t\colon Y \to Z$ in $\mathcal{S}$ such that $t \circ f = t \circ g$. This is visualized by the diagram:*

$$
W \xrightarrow{\ s\ } X \overset{f}{\underset{g}{\rightrightarrows}} Y \dashrightarrow[]{\ t\ } Z \,.
$$

Remark 7.1.6. Axioms S1–S2 asserts that there is a half-full subcategory $\widetilde{\mathcal{S}}$ of $\mathcal{C}$ with $\mathrm{Ob}(\widetilde{\mathcal{S}}) = \mathrm{Ob}(\mathcal{C})$ and $\mathrm{Mor}(\widetilde{\mathcal{S}}) = \mathcal{S}$. With these axioms, the notion of a right multiplicative system is stable by equivalence of categories.

Remark 7.1.7. The notion of a *left multiplicative* system is defined similarly by reversing the arrows. This means that the condition S3 and S4 are replaced by the conditions S'3 and S'4 below.

S'3 Given two morphisms $f\colon X \to Y$ and $t\colon Y' \to Y$ with $t \in \mathcal{S}$, there exist $s\colon X' \to X$ and $g\colon X' \to Y'$ with $s \in \mathcal{S}$ and $t \circ g = f \circ s$. This is visualized by the diagram:

$$
\begin{array}{ccc}
X' & \dashrightarrow[]{\ g\ } & Y' \\
{\scriptstyle s}\downarrow & & \downarrow{\scriptstyle t} \\
X & \xrightarrow{\ f\ } & Y \,.
\end{array}
$$

S'4 Let $f, g\colon X \rightrightarrows Y$ be two parallel morphisms. If there exists $t\colon Y \to Z$ in $\mathcal{S}$ such that $t \circ f = t \circ g$ then there exists $s\colon W \to X$ in $\mathcal{S}$ such that $f \circ s = g \circ s$. This is visualized by the diagram

$$
W \dashrightarrow[]{\ s\ } X \overset{f}{\underset{g}{\rightrightarrows}} Y \xrightarrow{\ t\ } Z \,.
$$

Remark 7.1.8. In the literature, "a multiplicative system" often means a system which is both right and left multiplicative. Moreover, some authors, in particular [24], call "right" what we call "left" and conversely. In [24], they call our "right multiplicative system" a left multiplicative system since, as we will see later, any morphism in the localization $\mathcal{C}_{\mathcal{S}}$ is written as $Q(s)^{-1} \circ Q(f)$ for some $s \in \mathcal{S}$ and $f \in \mathrm{Mor}(\mathcal{C})$. In this book we call it a right multiplicative system since $\mathrm{Hom}_{\mathcal{C}_{\mathcal{S}}}(Q(X), Q(Y))$ is expressed as the inductive limit

$\varinjlim_{Y \to Y'} \mathrm{Hom}_{\mathcal{C}}(X, Y')$ over the right arrows $Y \to Y'$ in $\mathcal{S}$. The terminology "right localization of functors" (Definition 7.3.1) comes from the same reason. Its particular case, "right derived functor" is widely used.

Definition 7.1.9. *Assume that $\mathcal{S}$ satisfies the axioms S1–S2 and let $X \in \mathcal{C}$. The categories $\mathcal{S}^X$, $\mathcal{S}_X$ and the functors $\alpha^X \colon \mathcal{S}^X \to \mathcal{C}$, $\alpha_X \colon \mathcal{S}_X \to \mathcal{C}$ are defined as follows.*

$$\mathrm{Ob}(\mathcal{S}^X) = \left\{ s \colon X \to X' \, ; \, s \in \mathcal{S} \right\} ,$$

$$\mathrm{Hom}_{\mathcal{S}^X}((s \colon X \to X'), (s' \colon X \to X'')) = \left\{ h \in \mathrm{Hom}_{\mathcal{C}}(X', X'') \, ; \, h \circ s = s' \right\} ,$$

$$\mathrm{Ob}(\mathcal{S}_X) = \left\{ s \colon X' \to X \, ; \, s \in \mathcal{S} \right\} ,$$

$$\mathrm{Hom}_{\mathcal{S}_X}((s \colon X' \to X), (s' \colon X'' \to X)) = \left\{ h \in \mathrm{Hom}_{\mathcal{C}}(X', X'') \, ; \, s' \circ h = s \right\} ,$$

$$\alpha^X((s \colon X \to X')) = X' ,$$

$$\alpha_X((s \colon X' \to X)) = X' .$$

One should be aware that we do not ask $h \in \mathcal{S}$ in the definition of the categories $\mathcal{S}^X$ and $\mathcal{S}_X$. Therefore $\mathcal{S}^X$ is a full subcategory of $\mathcal{C}^X$ and $\mathcal{S}_X$ is a full subcategory of $\mathcal{C}_X$ (see Definition 3.4.1).

In the sequel we shall concentrate on right multiplicative systems.

Proposition 7.1.10. *Assume that $\mathcal{S}$ is a right multiplicative system. Then the category $\mathcal{S}^X$ is filtrant.*

Proof. (a) Let $s \colon X \to X'$ and $s' \colon X \to X''$ belong to $\mathcal{S}$. By S3, there exist $t \colon X' \to X'''$ and $t' \colon X'' \to X'''$ such that $t' \circ s' = t \circ s$, and $t \in \mathcal{S}$. Hence, $t \circ s \in \mathcal{S}$ by S2 and $(X \to X''')$ belongs to $\mathcal{S}^X$.
(b) Let $s \colon X \to X'$ and $s' \colon X \to X''$ belong to $\mathcal{S}$, and consider two morphisms $f, g \colon X' \to X''$ with $f \circ s = g \circ s = s'$. By S4 there exists $t \colon X'' \to W$ in $\mathcal{S}$ such that $t \circ f = t \circ g$. Hence $t \circ s' \colon X \to W$ belongs to $\mathcal{S}^X$ and the two compositions $(X', s) \underset{g}{\overset{f}{\rightrightarrows}} (X'', s') \overset{t}{\longrightarrow} (W, t \circ s')$ coincide.   q.e.d.

Definition 7.1.11. *Let $\mathcal{S}$ be a right multiplicative system and let $X, Y \in \mathrm{Ob}(\mathcal{C})$. We set*

$$\mathrm{Hom}_{\mathcal{C}_{\mathcal{S}}^r}(X, Y) = \varinjlim_{(Y \to Y') \in \mathcal{S}^Y} \mathrm{Hom}_{\mathcal{C}}(X, Y') = \varinjlim \mathrm{Hom}_{\mathcal{C}}(X, \alpha^Y) .$$

Lemma 7.1.12. *Assume that $\mathcal{S}$ is a right multiplicative system. Let $Y \in \mathcal{C}$ and let $s \colon X \to X' \in \mathcal{S}$. Then s induces an isomorphism*

$$\mathrm{Hom}_{\mathcal{C}_{\mathcal{S}}^r}(X', Y) \underset{\circ s}{\overset{\sim}{\to}} \mathrm{Hom}_{\mathcal{C}_{\mathcal{S}}^r}(X, Y) .$$

Proof. (i) The map $\circ s$ is surjective. This follows from S3, as visualized by the diagram in which $s, t, t' \in \mathcal{S}$:

$$X \xrightarrow{\;f\;} Y' \xleftarrow{\;t\;} Y \;.$$
$$\downarrow s \qquad \qquad \vdots \, t'$$
$$X' \cdots\cdots> Y''$$

(ii) The map $\circ s$ is injective. This follows from S4, as visualized by the diagram in which $s, t, t' \in \mathcal{S}$:

$$X \xrightarrow{\;s\;} X' \underset{g}{\overset{f}{\rightrightarrows}} Y' \cdots\overset{t'}{\cdots}> Y'' \;.$$
$$\uparrow t$$
$$Y$$

q.e.d.

Using Lemma 7.1.12, we define the composition

(7.1.2) $\mathrm{Hom}_{\mathcal{C}_{\mathcal{S}}^r}(X, Y) \times \mathrm{Hom}_{\mathcal{C}_{\mathcal{S}}^r}(Y, Z) \to \mathrm{Hom}_{\mathcal{C}_{\mathcal{S}}^r}(X, Z)$

as

$$\varinjlim_{Y \to Y'} \mathrm{Hom}_{\mathcal{C}}(X, Y') \times \varinjlim_{Z \to Z'} \mathrm{Hom}_{\mathcal{C}}(Y, Z')$$
$$\simeq \varinjlim_{Y \to Y'} \left(\mathrm{Hom}_{\mathcal{C}}(X, Y') \times \varinjlim_{Z \to Z'} \mathrm{Hom}_{\mathcal{C}}(Y, Z') \right)$$
$$\xleftarrow{\sim} \varinjlim_{Y \to Y'} \left(\mathrm{Hom}_{\mathcal{C}}(X, Y') \times \varinjlim_{Z \to Z'} \mathrm{Hom}_{\mathcal{C}}(Y', Z') \right)$$
$$\to \varinjlim_{Y \to Y'} \varinjlim_{Z \to Z'} \mathrm{Hom}_{\mathcal{C}}(X, Z')$$
$$\simeq \varinjlim_{Z \to Z'} \mathrm{Hom}_{\mathcal{C}}(X, Z') \;.$$

Lemma 7.1.13. *The composition* (7.1.2) *is associative.*

The verification is left to the reader.

Hence we get a big category $\mathcal{C}_{\mathcal{S}}^r$ whose objects are those of $\mathcal{C}$ and morphisms are given by Definition 7.1.11.

Remark 7.1.14. One should be aware that $\mathcal{C}_{\mathcal{S}}^r$ is not necessarily a $\mathcal{U}$-category. It is a $\mathcal{U}$-category if $\mathcal{S}^X$ is cofinally small for every $X \in \mathcal{C}$.

Let us denote by $Q_{\mathcal{S}}^r \colon \mathcal{C} \to \mathcal{C}_{\mathcal{S}}^r$ the natural functor associated with

$$\mathrm{Hom}_{\mathcal{C}}(X, Y) \to \varinjlim_{(Y \to Y') \in \mathcal{S}^Y} \mathrm{Hom}_{\mathcal{C}}(X, Y') \;.$$

If there is no risk of confusion, we denote this functor simply by Q.

Lemma 7.1.15. *If $s: X \to Y$ belongs to $\mathcal{S}$, then $Q(s)$ is invertible.*

Proof. For any $Z \in \mathcal{C}_{\mathcal{S}}^r$, the map $\operatorname{Hom}_{\mathcal{C}_{\mathcal{S}}^r}(Y, Z) \to \operatorname{Hom}_{\mathcal{C}_{\mathcal{S}}^r}(X, Z)$ is bijective by Lemma 7.1.12.   q.e.d.

A morphism $f: Q(X) \to Q(Y)$ in $\mathcal{C}_{\mathcal{S}}^r$ is thus given by an equivalence class of triplets (Y', t, f') with $t: Y \to Y', t \in \mathcal{S}$ and $f': X \to Y'$, that is:

$$X \xrightarrow{\ f'\ } Y' \xleftarrow{\ t\ } Y \ ,$$

the equivalence relation being defined as follows: $(Y', t, f') \sim (Y'', t', f'')$ if there exist (Y''', t''', f''') $(t, t', t'' \in \mathcal{S})$ and a commutative diagram:

(7.1.3)

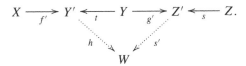

Note that the morphism (Y', t, f') in $\mathcal{C}_{\mathcal{S}}^r$ is $Q(t)^{-1} \circ Q(f')$, that is,

(7.1.4) $$f = Q(t)^{-1} \circ Q(f') \ .$$

For two parallel arrows $f, g: X \rightrightarrows Y$ in $\mathcal{C}$ we have the equivalence

(7.1.5)
$Q(f) = Q(g)$ holds in $\operatorname{Mor}(\mathcal{C}_{\mathcal{S}}^r)$
$\Longleftrightarrow$ there exits $s: Y \to Y'$ in $\mathcal{S}$ such that $s \circ f = s \circ g$.

The composition of two morphisms $(Y', t, f'): X \to Y$ and $(Z', s, g'): Y \to Z$ is defined by the diagram below with $t, s, s' \in \mathcal{S}$:

$$X \xrightarrow{\ f'\ } Y' \xleftarrow{\ t\ } Y \xrightarrow{\ g'\ } Z' \xleftarrow{\ s\ } Z.$$

$$h \searrow \qquad \swarrow s'$$
$$W$$

Theorem 7.1.16. *Assume that $\mathcal{S}$ is a right multiplicative system. Then the big category $\mathcal{C}_{\mathcal{S}}^r$ and the functor Q define a localization of $\mathcal{C}$ by $\mathcal{S}$.*

Proof. Let us check the conditions of Definition 7.1.1.
(a) follows from Lemma 7.1.15.
(b) For $X \in \operatorname{Ob}(\mathcal{C}_{\mathcal{S}}) = \operatorname{Ob}(\mathcal{C})$, set $F_{\mathcal{S}}(X) = F(X)$. For $X, Y \in \mathcal{C}$, we have a chain of morphisms

$$\text{Hom}_{\mathcal{C}_{\mathcal{S}}}(X, Y) = \varinjlim_{(Y \to Y') \in \mathcal{S}^Y} \text{Hom}_{\mathcal{C}}(X, Y')$$

$$\to \varinjlim_{(Y \to Y') \in \mathcal{S}^Y} \text{Hom}_{\mathcal{A}}(F(X), F(Y'))$$

$$\simeq \varinjlim_{(Y \to Y') \in \mathcal{S}^Y} \text{Hom}_{\mathcal{A}}(F(X), F(Y))$$

$$\simeq \text{Hom}_{\mathcal{A}}(F_{\mathcal{S}}(X), F_{\mathcal{S}}(Y)) \,.$$

This defines the functor $F_{\mathcal{S}} \colon \mathcal{C}_{\mathcal{S}} \to \mathcal{A}$.

(c) follows from Lemma 7.1.3. Indeed, with the notations of this Lemma, choose $X = Q(Y)$ and $s = \text{id}_{Q(Y)}$. Any morphism $t \colon Q(Y) \to Q(Y')$ is given by morphisms $Y \xrightarrow{t'} Y'' \xleftarrow{s'} Y'$ with $s' \in \mathcal{S}$, and the diagram in Lemma 7.1.3 (b) commutes.   q.e.d.

Notation 7.1.17. From now on, we shall write $\mathcal{C}_{\mathcal{S}}$ instead of $\mathcal{C}_{\mathcal{S}}^r$. This is justified by Theorem 7.1.16.

Remark 7.1.18. (i) In the above construction, we have used the property of $\mathcal{S}$ of being a right multiplicative system. If $\mathcal{S}$ is a left multiplicative system, we set

$$\text{Hom}_{\mathcal{C}_{\mathcal{S}}^l}(X, Y) = \varinjlim_{(X' \to X) \in \mathcal{S}_X} \text{Hom}_{\mathcal{C}}(X', Y) = \varinjlim \text{Hom}_{\mathcal{C}}(\alpha_X, Y) \,.$$

Then $\mathcal{C}_{\mathcal{S}}^l$ is a localization of $\mathcal{C}$ by $\mathcal{S}$.

(ii) When $\mathcal{S}$ is both a right and left multiplicative system, the two constructions give equivalent categories. Hence, we have

$$\text{Hom}_{\mathcal{C}_{\mathcal{S}}}(X, Y) \simeq \varinjlim_{(X' \to X) \in \mathcal{S}_X} \text{Hom}_{\mathcal{C}}(X', Y)$$

$$\xrightarrow{\sim} \varinjlim_{(X' \to X) \in \mathcal{S}_X, (Y \to Y') \in \mathcal{S}^Y} \text{Hom}_{\mathcal{C}}(X', Y')$$

$$\xleftarrow{\sim} \varinjlim_{(Y \to Y') \in \mathcal{S}^Y} \text{Hom}_{\mathcal{C}}(X, Y') \,.$$

Definition 7.1.19. *We say that a right multiplicative system $\mathcal{S}$ is right saturated, or simply saturated, if it satisfies*

S5 *for any morphisms $f \colon X \to Y$, $g \colon Y \to Z$ and $h \colon Z \to W$ such that $g \circ f$ and $h \circ g$ belong to $\mathcal{S}$, the morphism f belongs to $\mathcal{S}$.*

Proposition 7.1.20. *Let $\mathcal{S}$ be a right multiplicative system.*

(i) *For a morphism $f \colon X \to Y$, $Q(f)$ is an isomorphism in $\mathcal{C}_{\mathcal{S}}$ if and only if there exist $g \colon Y \to Z$ and $h \colon Z \to W$ such that $g \circ f \in \mathcal{S}$ and $h \circ g \in \mathcal{S}$.*

(ii) *The right multiplicative system $\mathcal{S}$ is right saturated if and only if $\mathcal{S}$ coincides with the family of morphisms f such that $Q(f)$ is an isomorphism.*

Proof. (i)-(a) Let $f\colon X \to Y$ be a morphism in $\mathcal{C}$ and assume that $Q(f)$ is an isomorphism. Let (X', s, g) be the inverse of $Q(f)$ in $\mathcal{C}_\mathcal{S}$. Hence we get $g\colon Y \to X'$ and $s\colon X \to X'$ such that $s \in \mathcal{S}$ and $Q(s)^{-1} \circ Q(g)$ is the inverse of $Q(f)$. Since $Q(g) \circ Q(f) = Q(s)$, there exists $t\colon X' \to X''$ in $\mathcal{S}$ such that $t \circ g \circ f = t \circ s$ (see (7.1.5)). This is visualized by the diagram

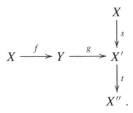

Since $t \circ s \in \mathcal{S}$, we have thus proved that, for $f\colon X \to Y$ in $\mathcal{C}$, if $Q(f)$ is an isomorphism, then there exists $g\colon Y \to Z$ such that $g \circ f \in \mathcal{S}$. Then $Q(g) \circ Q(f)$ is an isomorphism, and hence $Q(g)$ is an isomorphism. Therefore, there exists $h\colon Z \to W$ such that $h \circ g \in \mathcal{S}$.
(i)-(b) Conversely, assume that $g \circ f$ and $h \circ g$ belong to $\mathcal{S}$. Then $Q(g)$ has a right inverse and a left inverse, hence is an isomorphism. Since $Q(g) \circ Q(f)$ is an isomorphism, it follows that $Q(f)$ is an isomorphism.
(ii) follows from (i).   q.e.d.

Assume that $\mathcal{S}$ is a right multiplicative system and let $X \in \mathcal{C}$. The functor

$$(7.1.6) \qquad\qquad \theta\colon \mathcal{S}^X \to \mathcal{C}_{Q(X)}$$

is defined as follows. To $s\colon X \to Y \in \mathcal{S}^X$, associate $Q(s)^{-1}\colon Q(Y) \to Q(X)$ in $\mathcal{C}_{Q(X)}$.

Lemma 7.1.21. *Assume that $\mathcal{S}$ is a right multiplicative system and let $X \in \mathcal{C}$. The functor θ in (7.1.6) is cofinal.*

Proof. Recall that an object $(Y, f) \in \mathcal{C}_{Q(X)}$ is a pair of $Y \in \mathcal{C}$ and $f\colon Q(Y) \to Q(X) \in \mathrm{Mor}(\mathcal{C}_\mathcal{S})$, and a morphism $(Y, f) \to (Z, g)$ in $\mathcal{C}_{Q(X)}$ is a morphism $h\colon Y \to Z$ in $\mathcal{C}$ such that $g \circ Q(h) = f$. An object $(s, X') \in \mathcal{S}^X$ is a morphism $X \xrightarrow{s} X' \in \mathcal{S}$. Also recall that $\mathcal{S}^X$ is filtrant.
Let us check that θ in (7.1.6) satisfies the conditions in Proposition 3.2.2 (iii).
(a) Let $(Y, f) \in \mathcal{C}_{Q(X)}$. There exist morphisms $Y \xrightarrow{f'} Y' \xleftarrow{t} X$ in $\mathcal{C}$ such that $t \in \mathcal{S}$ and $f = Q(t)^{-1} \circ Q(f')$. Therefore f' defines a morphism $(Y, f) \to \theta((t, Y'))$.
(b) Let (s, X') be an object of $\mathcal{S}^X$, (Y, f) an object of $\mathcal{C}_{Q(X)}$, and let $h, h'\colon Y \rightrightarrows X'$ be a pair of parallel morphisms in $\mathcal{C}$ such that $f = Q(s)^{-1} \circ Q(h) = Q(s)^{-1} \circ Q(h')$. Since $Q(h) = Q(h')$, there exists a morphism $t\colon X' \to X''$ in $\mathcal{S}$ such that $t \circ h = t \circ h'$. Then t defines a morphism $\varphi\colon (s, X') \to (t \circ s, X'')$ in $\mathcal{S}^X$ and $\theta(\varphi) \circ h = \theta(\varphi) \circ h'$.   q.e.d.

Let us give some easy properties of the localization functor Q.

Proposition 7.1.22. *Let S be a right multiplicative system.*

(i) *The functor $Q \colon C \to C_S$ is right exact.*

(ii) *Let $\alpha \colon I \to C$ be an inductive system in C indexed by a finite category I. Assume that $\varinjlim \alpha$ exists in C. Then $\varinjlim (Q \circ \alpha)$ exists in C_S and is isomorphic to $Q(\varinjlim \alpha)$.*

(iii) *Assume that C admits cokernels. Then C_S admits cokernels and Q commutes with cokernels.*

(iv) *Assume that C admits finite coproducts. Then C_S admits finite coproducts and Q commutes with finite coproducts.*

(v) *If C admits finite inductive limits, then so does C_S.*

Proof. (i) Recall that Q is right exact if for any $X \in C$, the category $C_{Q(X)}$ is filtrant. Therefore the result follows from Lemma 7.1.21, Proposition 7.1.10 and Proposition 3.2.2.

(ii) follows from (i) and Proposition 3.3.2.

(iii) By (ii), it is enough to remark that any pair of parallel arrows in C_S is isomorphic to the image by Q of a pair of parallel arrows in C.

(iv) By (ii), it is enough to remark that a finite family of objects in C_S is the image by Q of a finite family of objects in C.

(v) follows from (iii) and (iv).   q.e.d.

7.2 Localization of Subcategories

Proposition 7.2.1. *Let C be a category, I a full subcategory, S a right multiplicative system in C, and let T be the family of morphisms in I which belong to S.*

(i) *Assume that T is a right multiplicative system in I. Then $I_T \to C_S$ is well defined.*

(ii) *Assume that for every $f \colon X \to Y$ with $f \in S$, $X \in I$, there exist $g \colon Y \to W$ with $W \in I$ and $g \circ f \in S$. Then T is a right multiplicative system and $I_T \to C_S$ is fully faithful.*

Proof. (i) is obvious.

(ii) It is left to the reader to check that T is a right multiplicative system. For $X \in I$ define the category T^X as the full subcategory of S^X whose objects are the morphisms $s \colon X \to Y$ with $Y \in I$. Then the functor $T^X \to S^X$ is cofinal by Propositions 7.1.10 and 3.2.4, and the result follows from Definition 7.1.11 and Proposition 2.5.2.   q.e.d.

Corollary 7.2.2. *Let C be a category, I a full subcategory, S a right multiplicative system in C, T the family of morphisms in I which belong to S. Assume that for any $X \in C$ there exists $s \colon X \to W$ with $W \in I$ and $s \in S$. Then T is a right multiplicative system and I_T is equivalent to C_S.*

Proof. The natural functor $\mathcal{I}_T \to \mathcal{C}_S$ is fully faithful by Proposition 7.2.1 and is essentially surjective by the assumption.   q.e.d.

7.3 Localization of Functors

Let $\mathcal{C}$ be a category, $\mathcal{S}$ a right multiplicative system in $\mathcal{C}$ and $F: \mathcal{C} \to \mathcal{A}$ a functor. In general, F does not send morphisms in $\mathcal{S}$ to isomorphisms in $\mathcal{A}$. In other words, F does not factorize through $\mathcal{C}_S$. It is however possible in some cases to define a localization of F as follows.

Definition 7.3.1. *Let $\mathcal{S}$ be a family of morphisms in $\mathcal{C}$ and assume that the localization $Q: \mathcal{C} \to \mathcal{C}_S$ exists.*

(i) *We say that F is* right *localizable if the functor $Q^\dagger F$ (see Definition 2.3.2) exists. In such a case, we say that $Q^\dagger F$ is a* right localization *of F and we denote it by $R_S F$. In other words, the right localization of F is a functor $R_S F: \mathcal{C}_S \to \mathcal{A}$ together with a morphism of functors $\tau: F \to R_S F \circ Q$ such that for any functor $G: \mathcal{C}_S \to \mathcal{A}$ the map*

$$(7.3.1) \qquad \mathrm{Hom}_{\mathrm{Fct}(\mathcal{C}_S, \mathcal{A})}(R_S F, G) \to \mathrm{Hom}_{\mathrm{Fct}(\mathcal{C}, \mathcal{A})}(F, G \circ Q)$$

is bijective. (This map is the composition $\mathrm{Hom}_{\mathrm{Fct}(\mathcal{C}_S, \mathcal{A})}(R_S F, G) \to$ $\mathrm{Hom}_{\mathrm{Fct}(\mathcal{C}, \mathcal{A})}(R_S F \circ Q, G \circ Q) \xrightarrow{\circ \tau} \mathrm{Hom}_{\mathrm{Fct}(\mathcal{C}, \mathcal{A})}(F, G \circ Q).)$
(ii) *We say that F is* universally right *localizable if for any functor $K: \mathcal{A} \to \mathcal{A}'$, the functor $K \circ F$ is localizable and $R_S(K \circ F) \xrightarrow{\sim} K \circ R_S F$.*

Note that if $(R_S F, \tau)$ exists, it is unique up to a unique isomorphism.

The notion of a (universally) left localizable functor is similarly defined. The left localization of F is $Q^\ddagger F$, that is, a functor $L_S F: \mathcal{C}_S \to \mathcal{A}$ together with $\sigma: L_S F \circ Q \to F$ such that for any functor $G: \mathcal{C}_S \to \mathcal{A}$, σ induces a bijection

$$(7.3.2) \qquad \mathrm{Hom}_{\mathrm{Fct}(\mathcal{C}_S, \mathcal{A})}(G, L_S F) \xrightarrow{\sim} \mathrm{Hom}_{\mathrm{Fct}(\mathcal{C}, \mathcal{A})}(G \circ Q, F) \, .$$

One shall be aware that even if F admits both a right and a left localization, the two localizations are not isomorphic in general. However, when the localization $Q: \mathcal{C} \to \mathcal{C}_S$ exists and F is right and left localizable, the canonical morphisms of functors $L_S F \circ Q \to F \to R_S F \circ Q$ together with the isomorphism $\mathrm{Hom}(L_S F \circ Q, R_S F \circ Q) \simeq \mathrm{Hom}(L_S F, R_S F)$ in (7.1.1) gives the canonical morphism of functors

$$(7.3.3) \qquad\qquad L_S F \to R_S F \, .$$

From now on, we shall concentrate on right localizations.

Proposition 7.3.2. *Let C be a category, $\mathcal{I}$ a full subcategory, $\mathcal{S}$ a right multiplicative system in C, $\mathcal{T}$ the family of morphisms in $\mathcal{I}$ which belong to $\mathcal{S}$. Let $F: C \to \mathcal{A}$ be a functor. Assume that*

(i) *for any $X \in C$, there exists $s: X \to W$ with $W \in \mathcal{I}$ and $s \in \mathcal{S}$,*
(ii) *for any $t \in \mathcal{T}$, $F(t)$ is an isomorphism.*

Then F is universally right localizable and the composition $\mathcal{I} \to C \xrightarrow{Q} C_{\mathcal{S}} \xrightarrow{R_{\mathcal{S}}F} \mathcal{A}$ is isomorphic to the restriction of F to $\mathcal{I}$.

Proof. Denote by $\iota: \mathcal{I} \to C$ the natural functor. By hypothesis (i) and Corollary 7.2.2, $\iota_Q: \mathcal{I}_{\mathcal{T}} \to C_{\mathcal{S}}$ is an equivalence. By hypothesis (ii) the localization $F_{\mathcal{T}}$ of $F \circ \iota$ exists. Consider the solid diagram:

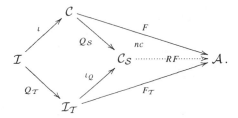

Denote by ι_Q^{-1} a quasi-inverse of ι_Q and set $RF := F_{\mathcal{T}} \circ \iota_Q^{-1}$. Then the diagram above is commutative, except the triangle $(C, C_{\mathcal{S}}, \mathcal{A})$ labeled by nc. Let us show that RF is the right localization of F. Let $G: C_{\mathcal{S}} \to \mathcal{A}$ be a functor. We have the chain of a morphism and isomorphisms:

$$
\begin{aligned}
\mathrm{Hom}_{\mathrm{Fct}(C,\mathcal{A})}(F, G \circ Q_{\mathcal{S}}) &\xrightarrow{\lambda} \mathrm{Hom}_{\mathrm{Fct}(\mathcal{I},\mathcal{A})}(F \circ \iota, G \circ Q_{\mathcal{S}} \circ \iota) \\
&\simeq \mathrm{Hom}_{\mathrm{Fct}(\mathcal{I},\mathcal{A})}(F_{\mathcal{T}} \circ Q_{\mathcal{T}}, G \circ \iota_Q \circ Q_{\mathcal{T}}) \\
&\simeq \mathrm{Hom}_{\mathrm{Fct}(\mathcal{I}_{\mathcal{T}},\mathcal{A})}(F_{\mathcal{T}}, G \circ \iota_Q) \\
&\simeq \mathrm{Hom}_{\mathrm{Fct}(C_{\mathcal{S}},\mathcal{A})}(F_{\mathcal{T}} \circ \iota_Q^{-1}, G) \\
&\simeq \mathrm{Hom}_{\mathrm{Fct}(C_{\mathcal{S}},\mathcal{A})}(RF, G).
\end{aligned}
$$

(7.3.4)

The second isomorphism follows from the fact that $Q_{\mathcal{T}}$ satisfies the hypothesis (c) of Definition 7.1.1 by Theorem 7.1.16. To conclude, it remains to prove that the morphism λ is bijective. Let us check that Lemma 7.1.3 applies to $\mathcal{I} \xrightarrow{\iota} C \xrightarrow{Q_{\mathcal{S}}} C_{\mathcal{S}}$ and hence to $\mathcal{I} \xrightarrow{\iota} C \xrightarrow{G \circ Q_{\mathcal{S}}} \mathcal{A}$. Let $X \in C$. By the hypothesis, there exist $Y \in \mathcal{I}$ and $s: X \to \iota(Y)$ with $s \in \mathcal{S}$. Therefore, $F(s)$ is an isomorphism and condition (a) in Lemma 7.1.3 is satisfied. Condition (b) follows from the fact that ι is fully faithful together with axiom S3 of right multiplicative systems.

Hence F is localizable and $R_{\mathcal{S}}F \simeq F_{\mathcal{T}} \circ \iota_Q^{-1}$.

If $K: \mathcal{A} \to \mathcal{A}'$ is another functor, $K \circ F(t)$ will be an isomorphism for any $t \in \mathcal{T}$. Hence, $K \circ F$ is localizable and we have

$$R_S(K \circ F) \simeq (K \circ F)_T \circ \iota_Q^{-1} \simeq K \circ F_T \circ \iota_Q^{-1} \simeq K \circ R_S F .$$

q.e.d.

Under suitable hypotheses, all functors from $\mathcal{C}$ to $\mathcal{A}$ are localizable. The functor $Q \colon \mathcal{C} \to \mathcal{C}_S$ defines the functor

(7.3.5)
$$Q_* \colon \mathrm{Fct}(\mathcal{C}_S, \mathcal{A}) \to \mathrm{Fct}(\mathcal{C}, \mathcal{A}) ,$$
$$G \mapsto G \circ Q.$$

Proposition 7.3.3. *We make the hypotheses*

(7.3.6)
$$\begin{cases} \mathcal{A} \text{ admits small filtrant inductive limits,} \\ \mathcal{S} \text{ is a right multiplicative system,} \\ \text{for each } X \in \mathcal{C}, \text{ the category } \mathcal{S}^X \text{ is cofinally small.} \end{cases}$$

Then

(i) $\mathcal{C}_S$ *is a $\mathcal{U}$-category,*
(ii) *the functor Q_* in (7.3.5) admits a left adjoint functor $Q^\dagger \colon \mathrm{Fct}(\mathcal{C}, \mathcal{A}) \to \mathrm{Fct}(\mathcal{C}_S, \mathcal{A})$,*
(iii) *any functor $F \colon \mathcal{C} \to \mathcal{A}$ is right localizable and*

(7.3.7) $$R_S F(Q(X)) = \varinjlim_{(X \to X') \in \mathcal{S}^X} F(X') \quad \text{for any } X \in \mathcal{C} .$$

Proof. (i) is obvious.
(ii) and (iii) By Lemma 7.1.21, $\mathcal{S}^X$ is cofinal to $\mathcal{C}_{Q(X)}$. Hence, this last category is cofinally small and filtrant, and we may apply Theorem 2.3.3.   q.e.d.

If the category $\mathcal{A}$ does not admit small filtrant inductive limits, one method would be to embed it in the category of its ind-objects. However, one shall be aware that this embedding does not commute with small filtrant inductive limits. We discuss this point in the subsequent section.

7.4 Indization and Localization

Let $\mathcal{C}$ be a category and $\mathcal{S}$ a right multiplicative system. In this section we shall assume that for every $X \in \mathcal{C}$, the category $\mathcal{S}^X$ is cofinally small. We shall make the link between localization and ind-objects.

As in Chap. 6, let us denote by $\iota_{\mathcal{A}}$ the natural functor $\mathcal{A} \to \mathrm{Ind}(\mathcal{A})$ and similarly with $\iota_{\mathcal{C}}$.

The natural isomorphism

$$\varinjlim_{(Y \to Y') \in \mathcal{S}^Y} \mathrm{Hom}_{\mathcal{C}}(X, Y') \xrightarrow{\sim} \varprojlim_{(X \to X') \in \mathcal{S}^X} \varinjlim_{(Y \to Y') \in \mathcal{S}^Y} \mathrm{Hom}_{\mathcal{C}}(X', Y')$$

defines the isomorphism

(7.4.1) $\text{Hom}_{\mathcal{C}_\mathcal{S}}(X, Y) \to \text{Hom}_{\text{Ind}(\mathcal{C})}(\text{"}\varinjlim\text{"}\,\alpha^X, \text{"}\varinjlim\text{"}\,\alpha^Y)$.

Recall that $\alpha^X \colon \mathcal{S}^X \to \mathcal{C}$ is the forgetful functor $(X \to X') \mapsto X'$. It is easily checked that the isomorphism (7.4.1) commutes with the composition. Therefore

Proposition 7.4.1. *Assume that $\mathcal{S}^X$ is cofinally small for any $X \in \mathcal{C}$. The functor*

$$\alpha_\mathcal{S} \colon \mathcal{C}_\mathcal{S} \to \text{Ind}(\mathcal{C}), \ X \mapsto \text{"}\varinjlim\text{"}\,\alpha^X = \varinjlim_{(X \to X') \in \mathcal{S}^X} X'$$

is well defined and fully faithful.

One shall be aware that the diagram

(where $\iota_\mathcal{C}$ denotes the natural functor) is *not* commutative in general. However, there is a natural morphism of functors:

(7.4.2) $\iota_\mathcal{C} \to \alpha_\mathcal{S} \circ Q$ given by $\iota_\mathcal{C}(X) \to \text{"}\varinjlim\text{"}\,\alpha^X \simeq (\alpha_\mathcal{S} \circ Q)(X)$.

Let $F \colon \mathcal{C} \to \mathcal{A}$ be a functor. Consider the diagram

$$\begin{array}{ccc} \mathcal{C} & \xrightarrow{\ \ F\ \ } & \mathcal{A} \\ {\scriptstyle Q}\downarrow & & \downarrow{\scriptstyle \iota_\mathcal{A}} \\ \mathcal{C}_\mathcal{S} & \xrightarrow[\alpha_\mathcal{S}]{} \text{Ind}(\mathcal{C}) \xrightarrow[IF]{} & \text{Ind}(\mathcal{A}) \end{array}$$

By (7.3.7), we have

(7.4.3) $R_\mathcal{S}(\iota_\mathcal{A} \circ F) \simeq IF \circ \alpha_\mathcal{S}$.

Definition 7.4.2. *The functor F is right localizable at $X \in \mathcal{C}$ if $\text{"}\varinjlim\text{"}(F \circ \alpha^X) = \varinjlim_{(X \to X') \in \mathcal{S}^X} F(X')$ is representable by an object of $\mathcal{A}$.*

Lemma 7.4.3. *If $G \colon \mathcal{A} \to \mathcal{A}'$ is a functor and F is right localizable at X, then $G \circ F$ is right localizable at X.*

Proof. This follows from the fact that $IG \colon \text{Ind}(\mathcal{A}) \to \text{Ind}(\mathcal{A}')$ sends $\mathcal{A}$ to $\mathcal{A}'$.
                                                                                                     q.e.d.

Proposition 7.4.4. *Let* $F \colon \mathcal{C} \to \mathcal{A}$ *be a functor,* $\mathcal{S}$ *a right multiplicative system in* $\mathcal{C}$. *We assume that the category* $\mathcal{S}^X$ *is cofinally small for any* $X \in \mathcal{C}$. *The two conditions below are equivalent:*

(i) F *is right localizable at each* $X \in \mathcal{C}$,
(ii) F *is universally right localizable.*

Proof. (i) $\Rightarrow$ (ii). If F is localizable at each $X \in \mathcal{C}$, then for any functor $K \colon \mathcal{A} \to \mathcal{A}'$, $K \circ F$ is localizable at each $X \in \mathcal{C}$ by Lemma 7.4.3. Hence it is enough to prove that F is right localizable. By (7.4.3) and the hypothesis, there exists a functor $H \colon \mathcal{C}_{\mathcal{S}} \to \mathcal{A}$ such that $R_{\mathcal{S}}(\iota_{\mathcal{A}} \circ F) \simeq \iota_{\mathcal{A}} \circ H$. To check that H is a right localization of F, consider a functor $G \colon \mathcal{C}_{\mathcal{S}} \to \mathcal{A}$. We have the chain of isomorphisms

$$\operatorname{Hom}_{\operatorname{Fct}(\mathcal{C}_{\mathcal{S}}, \mathcal{A})}(H, G) \simeq \operatorname{Hom}_{\operatorname{Fct}(\mathcal{C}_{\mathcal{S}}, \operatorname{Ind}(\mathcal{A}))}(R_{\mathcal{S}}(\iota_{\mathcal{A}} \circ F), \iota_{\mathcal{A}} \circ G)$$
$$\simeq \operatorname{Hom}_{\operatorname{Fct}(\mathcal{C}, \operatorname{Ind}(\mathcal{A}))}(\iota_{\mathcal{A}} \circ F, \iota_{\mathcal{A}} \circ G \circ Q)$$
$$\simeq \operatorname{Hom}_{\operatorname{Fct}(\mathcal{C}, \mathcal{A})}(F, G \circ Q) \,.$$

(ii) $\Rightarrow$ (i). The hypothesis implies $R_{\mathcal{S}}(\iota_{\mathcal{A}} \circ F) \simeq \iota_{\mathcal{A}} \circ R_{\mathcal{S}} F$. Therefore, $R_{\mathcal{S}}(\iota_{\mathcal{A}} \circ F)(X) \simeq \underset{(X \to X') \in \mathcal{S}^X}{\text{``}\varinjlim\text{''}} F(X') \in \mathcal{A}$.   q.e.d.

Remark 7.4.5. Let $\mathcal{C}$ (resp. $\mathcal{C}'$) be a category and $\mathcal{S}$ (resp. $\mathcal{S}'$) a right multiplicative system in $\mathcal{C}$ (resp. $\mathcal{C}'$). It is immediately checked that $\mathcal{S} \times \mathcal{S}'$ is a right multiplicative system in the category $\mathcal{C} \times \mathcal{C}'$ and $(\mathcal{C} \times \mathcal{C}')_{\mathcal{S} \times \mathcal{S}'}$ is equivalent to $\mathcal{C}_{\mathcal{S}} \times \mathcal{C}'_{\mathcal{S}'}$. Since a bifunctor is a functor on the product $\mathcal{C} \times \mathcal{C}'$, we may apply the preceding results to the case of bifunctors. For example, let $(X, Y) \in \mathcal{C}_{\mathcal{S}} \times \mathcal{C}'_{\mathcal{S}'}$. Then F is right localizable at (X, Y) if

$$\underset{(X \to X') \in \mathcal{S}^X, (Y \to Y') \in \mathcal{S}'^Y}{\text{``}\varinjlim\text{''}} F(X', Y')$$

is representable.

Exercises

Exercise 7.1. Let $\mathcal{C}$ be a category, $\mathcal{S}$ a right multiplicative system. Let $\mathcal{T}$ be the set of morphisms $f \colon X \to Y$ in $\mathcal{C}$ such that there exist $g \colon Y \to Z$ and $h \colon Z \to W$, with $h \circ g$ and $g \circ f$ in $\mathcal{S}$.

Prove that $\mathcal{T}$ is the smallest right saturated multiplicative system containing $\mathcal{S}$ and that the natural functor $\mathcal{C}_{\mathcal{S}} \to \mathcal{C}_{\mathcal{T}}$ is an equivalence.

Exercise 7.2. Let $\mathcal{C}$ be a category, $\mathcal{S}$ a right and left multiplicative system. Prove that $\mathcal{S}$ is right saturated if and only if for any $f \colon X \to Y$, $g \colon Y \to Z$, $h \colon Z \to W$, $h \circ g \in \mathcal{S}$ and $g \circ f \in \mathcal{S}$ imply $g \in \mathcal{S}$.

Exercise 7.3. Let $\mathcal{C}$ be a category with a zero object 0, $\mathcal{S}$ a right multiplicative system.
(i) Show that $\mathcal{C}_{\mathcal{S}}$ has a zero object (still denoted by 0).
(ii) Prove that $Q(X) \simeq 0$ if and only if there exists $Y \in \mathcal{C}$ such that $0\colon X \to Y$ belongs to $\mathcal{S}$.

Exercise 7.4. Let $\mathcal{C}$ be a category, $\mathcal{S}$ a right multiplicative system. Consider morphisms $f\colon X \to Y$ and $f'\colon X' \to Y'$ in $\mathcal{C}$ and morphisms $u\colon X \to X'$ and $v\colon Y \to Y'$ in $\mathcal{C}_{\mathcal{S}}$, and assume that $Q(f') \circ u = v \circ Q(f)$. Prove that there exists a commutative diagram in $\mathcal{C}$

$$
\begin{array}{ccccc}
X & \xrightarrow{\;u'\;} & X_1 & \xleftarrow{\;s\;} & X' \\
\big\downarrow{f} & & \big\downarrow & & \big\downarrow{f'} \\
Y & \xrightarrow{\;v'\;} & Y_1 & \xleftarrow{\;t\;} & Y'
\end{array}
$$

with s and t in $\mathcal{S}$, $u = Q(s)^{-1} \circ Q(u')$ and $v = Q(t)^{-1} \circ Q(v')$.

Exercise 7.5. Let $F\colon \mathcal{C} \to \mathcal{A}$ be a functor and assume that $\mathcal{C}$ admits finite inductive limits and F is right exact. Let $\mathcal{S}$ denote the set of morphisms s in $\mathcal{C}$ such that $F(s)$ is an isomorphism.
(i) Prove that $\mathcal{S}$ is a right saturated multiplicative system.
(ii) Prove that the localized functor $F_{\mathcal{S}}\colon \mathcal{C}_{\mathcal{S}} \to \mathcal{A}$ is faithful.

Exercise 7.6. Let $\mathcal{C} \underset{R}{\overset{L}{\rightleftarrows}} \mathcal{C}'$ be functors and let ε and η be two morphisms of functors as in (1.5.4) and (1.5.5). Assume that $\langle L, R, \eta, \varepsilon \rangle$ is an adjunction (see § 1.5) and that R is fully faithful (or, equivalently, $\eta\colon L \circ R \to \mathrm{id}_{\mathcal{C}'}$ is an isomorphism). Set $\mathcal{S} = \{u \in \mathrm{Mor}(\mathcal{C})\,;\, L(u) \text{ is an isomorphism}\}$.
(i) Prove that $\varepsilon(X)\colon X \to RL(X)$ belongs to $\mathcal{S}$ for every $X \in \mathcal{C}$.
(ii) Prove that $\mathcal{S}$ is a right saturated multiplicative system.
(iii) Prove that the functor $\iota\colon \mathcal{C}_{\mathcal{S}} \to \mathcal{C}'$ induced by L is an equivalence of categories.
(iv) Prove that any functor $F\colon \mathcal{C} \to \mathcal{A}$ is universally right localizable with respect to $\mathcal{S}$ and $R_{\mathcal{S}}F \simeq F \circ R \circ \iota$.

Exercise 7.7. Let $\mathcal{C}$ be a category and $\mathcal{S}$ a right saturated multiplicative system. Assume that $\mathrm{id}_{\mathcal{C}}\colon \mathcal{C} \to \mathcal{C}$ is universally right localizable with respect to $\mathcal{S}$. Prove that $R_{\mathcal{S}}\,\mathrm{id}_{\mathcal{C}}\colon \mathcal{C}_{\mathcal{S}} \to \mathcal{C}$ is fully faithful and is a right adjoint of the localization functor $Q\colon \mathcal{C} \to \mathcal{C}_{\mathcal{S}}$.

Exercise 7.8. Give an alternative proof of Lemma 7.1.3 by showing that
$$G(X) \xrightarrow{\;\sim\;} \varprojlim_{(X \to Q(Y)) \in \mathcal{C}^X} G(Q(Y)) \quad \text{for any } X \in \mathcal{C}'.$$

Exercise 7.9. Consider three categories $\mathcal{C}, \mathcal{C}', \mathcal{A}$ and a functor $Q\colon \mathcal{C} \to \mathcal{C}'$. Assume

(i) Q is essentially surjective,
(ii) for any $X, Y \in \mathcal{C}$ and any morphism $f: Q(X) \to Q(Y)$, there exist $Y' \in \mathcal{C}$ and morphisms $t: Y \to Y'$, $s: X \to Y'$ such that $Q(s) = Q(t) \circ f$ and $Q(t)$ is an isomorphism.

Prove that the functor $Q_*: \mathrm{Fct}(\mathcal{C}', \mathcal{A}) \to \mathrm{Fct}(\mathcal{C}, \mathcal{A})$ is fully faithful.

8

Additive and Abelian Categories

Many results or constructions in the category $\text{Mod}(R)$ of modules over a ring R have their counterparts in other contexts, such as finitely generated R-modules, or graded modules over a graded ring, or sheaves of R-modules, etc. Hence, it is natural to look for a common language which avoids to repeat the same arguments. This is the language of additive and abelian categories.

In this chapter, we begin by explaining the notion of additive categories. Then, we give the main properties of abelian categories and the basic results on exact sequences, injective objects, etc. in such categories. In particular, we introduce the important notion of a Grothendieck category, an abelian category which admits exact small filtrant inductive limits and a generator.

Then we study the action of a ring on an abelian category and prove the Gabriel-Popescu theorem (see [54]) which asserts that a Grothendieck category is embedded in the category of modules over the ring of endomorphisms of a generator.

We study with some details the abelian category $\text{Ind}(\mathcal{C})$ of ind-objects of an abelian category $\mathcal{C}$ and show in particular that the category $\text{Ind}(\mathcal{C})$ is abelian and the natural functor $\mathcal{C} \to \text{Ind}(\mathcal{C})$ is exact.

Finally we prove that under suitable hypotheses, the Kan extension of a right (or left) exact functor defined on an additive subcategory of an abelian category remains exact.

Complementary results on abelian categories will be given in the Exercises as well as in Sect. 9.6.

8.1 Group Objects

The notion of representable functor allows us to extend various algebraic notions to categories. Let us simply give one example.

We denote by **Group** the category of groups and we denote by $for\colon$ **Group** $\to$ **Set** the forgetful functor.

Definition 8.1.1. *Let $\mathcal{C}$ be a category. An object G in $\mathcal{C}$ is called a group object if there is given a functor $\widetilde{G} \colon \mathcal{C}^{\mathrm{op}} \to$ **Group** such that G represents for $\circ \widetilde{G}$.*

In other words, a group object structure on G is a decomposition of the functor $\mathrm{Hom}_{\mathcal{C}}(\,\cdot\,, G) \colon \mathcal{C}^{\mathrm{op}} \to$ **Set** into $\mathcal{C}^{\mathrm{op}} \to$ **Group** $\to$ **Set**.

Let us identify G with $\widetilde{G}$. For $X \in \mathcal{C}$, we shall write $G(X)$ instead of $\mathrm{Hom}_{\mathcal{C}}(X, G)$.

(i) Denote by $\mu_X \colon G(X) \times G(X) \to G(X)$ the multiplication map of the group $G(X)$. This map is functorial with respect to X, that is, if $f \colon X \to Y$ is a morphism in $\mathcal{C}$, the diagram below commutes:

$$
\begin{array}{ccc}
G(X) \times G(X) & \xrightarrow{\ \mu_X\ } & G(X) \\
{\scriptstyle G(f) \times G(f)} \downarrow & & \downarrow {\scriptstyle G(f)} \\
G(Y) \times G(Y) & \xrightarrow{\ \mu_Y\ } & G(Y)\,.
\end{array}
$$

Since there is a functorial isomorphism $G(X) \times G(X) \simeq (G \times G)(X)$, we get a morphism in $\mathcal{C}^\wedge$:

$$
\mu \colon G \times G \to G\,.
$$

The associativity of the multiplication in groups implies that the diagram below in $\mathcal{C}^\wedge$ is commutative

$$
\begin{array}{ccc}
G \times G \times G & \xrightarrow{\ \mathrm{id} \times \mu\ } & G \times G \\
{\scriptstyle \mu \times \mathrm{id}} \downarrow & & \downarrow {\scriptstyle \mu} \\
G \times G & \xrightarrow{\quad \mu \quad} & G\,.
\end{array}
$$

We shall say that the morphism μ is "associative".

(ii) Denote by e the neutral element in $G(X)$. It gives a map $\{\mathrm{pt}\} \to G(X)$, functorial with respect to X. Hence we get a morphism (that we denote by the same letter) $e \colon \mathrm{pt}_{\mathcal{C}^\wedge} \to G$. Here, $\mathrm{pt}_{\mathcal{C}^\wedge}$ is the terminal object of $\mathcal{C}^\wedge$. The identities $x \cdot e = x$ and $e \cdot x = x$ are translated into the commutative diagrams

(iii) Denote by $a_X \colon G(X) \to G(X)$ the map $x \mapsto x^{-1}$. These maps are functorial with respect to X and define a morphism $a \colon G \to G$. The identities $x \cdot x^{-1} = e$ and $x^{-1} \cdot x = e$ are translated into the commutative diagrams

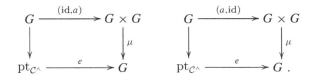

Conversely, the data of (G, μ, e, a) satisfying the above properties endows G with a structure of a group object.

If $\mathcal{C}$ admits finite products, then these diagrams are well defined in $\mathcal{C}$.

Now assume that G represents a functor with values in $\mathrm{Mod}(\mathbb{Z})$. In such a case, we say that G is a *commutative group object*.

Let us denote by

$$(8.1.1) \qquad\qquad v \colon G \times G \to G \times G$$

the morphism associated to the map $(a, b) \mapsto (b, a)$. Then the condition for the group object to be commutative is $\mu \circ v = \mu$. In other words, the diagram below commutes

$$
\begin{array}{ccc}
G \times G & \xrightarrow{\;v\;} & G \times G \\
& \searrow{\scriptstyle \mu} & \downarrow{\scriptstyle \mu} \\
& & G \;.
\end{array}
$$

Lemma 8.1.2. *Let $F \colon \mathcal{C} \to \mathcal{C}'$ be a functor and assume one of the following conditions*

  (i) *$\mathcal{C}$ admits finite products and F commutes with such products,*

  (ii) *F is left exact.*

If an object $X \in \mathcal{C}$ has a structure of a group object, then so does $F(X)$.

Proof. The case (i) is obvious. In case (ii), the functor $\widehat{F} \colon \mathcal{C}^\wedge \to (\mathcal{C}')^\wedge$ is exact by Corollary 3.3.19. In particular, $\widehat{F}$ commutes with finite products.   q.e.d.

8.2 Additive Categories

Definition 8.2.1. *A pre-additive category is a category $\mathcal{C}$ such that for any $X, Y \in \mathcal{C}$, $\mathrm{Hom}_{\mathcal{C}}(X, Y)$ is endowed with a structure of an additive group and the composition map $\circ$ is bilinear.*

Example 8.2.2. $\mathrm{Mod}(\mathbb{Z})$ is a pre-additive category.

Lemma 8.2.3. *Let X_1 and X_2 be objects of a pre-additive category $\mathcal{C}$.*

(i) *Assume that $X_1 \times X_2$ exists in C and denote by $p_k \colon X_1 \times X_2 \to X_k$ the projection $(k = 1, 2)$. Let $i_k \colon X_k \to X_1 \times X_2$ be the morphism defined by*

(8.2.1)
$$p_j \circ i_k = \begin{cases} \mathrm{id}_{X_k} & \text{if } j = k \,, \\ 0 & \text{if } j \neq k. \end{cases}$$

Then, we have

(8.2.2)
$$i_1 \circ p_1 + i_2 \circ p_2 = \mathrm{id}_{X_1 \times X_2} \,.$$

(ii) *Conversely, let $Z \in C$ and let $p_k \colon Z \to X_k$ and $i_k \colon X_k \to Z$ be morphisms $(k = 1, 2)$ satisfying (8.2.1) and (8.2.2). Then Z is a product of X_1 and X_2 by (p_1, p_2) and a coproduct by (i_1, i_2).*

Proof. (i) We have

$$p_1 \circ (i_1 \circ p_1 + i_2 \circ p_2) = (p_1 \circ i_1) \circ p_1 + (p_1 \circ i_2) \circ p_2 = p_1 = p_1 \circ \mathrm{id}_{X_1 \times X_2} \,.$$

Similarly,

$$p_2 \circ (i_1 \circ p_1 + i_2 \circ p_2) = p_2 \circ \mathrm{id}_{X_1 \times X_2} \,.$$

Hence, $i_1 \circ p_1 + i_2 \circ p_2 = \mathrm{id}_{X_1 \times X_2}$.

(ii) For any $Y \in C$, write

$$\widetilde{Z} := \mathrm{Hom}_C(Y, Z) \in \mathrm{Mod}(\mathbb{Z}) \,,$$
$$\widetilde{X}_k := \mathrm{Hom}_C(Y, X_k) \in \mathrm{Mod}(\mathbb{Z}), \quad k = 1, 2 \,.$$

The morphisms $\widetilde{X}_k \xrightarrow{\widetilde{i}_k} \widetilde{Z} \xrightarrow{\widetilde{p}_k} \widetilde{X}_k$ satisfy the conditions similar to (8.2.1) and (8.2.2) and we get an isomorphism $\widetilde{Z} \xrightarrow{\sim} \widetilde{X}_1 \times \widetilde{X}_2$ by a classical result of additive groups. Hence, Z is a product of X_1 and X_2. By reversing the arrows, we find that Z is a coproduct of X_1 and X_2.   q.e.d.

We can reformulate Lemma 8.2.3.

Corollary 8.2.4. *Let C be a pre-additive category and let X_1, $X_2 \in C$. If $X_1 \times X_2$ exists in C, then $X_1 \sqcup X_2$ also exists. Moreover denoting by $i_j \colon X_j \to X_1 \sqcup X_2$ and $p_j \colon X_1 \times X_2 \to X_j$ the j-th co-projection and projection, the morphism*
$$r \colon X_1 \sqcup X_2 \to X_1 \times X_2$$

given by
$$p_j \circ r \circ i_k = \begin{cases} \mathrm{id}_{X_k} & \text{if } j = k \,, \\ 0 & \text{if } j \neq k \,. \end{cases}$$

is an isomorphism

Notation 8.2.5. (i) The object Z in Lemma 8.2.3 (ii) is denoted by $X_1 \oplus X_2$ and is called a direct sum of X_1 and X_2. Note that a direct sum of X_1 and X_2 is also a product as well as a coproduct of X_1 and X_2.
(ii) For historical reasons, if $\{X_i\}_{i \in I}$ is a small family of objects of $\mathcal{C}$ and the coproduct $\coprod_{i \in I} X_i$ exists in $\mathcal{C}$, it is denoted by $\bigoplus_{i \in I} X_i$ and still called the direct sum of the X_i's.

Corollary 8.2.6. *Let $\mathcal{C}$ be a pre-additive category, $X, Y \in \mathcal{C}$ and $f_1, f_2 \in \mathrm{Hom}_{\mathcal{C}}(X, Y)$. Assume that the direct sums $X \oplus X$ and $Y \oplus Y$ exist. Then $f_1 + f_2 \in \mathrm{Hom}_{\mathcal{C}}(X, Y)$ coincides with the composition*

$$X \xrightarrow{\delta_X} X \oplus X \xrightarrow{f_1 \oplus f_2} Y \oplus Y \xrightarrow{\sigma_Y} Y \ .$$

Here $\delta_X \colon X \to X \times X \simeq X \oplus X$ is the diagonal morphism and $\sigma_Y \colon Y \oplus Y \simeq Y \sqcup Y \to Y$ is the codiagonal morphism.

Proof. Let $i_j \colon X \to X \oplus X$ and $p_j \colon X \oplus X \to X$ be the j-th co-projection and projection. Then we have $p_1 \circ (i_1 + i_2) = p_1 \circ i_1 + p_1 \circ i_2 = \mathrm{id}_X = p_1 \circ \delta_X$ and similarly $p_2 \circ (i_1 + i_2) = p_2 \circ \delta_X$. Hence we obtain $i_1 + i_2 = \delta_X$. On the other hand we have $\sigma_Y \circ (f_1 \sqcup f_2) \circ i_j = f_j$, which implies

$$\begin{aligned}
\sigma_Y \circ (f_1 \oplus f_2) \circ \delta_X &= \sigma_Y \circ (f_1 \sqcup f_2) \circ (i_1 + i_2) \\
&= \sigma_Y \circ (f_1 \sqcup f_2) \circ i_1 + \sigma_Y \circ (f_1 \sqcup f_2) \circ i_2 \\
&= f_1 + f_2 \ .
\end{aligned}$$

q.e.d.

Definition 8.2.7. *Let $F \colon \mathcal{C} \to \mathcal{C}'$ be a functor of pre-additive categories. We say that F is* additive *if the map $\mathrm{Hom}_{\mathcal{C}}(X, Y) \to \mathrm{Hom}_{\mathcal{C}'}(F(X), F(Y))$ is additive for any $X, Y \in \mathcal{C}$.*

Definition 8.2.8. *An* additive category *is a category $\mathcal{C}$ satisfying the conditions (i)–(iv) below.*

  (i) *$\mathcal{C}$ has a zero object, denoted by 0.*
 (ii) *For any $X_1, X_2 \in \mathcal{C}$, the product $X_1 \times X_2$ and the coproduct $X_1 \sqcup X_2$ exist.*
(iii) *For any $X_1, X_2 \in \mathcal{C}$, define the morphism $r \colon X_1 \sqcup X_2 \to X_1 \times X_2$ as follows: the composition $X_k \to X_1 \sqcup X_2 \xrightarrow{r} X_1 \times X_2 \to X_j$ is 0 if $j \neq k$ and is id_{X_k} if $j = k$. Then r is an isomorphism.*
      (Recall that, for $X, Y \in \mathcal{C}$, the zero morphism $0 \colon X \to Y$ is the composition $X \to 0 \to Y$).
 (iv) *For any $X \in \mathcal{C}$, there exists $a \in \mathrm{Hom}_{\mathcal{C}}(X, X)$ such that the composition*

$$X \xrightarrow{\delta_X} X \times X \xrightarrow{(a, \mathrm{id}_X)} X \times X \xleftarrow[r]{\sim} X \sqcup X \xrightarrow{\sigma_X} X$$

is the zero morphism. Here, δ_X is the diagonal morphism and σ_X is the codiagonal morphism.

Note that if $\mathcal{C}$ is additive, then so is $\mathcal{C}^{\mathrm{op}}$.

Lemma 8.2.9. *Let $\mathcal{C}$ be a pre-additive category which admits finite products. Then $\mathcal{C}$ is additive.*

Proof. This follows from Lemma 8.2.3 and Corollary 8.2.6. Note that the morphism a in Definition 8.2.8 (iv) is given by $-\mathrm{id}_X$.   q.e.d.

Lemma 8.2.10. *Let $\mathcal{C}$ be an additive category. Then any $X \in \mathcal{C}$ has a structure of a commutative group object.*

Proof. We define the composition morphism $\mu\colon X \times X \to X$ by the composition

$$X \times X \xleftarrow[\ r\ ]{\sim} X \sqcup X \xrightarrow{\ \sigma_X\ } X \ .$$

Then μ satisfies the associative law thanks to the commutative diagram below:

$$
\begin{array}{ccccc}
X \times X \times X & \xleftarrow[r \times X]{\sim} & (X \sqcup X) \times X & \xrightarrow{\sigma_X \times X} & X \times X \\
{\scriptstyle\sim}\big\uparrow{\scriptstyle X \times r} & & {\scriptstyle\sim}\big\uparrow & & {\scriptstyle\sim}\big\uparrow{\scriptstyle r} \\
X \times (X \sqcup X) & \xleftarrow{\ \sim\ } & X \sqcup X \sqcup X & \xrightarrow{\sigma_X \sqcup X} & X \sqcup X \\
\big\downarrow{\scriptstyle X \times \sigma_X} & & \big\downarrow{\scriptstyle X \sqcup \sigma_X} & & \big\downarrow{\scriptstyle \sigma_X} \\
X \times X & \xleftarrow[\ r\ ]{\sim} & X \sqcup X & \xrightarrow{\ \sigma_X\ } & X \ .
\end{array}
$$

The inverse morphism $a\colon X \to X$ is given by Definition 8.2.8 (iv). It is easily checked that these data give a commutative group structure on X.   q.e.d.

In the sequel, we shall denote by *for* the forgetful functor $\mathrm{Mod}(\mathbb{Z}) \to \mathbf{Set}$.

Lemma 8.2.11. *Let $\mathcal{C}$ be an additive category and let $F\colon \mathcal{C} \to \mathrm{Mod}(\mathbb{Z})$ be a functor commuting with finite products. For any $X \in \mathcal{C}$, the addition map $F(X) \times F(X) \to F(X)$ of the additive group $F(X)$ is given by the composition*

$$\xi\colon \quad F(X) \times F(X) \xleftarrow{\ \sim\ } F(X \times X) \xleftarrow[F(r)]{\sim} F(X \sqcup X) \xrightarrow{F(\sigma_X)} F(X) \ .$$

Proof. Let $i_\nu\colon F(X) \to F(X) \times F(X)$ ($\nu = 1, 2$) be the map given by $i_1(x) = (x, 0)$, $i_2(x) = (0, x)$. By the commutative diagram

$$
\begin{array}{ccccccc}
F(X) & \xleftarrow{\ \sim\ } & F(X \times 0) & \xleftarrow{\ \sim\ } & F(X \sqcup 0) & \longrightarrow & F(X) \\
{\scriptstyle i_1}\big\downarrow & & \big\downarrow & & \big\downarrow & & \big\downarrow{\scriptstyle \mathrm{id}} \\
F(X) \times F(X) & \xleftarrow{\ \sim\ } & F(X \times X) & \xleftarrow{\ \sim\ } & F(X \sqcup X) & \longrightarrow & F(X) \ ,
\end{array}
$$

we obtain $\xi \circ i_1 = \mathrm{id}_{F(X)}$. Similarly, $\xi \circ i_2 = \mathrm{id}_{F(X)}$. Since ξ is a morphism in $\mathrm{Mod}(\mathbb{Z})$, we obtain the result.   q.e.d.

Proposition 8.2.12. *Let C be an additive category and let $F, F' \colon C \to \mathrm{Mod}(\mathbb{Z})$ be functors commuting with finite products. Then*

$$\mathrm{Hom}_{\,\mathrm{Fct}(C, \mathrm{Mod}(\mathbb{Z}))}(F, F') \xrightarrow{\sim} \mathrm{Hom}_{\,\mathrm{Fct}(C, \mathbf{Set})}(\textit{for} \circ F, \textit{for} \circ F') \,.$$

Proof. The injectivity of the map is obvious since *for* is faithful. Let us prove the surjectivity.

Let $\varphi \colon \textit{for} \circ F \to \textit{for} \circ F'$ be a morphism of functors. By Lemma 8.2.11, the map $\textit{for} \circ F(X) \to \textit{for} \circ F'(X)$ commutes with the addition map, and hence it gives a morphism $\widetilde{\varphi}(X) \colon F(X) \to F'(X)$ in $\mathrm{Mod}(\mathbb{Z})$. It is easily checked that the family of morphisms $\{\widetilde{\varphi}(X)\}_{X \in C}$ defines a morphism $F \to F'$.   q.e.d.

Proposition 8.2.13. *Let C be an additive category and let $F \colon C \to \mathbf{Set}$ be a functor commuting with finite products. Then there is a functor $\widetilde{F} \colon C \to \mathrm{Mod}(\mathbb{Z})$ such that F is isomorphic to the composition $C \xrightarrow{\widetilde{F}} \mathrm{Mod}(\mathbb{Z}) \xrightarrow{\textit{for}} \mathbf{Set}$. Moreover, such an $\widetilde{F}$ is unique up to unique isomorphism.*

Proof. Any $X \in C$ has a structure of a commutative group object. Hence, by Lemma 8.1.2, $F(X)$ has a structure of a commutative group object, hence defines an object $\widetilde{F}(X) \in \mathrm{Mod}(\mathbb{Z})$. The uniqueness follows from Proposition 8.2.12.   q.e.d.

Theorem 8.2.14. *Let C be an additive category. Then C has a unique structure of a pre-additive category.*

Proof. Let $X \in C$. By applying Proposition 8.2.13 and 8.2.12 to the functor $F = \mathrm{Hom}_C(X, \cdot)$, we obtain that $\mathrm{Hom}_C(X, Y)$ has a structure of an additive group. For $f, g \in \mathrm{Hom}_C(X, Y)$, $f + g \in \mathrm{Hom}_C(X, Y)$ is given by the composition

(8.2.3)

$$
\begin{array}{ccccc}
X & \xrightarrow{\ \delta_X\ } & X \times X & \xrightarrow{\ f \times g\ } & Y \times Y \\
 & & \big\uparrow{\scriptstyle \sim} & & \big\uparrow{\scriptstyle \sim} \\
 & & X \sqcup X & \xrightarrow{\ f \sqcup g\ } & Y \sqcup Y \xrightarrow{\ \sigma_Y\ } Y \,.
\end{array}
$$

Hence, $+$ is symmetric by reversing the arrows.

For $h \in \mathrm{Hom}_C(W, X)$, $\mathrm{Hom}_C(X, \cdot) \xrightarrow{\circ h} \mathrm{Hom}_C(W, \cdot)$ is a morphism in $\mathrm{Fct}(C, \mathrm{Mod}(\mathbb{Z}))$ by Proposition 8.2.13. Hence, $(f + g) \circ h = f \circ h + g \circ h$ for $f, g \in \mathrm{Hom}_C(X, Y)$. By reversing the arrows we obtain $k \circ (f + g) = k \circ f + k \circ g$ for $k \in \mathrm{Hom}_C(Y, Z)$. Thus C has a structure of a pre-additive category.

Conversely, if C has a structure of a pre-additive category, then for f, $g \in \mathrm{Hom}_C(X, Y)$, $f + g$ is given by (8.2.3) in virtue of Corollary 8.2.6.   q.e.d.

Proposition 8.2.15. *Let C and C' be additive categories and let $F \colon C \to C'$ be a functor. Then F is an additive functor if and only if it commutes with finite products.*

Proof. For any $X \in C$, we have two functors $\alpha, \beta \colon C \to \mathrm{Mod}(\mathbb{Z})$ given by $C \ni Y \mapsto \alpha(Y) := \mathrm{Hom}_C(X, Y)$ and $C \ni Y \mapsto \beta(Y) := \mathrm{Hom}_{C'}(F(X), F(Y))$. Then α and β commute with finite products and hence the canonical morphism *for* $\circ \alpha \to$ *for* $\circ \beta$ lifts to a morphism $\alpha \to \beta$ by Proposition 8.2.12.   q.e.d.

Corollary 8.2.16. *Let C and C' be additive categories and let $F \colon C \to C'$ be a fully faithful functor. Then F is additive.*

Proof. Let $X, Y \in C$. We endow the set $\mathrm{Hom}_C(X, Y)$ with the additive group structure inherited from the bijection $\mathrm{Hom}_C(X, Y) \simeq \mathrm{Hom}_{C'}(F(X), F(Y))$. This defines a pre-additive structure on C, and this structure coincides with the original one by Theorem 8.2.14. Hence F is additive.   q.e.d.

Examples 8.2.17. (i) If R is a ring, $\mathrm{Mod}(R)$, $\mathrm{Mod}^f(R)$ and $\mathrm{Mod}^{fp}(R)$ (see Example 1.2.4 (iv)) are additive categories.
(ii) **Ban**, the category of $\mathbb{C}$-Banach spaces and linear continuous maps is additive.
(iii) Let I be a small category. If C is additive, the category $\mathrm{Fct}(I, C)$ of functors from I to C, is additive.

All along this book we shall encounter sequences of morphisms in additive categories.

Definition 8.2.18. *A complex $X^\bullet$ in an additive category C is a sequence of objects $\{X^j\}_{j \in \mathbb{Z}}$ and morphisms $d_X^j \colon X^j \to X^{j+1}$ such that $d_X^j \circ d_X^{j-1} = 0$ for all j.*

Remark 8.2.19. We shall also encounter finite sequences of morphisms

$$X^j \xrightarrow{d^j} X^{j+1} \xrightarrow{d^{j+1}} \cdots \xrightarrow{d^{k-1}} X^k$$

such that $d^n \circ d^{n-1} = 0$ when it is defined. In such a case we also call such a sequence a (finite) complex. We sometimes identify it with the complex

$$\cdots \to 0 \to X^j \xrightarrow{d^j} X^{j+1} \to \cdots \to X^k \to 0 \to \cdots .$$

In particular, $X' \xrightarrow{f} X \xrightarrow{g} X''$ is a complex if and only if $g \circ f = 0$.

In the subsequent chapters we shall often encounter diagrams in additive categories which commute up to sign.

Definition 8.2.20. *Let $\varepsilon = \pm 1$. A diagram in an additive category* $\begin{array}{ccc} X & \xrightarrow{f} & Y \\ h\downarrow & & g\downarrow \\ V & \xrightarrow{k} & Z \end{array}$

is ε-commutative if $g \circ f = \varepsilon(k \circ h)$. If it is (-1)-commutative, we say also that it is anti-commutative *(or anti-commutes).*

Convention 8.2.21. All along this book, a diagram in an additive category with horizontal and vertical arrows will be called *a diagram of complexes* if all rows and all columns are complexes.

8.3 Abelian Categories

From now on, $\mathcal{C}$, $\mathcal{C}'$ will denote additive categories.

Definition 8.3.1. *Let* $f: X \to Y$ *be a morphism in* $\mathcal{C}$.

(i) *The* kernel *of* f, *if it exists, is the fiber product of* $X \xrightarrow{f} Y \leftarrow 0$, *that is,* $X \times_Y 0$. *It is denoted by* $\operatorname{Ker} f$. *Equivalently,* $\operatorname{Ker} f$ *is the equalizer of the parallel arrows* $f, 0: X \rightrightarrows Y$.

(ii) *The* cokernel *of* f, *if it exists, is the kernel of* f *in* $\mathcal{C}^{\mathrm{op}}$. *It is denoted by* $\operatorname{Coker} f$. *Equivalently,* $\operatorname{Coker} f$ *is the co-equalizer of the parallel arrows* $f, 0: X \rightrightarrows Y$.

Note that for a pair of parallel arrows $f, g: X \rightrightarrows Y$, we have $\operatorname{Ker}(f, g) = \operatorname{Ker}(f - g)$ and $\operatorname{Coker}(f, g) = \operatorname{Coker}(f - g)$.

By its definition, $\operatorname{Ker} f$ is a representative of the contravariant functor

$$\operatorname{Ker}(\operatorname{Hom}_{\mathcal{C}}(\cdot, f)): Z \mapsto \operatorname{Ker}\big(\operatorname{Hom}_{\mathcal{C}}(Z, X) \to \operatorname{Hom}_{\mathcal{C}}(Z, Y)\big).$$

Here, Ker on the right hand side is the kernel in the category of additive groups, that is, the inverse image of $\{0\}$.

Hence, if $\operatorname{Ker} f$ exists, it is unique up to a unique isomorphism, and there is a morphism $h: \operatorname{Ker} f \to X$ with $f \circ h = 0$ and such that any $g: W \to X$ with $f \circ g = 0$ factorizes uniquely through h. This can be visualized by the diagram:

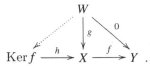

Recall that

(8.3.1) h is a monomorphism.

Hence $\operatorname{Ker} h \simeq 0$. Also note that $\operatorname{Ker} f \simeq 0$ if and only if f is a monomorphism. Finally, note that $\operatorname{Ker} f \xrightarrow{\sim} X$ if and only if f is the zero morphism.

Similarly, $\operatorname{Coker} f$ is a representative of the functor

$$\operatorname{Ker}(\operatorname{Hom}_{\mathcal{C}}(f, \cdot)): Z \mapsto \operatorname{Ker}\big(\operatorname{Hom}_{\mathcal{C}}(Y, Z) \to \operatorname{Hom}_{\mathcal{C}}(X, Z)\big).$$

If $\operatorname{Coker} f$ exists, it is unique up to a unique isomorphism, and there is a morphism $k: Y \to \operatorname{Coker} f$ with $k \circ f = 0$ and such that any $g: Y \to W$ with $g \circ f = 0$ factorizes uniquely through k. The cokernel may be visualized by the diagram:

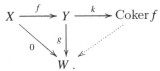

Note that

(8.3.2) k is an epimorphism,

and $\operatorname{Coker} f \simeq 0$ if and only if f is an epimorphism.

Example 8.3.2. Let R be a ring. The kernel of a morphism $f \colon M \to N$ in $\operatorname{Mod}(R)$ is the R-module $f^{-1}(0)$ and the cokernel of f is the quotient R-module $N/f(M)$. Let I be a left ideal which is not finitely generated and let $M = R/I$. Then the natural morphism $R \to M$ has no kernel in $\operatorname{Mod}^{\mathrm{f}}(R)$.

Let $\mathcal{C}$ be an additive category in which every morphism admits a kernel and a cokernel. Recall that (see Proposition 2.2.4):

$$Y_0 \sqcup_X Y_1 \simeq \operatorname{Coker}(i_0 \circ f_0 - i_1 \circ f_1 \colon X \to Y_0 \oplus Y_1)$$
$$\text{for morphisms } f_0 \colon X \to Y_0 \text{ and } f_1 \colon X \to Y_1 \,,$$
$$X_0 \times_Y X_1 \simeq \operatorname{Ker}(g_0 \circ p_0 - g_1 \circ p_1 \colon X_0 \oplus X_1 \to Y)$$
$$\text{for morphisms } g_0 \colon X_0 \to Y \text{ and } g_1 \colon X_1 \to Y \,.$$

Here, $i_\nu \colon Y_\nu \to Y_0 \oplus Y_1$ is the co-projection and $p_\nu \colon X_0 \oplus X_1 \to X_\nu$ the projection ($\nu = 0, 1$).

Notation 8.3.3. We shall often write $Y_0 \oplus_X Y_1$ instead of $Y_0 \sqcup_X Y_1$.

Also recall the image and coimage of a morphism given in Definition 5.1.1:

$$\operatorname{Coim} f = \operatorname{Coker}(X \times_Y X \rightrightarrows X) \,,$$
$$\operatorname{Im} f = \operatorname{Ker}(Y \rightrightarrows Y \oplus_X Y) \,.$$

Proposition 8.3.4. *Let $\mathcal{C}$ be an additive category which admits kernels and cokernels. Let $f \colon X \to Y$ be a morphism in $\mathcal{C}$. We have*

$$\operatorname{Coim} f \simeq \operatorname{Coker} h, \text{ where } h \colon \operatorname{Ker} f \to X \,,$$
$$\operatorname{Im} f \simeq \operatorname{Ker} k, \text{ where } k \colon Y \to \operatorname{Coker} f \,.$$

Proof. It is enough to treat Coim. Recall that $p_1, p_2 \colon X \times_Y X \rightrightarrows X$ denote the two canonical morphisms. Let $Z \in \mathcal{C}$. By the definition of Coim, we have

$$\operatorname{Hom}_{\mathcal{C}}(\operatorname{Coim} f, Z) \simeq \{u \colon X \to Z; u \circ p_1 = u \circ p_2\} \,.$$

Using the definition of $X \times_Y X$, we also have

$$\operatorname{Hom}_{\mathcal{C}}(\operatorname{Coim} f, Z) \simeq \{u \colon X \to Z; u \circ \varphi_1 = u \circ \varphi_2 \text{ for any } W \in \mathcal{C} \text{ and}$$
$$\text{any } \varphi_1, \varphi_2 \in \operatorname{Hom}_{\mathcal{C}}(W, X) \text{ with } f \circ \varphi_1 = f \circ \varphi_2\} \,.$$

The condition on u is equivalent to

$$u \circ \varphi = 0 \text{ for any } W \in \mathcal{C} \text{ and any } \varphi \in \operatorname{Hom}_{\mathcal{C}}(W, X) \text{ with } f \circ \varphi = 0 \,.$$

Since such a φ factors uniquely through $h\colon \mathrm{Ker}\, f \to X$, we obtain

$$\mathrm{Hom}_{\mathcal{C}}(\mathrm{Coim}\, f, Z) \simeq \{u\colon X \to Z; u \circ h = 0\}$$
$$\simeq \mathrm{Hom}_{\mathcal{C}}(\mathrm{Coker}\, h, Z)$$

functorially in Z. Hence, $\mathrm{Coim}\, f \simeq \mathrm{Coker}\, h$.   q.e.d.

Applying Propositions 8.3.4 and 5.1.2, we get a natural morphism $\mathrm{Coim}\, f \xrightarrow{u} \mathrm{Im}\, f$. This morphism is described by the diagram (see Proposition 5.1.2):

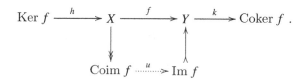

Definition 8.3.5. *An additive category $\mathcal{C}$ is* abelian *if:*

(i) *any morphism admits a kernel and a cokernel,*
(ii) *any morphism f in $\mathcal{C}$ is strict (see Definition 5.1.4), i.e., the natural morphism $\mathrm{Coim}\, f \to \mathrm{Im}\, f$ is an isomorphism.*

Recall that in an additive category, a morphism f is a monomorphism (resp. an epimorphism) if and only if $\mathrm{Ker}\, f \simeq 0$ (resp. $\mathrm{Coker}\, f \simeq 0$). In an abelian category, a morphism which is both a monomorphism and an epimorphism is an isomorphism (see Proposition 5.1.5 (ii)).

Note that abelian categories admit finite inductive limits and finite projective limits.

Remark 8.3.6. The following assertions are easily checked.

(i) If $\{\mathcal{C}_i\}_{i \in I}$ is a small family of abelian categories, then the product category $\prod_{i \in I} \mathcal{C}_i$ is abelian.
(ii) Let I be a small category. If $\mathcal{C}$ is abelian, the category $\mathcal{C}^I$ of functors from I to $\mathcal{C}$ is abelian. For example, if F, $G\colon I \to \mathcal{C}$ are two functors and $\varphi\colon F \to G$ is a morphism of functors, define the functor N by $N(X) := \mathrm{Ker}(F(X) \to G(X))$. Clearly, N is a kernel of φ.
(iii) If $\mathcal{C}$ is abelian, then the opposite category $\mathcal{C}^{\mathrm{op}}$ is abelian. Note that for a morphism $f\colon X \to Y$ in $\mathcal{C}$, we have $\mathrm{Ker}(f^{\mathrm{op}}) \simeq (\mathrm{Coker}(f))^{\mathrm{op}}$, $\mathrm{Coker}(f^{\mathrm{op}}) \simeq (\mathrm{Ker}(f))^{\mathrm{op}}$, $\mathrm{Im}(f^{\mathrm{op}}) \simeq (\mathrm{Coim}(f))^{\mathrm{op}}$ and $\mathrm{Coim}(f^{\mathrm{op}}) \simeq (\mathrm{Im}(f))^{\mathrm{op}}$.

Examples 8.3.7. (i) If R is a ring, $\mathrm{Mod}(R)$ is an abelian category.

(ii) The category $\mathrm{Mod}^{\mathrm{f}}(R)$ is abelian if and only if R is a Noether ring. If $\mathrm{Mod}^{\mathrm{fp}}(R)$ is abelian, we say that R is coherent .

(iii) The category **Ban** of Banach spaces over $\mathbb{C}$ admits kernels and cokernels. If $f\colon X \to Y$ is a morphism of Banach spaces, then $\mathrm{Ker}\, f = f^{-1}(0)$ and

Coker $f = Y/\overline{\mathrm{Im}\, f}$ where $\overline{\mathrm{Im}\, f}$ denotes the closure of the vector space $\mathrm{Im}\, f$. It is well-known that there exist continuous linear maps f which are injective, with dense and non closed image. For such an f, $\mathrm{Ker}\, f = \mathrm{Coker}\, f = 0$, $\mathrm{Coim}\, f \simeq X$ and $\mathrm{Im}\, f \simeq Y$, but $\mathrm{Coim}\, f \to \mathrm{Im}\, f$ is not an isomorphism. Thus **Ban** is not abelian. However, **Ban** is a quasi-abelian category in the sense of J-P. Schneiders [61].

Unless otherwise stated, $\mathcal{C}$ is assumed to be *abelian until the end of this section.*

Consider a complex

(8.3.3) $$ X' \xrightarrow{f} X \xrightarrow{g} X'' \quad (\text{hence } g \circ f = 0). $$

Since $\mathrm{Im}\, f \to X \to X''$ is zero, $\mathrm{Im}\, f \to X$ factors through $\mathrm{Ker}\, g$. Similarly, $X \to \mathrm{Im}\, g$ factors through $\mathrm{Coker}\, f$. We thus have a commutative diagram

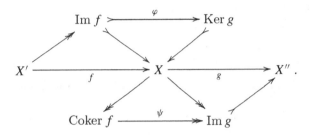

Note that φ is a monomorphism and ψ is an epimorphism. Let $u \colon \mathrm{Ker}\, g \to \mathrm{Coker}\, f$ be the composition $\mathrm{Ker}\, g \to X \to \mathrm{Coker}\, f$. We have the morphisms $\mathrm{Im}\, f \rightarrowtail \mathrm{Ker}\, u \rightarrowtail \mathrm{Ker}(X \to \mathrm{Coker}\, f) \simeq \mathrm{Im}\, f$. Therefore $\mathrm{Ker}\, u \simeq \mathrm{Im}\, f$. Similarly, $\mathrm{Coker}\, u \simeq \mathrm{Im}\, g$. Since $\mathrm{Im}\, u \simeq \mathrm{Coim}\, u$, we get the isomorphisms

(8.3.4)
$$ \mathrm{Im}\, u \simeq \mathrm{Coker}(\mathrm{Im}\, f \to \mathrm{Ker}\, g) \simeq \mathrm{Coker}(X' \to \mathrm{Ker}\, g) $$
$$ \simeq \mathrm{Ker}(\mathrm{Coker}\, f \to \mathrm{Im}\, g) \simeq \mathrm{Ker}(\mathrm{Coker}\, f \to X''). $$

Therefore the conditions below are equivalent

(8.3.5)
$$ u = 0 \iff \mathrm{Im}\, f \xrightarrow{\sim} \mathrm{Ker}\, g \iff X' \twoheadrightarrow \mathrm{Ker}\, g $$
$$ \iff \mathrm{Coker}\, f \xrightarrow{\sim} \mathrm{Im}\, g \iff \mathrm{Coker}\, f \rightarrowtail X''. $$

Definition 8.3.8. *Consider a complex* $X' \xrightarrow{f} X \xrightarrow{g} X''$ *as in* (8.3.3).

(i) *We shall denote by* $H(X' \xrightarrow{f} X \xrightarrow{g} X'')$ *any of the isomorphic objects in* (8.3.4) *and call it the* cohomology *of the complex* (8.3.3).

(ii) *The complex* (8.3.3) *is* exact *if the equivalent conditions in* (8.3.5) *are satisfied, that is, if* $H(X' \xrightarrow{f} X \xrightarrow{g} X'') \simeq 0$.

(iii) *More generally, a complex* $X^j \to \cdots \to X^k$ *is exact if any sequence* $X^{n-1} \to X^n \to X^{n+1}$ *extracted from this complex is exact.*
An exact complex is also often called an exact sequence.

Convention 8.3.9. All along this book, a diagram of complexes in an abelian category (see Convention 8.2.21) will be called *an exact diagram* if all rows and all columns are exact.

Note that the complex (8.3.3) is exact if and only if it is exact in $\mathcal{C}^{\mathrm{op}}$. Indeed, we have (see Remark 8.3.6 (iii))

$$H(X''^{\mathrm{op}} \xrightarrow{g^{\mathrm{op}}} X^{\mathrm{op}} \xrightarrow{f^{\mathrm{op}}} X'^{\mathrm{op}}) \simeq H(X' \xrightarrow{f} X \xrightarrow{g} X'')^{\mathrm{op}} .$$

A complex $0 \to X' \xrightarrow{f} X$ (resp. $X \xrightarrow{g} X'' \to 0$) is exact if and only if f is a monomorphism (resp. g is an epimorphism).

Note that a complex $X \xrightarrow{u} Y \overset{v}{\underset{w}{\rightrightarrows}} Z$ is exact in the sense of Definition 2.2.2 if and only if the sequence $0 \to X \xrightarrow{u} Y \xrightarrow{v-w} Z$ is exact.

Hence, a complex $0 \to X' \xrightarrow{f} X \xrightarrow{g} X''$ (resp. $X' \xrightarrow{f} X \xrightarrow{g} X'' \to 0$) is exact if and only if $X' \to \operatorname{Ker} g$ is an isomorphism (resp. $\operatorname{Coker} f \to X''$ is an isomorphism). A complex

$$0 \to X' \xrightarrow{f} X \xrightarrow{g} X'' \to 0$$

is exact if and only if $X' \to \operatorname{Ker} g$ and $\operatorname{Coker} f \to X''$ are isomorphisms. Such an exact complex is called *a short exact sequence.*
Any morphism $f : X \to Y$ may be decomposed into short exact sequences:

(8.3.6)
$$0 \to \operatorname{Ker} f \to X \to \operatorname{Im} f \to 0 ,$$
$$0 \to \operatorname{Im} f \to Y \to \operatorname{Coker} f \to 0.$$

Recalling Definition 2.2.7, we see that a square

(8.3.7)
$$\begin{array}{ccc} X' & \xrightarrow{f'} & Y' \\ {\scriptstyle g'}\downarrow & & \downarrow{\scriptstyle g} \\ X & \xrightarrow{f} & Y \end{array}$$

is Cartesian if and only if the sequence $0 \to X' \xrightarrow{(g',f')} X \oplus Y' \xrightarrow{(f,-g)} Y$ is exact. The square is co-Cartesian if and only if the sequence $X' \xrightarrow{(g',f')} X \oplus Y' \xrightarrow{(f,-g)} Y \to 0$ is exact.

Notations 8.3.10. Familiar notions for the categories of vector spaces are naturally extended to abelian categories. Let $Y \rightarrowtail X$ be a monomorphism. We sometimes identify Y with the isomorphism class of such monomorphisms, say abusively that Y is a subobject of X (see Definition 1.2.18), and write $Y \subset X$. Similarly, we sometimes abusively call the cokernel of $Y \to X$ a quotient of X and denote it by X/Y.

If X_1 and X_2 are subobjects of X, we sometimes set $X_1 \cap X_2 = X_1 \times_X X_2$, and $X_1 + X_2 = \mathrm{Im}(X_1 \oplus X_2 \to X)$. For a finite family of subobjects $\{X_i\}_{i \in I}$ of X we define similarly the subobjects $\bigcap_{i \in I} X_i$ and $\sum_i X_i$.

If $f \colon X \to Y$ is a morphism and Z is a subobject of Y, we set $f^{-1}(Z) = X \times_Y Z$.

We shall now prove some lemmas of constant use.

Lemma 8.3.11. *Consider the square* (8.3.7).

(a) *Assume that* (8.3.7) *is Cartesian.*
   (i) *We have* $\mathrm{Ker}\, f' \xrightarrow{\sim} \mathrm{Ker}\, f$,
   (ii) *if f is an epimorphism, then* (8.3.7) *is co-Cartesian and f' is an epimorphism.*
(b) *Assume that* (8.3.7) *is co-Cartesian.*
   (i) *We have* $\mathrm{Coker}\, f \xrightarrow{\sim} \mathrm{Coker}\, f'$,
   (ii) *if f' is a monomorphism, then* (8.3.7) *is Cartesian and f is a monomorphism.*

Proof. (a) (i) Let $S \in \mathcal{C}$. There is a chain of isomorphisms

$$
\begin{aligned}
\mathrm{Hom}\,(S, \mathrm{Ker}\, f') &\simeq \mathrm{Ker}\big(\mathrm{Hom}\,(S, X') \to \mathrm{Hom}\,(S, Y')\big) \\
&\simeq \mathrm{Ker}\big(\mathrm{Hom}\,(S, X) \times_{\mathrm{Hom}\,(S,Y)} \mathrm{Hom}\,(S, Y') \to \mathrm{Hom}\,(S, Y')\big) \\
&\simeq \mathrm{Ker}\big(\mathrm{Hom}\,(S, X) \to \mathrm{Hom}\,(S, Y)\big) \\
&\simeq \mathrm{Hom}\,(S, \mathrm{Ker}\, f)\,.
\end{aligned}
$$

(ii) If f is an epimorphism, then the sequence $0 \to X' \to X \oplus Y' \to Y \to 0$ is exact, hence the square is both Cartesian and co-Cartesian. Therefore $\mathrm{Coker}\, f' \simeq \mathrm{Coker}\, f$ by applying (i) with the arrows reversed.
(b) follows from (a) by reversing the arrows.   q.e.d.

Lemma 8.3.12. *Let $X' \xrightarrow{f} X \xrightarrow{g} X''$ be a complex (i.e., $g \circ f = 0$). Then the conditions below are equivalent:*

(i) *the complex $X' \xrightarrow{f} X \xrightarrow{g} X''$ is exact,*
(ii) *for any morphism $h \colon S \to X$ such that $g \circ h = 0$, there exist an epimorphism $f' \colon S' \twoheadrightarrow S$ and a commutative diagram*

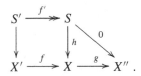

Proof. (i) $\Rightarrow$ (ii). It is enough to choose $X' \times_{\mathrm{Ker}\, g} S$ as S'. Since $X' \to \mathrm{Ker}\, g$ is an epimorphism, $S' \to S$ is an epimorphism by Lemma 8.3.11.

(ii) $\Rightarrow$ (i). Choose $S = \mathrm{Ker}\, g$. Then the composition $S' \to X' \to \mathrm{Ker}\, g$ is an epimorphism. Hence $X' \to \mathrm{Ker}\, g$ is an epimorphism.   q.e.d.

Lemma 8.3.13. [The "five lemma"] *Consider a commutative diagram whose rows are complexes*

$$
\begin{array}{ccccccc}
X^0 & \longrightarrow & X^1 & \longrightarrow & X^2 & \longrightarrow & X^3 \\
{\scriptstyle f^0}\downarrow & & {\scriptstyle f^1}\downarrow & & {\scriptstyle f^2}\downarrow & & {\scriptstyle f^3}\downarrow \\
Y^0 & \longrightarrow & Y^1 & \longrightarrow & Y^2 & \longrightarrow & Y^3 ,
\end{array}
$$

and assume that $X^1 \to X^2 \to X^3$ and $Y^0 \to Y^1 \to Y^2$ are exact sequences.

(i) *If f^0 is an epimorphism and f^1, f^3 are monomorphisms, then f^2 is a monomorphism.*

(ii) *If f^3 is a monomorphism and f^0, f^2 are epimorphisms, then f^1 is an epimorphism.*

The classical "five lemma" corresponds to the case of five morphisms $f^j : X^j \to Y^j$, $j = 0, \ldots, 4$ and exact complexes. It asserts that if f^0, f^1, f^3, f^4 are isomorphisms, then f^2 is also an isomorphism. Clearly, this is a consequence of Lemma 8.3.13.

Proof. (ii) is deduced from (i) by reversing the arrows. Hence, it is enough to prove (i). Let $h : S \to X^2$ be a morphism such that $h \circ f^2 = 0$. We shall prove that $h = 0$. The composition $S \to X^2 \to X^3 \xrightarrow{f^3} Y^3$ vanishes. Since f^3 is a monomorphism by the hypothesis, the composition $S \to X^2 \to X^3$ vanishes. Applying Lemma 8.3.12, there exist an epimorphism $S^1 \twoheadrightarrow S$ and a commutative solid diagram

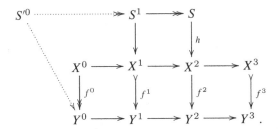

Since the composition $S^1 \to X^1 \to Y^1 \to Y^2$ vanishes, we find by applying again Lemma 8.3.12 that there exists an epimorphism $S'^0 \twoheadrightarrow S^1$ such that $S'^0 \to S^1 \to X^1 \to Y^1$ factors as $S'^0 \to Y^0 \to Y^1$.

Since $f^0 : X^0 \to Y^0$ is an epimorphism, there exists an epimorphism $S^0 \twoheadrightarrow S'^0$ such that $S^0 \to S'^0 \to Y^0$ factors through $S^0 \to X^0 \to Y^0$. We get a diagram

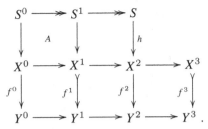

Note that the square diagram labeled "A" commutes since the two composi-
tions $S^0 \to S^1 \to X^1 \xrightarrow{f^1} Y^1$ and $S^0 \to X^0 \to X^1 \xrightarrow{f^1} Y^1$ coincide and f^1 is
a monomorphism. Therefore the composition $S^0 \to S^1 \to S \to X^2$ vanishes.
Since $S^0 \to S^1 \to S$ is an epimorphism, $S \to X^2$ vanishes. This shows that
$f^2 \colon X^2 \to Y^2$ is a monomorphism.   q.e.d.

Proposition 8.3.14. *Let* $0 \to X' \xrightarrow{f} X \xrightarrow{g} X'' \to 0$ *be a short exact sequence
in* $\mathcal{C}$. *Then the conditions below are equivalent:*

(i) *there exists* $h \colon X'' \to X$ *such that* $g \circ h = \mathrm{id}_{X''}$,
(ii) *there exists* $k \colon X \to X'$ *such that* $k \circ f = \mathrm{id}_{X'}$,
(iii) *there exist* $h \colon X'' \to X$ *and* $k \colon X \to X'$ *such that* $\mathrm{id}_X = f \circ k + h \circ g$,
(iv) *there exist* $\varphi = (k, g)$ *and* $\psi = (f, h)$ *such that* $X \xrightarrow{\varphi} X' \oplus X''$ *and*
$X' \oplus X'' \xrightarrow{\psi} X$ *are isomorphisms inverse to each other,*
(v) *for any* $Y \in \mathcal{C}$, *the map* $\mathrm{Hom}_{\mathcal{C}}(Y, X) \xrightarrow{g \circ} \mathrm{Hom}_{\mathcal{C}}(Y, X'')$ *is surjective,*
(vi) *for any* $Y \in \mathcal{C}$, *the map* $\mathrm{Hom}_{\mathcal{C}}(X, Y) \xrightarrow{\circ f} \mathrm{Hom}_{\mathcal{C}}(X', Y)$ *is surjective.*

Proof. (i) $\Rightarrow$ (iii). Since $g = g \circ h \circ g$, we get $g \circ (\mathrm{id}_X - h \circ g) = 0$, which implies
that $\mathrm{id}_X - h \circ g$ factors through $\mathrm{Ker}\, g$, that is, through X'. Hence, there exists
$k \colon X \to X'$ such that $\mathrm{id}_X - h \circ g = f \circ k$.
(iii) $\Rightarrow$ (i). Since $g \circ f = 0$, we find $g = g \circ h \circ g$, that is $(g \circ h - \mathrm{id}_{X''}) \circ g = 0$.
Since g is an epimorphism, this implies $g \circ h - \mathrm{id}_{X''} = 0$.
(iii) $\Leftrightarrow$ (ii) follows by reversing the arrows.
(iv) $\Leftrightarrow$ (iii) is obvious, as well as (i) $\Leftrightarrow$ (v) and (ii) $\Leftrightarrow$ (vi).   q.e.d.

Definition 8.3.15. *If the equivalent conditions of Proposition* 8.3.14 *are satisfied, we say that the short exact sequence* splits.

Note that an additive functor of abelian categories sends split short exact
sequences to split short exact sequences.

Definition 8.3.16. *An abelian category is called* semisimple *if all short exact
sequences split.*

For another characterization of semisimplicity, see §13.1.

Examples 8.3.17. (i) In the category $\mathrm{Mod}(\mathbb{Z})$, the exact sequence $0 \to \mathbb{Z} \xrightarrow{2 \cdot}$
$\mathbb{Z} \to \mathbb{Z}/2\mathbb{Z} \to 0$ does not split.
(ii) If k is a field, then $\mathrm{Mod}(k)$ is semisimple.

Proposition 8.3.18. *Let* $F \colon \mathcal{C} \to \mathcal{C}'$ *be an additive functor of abelian categories. Then* F *is left exact if and only if it commutes with kernels, that is, if and only if, for any exact sequence* $0 \to X' \to X \to X''$ *in* $\mathcal{C}$, *the sequence* $0 \to F(X') \to F(X) \to F(X'')$ *is exact.*

Proof. Applying Proposition 3.3.3, we find that $F \colon \mathcal{C} \to \mathcal{C}'$ is left exact if and only if it commutes with finite projective limits. Since F is additive, it commutes with finite products. Therefore, F commutes with finite projective limits if and only if it commutes with kernels, by Proposition 2.2.9.   q.e.d.

Similarly, an additive functor F is right exact if and only if it commutes with cokernels, that is, if and only if if for any exact sequence $X' \to X \to X'' \to 0$ in $\mathcal{C}$, the sequence $F(X') \to F(X) \to F(X'') \to 0$ is exact.

Recall that a contravariant functor $G \colon \mathcal{C} \to \mathcal{C}'$ is a functor from $\mathcal{C}^{\mathrm{op}}$ to $\mathcal{C}'$. Hence a contravariant functor G is left (resp. right) exact if and only if it sends an exact sequence $X' \to X \to X'' \to 0$ (resp. $0 \to X' \to X \to X''$) to an exact sequence $0 \to G(X'') \to G(X) \to G(X')$ (resp. $G(X'') \to G(X) \to G(X') \to 0$).

Note that F is left exact if and only if for any exact sequence $0 \to X' \to X \to X'' \to 0$ in $\mathcal{C}$, the sequence $0 \to F(X') \to F(X) \to F(X'')$ is exact, and similarly for right exact functors. Moreover F is exact if and only if for any exact sequence $X' \to X \to X''$ in $\mathcal{C}$, the sequence $F(X') \to F(X) \to F(X'')$ is exact. (See Exercise 8.17.)

Recall (see Proposition 3.3.7) that the functor $\mathrm{Hom}_{\mathcal{C}} \colon \mathcal{C}^{\mathrm{op}} \times \mathcal{C} \to \mathrm{Mod}(\mathbb{Z})$ is left exact with respect to each of its arguments. Moreover, if $F \colon \mathcal{C} \to \mathcal{C}'$ and $G \colon \mathcal{C}' \to \mathcal{C}$ are two functors, and F is a left adjoint to G, then F is right exact and G is left exact.

Example 8.3.19. Let k be a field and let $A = k[x]$. Consider the additive functor $F \colon \mathrm{Mod}(A) \to \mathrm{Mod}(A)$ given by $M \mapsto x \cdot M$. Then F sends a monomorphism to a monomorphism and an epimorphism to an epimorphism. On the other-hand, consider the exact sequence $0 \to x \cdot A \to A \to A/(x \cdot A) \to 0$. Applying the functor F, we get the sequence $0 \to x^2 \cdot A \to x \cdot A \to 0 \to 0$. Neither the sequence $0 \to x^2 \cdot A \to x \cdot A \to 0$ nor the sequence $x^2 \cdot A \to x \cdot A \to 0 \to 0$ is exact. Hence, the functor F is neither left nor right exact. (See Exercise 8.33.)

Example 8.3.20. Let R be a k-algebra.
(i) The bifunctor $\mathrm{Hom}_R \colon \mathrm{Mod}(R)^{\mathrm{op}} \times \mathrm{Mod}(R) \to \mathrm{Mod}(k)$ is left exact with respect to each of its argument. If R is a field, this functor is exact.
(ii) The bifunctor $\cdot \otimes_R \cdot \colon \mathrm{Mod}(R^{\mathrm{op}}) \times \mathrm{Mod}(R) \to \mathrm{Mod}(k)$ is right exact with respect to each of its argument. If R is a field, this functor is exact.
(iii) Recall that the category $\mathrm{Mod}(R)$ admits small inductive and projective limits. Moreover, if I is small and filtrant, the functor $\varinjlim \colon \mathrm{Mod}(R)^I \to \mathrm{Mod}(R)$ is exact. If I is discrete, then $\varinjlim$ and $\varprojlim$ are exact.

By Proposition 2.2.9, the abelian category $\mathcal{C}$ admits small projective (resp. inductive) limits if and only if it admits small products (resp. direct sums).

We shall introduce several notions concerning subcategories, which will be frequently used later.

Definition 8.3.21. *Let $\mathcal{J}$ be a full subcategory of $\mathcal{C}$. Denote by $\mathcal{J}'$ the full subcategory of $\mathcal{C}$ defined as follows: $X \in \mathcal{J}'$ if and only if there exist $Y \in \mathcal{J}$ and an isomorphism $X \simeq Y$.*

(i) *We say that $\mathcal{J}$ is closed by subobjects (resp. by quotients) if for any monomorphism $X \rightarrowtail Y$ (resp. epimorphism $Y \twoheadrightarrow X$) with $Y \in \mathcal{J}$, we have $X \in \mathcal{J}'$.*

(ii) *We say that $\mathcal{J}$ is closed by kernels (resp. cokernels) if for any morphism $f \colon X \to Y$ in $\mathcal{J}$, $\mathrm{Ker}\, f$ (resp. $\mathrm{Coker}\, f$) belongs to $\mathcal{J}'$.*

(iii) *We say that $\mathcal{J}$ is closed by extensions in $\mathcal{C}$ if for any exact sequence $0 \to X' \to X \to X'' \to 0$ in $\mathcal{C}$ with X', X'' in $\mathcal{J}$, we have $X \in \mathcal{J}'$.*

(iv) *We say that $\mathcal{J}$ is thick in $\mathcal{C}$ if it is closed by kernels, cokernels and extensions.*

(v) *We say that $\mathcal{J}$ is cogenerating in $\mathcal{C}$ if for any $X \in \mathcal{C}$ there exist $Y \in \mathcal{J}$ and a monomorphism $X \rightarrowtail Y$.*

(vi) *We say that $\mathcal{J}$ is generating in $\mathcal{C}$ if $\mathcal{J}^{\mathrm{op}}$ is cogenerating in $\mathcal{C}^{\mathrm{op}}$. This is equivalent to saying that for any $X \in \mathcal{C}$ there exist $Y \in \mathcal{J}$ and an epimorphism $Y \twoheadrightarrow X$.*

(vii) *We say that $\mathcal{J}$ is a fully abelian subcategory of $\mathcal{C}$ if $\mathcal{J}$ is an abelian full subcategory of $\mathcal{C}$ and the embedding functor is exact.*

Remark 8.3.22. (i) A full subcategory $\mathcal{J}$ of $\mathcal{C}$ is additive if and only if $0 \in \mathcal{J}$ and $X \oplus Y \in \mathcal{J}$ for any $X, Y \in \mathcal{J}$ (see Corollary 8.2.16).
(ii) A full additive subcategory $\mathcal{J}$ of $\mathcal{C}$ is a fully abelian subcategory if and only if $\mathcal{J}$ is closed by kernels and cokernels.
(iii) A full additive subcategory $\mathcal{J}$ of $\mathcal{C}$ is thick if and only if for any exact sequence $X_0 \to X_1 \to X_2 \to X_3 \to X_4$ in $\mathcal{C}$, $X_\nu \in \mathcal{J}$ for $\nu = 0, 1, 3, 4$ implies that X_2 is isomorphic to an object of $\mathcal{J}$.

Let us give a criterion for a fully abelian subcategory to be thick.

Lemma 8.3.23. *Let $\mathcal{C}$ be an abelian category and $\mathcal{J}$ a fully abelian subcategory. Assume that*

(8.3.8)
$$\begin{cases} \textit{for any epimorphism } X \to Y \textit{ with } Y \in \mathcal{J}, \textit{ there exists a morphism } Y' \to X \textit{ with } Y' \in \mathcal{J} \textit{ such that the composition } Y' \to X \to \\ Y \textit{ is an epimorphism,} \end{cases}$$

Then $\mathcal{J}$ is thick in $\mathcal{C}$.

Proof. We may assume that $\mathcal{J}$ is saturated. Consider an exact sequence $0 \to Y' \to X \to Y'' \to 0$ in $\mathcal{C}$ with Y', Y'' in $\mathcal{J}$. We shall show that $X \in \mathcal{J}$. By the hypothesis, there exists an exact commutative diagram with $Y \in \mathcal{J}$:

$$Y \xrightarrow{\ u\ } Y'' \longrightarrow 0$$
$$\downarrow \qquad \downarrow \text{id}$$
$$X \longrightarrow Y'' \longrightarrow 0 \ .$$

Consider the commutative exact diagram:

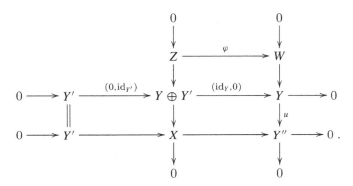

Then, φ is an isomorphism by Exercise 8.19. Hence, $Z \in \mathcal{J}$ and this implies $X \in \mathcal{J}$.   q.e.d.

If $\mathcal{J}$ is cogenerating in $\mathcal{C}$, then for each $X \in \mathcal{C}$ there exists an exact sequence

(8.3.9) $0 \to X \to Y^0 \to Y^1 \to \cdots$

with the Y^j's in $\mathcal{J}$. Indeed, the Y^j's are constructed by induction by embedding $\mathrm{Coker}(Y^{n-1} \to Y^n)$ into $Y^{n+1} \in \mathcal{J}$. Similarly, if $\mathcal{J}$ is generating, there is an exact sequence

(8.3.10) $\cdots \to Y^{-1} \to Y^0 \to X \to 0$

with the Y^j's in $\mathcal{J}$.

Recall (see Proposition 5.2.4) that in an abelian category, the conditions below are equivalent:

(i) G is a generator, that is, the functor $\varphi_G = \mathrm{Hom}_\mathcal{C}(G, \bullet)$ is conservative,
(ii) The functor φ_G is faithful.

(See Exercise 8.27 for an example in which φ_G is not conservative although $\varphi_G(X) \simeq 0$ implies $X \simeq 0$.)

Moreover, if $\mathcal{C}$ admits small inductive limits, the conditions above are equivalent to:

(iii) for any $X \in \mathcal{C}$, there exist a small set I and an epimorphism $G^{\sqcup I} \twoheadrightarrow X$.

Let us introduce a class of abelian categories which is extremely useful in practice and to which we shall come back in Sect. 9.6.

Definition 8.3.24. *A Grothendieck $\mathcal{U}$-category $\mathcal{C}$ is an abelian $\mathcal{U}$-category such that $\mathcal{C}$ admits a generator and $\mathcal{U}$-small inductive limits, and $\mathcal{U}$-small filtrant inductive limits are exact.*

Hence, the definition depends on the choice of a universe $\mathcal{U}$. However, if there is no risk of confusion, we do not mention $\mathcal{U}$.

Examples 8.3.25. (i) Let R be a ring. Then $\mathrm{Mod}(R)$ is a Grothendieck category.
(ii) Let $\mathcal{C}$ be a small abelian category. We shall prove in Theorem 8.6.5 below that $\mathrm{Ind}(\mathcal{C})$ is a Grothendieck category.

Corollary 8.3.26. *Let $\mathcal{C}$ be a Grothendieck category and let $X \in \mathcal{C}$. Then the family of quotients of X and the family of subobjects of X are small sets.*

Proof. Apply Proposition 5.2.9.   q.e.d.

Proposition 8.3.27. *Let $\mathcal{C}$ be a Grothendieck category. Then $\mathcal{C}$ satisfies the following properties.*

(i) *$\mathcal{C}$ admits small projective limits,*
(ii) *if a functor $F \colon \mathcal{C}^{\mathrm{op}} \to \mathbf{Set}$ commutes with small projective limits, then F is representable,*
(iii) *if a functor $F \colon \mathcal{C} \to \mathcal{C}'$ commutes with small inductive limits, then F admits a right adjoint.*

Proof. Apply Corollary 5.2.10 and Proposition 5.2.8.   q.e.d.

8.4 Injective Objects

Let $\mathcal{C}$ be an abelian category.

Definition 8.4.1.   (i) *An object I of $\mathcal{C}$ is* injective *if the functor $\mathrm{Hom}_{\mathcal{C}}(\,\cdot\,, I)$ is exact. The category $\mathcal{C}$ has* enough injectives *if the full subcategory of injective objects is cogenerating, i.e., for any $X \in \mathcal{C}$ there exists a monomorphism $X \rightarrowtail I$ with I injective.*

(ii) *An object P is* projective *in $\mathcal{C}$ if it is injective in $\mathcal{C}^{\mathrm{op}}$, i.e., if the functor $\mathrm{Hom}_{\mathcal{C}}(P, \,\cdot\,)$ is exact. The category $\mathcal{C}$ has* enough projectives *if the full subcategory of projective objects is generating, i.e., for any $X \in \mathcal{C}$ there exists an epimorphism $P \twoheadrightarrow X$ with P projective.*

Example 8.4.2. (i) Let R be a ring. Free R-modules are projective. It follows immediately that the category $\mathrm{Mod}(R)$ has enough projectives. It is a classical result (see Exercise 8.24) that the category $\mathrm{Mod}(R)$ has enough injectives. We shall prove later that any Grothendieck category has enough injectives.
(ii) If k is a field, then any object of $\mathrm{Mod}(k)$ is both injective and projective.

Proposition 8.4.3. *An object $I \in \mathcal{C}$ is injective if and only if, for any $X, Y \in \mathcal{C}$ and any solid diagram in which the row is exact*

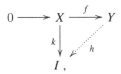

the dotted arrow may be completed, making the whole diagram commutative.

Proof. Consider an exact sequence $0 \to X \xrightarrow{f} Y \xrightarrow{g} Z \to 0$ and apply the functor $\mathrm{Hom}_{\mathcal{C}}(\cdot, I)$. Since this functor is left exact, $\mathrm{Hom}_{\mathcal{C}}(\cdot, I)$ is exact if and only if the map $\mathrm{Hom}_{\mathcal{C}}(Y, I) \xrightarrow{f \circ} \mathrm{Hom}_{\mathcal{C}}(X, I)$ is surjective.   q.e.d.

Lemma 8.4.4. *Let $0 \to X' \xrightarrow{f} X \xrightarrow{g} X'' \to 0$ be an exact sequence in $\mathcal{C}$, and assume that X' is injective. Then the sequence splits.*

Proof. Applying the preceding result with $k = \mathrm{id}_{X'}$, we find $h \colon X \to X'$ such that $h \circ f = \mathrm{id}_{X'}$. Then apply Proposition 8.3.14.   q.e.d.

It follows that if $F \colon \mathcal{C} \to \mathcal{C}'$ is an additive functor of abelian categories and the hypotheses of the lemma are satisfied, then the sequence $0 \to F(X') \to F(X) \to F(X'') \to 0$ splits and in particular is exact.

Lemma 8.4.5. *Let X', X'' belong to $\mathcal{C}$. Then $X' \oplus X''$ is injective if and only if X' and X'' are injective.*

Proof. It is enough to remark that for two additive functors of abelian categories F and G, $X \mapsto F(X) \oplus G(X)$ is exact if and only if F and G are exact.
                    q.e.d.

Applying Lemmas 8.4.4 and 8.4.5, we get:

Proposition 8.4.6. *Let $0 \to X' \to X \to X'' \to 0$ be an exact sequence in $\mathcal{C}$ and assume that X' and X are injective. Then X'' is injective.*

Proposition 8.4.7. *Let $\mathcal{C}$ denote a Grothendieck category and let $\{G_i\}_{i \in I}$ be a system of generators. Then an object $Z \in \mathcal{C}$ is injective if and only if for any $i \in I$ and any subobject $W \subset G_i$, the natural map $\mathrm{Hom}_{\mathcal{C}}(G_i, Z) \to \mathrm{Hom}_{\mathcal{C}}(W, Z)$ is surjective.*

Proof. The necessity of the condition is clear. Let us prove that it is sufficient. Let $f \colon X' {\rightarrowtail} X$ be a monomorphism and let $h \colon X' \to Z$ a morphism. Consider a commutative diagram D

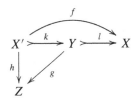

with $l \circ k = f$ and l is a monomorphism.

In the sequel, we shall write for short $D = (Y, g, l)$. Such diagrams form a category Δ, a morphism $D = (Y, g, l) \to D' = (Y', g', l')$ being a commutative diagram

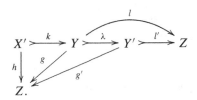

Denote by Σ the set of isomorphism classes of Δ. Since $\mathrm{card}(\mathrm{Hom}_\Delta(D, D')) \leq 1$ for any $D, D' \in \Delta$, Σ is a small ordered set. Moreover Δ is equivalent to the category associated with the ordered set Σ.

Since filtrant inductive limits are exact, Σ is inductively ordered. Let $D_0 = (Y_0, g_0, l_0)$ be a maximal element. By the definition of a system of generators, in order to prove that $Y_0 = X$, it is enough to check that, for each $i \in I$, the monomorphism $\mathrm{Hom}_\mathcal{C}(G_i, Y_0) \rightarrowtail \mathrm{Hom}_\mathcal{C}(G_i, X)$ is surjective. Let $\varphi \colon G_i \to X$ be a morphism. Define $Y := Y_0 \times_X G_i$. Since $Y_0 \to X$ is a monomorphism, $Y \to G_i$ is a monomorphism. Define $Y_1 := Y_0 \oplus_Y G_i$. Since we have an exact sequence $0 \to Y \to Y_0 \oplus G_i \xrightarrow{u} X$, we get $Y_1 \simeq \mathrm{Im}\, u \subset X$. By the assumption on Z, the composition $Y \to Y_0 \to Z$ factorizes through $Y \rightarrowtail G_i$. The morphism $G_i \to Z$ factorizes through Y_1, as in the diagram:

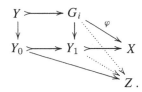

Since Y_0 is maximal, $Y_0 \simeq Y_1$ and $G_i \to X$ factorizes through Y_0.   q.e.d.

8.5 Ring Action

Let k denote a commutative ring.

A category $\mathcal{C}$ is a k-pre-additive category if for all X and Y in $\mathcal{C}$, $\mathrm{Hom}_\mathcal{C}(X, Y)$ is endowed with the structure of a k-module and the composition of morphisms is k-bilinear. The notion of k-additive functor between

k-pre-additive categories and that of k-additive category, k-abelian category are naturally defined. Note that additive categories are $\mathbb{Z}$-additive. Also note that for $X \in \mathcal{C}$, $\mathrm{End}_{\mathcal{C}}(X) := \mathrm{Hom}_{\mathcal{C}}(X, X)$ is a k-algebra.

There is an alternative definition. Let $\mathcal{C}$ be an additive category. Recall that $\mathrm{End}\,(\mathrm{id}_{\mathcal{C}})$ denotes the set of endomorphisms of the functor $\mathrm{id}_{\mathcal{C}}$. Then $\mathrm{End}\,(\mathrm{id}_{\mathcal{C}})$ has a structure of a ring with unit, and it follows from Lemma 1.3.8 that this ring is commutative. Clearly, a structure of a k-additive category on $\mathcal{C}$ is equivalent to the data of a morphism of rings $k \to \mathrm{End}\,(\mathrm{id}_{\mathcal{C}})$.

Definition 8.5.1. *Let R be a k-algebra and $\mathcal{C}$ a k-additive category. The category $\mathrm{Mod}(R, \mathcal{C})$ is defined as follows.*

$$\mathrm{Ob}(\mathrm{Mod}(R, \mathcal{C})) = \big\{(X, \xi_X)\,;\, X \in \mathcal{C} \text{ and } \xi_X : R \to \mathrm{End}_{\mathcal{C}}(X) \text{ is a morphism}$$
$$\text{of } k\text{-algebras}\big\}\,,$$
$$\mathrm{Hom}_{\mathrm{Mod}(R,\mathcal{C})}((X, \xi_X), (Y, \xi_Y)) = \big\{f : X \to Y\,;\, f \circ \xi_X(a) = \xi_Y(a) \circ f$$
$$\text{for all } a \in R\big\}\,.$$

Clearly, $\mathrm{Mod}(R, \mathcal{C})$ is k-additive and the functor *for*: $\mathrm{Mod}(R, \mathcal{C}) \to \mathcal{C}$ given by $(X, \xi_X) \mapsto X$ is k-additive and faithful. If R is commutative, $\mathrm{Mod}(R, \mathcal{C})$ is an R-additive category. More generally, $\mathrm{Mod}(R, \mathcal{C})$ is a $Z(R)$-additive category, where $Z(R)$ denotes the center of R.

Note that if $X \in \mathrm{Mod}(R, \mathcal{C})$ and $Y \in \mathcal{C}$, then $\mathrm{Hom}_{\mathcal{C}}(Y, X) \in \mathrm{Mod}(R)$ and $\mathrm{Hom}_{\mathcal{C}}(X, Y) \in \mathrm{Mod}(R^{\mathrm{op}})$.

If $F : \mathcal{C} \to \mathcal{C}'$ is a k-additive functor, it induces a functor $F_R : \mathrm{Mod}(R, \mathcal{C}) \to \mathrm{Mod}(R, \mathcal{C}')$ and the diagram below quasi-commutes

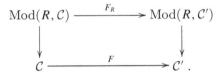

Proposition 8.5.2. (i) *Let $\mathcal{C}$ be a k-abelian category. Then $\mathrm{Mod}(R, \mathcal{C})$ is k-abelian and the natural functor for: $\mathrm{Mod}(R, \mathcal{C}) \to \mathcal{C}$ is faithful and exact.*

 (ii) *Let $F : \mathcal{C} \to \mathcal{C}'$ be a right (resp. left) exact functor of k-abelian categories. Then $F_R : \mathrm{Mod}(R, \mathcal{C}) \to \mathrm{Mod}(R, \mathcal{C}')$ is right (resp. left) exact.*

The proof is obvious.

Notation 8.5.3. Let *for* denote the forgetful functor $\mathrm{Mod}(R, \mathcal{C}) \to \mathcal{C}$. Clearly, *for* is faithful, but not fully faithful in general. However, we shall often denote by the same symbol X an object of $\mathrm{Mod}(R, \mathcal{C})$ and its image by *for* in $\mathcal{C}$. If $F : \mathcal{C} \to \mathcal{C}'$ is a functor, we shall often write F instead of F_R.

Example 8.5.4. We have $\mathrm{Mod}(R, \mathrm{Mod}(k)) \simeq \mathrm{Mod}(R)$.

Note that
$$\mathrm{Mod}(R, \mathcal{C})^{\mathrm{op}} \simeq \mathrm{Mod}(R^{\mathrm{op}}, \mathcal{C}^{\mathrm{op}}) ,$$

where R^{op} denotes the opposite ring of R.

Proposition 8.5.5. *Let $\mathcal{C}$ be a k-abelian category and R a k-algebra.*

(a) *Assume that $\mathcal{C}$ admits small inductive limits. Then*
   (i) *for any $X \in \mathrm{Mod}(R, \mathcal{C})$ and $N \in \mathrm{Mod}(R^{\mathrm{op}})$, the functor $Y \mapsto \mathrm{Hom}_{R^{\mathrm{op}}}(N, \mathrm{Hom}_{\mathcal{C}}(X, Y))$ is representable,*
   (ii) *denoting by $N \otimes_R X$ its representative, the functor*

$$\bullet \otimes_R \bullet : \mathrm{Mod}(R^{\mathrm{op}}) \times \mathrm{Mod}(R, \mathcal{C}) \to \mathcal{C}$$

   is additive and right exact in each variable.
(b) *Assume that $\mathcal{C}$ admits small projective limits. Then*
   (i) *for any $X \in \mathrm{Mod}(R, \mathcal{C})$ and $M \in \mathrm{Mod}(R)$, the functor $Y \mapsto \mathrm{Hom}_R(M, \mathrm{Hom}_{\mathcal{C}}(Y, X))$ is representable,*
   (ii) *denoting by $\mathrm{Hom}_R(M, X)$ its representative, the functor*

$$\mathrm{Hom}_R(\bullet, \bullet) \colon (\mathrm{Mod}(R))^{\mathrm{op}} \times \mathrm{Mod}(R, \mathcal{C}) \to \mathcal{C}$$

   is additive and left exact in each variable.

Proof. (a) (i) First, assume that $N = R^{\oplus I}$ for a small set I. The hypothesis implies the isomorphism, functorial with respect to $Y \in \mathcal{C}$:

$$\mathrm{Hom}_R(R^{\oplus I}, \mathrm{Hom}_{\mathcal{C}}(X, Y)) \simeq \mathrm{Hom}_{\mathcal{C}}(X, Y)^I \simeq \mathrm{Hom}_{\mathcal{C}}(X^{\oplus I}, Y) .$$

In the general case, we may find an exact sequence $R^{\oplus J} \to R^{\oplus I} \to N \to 0$, with I and J small. The sequence

$$0 \to \mathrm{Hom}_R(N, \mathrm{Hom}_{\mathcal{C}}(X, Y)) \to \mathrm{Hom}_R(R^{\oplus I}, \mathrm{Hom}_{\mathcal{C}}(X, Y))$$
$$\to \mathrm{Hom}_R(R^{\oplus J}, \mathrm{Hom}_{\mathcal{C}}(X, Y))$$

is exact. Hence, $\mathrm{Coker}(X^{\oplus J} \to X^{\oplus I})$ represents $N \otimes_R X$.
(a) (ii) is obvious.
(b) Apply the result (a) to the category $\mathcal{C}^{\mathrm{op}}$.   q.e.d.

Remark 8.5.6. In the situation of Proposition 8.5.5, if R is a k-algebra consider another k-algebra S, and assume that M is an $(S \otimes_k R^{\mathrm{op}})$-module. Then $M \otimes_R X$ belongs to $\mathrm{Mod}(S, \mathcal{C})$. For an $(R \otimes_k S^{\mathrm{op}})$-module N, $\mathrm{Hom}_R(N, X)$ belongs to $\mathrm{Mod}(S, \mathcal{C})$.

Remark 8.5.7. If $M \in \mathrm{Mod}(R^{\mathrm{op}})$ or $M \in \mathrm{Mod}(R)$ is of finite presentation, the above construction shows that $M \otimes_R X$ and $\mathrm{Hom}_R(M, X)$ are well defined without assuming that $\mathcal{C}$ admits small inductive or projective limits.

To end this section, let us recall a result of Gabriel-Popescu (see [54]). Let $\mathcal{C}$ be a Grothendieck category and G a generator. Set $R = (\mathrm{End}_{\mathcal{C}}(G))^{\mathrm{op}}$. Hence, G belongs to $\mathrm{Mod}(R^{\mathrm{op}}, \mathcal{C})$. Define the functors

$$\varphi_G \colon \mathcal{C} \to \mathrm{Mod}(R), \quad \varphi_G(X) = \mathrm{Hom}_{\mathcal{C}}(G, X) ,$$
$$\psi_G \colon \mathrm{Mod}(R) \to \mathcal{C}, \quad \psi_G(M) = G \otimes_R M .$$

Theorem 8.5.8. [Gabriel-Popescu] *Let $\mathcal{C}$ be a Grothendieck category and G a generator.*

(i) *The pair (ψ_G, φ_G) is a pair of adjoint functors,*
(ii) *$\psi_G \circ \varphi_G \to \mathrm{id}_{\mathcal{C}}$ is an isomorphism,*
(iii) *φ_G is fully faithful,*
(iv) *ψ_G is exact.*

Proof. We shall write φ and ψ instead of φ_G and ψ_G, respectively.

(i) is obvious, since for $X \in \mathcal{C}$ and $M \in \mathrm{Mod}(R)$ we have

$$\mathrm{Hom}_{\mathcal{C}}(\psi(M), X) = \mathrm{Hom}_{\mathcal{C}}(M \otimes_R G, X)$$
$$\simeq \mathrm{Hom}_R(M, \mathrm{Hom}_{\mathcal{C}}(G, X)) = \mathrm{Hom}_R(M, \varphi(X)) .$$

(ii) is equivalent to (iii) by Proposition 1.5.6.

(iii) Denote by $\mathcal{F}$ the full subcategory of $\mathrm{Mod}(R)$ consisting of the products of finite copies of R. Then $\psi|_{\mathcal{F}} \colon \mathcal{F} \to \mathcal{C}$ is fully faithful. By Theorem 5.3.6, the functor $\lambda \colon \mathcal{C} \to \mathcal{F}^{\wedge}$ (denoted by φ in this theorem) is fully faithful. On the other hand, the functor $\lambda' \colon \mathrm{Mod}(R) \to \mathcal{F}^{\wedge}$ is fully faithful, again by Theorem 5.3.6. Then the result follows from the commutative diagram of categories:

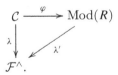

(iv) Since ψ is right exact, in order to prove that it is exact, it remains to prove that it sends a monomorphism $M \rightarrowtail N$ in $\mathrm{Mod}(R)$ to a monomorphism $\psi(M) \rightarrowtail \psi(N)$ in $\mathcal{C}$. We decompose the proof into several steps.

(iv) (a) Assume that M is finitely generated and $N = R^{\oplus J}$ for a small set J. There exist a finite set I and an epimorphism $R^{\oplus I} \twoheadrightarrow M$. Since ψ is right exact, $\psi(R^{\oplus I}) \to \psi(M)$ is still an epimorphism. Hence, it is enough to prove that the composition $\mathrm{Ker}(\psi(R^{\oplus I}) \to \psi(R^{\oplus J})) \to \psi(R^{\oplus J})$ vanishes. Since φ is faithful and left exact, we are reduced to prove the vanishing of the morphism

(8.5.1) $$\mathrm{Ker}(\varphi\psi(R^{\oplus I}) \to \varphi\psi(R^{\oplus J})) \to \varphi\psi(R^{\oplus J}).$$

Consider the diagram:

$$
\begin{array}{ccc}
R^{\oplus I} & \longrightarrow & R^{\oplus J} \\
\scriptstyle\sim\downarrow & & \downarrow \quad\searrow \\
\varphi\psi(R^{\oplus I}) & \longrightarrow \varphi\psi(R^{\oplus J}) & \longrightarrow R^{J}\,.
\end{array}
$$

Its commutativity implies the isomorphism

$$\mathrm{Ker}(R^{\oplus I} \to R^{\oplus J}) \xrightarrow{\sim} \mathrm{Ker}\big(\varphi\psi(R^{\oplus I}) \to \varphi\psi(R^{\oplus J})\big)\,.$$

Then the vanishing of the morphism in (8.5.1) follows from the commutative diagram:

$$
\begin{array}{ccc}
\mathrm{Ker}(R^{\oplus I} \to R^{\oplus J}) & \xrightarrow{\ 0\ } & R^{\oplus J} \\
\scriptstyle\sim\downarrow & & \downarrow \\
\mathrm{Ker}\big(\varphi\psi(R^{\oplus I}) \to \varphi\psi(R^{\oplus J})\big) & \longrightarrow & \varphi\psi(R^{\oplus J}).
\end{array}
$$

(iv) (b) Assume that $N = R^{\oplus J}$ for a small set J and M is an arbitrary R-submodule of N. For any finitely generated submodule M' of M, the morphism $\psi(M') \to \psi(N)$ is a monomorphism by (iv) (a). Since $M \simeq \varinjlim M'$ where M' ranges over the filtrant family of finitely generated submodules of M and ψ commutes with small inductive limits, the result follows. (Recall that by the hypotheses, filtrant inductive limits are exact in $\mathcal{C}$.)

(iv) (c) Finally, we treat the general case. We choose an epimorphism $R^{\oplus J} \twoheadrightarrow N$, where J is a small set. We set $K := M \times_N (R^{\oplus J})$ and $L := \mathrm{Ker}(K \to M)$. We get the exact commutative diagram

$$
\begin{array}{ccccccccc}
& & 0 & & 0 & & & & \\
& & \downarrow & & \downarrow & & & & \\
0 & \longrightarrow & L & \longrightarrow & K & \longrightarrow & M & \longrightarrow & 0 \\
& & \| & & \downarrow & & \downarrow & & \\
0 & \longrightarrow & L & \longrightarrow & R^{\oplus J} & \longrightarrow & N & \longrightarrow & 0\,.
\end{array}
$$

Applying the right exact functor ψ, we get the commutative diagram with exact rows

$$
\begin{array}{ccccccc}
& & 0 & & 0 & & \\
& & \downarrow & & \downarrow & & \\
\psi(L) & \longrightarrow & \psi(K) & \longrightarrow & \psi(M) & \longrightarrow & 0 \\
\| & & \downarrow & & \downarrow & & \\
\psi(L) & \longrightarrow & \psi(R^{\oplus J}) & \longrightarrow & \psi(N) & \longrightarrow & 0\,.
\end{array}
$$

By the result of (iv) (b), the middle column is exact. Hence, the right column is exact.   q.e.d.

8.6 Indization of Abelian Categories

Let $\mathcal{C}$ be an abelian $\mathcal{U}$-category. Then the big category $\mathcal{C}^{\wedge,add}$ of additive functors from $\mathcal{C}^{op}$ to $\mathrm{Mod}(\mathbb{Z})$ is abelian. By Proposition 8.2.12, we may regard $\mathcal{C}^{\wedge,add}$ as a full subcategory of $\mathcal{C}^{\wedge}$. Recall that $\mathcal{C}$ and $\mathrm{Mod}(\mathbb{Z})$ are $\mathcal{U}$-categories by the hypothesis and notice that $\mathcal{C}^{\wedge,add}$ may not be a $\mathcal{U}$-category.

Notation 8.6.1. Recall that if $\mathcal{C}$ is a category, we denote by "$\varinjlim$" the inductive limit in $\mathcal{C}^{\wedge}$. If $\{X_i\}_{i\in I}$ is a small family of objects of an additive category $\mathcal{C}$ indexed by a set I, we write "$\bigoplus_{i\in I}$" X_i for "$\varinjlim_{J}$" $(\bigoplus_{i\in J} X_i)$, where J ranges over the set of finite subsets of I. Hence,

$$\mathrm{Hom}_{\mathcal{C}^{\wedge}}(Z, \text{``}\bigoplus_{i\in I}\text{''} X_i) \simeq \bigoplus_{i\in I} \mathrm{Hom}_{\mathcal{C}^{\wedge}}(Z, X_i)$$

for $Z \in \mathcal{C}$.

Note that the functor

$$h_{\mathcal{C}}: \mathcal{C} \to \mathcal{C}^{\wedge,add}, \quad X \mapsto \mathrm{Hom}_{\mathcal{C}}(\cdot, X)$$

makes $\mathcal{C}$ a full subcategory of $\mathcal{C}^{\wedge,add}$ and this functor is left exact, but not exact in general.

Recall that an ind-object in $\mathcal{C}$ is an object $A \in \mathcal{C}^{\wedge}$ which is isomorphic to "$\varinjlim$" α for some functor $\alpha: I \to \mathcal{C}$ with I filtrant and small. Hence, $\mathrm{Ind}(\mathcal{C})$ is a full pre-additive subcategory of $\mathcal{C}^{\wedge,add}$. Recall that $\mathrm{Ind}(\mathcal{C})$ is a $\mathcal{U}$-category.

Proposition 8.6.2. *Let $A \in \mathcal{C}^{\wedge,add}$. Then the two conditions below are equivalent.*

(i) *The functor A belongs to $\mathrm{Ind}(\mathcal{C})$.*
(ii) *The functor A is left exact and $\mathcal{C}_A$ is cofinally small.*

Proof. This follows from Proposition 6.1.7.   q.e.d.

Corollary 8.6.3. *Let $\mathcal{C}$ be a small abelian category. Then $\mathrm{Ind}(\mathcal{C})$ is equivalent to the full additive subcategory $\mathcal{C}^{\wedge,add,l}$ of $\mathcal{C}^{\wedge,add}$ consisting of left exact functors.*

Lemma 8.6.4.   (i) *The category $\mathrm{Ind}(\mathcal{C})$ is additive and admits kernels and cokernels.*

(ii) *Let I be small and filtrant, let $\alpha, \beta \colon I \to \mathcal{C}$ be two functors, and let $\varphi \colon \alpha \to \beta$ be a morphism of functors. Let $f = \text{"}\varinjlim\text{"}\,\varphi$. Then $\operatorname{Ker} f \simeq \text{"}\varinjlim\text{"}(\operatorname{Ker}\varphi)$ and $\operatorname{Coker} f \simeq \text{"}\varinjlim\text{"}(\operatorname{Coker}\varphi)$.*

(iii) *If $\varphi \colon A \to B$ is a morphism in $\operatorname{Ind}(\mathcal{C})$, the kernel of φ in $\mathcal{C}^{\wedge,add}$ is its kernel in $\operatorname{Ind}(\mathcal{C})$.*

This is a particular case of Propositions 6.1.16 and 6.1.18.

Theorem 8.6.5. (i) *The category $\operatorname{Ind}(\mathcal{C})$ is abelian.*
(ii) *The natural functor $\mathcal{C} \to \operatorname{Ind}(\mathcal{C})$ is fully faithful and exact, and the natural functor $\operatorname{Ind}(\mathcal{C}) \to \mathcal{C}^{\wedge,add}$ is fully faithful and left exact.*
(iii) *The category $\operatorname{Ind}(\mathcal{C})$ admits small inductive limits. Moreover, inductive limits over small filtrant categories are exact.*
(iv) *Assume that $\mathcal{C}$ admits small projective limits. Then $\operatorname{Ind}(\mathcal{C})$ admits small projective limits.*
(v) *"$\bigoplus$" is a coproduct in $\operatorname{Ind}(\mathcal{C})$.*
(vi) *Assume that $\mathcal{C}$ is essentially small. Then $\operatorname{Ind}(\mathcal{C})$ admits a generator, and hence is a Grothendieck category.*

Proof. (i) We know by Lemma 8.6.4 that $\operatorname{Ind}(\mathcal{C})$ admits kernels and cokernels. Let f be a morphism in $\operatorname{Ind}(\mathcal{C})$. We may assume $f = \text{"}\varinjlim\text{"}\,\varphi$ as in Lemma 8.6.4 (ii). Then $\operatorname{Coim} f \simeq \text{"}\varinjlim\text{"} \operatorname{Coim}\varphi$ and $\operatorname{Im} f \simeq \text{"}\varinjlim\text{"} \operatorname{Im}\varphi$, by Lemma 8.6.4. Hence $\operatorname{Coim} f \simeq \operatorname{Im} f$.
(ii) follows from Lemma 8.6.4.
(iii) follows from Proposition 6.1.19.
(iv) follows from Corollary 6.1.17 (ii).
(v) is obvious.
(vi) Let $\{X_i\}_{i\in I}$ be a small set of objects of $\mathcal{C}$ such that any object of $\mathcal{C}$ is isomorphic to some X_i. Then this family is a system of $\mathcal{U}$-generators in $\operatorname{Ind}(\mathcal{C})$ and "$\bigoplus_{i\in I}$" X_i is a generator.   q.e.d.

Proposition 8.6.6. *Let $0 \to A' \xrightarrow{f} A \xrightarrow{g} A'' \to 0$ be an exact sequence in $\operatorname{Ind}(\mathcal{C})$ and let $\mathcal{J}$ be a full additive subcategory of $\mathcal{C}$.*

(a) *There exist a small filtrant category I and an exact sequence of functors from I to $\mathcal{C}$, $0 \to \alpha' \xrightarrow{\varphi} \alpha \xrightarrow{\psi} \alpha'' \to 0$ such that $f \simeq \text{"}\varinjlim\text{"}\,\varphi$ and $g \simeq \text{"}\varinjlim\text{"}\,\psi$.*

(b) *Assume that A belongs to $\operatorname{Ind}(\mathcal{J})$. Then we may choose the functor α in (a) with values in $\mathcal{J}$.*
(c) *Assume that A' belongs to $\operatorname{Ind}(\mathcal{J})$. Then we may choose the functor α' in (a) with values in $\mathcal{J}$.*

Proof. (a) By Proposition 6.1.13, we may assume that there exist I filtrant and small, functors $\alpha, \beta \colon I \to \mathcal{C}$ and a morphism of functor $\lambda \colon \alpha \to \beta$ such that $A \simeq \text{``}\varinjlim\text{''}\,\alpha$, $A'' \simeq \text{``}\varinjlim\text{''}\,\beta$, and $g \simeq \text{``}\varinjlim\text{''}\,\lambda$.

Set $\alpha'(i) = \operatorname{Ker}\lambda(i)$, denote by $\varphi(i) \colon \alpha'(i) \to \alpha(i)$ the natural morphism, and set $\alpha''(i) = \operatorname{Coker}\varphi(i)$. Since the sequence of functors $0 \to \alpha' \to \alpha \to \beta$ is exact, we get $A' \simeq \text{``}\varinjlim\text{''}\,\alpha'$. Since the sequences $0 \to \alpha'(i) \to \alpha(i) \to \alpha''(i) \to 0$ are exact, the sequence $0 \to \text{``}\varinjlim\text{''}\,\alpha' \to \text{``}\varinjlim\text{''}\,\alpha \to \text{``}\varinjlim\text{''}\,\alpha'' \to 0$ is exact. Hence, $\text{``}\varinjlim\text{''}\,\alpha'' \simeq A''$.

(b) The proof in (a) shows that if $A \in \operatorname{Ind}(\mathcal{J})$, we may assume α with values in $\mathcal{J}$.

(c) The result will follow from Lemma 8.6.7 below.   q.e.d.

Lemma 8.6.7. *Let I be a small and filtrant category, $\alpha \colon I \to \mathcal{C}$ a functor, $A = \text{``}\varinjlim\text{''}\,\alpha$ and let $f \colon A \rightarrowtail B$ be a monomorphism in $\operatorname{Ind}(\mathcal{C})$. Then there exist a small and filtrant category K, a cofinal functor $p \colon K \to I$, a functor $\beta \colon K \to \mathcal{C}$ and a monomorphism of functor $\varphi \colon \alpha \circ p \rightarrowtail \beta$ such that $f \simeq \text{``}\varinjlim\text{''}\,\varphi$.*

Proof. By Proposition 8.6.6 (a), there exist a small filtrant category J, functors $\alpha', \beta' \colon J \to \mathcal{C}$, and a monomorphism of functors $\varphi \colon \alpha' \to \beta'$ such that $f \colon A \to B$ is isomorphic to $\text{``}\varinjlim\text{''}\,\varphi \colon \text{``}\varinjlim\text{''}\,\alpha' \to \text{``}\varinjlim\text{''}\,\beta'$.

By Proposition 6.1.13 applied to $\operatorname{id}_A \colon \text{``}\varinjlim\text{''}\,\alpha' \to \text{``}\varinjlim\text{''}\,\alpha$, there exist a small and filtrant category K, cofinal functors $p_I \colon K \to I$ and $p_J \colon K \to J$, and a morphism of functors $\psi \colon \alpha' \circ p_J \to \alpha \circ p_I$ such that $\text{``}\varinjlim\text{''}\,\psi \simeq \operatorname{id}_A$.

For $k \in K$, define $\beta(k)$ as the coproduct of $\alpha(p_I(k))$ and $\beta'(p_J(k))$ over $\alpha'(p_J(k))$. In other words, the square below is co-Cartesian:

$$
\begin{array}{ccc}
\alpha'(p_J(k)) & \xrightarrow{\ \varphi(p_J(k))\ } & \beta'(p_J(k)) \\[2pt]
{\scriptstyle\psi(k)}\big\downarrow & & \big\downarrow \\[2pt]
\alpha(p_I(k)) & \xrightarrow[\ \ \xi(k)\ \ ]{} & \beta(k)\,.
\end{array}
$$

It follows that the arrow $\xi(k) \colon \alpha(p_I(k)) \to \beta(k)$ is a monomorphism by Lemma 8.3.11. Passing to the inductive limit with respect to $k \in K$, the square remains co-Cartesian and it follows that $B \simeq \text{``}\varinjlim\text{''}\,\beta$, $f \simeq \text{``}\varinjlim\text{''}\,\xi$. q.e.d.

Corollary 8.6.8. *Let $F \colon \mathcal{C} \to \mathcal{C}'$ be an additive functor of abelian categories, $IF \colon \operatorname{Ind}(\mathcal{C}) \to \operatorname{Ind}(\mathcal{C}')$ the associated functor. If F is left (resp. right) exact, then IF is left (resp. right) exact.*

Proof. Apply Proposition 8.6.6 (a).   q.e.d.

Proposition 8.6.9. *Let $f \colon A \to B$ be a morphism in $\operatorname{Ind}(\mathcal{C})$. The two conditions below are equivalent.*

(i) f *is an epimorphism,*

(ii) *for any solid diagram* $X \cdots\overset{g}{\cdots}\!\!> Y$ *in* $\mathrm{Ind}(\mathcal{C})$ *with* $Y \in \mathcal{C}$*, the dotted arrows*

$$\begin{array}{ccc} X & \overset{g}{\dashrightarrow} & Y \\ \downarrow & & \downarrow \\ A & \overset{f}{\longrightarrow} & B \end{array}$$

may be completed to a commutative diagram with $X \in \mathcal{C}$ *such that* g *is an epimorphism.*

Proof. We may assume from the beginning that there exist a small and filtrant category I and a morphism of functors $\varphi\colon \alpha \to \beta$ such that $f = \text{``}\underrightarrow{\lim}\text{''}\,\varphi$.

(i) $\Rightarrow$ (ii). Assume that f is an epimorphism. By Proposition 8.6.6 we may assume that $\varphi(i)\colon \alpha(i) \to \beta(i)$ is an epimorphism for all $i \in I$. The morphism $Y \to B$ factors through $Y \to \beta(i)$ for some $i \in I$. Hence the result follows from the corresponding one when replacing $\mathrm{Ind}(\mathcal{C})$ by $\mathcal{C}$ (see Proposition 8.3.12).

(ii) $\Rightarrow$ (i). For each $i \in I$ we shall apply the hypothesis with $Y = \beta(i)$. We find a commutative diagram

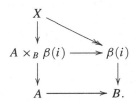

Hence, $A \times_B \beta(i) \to \beta(i)$ is an epimorphism. Applying the functor $\text{``}\underrightarrow{\lim}\text{''}$, we find that $A \simeq \text{``}\underset{i}{\underrightarrow{\lim}}\text{''}\,(A \times_B \beta(i)) \to \text{``}\underset{i}{\underrightarrow{\lim}}\text{''}\,\beta(i) \simeq B$ is an epimorphism.   q.e.d.

Corollary 8.6.10. *A complex* $A \overset{f}{\to} B \overset{g}{\to} C$ *in* $\mathrm{Ind}(\mathcal{C})$ *is exact if and only if for any solid commutative diagram in* $\mathrm{Ind}(\mathcal{C})$ *with* $Y \in \mathcal{C}$

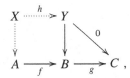

the dotted arrows may be completed to a commutative diagram with $X \in \mathcal{C}$ *such that h is an epimorphism.*

Proof. Apply Proposition 8.6.9 to the morphism $A \to \mathrm{Ker}\,g$.   q.e.d.

Proposition 8.6.11. *The category $\mathcal{C}$ is thick in* $\mathrm{Ind}(\mathcal{C})$.

This follows from Proposition 8.6.9 and Lemma 8.3.23.

Proposition 8.6.12. *Let $\mathcal{C}$ be an abelian category, $\mathcal{J} \subset \mathcal{C}$ an additive subcategory closed by extension in $\mathcal{C}$. Then* $\mathrm{Ind}(\mathcal{J})$ *is closed by extension in* $\mathrm{Ind}(\mathcal{C})$.

Proof. Let $A \in \mathrm{Ind}(\mathcal{C})$. Remark first that $A \in \mathrm{Ind}(\mathcal{J})$ if and only if any morphism $X \to A$ with $X \in \mathcal{C}$ factorizes through an object $Y \in \mathcal{J}$ (see Exercise 6.11). Now consider an exact sequence in $\mathrm{Ind}(\mathcal{C})$: $0 \to A' \to A \to A'' \to 0$ and assume that A', A'' belong to $\mathrm{Ind}(\mathcal{J})$. Consider a morphism $X \to A$ with $X \in \mathcal{C}$. The composition $X \to A \to A''$ factorizes through an object $Y'' \in \mathcal{J}$. Since $A \to A''$ is an epimorphism, there exists an epimorphism $X_1 \to Y''$ in $\mathcal{C}$ such that the composition $X_1 \to Y'' \to A''$ factorizes through $A \to A''$. Hence we get the commutative diagram

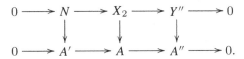

Set $X_2 = X \oplus X_1$ and define $N = \mathrm{Ker}(X_2 \to Y'') \in \mathcal{C}$. We get the commutative exact diagram:

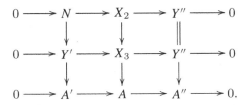

The morphism $N \to A'$ factorizes through an object $Y' \in \mathcal{J}$. Set $X_3 = Y' \oplus_N X_2$. We get the commutative diagram

$$
\begin{array}{ccccccccc}
0 & \longrightarrow & N & \longrightarrow & X_2 & \longrightarrow & Y'' & \longrightarrow & 0 \\
& & \downarrow & & \downarrow & & \| & & \\
0 & \longrightarrow & Y' & \longrightarrow & X_3 & \longrightarrow & Y'' & \longrightarrow & 0 \\
& & \downarrow & & \downarrow & & \downarrow & & \\
0 & \longrightarrow & A' & \longrightarrow & A & \longrightarrow & A'' & \longrightarrow & 0.
\end{array}
$$

Since the top square on the left is co-Cartesian, the middle row is exact. Since Y' and Y'' belong to $\mathcal{J}$, we get $X_3 \in \mathcal{J}$. Hence, $X \to A$ factors through $X_3 \in \mathcal{J}$.   q.e.d.

8.7 Extension of Exact Functors

Let $\mathcal{C}$ be an abelian category, $\mathcal{J}$ a full additive subcategory of $\mathcal{C}$, and let $j: \mathcal{J} \to \mathcal{C}$ be the embedding. Let $\mathcal{A}$ be another abelian category. Recall that the functor $j_*: \mathrm{Fct}(\mathcal{C}, \mathcal{A}) \to \mathrm{Fct}(\mathcal{J}, \mathcal{A})$ is defined by $j_* G = G \circ j$.

Notation 8.7.1. (i) We denote by $\mathrm{Fct}^r(\mathcal{C}, \mathcal{A})$ the full additive subcategory of $\mathrm{Fct}(\mathcal{C}, \mathcal{A})$ consisting of additive right exact functors.
(ii) We denote by $\mathrm{Fct}^r(\mathcal{J}, \mathcal{A})$ the full additive subcategory of $\mathrm{Fct}(\mathcal{J}, \mathcal{A})$ consisting of additive functors F which satisfy: for any exact sequence $Y' \to Y \to Y'' \to 0$ in $\mathcal{C}$ with Y', Y, Y'' in $\mathcal{J}$, the sequence $F(Y') \to F(Y) \to F(Y'') \to 0$ is exact in $\mathcal{A}$.

Note that j_* induces a functor (we keep the same notation)

(8.7.1) $j_* \colon \mathrm{Fct}^r(\mathcal{C}, \mathcal{A}) \to \mathrm{Fct}^r(\mathcal{J}, \mathcal{A})$.

Theorem 8.7.2. *Assume that $\mathcal{J}$ is generating in $\mathcal{C}$. Then the functor j_* induces an equivalence of categories $\mathrm{Fct}^r(\mathcal{C}, \mathcal{A}) \xrightarrow{\sim} \mathrm{Fct}^r(\mathcal{J}, \mathcal{A})$.*

Of course, one can deduce similar results for left exact or for contravariant functors. We leave the precise formulation to the reader.

The proof here is a toy model of the construction of derived categories studied in Chapter 11–13. Let us explain the idea of the proof. For $A \in \mathrm{Fct}^r(\mathcal{J}, \mathcal{A})$, we construct $A^+ \in \mathrm{Fct}^r(\mathcal{C}, \mathcal{A})$ whose image by j_* is isomorphic to A as follows. We can define the functor $K_0 \colon \mathrm{Mor}(\mathcal{J}) \to \mathcal{C}$ by $u \mapsto \mathrm{Coker}\,u$. On the other hand, we have the functor $A' \colon \mathrm{Mor}(\mathcal{J}) \to \mathcal{A}$ given by $u \mapsto \mathrm{Coker}(A(u))$. We will show that the diagram below can be completed with a dotted arrow:

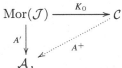

and then prove that A^+ belongs to $\mathrm{Fct}^r(\mathcal{C}, \mathcal{A})$ and its image by j_* is isomorphic to A.

We set

$\mathcal{D} := \mathrm{Mor}(\mathcal{C})$,

$K \colon \mathcal{D} \to \mathcal{C}$ the functor which associates $\mathrm{Coker}(u)$ to $u \in \mathcal{D}$.

Note that $\mathcal{D}$ is an abelian category.

Lemma 8.7.3. *For any $u, v \in \mathcal{D}$ and any morphism $f \colon K(u) \to K(v)$, there exist $w \in \mathcal{D}$ and morphisms $\alpha \colon w \to u$ and $\beta \colon w \to v$ such that $K(\alpha)$ is an isomorphism in $\mathcal{C}$ and $f \circ K(\alpha) = K(\beta)$.*

Proof. Let $u \colon Y \to X$ and $v \colon Y' \to X'$. Then construct X_1, Y_1, $Y_2 \in \mathcal{C}$ such that we have a commutative diagram with the three Cartesian squares marked by $\square$:

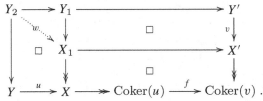

Then $w \colon Y_2 \to X_1$ satisfies the desired condition.   q.e.d.

Let $\mathcal{D}'$ be the category defined as follows:

$$\mathrm{Ob}(\mathcal{D}') = \mathrm{Ob}(\mathcal{D}),$$
$$\mathrm{Hom}_{\mathcal{D}'}(u, v) = \mathrm{Im}\big(\mathrm{Hom}_{\mathcal{D}}(u, v) \to \mathrm{Hom}_{\mathcal{C}}(K(u), K(v))\big) \quad \text{for } u, v \in \mathcal{D}.$$

Then $\mathcal{D}'$ is an additive category. The functor K decomposes as

$$\mathcal{D} \to \mathcal{D}' \xrightarrow{\ Q\ } \mathcal{C},$$

and Q is a faithful additive functor.

Let $\mathcal{S}$ be the set of morphisms s in $\mathcal{D}'$ such that $Q(s)$ is an isomorphism. Since Q is faithful, any morphism in $\mathcal{S}$ is a monomorphism by Proposition 1.2.12.

Lemma 8.7.4. (i) $\mathcal{S}$ *is a left multiplicative system,*
  (ii) *the functor Q decomposes as $\mathcal{D}' \to \mathcal{D}'_{\mathcal{S}} \to \mathcal{C}$ and the functor $\mathcal{D}'_{\mathcal{S}} \to \mathcal{C}$ is an equivalence of categories.*

Proof. (i) Let us check the condition in Definition 7.1.5. The conditions S1 and S2 are obvious. Let us show S'3 (with the notations there). Applying Lemma 8.7.3 to $K(t)^{-1} \circ K(f) \colon K(X) \to K(Y')$, there exist $X' \in \mathcal{D}'$ and morphisms $s \colon X' \to X$ and $g \colon X' \to Y'$ such that $K(s)$ is an isomorphism and $K(g) = K(t)^{-1} \circ K(f) \circ K(s)$. The condition S'4 immediately follows from the fact that any morphism in $\mathcal{S}$ is a monomorphism.
(ii) Since $Q \colon \mathcal{D}' \to \mathcal{C}$ sends the morphisms in $\mathcal{S}$ to isomorphisms, Q decomposes as $\mathcal{D}' \to \mathcal{D}'_{\mathcal{S}} \to \mathcal{C}$. For $u, v \in \mathcal{D}'$, the map

$$\mathrm{Hom}_{\mathcal{D}'_{\mathcal{S}}}(u, v) \simeq \varinjlim_{(u' \to u) \in \mathcal{S}_u} \mathrm{Hom}_{\mathcal{D}'}(u', v) \to \mathrm{Hom}_{\mathcal{C}}(Q(u), Q(v))$$

is injective because Q is faithful, and is surjective by Lemma 8.7.3. The functor $\mathcal{D}'_{\mathcal{S}} \to \mathcal{C}$ is therefore fully faithful, and it is evidently essentially surjective. Hence the functor $\mathcal{D}'_{\mathcal{S}} \to \mathcal{C}$ is an equivalence of categories.   q.e.d.

Let us denote by $\mathcal{D}_0$ the category $\mathrm{Mor}(\mathcal{J})$ and by $\mathcal{D}'_0$ the full subcategory of $\mathcal{D}'$ such that $\mathrm{Ob}(\mathcal{D}'_0) = \mathrm{Ob}(\mathcal{D}_0)$. Note that $\mathcal{D}_0$ is a full additive subcategory of $\mathcal{D}$. We set $\mathcal{T} = \mathcal{S} \cap \mathrm{Mor}(\mathcal{D}'_0)$.

Lemma 8.7.5. (i) *For any $u \in \mathcal{D}'$, there exists a morphism $\alpha \colon v \to u$ in $\mathcal{S}$ such that $v \in \mathcal{D}'_0$.*
  (ii) *The family of morphisms $\mathcal{T}$ is a left multiplicative system in $\mathcal{D}'_0$ and the functor $(\mathcal{D}'_0)_{\mathcal{T}} \to \mathcal{D}'_{\mathcal{S}}$ is an equivalence of categories.*

Proof. (i) Let us represent u by an object $u \colon Y \to X$ in $\mathcal{D}$. Take an epimorphism $X' \twoheadrightarrow X$ with $X' \in \mathcal{J}$ and then take an epimorphism $Y' \twoheadrightarrow Y \times_X X'$ with $Y' \in \mathcal{J}$. Then $v \colon Y' \to X'$ belongs to $\mathcal{D}_0$ and the morphism $v \to u$ induces an isomorphism $\mathrm{Coker}(v) \xrightarrow{\sim} \mathrm{Coker}(u)$.
(ii) then follows from Corollary 7.2.2 (with the arrows reversed).   q.e.d.

Applying Lemmas 8.7.4 and 8.7.5, we obtain that $(\mathcal{D}_0')_{\mathcal{T}} \to \mathcal{C}$ is an equivalence of categories. Let $K_0: \mathcal{D}_0 \to \mathcal{C}$ be the functor $u \mapsto \mathrm{Coker}(u)$. Then we have proved that K_0 decomposes as

$$\mathcal{D}_0 \to \mathcal{D}_0' \to (\mathcal{D}_0')_{\mathcal{T}} \xrightarrow{\sim} \mathcal{C} \ .$$

For $A \in \mathrm{Fct}^r(\mathcal{J}, \mathcal{A})$, we shall first construct $A^+ \in \mathrm{Fct}^r(\mathcal{C}, \mathcal{A})$ such that $j_* A^+ \simeq A$. We need two lemmas.

Lemma 8.7.6. *Let $A': \mathcal{D}_0 \to \mathcal{A}$ be the functor which associates $\mathrm{Coker}(A(u))$ to $u \in \mathcal{D}_0$. Then A' decomposes as $\mathcal{D}_0 \to \mathcal{D}_0' \to (\mathcal{D}_0')_{\mathcal{T}} \xrightarrow{\sim} \mathcal{C} \to \mathcal{A}$.*

Proof. It is enough to show the following two statements:

(8.7.2)   if $\alpha: u \to v$ in $\mathrm{Mor}(\mathcal{D}_0)$ satisfies $K(\alpha) = 0$, then $A'(\alpha) = 0$,

(8.7.3)   for $\alpha: u \to v$ in $\mathrm{Mor}(\mathcal{D}_0)$, if $K(\alpha)$ is an isomorphism, then $A'(\alpha)$ is an isomorphism .

Let us first show (8.7.3). Let us represent $\alpha: u \to v$ by a commutative diagram in $\mathcal{J}$:

(8.7.4)

$$
\begin{array}{ccc}
Y & \xrightarrow{\alpha_1} & Y' \\
{\scriptstyle u}\downarrow & & \downarrow{\scriptstyle v} \\
X & \xrightarrow{\alpha_0} & X'.
\end{array}
$$

The condition that $K(\alpha)$ is an isomorphism is equivalent to the fact that the sequence $Y \to X \oplus Y' \to X' \to 0$ is exact. This complex remains exact after applying A. Hence $\mathrm{Coker}(A(u)) \to \mathrm{Coker}(A(v))$ is an isomorphism.

Let us show (8.7.2). Let us represent α as in (8.7.4). The condition $K(\alpha) = 0$ implies that $X_1 := X \times_{X'} Y' \to X$ is an epimorphism. Set $Y_1 = X_1 \times_X Y$ and let $u_1: Y_1 \to X_1$ be the first projection. Then $u_1 \in \mathcal{D}$ and the morphism $\beta: u_1 \to u$ belongs to $\mathcal{S}$. By Lemma 8.7.5, there exists a morphism $\gamma: w \to u_1$ such that γ belongs to $\mathcal{S}$ and $w \in \mathcal{D}_0$. Thus we obtain a commutative diagram in $\mathcal{C}$:

$$
\begin{array}{ccccccc}
Y_2 & \xrightarrow{\gamma_1} & Y_1 & \xrightarrow{\beta_1} & Y & \xrightarrow{\alpha_1} & Y' \\
{\scriptstyle w}\downarrow & & {\scriptstyle u_1}\downarrow & & {\scriptstyle u}\downarrow & & \downarrow{\scriptstyle v} \\
X_2 & \xrightarrow{\gamma_0} & X_1 & \xrightarrow{\beta_0} & X & \xrightarrow{\alpha_0} & X'.
\end{array}
$$

Since $w \to u$ belongs to $\mathcal{S}$, (8.7.3) implies that $A'(w) \to A'(u)$ is an isomorphism. Since the composition $X_2 \xrightarrow{\beta_0 \circ \gamma_0} X \xrightarrow{\alpha_0} X'$ in $\mathcal{J}$ decomposes through $v: Y' \to X'$ as seen by the diagram above, the composition $A'(w) \xrightarrow{\sim} A'(u) \xrightarrow{A'(\alpha)} A'(v)$ vanishes. Hence we obtain the desired result, $A'(\alpha) = 0$.   q.e.d.

By Lemma 8.7.6, the functor $A': \mathcal{D}_0 \to \mathcal{A}$ decomposes through $\mathcal{C}$. Let $A^+: \mathcal{C} \to \mathcal{A}$ be the functor thus obtained. By the construction, it is obvious that A^+ commutes with finite products, and hence it is an additive functor by Proposition 8.2.15.

Lemma 8.7.7. *The functor A^+ is right exact.*

Proof. For an exact sequence $X_1 \to X_2 \to X_3 \to 0$ in $\mathcal{C}$, we can construct a commutative diagram in $\mathcal{C}$

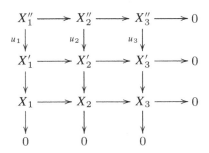

with $X'_j, X''_j \in \mathcal{J}$ ($j = 1, 2, 3$). Then $A'(u_j) \simeq \operatorname{Coker} A(u_j) \simeq A^+(X_j)$. Since A sends the first two rows to exact sequences, we obtain that $A^+(X_1) \to A^+(X_2) \to A^+(X_3) \to 0$ is exact.   q.e.d.

We can now complete the proof of the theorem. We have obtained the functor $\operatorname{Fct}^r(\mathcal{J}, \mathcal{A}) \to \operatorname{Fct}^r(\mathcal{C}, \mathcal{A})$ which associates $A^+ \in \operatorname{Fct}^r(\mathcal{C}, \mathcal{A})$ to $A \in \operatorname{Fct}^r(\mathcal{J}, \mathcal{A})$. It is obvious that this functor is a quasi-inverse to the functor j_* in (8.7.1).

Exercises

Exercise 8.1. Let $\mathcal{C}$ be a category admitting finite products, let $Gr(\mathcal{C})$ denote the category of group objects in $\mathcal{C}$ and denote by $for: Gr(\mathcal{C}) \to \mathcal{C}$ the forgetful functor. Prove that $Gr(\mathcal{C})$ admits finite products and that for commutes with finite products.

Exercise 8.2. Recall that **Group** denotes the category of groups and that $for:$ **Group** $\to$ **Set** is the forgetful functor.
(i) Prove that **Group** admits small projective limits and for commutes with such limits.
(ii) Let $X \in$ **Group** and assume that X is a group object in **Group**. Prove that the group structure on $for(X)$ induced by the group object structure in **Set** (see Lemma 8.1.2) is commutative and coincides with the group structure on $for(X)$ induced by the fact that $X \in$ **Group**.

Exercise 8.3. Let $\mathcal{C}$ and $\mathcal{C}'$ be additive categories.
(i) Let $X, Y \in \mathcal{C}$. Prove that if the first projection $X \times Y \to X$ is an isomorphism, then $Y \simeq 0$.
(ii) Prove that a functor $F : \mathcal{C} \to \mathcal{C}'$ is additive as soon as $F(X \times Y) \overset{\sim}{\to} F(X) \times F(Y)$ for any $X, Y \in \mathcal{C}$.

Exercise 8.4. Let $\mathcal{C}$ be an additive category and let $\mathcal{S}$ be a right multiplicative system. Prove that the localization $\mathcal{C}_{\mathcal{S}}$ is an additive category and $Q : \mathcal{C} \to \mathcal{C}_{\mathcal{S}}$ is an additive functor.

Exercise 8.5. Let $\mathcal{C}$ be an additive category and assume that $\mathcal{C}$ is idempotent complete. Let $X \in \mathcal{C}$.
(i) Let $p : X \to X$ be an idempotent (i.e., $p^2 = p$). Prove that there exists an isomorphism $X \simeq Y \oplus Z$ such that $p = g \circ f$ where $f : X \to Y$ is the projection and $g : Y \to X$ is the embedding.
(ii) Assume that $\mathrm{id}_X = \sum_{i \in I} e_i$, where $e_i \in \mathrm{Hom}_{\mathcal{C}}(X, X)$, I is finite, $e_i e_j = 0$ if $i \neq j$. Prove that $X \simeq \oplus_i X_i$ with $X_i \simeq \mathrm{Im}\, e_i$.

Exercise 8.6. Let $\mathcal{C}$ be an additive category and $\mathcal{N}$ a full additive subcategory of $\mathcal{C}$. For $X, Y \in \mathcal{C}$, define $\mathcal{N}(X, Y)$ as the set of morphisms $f : X \to Y$ in $\mathcal{C}$ which factorize through some $Z \in \mathcal{N}$.
(i) Prove that $\mathcal{N}(X, Y)$ is an additive subgroup of $\mathrm{Hom}_{\mathcal{C}}(X, Y)$.
(ii) Define the category $\mathcal{C}_{\mathcal{N}}$ by setting $\mathrm{Ob}(\mathcal{C}_{\mathcal{N}}) = \mathrm{Ob}(\mathcal{C})$ and $\mathrm{Hom}_{\mathcal{C}_{\mathcal{N}}}(X, Y) = \mathrm{Hom}_{\mathcal{C}}(X, Y)/\mathcal{N}(X, Y)$. Prove that $\mathcal{C}_{\mathcal{N}}$ is a well-defined additive category.
(iii) Assume that $\mathcal{N}$ is idempotent complete. Prove that a pair of objects X, Y in $\mathcal{C}$ are isomorphic in $\mathcal{C}_{\mathcal{N}}$ if and only if there exist $Z_1, Z_2 \in \mathcal{N}$ and an isomorphism $X \oplus Z_1 \simeq Y \oplus Z_2$. (Hint: if $f : X \to Y$ and $g : Y \to X$ satisfy $g \circ f = \mathrm{id}_X$, and if there exists $Y \overset{u}{\to} Z \overset{v}{\to} Y$ such that $\mathrm{id}_Y = f \circ g + v \circ u$ and $Z \in \mathcal{N}$, then $p := (u \circ v)^2 \in \mathrm{End}_{\mathcal{C}}(Z)$ satisfies $p = p^2$ and $v \circ u = v \circ p \circ u$.)

Exercise 8.7. Let I be a small set and let $\mathcal{C}$ be an additive category admitting coproducts indexed by I. Let $\mathcal{N}$ be a full additive subcategory of $\mathcal{C}$ closed by coproducts indexed by I. Prove that the category $\mathcal{C}_{\mathcal{N}}$ defined in Exercise 8.6 admits coproducts indexed by I and the functor $\mathcal{C} \to \mathcal{C}_{\mathcal{N}}$ commutes with such coproducts.

Exercise 8.8. Let $F : \mathcal{C} \to \mathcal{C}'$ be an additive functor of additive categories and assume that F admits a left (or right) adjoint G. Prove that G is additive.

Exercise 8.9. Let $\mathcal{C}$ be a small category and denote by $\mathcal{C}^{\wedge, ab}$ the category of functors from $\mathcal{C}^{\mathrm{op}}$ to $\mathrm{Mod}(\mathbb{Z})$. Denote by $\varphi : \mathcal{C} \to \mathcal{C}^{\wedge, ab}$ the functor

$$\varphi(X) : Y \mapsto \mathbb{Z}^{\oplus \mathrm{Hom}_{\mathcal{C}}(Y, X)} .$$

(i) Prove that φ is faithful.
(ii) Let $\mathcal{C}^+$ denote the full subcategory of $\mathcal{C}^{\wedge, ab}$ consisting of objects which are finite products of objects of the form $\varphi(X)$ with $X \in \mathcal{C}$. Prove that $\mathcal{C}^+$ is an additive category and

$$\mathrm{Hom}_{\mathcal{C}^+}(\varphi(X), \varphi(Y)) \simeq \mathbb{Z}^{\oplus \mathrm{Hom}_{\mathcal{C}}(X,Y)}.$$

(iii) Let $\mathcal{A}$ be an additive category and $F \colon \mathcal{C} \to \mathcal{A}$ a functor. Prove that there exists an additive functor $F' \colon \mathcal{C}^+ \to \mathcal{A}$ such that F is isomorphic to the composition $F' \circ \varphi$.

Exercise 8.10. Let k be a commutative ring and let $\mathcal{C}_1, \mathcal{C}_2$ be k-additive categories. Let $\Phi \colon \mathcal{C}_1 \times \mathcal{C}_2 \to (\mathcal{C}_1 \times \mathcal{C}_2)^\wedge$ be the functor given by

$$\Phi\big((X_1, X_2)\big)(Y_1, Y_2) = \mathrm{Hom}_{\mathcal{C}_1}(Y_1, X_1) \otimes_k \mathrm{Hom}_{\mathcal{C}_2}(Y_2, X_2) \,.$$

Let $\mathcal{C}_1 \otimes_k \mathcal{C}_2$ be the full subcategory of $(\mathcal{C}_1 \times \mathcal{C}_2)^\wedge$ consisting of objects isomorphic to finite products of images of objects of $\mathcal{C}_1 \times \mathcal{C}_2$ by Φ.
(i) Prove that $\mathcal{C}_1 \otimes_k \mathcal{C}_2$ is a k-additive category and prove that the functor $\varphi \colon \mathcal{C}_1 \times \mathcal{C}_2 \to \mathcal{C}_1 \otimes_k \mathcal{C}_2$ induced by Φ is k-bilinear, that is, k-additive with respect to each argument.
(ii) Let $\mathcal{A}$ be a k-additive category and let $F \colon \mathcal{C}_1 \times \mathcal{C}_2 \to \mathcal{A}$ be a k-bilinear functor. Prove that F decomposes as $\mathcal{C}_1 \times \mathcal{C}_2 \xrightarrow{\varphi} \mathcal{C}_1 \otimes_k \mathcal{C}_2 \xrightarrow{G} \mathcal{A}$ where G is unique up to unique isomorphism.
(iii) Prove that $\mathcal{C}_1^{\mathrm{op}} \otimes_k \mathcal{C}_2^{\mathrm{op}}$ is equivalent to $(\mathcal{C}_1 \otimes_k \mathcal{C}_2)^{\mathrm{op}}$.

Exercise 8.11. Let $\mathcal{C}$ be an abelian category, $\mathcal{S}$ a right and left multiplicative system. Prove that the localization $\mathcal{C}_\mathcal{S}$ is abelian and the functor $\mathcal{C} \to \mathcal{C}_\mathcal{S}$ is exact.

Exercise 8.12. Let $\mathcal{C}$ be an abelian category, $\mathcal{N}$ a full additive subcategory closed by subobjects, quotients and extensions (see Definition 8.3.21). Let $\mathcal{S}$ denote the family of morphisms in $\mathcal{C}$ defined by $f \in \mathcal{S}$ if and only if $\mathrm{Ker}\, f$ and $\mathrm{Coker}\, f$ belong to $\mathcal{N}$. Prove that $\mathcal{S}$ is a right and left saturated multiplicative system and that the localization $\mathcal{C}_\mathcal{S}$ (usually denoted by $\mathcal{C}/\mathcal{N}$) is an abelian category.

Exercise 8.13. We keep the notations of Exercise 8.12. Let $\mathcal{C}$ be a Grothendieck category, $\mathcal{N}$ a full additive subcategory closed by subobjects, quotients, extensions and small inductive limits.
(i) Prove that for any object $X \in \mathcal{C}$, there exists a maximal subobject Y of X with $Y \in \mathcal{N}$, and prove the isomorphism $\mathrm{Hom}_\mathcal{C}(Z, X/Y) \simeq \mathrm{Hom}_{\mathcal{C}/\mathcal{N}}(Z, X)$ for any $Z \in \mathcal{C}$.
(ii) Prove that the functor $\mathcal{C} \to \mathcal{C}/\mathcal{N}$ admits a right adjoint.
(iii) Prove that $\mathcal{C}/\mathcal{N}$ is a Grothendieck category and the localization functor $\mathcal{C} \to \mathcal{C}/\mathcal{N}$ commutes with small inductive limits.

Exercise 8.14. Recall that, for an additive category $\mathcal{C}$, $\mathrm{End}\,(\mathrm{id}_\mathcal{C})$ denotes the commutative ring of morphisms of the identity functor on $\mathcal{C}$.
(i) Let R be a ring. Prove that $\mathrm{End}\,(\mathrm{id}_{\mathrm{Mod}(R)})$ is isomorphic to the center of R.

(ii) Let $\mathcal{C}$ be the category of finite abelian groups. Prove that $\mathrm{End}\,(\mathrm{id}_{\mathcal{C}}) \simeq \prod_p \mathbb{Z}_p$ where p ranges over the set of prime integers and $\mathbb{Z}_p := \varprojlim_n (\mathbb{Z}/\mathbb{Z}p^n)$ is the ring of p-adic integers.

Exercise 8.15. Let $\mathcal{C}$ be an abelian category.
(i) Prove that a complex $0 \to X \to Y \to Z$ is exact if and only if the complex of abelian groups $0 \to \mathrm{Hom}_{\mathcal{C}}(W, X) \to \mathrm{Hom}_{\mathcal{C}}(W, Y) \to \mathrm{Hom}_{\mathcal{C}}(W, Z)$ is exact for any object $W \in \mathcal{C}$.
(ii) By reversing the arrows, state and prove a similar statement for a complex $X \to Y \to Z \to 0$.

Exercise 8.16. Let $\mathcal{C}$ be an abelian category and let $f : X \to Y$ and $g : Y \to Z$ be morphisms in $\mathcal{C}$. Prove that there exists an exact complex

$$0 \to \mathrm{Ker}(f) \to \mathrm{Ker}(g \circ f) \to \mathrm{Ker}(g)$$
$$\to \mathrm{Coker}(f) \to \mathrm{Coker}(g \circ f) \to \mathrm{Coker}(g) \to 0.$$

Here, $\mathrm{Ker}(g) \to \mathrm{Coker}(f)$ is given by the composition $\mathrm{Ker}(g) \to Y \to \mathrm{Coker}(f)$.

Exercise 8.17. Let $F : \mathcal{C} \to \mathcal{C}'$ be an additive functor of abelian categories.
(i) Prove that F is left exact if and only if for any exact sequence $0 \to X' \to X \to X'' \to 0$ in $\mathcal{C}$, the sequence $0 \to F(X') \to F(X) \to F(X'')$ is exact.
(ii) Prove that the conditions (a)–(c) below are equivalent:

(a) F is exact,
(b) for any exact sequence $0 \to X' \to X \to X'' \to 0$ in $\mathcal{C}$, the sequence $0 \to F(X') \to F(X) \to F(X'') \to 0$ is exact,
(c) for any exact sequence $X' \to X \to X''$ in $\mathcal{C}$, the sequence $F(X') \to F(X) \to F(X'')$ is exact.

Exercise 8.18. Let $F : \mathcal{C} \to \mathcal{C}'$ be an additive functor of abelian categories.
(i) Prove that F is left exact if and only if for any monomorphism $X \rightarrowtail Y$ in $\mathcal{C}$, the sequence $F(X) \to F(Y) \rightrightarrows F(Y \oplus_X Y)$ is exact.
(ii) Similarly, F is right exact if and only if for any epimorphism $X \twoheadrightarrow Y$ in $\mathcal{C}$, the sequence $F(X \times_Y X) \rightrightarrows F(X) \to F(Y)$ is exact.

Exercise 8.19. Let $\mathcal{C}$ be an abelian category and consider a commutative diagram of complexes

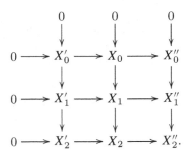

Assume that all rows are exact as well as the middle and right columns. Prove that all columns are exact.

Exercise 8.20. Let $\mathcal{C}$ be an abelian category. An object $X \in \mathcal{C}$ is *simple* if it is not isomorphic to 0 and any subobject of X is either X or 0. In this exercise, we write $X \supset Y$ or $Y \subset X$ to denote a subobject Y of X. A sequence $X = X_0 \supset X_1 \supset \cdots \supset X_{n-1} \supset X_n = 0$ is *a composition series* if X_i/X_{i+1} is simple for all i with $0 \le i < n$.

(i) Prove that the conditions below are equivalent:

(a) there exists a composition series $X = X_0 \supset X_1 \supset \cdots \supset X_n = 0$,
(b) there exists an integer n such that for any sequence $X = X_0 \supsetneq X_1 \supsetneq \cdots \supsetneq X_m = 0$, we have $m \le n$,
(c) any decreasing sequence $X = X_0 \supset X_1 \supset \cdots \supset X_m \supset \cdots$ is stationary (*i.e.* $X_m = X_{m+1}$ for $m \gg 0$) and any increasing sequence $X_0 \subset \cdots \subset X_m \subset \cdots \subset X$ is stationary,
(d) for any set $\mathcal{S}$ of subobjects of X ordered by inclusion, if $\mathcal{S}$ is filtrant then $\mathcal{S}$ has a largest element, and if $\mathcal{S}$ is cofiltrant then $\mathcal{S}$ has a smallest element.

(ii) Prove that the integer n in (a) depends only on X.
If the equivalent conditions above are satisfied, we say that X has finite length and the integer n in (a) is called the length of X.

Exercise 8.21. Let $\mathcal{C}$ be an abelian category and consider a commutative exact diagram:

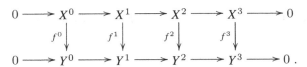

Prove that the following two conditions are equivalent:

(a) the middle square (X^1, X^2, Y^1, Y^2) is Cartesian,
(b) f^0 is an isomorphism and f^3 is a monomorphism.

Exercise 8.22. Let $\mathcal{C}$ be an abelian category and consider the diagram of complexes that we assume to be commutative except the two squares marked by "nc":

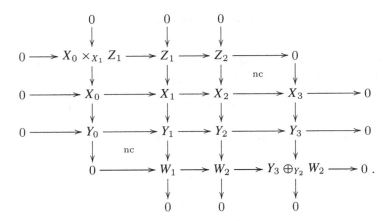

Assume that the second and third rows, as well as the second and third columns are exact. Prove that the conditions below are equivalent:

(a) the whole diagram (including the squares marked by "nc") is commutative and all rows and columns are exact,
(b) $X_1 \to X_2 \times_{Y_2} Y_1$ is an epimorphism,
(c) $X_2 \oplus_{X_1} Y_1 \to Y_2$ is a monomorphism.

Exercise 8.23. Let $\mathcal{C}$ be an abelian category and $\mathcal{J}$ a full additive subcategory. Let $X \in \mathcal{C}$. We say that

(a) X is $\mathcal{J}$-finite if there exists an epimorphism $Y \twoheadrightarrow X$ with $Y \in \mathcal{J}$,
(b) X is $\mathcal{J}$-pseudo-coherent if for any morphism $Y \xrightarrow{\varphi} X$ with $Y \in \mathcal{J}$, $\mathrm{Ker}\,\varphi$ is $\mathcal{J}$-finite,
(c) X is $\mathcal{J}$-coherent if X is $\mathcal{J}$-finite and $\mathcal{J}$-pseudo-coherent.

We denote by $\mathrm{coh}(\mathcal{J})$ the full subcategory of $\mathcal{C}$ consisting of $\mathcal{J}$-coherent objects.

(i) Consider an exact sequence $0 \to W \xrightarrow{f} X \xrightarrow{g} Y$ in $\mathcal{C}$ and assume that X is $\mathcal{J}$-finite and Y is $\mathcal{J}$-pseudo-coherent. Prove that W is $\mathcal{J}$-finite. (Hint: choose an epimorphism $\psi \colon Z \twoheadrightarrow X$ with $Z \in \mathcal{J}$, then construct an exact commutative diagram as below with $V \in \mathcal{J}$:

$$
\begin{array}{ccccc}
V & \xrightarrow{\ h\ } & Z & \longrightarrow & Y \\
\downarrow{\varphi} & & \downarrow{\psi} & & \| \\
0 \longrightarrow W & \xrightarrow{\ f\ } & X & \xrightarrow{\ g\ } & Y
\end{array}
$$

and prove, using Lemma 8.3.13, that φ is an epimorphism.)
(ii) Deduce from (i) that $\mathrm{coh}(\mathcal{J})$ is closed by kernels.
(iii) Prove that $\mathrm{coh}(\mathcal{J})$ is closed by extensions. (Hint: for an exact sequence $0 \to Y' \to Y \xrightarrow{v} Y'' \to 0$ and $u \colon X \to Y$, there is an exact sequence $0 \to \mathrm{Ker}\,u \to \mathrm{Ker}(v \circ u) \to Y'$.)

(iv) Assume that for any exact sequence $0 \to X' \to X \to X'' \to 0$, if X'' belongs to $\mathcal{J}$ and X' is $\mathcal{J}$-coherent, then X is $\mathcal{J}$-finite. Prove that $\mathrm{coh}(\mathcal{J})$ is abelian and the inclusion functor $\mathrm{coh}(\mathcal{J}) \hookrightarrow \mathcal{C}$ is exact. (Hint: it is enough to check that $\mathrm{coh}(\mathcal{J})$ is closed by cokernels. Consider an exact sequence $0 \to X' \to X \to X'' \to 0$ and assume that X' and X are $\mathcal{J}$-coherent. Clearly, X'' is $\mathcal{J}$-finite. Let $\varphi : S \to X''$ be a morphism with $S \in \mathcal{J}$, set $Y := X \times_{X''} S$ and consider the commutative exact diagram

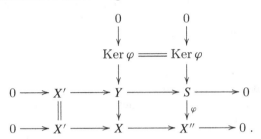

Show that $\mathrm{Ker}\,\varphi$ is $\mathcal{J}$-finite.)
When R is a ring, $\mathcal{C} = \mathrm{Mod}(R)$ and $\mathcal{J}$ is the full subcategory of free modules of finite rank, the $\mathcal{J}$-coherent objects of $\mathcal{C}$ are called coherent R-modules. Recall that a ring R is left coherent if it is coherent as a left R-module.

Exercise 8.24. In this exercise, we shall prove that the category $\mathrm{Mod}(R)$ of modules over a ring R admits enough injective objects. If M is a $\mathbb{Z}$-module, set $M^{\vee} = \mathrm{Hom}_{\mathbb{Z}}(M, \mathbb{Q}/\mathbb{Z})$.
(i) Prove that a $\mathbb{Z}$-module M is injective in $\mathrm{Mod}(\mathbb{Z})$ if and only if $nM = M$ for every positive integer n.
(ii) Prove that $\mathbb{Q}/\mathbb{Z}$ is injective in $\mathrm{Mod}(\mathbb{Z})$.
(iii) Define a natural morphism $M \to M^{\vee\vee}$ and prove that this morphism is injective.
(iv) Prove that for $M, N \in \mathrm{Mod}(\mathbb{Z})$, the map $\mathrm{Hom}_{\mathbb{Z}}(M, N) \to \mathrm{Hom}_{\mathbb{Z}}(N^{\vee}, M^{\vee})$ is injective.
(iv) Prove that if P is a projective R-module, then $P^{\vee}$ is R^{op}-injective.
(v) Let M be an R-module. Prove that there exist an injective R-module Z and a monomorphism $M \to Z$.

Exercise 8.25. Let $F : \mathcal{C} \to \mathcal{C}'$ be an additive functor of abelian categories. Consider the conditions

(a) F is faithful,
(b) F is conservative,
(c) $F(X) \simeq 0 \Rightarrow X \simeq 0$ for any $X \in \mathcal{C}$.

Prove that (a) $\Rightarrow$ (b) $\Rightarrow$ (c) and that these three conditions are equivalent when assuming that F is exact. (Hint: use Proposition 2.2.3.)
(See Exercise 8.27 for an example which shows that (c) does not imply (b) and see Exercise 8.26 for an example which shows that (b) does not imply (a).)

Exercise 8.26. Let k be a field and let $A = k[x]$. Let $F \colon \mathrm{Mod}(A) \to \mathrm{Mod}(A)$ be the functor which associates $xM \oplus (M/xM)$ to $M \in \mathrm{Mod}(A)$. Prove that F is conservative but is not faithful.

Exercise 8.27. Let k be a field, let $A = k[x, y]$ be the k-algebra generated by indeterminates x, y, and let $\mathfrak{a}$ be the ideal $Ax + Ay$.
(i) Prove that $\mathfrak{a}$ is not a generator of the category $\mathrm{Mod}(A)$.
(ii) Prove that for any $X \in \mathrm{Mod}(A)$, $\mathrm{Hom}_A(\mathfrak{a}, X) \simeq 0$ implies $X \simeq 0$. (Hint: reduce to the case $X = A/I$ and use $\mathrm{Hom}_A(\mathfrak{a}, A/\mathfrak{a}) \simeq k \oplus k$.)
(iii) Prove that the functor $\mathrm{Hom}_A(\mathfrak{a}, \cdot) \colon \mathrm{Mod}(A) \to \mathbf{Set}$ is neither conservative nor faithful. (Hint: consider $\mathfrak{a} \to A \to A/\mathfrak{a}$.)

Exercise 8.28. In this exercise and the next one, we shall give an example of an additive category $\mathcal{C}_0$ which is not abelian, although it admits kernels and cokernels, and any morphism which is both a monomorphism and an epimorphism is an isomorphism.
Let $\mathcal{C}$ be an abelian category which admits small inductive limits, let $G \in \mathcal{C}$ and denote by $\mathcal{C}_0$ the full subcategory of $\mathcal{C}$ consisting of objects X such that there exists an epimorphism $G^{\oplus I} \twoheadrightarrow X$ for some small set I. Let $\alpha \colon \mathcal{C}_0 \to \mathcal{C}$ denote the canonical functor.
(i) Prove that the functor α admits a right adjoint functor $\beta \colon \mathcal{C} \to \mathcal{C}_0$ and that for any $X \in \mathcal{C}$, $\alpha\beta(X) \to X$ is a monomorphism and $\mathrm{id}_{\mathcal{C}_0} \to \beta \circ \alpha$ is an isomorphism of functors.
(ii) Prove that $\mathcal{C}_0$ is an additive category which admits kernels and cokernels.
We shall denote by Ker_0, Coker_0, Im_0, and Coim_0 the kernel, cokernel, image and coimage in $\mathcal{C}_0$.
(iii) Let f be a morphism in $\mathcal{C}_0$. Prove that $\mathrm{Coker}_0 f \simeq \mathrm{Coker} f$ and $\mathrm{Ker}_0 f \simeq \beta(\mathrm{Ker} f)$.
(iv) Let f be a morphism in $\mathcal{C}_0$. Prove that f is a strict morphism if and only if $\mathrm{Ker} f$ belongs to $\mathcal{C}_0$.

Exercise 8.29. Let k be a field, let $A = k[x, y]$ and let $\mathcal{C} = \mathrm{Mod}(A)$. Let $\mathfrak{a}$ denote the ideal $\mathfrak{a} = Ax + Ay$ and let $\mathfrak{a}^2$ denote its square, the ideal generated by ab with $a, b \in \mathfrak{a}$. We define $\mathcal{C}_0$, $\alpha \colon \mathcal{C}_0 \to \mathcal{C}$ and $\beta \colon \mathcal{C} \to \mathcal{C}_0$ as in Exercise 8.28, taking $\mathfrak{a}$ as G. (See also Exercise 5.8.)
(i) Let X be an A-module such that $\mathfrak{a} X = 0$. Prove that X belongs to $\mathcal{C}_0$.
(ii) Prove that $\beta(X) \simeq 0$ implies $X \simeq 0$ (see Exercise 8.27) and prove that $\beta(A) \simeq \mathfrak{a}$.
(iii) Prove that a morphism $u \colon X \to Y$ in $\mathcal{C}_0$ is a monomorphism (resp. an epimorphism) if and only if $\alpha(u)$ is a monomorphism (resp. an epimorphism) in $\mathcal{C}$. (Hint: use (ii).)
(iv) Prove that any monomorphism in $\mathcal{C}_0$ is strict.
(v) Prove that a morphism in $\mathcal{C}_0$ which is both a monomorphism and an epimorphism is an isomorphism.
(vi) Prove that the canonical morphism $u \colon \mathfrak{a} \to \mathfrak{a}/Ax$ is a morphism in $\mathcal{C}_0$. Prove also that

$$\mathrm{Ker}_0\, u \simeq \mathfrak{a}x, \quad \mathrm{Coim}_0\, u \simeq \mathfrak{a}/(\mathfrak{a}x), \quad \mathrm{Im}_0\, u \simeq \mathfrak{a}/(Ax) \,.$$

In particular, $\mathrm{Coim}_0\, u \to \mathrm{Im}_0\, u$ is not an isomorphism.

(vii) We shall show that strict epimorphisms are not stable by base changes in $\mathcal{C}_0$. Let $v\colon \mathfrak{a} \to \mathfrak{a}/\mathfrak{a}^2$ be the canonical morphism, and let $w\colon A/\mathfrak{a} \to \mathfrak{a}/\mathfrak{a}^2$ be the morphism that sends $1 \bmod \mathfrak{a}$ to $x \bmod \mathfrak{a}^2$.

$$\begin{array}{ccc} X & \longrightarrow & A/\mathfrak{a} \\ \downarrow & \square & \downarrow{\scriptstyle w} \\ \mathfrak{a} & \xrightarrow{\ v\ } & \mathfrak{a}/\mathfrak{a}^2 \end{array}$$

Denote by X the fiber product in $\mathcal{C}_0$ of $\mathfrak{a}$ and $A/\mathfrak{a}$ over $\mathfrak{a}/\mathfrak{a}^2$, that is, X is the kernel in $\mathcal{C}_0$ of $v\oplus w\colon \mathfrak{a}\oplus(A/\mathfrak{a}) \to \mathfrak{a}/\mathfrak{a}^2$. Prove that v is a strict epimorphism, prove that $X \simeq \mathfrak{a}^2$ and prove that the canonical morphism $X \to A/\mathfrak{a}$ is the zero morphism.

This shows that the category $\mathcal{C}_0$ is not quasi-abelian in the sense of Schneiders [61], since strict epimorphisms are stable by base change in quasi-abelian categories.

Exercise 8.30. Let $\mathcal{C}$ be an abelian category which admits small inductive limits and such that small filtrant inductive limits are exact. Prove that any object of finite length (see Exercise 8.20) is of finite type (see Exercise 6.7).

Exercise 8.31. Let k be a field, t an indeterminate, and denote by $\mathcal{C}$ the abelian category $\mathrm{Mod}(k[t])$. Denote by $\mathcal{C}_0$ the fully abelian subcategory consisting of $k[t]$-modules M for which there exists some $n \geq 0$ with $t^n M = 0$. Set $Z_n := k[t]t^{-n} \subset k[t, t^{-1}]$, and let $X_n := Z_n/k[t] \in \mathcal{C}_0$.
(i) Show that $X_n \to X_{n+1}$ is a monomorphism but $\varinjlim_n X_n \simeq 0$ in $\mathcal{C}_0$ (the limit is calculated in $\mathcal{C}_0$).
(ii) Set $Y_n = k[t]/k[t]t^{n+1}$. Show that $Y_n \to Y_{n-1}$ is an epimorphism but $\varprojlim_n Y_n \simeq 0$ in $\mathcal{C}_0$ (the limit is calculated in $\mathcal{C}_0$).
Note that neither $\varinjlim_n X_n$ nor $\varprojlim_n Y_n$ vanishes in $\mathcal{C}$ when the limits are calculated in $\mathcal{C}$.

Exercise 8.32. Let $\mathcal{C}$ be an abelian category with small projective limits and let $\{X_n\}_{n\in\mathbb{N}}$ be objects of $\mathcal{C}$.
(i) Prove that $\varprojlim_{n\in\mathbb{N}} \text{``}\bigoplus\text{''}_{m\geq n} X_m \simeq 0$ in $\mathrm{Ind}(\mathcal{C})$. (Recall that "$\bigoplus$" denotes the coproduct in $\mathrm{Ind}(\mathcal{C})$.)
(ii) Deduce that if $\mathcal{C} \neq \mathbf{Pt}$, filtrant projective limits in $\mathrm{Ind}(\mathcal{C})$ are not exact. (Hint: consider the exact sequences $0 \to \text{``}\bigoplus\text{''}_{m\geq n} X_m \to \text{``}\bigoplus\text{''}_{m\geq 0} X_m \to \text{``}\bigoplus\text{''}_{0\leq m<n} X_m \to 0$.)
(iii) Deduce that if $\mathcal{C} \neq \mathbf{Pt}$, the abelian categories $\mathrm{Ind}(\mathrm{Pro}(\mathcal{C}))$ and $\mathrm{Pro}(\mathrm{Ind}(\mathcal{C}))$ are not equivalent.

Exercise 8.33. Let A be a commutative ring and let $a \in A$. Consider the additive functor $F_a \colon \mathrm{Mod}(A) \to \mathrm{Mod}(A)$ which associates to a module M the submodule aM of M.
(i) Prove that F_a sends monomorphisms to monomorphisms and epimorphisms to epimorphisms.
(ii) Show that the conditions (a)–(d) below are equivalent.

(a) F_a is left exact,
(b) F_a is right exact,
(c) $a \in Aa^2$,
(d) there exists $c \in A$ such that $c = c^2$ and $Ac = Aa$.

(Hint: if $a = ba^2$, then ba is an idempotent.)

Exercise 8.34. In this exercise, we shall generalize the notion of split exact sequences in an abelian category (Definition 8.3.15) to the one in an arbitrary additive category. Let $\mathcal{C}$ be an additive category.
(i) Let $0 \to X \xrightarrow{f} Y \xrightarrow{g} Z \to 0$ be a complex in $\mathcal{C}$. Prove that the following conditions are equivalent:

(a) there exist $h \colon Z \to Y$ and $k \colon Y \to X$ such that $\mathrm{id}_Y = f \circ k + h \circ g$, $g \circ h = \mathrm{id}_Z$ and $k \circ f = \mathrm{id}_X$,
(b) there exits an isomorphism of complexes

$$
\begin{array}{ccccccccc}
0 & \longrightarrow & X & \longrightarrow & Y & \longrightarrow & Z & \longrightarrow & 0 \\
& & \downarrow{\scriptstyle \mathrm{id}_X} & & \downarrow{\scriptstyle \sim} & & \downarrow{\scriptstyle \mathrm{id}_Z} & & \\
0 & \longrightarrow & X & \longrightarrow & X \oplus Z & \longrightarrow & Z & \longrightarrow & 0\,,
\end{array}
$$

where the bottom row is the canonical complex,
(c) for any $W \in \mathcal{C}$, the complex

$$0 \to \mathrm{Hom}_{\mathcal{C}}(W, X) \to \mathrm{Hom}_{\mathcal{C}}(W, Y) \to \mathrm{Hom}_{\mathcal{C}}(W, Z) \to 0$$

in $\mathrm{Mod}(\mathbb{Z})$ is exact,
(d) for any $W \in \mathcal{C}$, the complex

$$0 \to \mathrm{Hom}_{\mathcal{C}}(Z, W) \to \mathrm{Hom}_{\mathcal{C}}(Y, W) \to \mathrm{Hom}_{\mathcal{C}}(X, W) \to 0$$

in $\mathrm{Mod}(\mathbb{Z})$ is exact.

If these equivalent conditions are satisfied, we say that the complex $0 \to X \to Y \to Z \to 0$ *splits*.
(ii) Assume that $\mathcal{C}$ is abelian. Prove that the above notion coincides with that of Definition 8.3.15.

Exercise 8.35. Let $\mathcal{C}$ be an abelian category which admits small inductive limits and such that small filtrant inductive limits are exact. Let I be a small set and $f\colon X \to \oplus_i Y_i$ a morphism in $\mathcal{C}$. For $i_0 \in I$, denote by f_{i_0} the composition $X \to \oplus_i Y_i \to Y_{i_0}$. Let $K \subset I$ and assume that $f_i = 0$ for all $i \in I \setminus K$. Prove that f decomposes as $X \xrightarrow{g} \oplus_{i \in K} Y_i \hookrightarrow \oplus_{i \in I} Y_i$. (Hint: use the isomorphism $\oplus_{i \in I} Y_i \simeq \varinjlim_J \oplus_{i \in J} Y_i$ where J ranges over the filtrant family of finite subsets of I and write $X \simeq \varinjlim_J X_J$.)

Exercise 8.36. Let $\mathcal{C}$ be an abelian category and let $\{G_i\}_{i \in I}$ be a small family of objects of $\mathcal{C}$. Consider the conditions:

(a) $\{G_i\}_{i \in I}$ is a system of generators,
(b) for any $X \in \mathcal{C}$ and any monomorphism $f\colon Z \hookrightarrow X$, if $\mathrm{Hom}_{\mathcal{C}}(G_i, Z) \to \mathrm{Hom}_{\mathcal{C}}(G_i, X)$ is surjective for all i, then f is an isomorphism,
(c) if $X \in \mathcal{C}$ satisfies $\mathrm{Hom}_{\mathcal{C}}(G_i, X) = 0$ for all $i \in I$, then $X \simeq 0$.

Prove that (a) $\Leftrightarrow$ (b) $\Rightarrow$ (c).
(Note that (c) does not implies (a), see Exercise 8.27.)

Exercise 8.37. Let $\mathcal{C}$ be an abelian category and let $\{X_n\}_{n \in \mathbb{N}}$ be an inductive system in $\mathcal{C}$ indexed by the ordered set $\mathbb{N}$.
(i) Assume that $\mathcal{C}$ admits countable coproducts and countable filtrant inductive limits are exact. Let $\mathrm{sh}\colon \oplus_{n \geq 0} X_n \to \oplus_{n \geq 0} X_n$ be the morphism in $\mathcal{C}$ associated with $X_n \to X_{n+1}$. Prove that the sequence

$$(8.7.5) \qquad 0 \to \bigoplus_{n \geq 0} X_n \xrightarrow{\mathrm{id}-\mathrm{sh}} \bigoplus_{n \geq 0} X_n \to \varinjlim_n X_n \to 0$$

is exact in $\mathcal{C}$. (Hint: $0 \to \bigoplus_{n=0}^{m} X_n \xrightarrow{\mathrm{id}-\mathrm{sh}} \bigoplus_{n=0}^{m+1} X_n \to X_{m+1} \to 0$ is exact.)
(ii) Prove that the sequence $0 \to \text{"}\bigoplus_{n \geq 0}\text{"} X_n \xrightarrow{\mathrm{id}-\mathrm{sh}} \text{"}\bigoplus_{n \geq 0}\text{"} X_n \to \text{"}\varinjlim_n\text{"} X_n \to 0$ is exact in $\mathrm{Ind}(\mathcal{C})$.
(iii) Assume that $\mathcal{C}$ admits countable coproducts. Prove that the sequence (8.7.5) is exact when assuming that $\text{"}\varinjlim_n\text{"} X_n$ belongs to $\mathcal{C}$.

Exercise 8.38. Let $\mathcal{C}$ be an abelian category which admits small projective limits and small inductive limits. Assume that small filtrant inductive limits are exact. Prove that for any small family $\{X_i\}_{i \in I}$ of objects of $\mathcal{C}$, the natural morphism $\oplus_{i \in I} X_i \to \prod_{i \in I} X_i$ is a monomorphism.

Exercise 8.39. Let k be a field of characteristic 0. The Weyl algebra $W := W_n(k)$ in n variables over k, is the k-algebra generated by x_i, ∂_i ($1 \leq i \leq n$) with the defining relations:

$$[x_i, x_j] = 0, \quad [\partial_i, \partial_j] = 0, \quad [\partial_i, x_j] = \delta_{ij} .$$

We set $\mathcal{A} = \mathrm{Mod}(k)$. We endow W with the increasing filtration for which each x_i and each ∂_j is of order one.

Let M be a finitely generated W-module, and let us endow it with a good filtration $M = \bigcup_{m \in \mathbb{Z}} M_m$, i.e., each M_m is a finite-dimensional vector space, $x_i M_m \subset M_{m+1}$, $\partial_i M_m \subset M_{m+1}$ for any m and i, and $M_m = 0$ for $m \ll 0$, $M_{m+1} = M_m + \sum_{i=1}^n (x_i M_m + \partial_i M_m)$ for $m \gg 0$.

(i) Show that "$\varinjlim_m$" M_m belongs to $\mathrm{Mod}(W, \mathrm{Ind}(\mathcal{A}))$.

(ii) Show that "$\varinjlim_m$" M_m does not depend on the choice of good filtrations.

(iii) Show that if $M \neq 0$, "$\varinjlim_m$" M_m is not isomorphic to the image of any object of $\mathrm{Ind}(\mathrm{Mod}(W, \mathcal{A}))$ by the natural functor $\mathrm{Ind}(\mathrm{Mod}(W, \mathcal{A})) \to \mathrm{Mod}(W, \mathrm{Ind}(\mathcal{A}))$. (Hint: otherwise, "$\varinjlim_m$" $M_m \simeq$ "$\varinjlim_i$" V_i with $V_i \in \mathrm{Mod}(W)$, and the W-linear morphisms $V_i \to M$ would then factorize through finite-dimensional vector spaces, and this implies they are zero.)

Exercise 8.40. Let $\mathcal{C}$ be an abelian category. Assume that finite inductive limits are stable by base change (see Definition 2.2.6). Prove that $\mathcal{C}$ is equivalent to **Pt**. (Hint: use Exercise 2.26.)

Exercise 8.41. Let k be a field, t an indeterminate, and denote by $\mathcal{C}$ the abelian category $\mathrm{Mod}(k[t])$ (see Exercise 8.31). Denote by $\mathcal{C}_1$ the fully abelian subcategory consisting of $k[t]$-modules M such that for any $u \in M$, there exists some $n \geq 0$ such that $t^n u = 0$.
(i) Prove that $\mathcal{C}_1$ admits small inductive limits and that the inclusion functor $\mathcal{C}_1 \hookrightarrow \mathcal{C}$ commutes with such limits.
(ii) Prove that $\mathcal{C}_1$ admits small projective limits and that the inclusion functor $\mathcal{C}_1 \hookrightarrow \mathcal{C}$ does not commute with such limits.
(iii) Prove that $\mathcal{C}_1$ is a Grothendieck category.
(iv) Prove that $k[t, t^{-1}]/k[t]$ is an injective cogenerator of $\mathcal{C}_1$.

Exercise 8.42. Let k be a commutative ring, R a k-algebra. Set $\mathcal{C} = \mathrm{Mod}(R)$. Prove that the object $F \in \mathcal{C}^\wedge$ given by $\mathcal{C} \ni M \mapsto \mathrm{Hom}_k(M, k)$ is representable and give its representative.

Exercise 8.43. Let $\mathcal{U}$ be a universe, k a $\mathcal{U}$-small field and let I be a set which is not $\mathcal{U}$-small. Let A be the polynomial ring $k[X_i; i \in I]$ where the X_i's are indeterminates. (Hence, A is not small.) Let $\mathcal{C}$ be the category of A-modules which are $\mathcal{U}$-small as sets.
(i) Prove that $\mathcal{C}$ is an abelian $\mathcal{U}$-category and that $\mathcal{C}$ admits $\mathcal{U}$-small inductive and projective limits.
(ii) Prove that $\mathcal{U}$-small filtrant inductive limits are exact.
(iii) Prove that the set of subobjects of any object of $\mathcal{C}$ is $\mathcal{U}$-small.
(iv) Prove that $\mathcal{C}$ has no generator.
(v) Prove that any projective object of $\mathcal{C}$ is isomorphic to zero.
(vi) Prove that any injective object of $\mathcal{C}$ is isomorphic to zero.

(vii) Prove that the object $F \in \mathcal{C}^\wedge$ given by $M \mapsto \mathrm{Hom}_k(M, k)$ commutes with small projective limits but is not representable.

(Hint: for (iv)–(vii), use the fact that the map $A \to \mathrm{End}_k(M)$ is not injective for any $M \in \mathcal{C}$.)

9

π-accessible Objects and $\mathcal{F}$-injective Objects

We introduce the notion of π-filtrant categories, where π is an infinite cardinal. When π is the countable cardinal $\aleph_0$, we recover the notion of filtrant categories. Then we generalize previous results concerning inductive limits over small and filtrant categories to π-filtrant categories. For example, we prove that inductive limits in **Set** over π-filtrant categories commute with projective limits over a category J as soon as the cardinal of $\mathrm{Mor}(J)$ is smaller than π. We define the full subcategory $\mathrm{Ind}^\pi(\mathcal{C})$ of $\mathrm{Ind}(\mathcal{C})$ of objects which are inductive limits over π-filtrant categories of objects of $\mathcal{C}$ and the full subcategory $\mathcal{C}_\pi$ of $\mathcal{C}$ of π-accessible objects, that is, objects X such that $\mathrm{Hom}_{\mathcal{C}}(X, \cdot)$ commutes with π-filtrant inductive limits. Then we give sufficient conditions which ensure that $\mathcal{C}_\pi$ is small and that $\varinjlim$ induces an equivalence $\mathrm{Ind}^\pi(\mathcal{C}_\pi) \xrightarrow{\sim} \mathcal{C}$. References are made to [64].

Next, given a family $\mathcal{F}$ of morphisms in $\mathcal{C}$, we define the notion of "$\mathcal{F}$-injective objects" and prove under suitable hypotheses the existence of "enough $\mathcal{F}$-injective objects". Some arguments used here were initiated in Grothendieck's paper [28] and play an essential role in the theory of model categories (see [32, 56]). Accessible objects are also discussed in [23], [1] and [49].

In the course of an argument, we need to prove a categorical version of Zorn's lemma which asserts that a small category which admits small filtrant inductive limits has what we call "a quasi-terminal object". We treat this technical result in a separate section.

We apply these results to abelian categories. A particular case is the Grothendieck theorem [28] on the existence of "enough injectives" in Grothendieck categories. We shall also apply these techniques in Chap. 13 to prove the existence of "enough homotopically injective complexes" in the category of unbounded complexes of a Grothendieck category.

To conclude, we prove the Freyd-Mitchell theorem which asserts that any small abelian category is equivalent to a full abelian subcategory of the category of modules over a suitable ring R. This justifies in some sense the common practice which consists in replacing an abelian category by $\mathrm{Mod}(\mathbb{Z})$ when chasing diagrams.

This chapter makes an intensive use of the notion of cardinals that we recall first.

9.1 Cardinals

In this chapter we fix a universe $\mathcal{U}$. As usual, $\mathcal{U}$-**Set** or simply **Set**, denotes the category of $\mathcal{U}$-sets.

Definition 9.1.1. *An ordered set I is* well ordered *if for any non empty subset $A \subset I$, there exists $a \in A$ such that $a \leq b$ for all $b \in A$ (that is, A admits a smallest element).*

In particular, if I is well ordered, then I is totally ordered. Let I be a well ordered set and let $a \in I$. Let $A_a = \{x \in I ; a < x\}$. Assuming A_a non empty (i.e., a is not a largest element of I), there exists a smallest element x in A_a. It is called the *successor* of a and denoted by $a + 1$.

Let I be a well ordered set and let $x \in I$. Then one and only one of the properties (a)–(c) below is satisfied.

(a) x is the smallest element in I,
(b) there exists $y \in I$ such that $x = y + 1$,
(c) x is not the smallest element of I and $x = \sup\{y ; y < x\}$.

Moreover, the element y given in (b) is unique. Indeed, $y = \sup\{z ; z < x\}$.

By the axiom of choice, any set may be well ordered.

Definition 9.1.2. *A $\mathcal{U}$-cardinal ω or simply a cardinal ω is an equivalence class of small sets with respect to the relation $X \sim Y$. Here, $X \sim Y$ if and only if X and Y are isomorphic in $\mathcal{U}$-**Set**.*

For a small set A, we denote by $\mathrm{card}(A)$ the associated cardinal.

We denote by $\aleph_0$ the countable cardinal, that is, $\mathrm{card}(\mathbb{N}) = \aleph_0$.

If $\pi_1 = \mathrm{card}(X_1)$ and $\pi_2 = \mathrm{card}(X_2)$ are two cardinals, we define

$$\pi_1 \cdot \pi_2 = \mathrm{card}(X_1 \times X_2) \,,$$
$$\pi_1^{\pi_2} = \mathrm{card}(\mathrm{Hom}_{\mathbf{Set}}(X_2, X_1)) \,.$$

These cardinals are well defined. Let us list some properties of cardinals.

(i) The set (which is no more small) of cardinals is well ordered. We denote as usual by $\leq$ this order. Recall that $\pi_1 \leq \pi_2$ if and only if there exist small sets A_1 and A_2 and an injective map $f \colon A_1 \to A_2$ such that $\mathrm{card}(A_1) = \pi_1$ and $\mathrm{card}(A_2) = \pi_2$.
(ii) For any cardinal π, $2^\pi > \pi$.
(iii) Let π be an infinite cardinal. Then $\pi \cdot \pi = \pi$.

(iv) An infinite cardinal π is said to be *regular* if it satisfies the condition that for any family of small sets $\{B_i\}_{i\in I}$ indexed by a small set I such that $\mathrm{card}(I) < \pi$ and $\mathrm{card}(B_i) < \pi$, we have $\mathrm{card}(\bigsqcup_i B_i) < \pi$. If π is an infinite cardinal, then its successor π' is a regular cardinal. Indeed, if $\mathrm{card}(I) < \pi'$ and $\mathrm{card}(B_i) < \pi'$, then $\mathrm{card}(I) \leq \pi$, $\mathrm{card}(B_i) \leq \pi$, and hence $\mathrm{card}(\bigsqcup_i B_i) \leq \pi \cdot \pi = \pi < \pi'$.
(v) Let π_0 be an infinite cardinal. There exists $\pi > \pi_0$ such that $\pi^{\pi_0} = \pi$. Indeed, $\pi := 2^{\pi_0}$ satisfies $\pi > \pi_0$ and $\pi^{\pi_0} = 2^{\pi_0\pi_0} = 2^{\pi_0} = \pi$.

In this chapter, π denotes an infinite cardinal.

9.2 π-filtrant Categories and π-accessible Objects

Some results of Chap. 3 will be generalized here.

Proposition 9.2.1. *Let I be a category. The following conditions* (i) *and* (ii) *are equivalent:*

(i) *The following two conditions hold:*
  (a) *for any $A \subset \mathrm{Ob}(I)$ such that $\mathrm{card}(A) < \pi$, there exists $j \in I$ such that for any $a \in A$ there exists a morphism $a \to j$ in I,*
  (b) *for any $i, j \in I$ and for any $B \subset \mathrm{Hom}_I(i, j)$ such that $\mathrm{card}(B) < \pi$, there exists a morphism $j \to k$ in I such that the composition $i \xrightarrow{s} j \to k$ does not depend on $s \in B$.*
(ii) *For any category J such that $\mathrm{card}(\mathrm{Mor}(J)) < \pi$ and any functor $\varphi\colon J \to I$, there exists $i \in I$ such that $\varprojlim_{j\in J} \mathrm{Hom}_I(\varphi(j), i) \neq \emptyset$.*

Note that condition (i) (a) implies that I is non empty. Indeed, apply this condition with $A = \emptyset$.

The proof is a variation of that of Lemma 3.1.2.

Proof. (i) $\Rightarrow$ (ii). Let J and φ be as in the statement (ii). Applying (a) to the family $\{\varphi(j)\}_{j\in J}$, there exist $i_0 \in I$ and morphisms $s(j)\colon \varphi(j) \to i_0$. Moreover, there exist $k(j) \in I$ and a morphism $\lambda(j)\colon i_0 \to k(j)$ such that the composition

$$\varphi(j) \xrightarrow{\varphi(t)} \varphi(j') \xrightarrow{s(j')} i_0 \xrightarrow{\lambda(j)} k(j)$$

does not depend on $t\colon j \to j'$. Now, there exist $i_1 \in I$ and morphisms $\xi(j)\colon k(j) \to i_1$.

Finally, take a morphism $t\colon i_1 \to i_2$ such that $t \circ \xi(j) \circ \lambda(j)$ does not depend on j. The family of morphisms $u_j\colon \varphi(j) \to i_0 \to k(j) \to i_1 \to i_2$ defines an element of $\varprojlim_{j\in J} \mathrm{Hom}_I(\varphi(j), i_2)$.

(ii) $\Rightarrow$ (i). By taking as J the discrete category A, we obtain (a). By taking as J the category with two objects a and b and morphisms id_a, id_b and a family of arrows $a \to b$ indexed by B, we obtain (b). q.e.d.

Definition 9.2.2. *A category I is* π-filtrant *if the equivalent conditions in* Proposition 9.2.1 *are satisfied.*

Note that for $\pi = \aleph_0$, a category is π-filtrant if and only if it is filtrant.

Also note that if π' is an infinite cardinal with $\pi' \leq \pi$, then any π-filtrant category is π'-filtrant.

Example 9.2.3. Let π be an infinite regular cardinal and let J be a well ordered set such that $\mathrm{card}(J) \geq \pi$. For $x \in J$, set $J_x = \{y \in J; y < x\}$. Define the subset I of J by

$$I = \{x \in J; \mathrm{card}(J_x) < \pi\} \,.$$

It is obvious that $x \leq y$ and $y \in I$ implies $x \in I$. Then

(i) I is well ordered,
(ii) I is π-filtrant,
(iii) $\sup(I)$ does not exist in I.

In order to prove (ii), let us check condition (i) of Proposition 9.2.1. Condition (i) (b) is obviously satisfied. Let $A \subset I$ with $\mathrm{card}(A) < \pi$. Set $A' = \bigcup_{a \in A} J_a$. Since π is regular, $\mathrm{card}(A') < \pi$. Hence $A' \neq J$ and $b := \inf(J \setminus A)$ exists. Then $J_b \subset A'$, and $b \in I$. This shows (i) (a) and I is π-filtrant. Let us check (iii). If $\sup(I)$ exists in I, then $\mathrm{card}(I) < \pi$ and $\sup(I) + 1$ exists in J and belongs to I, which is a contradiction.

Example 9.2.4. Let π be an infinite regular cardinal and let A be a small set. Set

$$I = \{B \subset A \,;\, \mathrm{card}(B) < \pi\} \,.$$

The inclusion relation defines an order on I. Regarding I as a category, I is π-filtrant. Indeed, condition (i) (b) in Proposition 9.2.1 is obviously satisfied. On the other hand, for any $S \subset I$ with $\mathrm{card}(S) < \pi$, we have $\mathrm{card}(\bigcup_{B \in S} B) < \pi$. This implies (i) (a).

Lemma 9.2.5. *Let $\varphi \colon J \to I$ be a functor. Assume that J is π-filtrant and φ is cofinal. Then I is π-filtrant.*

Proof. By Proposition 3.2.4, I is filtrant. Let us check property (i) of Proposition 9.2.1.
(a) Let $A \subset \mathrm{Ob}(I)$ with $\mathrm{card}(A) < \pi$. For any $a \in A$, there exist $j(a) \in J$ and a morphism $a \to \varphi(j(a))$. Let $A' = \{j(a); a \in A\}$. There exist $j \in J$ and morphisms $j(a) \to j$ in J. Therefore there exist morphisms $a \to \varphi(j(a)) \to \varphi(j)$ in I.
(b) Let $i_1, i_2 \in I$ and let $B \subset \mathrm{Hom}_I(i_1, i_2)$ with $\mathrm{card}(B) < \pi$. There exist $j_1 \in J$ and a morphism $i_1 \to \varphi(j_1)$. For each $s \in B$, there exist $j(s) \in J$, a morphism $t(s) \colon j_1 \to j(s)$ in J, and a commutative diagram

$$
\begin{array}{ccc}
i_1 & \xrightarrow{\quad s \quad} & i_2 \\
\downarrow & & \downarrow \\
\varphi(j_1) & \xrightarrow{\varphi(t(s))} & \varphi(j(s)).
\end{array}
$$

By property (i) (a), there exist $j_2 \in J$ and morphisms $j(s) \to j_2$. By property (i) (b), there exists a morphism $j_2 \to j_3$ such that the composition $j_1 \to j(s) \to j_2 \to j_3$ does not depend on s. Hence the composition $i_1 \xrightarrow{s} i_2 \to \varphi(j(s)) \to \varphi(j_3)$ does not depend on $s \in B$.   q.e.d.

Remark 9.2.6. Let I be a cofinally small π-filtrant category. Then there exist a small π-filtrant category I' and a cofinal functor $I' \to I$. Indeed, there exists a small subset $S \subset \mathrm{Ob}(I)$ such that for any $i \in I$ there exists a morphism from i to an element of S. Define I' as the full subcategory of I such that $\mathrm{Ob}(I') = S$. Then I' is π-filtrant and cofinal to I by Proposition 3.2.4.

Definition 9.2.7. *Let π be an infinite cardinal and let $\mathcal{C}$ be a category which admits π-filtrant small inductive limits.*

(i) *An object $X \in \mathcal{C}$ is π-accessible if for any π-filtrant small category I and any functor $\alpha: I \to \mathcal{C}$, the natural map*

$$
\varinjlim_{i \in I} \mathrm{Hom}_{\mathcal{C}}(X, \alpha(i)) \to \mathrm{Hom}_{\mathcal{C}}(X, \varinjlim_{i \in I} \alpha(i))
$$

is an isomorphism.
(ii) *We denote by $\mathcal{C}_{\pi}$ the full subcategory of $\mathcal{C}$ consisting of π-accessible objects.*

Remark 9.2.8. (i) If $\mathcal{C}$ is discrete, then $\mathcal{C} = \mathcal{C}_{\pi}$, since any functor $I \to \mathcal{C}$ with I filtrant, is a constant functor.
(ii) If $\pi' \le \pi$, then $\mathcal{C}_{\pi'} \subset \mathcal{C}_{\pi}$.
(iii) If $\pi = \aleph_0$, $\mathcal{C}_{\pi}$ is the category of objects of finite presentation (see Definition 6.3.3).
(iv) We shall give later a condition which ensures that $\mathrm{Ob}(\mathcal{C}) \simeq \bigcup_{\pi} \mathrm{Ob}(\mathcal{C}_{\pi})$.

Proposition 9.2.9. *Let π be an infinite cardinal. Let J be a category such that $\mathrm{card}(\mathrm{Mor}(J)) < \pi$ and let I be a small π-filtrant category. Consider a functor $\alpha: J^{\mathrm{op}} \times I \to \mathbf{Set}$. Then the natural map λ below is bijective:*

(9.2.1) $$\lambda : \varinjlim_{i \in I} \varprojlim_{j \in J} \alpha(j, i) \to \varprojlim_{j \in J} \varinjlim_{i \in I} \alpha(j, i) .$$

Proof. (i) Injectivity. Let $s, t \in \varinjlim_{i \in I} \varprojlim_{j \in J} \alpha(j, i)$ such that $\lambda(s) = \lambda(t)$. We may assume from the beginning that $s, t \in \varprojlim_{j \in J} \alpha(j, i_0)$ for some $i_0 \in I$. Let $s = \{s(j)\}_{j \in J}$, $t = \{t(j)\}_{j \in J}$ with $s(j), t(j) \in \alpha(j, i_0)$. By the hypothesis, for each

$j \in J$ there exists a morphism $i_0 \to i(j)$ in I such that $s(j) \in \alpha(j, i_0)$ and $t(j) \in \alpha(j, i_0)$ have the same image in $\alpha(j, i(j))$. Since I is π-filtrant and $\text{card}(J) < \pi$, there exist $i_1 \in I$ and morphisms $i(j) \to i_1$ such that the composition $i_0 \to i(j) \to i_1$ does not depend on j. Since $s(j) \in \alpha(j, i_0)$ and $t(j) \in \alpha(j, i_0)$ have the same image in $\alpha(j, i_1)$ for any $j \in J$, $s, t \in \varprojlim_{j \in J} \alpha(j, i_0)$ have the same image in $\varprojlim_{j \in J} \alpha(j, i_1)$.

(ii) Surjectivity. Let $s \in \varprojlim_{j \in J} \varinjlim_{i \in I} \alpha(j, i)$. Then $s = \{s(j)\}_{j \in J}$, $s(j) \in \varinjlim_{i \in I} \alpha(j, i)$. For each $j \in J$, there exist $i(j) \in I$ and $\tilde{s}(j) \in \alpha(j, i(j))$ whose image is $s(j)$. Since $\text{card}(J) < \pi$, there exist $i_0 \in I$ and morphisms $i(j) \to i_0$. Hence we may assume from the beginning that $s(j) \in \alpha(j, i_0)$ for some i_0 which does not depend on j.

Let $u: j \to j'$ be a morphism in J^{op}. It defines a morphism

$$u_{i_0}: \alpha(j, i_0) \to \alpha(j', i_0)$$

and $u_{i_0}(s(j)) = s(j')$ in $\varinjlim_{i \in I} \alpha(j', i)$. There exist $j(u) \in I$ and a morphism $i_0 \to j(u)$ such that $u_{i_0}(s(j))$ and $s(j')$ have the same image in $\alpha(j', j(u))$. Since $\text{card}(\text{Mor}(J)) < \pi$, there exist $i_1 \in I$ and morphisms $j(u) \to i_1$ such that the composition $i_0 \to j(u) \to i_1$ does not depend on u.

We now define $t(j) \in \alpha(j, i_1)$ as the image of $s(j)$. For any morphism $u: j \to j'$ in J^{op}, we have $u_{i_1}(t(j)) = t(j')$. Therefore $\{t(j)\}_{j \in J}$ defines an element t of $\varprojlim_{j \in J} \alpha(j, i_1)$ whose image in $\varprojlim_{j \in J} \varinjlim_{i \in I} \alpha(j, i)$ is s.   q.e.d.

Proposition 9.2.10. *Assume that $\mathcal{C}$ admits small π-filtrant inductive limits. Let J be a category such that $\text{card}(\text{Mor}(J)) < \pi$. Let $\beta: J \to \mathcal{C}_\pi$ be a functor. If $\varinjlim_{j \in J} \beta(j)$ exists in $\mathcal{C}$, then it belongs to $\mathcal{C}_\pi$.*

Proof. Let I be a π-filtrant category and let $\alpha: I \to \mathcal{C}$ be a functor. There is a chain of isomorphisms

$$\varinjlim_{i \in I} \text{Hom}_{\mathcal{C}}(\varinjlim_{j \in J} \beta(j), \alpha(i)) \simeq \varinjlim_{i \in I} \varprojlim_{j \in J} \text{Hom}_{\mathcal{C}}(\beta(j), \alpha(i)) \,,$$

$$\text{Hom}_{\mathcal{C}}(\varinjlim_{j \in J} \beta(j), \varinjlim_{i \in I} \alpha(i)) \simeq \varprojlim_{j \in J} \text{Hom}_{\mathcal{C}}(\beta(j), \varinjlim_{i \in I} \alpha(i))$$

$$\simeq \varprojlim_{j \in J} \varinjlim_{i \in I} \text{Hom}_{\mathcal{C}}(\beta(j), \alpha(i)) \,,$$

where the last isomorphism follows from the fact that $\beta(j) \in \mathcal{C}_\pi$. Hence, the result follows from Proposition 9.2.9.   q.e.d.

Corollary 9.2.11. *Assume that $\mathcal{C}$ admits small inductive limits. Then $\mathcal{C}_\pi$, as well as $(\mathcal{C}_\pi)_X$ for any $X \in \mathcal{C}$, is π-filtrant.*

Proof. Let us check condition (ii) of Proposition 9.2.1. Let J be a category such that $\mathrm{card}(\mathrm{Mor}(J)) < \pi$ and let $\varphi\colon J \to \mathcal{C}_\pi$ be a functor. By the hypothesis the object $Z := \varinjlim \varphi$ exists in $\mathcal{C}$ and by Proposition 9.2.10 it belongs to $\mathcal{C}_\pi$. Then

$$\varprojlim \mathrm{Hom}_{\mathcal{C}_\pi}(\varphi, Z) \simeq \mathrm{Hom}_{\mathcal{C}}(\varinjlim \varphi, Z) \simeq \mathrm{Hom}_{\mathcal{C}}(Z, Z) \neq \emptyset \,.$$

The case of $(\mathcal{C}_\pi)_X$ is similar.   q.e.d.

Corollary 9.2.12. *Let π be an infinite regular cardinal. Then*

$$\mathbf{Set}_\pi = \{A \in \mathbf{Set}; \mathrm{card}(A) < \pi\} \,.$$

Proof. (i) Let $A \in \mathbf{Set}$ with $\mathrm{card}(A) < \pi$. The set $\{\mathrm{pt}\}$ clearly belongs to $\mathbf{Set}_\pi$. By Proposition 9.2.10, $A \simeq \{\mathrm{pt}\}^{\sqcup A}$ belongs to $\mathbf{Set}_\pi$.
(ii) Conversely, let $A \in \mathbf{Set}_\pi$. Set $I = \{B \subset A; \mathrm{card}(B) < \pi\}$. Then I is π-filtrant as seen in Example 9.2.4. Consider the functor $\alpha\colon I \to \mathbf{Set}$, $\alpha(B) = B$. Since A is π-accessible, we obtain

$$\varinjlim_{B \in I} \mathrm{Hom}_{\mathbf{Set}}(A, B) \xrightarrow{\sim} \mathrm{Hom}_{\mathbf{Set}}(A, \varinjlim_{B \in I} B)$$
$$\simeq \mathrm{Hom}_{\mathbf{Set}}(A, A) \,.$$

Hence $\mathrm{id}_A \in \mathrm{Hom}_{\mathbf{Set}}(A, \varinjlim_{B \in I} B)$ belongs to $\mathrm{Hom}_{\mathbf{Set}}(A, B)$ for some $B \in I$, that is, $\mathrm{id}_A\colon A \to A$ factors through $B \rightarrowtail A$, which implies $A = B$.   q.e.d.

Definition 9.2.13. *Let $\mathcal{C}$ be a category (we do not assume that $\mathcal{C}$ admits small inductive limits). We set*

$$\mathrm{Ind}^\pi(\mathcal{C}) = \{A \in \mathrm{Ind}(\mathcal{C}); \mathcal{C}_A \text{ is } \pi\text{-filtrant}\} \,.$$

Remark 9.2.14. (i) If $\pi = \aleph_0$, then $\mathrm{Ind}^\pi(\mathcal{C}) = \mathrm{Ind}(\mathcal{C})$.
(ii) For $\pi' \leq \pi$, we have $\mathrm{Ind}^{\pi'}(\mathcal{C}) \supset \mathrm{Ind}^\pi(\mathcal{C})$.

Lemma 9.2.15. *Let $A \in \mathcal{C}$. Then $A \in \mathrm{Ind}^\pi(\mathcal{C})$ if and only if there exist a small π-filtrant category I and a functor $\alpha\colon I \to \mathcal{C}$ with $A \simeq \text{“}\varinjlim\text{”}\, \alpha$.*

Proof. (i) The condition is necessary since $A \simeq \text{“}\varinjlim\text{”}_{X \in \mathcal{C}_A} X$ by Proposition 2.6.3 and $\mathcal{C}_A$ is cofinally small by Proposition 6.1.5 (see Remark 9.2.6).
(ii) Conversely, assume that $A \simeq \text{“}\varinjlim\text{”}\, \alpha$ with $\alpha\colon I \to \mathcal{C}$ for a small and π-filtrant category I. Then the natural functor $I \to \mathcal{C}_A$ is cofinal by Proposition 2.6.3. To conclude, apply Lemma 9.2.5.   q.e.d.

Lemma 9.2.16. *The category $\mathrm{Ind}^\pi(\mathcal{C})$ is closed by π-filtrant inductive limits.*

Proof. Let $\alpha\colon I \to \mathrm{Ind}^\pi(\mathcal{C})$ be a functor with I small and π-filtrant and let $A = \text{``}\varinjlim\text{''}\,\alpha \in \mathrm{Ind}(\mathcal{C})$. We can prove that $\mathcal{C}_A$ is π-filtrant as in the proof of Theorem 6.1.8.   q.e.d.

Proposition 9.2.17. *Let $\mathcal{C}$ be a category and assume that $\mathcal{C}$ admits inductive limits indexed by any category J such that $\mathrm{card}(\mathrm{Mor}(J)) < \pi$. Let $A \in \mathrm{Ind}(\mathcal{C})$. Then the conditions (i)–(iii) below are equivalent.*

(i) $A \in \mathrm{Ind}^\pi(\mathcal{C})$,

(ii) *for any category J such that $\mathrm{card}(\mathrm{Mor}(J)) < \pi$ and any functor $\varphi\colon J \to \mathcal{C}$, the natural map*

$$(9.2.2) \qquad A(\varinjlim_{j \in J}\varphi(j)) \to \varprojlim_{j \in J} A(\varphi(j))$$

is surjective,

(iii) *for any category J and any functor φ as in (ii), the natural map (9.2.2) is bijective.*

Proof. (ii) $\Rightarrow$ (i). By the hypothesis, any functor $\varphi\colon J \to \mathcal{C}_A$ factorizes through the constant functor $\Delta_{\varinjlim\varphi}$. Hence $\mathcal{C}_A$ is π-filtrant by Proposition 9.2.1 (ii).

(iii) $\Rightarrow$ (ii) is obvious.

(i) $\Rightarrow$ (iii). Let $A \simeq \text{``}\varinjlim_i\text{''}\,\alpha(i)$, where α is a functor $I \to \mathcal{C}$ and I is small and π-filtrant. We get by Proposition 9.2.9

$$\varprojlim_{j \in J} A(\varphi(j)) \simeq \varprojlim_{j \in J}\varinjlim_{i \in I} \mathrm{Hom}_{\mathcal{C}}(\varphi(j), \alpha(i)) \simeq \varinjlim_{i \in I}\varprojlim_{j \in J} \mathrm{Hom}_{\mathcal{C}}(\varphi(j), \alpha(i))$$

$$\simeq \varinjlim_{i \in I} \mathrm{Hom}_{\mathcal{C}}(\varinjlim_{j \in J}\varphi(j), \alpha(i)) \simeq A(\varinjlim_{j \in J}\varphi(j))\,.$$

                                                          q.e.d.

Proposition 9.2.18. *Assume that $\mathcal{C}$ admits small π-filtrant inductive limits. Then the functor $\sigma_\pi\colon \mathrm{Ind}^\pi(\mathcal{C}_\pi) \to \mathcal{C}$, $\text{``}\varinjlim\text{''}\,\alpha \mapsto \varinjlim\alpha$, is fully faithful.*

The proof below is similar to that of Proposition 6.3.4.

Proof. Let I, J be small π-filtrant categories, let $\alpha\colon I \to \mathcal{C}_\pi$, $\beta\colon J \to \mathcal{C}_\pi$ be functors and let $A = \text{``}\varinjlim_{i \in I}\text{''}\,\alpha(i)$, $B = \text{``}\varinjlim_{j \in J}\text{''}\,\beta(j)$. There is a chain of isomorphisms

$$\mathrm{Hom}_{\mathrm{Ind}^\pi(\mathcal{C}_\pi)}(A, B) \simeq \varprojlim_{i \in I}\varinjlim_{j \in J} \mathrm{Hom}_{\mathcal{C}}(\alpha(i), \beta(j))$$

$$\simeq \varprojlim_{i \in I} \mathrm{Hom}_{\mathcal{C}}(\alpha(i), \varinjlim_{j \in J}\beta(j))$$

$$\simeq \mathrm{Hom}_{\mathcal{C}}(\varinjlim_{i \in I}\alpha(i), \varinjlim_{j \in J}\beta(j))\,.$$

                                                          q.e.d.

Proposition 9.2.19. *Let C be a category and assume that*

(i) *C admits small inductive limits,*
(ii) *C admits a system of generators $\{G_\nu\}_\nu$ such that each G_ν is π-accessible,*
(iii) *for any object $X \in C$, the category $(C_\pi)_X$ is cofinally small.*

Then the functor $\sigma_\pi : \mathrm{Ind}^\pi(C_\pi) \to C$ is an equivalence.

Proof. By Proposition 9.2.18 it remains to show that σ_π is essentially surjective. For any $X \in C$, $(C_\pi)_X$ is π-filtrant by Corollary 9.2.11. Hence the object $\underset{(X' \to X) \in (C_\pi)_X}{\text{"}\varinjlim\text{"}} X'$ belongs to $\mathrm{Ind}^\pi(C_\pi)$. It is then enough to check that the morphism

$$\lambda : \varinjlim_{(X' \to X) \in (C_\pi)_X} X' \to X$$

is an isomorphism. Since $\{G_\nu\}$ is a system of generators, this is equivalent to saying that the morphisms

$$(9.2.3) \quad \lambda_\nu : \varinjlim_{(X' \to X) \in (C_\pi)_X} (X'(G_\nu)) \simeq (\varinjlim_{(X' \to X) \in (C_\pi)_X} X')(G_\nu) \to X(G_\nu)$$

are isomorphisms for all ν.
(i) λ_ν is surjective since $(G_\nu \overset{u}{\to} X) \in (C_\pi)_X$ for all $u \in X(G_\nu)$.
(ii) Let us show that λ_ν is injective. Let $f, g \in X'(G_\nu)$, that is, $f, g : G_\nu \rightrightarrows X'$, and assume that their compositions with $X' \to X$ coincide. Let $X'' = \mathrm{Coker}(f, g)$. By Proposition 9.2.10, $X'' \in C_\pi$. Then the two compositions $G_\nu \rightrightarrows X' \to X''$ coincide, which implies that f and g have same image in $\varinjlim_{(X' \to X) \in (C_\pi)_X} X'(G_\nu)$.   q.e.d.

9.3 π-accessible Objects and Generators

Recall that π is an infinite cardinal. We shall assume that π is regular. Now we consider the following hypotheses on C.

$$(9.3.1) \quad \begin{cases} \text{(i) } C \text{ admits small inductive limits,} \\ \text{(ii) } C \text{ admits finite projective limits,} \\ \text{(iii) small filtrant inductive limits are exact,} \\ \text{(iv) there exists a generator } G, \\ \text{(v) any epimorphism in } C \text{ is strict.} \end{cases}$$

Note that under the assumption (9.3.1), C admits small projective limits by Corollary 5.2.10 and Proposition 5.2.8.

Lemma 9.3.1. *Assume* (9.3.1) *and let* π *be an infinite regular cardinal. Let I be a π-filtrant small category, $\alpha\colon I \to \mathcal{C}$ a functor, and let $\varinjlim \alpha \to Y$ be an epimorphism in $\mathcal{C}$. Assume either* $\mathrm{card}(Y(G)) < \pi$ *or* $Y \in \mathcal{C}_\pi$. *Then there exists* $i_0 \in I$ *such that* $\alpha(i_0) \to Y$ *is an epimorphism.*

Proof. Set $X_i = \alpha(i)$ and $Y_i = \mathrm{Im}(X_i \to Y) = \mathrm{Ker}(Y \rightrightarrows Y \sqcup_{X_i} Y)$. Since small filtrant inductive limits are exact,

$$\varinjlim_i Y_i \simeq \mathrm{Ker}(Y \rightrightarrows Y \underset{\varinjlim_i X_i}{\sqcup} Y) \simeq \mathrm{Im}(\varinjlim_i X_i \to Y) \simeq Y \, ,$$

where the last isomorphism follows from the hypothesis that $\varinjlim_i X_i \to Y$ is an epimorphism together with Proposition 5.1.2 (iv).

(a) Assume that $\mathrm{card}(Y(G)) < \pi$. Set $S = \varinjlim_i Y_i(G) \subset Y(G)$. Then $\mathrm{card}(S) \le \mathrm{card}(Y(G)) < \pi$. By Corollary 9.2.12, $S \in \mathbf{Set}_\pi$ and this implies

$$\varinjlim_{i \in I} \mathrm{Hom}_{\mathbf{Set}}(S, Y_i(G)) \simeq \mathrm{Hom}_{\mathbf{Set}}(S, S) \, .$$

Hence, there exist i_0 and a morphism $S \to Y_{i_0}(G)$ such that the composition $S \to Y_{i_0}(G) \to S$ is the identity. Therefore $Y_{i_0}(G) = S$ and hence, $Y_{i_0}(G) \to Y_i(G)$ is bijective for any $i_0 \to i$. Hence $Y_{i_0} \to Y_i$ is an isomorphism, which implies that $Y_{i_0} \to Y$ is an isomorphism. Applying Proposition 5.1.2 (iv), we find that $X_{i_0} \to Y$ is an epimorphism.

(b) Assume that $Y \in \mathcal{C}_\pi$. Then $\varinjlim_{i \in I} \mathrm{Hom}_{\mathcal{C}}(Y, Y_i) \to \mathrm{Hom}_{\mathcal{C}}(Y, Y)$ is an isomorphism. This shows that id_Y decomposes as $Y \to Y_i \to Y$ for some $i \in I$. Hence, $Y_i \simeq Y$ (see Exercise 1.7) and $X_i \to Y$ is an epimorphism by Proposition 5.1.2 (iv). q.e.d.

Proposition 9.3.2. *Assume* (9.3.1) *and let* π *be an infinite regular cardinal. Let* $A \in \mathcal{C}$ *and assume that* $\mathrm{card}(A(G)) < \pi$ *and* $\mathrm{card}(G^{\sqcup A(G)}(G)) < \pi$. *Then* $A \in \mathcal{C}_\pi$.

Proof. First, note that $\mathbf{Set} \ni E \mapsto G^{\sqcup E} \in \mathcal{C}$ is a well-defined covariant functor.

Also note that $\mathrm{card}(G^{\sqcup S}(G)) < \pi$ for any $S \subset A(G)$. Indeed, there exist maps $S \to A(G) \to S$ whose composition is the identity. Hence, the composition $G^{\sqcup S}(G) \to G^{\sqcup A(G)}(G) \to G^{\sqcup S}(G)$ is the identity.

Let I be a small π-filtrant category and let $\alpha\colon I \to \mathcal{C}$ be a functor. Set $X_i = \alpha(i)$ and $X_\infty = \varinjlim_{i \in I} \alpha(i)$. We shall show that the map λ below is bijective:

(9.3.2) $$\lambda\colon \varinjlim_{i \in I} \mathrm{Hom}_{\mathcal{C}}(A, X_i) \to \mathrm{Hom}_{\mathcal{C}}(A, X_\infty) \, .$$

(i) λ is injective. (Here, we shall only use $\mathrm{card}(A(G)) < \pi$.)
Let $f, g\colon A \rightrightarrows X_{i_0}$ and assume that the two compositions $A \rightrightarrows X_{i_0} \to X_\infty$ coincide. For each morphism $s\colon i_0 \to i$, set $N_s = \mathrm{Ker}(A \rightrightarrows X_i)$. Since $\varinjlim_{s \in I^{i_0}}$ is exact in $\mathcal{C}$ by the hypothesis, we obtain

$$\varinjlim_{s \in I^{i_0}} N_s \simeq \mathrm{Ker}(A \rightrightarrows \varinjlim_{s \in I^{i_0}} X_i)$$

$$\simeq \mathrm{Ker}(A \rightrightarrows X_\infty) \simeq A .$$

Since N_s is a subobject of A and $\mathrm{card}(A(G)) < \pi$, we may apply Lemma 9.3.1 and conclude that there exists $i_0 \to i_1$ such that $N_{i_1} \to A$ is an epimorphism. Hence, the two compositions $A \rightrightarrows X_{i_0} \to X_{i_1}$ coincide.

(ii) λ is surjective.
Let $f \in \mathrm{Hom}_{\mathcal{C}}(A, X_\infty)$. For each $i \in I$ define $Y_i = X_i \times_{X_\infty} A$. Since $\varinjlim_i$ is exact, $\varinjlim_i Y_i \simeq A$. Since $\mathrm{card}(A(G)) < \pi$, Lemma 9.3.1 implies that there exists $i_0 \in I$ such that $Y_{i_0} \to A$ is an epimorphism. Set

$$K = \mathrm{Im}(Y_{i_0}(G) \to A(G)) \subset A(G) .$$

Consider the commutative diagram below:

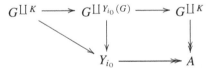

The left vertical arrow is an epimorphism by Proposition 5.2.3 (iv). Hence $G^{\amalg K} \to A$ is an epimorphism. Choosing a section $K \to Y_{i_0}(G)$, we get a commutative diagram

$$
\begin{array}{ccccc}
G^{\amalg K} & \longrightarrow & G^{\amalg Y_{i_0}(G)} & \longrightarrow & G^{\amalg K} \\
& \searrow & \downarrow & & \downarrow \\
& & Y_{i_0} & \longrightarrow\!\!\!\!\!\rightarrow & A
\end{array}
$$

such that the composition $G^{\amalg K} \to G^{\amalg Y_{i_0}(G)} \to G^{\amalg K}$ is the identity. Set $B = G^{\amalg K}$. The composition $B \to Y_{i_0} \to A$ is an epimorphism. We obtain a commutative diagram

$$
\begin{array}{ccc}
B & \longrightarrow\!\!\!\!\!\rightarrow & A \\
\downarrow & & \downarrow f \\
X_{i_0} & \longrightarrow & X_\infty .
\end{array}
$$

Since all epimorphisms are strict, the sequence

(9.3.3) $B \times_A B \rightrightarrows B \to A$

is exact. On the other hand, since $\mathrm{card}(B(G)) = \mathrm{card}(G^{\amalg K}(G)) < \pi$, we have $\mathrm{card}(B \times_A B)(G) \leq \mathrm{card}(B(G))^2 < \pi$. Then, applying part (i) to $B \times_A B$, the natural map

$$\varinjlim_{i \in I} \mathrm{Hom}_{\mathcal{C}}(B \times_A B, X_i) \to \mathrm{Hom}_{\mathcal{C}}(B \times_A B, X_\infty)$$

is injective. Consider the diagram

$$
\begin{array}{ccc}
B \times_A B \rightrightarrows B & \longrightarrow & A \\
\downarrow & & \downarrow f \\
X_{i_0} & \longrightarrow & X_\infty \, .
\end{array}
$$

Since the two compositions $B \times_A B \rightrightarrows B \to X_{i_0} \to X_\infty$ are equal, there exists an arrow $i_0 \to i$ such that the two compositions $B \times_A B \rightrightarrows B \to X_{i_0} \to X_i$ are equal. Hence, the exactness of (9.3.3) implies that $B \to X_{i_0} \to X_i$ decomposes into $B \to A \to X_i$. Since $B \to A$ is an epimorphism, the composition $A \to X_i \to X_\infty$ coincides with f.   q.e.d.

We keep hypothesis (9.3.1) and choose an infinite regular cardinal π_0 such that

$$\mathrm{card}(G(G)) < \pi_0, \quad \mathrm{card}(G^{\sqcup G(G)}(G)) < \pi_0 \, .$$

By Proposition 9.3.2, we get $G \in \mathcal{C}_{\pi_0}$. Now choose a cardinal $\pi_1 \geq \pi_0$ such that if X is a quotient of $G^{\sqcup A}$ for a set A with $\mathrm{card}(A) < \pi_0$, then $\mathrm{card}(X(G)) < \pi_1$. (Since the set of quotients of $G^{\sqcup A}$ is small, such a cardinal π_1 exists.)

Let π be the successor of 2^{π_1}. The cardinals π and π_0 satisfy

(9.3.4)
$$
\begin{cases}
\text{(a) } \pi \text{ and } \pi_0 \text{ are infinite regular cardinals,} \\
\text{(b) } G \in \mathcal{C}_{\pi_0} \, , \\
\text{(c) } \pi'^{\pi_0} < \pi \text{ for any } \pi' < \pi \, , \\
\text{(d) if } X \text{ is a quotient of } G^{\sqcup A} \text{ for a set } A \text{ with } \mathrm{card}(A) < \pi_0, \\
\qquad \text{then } \mathrm{card}(X(G)) < \pi.
\end{cases}
$$

Indeed (c) is proved as follows: if $\pi' < \pi$, then $\pi' \leq 2^{\pi_1}$ and $\pi'^{\pi_0} \leq (2^{\pi_1})^{\pi_0} = 2^{\pi_0 \cdot \pi_1} = 2^{\pi_1} < \pi$.

Lemma 9.3.3. *Assume* (9.3.1) *and* (9.3.4). *Let* $A \in \mathbf{Set}$ *with* $\mathrm{card}(A) < \pi$ *and let* $X \in \mathcal{C}$ *be a quotient of* $G^{\sqcup A}$. *Then* $\mathrm{card}(X(G)) < \pi$.

Proof. Let $u \colon G^{\sqcup A} \to X$ be an epimorphism. Set $I = \{B \subset A; \mathrm{card}(B) < \pi_0\}$. Then I is a π_0-filtrant ordered set by Example 9.2.4.

By 9.3.4 (c), we have

$$\mathrm{card}(I) \leq \mathrm{card}(A)^{\pi_0} < \pi \, .$$

For $B \in I$, set

$$X_B = \mathrm{Coim}(G^{\sqcup B} \to G^{\sqcup A} \to X) \, .$$

Then $\{X_B\}_{B \in I}$ is a π_0-filtrant inductive system of subobjects of X by Proposition 5.1.7. Since small filtrant inductive limits are exact, $\varinjlim_{B \in I} X_B \to X$ is a monomorphism. Since $\varinjlim_{B \in I} G^{\sqcup B} \simeq G^{\sqcup A}$, the morphism $\varinjlim_{B \in I} X_B \to X$ is also an epimorphism. It is thus an isomorphism by the hypothesis (9.3.1) (v) and Proposition 5.1.5 (ii). Since $G \in \mathcal{C}_{\pi_0}$ and I is π_0-filtrant,

$$X(G) \simeq \varinjlim_{B \in I} X_B(G) \ .$$

Since $\mathrm{card}(X_B(G)) < \pi$ by (9.3.4) (d) and $\mathrm{card}(I) < \pi$, we obtain

$$\mathrm{card}(X(G)) \leq \mathrm{card}(\bigsqcup_{B \in I} X_B(G)) < \pi \ .$$

q.e.d.

Theorem 9.3.4. *Assume* (9.3.1) *and* (9.3.4). *Then*

$$\mathcal{C}_\pi \simeq \{X \in \mathcal{C} \, ; \, \mathrm{card}(X(G)) < \pi\} \ .$$

Proof. Set $S_\pi = \{X \in \mathcal{C} \, ; \, \mathrm{card}(X(G)) < \pi\}$.
(i) $S_\pi \subset \mathcal{C}_\pi$. If $X \in S_\pi$, then Lemma 9.3.3 implies that $\mathrm{card}(G^{\sqcup X(G)}(G)) < \pi$. Then Proposition 9.3.2 implies $X \in \mathcal{C}_\pi$.
(ii) $\mathcal{C}_\pi \subset S_\pi$. Let $X \in \mathcal{C}_\pi$. Set $I = \{A \subset X(G) \, ; \, \mathrm{card}(A) < \pi\}$. Then I is π-filtrant. For $A \in I$ we get the morphisms

$$G^{\sqcup A} \to G^{\sqcup X(G)} \twoheadrightarrow X \ .$$

Since

$$\varinjlim_{A \in I} G^{\sqcup A} \xrightarrow{\sim} G^{\sqcup X(G)} \to X$$

is an epimorphism, Lemma 9.3.1 implies that $G^{\sqcup A} \to X$ is an epimorphism for some $A \in I$. Then Lemma 9.3.3 implies that $\mathrm{card}(X(G)) < \pi$.   q.e.d.

Corollary 9.3.5. *Assume* (9.3.1) *and* (9.3.4). *Then*

(i) $\mathcal{C}_\pi$ *is essentially small,*
(ii) *if* $f \colon X \twoheadrightarrow Y$ *is an epimorphism and* $X \in \mathcal{C}_\pi$, *then* $Y \in \mathcal{C}_\pi$,
(iii) *if* $f \colon X \rightarrowtail Y$ *is a monomorphism and* $Y \in \mathcal{C}_\pi$, *then* $X \in \mathcal{C}_\pi$,
(iv) $\mathcal{C}_\pi$ *is closed by finite projective limits.*

Proof. (i) Let $X \in \mathcal{C}_\pi$. There exist a set I with $\mathrm{card}(I) < \pi$ and an epimorphism $G^{\sqcup I} \twoheadrightarrow X$. (Take $X(G)$ as I.) Since the set of quotients of $G^{\sqcup I}$ is small, $\mathcal{C}_\pi$ is essentially small.
(ii) Since $G^{\sqcup X(G)} \to X$ is an epimorphism, we obtain an epimorphism $G^{\sqcup X(G)} \to Y$. Since $\mathrm{card}(X(G)) < \pi$ by Theorem 9.3.4, the result follows from Lemma 9.3.3 and Theorem 9.3.4.

(iii) Since $X(G) \subset Y(G)$, we have $\mathrm{card}(X(G)) \leq \mathrm{card}(Y(G))$, and the result follows from Theorem 9.3.4.

(iv) Let $\{X_i\}_{i \in I}$ be a finite projective system in $\mathcal{C}$ and set $X = \varprojlim_i X_i$. We have

$$X(G) = \varprojlim_i (X_i(G)) \subset \prod_i X_i(G) ,$$

and $\mathrm{card}(X(G)) \leq \mathrm{card}(\prod_i X_i(G)) < \pi$.   q.e.d.

Corollary 9.3.6. *Assume* (9.3.1) *and* (9.3.4). *Then the functor* $\varinjlim \colon \mathrm{Ind}^\pi(\mathcal{C}_\pi)$ *$\to \mathcal{C}$ is an equivalence.*

Proof. Apply Proposition 9.2.19.   q.e.d.

Corollary 9.3.7. *Assume* (9.3.1). *Then for any small subset S of* $\mathrm{Ob}(\mathcal{C})$, *there exists an infinite cardinal π such that $S \subset \mathrm{Ob}(\mathcal{C}_\pi)$.*

Corollary 9.3.8. *Assume* (9.3.1) *and let κ be a cardinal. Then there exist a full subcategory $S \subset \mathcal{C}$ and an infinite regular cardinal $\pi > \kappa$ such that*

  (i) *S is essentially small,*
  (ii) *if $X \twoheadrightarrow Y$ is an epimorphism and $X \in S$, then $Y \in S$,*
  (iii) *if $X \rightarrowtail Y$ is a monomorphism and $Y \in S$, then $X \in S$,*
  (iv) *there exists an object $G \in S$ which is a generator in $\mathcal{C}$,*
  (v) *for any epimorphism $f \colon X \twoheadrightarrow Y$ in $\mathcal{C}$ with $Y \in S$, there exist $Z \in S$ and a monomorphism $g \colon Z \to X$ such that $f \circ g \colon Z \to Y$ is an epimorphism,*
  (vi) *any $X \in S$ is π-accessible in $\mathcal{C}$,*
  (vii) *S is closed by inductive limits indexed by categories J which satisfy* $\mathrm{card}(\mathrm{Mor}(J)) < \pi$.

Proof. Choose cardinals $\pi_0 > \kappa$ and π as in (9.3.4) and set $S = \mathcal{C}_\pi$. We only have to check (v). Consider the epimorphisms $G^{\sqcup X(G)} \twoheadrightarrow X \twoheadrightarrow Y$ and set $I = \{A \subset X(G) \,;\, \mathrm{card}(A) < \pi\}$. Then I is π-filtrant. Since $\varinjlim_{A \in I} G^{\sqcup A} \simeq G^{\sqcup X(G)}$, Lemma 9.3.1 implies that there exists $A \in I$ such that $G^{\sqcup A} \to Y$ is an epimorphism. Hence, it is enough to set $Z = \mathrm{Im}(G^{\sqcup A} \to X)$.   q.e.d.

9.4 Quasi-Terminal Objects

Definition 9.4.1. *Let $\mathcal{C}$ be a category. An object $X \in \mathcal{C}$ is* quasi-terminal *if any morphism $u \colon X \to Y$ in $\mathcal{C}$ admits a left inverse, that is, there exists $v \colon Y \to X$ such that $v \circ u = \mathrm{id}_X$.*

Hence, any endomorphism of a quasi-terminal object is an automorphism.

The aim of this section is to prove Theorem 9.4.2 below, a categorical variant of Zorn's lemma.

Theorem 9.4.2. *Let C be an essentially small non empty category which admits small filtrant inductive limits. Then C has a quasi-terminal object.*

The proof decomposes into several steps. We may assume that C is small.

Sublemma 9.4.3. *There exists an object $X \in C$ such that if there exists a morphism $X \to Y$, then there exists a morphism $Y \to X$.*

Proof of Sublemma 9.4.3. Let $\mathcal{F}$ be the set of filtrant subcategories of C. Since C is non empty, $\mathcal{F}$ is non empty, and $\mathcal{F}$ is clearly inductively ordered. Let $\mathcal{S}$ be a maximal element of $\mathcal{F}$. Since $\mathcal{S}$ is small, $X := \varinjlim_{S \in \mathcal{S}} S$ exists in C.

We shall prove that X satisfies the condition of the statement. For $S \in \mathcal{S}$, let us denote by $a_S \colon S \to X$ the canonical morphism. Let $u \colon X \to Y$ be a morphism in C.
(i) $Y \in \mathcal{S}$. Otherwise, define the subcategory $\widetilde{\mathcal{S}}$ of C by setting

$$\mathrm{Ob}(\widetilde{\mathcal{S}}) = \mathrm{Ob}(\mathcal{S}) \sqcup \{Y\}\,,$$
$$\mathrm{Mor}(\widetilde{\mathcal{S}}) = \mathrm{Mor}(\mathcal{S}) \sqcup \{\mathrm{id}_Y\} \sqcup \{u \circ a_S \in \mathrm{Hom}_C(S, Y); \ S \in \mathcal{S}\}\,.$$

It is easily checked that $\widetilde{\mathcal{S}}$ is a subcategory of C and Y is a terminal object of $\widetilde{\mathcal{S}}$. Hence $\widetilde{\mathcal{S}}$ is a filtrant subcategory containing $\mathcal{S}$. This contradicts the fact that $\mathcal{S}$ is maximal.
(ii) Since $Y \in \mathcal{S}$, there exists a morphism $Y \to X$, namely the morphism a_Y.
q.e.d.

Sublemma 9.4.4. *For any $X \in C$, there exists a morphism $f \colon X \to Y$ satisfying the property:*
$P(f)$: for any morphism $u \colon Y \to Z$, there exists a morphism $v \colon Z \to Y$ such that $v \circ u \circ f = f$.

Proof of Sublemma 9.4.4. The category C^X is non empty, essentially small and admits small filtrant inductive limits. Applying Sublemma 9.4.3, we find an object $(X \xrightarrow{f} Y) \in C^X$ such that for any object $(X \to Z)$ and morphism $u \colon (X \to Y) \to (X \xrightarrow{u \circ f} Z)$ in C^X, there exists a morphism $v \colon (X \to Z) \to (X \to Y)$ in C^X.
q.e.d.

Let us choose an infinite regular cardinal π such that $\mathrm{card}(\mathrm{Mor}(C)) < \pi$.

Sublemma 9.4.5. *Let I be a π-filtrant small category and let $\{X_i\}_{i \in I}$ be an inductive system in C indexed by I. Then there exists $i_0 \in I$ such that $X_{i_0} \to \varinjlim_i X_i$ is an epimorphism.*

Proof of Sublemma 9.4.5. The proof is similar to that of Lemma 9.3.1. Set $X = \varinjlim_i X_i$ and let $a_i \colon X_i \to X$ denote the canonical morphism. Let $F \in \mathcal{C}^\wedge$ denote the image of "$\varinjlim_i$" $X_i \to X$, that is,

$$F(Y) = \mathrm{Im}\big(\varinjlim_i \mathrm{Hom}_{\mathcal{C}}(Y, X_i) \to \mathrm{Hom}_{\mathcal{C}}(Y, X)\big) .$$

Since $\mathrm{card}(\mathrm{Hom}_{\mathcal{C}}(Y, X)) < \pi$, we have $F(Y) \in \mathbf{Set}_\pi$. Therefore, there exists $i_Y \in I$ such that $\mathrm{Hom}_{\mathcal{C}}(Y, X_{i_Y}) \to F(Y)$ is surjective (apply Lemma 9.3.1 with $\mathcal{C} = \mathbf{Set}$). Since $\mathrm{card}(\{i_Y; Y \in \mathrm{Ob}(\mathcal{C})\}) < \pi$ and I is π-filtrant, there exists $i_0 \in I$ such that there exists a morphism $i_Y \to i_0$ for any $Y \in \mathrm{Ob}(\mathcal{C})$. Hence $\mathrm{Hom}_{\mathcal{C}}(Y, X_{i_0}) \to F(Y)$ is surjective for any $Y \in \mathcal{C}$. In particular, for any $i \in I$, there exists a morphism $h_i \colon X_i \to X_{i_0}$ such that $a_{i_0} \circ h_i = a_i$.

Let us show that $a_{i_0} \colon X_{i_0} \to X$ is an epimorphism. Let $f_1, f_2 \colon X \rightrightarrows Y$ be a pair of parallel arrows such that $f_1 \circ a_{i_0} = f_2 \circ a_{i_0}$. Then, for any $i \in I$, we have

$$f_1 \circ a_i = f_1 \circ a_{i_0} \circ h_i = f_2 \circ a_{i_0} \circ h_i = f_2 \circ a_i .$$

Hence, $f_1 = f_2$.   q.e.d.

Proof of Theorem 9.4.2. As in Example 9.2.3, let us choose a small π-filtrant well ordered set I such that $\sup(I)$ does not exist. Let us define an inductive system $\{X_i\}_{i \in I}$ in $\mathcal{C}$ by transfinite induction. For the smallest element $0 \in I$, we choose an arbitrary object $X_0 \in \mathcal{C}$. Let $i > 0$ and assume that X_j and $u_{jk} \colon X_k \to X_j$ have been constructed for $k \le j < i$.

(a) If $i = j + 1$ for some j, then take $u_{ij} \colon X_j \to X_i$ with the property $P(u_{ij})$ in Sublemma 9.4.4. Then define $u_{ik} = u_{ij} \circ u_{jk}$ for any $k \le j$.
(b) If $i = \sup\{j; j < i\}$, set $X_i = \varinjlim_{j<i} X_j$ and define u_{ij} as the canonical morphism $X_j \to X_i$.

We shall prove that $X := \varinjlim_{i \in I} X_i$ is a quasi-terminal object. Let $a_i \colon X_i \to X$ denote the canonical morphism. Hence, $a_j = a_i \circ u_{ij}$ for $j \le i$. By Sublemma 9.4.5, there exists $i_0 \in I$ such that $a_{i_0} \colon X_{i_0} \to X$ is an epimorphism. Let $u \colon X \to Y$ be a morphism. By the property $P(u_{i_0+1,i_0})$ applied to $u \circ a_{i_0+1} \colon X_{i_0+1} \to Y$, we find a morphism $w \colon Y \to X_{i_0+1}$ such that $w \circ u \circ a_{i_0+1} \circ u_{i_0+1,i_0} = u_{i_0+1,i_0}$. Set $v = a_{i_0+1} \circ w \in \mathrm{Hom}_{\mathcal{C}}(Y, X)$. Then

$$(v \circ u) \circ a_{i_0} = (a_{i_0+1} \circ w) \circ u \circ (a_{i_0+1} \circ u_{i_0+1,i_0})$$
$$= a_{i_0+1} \circ u_{i_0+1,i_0} = a_{i_0} = \mathrm{id}_X \circ a_{i_0} .$$

Since a_{i_0} is an epimorphism, we conclude that $v \circ u = \mathrm{id}_X$.   q.e.d.

9.5 $\mathcal{F}$-injective Objects

Let $\mathcal{C}$ denote a $\mathcal{U}$-category.

Definition 9.5.1. (i) *Let $\mathcal{F} \subset \mathrm{Mor}(\mathcal{C})$ be a family of morphisms in $\mathcal{C}$. An object $I \in \mathcal{C}$ is $\mathcal{F}$-injective if for any solid diagram*

(9.5.1)
$$
\begin{array}{ccc}
X & \longrightarrow & I \\
{\scriptstyle f}\downarrow & \nearrow & \\
Z & &
\end{array}
$$

with $f \in \mathcal{F}$, there exists a dotted arrow making the whole diagram commutative.

In other words, I is $\mathcal{F}$-injective if the map $\mathrm{Hom}_{\mathcal{C}}(Z, I) \xrightarrow{\circ f} \mathrm{Hom}_{\mathcal{C}}(X, I)$ is surjective for any $f \colon X \to Z$ in $\mathcal{F}$.

(ii) *An object is $\mathcal{F}$-projective in $\mathcal{C}$ if it is $\mathcal{F}^{\mathrm{op}}$-injective in $\mathcal{C}^{\mathrm{op}}$.*

Example 9.5.2. Let $\mathcal{C}$ be an abelian category and let $\mathcal{F} \subset \mathrm{Mor}(\mathcal{C})$ be the family of monomorphisms. Then the $\mathcal{F}$-injective objects are the injective objects.

We shall consider a subcategory $\mathcal{C}_0$ of $\mathcal{C}$ and we shall make the hypotheses below

(9.5.2)
$$
\left\{
\begin{array}{l}
\text{(i) } \mathcal{C}_0 \text{ admits small filtrant inductive limits and } \mathcal{C}_0 \to \mathcal{C} \text{ com-} \\
\quad \text{mutes with such limits,} \\
\text{(ii) for any } X, Y, X' \in \mathcal{C}_0, \text{ any } u \colon X \to Y \text{ in } \mathrm{Mor}(\mathcal{C}_0) \text{ and any} \\
\quad f \colon X \to X' \text{ in } \mathrm{Mor}(\mathcal{C}), \text{ there exists a commutative diagram} \\
\quad \begin{array}{ccc} X & \xrightarrow{u} & Y \\ {\scriptstyle f}\downarrow & & \downarrow{\scriptstyle g} \\ X' & \xrightarrow{u'} & Y' \end{array} \quad \text{with } u' \in \mathrm{Mor}(\mathcal{C}_0) \text{ and } g \in \mathrm{Mor}(\mathcal{C}).
\end{array}
\right.
$$

Lemma 9.5.3. *Assume* (9.5.2). *Then for any $X' \in \mathcal{C}_0$, any small family $\{u_i \colon X_i \to Y_i\}_{i \in I}$ in $\mathrm{Mor}(\mathcal{C}_0)$ and any family $\{f_i \colon X_i \to X'\}_{i \in I}$ in $\mathrm{Mor}(\mathcal{C})$, there exist $u' \colon X' \to Y'$ in $\mathrm{Mor}(\mathcal{C}_0)$ and $\{g_i \colon Y_i \to Y'\}_{i \in I}$ in $\mathrm{Mor}(\mathcal{C})$ such that the diagrams* $\begin{array}{ccc} X_i & \xrightarrow{u_i} & Y_i \\ {\scriptstyle f_i}\downarrow & & \downarrow{\scriptstyle g_i} \\ X' & \xrightarrow{u'} & Y' \end{array}$ *commute for all $i \in I$.*

Proof. When I is empty, it is enough to take $\mathrm{id}_{X'}$ as u'. Assume that I is non empty. We may assume that I is well ordered. We shall construct an inductive system $\{X' \to Y'_i\}_{i \in I}$ in $\mathcal{C}_0$ and morphisms $Y_i \to Y'_i$ by transfinite induction.

If $i = 0$ is the smallest element of I, let us take a morphism $X' \to Y'_0$ in $\mathcal{C}_0$ such that $X_0 \to X' \to Y'_0$ factors through $X_0 \xrightarrow{u_0} Y_0$.

Let $i > 0$. By (9.5.2) (i), $Y'_{<i} := \varinjlim_{j<i} Y'_j$ exists in $\mathcal{C}_0$. Then we take $Y'_i \in \mathcal{C}_0$ and a morphism $Y'_{<i} \to Y'_i$ in $\mathcal{C}_0$ such that there exists a commutative diagram

$$
\begin{array}{ccc}
X_i & \xrightarrow{\;\;u_i\;\;} & Y_i \\
{\scriptstyle f_i}\downarrow & \searrow & \downarrow \\
X' \longrightarrow & Y'_{<i} \longrightarrow & Y'_i \;.
\end{array}
$$

We have thus constructed an inductive system $\{Y'_i\}_{i \in I}$ in $\mathcal{C}_0$. Set $Y' := \varinjlim_i Y'_i$. Then $X' \to Y'$ satisfies the desired properties.   q.e.d.

Theorem 9.5.4. *Assume (9.5.2). Let $\mathcal{F} \subset \mathrm{Mor}(\mathcal{C}_0)$ be a small set and assume that there exists an infinite cardinal π such that for any $u \colon X \to Z$ in $\mathcal{F}$, $X \in \mathcal{C}_\pi$. Then, for any $X \in \mathcal{C}_0$, there exists a morphism $f \colon X \to Y$ such that $f \in \mathrm{Mor}(\mathcal{C}_0)$ and Y is $\mathcal{F}$-injective in $\mathcal{C}$.*

Proof. We may assume from the beginning that π is an infinite regular cardinal. Choose a well ordered π-filtrant set I such that $\sup(I)$ does not exist, as in Example 9.2.3. For $i, j \in I$ with $j \le i$, we shall define by transfinite induction on i:

(9.5.3) Y_i in $\mathcal{C}_0$ and $u_{ij} \colon Y_j \to Y_i$ in $\mathrm{Mor}(\mathcal{C}_0)$

such that $u_{ii} = \mathrm{id}_{Y_i}$, $u_{ij} \circ u_{jk} = u_{ik}$ for $k \le j \le i$.

Denote by 0 the smallest element in I. Set $Y_0 = X$, $u_{00} = \mathrm{id}_X$. For $i > 0$, assume that Y_k and u_{jk} are constructed for $k \le j < i$.
(a) If $\sup\{j; j < i\} = i$, set

$$
Y_i = \varinjlim_{j<i} Y_j
$$

and define the morphisms $Y_j \to Y_i$ as the natural ones.
(b) If $i = j + 1$ for some j, define the set

$$
S_j = \{B \xleftarrow{v} A \xrightarrow{u} Y_j ; v \in \mathcal{F}\} \;.
$$

Then S_j is a small set. For $s \in S_j$, we denote by $B_s \xleftarrow{v_s} A_s \xrightarrow{u_s} Y_j$ the corresponding diagram. It follows from Lemma 9.5.3 that there exist a morphism $u_{ij} \colon Y_j \to Y_i$ in $\mathcal{C}_0$ and a commutative diagram

$$
\begin{array}{ccc}
A_s & \xrightarrow{\;\;u_s\;\;} & Y_j \\
{\scriptstyle v_s}\downarrow & & \downarrow{\scriptstyle u_{ij}} \\
B_s & \xrightarrow{\;\;w_s\;\;} & Y_i
\end{array}
$$

for every $s \in S_j$. For $k \leq j$, we define u_{ik} as the composition $u_{ij} \circ u_{jk}$.

We have thus constructed an inductive system $\{Y_i\}_{i \in I}$ in $\mathcal{C}_0$.

Set $Y = \varinjlim\limits_{i \in I} Y_i$. Then the morphism $X \to Y$ belongs to $\mathrm{Mor}(\mathcal{C}_0)$. Let us show that Y is $\mathcal{F}$-injective. Consider a diagram

(9.5.4)
$$
\begin{array}{ccc}
Z_1 & \xrightarrow{w} & Y \\
{\scriptstyle f}\downarrow & & \\
Z_2 & &
\end{array}
$$

with $f \in \mathcal{F}$ and $w \in \mathrm{Mor}(\mathcal{C})$. Since $Z_1 \in \mathcal{C}_\pi$, there is an isomorphism

$$
\mathrm{Hom}_{\mathcal{C}}(Z_1, Y) \simeq \varinjlim_{i \in I} \mathrm{Hom}_{\mathcal{C}}(Z_1, Y_i)
$$

and there exists $j \in I$ such that diagram (9.5.4) decomposes into

$$
\begin{array}{ccc}
Z_1 & \longrightarrow Y_j \longrightarrow Y \ . \\
{\scriptstyle f}\downarrow & \\
Z_2 &
\end{array}
$$

Now $(Z_2 \leftarrow Z_1 \to Y_j)$ is equal to $(B_s \leftarrow A_s \to Y_j)$ for some $s \in S_j$ and we get the commutative diagram

$$
\begin{array}{ccccccc}
Z_1 & \xrightarrow{\sim} & A_s & \xrightarrow{u_s} & Y_j & \longrightarrow & Y \ . \\
{\scriptstyle f}\downarrow & & {\scriptstyle v_s}\downarrow & & \downarrow & \nearrow & \\
Z_2 & \xrightarrow{\sim} & B_s & \xrightarrow{w_s} & Y_{j+1} & &
\end{array}
$$

This completes the proof.   q.e.d.

Let $\mathcal{C}$ be a category, $\mathcal{C}_0$ a subcategory and $\mathcal{F} \subset \mathrm{Mor}(\mathcal{C}_0)$ a family of morphisms in $\mathcal{C}_0$. We introduce the following condition on a morphism $f \colon X \to Y$ in $\mathcal{C}_0$.

(9.5.5) $\left\{ \begin{array}{l} \text{Any Cartesian square in } \mathcal{C} \quad \begin{array}{ccc} U & \xrightarrow{s} & V \\ {\scriptstyle u}\downarrow & {\scriptstyle \xi}\nearrow & \downarrow{\scriptstyle v} \\ X & \xrightarrow{f} & Y \end{array} \quad \text{(without the dotted} \\[2em] \text{arrow) such that } s \in \mathcal{F} \text{ can be completed to a commutative} \\ \text{diagram in } \mathcal{C} \text{ with a dotted arrow } \xi. \end{array} \right.$

Theorem 9.5.5. *Let $\mathcal{C}$ be a category, $\mathcal{C}_0$ a subcategory and $\mathcal{F} \subset \mathrm{Mor}(\mathcal{C}_0)$ a family of morphisms in $\mathcal{C}_0$. Assume (9.5.2) and also*

(9.5.6) *for any $X \in \mathcal{C}_0$, $(\mathcal{C}_0)_X$ is essentially small,*

(9.5.7) *any Cartesian square* $\begin{array}{ccc} X' & \xrightarrow{f'} & Y' \\ {\scriptstyle u}\downarrow & & \downarrow{\scriptstyle v} \\ X & \xrightarrow{f} & Y \end{array}$ *in $\mathcal{C}$ with $f, f' \in \mathrm{Mor}(\mathcal{C}_0)$*

decomposes into a commutative diagram 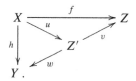 *such*

that the square labeled by A is co-Cartesian, $g, h \in \mathrm{Mor}(\mathcal{C}_0)$ and $f = h \circ g$,

(9.5.8) *if a morphism $f: X \to Y$ in $\mathcal{C}_0$ satisfies condition (9.5.5), then f is an isomorphism.*

Then any $Y \in \mathcal{C}$ which is $\mathcal{F}$-injective is $\mathrm{Mor}(\mathcal{C}_0)$-injective.

Proof. Let Y be $\mathcal{F}$-injective and consider morphisms $Y \xleftarrow{h} X \xrightarrow{f} Z$ with $f \in \mathrm{Mor}(\mathcal{C}_0)$. We shall show that h factorizes through f.

Let us denote by $D(u, v, w, Z')$ a commutative diagram in $\mathcal{C}$ with $Z' \in \mathcal{C}_0$, $u, v \in \mathrm{Mor}(\mathcal{C}_0)$:

$$\begin{array}{ccc} X & \xrightarrow{\quad f \quad} & Z \\ {\scriptstyle h}\downarrow & \searrow{\scriptstyle u} \quad \nearrow{\scriptstyle v} & \\ & Z' & \\ Y\,. & \nwarrow{\scriptstyle w} & \end{array}$$

Denote by I the category of such diagrams, a morphism $D(u_1, v_1, w_1, Z_1') \to D(u_2, v_2, w_2, Z_2')$ being a morphism $Z_1' \to Z_2'$ in $\mathcal{C}_0$ which satisfies the natural commutation relations.

By hypothesis (9.5.6), I is essentially small. By hypothesis (9.5.2), I admits small filtrant inductive limits. Applying Theorem 9.4.2, I has a quasi-terminal object, which we denote by $D(u_0, v_0, w_0, Z_0)$.

It remains to show that v_0 is an isomorphism. For that purpose, we shall use (9.5.8).

Consider a Cartesian square

$$\begin{array}{ccc} U & \xrightarrow{\ s\ } & V \\ {\scriptstyle \alpha}\downarrow & & \downarrow{\scriptstyle \gamma} \\ Z_0 & \xrightarrow{\ v_0\ } & Z \end{array}$$

with $s \in \mathcal{F}$. Then by (9.5.7), it decomposes into a commutative diagram

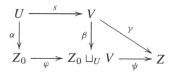

with φ, $\psi \in \mathrm{Mor}(\mathcal{C}_0)$ such that $v_0 = \psi \circ \varphi$. Since Y is $\mathcal{F}$-injective, the composition $U \xrightarrow{\alpha} Z_0 \xrightarrow{w_0} Y$ factors as $U \xrightarrow{s} V \to Y$, which induces a morphism $w_1 \colon Z_0 \sqcup_U V \to Y$. We thus obtain a commutative diagram

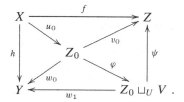

This defines a morphism ξ in I:

$$\xi \colon D(u_0, v_0, w_0, Z_0) \to D(\varphi \circ u_0, \psi, w_1, Z_0 \sqcup_U V).$$

Hence, ξ admits a left inverse. We get a morphism $\eta \colon Z_0 \sqcup_U V \to Z_0$ in $\mathcal{C}_0$ such that the two triangles in the diagram below commute

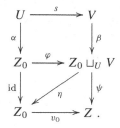

Therefore, the whole diagram commutes, and v_0 satisfies condition (9.5.5), hence is an isomorphism by assumption (9.5.8).   q.e.d.

9.6 Applications to Abelian Categories

We shall apply some of the preceding results to abelian categories. Note that a Grothendieck category (see Definition 8.3.24) satisfies hypothesis (9.3.1).

Let us summarize the principal results that we shall use in the sequel. These results follow from Corollaries 9.3.7 and 9.3.8.

Theorem 9.6.1. *Let $\mathcal{C}$ be a Grothendieck category. Then, for any small subset E of $\mathrm{Ob}(\mathcal{C})$, there exist an infinite cardinal π and a full subcategory $\mathcal{S}$ of $\mathcal{C}$ satisfying the conditions:*

   (i) $\mathrm{Ob}(\mathcal{S})$ *contains* E,

  (ii) $\mathcal{S}$ *is a fully abelian subcategory of* $\mathcal{C}$,

 (iii) $\mathcal{S}$ *is essentially small*,

 (iv) $\mathcal{S}$ *contains a generator of* $\mathcal{C}$,

  (v) $\mathcal{S}$ *is closed by subobjects and quotients in* $\mathcal{C}$,

 (vi) *for any epimorphism* $f\colon X \twoheadrightarrow Y$ *in* $\mathcal{C}$ *with* $Y \in \mathcal{S}$, *there exist* $Z \in \mathcal{S}$ *and a monomorphism* $g\colon Z \to X$ *such that* $f \circ g\colon Z \to Y$ *is an epimorphism*,

 (vii) $\mathcal{S}$ *is closed by countable direct sums*,

(viii) *any object in* $\mathcal{S}$ *is* π-*accessible*.

On the Existence of Enough Injectives and Injective Cogenerators

The next result is due to Grothendieck [28].

Theorem 9.6.2. *Let* $\mathcal{C}$ *be a Grothendieck category. Then* $\mathcal{C}$ *admits enough injectives.*

Proof. We shall apply Theorem 9.5.4. Let G be a generator of $\mathcal{C}$. We take as $\mathcal{C}_0$ the category whose objects are those of $\mathcal{C}$, the morphisms in $\mathcal{C}_0$ being the monomorphisms in $\mathcal{C}$. Let $\mathcal{F} \subset \mathrm{Mor}(\mathcal{C}_0)$ be the set of monomorphisms $N \hookrightarrow G$. This is a small set by Corollary 8.3.26. By Proposition 8.4.7, an object of $\mathcal{C}$ is injective if it is $\mathcal{F}$-injective.

Hypothesis (9.5.2) is clearly satisfied (use Lemma 8.3.11). By Theorem 9.6.1, there exists an infinite cardinal π such that any subobject of G is π-accessible. Applying Theorem 9.5.4, we obtain that for any $X \in \mathcal{C}$ there exists a monomorphism $X \hookrightarrow Y$ such that Y is $\mathcal{F}$-injective.   q.e.d.

Theorem 9.6.3. *Let* $\mathcal{C}$ *be a Grothendieck category. Then* $\mathcal{C}$ *admits an injective cogenerator* K.

Proof. Let G be a generator. By the result of Proposition 5.2.9, there exists a family $\{G_j\}_{j \in J}$ indexed by a small set J such that any quotient of G is isomorphic to some G_j. Let $S = \bigoplus_{j \in J} G_j$. By Theorem 9.6.2, there exist an injective object K and a monomorphism $S \hookrightarrow K$. We shall show that K is a cogenerator.

First, let us show that

$$(9.6.1) \qquad\qquad \mathrm{Hom}_{\mathcal{C}}(X, K) \simeq 0 \text{ implies } X \simeq 0.$$

For any morphism $u\colon G \to X$, the map $\mathrm{Hom}_{\mathcal{C}}(X, K) \to \mathrm{Hom}_{\mathcal{C}}(\mathrm{Im}\,u, K)$ is surjective. Hence, $\mathrm{Hom}_{\mathcal{C}}(\mathrm{Im}\,u, K) \simeq 0$. Since $\mathrm{Im}\,u$ is isomorphic to some G_j, there exists a monomorphism $\mathrm{Im}\,u \hookrightarrow K$. Hence, $\mathrm{Im}\,u \simeq 0$. Thus $\mathrm{Hom}_{\mathcal{C}}(G, X) \simeq 0$ and $X \simeq 0$. This prove (9.6.1).

To conclude, consider a morphism $f\colon X \to Y$ such that $\mathrm{Hom}_{\mathcal{C}}(f, K)$ is bijective, and let us show that f is an isomorphism. It is enough to check that $\mathrm{Ker}\,f \simeq \mathrm{Coker}\,f \simeq 0$. This follows from (9.6.1) and the exact sequence

$$0 \to \mathrm{Hom}_{\mathcal{C}}(\mathrm{Coker}\, f, K) \to \mathrm{Hom}_{\mathcal{C}}(Y, K)$$
$$\to \mathrm{Hom}_{\mathcal{C}}(X, K) \to \mathrm{Hom}_{\mathcal{C}}(\mathrm{Ker}\, f, K) \to 0\,.$$

q.e.d.

Corollary 9.6.4. *Let $\mathcal{C}$ be a Grothendieck category.*

(i) *A functor $F \colon \mathcal{C} \to \mathbf{Set}$ is representable if F commutes with small projective limits.*

(ii) *Let $\mathcal{A}$ be another category. If a functor $R \colon \mathcal{C} \to \mathcal{A}$ commutes with small projective limits, then R admits a left adjoint.*

Proof. It is enough to apply Theorem 9.6.3, Theorem 5.2.6 and Proposition 5.2.8 (with the arrows reversed). q.e.d.

Corollary 9.6.5. *Let $\mathcal{C}$ be a small abelian category. Then $\mathrm{Ind}(\mathcal{C})$ admits an injective cogenerator.*

Proof. Apply Theorem 8.6.5 and Proposition 9.6.3. q.e.d.

Let us give an important application of Theorem 9.6.3.

Corollary 9.6.6. *Let $\mathcal{C}$ be a Grothendieck category. Denote by $\mathcal{I}_{inj}$ the full additive subcategory of $\mathcal{C}$ consisting of injective objects and by $\iota \colon \mathcal{I}_{inj} \to \mathcal{C}$ the inclusion functor. Then there exist a (not necessarily additive) functor $\Psi \colon \mathcal{C} \to \mathcal{I}_{inj}$ and a morphism of functor $\mathrm{id}_{\mathcal{C}} \to \iota \circ \Psi$ such that $X \to \Psi(X)$ is a monomorphism for any $X \in \mathcal{C}$.*

Proof. The category $\mathcal{C}$ admits an injective cogenerator K by Proposition 9.6.3 and admits small products by Proposition 8.3.27. Consider the (non additive) functor

$$\Psi \colon \mathcal{C} \to \mathcal{C}, \quad X \mapsto K^{\mathrm{Hom}\,(X,K)}\,.$$

The identity of

$$\mathrm{Hom}_{\mathbf{Set}}(\mathrm{Hom}_{\mathcal{C}}(X, K), \mathrm{Hom}_{\mathcal{C}}(X, K)) \simeq \mathrm{Hom}_{\mathcal{C}}(X, K^{\mathrm{Hom}\,(X,K)})$$

defines a morphism $X \to \Psi(X) = K^{\mathrm{Hom}\,(X,K)}$, and this morphism is a monomorphism by Proposition 5.2.3 (iv). q.e.d.

Corollary 9.6.7. *Let $\mathcal{C}$ be a Grothendieck category and I a small category. Let $\alpha \colon I \to \mathcal{C}$ be a functor. Denote by $\mathcal{I}_{inj}$ the full additive subcategory of $\mathcal{C}$ consisting of injective objects. Then there exist a functor $\beta \colon I \to \mathcal{I}_{inj}$ and a monomorphism $\alpha \rightarrowtail \beta$ in $\mathrm{Fct}(I, \mathcal{C})$.*

Proof. Take $\beta = \Psi \circ \alpha$, where Ψ is the functor given by Corollary 9.6.6. q.e.d.

The Freyd-Mitchell Theorem

Lemma 9.6.8. *Let $\mathcal{C}$ be an abelian category which admits a projective generator G and small inductive limits. Let $J \subset \mathrm{Ob}(\mathcal{C})$ be a small set. Then there exists a projective generator P such that any $X \in J$ is isomorphic to a quotient of P.*

Proof. For $X \in \mathcal{C}$, the morphism $G^{\oplus \mathrm{Hom}\,(G,X)} \to X$ is an epimorphism. Hence, it is enough to set $P = \bigoplus_{X \in J} G^{\oplus \mathrm{Hom}\,(G,X)}$. \hfill q.e.d.

We set

$$
(9.6.2) \quad \begin{cases} R_G := \text{ the opposite ring of the ring } \mathrm{Hom}_{\mathcal{C}}(G,G)\,, \\ \varphi_G := \text{ the functor } \mathrm{Hom}_{\mathcal{C}}(G,\,\cdot\,) \colon \mathcal{C} \to \mathrm{Mod}(R_G)\,. \end{cases}
$$

Lemma 9.6.9. *Let $\mathcal{C}$ be an abelian category which admits a projective generator G. Let $X \in \mathcal{C}$ be a quotient of a finite direct sum of copies of G. Then the map*

$$
(9.6.3) \qquad \mathrm{Hom}_{\mathcal{C}}(X,Y) \to \mathrm{Hom}_{R_G}(\varphi_G(X), \varphi_G(Y))
$$

is bijective for all $Y \in \mathcal{C}$.

Proof. For short, we shall write R and φ instead of R_G and φ_G, respectively. By the assumption, φ is an exact functor. By Proposition 5.2.3, the functor φ is faithful. Taking an epimorphism $G^{\oplus m} \twoheadrightarrow X$, set $N = \mathrm{Ker}(G^{\oplus m} \to X)$.

Then $0 \to \varphi(N) \to \varphi(G^{\oplus m}) \to \varphi(X) \to 0$ is exact. Let us consider the exact commutative diagram (in this diagram, we write Hom instead of $\mathrm{Hom}_{\mathcal{C}}$ or Hom_R for short)

$$
\begin{array}{ccccccc}
0 & \longrightarrow & \mathrm{Hom}\,(X,Y) & \longrightarrow & \mathrm{Hom}\,(G^{\oplus m},Y) & \longrightarrow & \mathrm{Hom}\,(N,Y) \\
& & \downarrow & & \downarrow & & \downarrow \\
0 & \to & \mathrm{Hom}\,(\varphi(X),\varphi(Y)) & \to & \mathrm{Hom}\,(\varphi(G^{\oplus m}),\varphi(Y)) & \to & \mathrm{Hom}\,(\varphi(N),\varphi(Y)).
\end{array}
$$

Since $\mathrm{Hom}_{\mathcal{C}}(G,Y) = \varphi(Y) \simeq \mathrm{Hom}_R(\varphi(G),\varphi(Y))$, the middle vertical arrow is an isomorphism. Since φ is faithful, the right vertical arrow is a monomorphism. Therefore the left vertical arrow is an isomorphism. \hfill q.e.d.

The next theorem is due to Freyd and Mitchell.

Theorem 9.6.10. *Let $\mathcal{C}$ be a small abelian category. There exist a ring R and an exact fully faithful functor $\mathcal{C} \to \mathrm{Mod}(R)$. In other words, $\mathcal{C}$ is equivalent to a fully abelian subcategory of $\mathrm{Mod}(R)$.*

Proof. The category $\mathcal{C}^{\mathrm{op}}$ is abelian. Applying Corollary 9.6.5, the abelian category $\mathrm{Ind}(\mathcal{C}^{\mathrm{op}})$ admits an injective cogenerator. Hence $\mathrm{Pro}(\mathcal{C}) \simeq (\mathrm{Ind}(\mathcal{C}^{\mathrm{op}}))^{\mathrm{op}}$ admits a projective generator. We regard $\mathcal{C}$ as a full subcategory of $\mathrm{Pro}(\mathcal{C})$. Note that $\mathcal{C} \to \mathrm{Pro}(\mathcal{C})$ is an exact functor. By Lemmas 9.6.8 there exists a projective generator G of $\mathrm{Pro}(\mathcal{C})$ such that any object of $\mathcal{C}$ is a quotient of G. Then Lemma 9.6.9 implies that the functor $\varphi_G \colon \mathcal{C} \to \mathrm{Mod}(R_G)$ is fully faithful, and φ_G is obviously exact.   q.e.d.

Exercises

Exercise 9.1. Let $\mathcal{C}$ be an abelian category. A monomorphism $f \colon X \rightarrowtail Y$ is *essential* if for any subobject W of Y, $W \cap X \simeq 0$ implies $W \simeq 0$. Prove that f is essential if and only if a morphism $g \colon Y \to Z$ is a monomorphism as soon as $g \circ f$ is a monomorphism.

Exercise 9.2. Let $\mathcal{C}$ be a Grothendieck category and let $f \colon X \rightarrowtail Y$ be a monomorphism. Prove that there exists an essential monomorphism $h \colon X \rightarrowtail Z$ which factorizes as $X \xrightarrow{f} Y \to Z$.
(Hint: let Σ denote the set of subobjects W of Y satisfying $W \cap X = 0$. Then Σ is a small set and is inductively ordered.)

Exercise 9.3. Let $\mathcal{C}$ be a Grothendieck category and let $Z \in \mathcal{C}$. Prove that Z is injective if and only if any essential monomorphism $f \colon Z \rightarrowtail W$ is an isomorphism.

Exercise 9.4. Let $\mathcal{C}$ be a Grothendieck category and let $\{X \to Y_i\}_{i \in I}$ be an inductive system of morphisms in $\mathcal{C}$ indexed by a small and filtrant category I. Assume that all morphisms $X \to Y_i$ are essential monomorphisms. Prove that $f \colon X \to \varinjlim_i Y_i$ is an essential monomorphism.

Exercise 9.5. Let $\mathcal{C}$ be a Grothendieck category and let G be a generator. Set $R = (\mathrm{End}_{\mathcal{C}}(G))^{\mathrm{op}}$. Recall that the category $\mathrm{Mod}(R)$ admits enough injectives by the result of Exercise 8.24.
(i) Prove that if $f \colon X \to Y$ is an essential monomorphism in $\mathcal{C}$, then $\varphi_G(f)$ is an essential monomorphism in $\mathrm{Mod}(R)$.
(ii) Deduce another proof of Theorem 9.6.2.

Exercise 9.6. Let $\mathcal{C}$ be a Grothendieck category and let $X \in \mathcal{C}$.
(i) Prove that there exist an injective object I and an essential monomorphism $X \rightarrowtail I$.
(ii) Let $u \colon X \to I$ and $u' \colon X \to I'$ be two essential monomorphisms, with I and I' injectives. Prove that there exists an isomorphism $g \colon I \xrightarrow{\sim} I'$ such that $g \circ u = u'$. (Note that such a g is not unique in general.)

Exercise 9.7. Let π be an infinite cardinal and I a small category. Assume that for any category J such that $\text{card}(\text{Mor}(J)) < \pi$ and any functor $\alpha\colon J^{\text{op}} \times I \to \mathbf{Set}$, the morphism λ in (9.2.1) is an isomorphism. Prove that I is π-filtrant. (Hint: for any $\varphi\colon J \to I$, apply (9.2.1) to $\alpha(j, i) = \text{Hom}_{\mathcal{C}}(\varphi(j), i)$ and use Lemma 2.4.7.)

Exercise 9.8. Let π be an infinite cardinal and let $\mathcal{C}$ be a category which admits inductive limits indexed by any category J such that $\text{card}(\text{Mor}(J)) < \pi$. Let $F\colon \mathcal{C} \to \mathcal{C}'$ be a functor. Prove that F commutes with such inductive limits if and only if $\mathcal{C}_Y$ is π-filtrant for any $Y \in \mathcal{C}'$.

Exercise 9.9. Let X and Y be two quasi-terminal objects in a category $\mathcal{C}$. Prove that any morphism $f\colon X \to Y$ is an isomorphism.

Exercise 9.10. Let $\mathcal{C}$ be a category and $X \in \mathcal{C}$. Assume that $(X, \text{id}_X) \in \mathcal{C}^X$ is a terminal object of $\mathcal{C}^X$. Prove that X is a quasi-terminal object.

Exercise 9.11. Let A be a small set and let $\mathcal{C}$ be the category defined as follows.

$$\text{Ob}(\mathcal{C}) = \{x, y\}\,,$$
$$\text{Hom}_{\mathcal{C}}(x, x) = \{\text{id}_x\}\,,$$
$$\text{Hom}_{\mathcal{C}}(y, y) = \{\text{id}_y\} \sqcup \{p_a; a \in A\}\,,$$
$$\text{Hom}_{\mathcal{C}}(x, y) = \{u\}\,,$$
$$\text{Hom}_{\mathcal{C}}(y, x) = \{v_a; a \in A\}\,,$$

with the relations $p_a \circ p_b = p_b$, $v_a \circ p_b = v_b$ for any $a, b \in A$.
(i) Prove that $\mathcal{C}$ is a category.
(ii) Prove that there exists a fully faithful functor $F\colon \mathcal{C} \to \mathbf{Set}^A$.
(iii) Prove that $\mathcal{C} \to \text{Ind}(\mathcal{C})$ is an equivalence of categories.
(iv) Prove that $\mathcal{C}$ admits filtrant inductive limits.
(v) Prove that x is a quasi-terminal object of $\mathcal{C}$ (see Definition 9.4.1) and observe that a left inverse of $u\colon x \to y$ is not unique.

Exercise 9.12. Let $\mathcal{C}$ be an abelian category and G a projective object of $\mathcal{C}$. Assume that any object of $\mathcal{C}$ is a quotient of a direct sum of finite copies of G. Define R_G and φ_G as in (9.6.2). Prove that R_G is a left coherent ring (see Exercise 8.23) and that φ_G gives an equivalence $\mathcal{C} \xrightarrow{\sim} \text{Mod}^{\text{coh}}(R_G)$, where $\text{Mod}^{\text{coh}}(R_G)$ is the full subcategory of $\text{Mod}(R_G)$ consisting of coherent R_G-modules.

Exercise 9.13. Let $\mathcal{C}$ be a *small* category which admits small products. Prove that for any pair of objects X, Y in $\mathcal{C}$, $\text{Hom}_{\mathcal{C}}(X, Y)$ has at most one element and that $\mathcal{C}$ is equivalent to the category associated with an ordered set I such that for any subset J of I, $\inf(J)$ exists in I. (Hint: assume there exist $X, Y \in \mathcal{C}$ such that $\text{Hom}_{\mathcal{C}}(X, Y)$ has more than one element and set $M = \text{Ob}(\text{Mor}(\mathcal{C}))$, $\pi = \text{card}(M)$. By considering $\text{Hom}_{\mathcal{C}}(X, Y^M)$, find a contradiction.)
(The result of this exercise is due to Freyd [22].)

10

Triangulated Categories

Triangulated categories play an increasing role in mathematics and this subject deserves a whole book.

In this chapter we define and give the main properties of triangulated categories and cohomological functors and prove in particular that the localization of a triangulated category is still triangulated. We also show that under natural hypotheses, the Kan extension of a cohomological functor remains cohomological.

Then we study triangulated categories admitting small direct sums. Such categories are studied by many authors, in particular [6] and [53]. Here, we prove the so-called "Brown representability theorem" [11] in the form due to Neeman [53], more precisely, a variant due to [44], which asserts that any cohomological contravariant functor defined on a triangulated category admitting small direct sums and a suitable system of generators is representable as soon as it sends small direct sums to products. (The fact that Brown's theorem could be adapted to triangulated categories was also noticed by Keller [42].)

There also exist variants of the Brown representability theorem for triangulated categories which do not admit small direct sums. For results in this direction, we refer to [8].

We ask the reader to wait until Chap. 11 to encounter examples of triangulated categories. In fact, it would have been possible to formulate the important Theorem 11.3.8 below before defining triangulated categories, by listing the properties which become the axioms of these categories. We have chosen to give the axioms first in order to avoid repetitions, and also because the scope of triangulated categories goes much beyond the case of complexes in additive categories.

We do not treat here t-structures on triangulated categories and refer to the original paper [4] (see also [38] for an exposition). Another important closely related subject which is not treated here is the theory of A_∞-algebras (see [41, 43]).

10.1 Triangulated Categories

Definition 10.1.1. (i) *A category with* translation $(\mathcal{D}, T)$ *is a category* $\mathcal{D}$ *endowed with an equivalence of categories* $T \colon \mathcal{D} \xrightarrow{\sim} \mathcal{D}$. *The functor* T *is called the translation functor.*

(ii) *A functor of categories with translation* $F \colon (\mathcal{D}, T) \to (\mathcal{D}', T')$ *is a functor* $F \colon \mathcal{D} \to \mathcal{D}'$ *together with an isomorphism* $F \circ T \simeq T' \circ F$. *If* $\mathcal{D}$ *and* $\mathcal{D}'$ *are additive categories and* F *is additive, we say that* F *is a functor of additive categories with translation.*

(iii) *Let* $F, F' \colon (\mathcal{D}, T) \to (\mathcal{D}', T')$ *be two functors of categories with translation. A morphism* $\theta \colon F \to F'$ *of functors of categories with translation is a morphism of functors such that the diagram below commutes:*

$$
\begin{array}{ccc}
F \circ T & \xrightarrow{\ \theta \circ T\ } & F' \circ T \\
\sim \downarrow & & \downarrow \sim \\
T' \circ F & \xrightarrow{\ T' \circ \theta\ } & T' \circ F' .
\end{array}
$$

(iv) *A subcategory with translation* $(\mathcal{D}', T')$ *of* $(\mathcal{D}, T)$ *is a category with translation such that* $\mathcal{D}'$ *is a subcategory of* $\mathcal{D}$ *and the translation functor* T' *is the restriction of* T.

(v) *Let* $(\mathcal{D}, T)$, $(\mathcal{D}', T')$ *and* $(\mathcal{D}'', T'')$ *be additive categories with translation. A bifunctor of additive categories with translation* $F \colon \mathcal{D} \times \mathcal{D}' \to \mathcal{D}''$ *is an additive bifunctor endowed with functorial isomorphisms*

$$
\theta_{X,Y} \colon F(TX, Y) \xrightarrow{\sim} T'' F(X, Y) \ \text{and}\ \theta'_{X,Y} \colon F(X, T'Y) \xrightarrow{\sim} T'' F(X, Y)
$$

for $(X, Y) \in \mathcal{D} \times \mathcal{D}'$ *such that the diagram below anti-commutes (see Definition 8.2.20):*

$$
\begin{array}{ccc}
F(TX, T'Y) & \xrightarrow{\ \theta_{X,T'Y}\ } & T'' F(X, T'Y) \\
\theta'_{TX,Y} \downarrow & \text{ac} & \downarrow T'' \theta'_{X,Y} \\
T'' F(TX, Y) & \xrightarrow{\ T'' \theta_{X,Y}\ } & T''^2 F(X, Y) .
\end{array}
$$

Remark 10.1.2. The anti-commutativity of the diagram above will be justified in Chapter 11 (see Proposition 11.2.11 and Lemma 11.6.3).

Notations 10.1.3. (i) We shall denote by T^{-1} a quasi-inverse of T. Then T^n is well defined for $n \in \mathbb{Z}$. These functors are unique up to unique isomorphism. (ii) If there is no risk of confusion, we shall write $\mathcal{D}$ instead of $(\mathcal{D}, T)$ and TX instead of $T(X)$.

Definition 10.1.4. *Let* $(\mathcal{D}, T)$ *be an additive category with translation. A triangle in* $\mathcal{D}$ *is a sequence of morphisms*

$$(10.1.1) \qquad\qquad X \xrightarrow{f} Y \xrightarrow{g} Z \xrightarrow{h} TX .$$

A morphism of triangles is a commutative diagram:

$$
\begin{array}{ccccccc}
X & \xrightarrow{\ f\ } & Y & \xrightarrow{\ g\ } & Z & \xrightarrow{\ h\ } & TX \\
{\scriptstyle\alpha}\downarrow & & {\scriptstyle\beta}\downarrow & & {\scriptstyle\gamma}\downarrow & & \downarrow{\scriptstyle T(\alpha)} \\
X' & \xrightarrow{\ f'\ } & Y' & \xrightarrow{\ g'\ } & Z' & \xrightarrow{\ h'\ } & TX' .
\end{array}
$$

Remark 10.1.5. For $\varepsilon_1, \varepsilon_2, \varepsilon_3 = \pm 1$, the triangle $X \xrightarrow{\varepsilon_1 f} Y \xrightarrow{\varepsilon_2 g} Z \xrightarrow{\varepsilon_3 h} TX$ is isomorphic to the triangle (10.1.1) if $\varepsilon_1 \varepsilon_2 \varepsilon_3 = 1$, but if $\varepsilon_1 \varepsilon_2 \varepsilon_3 = -1$, it is not isomorphic to the triangle (10.1.1) in general.

Definition 10.1.6. *A* triangulated category *is an additive category* $(\mathcal{D}, T)$ *with translation endowed with a family of triangles, called* distinguished triangles (d.t. *for short*), *this family satisfying the axioms* TR0 – TR5 *below.*

TR0 *A triangle isomorphic to a* d.t. *is a* d.t.

TR1 *The triangle* $X \xrightarrow{\mathrm{id}_X} X \to 0 \to TX$ *is a* d.t.

TR2 *For all* $f : X \to Y$, *there exists a* d.t. $X \xrightarrow{f} Y \to Z \to TX$.

TR3 *A triangle* $X \xrightarrow{f} Y \xrightarrow{g} Z \xrightarrow{h} TX$ *is a* d.t. *if and only if* $Y \xrightarrow{-g} Z \xrightarrow{-h} TX \xrightarrow{-T(f)} TY$ *is a* d.t.

TR4 *Given two* d.t.'s $X \xrightarrow{f} Y \xrightarrow{g} Z \xrightarrow{h} TX$ *and* $X' \xrightarrow{f'} Y' \xrightarrow{g'} Z' \xrightarrow{h'} TX'$ *and morphisms* $\alpha : X \to X'$ *and* $\beta : Y \to Y'$ *with* $f' \circ \alpha = \beta \circ f$, *there exists a morphism* $\gamma : Z \to Z'$ *giving rise to a morphism of* d.t.'s:

$$
\begin{array}{ccccccc}
X & \xrightarrow{\ f\ } & Y & \xrightarrow{\ g\ } & Z & \xrightarrow{\ h\ } & TX \\
{\scriptstyle\alpha}\downarrow & & {\scriptstyle\beta}\downarrow & & {\scriptstyle\gamma}\downarrow & & \downarrow{\scriptstyle T(\alpha)} \\
X' & \xrightarrow{\ f'\ } & Y' & \xrightarrow{\ g'\ } & Z' & \xrightarrow{\ h'\ } & TX' .
\end{array}
$$

TR5 *Given three* d.t.'s

$$X \xrightarrow{f} Y \xrightarrow{h} Z' \to TX ,$$

$$Y \xrightarrow{g} Z \xrightarrow{k} X' \to TY ,$$

$$X \xrightarrow{g \circ f} Z \xrightarrow{l} Y' \to TX,$$

there exists a d.t. $Z' \xrightarrow{u} Y' \xrightarrow{v} X' \xrightarrow{w} TZ'$ *making the diagram below commutative:*

(10.1.2)

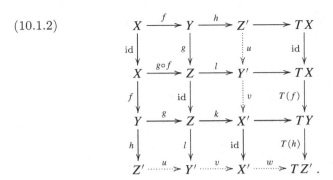

Diagram (10.1.2) is often called the *octahedron diagram*. Indeed, it can be written using the vertices of an octahedron.

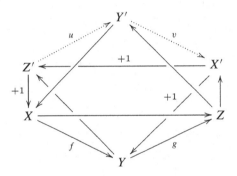

Here, for example, $X' \xrightarrow{+1} Y$ means a morphism $\dot{X}' \to TY$.

Notation 10.1.7. The translation functor T is called the suspension functor by the topologists.

Remark 10.1.8. The morphism γ in TR4 is not unique and this is the origin of many troubles. See the paper [7] for an attempt to overcome this difficulty.

Definition 10.1.9. (i) *A* triangulated functor *of triangulated categories* F : $(\mathcal{D}, T) \to (\mathcal{D}', T')$ *is a functor of additive categories with translation sending distinguished triangles to distinguished triangles. If moreover F is an equivalence of categories, F is called an equivalence of triangulated categories.*

(ii) *Let $F, F' : (\mathcal{D}, T) \to (\mathcal{D}', T')$ be triangulated functors. A morphism $\theta : F \to F'$ of triangulated functors is a morphism of functors of additive categories with translation.*

(iii) *A triangulated subcategory $(\mathcal{D}', T')$ of $(\mathcal{D}, T)$ is an additive subcategory with translation of $\mathcal{D}$ (i.e., the functor T' is the restriction of T) such that it is triangulated and that the inclusion functor is triangulated.*

Remark 10.1.10. (i) A triangle $X \xrightarrow{f} Y \xrightarrow{g} Z \xrightarrow{h} TX$ is *anti-distinguished* if
the triangle $X \xrightarrow{f} Y \xrightarrow{g} Z \xrightarrow{-h} TX$ is distinguished. Then $(\mathcal{D}, T)$ endowed
with the family of anti-distinguished triangles is triangulated. If we denote by
$(\mathcal{D}^{ant}, T)$ this triangulated category, then $(\mathcal{D}^{ant}, T)$ and $(\mathcal{D}, T)$ are equivalent
as triangulated categories (see Exercise 10.10).
(ii) Consider the contravariant functor op: $\mathcal{D} \to \mathcal{D}^{op}$, and define $T^{op} = op \circ$
$T^{-1} \circ op^{-1}$. Let us say that a triangle $X \xrightarrow{f} Y \xrightarrow{g} Z \xrightarrow{h} T^{op}(X)$ in $\mathcal{D}^{op}$
is distinguished if its image $Z^{op} \xrightarrow{g^{op}} Y^{op} \xrightarrow{f^{op}} X^{op} \xrightarrow{T(h^{op})} TZ^{op}$ by op is
distinguished. (Here, we write op instead of op^{-1} for short.) Then $(\mathcal{D}^{op}, T^{op})$
is a triangulated category.

Proposition 10.1.11. *If $X \xrightarrow{f} Y \xrightarrow{g} Z \to TX$ is a d.t. then $g \circ f = 0$.*

Proof. Applying TR1 and TR4 we get a commutative diagram:

$$\begin{array}{ccccccc}
X & \xrightarrow{id} & X & \longrightarrow & 0 & \longrightarrow & TX \\
{\scriptstyle id}\downarrow & & {\scriptstyle f}\downarrow & & \downarrow & & {\scriptstyle id}\downarrow \\
X & \xrightarrow{f} & Y & \xrightarrow{g} & Z & \longrightarrow & TX .
\end{array}$$

Then $g \circ f$ factorizes through 0.   q.e.d.

Definition 10.1.12. *Let $(\mathcal{D}, T)$ be a triangulated category and $\mathcal{C}$ an abelian
category. An additive functor $F: \mathcal{D} \to \mathcal{C}$ is* cohomological *if for any d.t.
$X \to Y \to Z \to TX$ in $\mathcal{D}$, the sequence $F(X) \to F(Y) \to F(Z)$ is exact in $\mathcal{C}$.*

Proposition 10.1.13. *For any $W \in \mathcal{D}$, the two functors $\mathrm{Hom}_{\mathcal{D}}(W, \cdot)$ and
$\mathrm{Hom}_{\mathcal{D}}(\cdot, W)$ are cohomological.*

Proof. Let $X \to Y \to Z \to TX$ be a d.t. and let $W \in \mathcal{D}$. We want to show
that
$$\mathrm{Hom}(W, X) \xrightarrow{f \circ} \mathrm{Hom}(W, Y) \xrightarrow{g \circ} \mathrm{Hom}(W, Z)$$
is exact, i.e. : for all $\varphi: W \to Y$ such that $g \circ \varphi = 0$, there exists $\psi: W \to X$
such that $\varphi = f \circ \psi$. This means that the dotted arrows below may be
completed, and this follows from the axioms TR4 and TR3.

$$\begin{array}{ccccccc}
W & \xrightarrow{id} & W & \longrightarrow & 0 & \longrightarrow & TW \\
\vdots & & {\scriptstyle \varphi}\downarrow & & \downarrow & & \vdots \\
X & \xrightarrow{f} & Y & \xrightarrow{g} & Z & \longrightarrow & TX .
\end{array}$$

By replacing $\mathcal{D}$ with $\mathcal{D}^{op}$, we obtain the assertion for $\mathrm{Hom}(\cdot, W)$.   q.e.d.

Remark 10.1.14. By TR3, a cohomological functor gives rise to a long exact
sequence:

(10.1.3) $\cdots \to F(T^{-1}Z) \to F(X) \to F(Y) \to F(Z) \to F(TX) \to \cdots$.

Proposition 10.1.15. *Consider a morphism of* d.t.'s:

$$
\begin{array}{ccccccc}
X & \xrightarrow{f} & Y & \xrightarrow{g} & Z & \xrightarrow{h} & TX \\
\downarrow{\alpha} & & \downarrow{\beta} & & \downarrow{\gamma} & & \downarrow{T(\alpha)} \\
X' & \xrightarrow{f'} & Y' & \xrightarrow{g'} & Z' & \xrightarrow{h'} & TX' .
\end{array}
$$

If α and β are isomorphisms, then so is γ.

Proof. Apply $\mathrm{Hom}\,(W, \cdot)$ to this diagram and write $\widetilde{X}$ instead of $\mathrm{Hom}\,(W, X)$, $\widetilde{\alpha}$ instead of $\mathrm{Hom}\,(W, \alpha)$, etc. We get the commutative diagram:

$$
\begin{array}{ccccccccccc}
\widetilde{X} & \xrightarrow{\widetilde{f}} & \widetilde{Y} & \xrightarrow{\widetilde{g}} & \widetilde{Z} & \xrightarrow{\widetilde{h}} & \widetilde{TX} & \xrightarrow{\widetilde{T(f)}} & \widetilde{TY} \\
\downarrow{\widetilde{\alpha}} & & \downarrow{\widetilde{\beta}} & & \downarrow{\widetilde{\gamma}} & & \downarrow{\widetilde{T(\alpha)}} & & \downarrow{\widetilde{T(\beta)}} \\
\widetilde{X}' & \xrightarrow{\widetilde{f}'} & \widetilde{Y}' & \xrightarrow{\widetilde{g}'} & \widetilde{Z}' & \xrightarrow{\widetilde{h}'} & \widetilde{TX}' & \xrightarrow{\widetilde{T(f)}} & \widetilde{TY}' .
\end{array}
$$

The rows are exact in view of the Proposition 10.1.13, and $\widetilde{\alpha}$, $\widetilde{\beta}$, $\widetilde{T(\alpha)}$ and $\widetilde{T(\beta)}$ are isomorphisms. Therefore $\widetilde{\gamma} = \mathrm{Hom}\,(W, \gamma) \colon \mathrm{Hom}\,(W, Z) \to \mathrm{Hom}\,(W, Z')$ is an isomorphism by Lemma 8.3.13. This implies that γ is an isomorphism by Corollary 1.4.7.   q.e.d.

Corollary 10.1.16. *Let $\mathcal{D}'$ be a full triangulated subcategory of $\mathcal{D}$.*

(i) *Consider a triangle $X \xrightarrow{f} Y \to Z \to TX$ in $\mathcal{D}'$ and assume that this triangle is distinguished in $\mathcal{D}$. Then it is distinguished in $\mathcal{D}'$.*

(ii) *Consider a d.t. $X \to Y \to Z \to TX$ in $\mathcal{D}$ with X and Y in $\mathcal{D}'$. Then Z is isomorphic to an object of $\mathcal{D}'$.*

Proof. There exists a d.t. $X \xrightarrow{f} Y \to Z' \to TX$ in $\mathcal{D}'$. Then $X \xrightarrow{f} Y \to Z \to TX$ is isomorphic to $X \xrightarrow{f} Y \to Z' \to TX$ in $\mathcal{D}$ by TR4 and Proposition 10.1.15.   q.e.d.

By Proposition 10.1.15, we obtain that the object Z given in TR2 is unique up to isomorphism. As already mentioned, the fact that this isomorphism is not unique is the source of many difficulties (e.g., gluing problems in sheaf theory). Let us give a criterion which ensures, in some very special cases, the uniqueness of the third term of a d.t.

Proposition 10.1.17. *In the situation of TR4 assume that $\mathrm{Hom}_{\mathcal{D}}(Y, X') = 0$ and $\mathrm{Hom}_{\mathcal{D}}(TX, Y') = 0$. Then γ is unique.*

Proof. We may replace α and β by the zero morphisms and prove that in this case, γ is zero.

$$
\begin{array}{ccccccc}
X & \xrightarrow{f} & Y & \xrightarrow{g} & Z & \xrightarrow{h} & TX \\
\downarrow 0 & & \downarrow 0 & & \downarrow \gamma & & \downarrow 0 \\
X' & \xrightarrow{f'} & Y' & \xrightarrow{g'} & Z' & \xrightarrow{h'} & TX' .
\end{array}
$$

We shall apply Proposition 10.1.13. Since $h' \circ \gamma = 0$, γ factorizes through g', i.e., there exists $u \colon Z \to Y'$ with $\gamma = g' \circ u$. Similarly, since $\gamma \circ g = 0$, γ factorizes through h, i.e., there exists $v \colon TX \to Z'$ with $\gamma = v \circ h$.

By TR4, there exists a morphism w defining a morphism of d.t.'s:

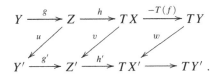

By the hypothesis, $w = 0$. Hence v factorizes through Y', and by the hypothesis this implies that $v = 0$. Therefore, $\gamma = 0$.   q.e.d.

Proposition 10.1.18. *Let $\mathcal{T}$ and $\mathcal{D}$ be triangulated categories and let $F \colon \mathcal{T} \to \mathcal{D}$ be a triangulated functor. Then F is exact (see Definition 3.3.1).*

Proof. (i) Let us show that F is right exact, that is, for any $X \in \mathcal{D}$, the category $\mathcal{T}_X$ is filtrant.

(a) The category $\mathcal{T}_X$ is non empty since it contains the object $0 \to X$.

(b) Let (Y_0, s_0) and (Y_1, s_1) be two objects in $\mathcal{T}_X$ with $Y_i \in \mathcal{T}$ and $s_i \colon F(Y_i) \to X$, $i = 0, 1$. The morphisms s_0 and s_1 define $s \colon F(Y_0 \oplus Y_1) \to X$. Hence, we obtain morphisms $(Y_i, s_i) \to (Y_0 \oplus Y_1, s)$ for $i = 0, 1$.

(c) Consider a pair of parallel arrows $f, g \colon (Y_0, s_0) \rightrightarrows (Y_1, s_1)$ in $\mathcal{T}_X$. Let us embed $f - g \colon Y_0 \to Y_1$ in a d.t. $Y_0 \xrightarrow{f-g} Y_1 \xrightarrow{h} Y \to TY_0$. Since $s_1 \circ F(f) = s_1 \circ F(g)$, Proposition 10.1.13 implies that the morphism $s_1 \colon F(Y_1) \to X$ factorizes as $F(Y_1) \to F(Y) \xrightarrow{t} X$. Hence, the two compositions $(Y_0, s_0) \rightrightarrows (Y_1, s_1) \to (Y, t)$ coincide.

(ii) Replacing $F \colon \mathcal{T} \to \mathcal{D}$ with $F^{\mathrm{op}} \colon \mathcal{T}^{\mathrm{op}} \to \mathcal{D}^{\mathrm{op}}$, we find that F is left exact.
                    q.e.d.

Proposition 10.1.19. *Let $\mathcal{D}$ be a triangulated category which admits direct sums indexed by a set I. Then direct sums indexed by I commute with the translation functor T, and a direct sum of distinguished triangles indexed by I is a distinguished triangle.*

Proof. The first assertion is obvious since T is an equivalence of categories.

Let $D_i \colon X_i \to Y_i \to Z_i \to TX_i$ be a family of d.t.'s indexed by $i \in I$. Let D be the triangle

$$
\oplus_{i \in I} D_i \colon \ \oplus_i X_i \to \oplus_i Y_i \to \oplus_i Z_i \to \oplus_i TX_i .
$$

By TR2 there exists a d.t. $D' : \oplus_i X_i \to \oplus_i Y_i \to Z \to T(\oplus_i X_i)$. By TR3 there exist morphisms of triangles $D_i \to D'$ and they induce a morphism $D \to D'$. Let $W \in \mathcal{D}$ and let us show that the morphism $\mathrm{Hom}_{\mathcal{D}}(D', W) \to \mathrm{Hom}_{\mathcal{D}}(D, W)$ is an isomorphism. This will imply the isomorphism $D \xrightarrow{\sim} D'$ by Corollary 1.4.7. Consider the commutative diagram of complexes

$$
\begin{array}{ccccc}
\mathrm{Hom}_{\mathcal{D}}(T(\oplus_i Y_i), W) & \to & \mathrm{Hom}_{\mathcal{D}}(T(\oplus_i X_i), W) & \longrightarrow & \mathrm{Hom}_{\mathcal{D}}(Z, W) \longrightarrow \\
\downarrow & & \downarrow & & \downarrow \\
\mathrm{Hom}_{\mathcal{D}}(\oplus_i T Y_i, W) & \longrightarrow & \mathrm{Hom}_{\mathcal{D}}(\oplus_i T X_i, W) & \longrightarrow & \mathrm{Hom}_{\mathcal{D}}(\oplus_i Z_i, W) \longrightarrow
\end{array}
$$

$$
\begin{array}{ccc}
\to & \mathrm{Hom}_{\mathcal{D}}(\oplus_i Y_i, W) & \to & \mathrm{Hom}_{\mathcal{D}}(\oplus_i X_i, W) \\
& \downarrow & & \downarrow \\
\to & \mathrm{Hom}_{\mathcal{D}}(\oplus_i Y_i, W) & \to & \mathrm{Hom}_{\mathcal{D}}(\oplus_i X_i, W).
\end{array}
$$

The first row is exact since the functor $\mathrm{Hom}_{\mathcal{D}}$ is cohomological. The second row is isomorphic to

$$
\prod_i \mathrm{Hom}_{\mathcal{D}}(T Y_i, W) \to \prod_i \mathrm{Hom}_{\mathcal{D}}(T X_i, W) \to \prod_i \mathrm{Hom}_{\mathcal{D}}(Z_i, W)
$$
$$
\to \prod_i \mathrm{Hom}_{\mathcal{D}}(Y_i, W) \to \prod_i \mathrm{Hom}_{\mathcal{D}}(X_i, W) \ .
$$

Since the functor $\prod_i$ is exact on $\mathrm{Mod}(\mathbb{Z})$, this complex is exact. Since the vertical arrows except the middle one are isomorphisms, the middle one is an isomorphism by Lemma 8.3.13.   q.e.d.

As particular cases of Proposition 10.1.19, we get:

Corollary 10.1.20. *Let $\mathcal{D}$ be a triangulated category.*

(i) *Let $X_1 \to Y_1 \to Z_1 \to TX_1$ and $X_2 \to Y_2 \to Z_2 \to TX_2$ be two d.t.'s. Then $X_1 \oplus X_2 \to Y_1 \oplus Y_2 \to Z_1 \oplus Z_2 \to TX_1 \oplus TX_2$ is a d.t.*
(ii) *Let $X, Y \in \mathcal{D}$. Then $X \to X \oplus Y \to Y \xrightarrow{0} TX$ is a d.t.*

10.2 Localization of Triangulated Categories

Let $\mathcal{D}$ be a triangulated category, $\mathcal{N}$ a full saturated subcategory. (Recall that $\mathcal{N}$ is saturated if $X \in \mathcal{D}$ belongs to $\mathcal{N}$ whenever X is isomorphic to an object of $\mathcal{N}$.)

Lemma 10.2.1. (a) *Let $\mathcal{N}$ be a full saturated triangulated subcategory of $\mathcal{D}$. Then $\mathrm{Ob}(\mathcal{N})$ satisfies conditions N1–N3 below.*
  N1 $0 \in \mathcal{N}$,
  N2 $X \in \mathcal{N}$ *if and only if* $TX \in \mathcal{N}$,
  N3 *if* $X \to Y \to Z \to TX$ *is a d.t. in $\mathcal{D}$ and $X, Z \in \mathcal{N}$ then $Y \in \mathcal{N}$.*

(b) *Conversely, let $\mathcal{N}$ be a full saturated subcategory of $\mathcal{D}$ and assume that $\mathrm{Ob}(\mathcal{N})$ satisfies conditions N1–N3 above. Then the restriction of T and the collection of* d.t.'s $X \to Y \to Z \to TX$ *in $\mathcal{D}$ with X, Y, Z in $\mathcal{N}$ make $\mathcal{N}$ a full saturated triangulated subcategory of $\mathcal{D}$. Moreover it satisfies*

  N'3 *if $X \to Y \to Z \to TX$ is a d.t. in $\mathcal{D}$ and two objects among X, Y, Z belong to $\mathcal{N}$, then so does the third one.*

Proof. (a) Assume that $\mathcal{N}$ is a full saturated triangulated subcategory of $\mathcal{D}$. Then N1 and N2 are clearly satisfied. Moreover N3 follows from Corollary 10.1.16 and the hypothesis that $\mathcal{N}$ is saturated.

(b) Let $\mathcal{N}$ be a full subcategory of $\mathcal{D}$ satisfying N1–N3. Then N'3 follows from N2 and N3.

(i) Let us prove that $\mathcal{N}$ is saturated. Let $f : X \xrightarrow{\sim} Y$ be an isomorphism with $X \in \mathcal{N}$. The triangle $X \xrightarrow{f} Y \to 0 \to TX$ being isomorphic to the d.t. $X \xrightarrow{\mathrm{id}_X} X \to 0 \to TX$, it is itself a d.t. Hence, $Y \in \mathcal{N}$.

(ii) Let $X, Y \in \mathcal{N}$. Since $X \to X \oplus Y \to Y \xrightarrow{0} TX$ is a d.t., we find that $X \oplus Y$ belongs to $\mathcal{N}$, and it follows that $\mathcal{N}$ is a full additive subcategory of $\mathcal{D}$.

(iii) The axioms of triangulated categories are then easily checked.   q.e.d.

Definition 10.2.2. *A* null system *in $\mathcal{D}$ is a full saturated subcategory $\mathcal{N}$ such that $\mathrm{Ob}(\mathcal{N})$ satisfies the conditions N1–N3 in Lemma 10.2.1 (a).*

We associate a family of morphisms to a null system as follows. Define:

(10.2.1)
$$\mathcal{N}Q := \{f : X \to Y; \text{ there exists a d.t. } X \to Y \to Z \to TX \text{ with } Z \in \mathcal{N}\}.$$

Theorem 10.2.3.   (i) *$\mathcal{N}Q$ is a right and left multiplicative system.*

  (ii) *Denote by $\mathcal{D}_{\mathcal{N}Q}$ the localization of $\mathcal{D}$ by $\mathcal{N}Q$ and by $Q : \mathcal{D} \to \mathcal{D}_{\mathcal{N}Q}$ the localization functor. Then $\mathcal{D}_{\mathcal{N}Q}$ is an additive category endowed with an automorphism (the image of T, still denoted by T).*

  (iii) *Define a d.t. in $\mathcal{D}_{\mathcal{N}Q}$ as being isomorphic to the image of a d.t. in $\mathcal{D}$ by Q. Then $\mathcal{D}_{\mathcal{N}Q}$ is a triangulated category and Q is a triangulated functor.*

  (iv) *If $X \in \mathcal{N}$, then $Q(X) \simeq 0$.*

  (v) *Let $F : \mathcal{D} \to \mathcal{D}'$ be a triangulated functor of triangulated categories such that $F(X) \simeq 0$ for any $X \in \mathcal{N}$. Then F factors uniquely through Q.*

One shall be aware that $\mathcal{D}_{\mathcal{N}Q}$ is a big category in general.

Notation 10.2.4. We will write $\mathcal{D}/\mathcal{N}$ instead of $\mathcal{D}_{\mathcal{N}Q}$.

Proof. (i) Since the opposite category of $\mathcal{D}$ is again triangulated and $\mathcal{N}^{\mathrm{op}}$ is a null system in $\mathcal{D}^{\mathrm{op}}$, it is enough to check that $\mathcal{N}Q$ is a right multiplicative system. Let us check the conditions S1–S4 in Definition 7.1.5.

S1: if $f : X \to Y$ is an isomorphism, the triangle $X \xrightarrow{f} Y \to 0 \to TX$ is a d.t. and we deduce $f \in \mathcal{N}Q$.

S2: Let $f\colon X \to Y$ and $g\colon Y \to Z$ be in $\mathcal{N}Q$. By TR3, there are d.t.'s $X \xrightarrow{f} Y \to Z' \to TX$, $Y \xrightarrow{g} Z \to X' \to TY$, and $X \xrightarrow{g \circ f} Z \to Y' \to TX$. By TR5, there exists a d.t. $Z' \to Y' \to X' \to TZ'$. Since Z' and X' belong to $\mathcal{N}$, so does Y'.

S3: Let $f\colon X \to Y$ and $s\colon X \to X'$ be two morphisms with $s \in \mathcal{N}Q$. By the hypothesis, there exists a d.t. $W \xrightarrow{h} X \xrightarrow{s} X' \to TW$ with $W \in \mathcal{N}$. By TR2, there exists a d.t. $W \xrightarrow{f \circ h} Y \xrightarrow{t} Z \to TW$, and by TR4, there exists a commutative diagram

$$
\begin{array}{ccccccc}
W & \xrightarrow{\ h\ } & X & \xrightarrow{\ s\ } & X' & \longrightarrow & TW \\
{\scriptstyle \mathrm{id}}\big\downarrow & & {\scriptstyle f}\big\downarrow & & \big\downarrow & & \big\downarrow \\
W & \xrightarrow[f \circ h]{} & Y & \xrightarrow[t]{} & Z & \longrightarrow & TW \ .
\end{array}
$$

Since $W \in \mathcal{N}$, we get $t \in \mathcal{N}Q$.

S4: Replacing f by $f - g$, it is enough to check that if there exists $s \in \mathcal{N}Q$ with $f \circ s = 0$, then there exists $t \in \mathcal{N}Q$ with $t \circ f = 0$. Consider the diagram

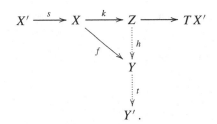

Here, the row is a d.t. with $Z \in \mathcal{N}$. Since $s \circ f = 0$, the arrow h, making the diagram commutative, exists by Proposition 10.1.13. There exists a d.t. $Z \to Y \xrightarrow{t} Y' \to TZ$ by TR2. We thus obtain $t \in \mathcal{N}Q$ since $Z \in \mathcal{N}$. Finally, $t \circ h = 0$ implies that $t \circ f = t \circ h \circ k = 0$.

(ii) follows from the result of Exercise 8.4.

(iii) Axioms TR0–TR3 are obviously satisfied. Let us prove TR4. With the notations of TR4, and using the result of Exercise 7.4, we may assume that there exists a commutative diagram in $\mathcal{D}$ of solid arrows, with s and t in $\mathcal{N}Q$

$$
\begin{array}{ccccccc}
X & \xrightarrow{\ f\ } & Y & \xrightarrow{\ g\ } & Z & \longrightarrow & TX \\
{\scriptstyle \alpha'}\big\downarrow & & {\scriptstyle \beta'}\big\downarrow & & {\scriptstyle \gamma'}\big\downarrow & & \big\downarrow{\scriptstyle T(\alpha')} \\
X_1 & \xrightarrow{\ f_1\ } & Y_1 & \dashrightarrow{\ g_1\ } & Z_1 & \dashrightarrow & TX_1 \\
{\scriptstyle s}\big\uparrow & & {\scriptstyle t}\big\uparrow & {\scriptstyle A} & \big\uparrow{\scriptstyle u} & {\scriptstyle B} & \big\uparrow{\scriptstyle T(s)} \\
X' & \xrightarrow[f']{} & Y' & \xrightarrow[g']{} & Z' & \longrightarrow & TX' \ .
\end{array}
$$

After having embedded $f_1 \colon X_1 \to Y_1$ in a d.t., we construct the commutative squares labeled by A and B with $u \in \mathcal{N}Q$ by using the result of Exercise 10.6. (In diagram (10.5.5) of this exercise, if Z^0 and Z^1 are in $\mathcal{N}$, then so is Z^2.) Then we construct the morphism γ' using TR4.

Let us prove TR5. Consider two morphisms in $\mathcal{D}/\mathcal{N}$: $f \colon X \to Y$ and $g \colon Y \to Z$. We may represent them by morphisms in $\mathcal{D}$: $\tilde{f} \colon \tilde{X} \to \tilde{Y}$ and $\tilde{g} \colon \tilde{Y} \to \tilde{Z}$. Then apply TR5 (in $\mathcal{D}$) and take the image in $\mathcal{D}/\mathcal{N}$ of the octahedron diagram (10.1.2).

(iv) Consider a d.t. $0 \to X \to X \to T(0)$. The morphism $0 \to X$ belongs to $\mathcal{N}Q$. Hence, $Q(0) \to Q(X)$ is an isomorphism.

(v) is obvious.   q.e.d.

Let $\mathcal{N}$ be a null system and let $X \in \mathcal{D}$. The categories $\mathcal{N}Q_X$ and $\mathcal{N}Q^X$ attached to the multiplicative system $\mathcal{N}Q$ (see Sect. 7.1) are given by:

$$(10.2.2) \quad \mathrm{Ob}(\mathcal{N}Q^X) = \{s \colon X \to X'; s \in \mathcal{N}Q\} \,,$$

$$(10.2.3) \quad \mathrm{Hom}_{\mathcal{N}Q^X}((s \colon X \to X'), (s' \colon X \to X'')) = \{h \colon X' \to X''; h \circ s = s'\}$$

and similarly for $\mathcal{N}Q_X$.

Remark 10.2.5. It follows easily from TR5 that the morphism h in (10.2.3) belongs to $\mathcal{N}Q$. Therefore, by considering $\mathcal{N}Q$ as a subcategory of $\mathcal{D}$, the category $\mathcal{N}Q^X$ is the category given by Definition 1.2.16 (with respect to the identity functor $\mathrm{id} \colon \mathcal{N}Q \to \mathcal{N}Q$). The same result holds for $\mathcal{N}Q_X$.

By Lemma 7.1.10 the categories $(\mathcal{N}Q_X)^{\mathrm{op}}$ and $\mathcal{N}Q^X$ are filtrant, and by the definition of the localization functor we get

$$
\begin{aligned}
\mathrm{Hom}_{\mathcal{D}/\mathcal{N}}(X, Y) &\simeq \varinjlim_{(Y \to \tilde{Y}') \in \mathcal{N}Q} \mathrm{Hom}_{\mathcal{D}}(X, Y') \\
&\simeq \varinjlim_{(X' \to X) \in \mathcal{N}Q} \mathrm{Hom}_{\mathcal{D}}(X', Y) \\
&\simeq \varinjlim_{(Y \to Y') \in \mathcal{N}Q, (X' \to X) \in \mathcal{N}Q} \mathrm{Hom}_{\mathcal{D}}(X', Y') \,.
\end{aligned}
$$

Now consider a full triangulated subcategory $\mathcal{I}$ of $\mathcal{D}$. We shall write $\mathcal{N} \cap \mathcal{I}$ for the full subcategory whose objects are $\mathrm{Ob}(\mathcal{N}) \cap \mathrm{Ob}(\mathcal{I})$. This is clearly a null system in $\mathcal{I}$.

Proposition 10.2.6. *Let $\mathcal{D}$ be a triangulated category, $\mathcal{N}$ a null system, $\mathcal{I}$ a full triangulated subcategory of $\mathcal{D}$. Assume condition (i) or (ii) below:*

(i) *any morphism $Y \to Z$ with $Y \in \mathcal{I}$ and $Z \in \mathcal{N}$ factorizes as $Y \to Z' \to Z$ with $Z' \in \mathcal{N} \cap \mathcal{I}$,*

(ii) *any morphism $Z \to Y$ with $Y \in \mathcal{I}$ and $Z \in \mathcal{N}$ factorizes as $Z \to Z' \to Y$ with $Z' \in \mathcal{N} \cap \mathcal{I}$.*

Then $\mathcal{I}/(\mathcal{N} \cap \mathcal{I}) \to \mathcal{D}/\mathcal{N}$ is fully faithful.

Proof. We may assume (ii), the case (i) being deduced by considering $\mathcal{D}^{\mathrm{op}}$. We shall apply Proposition 7.2.1. Let $f\colon X \to Y$ is a morphism in $\mathcal{N}Q$ with $X \in \mathcal{I}$. We shall show that there exists $g\colon Y \to W$ with $W \in \mathcal{I}$ and $g \circ f \in \mathcal{N}Q$. The morphism f is embedded in a d.t. $X \to Y \to Z \to TX$ with $Z \in \mathcal{N}$. By the hypothesis, the morphism $Z \to TX$ factorizes through an object $Z' \in \mathcal{N} \cap \mathcal{I}$. We may embed $Z' \to TX$ in a d.t. in $\mathcal{I}$ and obtain a commutative diagram of d.t.'s by TR4:

$$
\begin{array}{ccccccc}
X & \xrightarrow{f} & Y & \longrightarrow & Z & \longrightarrow & TX \\
\downarrow{\scriptstyle \mathrm{id}} & & \downarrow{\scriptstyle g} & & \downarrow & & \downarrow{\scriptstyle \mathrm{id}} \\
X & \xrightarrow{g \circ f} & W & \longrightarrow & Z' & \longrightarrow & TX\,.
\end{array}
$$

Since Z' belongs to $\mathcal{N}$, we get that $g \circ f \in \mathcal{N}Q \cap \mathrm{Mor}(\mathcal{I})$.   q.e.d.

Proposition 10.2.7. *Let $\mathcal{D}$ be a triangulated category, $\mathcal{N}$ a null system, $\mathcal{I}$ a full triangulated subcategory of $\mathcal{D}$, and assume conditions* (i) *or* (ii) *below:*

(i) *for any $X \in \mathcal{D}$, there exists a morphism $X \to Y$ in $\mathcal{N}Q$ with $Y \in \mathcal{I}$,*
(ii) *for any $X \in \mathcal{D}$, there exists a morphism $Y \to X$ in $\mathcal{N}Q$ with $Y \in \mathcal{I}$.*

Then $\mathcal{I}/(\mathcal{N} \cap \mathcal{I}) \to \mathcal{D}/\mathcal{N}$ is an equivalence of categories.

Proof. Apply Corollary 7.2.2.   q.e.d.

Proposition 10.2.8. *Let $\mathcal{D}$ be a triangulated category admitting direct sums indexed by a set I and let $\mathcal{N}$ be a null system closed by such direct sums. Let $Q\colon \mathcal{D} \to \mathcal{D}/\mathcal{N}$ denote the localization functor. Then $\mathcal{D}/\mathcal{N}$ admits direct sums indexed by I and the localization functor $Q\colon \mathcal{D} \to \mathcal{D}/\mathcal{N}$ commutes with such direct sums.*

Proof. Let $\{X_i\}_{i \in I}$ be a family of objects in $\mathcal{D}$. It is enough to show that $Q(\oplus_i X_i)$ is the direct sum of the family $Q(X_i)$, i.e., the map

$$
\mathrm{Hom}_{\mathcal{D}/\mathcal{N}}(Q(\bigoplus_{i \in I} X_i), Y) \to \prod_{i \in I} \mathrm{Hom}_{\mathcal{D}/\mathcal{N}}(Q(X_i), Y)
$$

is bijective for any $Y \in \mathcal{D}$.
(i) Surjectivity. Let $u_i \in \mathrm{Hom}_{\mathcal{D}/\mathcal{N}}(Q(X_i), Y)$. The morphism u_i is represented by a morphism $u_i'\colon X_i' \to Y$ in $\mathcal{D}$ together with a d.t. $X_i' \xrightarrow{v_i} X_i \xrightarrow{w_i} Z_i \to TX_i'$ in $\mathcal{D}$ with $Z_i \in \mathcal{N}$. We get a morphism $\oplus_i X_i' \to Y$ and a d.t. $\oplus_i X_i' \to \oplus_i X_i \to \oplus_i Z_i \to T(\oplus_i X_i')$ in $\mathcal{D}$ with $\oplus_i Z_i \in \mathcal{N}$.

(ii) Injectivity. Assume that the composition $Q(X_i) \to Q(\oplus_{i'} X_{i'}) \xrightarrow{u} Q(Y)$ is zero for every $i \in I$. By the definition, the morphism u is represented by morphisms $u'\colon \oplus_i X_i \xrightarrow{u'} Y' \xleftarrow{s} Y$ with $s \in \mathcal{N}Q$. Using the result of Exercise 10.11, we can find $Z_i \in \mathcal{N}$ such that $v_i'\colon X_i \to Y'$ factorizes as $X_i \to Z_i \to Y'$. Then $\oplus_i X_i \to Y'$ factorizes as $\oplus_i X_i \to \oplus_i Z_i \to Y'$. Since $\oplus_i Z_i \in \mathcal{N}$, $Q(u) = 0$.   q.e.d.

10.3 Localization of Triangulated Functors

Let $F\colon \mathcal{D} \to \mathcal{D}'$ be a functor of triangulated categories, $\mathcal{N}$ and $\mathcal{N}'$ null systems in $\mathcal{D}$ and $\mathcal{D}'$, respectively. The right or left localization of F (when it exists) is defined by mimicking Definition 7.3.1, replacing "functor" by "triangulated functor".

In the sequel, $\mathcal{D}$ (resp. $\mathcal{D}'$, $\mathcal{D}''$) is a triangulated category and $\mathcal{N}$ (resp. $\mathcal{N}'$, $\mathcal{N}''$) is a null system in this category. We denote by $Q\colon \mathcal{D} \to \mathcal{D}/\mathcal{N}$ (resp. $Q'\colon \mathcal{D}' \to \mathcal{D}'/\mathcal{N}'$, $Q''\colon \mathcal{D}'' \to \mathcal{D}''/\mathcal{N}''$) the localization functor and by $\mathcal{N}'Q$ (resp. $\mathcal{N}''Q$) the family of morphisms in $\mathcal{D}'$ (resp. $\mathcal{D}''$) defined in (10.2.1).

Definition 10.3.1. *We say that a triangulated functor $F\colon \mathcal{D} \to \mathcal{D}'$ is right (resp. left) localizable with respect to $(\mathcal{N}, \mathcal{N}')$ if $Q' \circ F\colon \mathcal{D} \to \mathcal{D}'/\mathcal{N}'$ is universally right (resp. left) localizable with respect to the multiplicative system $\mathcal{N}Q$ (see Definition 7.3.1). Recall that it means that, for any $X \in \mathcal{D}$,*
$$\underset{(X\to Y)\in\mathcal{N}Q^X}{\text{"}\varinjlim\text{"}} Q'F(Y) \quad (resp. \quad \underset{(Y\to X)\in\mathcal{N}Q_X}{\text{"}\varprojlim\text{"}} Q'F(Y)) \ is \ representable \ in \ \mathcal{D}'/\mathcal{N}'. \ If$$
there is no risk of confusion, we simply say that F is right (resp. left) localizable or that RF exists.

Definition 10.3.2. *Let $F\colon \mathcal{D} \to \mathcal{D}'$ be a triangulated functor of triangulated categories, $\mathcal{N}$ and $\mathcal{N}'$ null systems in $\mathcal{D}$ and $\mathcal{D}'$, and $\mathcal{I}$ a full triangulated subcategory of $\mathcal{D}$. Consider the conditions (i), (ii), (iii) below.*

(i) *For any $X \in \mathcal{D}$, there exists a morphism $X \to Y$ in $\mathcal{N}Q$ with $Y \in \mathcal{I}$.*
(ii) *For any $X \in \mathcal{D}$, there exists a morphism $Y \to X$ in $\mathcal{N}Q$ with $Y \in \mathcal{I}$.*
(iii) *For any $Y \in \mathcal{N} \cap \mathcal{I}$, $F(Y) \in \mathcal{N}'$.*

Then

(a) *if conditions (i) and (iii) are satisfied, we say that the subcategory $\mathcal{I}$ is F-injective with respect to $\mathcal{N}$ and $\mathcal{N}'$,*
(b) *if conditions (ii) and (iii) are satisfied, we say that the subcategory $\mathcal{I}$ is F-projective with respect to $\mathcal{N}$ and $\mathcal{N}'$.*

If there is no risk of confusion, we omit "with respect to $\mathcal{N}$ and $\mathcal{N}'$".

Note that if $F(\mathcal{N}) \subset \mathcal{N}'$, then $\mathcal{D}$ is both F-injective and F-projective.

Proposition 10.3.3. *Let $F\colon \mathcal{D} \to \mathcal{D}'$ be a triangulated functor of triangulated categories, $\mathcal{N}$ and $\mathcal{N}'$ null systems in $\mathcal{D}$ and $\mathcal{D}'$, and $\mathcal{I}$ a full triangulated category of $\mathcal{D}$.*

(a) *If $\mathcal{I}$ is F-injective with respect to $\mathcal{N}$ and $\mathcal{N}'$, then F is right localizable and its right localization is a triangulated functor.*
(b) *If $\mathcal{I}$ is F-projective with respect to $\mathcal{N}$ and $\mathcal{N}'$, then F left localizable and its left localization is a triangulated functor.*

Proof. Apply Proposition 7.3.2.   q.e.d.

Notation 10.3.4. (i) We denote by $R_{\mathcal{N}}^{\mathcal{N}'} F \colon \mathcal{D}/\mathcal{N} \to \mathcal{D}'/\mathcal{N}'$ the right localization of F with respect to $(\mathcal{N}, \mathcal{N}')$. If there is no risk of confusion, we simply write RF instead of $R_{\mathcal{N}}^{\mathcal{N}'} F$.
(ii) We denote by $L_{\mathcal{N}}^{\mathcal{N}'} F \colon \mathcal{D}/\mathcal{N} \to \mathcal{D}'/\mathcal{N}'$ the left localization of F with respect to $(\mathcal{N}, \mathcal{N}')$. If there is no risk of confusion, we simply write LF instead of $L_{\mathcal{N}}^{\mathcal{N}'} F$.

If $\mathcal{I}$ is F-injective, $R_{\mathcal{N}}^{\mathcal{N}'} F$ may be defined by the diagram:

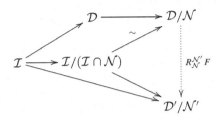

and

(10.3.1) $R_{\mathcal{N}}^{\mathcal{N}'} F(X) \simeq F(Y)$ for $(X \to Y) \in \mathcal{N}Q$ with $Y \in \mathcal{I}$.

Similarly, if $\mathcal{I}$ is F-projective, the diagram above defines $L_{\mathcal{N}}^{\mathcal{N}'} F$ and

(10.3.2) $L_{\mathcal{N}}^{\mathcal{N}'} F(X) \simeq F(Y)$ for $(Y \to X) \in \mathcal{N}Q$ with $Y \in \mathcal{I}$.

Proposition 10.3.5. *Let $F \colon \mathcal{D} \to \mathcal{D}'$ and $F' \colon \mathcal{D}' \to \mathcal{D}''$ be triangulated functors of triangulated categories and let $\mathcal{N}$, $\mathcal{N}'$ and $\mathcal{N}''$ be null systems in $\mathcal{D}$, $\mathcal{D}'$ and $\mathcal{D}''$, respectively.*

(i) *Assume that $R_{\mathcal{N}}^{\mathcal{N}'} F$, $R_{\mathcal{N}'}^{\mathcal{N}''} F'$ and $R_{\mathcal{N}}^{\mathcal{N}''} (F' \circ F)$ exist. Then there is a canonical morphism in $\mathrm{Fct}(\mathcal{D}/\mathcal{N}, \mathcal{D}''/\mathcal{N}'')$:*

(10.3.3) $R_{\mathcal{N}}^{\mathcal{N}''} (F' \circ F) \to R_{\mathcal{N}'}^{\mathcal{N}''} F' \circ R_{\mathcal{N}}^{\mathcal{N}'} F$.

(ii) *Let $\mathcal{I}$ and $\mathcal{I}'$ be full triangulated subcategories of $\mathcal{D}$ and $\mathcal{D}'$, respectively. Assume that $\mathcal{I}$ is F-injective with respect to $\mathcal{N}$ and $\mathcal{N}'$, $\mathcal{I}'$ is F'-injective with respect to $\mathcal{N}'$ and $\mathcal{N}''$, and $F(\mathcal{I}) \subset \mathcal{I}'$. Then $\mathcal{I}$ is $(F' \circ F)$-injective with respect to $\mathcal{N}$ and $\mathcal{N}''$, and (10.3.3) is an isomorphism.*

Proof. (i) By Definition 7.3.1, there are a bijection

$$\mathrm{Hom}\,(R_{\mathcal{N}}^{\mathcal{N}''} (F' \circ F), R_{\mathcal{N}'}^{\mathcal{N}''} F' \circ R_{\mathcal{N}}^{\mathcal{N}'} F)$$
$$\simeq \mathrm{Hom}\,(Q'' \circ F' \circ F, R_{\mathcal{N}'}^{\mathcal{N}''} F' \circ R_{\mathcal{N}}^{\mathcal{N}'} F \circ Q) ,$$

and natural morphisms of functors

$$Q'' \circ F' \to R_{\mathcal{N}'}^{\mathcal{N}''} F' \circ Q', \quad Q' \circ F \to R_{\mathcal{N}}^{\mathcal{N}'} F \circ Q .$$

We deduce the canonical morphisms

$$Q'' \circ F' \circ F \to R_{\mathcal{N}'}^{\mathcal{N}''} F' \circ Q' \circ F \to R_{\mathcal{N}'}^{\mathcal{N}''} F' \circ R_{\mathcal{N}}^{\mathcal{N}'} F \circ Q$$

and the result follows.

(ii) The fact that $\mathcal{I}$ is $(F' \circ F)$-injective follows immediately from the definition. Let $X \in \mathcal{D}$ and consider a morphism $X \to Y$ in $\mathcal{N}Q$ with $Y \in \mathcal{I}$. Then $R_{\mathcal{N}}^{\mathcal{N}'} F(X) \simeq F(Y)$ by (10.3.1) and $F(Y) \in \mathcal{I}'$ by the hypothesis. Hence $(R_{\mathcal{N}'}^{\mathcal{N}''} F')(F(Y)) \simeq F'F(Y)$ by (10.3.1) and we find

$$(R_{\mathcal{N}'}^{\mathcal{N}''} F')(R_{\mathcal{N}}^{\mathcal{N}'} F(X)) \simeq F'F(Y) .$$

On the other hand, $R_{\mathcal{N}}^{\mathcal{N}''}(F' \circ F)(X) \simeq F'F(Y)$ by (10.3.1) since $\mathcal{I}$ is $(F' \circ F)$-injective.   q.e.d.

Triangulated Bifunctors

Definition 10.3.6. *Let* $(\mathcal{D}, T)$, $(\mathcal{D}', T')$ *and* $(\mathcal{D}'', T'')$ *be triangulated categories. A triangulated bifunctor* $F \colon \mathcal{D} \times \mathcal{D}' \to \mathcal{D}''$ *is a bifunctor of additive categories with translation (see* Definition 10.1.1 (v)) *which sends d.t.'s in each argument to d.t.'s.*

Definition 10.3.7. *Let* $\mathcal{D}$, $\mathcal{D}'$ *and* $\mathcal{D}''$ *be triangulated categories,* $\mathcal{N}$, $\mathcal{N}'$ *and* $\mathcal{N}''$ *null systems in* $\mathcal{D}$, $\mathcal{D}'$ *and* $\mathcal{D}''$, *respectively. We say that a triangulated bifunctor* $F \colon \mathcal{D} \times \mathcal{D}' \to \mathcal{D}''$ *is right (resp. left) localizable with respect to* $(\mathcal{N} \times \mathcal{N}', \mathcal{N}'')$ *if* $Q'' \circ F \colon \mathcal{D} \times \mathcal{D}' \to \mathcal{D}''/\mathcal{N}''$ *is universally right (resp. left) localizable with respect to the multiplicative system* $\mathcal{N}Q \times \mathcal{N}'Q$ *(see* Remark 7.4.5). *If there is no risk of confusion, we simply say that* F *is right (resp. left) localizable.*

Notation 10.3.8. We denote by $R_{\mathcal{N} \times \mathcal{N}'}^{\mathcal{N}''} F \colon \mathcal{D}/\mathcal{N} \times \mathcal{D}'/\mathcal{N}' \to \mathcal{D}''/\mathcal{N}''$ the right localization of F with respect to $(\mathcal{N} \times \mathcal{N}', \mathcal{N}'')$, if it exists. If there is no risk of confusion, we simply write RF. We use similar notations for the left localization.

Definition 10.3.9. *Let* $\mathcal{D}$, $\mathcal{D}'$ *and* $\mathcal{D}''$ *be triangulated categories,* $\mathcal{N}$, $\mathcal{N}'$ *and* $\mathcal{N}''$ *null systems in* $\mathcal{D}$, $\mathcal{D}'$ *and* $\mathcal{D}''$, *respectively, and* $\mathcal{I}, \mathcal{I}'$ *full triangulated subcategories of* $\mathcal{D}$ *and* $\mathcal{D}'$, *respectively. Let* $F \colon \mathcal{D} \times \mathcal{D}' \to \mathcal{D}''$ *be a triangulated bifunctor. The pair* $(\mathcal{I}, \mathcal{I}')$ *is* F*-injective with respect to* $(\mathcal{N}, \mathcal{N}', \mathcal{N}'')$ *if*

(i) $\mathcal{I}'$ *is* $F(Y, \cdot)$*-injective with respect to* $\mathcal{N}'$ *and* $\mathcal{N}''$ *for any* $Y \in \mathcal{I}$,
(ii) $\mathcal{I}$ *is* $F(\cdot, Y')$*-injective with respect to* $\mathcal{N}$ *and* $\mathcal{N}''$ *for any* $Y' \in \mathcal{I}'$.

These two conditions are equivalent to saying that

(a) for any $X \in \mathcal{D}$, there exists a morphism $X \to Y$ in $\mathcal{N}Q$ with $Y \in \mathcal{I}$,
(b) for any $X' \in \mathcal{D}'$, there exists a morphism $X' \to Y'$ in $\mathcal{N}'Q$ with $Y' \in \mathcal{I}'$,

(c) $F(X, X')$ belongs to $\mathcal{N}''$ for $X \in \mathcal{I}$, $X' \in \mathcal{I}'$ as soon as X belongs to $\mathcal{N}$ or X' belongs to $\mathcal{N}'$.

The property for $(\mathcal{I}, \mathcal{I}')$ of being F-*projective* is defined similarly.

Proposition 10.3.10. *Let* $\mathcal{D}, \mathcal{N}, \mathcal{I}, \mathcal{D}', \mathcal{N}', \mathcal{I}', \mathcal{D}'', \mathcal{N}''$ *and* F *be as in* Definition 10.3.9. *Assume that* $(\mathcal{I}, \mathcal{I}')$ *is* F-*injective with respect to* $(\mathcal{N}, \mathcal{N}')$. *Then* F *is right localizable, its right localization* $R_{\mathcal{N}\mathcal{N}'}^{\mathcal{N}''} F$ *is a triangulated bifunctor*

$$R_{\mathcal{N}\mathcal{N}'}^{\mathcal{N}''} F : \mathcal{D}/\mathcal{N} \times \mathcal{D}'/\mathcal{N}' \to \mathcal{D}''/\mathcal{N}'' \, ,$$

and moreover

(10.3.4) $R_{\mathcal{N}\mathcal{N}'}^{\mathcal{N}''} F(X, X') \simeq F(Y, Y')$ *for* $(X \to Y) \in \mathcal{N}Q$ *and*
$$(X' \to Y') \in \mathcal{N}'Q \text{ with } Y \in \mathcal{I}, \ Y' \in \mathcal{I}'.$$

Of course, there exists a similar result by replacing "injective" with "projective" and reversing the arrows in (10.3.4).

Corollary 10.3.11. *Let* $\mathcal{D}$, $\mathcal{N}$, $\mathcal{I}$, $\mathcal{D}'$, $\mathcal{N}'$, *and* $\mathcal{D}''$, $\mathcal{N}''$ *be as in* Proposition 10.3.10. *Let* $F : \mathcal{D} \times \mathcal{D}' \to \mathcal{D}''$ *be a triangulated bifunctor. Assume that*

(i) $F(\mathcal{I}, \mathcal{N}') \subset \mathcal{N}''$,
(ii) *for any* $X' \in \mathcal{D}'$, $\mathcal{I}$ *is* $F(\cdot, X')$-*injective with respect to* $\mathcal{N}$.

Then F *is right localizable. Moreover,*

$$R_{\mathcal{N}\mathcal{N}'}^{\mathcal{N}''} F(X, X') \simeq R_{\mathcal{N}}^{\mathcal{N}''} F(\cdot, X')(X) \, .$$

Here again, there is a similar statement by replacing "injective" with "projective".

10.4 Extension of Cohomological Functors

In this section, we consider two triangulated categories $\mathcal{T}$ and $\mathcal{D}$, a triangulated functor $\varphi : \mathcal{T} \to \mathcal{D}$, an abelian category $\mathcal{A}$, and a cohomological functor $F : \mathcal{T} \to \mathcal{A}$. For $X \in \mathcal{D}$, we denote as usual by $\mathcal{T}_X$ the category whose objects are the pairs (Y, u) of objects $Y \in \mathcal{T}$ and morphisms $u : \varphi(Y) \to X$.
We make the hypotheses:

(10.4.1) $\begin{cases} \mathcal{A} \text{ admits small filtrant inductive limits and such limits are exact} , \\ \mathcal{T}_X \text{ is cofinally small for any } X \in \mathcal{D} . \end{cases}$

Note that the functor $\varphi : \mathcal{T} \to \mathcal{D}$ is exact by Proposition 10.1.18. Hence, Theorem 3.3.18 asserts that the functor $\varphi_* : \mathrm{Fct}(\mathcal{D}, \mathcal{A}) \to \mathrm{Fct}(\mathcal{T}, \mathcal{A})$ admits a left adjoint $\varphi^\dagger$ such that for $F : \mathcal{D} \to \mathcal{A}$ we have

(10.4.2) $$\varphi^\dagger F(X) = \varinjlim_{(\varphi(Y) \to X) \in \mathcal{T}_X} F(Y) \, ,$$

and there is a natural morphism of functors

$$(10.4.3) \qquad\qquad F \to (\varphi^\dagger F) \circ \varphi \, .$$

Theorem 10.4.1. *Let* $\varphi: \mathcal{T} \to \mathcal{D}$ *be a triangulated functor of triangulated categories, let* $\mathcal{A}$ *be an abelian category, and assume* (10.4.1). *Let* $F: \mathcal{T} \to \mathcal{A}$ *be a cohomological functor. Then the functor* $\varphi^\dagger F$ *is additive and cohomological.*

Proof. (i) Let us first show that $\varphi^\dagger F$ is additive. By Proposition 8.2.15, it is enough to show that $\varphi^\dagger F(X_1 \oplus X_2) \to \varphi^\dagger F(X_1) \oplus \varphi^\dagger F(X_2)$ is an isomorphism for any $X_1, X_2 \in \mathcal{D}$. Let $\xi: \mathcal{T}_{X_1} \times \mathcal{T}_{X_2} \to \mathcal{T}_{X_1 \oplus X_2}$ be the functor given by $((\varphi(Y_1) \to X_1), (\varphi(Y_2) \to X_2)) \mapsto (\varphi(Y_1 \oplus Y_2) \to X_1 \oplus X_2)$. Then ξ has a left adjoint $\eta: \mathcal{T}_{X_1 \oplus X_2} \to \mathcal{T}_{X_1} \times \mathcal{T}_{X_2}$ given by $(\varphi(Y) \to X_1 \oplus X_2) \mapsto ((\varphi(Y) \to X_1 \oplus X_2 \to X_1), (\varphi(Y) \to X_1 \oplus X_2 \to X_2))$. Hence ξ is a cofinal functor by Lemma 3.3.10. Moreover, the canonical functor $\mathcal{T}_{X_1} \times \mathcal{T}_{X_2} \to \mathcal{T}_{X_i}$ $(i = 1, 2)$ is cofinal. Hence we obtain

$$\begin{aligned}
\varphi^\dagger F(X_1 \oplus X_2) &\simeq \varinjlim_{Y \in \mathcal{T}_{X_1 \oplus X_2}} F(Y) \\
&\simeq \varinjlim_{(Y_1, Y_2) \in \mathcal{T}_{X_1} \oplus \mathcal{T}_{X_2}} F(Y_1 \oplus Y_2) \\
&\simeq \varinjlim_{(Y_1, Y_2) \in \mathcal{T}_{X_1} \oplus \mathcal{T}_{X_2}} F(Y_1) \oplus F(Y_2) \\
&\simeq \Big(\varinjlim_{Y_1 \in \mathcal{T}_{X_1}} F(Y_1) \Big) \oplus \Big(\varinjlim_{Y_2 \in \mathcal{T}_{X_2}} F(Y_2) \Big) \\
&\simeq \varphi^\dagger F(X_1) \oplus \varphi^\dagger F(X_2) \, .
\end{aligned}$$

(ii) Let us show that $\varphi^\dagger F$ is cohomological. We shall denote by X, Y, Z objects of $\mathcal{D}$ and by X_0, Y_0, Z_0 objects of $\mathcal{T}$.

By Proposition 10.1.18, the functor φ is exact. This result together with Corollary 3.4.6 implies that:

(a) for $X \in \mathcal{D}$ the category $\mathcal{T}_X$ is filtrant and cofinally small,
(b) for a morphism $g: Y \to Z$ in $\mathcal{D}$, the category $\mathrm{Mor}(\mathcal{T})_g$ is filtrant, cofinally small, and the two natural functors from $\mathrm{Mor}(\mathcal{T})_g$ to $\mathcal{T}_Y$ and $\mathcal{T}_Z$ are cofinal.

By (b), for a morphism $g: Y \to Z$ in $\mathcal{D}$, we get

$$\varphi^\dagger F(Y) \simeq \varinjlim_{(Y_0 \to Z_0) \in \mathrm{Mor}(\mathcal{T})_g} F(Y_0), \quad \varphi^\dagger F(Z) \simeq \varinjlim_{(Y_0 \to Z_0) \in \mathrm{Mor}(\mathcal{T})_g} F(Z_0) \, .$$

Moreover, since small filtrant inductive limits are exact in $\mathcal{A}$,

$$(10.4.4) \quad \mathrm{Ker}\,\varphi^\dagger F(g) \simeq \mathrm{Ker}\Big(\varinjlim_{g_0 \in \mathrm{Mor}(\mathcal{T})_g} F(g_0) \Big) \simeq \varinjlim_{g_0 \in \mathrm{Mor}(\mathcal{T})_g} (\mathrm{Ker}\,F(g_0)) .$$

Now consider a d.t. $X \xrightarrow{f} Y \xrightarrow{g} Z \to TX$ in $\mathcal{D}$. Let $(Y_0 \xrightarrow{g_0} Z_0) \in \mathrm{Mor}(\mathcal{T})_g$. Embed g_0 in a d.t. $X_0 \xrightarrow{f_0} Y_0 \xrightarrow{g_0} Z_0 \to TX_0$. In the diagram below, we may complete the dotted arrows in order to get a morphism of d.t.'s:

$$\varphi(X_0) \xrightarrow{\varphi(f_0)} \varphi(Y_0) \xrightarrow{\varphi(g_0)} \varphi(Z_0) \longrightarrow T(\varphi(X_0))$$

$$X \xrightarrow{\ f\ } Y \xrightarrow{\ g\ } Z \longrightarrow TX \ .$$

Applying the functor $\varphi^\dagger F$, and using the morphism of functors $F \to \varphi^\dagger F \circ \varphi$ (see (10.4.3)), we get a commutative diagram in $\mathcal{A}$ in which the row in the top is exact

$$F(X_0) \xrightarrow{F(f_0)} F(Y_0) \xrightarrow{F(g_0)} F(Z_0)$$

$$\varphi^\dagger F(X) \xrightarrow{\varphi^\dagger F(f)} \varphi^\dagger F(Y) \xrightarrow{\varphi^\dagger F(g)} \varphi^\dagger F(Z) \ .$$

We have a morphism $\mathrm{Coker}(F(f_0)) \to \mathrm{Coker}(\varphi^\dagger F(f))$. Since $F(X_0) \to F(Y_0) \to F(Z_0)$ is exact, the morphism $\mathrm{Ker}(F(g_0)) \to \mathrm{Coker}(F(f_0))$ vanishes and hence $\mathrm{Ker}(F(g_0)) \to \mathrm{Coker}(\varphi^\dagger F(f))$ vanishes. By (10.4.4), the morphism $\mathrm{Ker}(\varphi^\dagger F(g)) \to \mathrm{Coker}(\varphi^\dagger F(f))$ vanishes, which means that the sequence $\varphi^\dagger F(X) \xrightarrow{\varphi^\dagger F(f)} \varphi^\dagger F(Y) \xrightarrow{\varphi^\dagger F(g)} \varphi^\dagger F(Z)$ is exact.   q.e.d.

10.5 The Brown Representability Theorem

In this section we shall give a sufficient condition for the representability of contravariant cohomological functors on triangulated categories admitting small direct sums. Recall (Proposition 10.1.19) that in such categories, a small direct sum of d.t.'s is a d.t.

Definition 10.5.1. *Let $\mathcal{D}$ be a triangulated category admitting small direct sums. A* system of t-generators $\mathcal{F}$ *in $\mathcal{D}$ is a small family of objects of $\mathcal{D}$ satisfying conditions* (i) *and* (ii) *below.*

   (i) *$\mathcal{F}$ is a system of generators (see Definition 5.2.1), or equivalently, $\mathcal{F}$ is a small family of objects of $\mathcal{D}$ such that for any $X \in \mathcal{D}$ with $\mathrm{Hom}_{\mathcal{D}}(C, X) \simeq 0$ for all $C \in \mathcal{F}$, we have $X \simeq 0$.*
   (ii) *For any* countable *set I and any family $\{u_i \colon X_i \to Y_i\}_{i \in I}$ of morphisms in $\mathcal{D}$, the map $\mathrm{Hom}_{\mathcal{D}}(C, \oplus_i X_i) \xrightarrow{\oplus_i u_i} \mathrm{Hom}_{\mathcal{D}}(C, \oplus_i Y_i)$ vanishes for every $C \in \mathcal{F}$ as soon as $\mathrm{Hom}_{\mathcal{D}}(C, X_i) \xrightarrow{u_i} \mathrm{Hom}_{\mathcal{D}}(C, Y_i)$ vanishes for every $i \in I$ and every $C \in \mathcal{F}$.*

Note that the equivalence in (i) follows from the fact that, for a d.t. $X \xrightarrow{f} Y \to Z \to TX$, f is an isomorphism if and only if $Z \simeq 0$ (see Exercise 10.1).

Theorem 10.5.2. [The Brown representability Theorem] *Let $\mathcal{D}$ be a triangulated category admitting small direct sums and a system of t-generators $\mathcal{F}$.*

(i) *Let $H: \mathcal{D}^{\mathrm{op}} \to \mathrm{Mod}(\mathbb{Z})$ be a cohomological functor which commutes with small products (i.e., for any small family $\{X_i\}_{i \in I}$ in $\mathrm{Ob}(\mathcal{D})$, we have $H(\oplus_i X_i) \xrightarrow{\sim} \prod_i H(X_i)$). Then H is representable.*

(ii) *Let $\mathcal{K}$ be a full triangulated subcategory of $\mathcal{D}$ such that $\mathcal{F} \subset \mathrm{Ob}(\mathcal{K})$ and $\mathcal{K}$ is closed by small direct sums. Then the natural functor $\mathcal{K} \to \mathcal{D}$ is an equivalence.*

Similarly to the other representability theorems (see e.g. §5.2), this theorem implies the following corollary.

Corollary 10.5.3. *Let $\mathcal{D}$ be a triangulated category admitting small direct sums and a system of t-generators.*

(i) *$\mathcal{D}$ admits small products.*

(ii) *Let $F: \mathcal{D} \to \mathcal{D}'$ be a triangulated functor of triangulated categories. Assume that F commutes with small direct sums. Then F admits a right adjoint G, and G is triangulated.*

Proof. (i) For a small family $\{X_i\}_{i \in I}$ of objects in $\mathcal{D}$, the functor

$$Z \mapsto \prod_i \mathrm{Hom}_{\mathcal{D}}(Z, X_i)$$

is cohomological and commutes with small products. Hence it is representable.
(ii) For each $Y \in \mathcal{D}'$, the functor $X \mapsto \mathrm{Hom}_{\mathcal{D}'}(F(X), Y)$ is representable by Theorem 10.5.2. Hence F admits a right adjoint. Finally G is triangulated by the result of Exercise 10.3. q.e.d.

Remark 10.5.4. Condition (ii) in Definition 10.5.1 can be reformulated in many ways. Each of the following conditions is equivalent to (ii):

(ii)′ for any countable set I and any family $\{u_i: X_i \to Y_i\}_{i \in I}$ of morphisms in $\mathcal{D}$, the map $\mathrm{Hom}_{\mathcal{D}}(C, \oplus_i X_i) \xrightarrow{\oplus_i u_i} \mathrm{Hom}_{\mathcal{D}}(C, \oplus_i Y_i)$ is surjective for every $C \in \mathcal{F}$ as soon as $\mathrm{Hom}_{\mathcal{D}}(C, X_i) \xrightarrow{u_i} \mathrm{Hom}_{\mathcal{D}}(C, Y_i)$ is surjective for every $i \in I$ and every $C \in \mathcal{F}$.

(ii)″ for any countable set I and any family $\{u_i: X_i \to Y_i\}_{i \in I}$ of morphisms in $\mathcal{D}$, the map $\mathrm{Hom}_{\mathcal{D}}(C, \oplus_i X_i) \xrightarrow{\oplus_i u_i} \mathrm{Hom}_{\mathcal{D}}(C, \oplus_i Y_i)$ is injective for every $C \in \mathcal{F}$ as soon as $\mathrm{Hom}_{\mathcal{D}}(C, X_i) \xrightarrow{u_i} \mathrm{Hom}_{\mathcal{D}}(C, Y_i)$ is injective for every $i \in I$ and every $C \in \mathcal{F}$.

Indeed if we take a d.t. $X \to Y \to Z \to TX$, then we have an equivalence

$$\mathrm{Hom}_{\mathcal{D}}(C, X) \to \mathrm{Hom}_{\mathcal{D}}(C, Y) \text{ vanishes}$$
$$\Longleftrightarrow \mathrm{Hom}_{\mathcal{D}}(C, Y) \to \mathrm{Hom}_{\mathcal{D}}(C, Z) \text{ is injective}$$
$$\Longleftrightarrow \mathrm{Hom}_{\mathcal{D}}(C, T^{-1}Z) \to \mathrm{Hom}_{\mathcal{D}}(C, X) \text{ is surjective}.$$

Condition (ii) is also equivalent to the following condition:

(iii) for any countable set I, any family $\{X_i\}_{i \in I}$ in $\mathcal{D}$, any $C \in \mathcal{F}$ and any morphism $f : C \to \oplus_{i \in I} X_i$, there exists a family of morphisms $u_i : C_i \to X_i$ such that f decomposes into $C \to \oplus_i C_i \xrightarrow{\oplus u_i} \oplus_i X_i$ and each C_i is a small direct sum of objects in $\mathcal{F}$.

Indeed, let $\mathcal{S}$ be the full subcategory of $\mathcal{D}$ consisting of small direct sums of objects in $\mathcal{F}$. If a morphism $X \to Y$ in $\mathcal{D}$ satisfies the condition that $\operatorname{Hom}_{\mathcal{D}}(C, X) \to \operatorname{Hom}_{\mathcal{D}}(C, Y)$ vanishes for every $C \in \mathcal{F}$, then the same condition holds for every $C \in \mathcal{S}$. Hence it is easy to see that (iii) implies (ii). Conversely assume that (ii)$'$ is true. For a countable family of objects X_i in $\mathcal{D}$ set $C_i = \bigoplus_{C \in \mathcal{F}} C^{\oplus X_i(C)}$. Then $C_i \in \mathcal{S}$, and the canonical morphism $C_i \to X_i$ satisfies the condition that any morphism $C \to X_i$ with $C \in \mathcal{F}$ factors through $C_i \to X_i$. Hence (ii)$'$ implies that $\operatorname{Hom}_{\mathcal{D}}(C, \oplus_i C_i) \to \operatorname{Hom}_{\mathcal{D}}(C, \oplus_i X_i)$ is surjective. Hence any morphism $C \to \oplus_i X_i$ factors through $\oplus_i C_i \to \oplus_i X_i$.

Note that condition (iii) is a consequence of the following condition (iii)$'$, which is sufficient in most applications.

(iii)$'$ for any countable set I, any family $\{X_i\}_{i \in I}$ in $\mathcal{D}$, any $C \in \mathcal{F}$ and any morphism $f : C \to \oplus_{i \in I} X_i$, there exists a family of morphisms $u_i : C_i \to X_i$ with $C_i \in \mathcal{F}$ such that f decomposes into $C \to \oplus_i C_i \xrightarrow{\oplus u_i} \oplus_i X_i$.

Summing up, for a small family $\mathcal{F}$ of objects of $\mathcal{D}$, we have

$$\text{(ii)} \Leftrightarrow \text{(ii)}' \Leftrightarrow \text{(ii)}'' \Leftrightarrow \text{(iii)} \Leftarrow \text{(iii)}' \, .$$

The Brown representability theorem was proved by Neeman [53] under condition (iii)$'$, and later by Krause [44] under the condition (ii).

The rest of the section is devoted to the proof of the theorem.

Functors Commuting with Small Products

Let $\mathcal{S}$ be an additive $\mathcal{U}$-category which admits small direct sums. Let $\mathcal{S}^{\wedge, \mathrm{add}}$ be the category of additive functors from $\mathcal{S}^{\mathrm{op}}$ to $\operatorname{Mod}(\mathbb{Z})$. The category $\mathcal{S}^{\wedge, \mathrm{add}}$ is a big abelian category. By Proposition 8.2.12, $\mathcal{S}^{\wedge, \mathrm{add}}$ is regarded as a full subcategory of $\mathcal{S}^{\wedge}$.

A complex $F' \to F \to F''$ in $\mathcal{S}^{\wedge, \mathrm{add}}$ is exact if and only if $F'(X) \to F(X) \to F''(X)$ is exact for every $X \in \mathcal{S}$. Let $\mathcal{S}^{\wedge, \mathrm{prod}}$ be the full subcategory of $\mathcal{S}^{\wedge, \mathrm{add}}$ consisting of additive functors F commuting with small products, namely the canonical map $F(\oplus_i X_i) \to \prod_i F(X_i)$ is bijective for any small family $\{X_i\}_i$ of objects in $\mathcal{S}$.

Lemma 10.5.5. *The full category $\mathcal{S}^{\wedge, \mathrm{prod}}$ is a fully abelian subcategory of $\mathcal{S}^{\wedge, \mathrm{add}}$ closed by extension.*

Proof. It is enough to show that, for an exact complex $F_1 \to F_2 \to F_3 \to F_4 \to F_5$ in $\mathcal{S}^{\wedge, \mathrm{add}}$, if F_j belongs to $\mathcal{S}^{\wedge, \mathrm{prod}}$ for $j \neq 3$, then F_3 also belongs to

$\mathcal{S}^{\wedge,\mathrm{prod}}$ (see Remark 8.3.22). For a small family $\{X_i\}$ of objects in $\mathcal{S}$, we have an exact diagram in $\mathrm{Mod}(\mathbb{Z})$

$$
\begin{array}{ccccccccc}
F_1(\oplus_i X_i) & \longrightarrow & F_2(\oplus_i X_i) & \longrightarrow & F_3(\oplus_i X_i) & \longrightarrow & F_4(\oplus_i X_i) & \longrightarrow & F_5(\oplus_i X_i) \\
\downarrow{\scriptstyle\sim} & & \downarrow{\scriptstyle\sim} & & \downarrow & & \downarrow{\scriptstyle\sim} & & \downarrow{\scriptstyle\sim} \\
\prod_i F_1(X_i) & \longrightarrow & \prod_i F_2(X_i) & \longrightarrow & \prod_i F_3(X_i) & \longrightarrow & \prod_i F_4(X_i) & \longrightarrow & \prod_i F_5(X_i)\,.
\end{array}
$$

Since the vertical arrows are isomorphisms except the middle one, the five lemma (Lemma 8.3.13) implies that the middle arrow is an isomorphism. q.e.d.

Now assume that

(10.5.1)   there exists a small full subcategory $\mathcal{S}_0$ of $\mathcal{S}$ such that any object of $\mathcal{S}$ is a small direct sum of objects of $\mathcal{S}_0$.

Hence a complex $F' \to F \to F''$ in $\mathcal{S}^{\wedge,\mathrm{prod}}$ is exact if and only if $F'(X) \to F(X) \to F''(X)$ is exact for every $X \in \mathcal{S}_0$. In particular the restriction functor $\mathcal{S}^{\wedge,\mathrm{prod}} \to \mathcal{S}_0^{\wedge,\mathrm{add}}$ is exact, faithful and conservative. Hence, the category $\mathcal{S}^{\wedge,\mathrm{prod}}$ is a $\mathcal{U}$-category.

Let $\varphi \colon \mathcal{S} \to \mathcal{S}^{\wedge,\mathrm{prod}}$ be the functor which associates to $X \in \mathcal{S}$ the functor $\mathcal{S} \ni C \mapsto \mathrm{Hom}_{\mathcal{S}}(C, X)$. This functor commutes with small products. Since $\mathcal{S}^{\wedge,\mathrm{prod}} \to \mathcal{S}^{\wedge}$ is fully faithful, φ is a fully faithful additive functor by the Yoneda lemma.

Lemma 10.5.6. *Assume* (10.5.1). *Then, for any $F \in \mathcal{S}^{\wedge,\mathrm{prod}}$ we can find an object $X \in \mathcal{S}$ and an epimorphism $\varphi(X) \twoheadrightarrow F$.*

Proof. For any $C \in \mathcal{S}_0$, set $X_C = C^{\oplus F(C)}$. Then we have

$$
F(X_C) \simeq F(C)^{F(C)} = \mathrm{Hom}_{\mathbf{Set}}(F(C), F(C))\,.
$$

Hence $\mathrm{id}_{F(C)}$ gives an element $s_C \in F(X_C) \simeq \mathrm{Hom}_{\mathcal{S}^{\wedge,\mathrm{prod}}}(\varphi(X_C), F)$. Since the composition

$$
F(C) \to \mathrm{Hom}_{\mathcal{S}}(C, C) \times F(C) \to \mathrm{Hom}_{\mathcal{S}}(C, X_C) \simeq \varphi(X_C)(C) \to F(C)
$$

is the identity, the map $\varphi(X_C)(C) \to F(C)$ is surjective. Set $X = \oplus_{C \in \mathcal{S}_0} X_C$. Then $(s_C)_C \in \prod_C F(X_C) \simeq F(X)$ gives a morphism $\varphi(X) \to F$ and $\varphi(X)(C) \to F(C)$ is surjective for any $C \in \mathcal{S}_0$. Hence $\varphi(X) \to F$ is an epimorphism.   q.e.d.

Lemma 10.5.7. *Assume* (10.5.1).

(i) *The functor $\varphi \colon \mathcal{S} \to \mathcal{S}^{\wedge,\mathrm{prod}}$ commutes with small direct sums.*
(ii) *The abelian category $\mathcal{S}^{\wedge,\mathrm{prod}}$ admits small direct sums, and hence it admits small inductive limits.*

Proof. (i) For a small family $\{X_i\}_i$ of objects in $\mathcal{S}$ and $F \in \mathcal{S}^{\wedge,\text{prod}}$, we have

$$\text{Hom}_{\mathcal{S}^{\wedge,\text{prod}}}(\varphi(\oplus_i X_i), F) \simeq F(\oplus_i X_i)$$
$$\simeq \prod_i F(X_i) \simeq \prod_i \text{Hom}_{\mathcal{S}^{\wedge,\text{prod}}}(\varphi(X_i), F) \ .$$

(ii) For a small family $\{F_i\}_i$ of objects in $\mathcal{S}^{\wedge,\text{prod}}$, there exists an exact sequence $\varphi(X_i) \to \varphi(Y_i) \to F_i \to 0$ with $X_i, Y_i \in \mathcal{S}$ by Lemma 10.5.6. Since φ is fully faithful, there is a morphism $u_i \colon X_i \to Y_i$ which induces the morphism $\varphi(X_i) \to \varphi(Y_i)$. Then we have

$$\text{Coker}(\varphi(\oplus_i X_i) \xrightarrow{\oplus_i u_i} \varphi(\oplus_i Y_i)) \simeq \text{Coker}(\oplus_i \varphi(X_i) \to \oplus_i \varphi(Y_i))$$
$$\simeq \oplus_i \text{Coker}(\varphi(X_i) \to \varphi(Y_i)) \simeq \oplus_i F_i \ .$$

q.e.d.

Note that, for a small family $\{F_i\}_i$ of objects in $\mathcal{S}^{\wedge,\text{prod}}$ and $X \in \mathcal{S}$, the map $\oplus_i (F_i(X)) \to (\oplus_i F_i)(X)$ may be not bijective.

Proof of Theorem 10.5.2

Now let us come back to the original situation. Let $\mathcal{D}$ be a triangulated category admitting small direct sums and a system of t-generators $\mathcal{F}$. By replacing $\mathcal{F}$ with $\bigcup_{n \in \mathbb{Z}} T^n \mathcal{F}$, we may assume from the beginning that $T\mathcal{F} = \mathcal{F}$. Let $\mathcal{S}$ be the full subcategory of $\mathcal{D}$ consisting of small direct sums of objects in $\mathcal{F}$. Then $\mathcal{S}$ is an additive category which admits small direct sums. Moreover, $T\mathcal{S} = \mathcal{S}$, and T induces an automorphism $T \colon \mathcal{S}^{\wedge,\text{prod}} \to \mathcal{S}^{\wedge,\text{prod}}$ by $(TF)(C) = F(T^{-1}C)$ for $F \in \mathcal{S}^{\wedge,\text{prod}}$ and $C \in \mathcal{S}$. By its construction, $\mathcal{S}$ satisfies condition (10.5.1), and hence $\mathcal{S}^{\wedge,\text{prod}}$ is an abelian $\mathcal{U}$-category and Lemmas 10.5.5–10.5.7 hold. Note that a complex $F' \to F \to F''$ in $\mathcal{S}^{\wedge,\text{prod}}$ is exact if and only if $F'(C) \to F(C) \to F''(C)$ is exact for any $C \in \mathcal{F}$.

   We shall extend the functor $\varphi \colon \mathcal{S} \to \mathcal{S}^{\wedge,\text{prod}}$ to the functor $\widetilde{\varphi} \colon \mathcal{D} \to \mathcal{S}^{\wedge,\text{prod}}$ defined by $\widetilde{\varphi}(X)(C) = \text{Hom}_{\mathcal{D}}(C, X)$ for $X \in \mathcal{D}$ and $C \in \mathcal{S}$. Then $\widetilde{\varphi}$ commutes with T. Note that although $\varphi \colon \mathcal{S} \to \mathcal{S}^{\wedge,\text{prod}}$ is fully faithful, the functor $\widetilde{\varphi} \colon \mathcal{D} \to \mathcal{S}^{\wedge,\text{prod}}$ is not faithful in general.

   In the proof of the lemma below, we use the fact that $\mathcal{F}$ satisfies the condition (ii) in Definition 10.5.1.

Lemma 10.5.8.   (i) *The functor* $\widetilde{\varphi} \colon \mathcal{D} \to \mathcal{S}^{\wedge,\text{prod}}$ *is a cohomological functor.*
   (ii) *The functor* $\widetilde{\varphi} \colon \mathcal{D} \to \mathcal{S}^{\wedge,\text{prod}}$ *commutes with countable direct sums.*
   (iii) *Let* $\{X_i \to Y_i\}$ *be a countable family of morphisms in* $\mathcal{D}$. *If* $\widetilde{\varphi}(X_i) \to \widetilde{\varphi}(Y_i)$ *is an epimorphism for all* i, *then* $\widetilde{\varphi}(\oplus_i X_i) \to \widetilde{\varphi}(\oplus_i Y_i)$ *is an epimorphism.*

Proof. (i) is obvious.
Let us first prove (iii). For all $C \in \mathcal{F}$, the map $\text{Hom}_{\mathcal{D}}(C, X_i) \to \text{Hom}_{\mathcal{D}}(C, Y_i)$ is surjective. Hence Remark 10.5.4 (ii)$'$ implies that $\text{Hom}_{\mathcal{D}}(C, \oplus_i X_i) \to \text{Hom}_{\mathcal{D}}(C, \oplus_i Y_i)$ is surjective.

Finally let us prove (ii). Let $\{X_i\}_i$ be a countable family of objects of $\mathcal{D}$. Then we can find an epimorphism $\varphi(Y_i) \twoheadrightarrow \widetilde{\varphi}(X_i)$ in $\mathcal{S}^{\wedge,\mathrm{prod}}$ with $Y_i \in \mathcal{S}$ by Lemma 10.5.6. Let $W_i \to Y_i \to X_i \to TW_i$ be a d.t. Then take an epimorphism $\varphi(Z_i) \twoheadrightarrow \widetilde{\varphi}(W_i)$ with $Z_i \in \mathcal{S}$. Hence $\varphi(\oplus_i Z_i) \to \widetilde{\varphi}(\oplus_i W_i)$ and $\varphi(\oplus_i Y_i) \to \widetilde{\varphi}(\oplus_i X_i)$ are epimorphisms by (iii). On the other hand, $\oplus_i W_i \to \oplus_i Y_i \to \oplus_i X_i \to T(\oplus_i W_i)$ is a d.t., and hence $\widetilde{\varphi}(\oplus_i W_i) \to \varphi(\oplus_i Y_i) \to \widetilde{\varphi}(\oplus_i X_i)$ is exact by (i). Hence, $\varphi(\oplus_i Z_i) \to \varphi(\oplus_i Y_i) \to \widetilde{\varphi}(\oplus_i X_i) \to 0$ is exact. By Lemma 10.5.7, we have $\varphi(\oplus_i Z_i) \simeq \oplus_i \varphi(Z_i)$ and similarly for Y_i. Since $\varphi(Z_i) \to \varphi(Y_i) \to \widetilde{\varphi}(X_i) \to 0$ is exact for all i, $\oplus_i \varphi(Z_i) \to \oplus_i \varphi(Y_i) \to \oplus_i \widetilde{\varphi}(X_i) \to 0$ is also exact, from which we conclude that $\widetilde{\varphi}(\oplus_i X_i) \simeq \oplus_i \widetilde{\varphi}(X_i)$.   q.e.d.

Let $H \colon \mathcal{D}^{\mathrm{op}} \to \mathrm{Mod}(\mathbb{Z})$ be a cohomological functor commuting with small products. The restriction of H to $\mathcal{S}^{\mathrm{op}}$ defines $H_0 \in \mathcal{S}^{\wedge,\mathrm{prod}}$.

In the lemma below, we regard $\mathcal{D}$ as a full subcategory of $\mathcal{D}^\wedge$.

Lemma 10.5.9. *Let H and $\mathcal{K}$ be as in* Theorem 10.5.2. *Then there exists a commutative diagram in $\mathcal{D}^\wedge$*

(10.5.2)

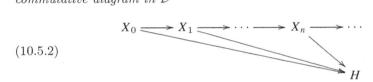

such that $X_n \in \mathcal{K}$ and $\mathrm{Im}\big(\widetilde{\varphi}(X_n) \to \widetilde{\varphi}(X_{n+1})\big) \xrightarrow{\sim} H_0$ in $\mathcal{S}^{\wedge,\mathrm{prod}}$.

Proof. We can take $X_0 \in \mathcal{S}$ and an epimorphism $\varphi(X_0) \twoheadrightarrow H_0$ in $\mathcal{S}^{\wedge,\mathrm{prod}}$ by Lemma 10.5.6. We shall construct $X_n \in \mathcal{K}$ inductively as follows. Assume that $X_0 \to X_1 \to \cdots \to X_n \to H$ has been constructed and $\mathrm{Im}\big(\widetilde{\varphi}(X_i) \to \widetilde{\varphi}(X_{i+1})\big) \xrightarrow{\sim} H_0$ for $0 \le i < n$. Let us take an exact sequence $\varphi(Z_n) \to \widetilde{\varphi}(X_n) \to H_0 \to 0$ with $Z_n \in \mathcal{S}$. Then take a d.t. $Z_n \to X_n \to X_{n+1} \to TZ_n$. Since Z_n and X_n belong to $\mathcal{K}$, X_{n+1} also belongs to $\mathcal{K}$. Since $Z_n \to X_n \to H$ vanishes and H is cohomological, $X_n \to H$ factors through $X_n \to X_{n+1}$. Since $\widetilde{\varphi}(Z_n) \to \widetilde{\varphi}(X_n) \to \widetilde{\varphi}(X_{n+1})$ is exact, we obtain that $\mathrm{Im}\big(\widetilde{\varphi}(X_n) \to \widetilde{\varphi}(X_{n+1})\big) \simeq \mathrm{Coker}\big(\widetilde{\varphi}(Z_n) \to \widetilde{\varphi}(X_n)\big) \simeq H_0$.   q.e.d.

Notation 10.5.10. Consider a functor $X \colon \mathbb{N} \to \mathcal{D}$, that is, a sequence of morphisms $X_0 \xrightarrow{f_0} X_1 \to \cdots \to X_n \xrightarrow{f_n} X_{n+1} \to \cdots$ in $\mathcal{D}$. Denote by

(10.5.3) $\mathrm{sh}_X \colon \bigoplus_{n \ge 0} X_n \to \bigoplus_{n \ge 0} X_n$

the morphism obtained as the composition

$$\bigoplus_{n \ge 0} X_n \xrightarrow{\oplus f_n} \bigoplus_{n \ge 0} X_{n+1} \simeq \bigoplus_{n \ge 1} X_n \hookrightarrow \bigoplus_{n \ge 0} X_n .$$

Consider a d.t.

$$(10.5.4) \qquad \oplus_{n \geq 0} X_n \xrightarrow{\mathrm{id} - \mathrm{sh}_X} \oplus_{n \geq 0} X_n \to Z \to T(\oplus_{n \geq 0} X_n) \,.$$

In the literature, Z is called the *homotopy colimit* of the inductive system $\{X_n, f_n\}_n$ and denoted by $\mathrm{hocolim}(X)$. Note that this object is unique up to isomorphism, but not up to unique isomorphism. Hence, $\{X_n, f_n\}_n \mapsto Z$ is not a functor.

Consider the functor $X \colon \mathbb{N} \to \mathcal{D}$ given by Lemma 10.5.9 and let sh_X be as in (10.5.3). Since $H(\oplus_{n \geq 0} X_n) \simeq \prod_{n \geq 0} H(X_n)$, the morphisms $X_n \to H$ define the morphism $\oplus_{n \geq 0} X_n \to H$. The commutativity of (10.5.2) implies that the composition $\oplus_{n \geq 0} X_n \xrightarrow{\mathrm{id} - \mathrm{sh}_X} \oplus_{n \geq 0} X_n \to H$ vanishes.

Lemma 10.5.11. *The sequence*

$$0 \to \widetilde{\varphi}(\oplus_{n \geq 0} X_n) \xrightarrow{\mathrm{id} - \mathrm{sh}_X} \widetilde{\varphi}(\oplus_{n \geq 0} X_n) \to H_0 \to 0 \,.$$

is exact in $\mathcal{S}^{\wedge, \mathrm{prod}}$.

Proof. Note that we have $\widetilde{\varphi}(\oplus_{n \geq 0} X_n) \simeq \oplus_{n \geq 0} \widetilde{\varphi}(X_n)$ by Lemma 10.5.8. Since $\mathrm{Im}\big(\widetilde{\varphi}(X_n) \to \widetilde{\varphi}(X_{n+1})\big) \simeq H_0$, we have "$\varinjlim_n$" $\widetilde{\varphi}(X_n) \simeq H_0$. Then $\varinjlim_n \widetilde{\varphi}(X_n) \simeq H_0$ and the the above sequence is exact by Exercise 8.37.   q.e.d.

Lemma 10.5.12. *There exist $Z \in \mathcal{K}$ and a morphism $Z \to H$ which induces an isomorphism $Z(C) \xrightarrow{\sim} H(C)$ for every $C \in \mathcal{F}$.*

Proof. Let Z be as in (10.5.4). Since H is cohomological, $\oplus_{n \geq 0} X_n \to H$ factors through Z. Set $X = \oplus_{n \geq 0} X_n$. Since $\widetilde{\varphi}$ is cohomological, we have an exact sequence in $\mathcal{S}^{\wedge, \mathrm{prod}}$:

$$\begin{array}{ccccccccc}
\widetilde{\varphi}(X) & \xrightarrow{\mathrm{id} - \mathrm{sh}_X} & \widetilde{\varphi}(X) & \longrightarrow & \widetilde{\varphi}(Z) & \longrightarrow & \widetilde{\varphi}(TX) & \xrightarrow{\widetilde{\varphi}(T(\mathrm{id} - \mathrm{sh}_X))} & \widetilde{\varphi}(TX) \\
& & & & & & \ \ \downarrow \wr & & \ \ \downarrow \wr \\
& & & & & & T(\widetilde{\varphi}(X)) & \xrightarrow{T(\widetilde{\varphi}(\mathrm{id} - \mathrm{sh}_X))} & T(\widetilde{\varphi}(X)) \,.
\end{array}$$

Applying Lemma 10.5.11, we find that the last right arrows are monomorphisms. Hence we have

$$\widetilde{\varphi}(Z) \simeq \mathrm{Coker}(\widetilde{\varphi}(X) \xrightarrow{\mathrm{id} - \mathrm{sh}_X} \widetilde{\varphi}(X)) \simeq H_0 \,,$$

where the last isomorphism follows from Lemma 10.5.11.   q.e.d.

Lemma 10.5.13. *The natural functor $\mathcal{K} \to \mathcal{D}$ is an equivalence.*

Proof. This functor being fully faithful, it remains to show that it is essentially surjective. Let $X \in \mathcal{D}$. Applying Lemma 10.5.12 to the functor $H = \mathrm{Hom}_{\mathcal{D}}(\cdot, X)$, we get $Z \in \mathcal{K}$ and a morphism $Z \to X$ which induces an isomorphism $Z(C) \xrightarrow{\sim} X(C)$ for all $C \in \mathcal{F}$. Since $\mathcal{F}$ is a system of generators, $Z \xrightarrow{\sim} X$.   q.e.d.

Lemma 10.5.14. *Let Z be as in* Lemma 10.5.12. *Then $Z \to H$ is an isomorphism.*

Proof. Let $\mathcal{K}$ denote the full subcategory of $\mathcal{D}$ consisting of objects Y such that $Z(T^n Y) \to H(T^n Y)$ is an isomorphism for any $n \in \mathbb{Z}$. Then $\mathcal{K}$ contains $\mathcal{F}$, is closed by small direct sums and is a triangulated subcategory of $\mathcal{D}$. Therefore $\mathcal{K} = \mathcal{D}$ by Lemma 10.5.13.   q.e.d.

The proof of Theorem 10.5.2 is complete.

Exercises

Exercise 10.1. Let $X \xrightarrow{f} Y \to Z \to TX$ be a d.t. in a triangulated category. Prove that f is an isomorphism if and only if Z is isomorphic to 0.

Exercise 10.2. Let $\mathcal{D}$ be a triangulated category and consider a commutative diagram in $\mathcal{D}$:

$$
\begin{array}{ccccccc}
X & \xrightarrow{\;f\;} & Y & \xrightarrow{\;g\;} & Z & \xrightarrow{\;h\;} & TX \\
\downarrow{\alpha} & & \downarrow{\beta} & & \downarrow{\gamma} & & \downarrow{T(\alpha)} \\
X' & \xrightarrow{\;f'\;} & Y' & \xrightarrow{\;g'\;} & Z' & \xrightarrow{\;h'\;} & TX' \, .
\end{array}
$$

Assume that α and β are isomorphisms, $T(f') \circ h' = 0$, and the first row is a d.t. Prove that the second row is also a d.t. under one of the hypotheses:
(i) for any $P \in \mathcal{D}$, the sequence below is exact:

$$\mathrm{Hom}\,(P, X') \to \mathrm{Hom}\,(P, Y') \to \mathrm{Hom}\,(P, Z') \to \mathrm{Hom}\,(P, TX')\,,$$

(ii) for any $P \in \mathcal{D}$, the sequence below is exact:

$$\mathrm{Hom}\,(TX', P) \to \mathrm{Hom}\,(Z', P) \to \mathrm{Hom}\,(Y', P) \to \mathrm{Hom}\,(X', P)\,.$$

Exercise 10.3. Let $F \colon \mathcal{D} \to \mathcal{D}'$ be a triangulated functor and assume that F admits an adjoint G. Prove that G is triangulated. (Hint: use Exercise 10.2.)

Exercise 10.4. Let $X \xrightarrow{f} Y \xrightarrow{g} Z \xrightarrow{h} TX$ be a d.t. in a triangulated category.
(i) Prove that if $h = 0$, this d.t. is isomorphic to $X \to X \oplus Z \to Z \xrightarrow{0} TX$.
(ii) Prove the same result by assuming now that there exists $k \colon Y \to X$ with $k \circ f = \mathrm{id}_X$.

Exercise 10.5. Let $f \colon X \to Y$ be a monomorphism in a triangulated category $\mathcal{D}$. Prove that there exist $Z \in \mathcal{D}$ and an isomorphism $h \colon Y \xrightarrow{\sim} X \oplus Z$ such that the composition $X \to Y \to X \oplus Z$ is the canonical morphism.

Exercise 10.6. In a triangulated category $\mathcal{D}$ consider the diagram of solid arrows

(10.5.5)

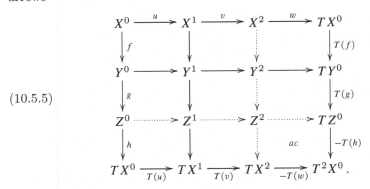

Assume that the two first rows and columns are d.t.'s. Show that the dotted arrows may be completed in order that all squares are commutative except the one labeled "ac" which is anti-commutative (see Definition 8.2.20), all rows and all columns are d.t.'s. (Hint: see [4], Proposition 1.1.11.)

Exercise 10.7. Let $\mathcal{D}$ be a triangulated category, $\mathcal{C}$ an abelian category, $F, G : \mathcal{D} \to \mathcal{C}$ two cohomological functors and $\theta : F \to G$ a morphism of functors. Define the full subcategory $\mathcal{T}$ of $\mathcal{D}$ consisting of objects $X \in \mathcal{D}$ such that $\theta(T^k(X)) : F(T^k(X)) \to G(T^k(X))$ is an isomorphism for all $k \in \mathbb{Z}$. Prove that $\mathcal{T}$ is triangulated. (Hint: use Lemma 8.3.13.)

Exercise 10.8. Let $\mathcal{D}$ be a triangulated category, $\mathcal{A}$ an abelian category and $F : \mathcal{D} \to \mathcal{A}$ a cohomological functor. Prove that F is exact.

Exercise 10.9. Let $\mathcal{D}$ be a triangulated category. Denote by $F : \mathcal{D} \to \mathcal{D}$ the translation functor T. By choosing a suitable isomorphism of functors $F \circ T \simeq T \circ F$, prove that F induces an equivalence of triangulated categories.

Exercise 10.10. Let $\mathcal{D}$ be a triangulated category and define the triangulated category $\mathcal{D}^{\mathrm{ant}}$ as follows: a triangle $X \xrightarrow{f} Y \xrightarrow{g} Z \xrightarrow{h} TX$ is distinguished in $\mathcal{D}^{\mathrm{ant}}$ if and only if $X \xrightarrow{f} Y \xrightarrow{g} Z \xrightarrow{-h} TX$ is distinguished in $\mathcal{D}$. Prove that $\mathcal{D}$ and $\mathcal{D}^{\mathrm{ant}}$ are equivalent as triangulated categories.

Exercise 10.11. Let $\mathcal{D}$ be a triangulated category, $\mathcal{N}$ a null system, and let $Q : \mathcal{D} \to \mathcal{D}/\mathcal{N}$ be the canonical functor.
(i) Let $f : X \to Y$ be a morphism in $\mathcal{D}$ and assume that $Q(f) = 0$ in $\mathcal{D}/\mathcal{N}$. Prove that there exists $Z \in \mathcal{N}$ such that f factorizes as $X \to Z \to Y$.
(ii) For $X \in \mathcal{D}$, prove that $Q(X) \simeq 0$ if and only if there exists Y such that $X \oplus Y \in \mathcal{N}$ and this last condition is equivalent to $X \oplus TX \in \mathcal{N}$.

Exercise 10.12. Let $F : \mathcal{D} \to \mathcal{D}'$ be a triangulated functor of triangulated categories. Let $\mathcal{N}$ be the full subcategory of $\mathcal{D}$ consisting of objects $X \in \mathcal{D}$ such that $F(X) \simeq 0$.

(i) Prove that $\mathcal{N}$ is a null system and F factorizes uniquely as $\mathcal{D} \to \mathcal{D}/\mathcal{N} \to \mathcal{D}'$.
(ii) Prove that if $X \oplus Y \in \mathcal{N}$, then $X \in \mathcal{N}$ and $Y \in \mathcal{N}$.

Exercise 10.13. Let $\mathcal{D}$ be a triangulated category admitting countable direct sums, let $X \in \mathcal{D}$ and let $p \colon X \to X$ be a projector (i.e., $p^2 = p$). Define the functor $\alpha \colon \mathbb{N} \to \mathcal{D}$ by setting $\alpha(n) = X$ and $\alpha(n \to n+1) = p$.
(i) Prove that $\varinjlim \alpha$ exists in $\mathcal{D}$ and is isomorphic to $\mathrm{hocolim}(\alpha)$. (See Notation 10.5.10.)
(ii) Deduce that $\mathcal{D}$ is idempotent complete. (See [53].)

Exercise 10.14. Let $\mathcal{D}$ be a triangulated category and let I be a filtrant category. Let $\alpha \xrightarrow{f} \beta \xrightarrow{g} \gamma \xrightarrow{h} T \circ \alpha$ be morphisms of functors from I to $\mathcal{D}$ such that $\alpha(i) \xrightarrow{f(i)} \beta(i) \xrightarrow{g(i)} \gamma(i) \xrightarrow{h(i)} T(\alpha(i))$ is a d.t. for all $i \in I$. Prove that if "$\varinjlim$" α and "$\varinjlim$" β are representable by objects of $\mathcal{D}$, then so is "$\varinjlim$" γ and the induced triangle "$\varinjlim$" $\alpha \to$ "$\varinjlim$" $\beta \to$ "$\varinjlim$" $\gamma \to T($"$\varinjlim$" $\alpha)$ is a d.t. (Hint: construct a morphism of d.t.'s

$$
\begin{array}{ccccccc}
\text{"}\varinjlim\text{"}\,\alpha & \longrightarrow & \text{"}\varinjlim\text{"}\,\beta & \longrightarrow & Z & \longrightarrow & T(\text{"}\varinjlim\text{"}\,\alpha) \\
\downarrow & & \downarrow & & \downarrow & & \downarrow \\
\alpha(i) & \longrightarrow & \beta(i) & \longrightarrow & \gamma(i) & \longrightarrow & T(\alpha(i))
\end{array}
$$

for some $i \in I$.)

Exercise 10.15. Let $\mathcal{D}$ be a triangulated category, $\mathcal{N}$ a null system, and let $\mathcal{N}^{\perp r}$ (resp. $\mathcal{N}^{\perp l}$) be the full subcategory of $\mathcal{D}$ consisting of objects Y such that $\mathrm{Hom}_{\mathcal{D}}(Z, Y) \simeq 0$ (resp. $\mathrm{Hom}_{\mathcal{D}}(Y, Z) \simeq 0$) for all $Z \in \mathcal{N}$.
(i) Prove that $\mathcal{N}^{\perp r}$ and $\mathcal{N}^{\perp l}$ are null systems in $\mathcal{D}$.
(ii) Prove that $\mathrm{Hom}_{\mathcal{D}}(X, Y) \xrightarrow{\sim} \mathrm{Hom}_{\mathcal{D}/\mathcal{N}}(X, Y)$ for any $X \in \mathcal{D}$ and any $Y \in \mathcal{N}^{\perp r}$.

In the sequel, we assume that $X \oplus Y \in \mathcal{N}$ implies $X \in \mathcal{N}$ and $Y \in \mathcal{N}$.
(iii) Prove that the following conditions are equivalent:

(a) $\mathcal{N}^{\perp r} \to \mathcal{D}/\mathcal{N}$ is an equivalence,
(b) $\mathcal{D} \to \mathcal{D}/\mathcal{N}$ has a right adjoint,
(c) $\iota \colon \mathcal{N} \to \mathcal{D}$ has a right adjoint R,
(d) for any $X \in \mathcal{D}$, there exist $X' \in \mathcal{N}$, $X'' \in \mathcal{N}^{\perp r}$ and a d.t. $X' \to X \to X'' \to TX'$,
(e) $\mathcal{N} \to \mathcal{D}/\mathcal{N}^{\perp r}$ is an equivalence,
(f) $\mathcal{D} \to \mathcal{D}/\mathcal{N}^{\perp r}$ has a left adjoint and $\mathcal{N} \simeq (\mathcal{N}^{\perp r})^{\perp l}$,
(g) $\iota' \colon \mathcal{N}^{\perp r} \to \mathcal{D}$ has a left adjoint L and $\mathcal{N} \simeq (\mathcal{N}^{\perp r})^{\perp l}$.

(iv) Assume that the equivalent conditions (a)–(g) in (iii) are satisfied. Let $L \colon \mathcal{D} \to \mathcal{N}^{\perp r}$, $R \colon \mathcal{D} \to \mathcal{N}$, $\iota \colon \mathcal{N} \to \mathcal{D}$ and $\iota' \colon \mathcal{N}^{\perp r} \to \mathcal{D}$ be as above.

(a) Prove that there exists a morphism of functors $\iota' \circ L \to T \circ \iota \circ R$ such that $\iota R(X) \to X \to \iota' L(X) \to T(\iota R(X))$ is a d.t. for all $X \in \mathcal{D}$.

(b) Let $\widetilde{\mathcal{D}}$ be the category whose objects are the triplets (X', X'', u) with $X' \in \mathcal{N}$, $X'' \in \mathcal{N}^{\perp r}$ and u is a morphism $X'' \to TX'$ in $\mathcal{D}$. A morphism $(X', X'', u) \to (Y', Y'', v)$ in $\widetilde{\mathcal{D}}$ is a pair $(w' \colon X' \to Y', w'' \colon X'' \to Y'')$ making the diagram below commutative

$$
\begin{array}{ccc}
X'' & \xrightarrow{\ u\ } & TX' \\
{\scriptstyle w''}\big\downarrow & & \big\downarrow{\scriptstyle T(w')} \\
Y'' & \xrightarrow{\ v\ } & TY' \ .
\end{array}
$$

Define an equivalence of categories $\mathcal{D} \xrightarrow{\sim} \widetilde{\mathcal{D}}$.

Exercise 10.16. (i) Let $\mathcal{D}$ be a triangulated category. Assume that $\mathcal{D}$ is abelian.

(a) Prove that $\mathcal{D}$ is a semisimple abelian category (see Definition 8.3.16). (Hint: use Exercise 10.5.)

(b) Prove that any triangle in $\mathcal{D}$ is a direct sum of three triangles

$$
\begin{array}{ccccccc}
X & \xrightarrow{\ \mathrm{id}_X\ } & X & \to & 0 & \to & TX \ , \\
0 & \to & Y & \xrightarrow{\ \mathrm{id}_Y\ } & Y & \to & T(0), \qquad \text{and} \\
T^{-1}Z & \to & 0 & \to & Z & \xrightarrow{\ \mathrm{id}_Z\ } & T(T^{-1}Z).
\end{array}
$$

(ii) Conversely let $(\mathcal{C}, T)$ be a category with translation and assume that $\mathcal{C}$ is a semisimple abelian category. We say that a triangle in $\mathcal{C}$ is distinguished if it is a direct sum of three triangles as in (i) (b). Prove that $\mathcal{C}$ is a triangulated category.

11

Complexes in Additive Categories

As already mentioned in the Introduction, one of the main ideas of homological algebra is to replace an object in an abelian category $\mathcal{C}$ by a complex of objects of the category, the components of these complexes having "good properties". For example, a module is replaced by a complex of projective modules.

In this chapter, we start by studying additive categories with translation, already encountered in Chapter 10. For such a category, there are natural notions of a differential object, a complex, the mapping cone of a morphism and of a morphism homotopic to zero. Identifying morphisms homotopic to zero with the zero morphism, we get the associated homotopy category.

One of the main result of this chapter is the fact that this homotopy category, endowed with the family of triangles isomorphic to those associated with a mapping cone, is a triangulated category.

We apply the preceding results to the category $\mathrm{Gr}(\mathcal{C})$ of sequences of objects of an additive category $\mathcal{C}$. The category $\mathrm{Gr}(\mathcal{C})$ is endowed with a natural translation functor, and we get the category $\mathrm{C}(\mathcal{C})$ of complexes of objects of $\mathcal{C}$ as well as the associated triangulated category $\mathrm{K}(\mathcal{C})$. We also introduce the simplicial category $\mathbf{\Delta}$, we construct complexes associated to it and give a criterion in order that such complexes are homotopic to zero.

If $F: \mathcal{C} \to \mathcal{C}'$ is an additive functor of additive categories, it defines naturally a triangulated functor $\mathrm{K}(F): \mathrm{K}(\mathcal{C}) \to \mathrm{K}(\mathcal{C}')$. Things become more delicate with bifunctors. Indeed, if $F: \mathcal{C} \times \mathcal{C}' \to \mathcal{C}''$ is an additive bifunctor, it defines naturally a functor from $\mathrm{C}(\mathcal{C}) \times \mathrm{C}(\mathcal{C}')$ to the category $\mathrm{C}^2(\mathcal{C}'')$ of double complexes in $\mathcal{C}''$, and it is necessary to construct (under suitable hypotheses) simple complexes associated with a double complex. As we shall see, signs should be treated with some care.

Finally, we apply these constructions to the bifunctor Hom.

11.1 Differential Objects and Mapping Cones

Definition 11.1.1. *Let* $(\mathcal{A}, T)$ *be an additive category with translation* (*see* Definition 10.1.1).

(i) *A* differential object *in* $(\mathcal{A}, T)$ *is an object* $X \in \mathcal{A}$ *endowed with a morphism* $d_X \colon X \to TX$, *called the* differential *of* X.

(ii) *A morphism* $f \colon X \to Y$ *of differential objects, also called a* differential morphism, *is a morphism* $f \colon X \to Y$ *such that the diagram below commutes:*

$$
\begin{array}{ccc}
X & \xrightarrow{\ d_X\ } & TX \\
\downarrow{\scriptstyle f} & & \downarrow{\scriptstyle T(f)} \\
Y & \xrightarrow{\ d_Y\ } & TY\,.
\end{array}
$$

We denote by $\mathcal{A}_d$ *the category of differential objects and differential morphisms.*

(iii) *A differential object* X *is a* complex *if* $T(d_X) \circ d_X = 0$. *We denote by* $\mathcal{A}_c$ *the full additive subcategory of* $\mathcal{A}_d$ *consisting of complexes. A differential morphism of complexes is also called a* morphism of complexes.

Clearly, if $F \colon \mathcal{A} \to \mathcal{A}'$ is a functor of additive categories with translation, it induces a functor $F \colon \mathcal{A}_d \to \mathcal{A}'_d$ and a functor $F \colon \mathcal{A}_c \to \mathcal{A}'_c$.

Definition 11.1.2. *Let* $(\mathcal{A}, T)$ *be an additive category with translation. For a differential object* $X \in \mathcal{A}_d$, *the differential object* TX *with the differential* $d_{TX} = -T(d_X)$ *is called the* shifted object *of* X.

Note that

- If X is a complex, then so is the shifted object TX.
- The pair $(\mathcal{A}_d, T)$ is an additive category with translation, as well as the pair $(\mathcal{A}_c, T)$.
- For a differential object X and an integer n, we have $d_{T^n X} = (-1)^n T^n(d_X)$.

Definition 11.1.3. *Let* $(\mathcal{A}, T)$ *be an additive category with translation, let* X *and* Y *be two differential objects and let* $f \colon X \to Y$ *be a morphism in* $\mathcal{A}$. *The* mapping cone $\mathrm{Mc}(f)$ *of* f *is the object* $TX \oplus Y$ *with differential*

$$
d_{\mathrm{Mc}(f)} := \begin{pmatrix} d_{TX} & 0 \\ T(f) & d_Y \end{pmatrix} = \begin{pmatrix} -T(d_X) & 0 \\ T(f) & d_Y \end{pmatrix}.
$$

Here we have used the column notation for morphisms between direct sums. Hence the composition $TX \to \mathrm{Mc}(f) \xrightarrow{d_{\mathrm{Mc}(f)}} T(\mathrm{Mc}(f)) \to TY$ in $\mathcal{A}$ is equal to $T(f)$.

Proposition 11.1.4. *Assume that* X *and* Y *are complexes. Then* $\mathrm{Mc}(f)$ *is a complex if and only if* f *is a morphism of complexes.*

Proof. Set

$$\begin{pmatrix} T(d_{T(X)}) & 0 \\ T^2(f) & T(d_Y) \end{pmatrix} \circ \begin{pmatrix} d_{TX} & 0 \\ T(f) & d_Y \end{pmatrix} = \begin{pmatrix} A & B \\ C & D \end{pmatrix} .$$

Then we have

$$A = T(d_{TX}) \circ d_{TX} = T(-T(d_X) \circ (-d_X)) = 0 ,$$
$$D = T(d_Y) \circ d_Y = 0 ,$$
$$B = 0 ,$$
$$C = T^2(f) \circ d_{TX} + T(d_Y) \circ T(f) = T(-T(f) \circ d_X + d_Y \circ f) .$$

Hence, $C = 0$ if and only if $-T(f) \circ d_X + d_Y \circ f = 0$, that is, if and only if f is a morphism of differential objects.   q.e.d.

We have:

- $\mathrm{Mc}(f)$ is not $TX \oplus Y$ in $\mathcal{A}_d$ unless f is the zero morphism,
- Mc is a functor from $\mathrm{Mor}(\mathcal{A}_d)$ to $\mathcal{A}_d$. Namely for a commutative diagram

$$\begin{array}{ccc} X & \xrightarrow{u} & X' \\ f \downarrow & & \downarrow f' \\ Y & \xrightarrow{v} & Y' \end{array} \quad \text{in } \mathcal{A}_d, \ T(u) \oplus v \text{ gives a morphism } \mathrm{Mc}(f) \to \mathrm{Mc}(f'),$$

- if $F \colon (\mathcal{A}, T) \to (\mathcal{A}', T')$ is a functor of additive categories with translation, then $F(\mathrm{Mc}(f)) \simeq \mathrm{Mc}(F(f))$.

Let $f \colon X \to Y$ be a morphism in $\mathcal{A}_d$. We introduce the differential morphisms

$$\alpha(f) \colon Y \to \mathrm{Mc}(f), \quad \alpha(f) = 0 \oplus \mathrm{id}_Y$$

and

$$\beta(f) \colon \mathrm{Mc}(f) \to TX, \quad \beta(f) = (\mathrm{id}_{TX}, 0) .$$

We get a triangle in $\mathcal{A}_d$:

(11.1.1) $$X \xrightarrow{f} Y \xrightarrow{\alpha(f)} \mathrm{Mc}(f) \xrightarrow{\beta(f)} TX .$$

We call such a triangle *a mapping cone triangle* in $\mathcal{A}_d$.

Remark 11.1.5. Consider a morphism $f \colon X \to Y$ in $\mathcal{A}_d$. We have a commutative diagram (the verification is left to the reader):

$$\begin{array}{ccccccc} TX & \xrightarrow{T(f)} & TY & \xrightarrow{T(\alpha(f))} & T(\mathrm{Mc}(f)) & \xrightarrow{-T(\beta(f))} & T^2X \\ \mathrm{id}\downarrow & & \mathrm{id}\downarrow & & \downarrow\wr & & \downarrow\mathrm{id} \\ TX & \xrightarrow{T(f)} & TY & \xrightarrow{\alpha(T(f))} & \mathrm{Mc}(T(f)) & \xrightarrow{\beta(T(f))} & T^2X . \end{array}$$

11.2 The Homotopy Category

Lemma 11.2.1. *Let $(\mathcal{A}, T)$ be an additive category with translation, let X and Y be two differential objects and let $u\colon X \to T^{-1}Y$ be a morphism in $\mathcal{A}$. (We do not ask u to be a differential morphism.) Set $f = T(u) \circ d_X + T^{-1}(d_Y) \circ u$. Then f is a differential morphism if and only if $d_Y \circ T^{-1}(d_Y) \circ u = T^2(u) \circ T(d_X) \circ d_X$. In particular, if X and Y are complexes, f is always a morphism of complexes.*

Proof. One has

$$d_Y \circ f = d_Y \circ T(u) \circ d_X + d_Y \circ T^{-1}(d_Y) \circ u \ ,$$
$$T(f) \circ d_X = T^2(u) \circ T(d_X) \circ d_X + T(T^{-1}(d_Y)) \circ T(u) \circ d_X.$$

q.e.d.

Definition 11.2.2. *Let $(\mathcal{A}, T)$ be an additive category with translation. A morphism $f\colon X \to Y$ in $\mathcal{A}_d$ is homotopic to zero if there exists a morphism $u\colon X \to T^{-1}Y$ in $\mathcal{A}$ such that:*

$$f = T(u) \circ d_X + T^{-1}(d_Y) \circ u.$$

Two morphisms $f, g\colon X \to Y$ are homotopic if $f - g$ is homotopic to zero.

A morphism homotopic to zero is visualized by the diagram (which is not commutative):

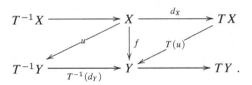

Note that a functor of additive categories with translation sends a morphism homotopic to zero to a morphism homotopic to zero.

Lemma 11.2.3. *Let $f\colon X \to Y$ and $g\colon Y \to Z$ be morphisms in $\mathcal{A}_d$. If f or g is homotopic to zero, then $g \circ f$ is homotopic to zero.*

Proof. If $f = T(u) \circ d_X + T^{-1}(d_Y) \circ u$ for some $u\colon X \to T^{-1}Y$, then we have

$$\begin{aligned}
g \circ f &= g \circ T(u) \circ d_X + g \circ T^{-1}(d_Y) \circ u \\
&= g \circ T(u) \circ d_X + T^{-1}(d_Z) \circ T^{-1}(g) \circ u \\
&= T(T^{-1}(g) \circ u) \circ d_X + T^{-1}(d_Z) \circ (T^{-1}(g) \circ u) \ .
\end{aligned}$$

Hence $g \circ f$ is homotopic to zero. The other assertion is similarly proved. q.e.d.

Set:

$$\mathrm{Ht}(X, Y) = \left\{ f \in \mathrm{Hom}_{\mathcal{A}_d}(X, Y) \,;\, f \text{ is homotopic to } 0 \right\} .$$

By the lemma above, the composition map induces a bilinear map:

$$(11.2.1) \quad \mathrm{Hom}_{\mathcal{A}_d}(X, Y)/\mathrm{Ht}(X, Y) \times \mathrm{Hom}_{\mathcal{A}_d}(Y, Z)/\mathrm{Ht}(Y, Z)$$
$$\to \mathrm{Hom}_{\mathcal{A}_d}(X, Z)/\mathrm{Ht}(X, Z) .$$

Definition 11.2.4. *The homotopy category* $\mathrm{K}_d(\mathcal{A})$ *is defined by:*

$$\mathrm{Ob}(\mathrm{K}_d(\mathcal{A})) = \mathrm{Ob}(\mathcal{A}_d) ,$$
$$\mathrm{Hom}_{\mathrm{K}_d(\mathcal{A})}(X, Y) = \mathrm{Hom}_{\mathcal{A}_d}(X, Y)/\mathrm{Ht}(X, Y) ,$$

and the composition of morphisms is given by (11.2.1).

In other words, a morphism homotopic to zero in $\mathcal{A}_d$ becomes the zero morphism in $\mathrm{K}_d(\mathcal{A})$.

The category $\mathrm{K}_d(\mathcal{A})$ is obviously additive and the translation functor T on $\mathcal{A}_d$ induces a translation functor (we keep the same notation) T on $\mathrm{K}_d(\mathcal{A})$. Hence, $(\mathrm{K}_d(\mathcal{A}), T)$ is an additive category with translation, and we have a functor of additive categories with translation $(\mathcal{A}_d, T) \to (\mathrm{K}_d(\mathcal{A}), T)$.

Two objects in $\mathcal{A}_d$ are called homotopic if they are isomorphic in $\mathrm{K}_d(\mathcal{A})$. Hence an object X in $\mathcal{A}_d$ is homotopic to 0 if and only if id_X is homotopic to zero.

Definition 11.2.5. *A distinguished triangle in* $(\mathrm{K}_d(\mathcal{A}), T)$ *is a triangle isomorphic in* $\mathrm{K}_d(\mathcal{A})$ *to a mapping cone triangle* (11.1.1).

Recall that we write "a d.t." instead of "distinguished triangle", for short.

Theorem 11.2.6. *The category* $\mathrm{K}_d(\mathcal{A})$ *endowed with the translation functor* T *and the family of* d.t.*'s is a triangulated category.*

Proof. The axioms TR0 and TR2 are obvious and TR1 follows from TR3 and the d.t. $0 \to X \to X \to T(0)$ associated with the mapping cone of $0 \to X$.

Proof of TR3. We shall construct a morphism $\varphi \colon TX \to \mathrm{Mc}(\alpha(f))$ in $\mathcal{A}_d$ such that:
(i) φ is an isomorphism in $\mathrm{K}_d(\mathcal{A})$,
(ii) the diagram below commutes in $\mathrm{K}_d(\mathcal{A})$:

$$
\begin{array}{ccccccc}
Y & \xrightarrow{\alpha(f)} & \mathrm{Mc}(f) & \xrightarrow{\beta(f)} & TX & \xrightarrow{-T(f)} & TY \\
\downarrow{\scriptstyle \mathrm{id}_Y} & & \downarrow{\scriptstyle \mathrm{id}_{\mathrm{Mc}(f)}} & & \downarrow{\scriptstyle \varphi} & & \downarrow{\scriptstyle \mathrm{id}_{TY}} \\
Y & \xrightarrow{\alpha(f)} & \mathrm{Mc}(f) & \xrightarrow{\alpha(\alpha(f))} & \mathrm{Mc}(\alpha(f)) & \xrightarrow{\beta(\alpha(f))} & TY .
\end{array}
$$

Define φ and ψ:

$$\varphi: TX \to \mathrm{Mc}(\alpha(f)) = TY \oplus TX \oplus Y,$$
$$\psi: \mathrm{Mc}(\alpha(f)) = TY \oplus TX \oplus Y \to TX$$

by:

$$\varphi = \begin{pmatrix} -T(f) \\ \mathrm{id}_{TX} \\ 0 \end{pmatrix}, \quad \psi = (0, \mathrm{id}_{TX}, 0).$$

We have to check that:
(a) φ and ψ are morphisms of differential objects,
(b) $\psi \circ \varphi = \mathrm{id}_{TX}$,
(c) $\varphi \circ \psi$ is homotopic to $\mathrm{id}_{\mathrm{Mc}(\alpha(f))}$,
(d) $\psi \circ \alpha(\alpha(f)) = \beta(f)$,
(e) $\beta(\alpha(f)) \circ \varphi = -T(f)$.
(Note that (c)+(d) $\Rightarrow$ (d'): $\varphi \circ \beta(f)$ is homotopic to $\alpha(\alpha(f))$.)

Let us prove (c), the other verifications being straightforward. Define $s: \mathrm{Mc}(\alpha(f)) \to T^{-1}(\mathrm{Mc}(\alpha(f)))$ by:

$$s = \begin{pmatrix} 0 & 0 & \mathrm{id}_Y \\ 0 & 0 & 0 \\ 0 & 0 & 0 \end{pmatrix}.$$

Then:

$$\mathrm{id}_{\mathrm{Mc}(\alpha(f))} - \varphi \circ \psi = T(s) \circ d_{\mathrm{Mc}(\alpha(f))} + T^{-1}(d_{\mathrm{Mc}(\alpha(f))}) \circ s.$$

Indeed:

$$d_{\mathrm{Mc}(\alpha(f))} = \begin{pmatrix} -T(d_Y) & 0 & 0 \\ 0 & -T(d_X) & 0 \\ \mathrm{id}_{TY} & T(f) & d_Y \end{pmatrix},$$

$$\varphi \circ \psi = \begin{pmatrix} 0 & -T(f) & 0 \\ 0 & \mathrm{id}_{TX} & 0 \\ 0 & 0 & 0 \end{pmatrix},$$

$$\mathrm{id}_{\mathrm{Mc}(\alpha(f))} - \varphi \circ \psi = \begin{pmatrix} \mathrm{id}_{TY} & T(f) & 0 \\ 0 & 0 & 0 \\ 0 & 0 & \mathrm{id}_Y \end{pmatrix}.$$

Proof of TR4. We may assume $Z = \mathrm{Mc}(f)$, $Z' = \mathrm{Mc}(f')$. Then saying that

commutes in $K_d(\mathcal{A})$ means that there exists a morphism $s\colon X \to T^{-1}(Y')$ in $\mathcal{A}$ with:
$$v \circ f - f' \circ u = T(s) \circ d_X + T^{-1}(d_{Y'}) \circ s.$$

Define:
$$w\colon \operatorname{Mc}(f) = TX \oplus Y \to \operatorname{Mc}(f') = TX' \oplus Y'$$

by
$$w = \begin{pmatrix} T(u) & 0 \\ T(s) & v \end{pmatrix}.$$

Then w is a morphism of differential objects and the diagram below commutes:

$$
\begin{array}{ccccccc}
X & \xrightarrow{\;f\;} & Y & \xrightarrow{\;\alpha(f)\;} & \operatorname{Mc}(f) & \xrightarrow{\;\beta(f)\;} & TX \\
{\scriptstyle u}\big\downarrow & & {\scriptstyle v}\big\downarrow & & {\scriptstyle w}\big\downarrow & & \big\downarrow{\scriptstyle T(u)} \\
X' & \xrightarrow{\;f'\;} & Y' & \xrightarrow{\;\alpha(f')\;} & \operatorname{Mc}(f') & \xrightarrow{\;\beta(f')\;} & TX' .
\end{array}
$$

Proof of TR5. We may assume that $Z' = \operatorname{Mc}(f)$, $X' = \operatorname{Mc}(g)$ and $Y' = \operatorname{Mc}(g \circ f)$. Let us define $u\colon Z' \to Y'$ and $v\colon Y' \to X'$ by

$$u\colon TX \oplus Y \to TX \oplus Z, \quad u = \begin{pmatrix} \operatorname{id}_{TX} & 0 \\ 0 & g \end{pmatrix},$$

$$v\colon TX \oplus Z \to TY \oplus Z, \quad v = \begin{pmatrix} T(f) & 0 \\ 0 & \operatorname{id}_Z \end{pmatrix}.$$

We define $w\colon X' \to TZ'$ as the composition $X' \xrightarrow{\beta(g)} TY \xrightarrow{T(\alpha(f))} TZ'$. Then the diagram in TR5 is commutative and it is enough to show that the triangle $Z' \xrightarrow{u} Y' \xrightarrow{v} X' \xrightarrow{w} TZ'$ is distinguished. For that purpose, we shall construct an isomorphism $\varphi\colon \operatorname{Mc}(u) \to X'$ and its inverse $\psi\colon X' \to \operatorname{Mc}(u)$ in $K_d(\mathcal{A})$ such that $\varphi \circ \alpha(u) = v$ and $\beta(u) \circ \psi = w$. We have

$$\operatorname{Mc}(u) = T(\operatorname{Mc}(f)) \oplus \operatorname{Mc}(g \circ f) = T^2 X \oplus TY \oplus TX \oplus Z$$

and $X' = \operatorname{Mc}(g) = TY \oplus Z$. We define φ and ψ by

$$\varphi = \begin{pmatrix} 0 & \operatorname{id}_{TY} & T(f) & 0 \\ 0 & 0 & 0 & \operatorname{id}_Z \end{pmatrix}, \quad \psi = \begin{pmatrix} 0 & 0 \\ \operatorname{id}_{TY} & 0 \\ 0 & 0 \\ 0 & \operatorname{id}_{TX} \end{pmatrix}.$$

It is easily checked that φ and ψ are morphisms of differential objects, and $\varphi \circ \alpha(u) = v$, $\beta(u) \circ \psi = w$ and $\varphi \circ \psi = \operatorname{id}_X$ hold in $\mathcal{A}_d$. Define a morphism in $\mathcal{A}$

$$s \colon \mathrm{Mc}(u) \to T^{-1}(\mathrm{Mc}(u)), \quad s = \begin{pmatrix} 0 & 0 & \mathrm{id}_{TX} & 0 \\ 0 & 0 & 0 & 0 \\ 0 & 0 & 0 & 0 \\ 0 & 0 & 0 & 0 \end{pmatrix}.$$

Then

$$\mathrm{id}_{\mathrm{Mc}(u)} - \psi \circ \varphi = T(s) \circ d_{\mathrm{Mc}(u)} + T^{-1}(d_{\mathrm{Mc}(u)}) \circ s.$$

Therefore $\psi \circ \varphi = \mathrm{id}_{\mathrm{Mc}(u)}$ holds in $\mathrm{K}_d(\mathcal{A})$.   q.e.d.

Remark 11.2.7. In proving Theorem 11.3.8, we have shown that some diagrams were commutative in $\mathrm{K}_d(\mathcal{A})$, that is, did commute in $\mathcal{A}_d$ up to homotopy. One should be aware that some of these diagrams did not commute in $\mathcal{A}_d$, and in fact, this last category is not triangulated in general.

Let $\mathrm{K}_c(\mathcal{A})$ be the full subcategory of $\mathrm{K}_d(\mathcal{A})$ consisting of complexes in $(\mathcal{A}, T)$. Then $\mathrm{K}_c(\mathcal{A})$ is an additive subcategory with translation. Since the mapping cone of a morphism of complexes is also a complex, we obtain the following proposition.

Proposition 11.2.8. *The category* $\mathrm{K}_c(\mathcal{A})$ *endowed with the translation functor* T *and the family of d.t.'s is a triangulated full subcategory of* $\mathrm{K}_d(\mathcal{A})$.

Proposition 11.2.9. *Let* $F \colon (\mathcal{A}, T) \to (\mathcal{A}', T')$ *be a functor of additive categories with translation. Then* F *defines naturally triangulated functors* $\mathrm{K}(F) \colon \mathrm{K}_d(\mathcal{A}) \to \mathrm{K}_d(\mathcal{A}')$ *and* $\mathrm{K}(F) \colon \mathrm{K}_c(\mathcal{A}) \to \mathrm{K}_c(\mathcal{A}')$.

Proof. As already noticed, F induces a functor $F \colon \mathcal{A}_d \to \mathcal{A}'_d$. Moreover F sends a morphism homotopic to zero in $\mathcal{A}_d$ to a morphism homotopic to zero in $\mathcal{A}'_d$, hence defines an additive functor from $\mathrm{K}_d(\mathcal{A})$ to $\mathrm{K}_d(\mathcal{A}')$. To conclude, notice that F sends a mapping cone triangle in $\mathcal{A}_d$ to a mapping cone triangle in $\mathcal{A}'_d$.   q.e.d.

When there is no risk of confusion, we shall simply denote by F the functor $\mathrm{K}(F)$.

Let $F \colon (\mathcal{A}, T) \times (\mathcal{A}', T') \to (\mathcal{A}'', T'')$ be a bifunctor of additive categories with translation. Then, $\theta_{X,Y} \colon F(TX, Y) \xrightarrow{\sim} T'' F(X, Y)$ in Definition 10.1.1 (v) induces a functorial isomorphism

$$_n\theta_{X,Y} \colon F(T^n X, Y) \xrightarrow{\sim} T''^n F(X, Y)$$

for $n \in \mathbb{Z}$. Similarly, $\theta'_{X,Y} \colon F(X, T'Y) \xrightarrow{\sim} T'' F(X, Y)$ induces a functorial isomorphism

$$_n\theta'_{X,Y} \colon F(X, T'^n Y) \xrightarrow{\sim} T''^n F(X, Y).$$

We can easily check that the diagram

$$F(T^n X, T'^m Y) \xrightarrow{\ _n\theta_{X,T'^m Y}\ } T''^n F(X, T'^m Y)$$

$$\left\downarrow{\scriptstyle {}_m\theta'_{T^n X,Y}}\right. \qquad\qquad (-1)^{nm} \qquad\qquad \left\downarrow{\scriptstyle T''^n({}_m\theta'_{X,Y})}\right.$$

$$T''^m F(T^n X, Y) \xrightarrow[\ T''^m({}_n\theta_{X,Y})\ ]{} T''^{n+m} F(X, Y) \,.$$

$(-1)^{nm}$-commutes (see Definition 8.2.20), i.e., it commutes or anti-commutes according that $(-1)^{nm} = 1$ or -1. For a differential object X in $(\mathcal{A}, T)$ and Y in $(\mathcal{A}', T')$, we have morphisms in $\mathcal{A}''$:

$$F(d_X, Y) : \ F(X, Y) \to F(TX, Y) \simeq T'' F(X, Y) \,,$$
$$F(X, d_Y) : \ F(X, Y) \to F(X, T'Y) \simeq T'' F(X, Y) \,.$$

We set

$$(11.2.2) \qquad d_{F(X,Y)} = F(d_X, Y) + F(X, d_Y) \colon F(X, Y) \to T'' F(X, Y) \,.$$

Thus we obtain a bifunctor of additive categories with translation $F \colon (\mathcal{A}_d, T) \times (\mathcal{A}'_d, T') \to (\mathcal{A}''_d, T'')$.

Lemma 11.2.10.   (i) *For a morphism $s \colon X \to T^n X'$ in $\mathcal{A}$ and a morphism $t \colon Y \to T'^m Y'$ in $\mathcal{A}'$, let us set*

$$F(s, Y) \colon F(X, Y) \to F(T^n X', Y) \xrightarrow[\ _n\theta_{X',Y}\ ]{\sim} T''^n F(X', Y) \,,$$

$$F(X, t) \colon F(X, Y) \to F(X, T'^m Y') \xrightarrow[\ _m\theta'_{X,Y'}\ ]{\sim} T''^m F(X, Y') \,.$$

Then one has

$$T''^m(F(s, Y')) \circ F(X, t) = (-1)^{nm} T''^n(F(X', t)) \circ F(s, Y) \,.$$

(ii) *We have*

$$T''(F(d_X, Y)) \circ F(X, d_Y) = -T''(F(X, d_Y)) \circ F(d_X, Y) \,,$$
$$T''(d_{F(X,Y)}) \circ d_{F(X,Y)} = F(T(d_X) \circ d_X, Y) + F(X, T'(d_Y) \circ d_Y) \,.$$

Proof. (i) We have the diagram in which all the squares commute except the right bottom square which $(-1)^{nm}$-commutes:

$$
\begin{array}{ccccc}
F(X, Y) & \xrightarrow{\ F(X,t)\ } & F(X, T'^m Y') & \xrightarrow[\ _m\theta'_{X,Y'}\ ]{\sim} & T''^m F(X, Y') \\
\downarrow{\scriptstyle F(s,Y)} & & \downarrow{\scriptstyle F(s,T'^m Y')} & & \downarrow{\scriptstyle T''^m(F(s,Y'))} \\
F(T^n X', Y) & \xrightarrow{\ F(T^n X',t)\ } & F(T^n X', T'^m Y') & \xrightarrow[\ _m\theta'_{T^n X',Y'}\ ]{\sim} & T''^m F(T^n X', Y') \\
{\scriptstyle\sim}\downarrow{\scriptstyle _n\theta_{X',Y}} & & {\scriptstyle _n\theta_{X',T'^m Y'}}\downarrow{\scriptstyle\sim} & (-1)^{nm} & {\scriptstyle\sim}\downarrow{\scriptstyle T''^m(_n\theta_{X',Y'})} \\
T''^n F(X', Y) & \xrightarrow[\ T''^n(F(X',t))\ ]{} & T''^n F(X', T'^m Y') & \xrightarrow[\ T''^n(_m\theta'_{X',Y'})\ ]{\sim} & T''^{n+m} F(X', Y') \,.
\end{array}
$$

(ii) The first equality follows from (i) and immediately implies the second one.
$$\text{q.e.d.}$$

Proposition 11.2.11. *Let* $F\colon (\mathcal{A}, T) \times (\mathcal{A}', T') \to (\mathcal{A}'', T'')$ *be a bifunctor of additive categories with translation. Then* F *defines naturally triangulated bifunctors* $\mathrm{K}(F)\colon \mathrm{K}_d(\mathcal{A}) \times \mathrm{K}_d(\mathcal{A}') \to \mathrm{K}_d(\mathcal{A}'')$ *and* $\mathrm{K}(F)\colon \mathrm{K}_c(\mathcal{A}) \times \mathrm{K}_c(\mathcal{A}') \to \mathrm{K}_c(\mathcal{A}'')$.

Proof. Let us show first that for a morphism $f\colon X \to X'$ in $\mathcal{A}_d$ homotopic to zero and $Y \in \mathcal{A}'_d$, the morphism $F(f, Y)\colon F(X, Y) \to F(X', Y)$ is homotopic to zero. By the assumption, there exists a morphism $s\colon X \to T^{-1}X'$ in $\mathcal{A}$ such that $f = T(s) \circ d_X + T^{-1}(d_{X'}) \circ s$. Set $s'' = F(s, Y)\colon F(X, Y) \to T''^{-1}F(X', Y)$. Then we have

$$
\begin{aligned}
T''(s) &\circ d_{F(X,Y)} + T''^{-1}(d_{F(X',Y)}) \circ s'' \\
&= F\big(T(s) \circ d_X + T^{-1}(d_{X'}) \circ s, Y\big) \\
&\quad + \Big(T'' F(s, Y) \circ F(X, d_Y) + T''^{-1}F(X, d_Y) \circ F(s, Y)\Big),
\end{aligned}
$$

in which the first term is equal to $F(f, Y)$ and the second term vanishes by Lemma 11.2.10 (i). Hence $F(f, Y)$ is homotopic to zero.

Similarly F sends the morphisms homotopic to zero in $\mathcal{A}'$ to morphisms homotopic to zero in $\mathcal{A}''$. Thus F induces a functor $\mathrm{K}(F)\colon \mathrm{K}_d(\mathcal{A}) \times \mathrm{K}_d(\mathcal{A}') \to \mathrm{K}_d(\mathcal{A}'')$. By Lemma 11.2.10 (ii), $\mathrm{K}(F)$ sends $\mathrm{K}_c(\mathcal{A}) \times \mathrm{K}_c(\mathcal{A}')$ to $\mathrm{K}_c(\mathcal{A}'')$. Finally note that $\mathrm{K}(F)$ sends the mapping cones to mapping cones.   q.e.d.

11.3 Complexes in Additive Categories

In this section, $\mathcal{C}$ denotes an additive category.

We introduced the notion of complexes in $\mathcal{C}$ in Definition 8.2.18. We reformulate this in the language of categories with translation.

Let $\mathbb{Z}_d$ denote the set $\mathbb{Z}$, considered as a discrete category. Recall that

- an object X of $\mathcal{C}^{\mathbb{Z}_d}$ is a family $\{X^n\}_{n \in \mathbb{Z}}$ of objects of $\mathcal{C}$,
- for $X = \{X^n\}_{n \in \mathbb{Z}}$ and $Y = \{Y^n\}_{n \in \mathbb{Z}}$ two objects of $\mathcal{C}^{\mathbb{Z}_d}$, a morphism $f\colon X \to Y$ is a family of morphisms $\{f^n\}_{n \in \mathbb{Z}}$, $f^n\colon X^n \to Y^n$.

Definition 11.3.1. *Let* $\mathcal{C}$ *be an additive category. The associated graded category* $(\mathrm{Gr}(\mathcal{C}), T)$ *is the additive category with translation given by* $\mathrm{Gr}(\mathcal{C}) = \mathcal{C}^{\mathbb{Z}_d}$ *and* $(TX)^n = X^{n+1}$ *for* $X = \{X^n\}_{n \in \mathbb{Z}} \in \mathrm{Gr}(\mathcal{C})$.

In $\mathrm{Gr}(\mathcal{C})$, a differential object X is thus a sequence of objects $X^n \in \mathcal{C}$ and morphisms $d_X^n\colon X^n \to X^{n+1}$ $(n \in \mathbb{Z})$. It is visualized as

$$(11.3.1) \qquad \cdots \to X^{n-1} \xrightarrow{d_X^{n-1}} X^n \xrightarrow{d_X^n} X^{n+1} \to \cdots .$$

A morphism of differential objects $f \colon X \to Y$ is a sequence of morphisms $f^n \colon X^n \to Y^n$ making the diagram below commutative:

$$
\begin{array}{ccccccccc}
\cdots & \longrightarrow & X^{n-1} & \xrightarrow{\ d_X^{n-1}\ } & X^n & \xrightarrow{\ d_X^n\ } & X^{n+1} & \longrightarrow & \cdots \\
& & \downarrow{\scriptstyle f^{n-1}} & & \downarrow{\scriptstyle f^n} & & \downarrow{\scriptstyle f^{n+1}} & & \\
\cdots & \longrightarrow & Y^{n-1} & \xrightarrow[\ d_Y^{n-1}\ ]{} & Y^n & \xrightarrow[\ d_Y^n\ ]{} & Y^{n+1} & \longrightarrow & \cdots .
\end{array}
$$

A complex in $\mathrm{Gr}(\mathcal{C})$ is thus a differential object X of $\mathrm{Gr}(\mathcal{C})$ such that

$$ d_X^n \circ d_X^{n-1} = 0 \text{ for all } n \in \mathbb{Z} . $$

It coincides with the notion introduced in Definition 8.2.18.

Notations 11.3.2. (i) For an additive category $\mathcal{C}$, we denote by $\mathrm{C}(\mathcal{C})$ the category consisting of complexes and morphisms of complexes in $\mathrm{Gr}(\mathcal{C})$. In other words, we set

(11.3.2) $$ \mathrm{C}(\mathcal{C}) := (\mathrm{Gr}(\mathcal{C}))_c . $$

An object of $\mathrm{C}(\mathcal{C})$ is often called "a complex in $\mathcal{C}$" and sometimes denoted by $X^\bullet$.

(ii) The translation functor T is also called the *the shift functor* and denoted by [1]. We write $X[n]$ instead of $T^n X$ $(n \in \mathbb{Z})$.

For $X \in \mathrm{C}(\mathcal{C})$ and $n \in \mathbb{Z}$, the object $X[n] \in \mathrm{C}(\mathcal{C})$ is thus given by:

$$
\begin{cases}
(X[n])^i = X^{i+n} , \\
d_{X[n]}^i = (-1)^n d_X^{i+n} .
\end{cases}
$$

Definition 11.3.3. *A complex $X^\bullet$ is* bounded *(resp.* bounded below, resp. bounded above*) if $X^n = 0$ for $|n| \gg 0$ (resp. $n \ll 0$, resp. $n \gg 0$).*

Notations 11.3.4. (i) We denote by $\mathrm{C}^*(\mathcal{C})$ $(* = \mathrm{b}, +, -)$ the full subcategory of $\mathrm{C}(\mathcal{C})$ consisting of bounded complexes (resp. bounded below, resp. bounded above).

(ii) We set $\mathrm{C}^{\mathrm{ub}}(\mathcal{C}) := \mathrm{C}(\mathcal{C})$. (Here, "ub" stands for "unbounded".)

(iii) Let $-\infty \leq a \leq b \leq +\infty$. We denote by $\mathrm{C}^{[a,b]}(\mathcal{C})$ the full additive subcategory of $\mathrm{C}(\mathcal{C})$ consisting of complexes whose j-th component is zero for $j \notin [a, b]$. We also write $\mathrm{C}^{\geq a}(\mathcal{C})$ (resp. $\mathrm{C}^{\leq a}(\mathcal{C})$) for $\mathrm{C}^{[a,\infty]}(\mathcal{C})$ (resp. $\mathrm{C}^{[-\infty,a]}(\mathcal{C})$).

Note that $\mathrm{C}^+(\mathcal{C})$ (resp. $\mathrm{C}^-(\mathcal{C})$, resp. $\mathrm{C}^{\mathrm{b}}(\mathcal{C})$) is the union of the $\mathrm{C}^{\geq a}(\mathcal{C})$'s (resp. $\mathrm{C}^{\leq b}(\mathcal{C})$'s, resp. $\mathrm{C}^{[a,b]}(\mathcal{C})$'s). All these categories are clearly additive.

We consider $\mathcal{C}$ as a full subcategory of $\mathrm{C}^{\mathrm{b}}(\mathcal{C})$ by identifying an object $X \in \mathcal{C}$ with the complex $X^\bullet$ "concentrated in degree 0":

$$ X^\bullet := \cdots \to 0 \to X \to 0 \to \cdots $$

where X stands in degree 0 in this complex.

Examples 11.3.5. (i) Let $f: X \to Y$ be a morphism in $\mathcal{C}$. We identify f with a morphism in $C(\mathcal{C})$. Then $\mathrm{Mc}(f)$ is the complex

$$\cdots \to 0 \to X \xrightarrow{f} Y \to 0 \to \cdots$$

where Y stands in degree 0.

(ii) Consider the morphism of complexes in which X^0 and Y^0 stand in degree 0:

$$
\begin{array}{ccccccc}
0 & \longrightarrow & X^0 & \xrightarrow{d_X} & X^1 & \longrightarrow & 0 \\
& & \downarrow{f^0} & & \downarrow{f^1} & & \\
0 & \longrightarrow & Y^0 & \xrightarrow{d_Y} & Y^1 & \longrightarrow & 0 \, .
\end{array}
$$

The mapping cone is the complex

$$0 \longrightarrow X^0 \xrightarrow{d^{-1}} X^1 \oplus Y^0 \xrightarrow{d^0} Y^1 \longrightarrow 0$$

where $X^1 \oplus Y^0$ stands in degree 0, $d^{-1} = (-d_X) \oplus f^0$ and $d^0 = (f^1, d_Y)$.

Applying Definition 11.2.2, we get the notion of a morphism of complexes homotopic to zero. Hence a morphism $f: X \to Y$ is homotopic to zero if there exist $s^n: X^n \to Y^{n-1}$ such that $f^n = s^{n+1} \circ d_X^n + d_Y^{n-1} \circ s^n$. Such a morphism is visualized by the diagram (which is not commutative):

$$
\begin{array}{ccccccc}
\cdots \longrightarrow & X^{n-1} & \longrightarrow & X^n & \xrightarrow{d_X^n} & X^{n+1} & \longrightarrow \cdots \\
& & \swarrow{s^n} & \downarrow{f^n} \,\, \swarrow{s^{n+1}} & & & \\
\cdots \longrightarrow & Y^{n-1} & \xrightarrow[d_Y^{n-1}]{} & Y^n & \longrightarrow & Y^{n+1} & \longrightarrow \cdots .
\end{array}
$$

Example 11.3.6. If $\mathcal{C}$ is abelian, a complex $0 \to X' \to X \to X'' \to 0$ is homotopic to zero if and only if it splits (see Definition 8.3.15).

Notations 11.3.7. (i) Let $\mathcal{C}$ be an additive category. We set

(11.3.3) $$\mathrm{K}(\mathcal{C}) := \mathrm{K}_c(\mathrm{Gr}(\mathcal{C})) \, .$$

Hence, an object of $\mathrm{K}(\mathcal{C})$ is a complex of objects of $\mathcal{C}$, and a morphism in $C(\mathcal{C})$ homotopic to zero becomes the zero morphism in $\mathrm{K}(\mathcal{C})$.

(ii) We define $\mathrm{K}^*(\mathcal{C})$ ($* = \mathrm{ub, b}, +, -, [a, b]$) as the full subcategory of $\mathrm{K}(\mathcal{C})$ such that $\mathrm{Ob}(\mathrm{K}^*(\mathcal{C})) = \mathrm{Ob}(C^*(\mathcal{C}))$ (see Notations 11.3.4).

Applying Theorem 11.2.6, we get:

Theorem 11.3.8. *The category* $\mathrm{K}(\mathcal{C})$ *endowed with the shift functor* [1] *and the family of* d.t.'s *is a triangulated category. Moreover, the categories* $\mathrm{K}^*(\mathcal{C})$ *($* = \mathrm{b}, +, -$) are full triangulated subcategories.*

The last assertion follows from the fact that $C^*(\mathcal{C})$ is closed by the mapping cones.

Notation 11.3.9. A d.t. $X \xrightarrow{f} Y \xrightarrow{g} Z \xrightarrow{h} X[1]$ is sometimes denoted by

$$X \xrightarrow{f} Y \xrightarrow{g} Z \xrightarrow{+1}$$

for short.

An additive functor of additive categories $F \colon \mathcal{C} \to \mathcal{C}'$ defines naturally an additive functor $C(F) \colon C(\mathcal{C}) \to C(\mathcal{C}')$, by setting

$$C(F)(X)^n = F(X^n), \quad d_{C(F)(X)}^n = F(d_X^n) \, .$$

Of course, $C(F)$ commutes with the shift functor. From now on, if there is no risk of confusion, we shall write F instead of $C(F)$. By Proposition 11.2.9, F induces a functor $K(F) \colon K(\mathcal{C}) \to K(\mathcal{C}')$. If there is no risk of confusion, we still denote this functor by F.

The next result is obvious.

Proposition 11.3.10. *Assume that an additive category $\mathcal{C}$ admits direct sums indexed by a set I. Then so do $C(\mathcal{C})$ and $K(\mathcal{C})$ and the natural functor $C(\mathcal{C}) \to K(\mathcal{C})$ commutes with such direct sums.*

Definition 11.3.11. *Let $\mathcal{C}$ be an additive category and let $n \in \mathbb{Z}$. The stupid truncation functors $\sigma^{\geq n} \colon C(\mathcal{C}) \to C^+(\mathcal{C})$ and $\sigma^{\leq n} \colon C(\mathcal{C}) \to C^-(\mathcal{C})$ are defined as follows. To $X^\bullet$ as in (11.3.1), we associate*

$$\sigma^{\geq n}(X^\bullet) = \cdots \to 0 \to 0 \to X^n \xrightarrow{d_X^n} X^{n+1} \to \cdots \, ,$$

$$\sigma^{\leq n}(X^\bullet) = \cdots \to X^{n-1} \xrightarrow{d_X^{n-1}} X^n \to 0 \to 0 \to \cdots \, .$$

We set $\sigma^{>n} = \sigma^{\geq n+1}$ and $\sigma^{<n} = \sigma^{\leq n-1}$.

See Exercise 11.12 for some properties of the stupid truncation functors.

As we shall see in the next chapter, there are other truncation functors when $\mathcal{C}$ is abelian, and the stupid truncation functors are in fact less useful.

Contravariant Functors

Let $\mathcal{C}$ be an additive category. We shall also encounter complexes with differentials which decrease the degree. We shall denote them using subscripts, as follows:

$$X_\bullet := \cdots \to X_{n+1} \xrightarrow{d_{n+1}^X} X_n \xrightarrow{d_n^X} X_{n-1} \to \cdots \, .$$

By setting $X^n = X_{-n}$ and $d_X^n = d_{-n}^X$, these two notions are equivalent.

Definition 11.3.12. *Let* $F \colon C^{\mathrm{op}} \to C'$ *be an additive functor. We define the functor* $C(F) \colon (C(C))^{\mathrm{op}} \to C(C')$ *by setting:*

$$C(F)(X^\bullet)^n = F(X^{-n}), \quad d^n_{C(F)(X)} = (-1)^n F(d_X^{-n-1}) \,.$$

With the convention of Definition 11.3.12, we get

$$F(X[1]) \simeq F(X)[-1] \,,$$

this isomorphism being given by

$$F(X[1])^n = F(X[1]^{-n})$$
$$= F(X^{1-n}) \xrightarrow{(-1)^{n-1}} F(X^{1-n}) = F(X)^{n-1} = (F(X)[-1])^n \,.$$

Indeed,

$$d^n_{F(X[1])} = (-1)^n F(d^{-n-1}_{X[1]}) = (-1)^n F(-d_X^{-n}) = (-1)^{n+1} F(d_X^{-n}) \,,$$

and

$$d^n_{F(X)[-1]} = -d^{n-1}_{F(X)} = -(-1)^{n-1} F(d_X^{-(n-1)-1}) = (-1)^n F(d_X^{-n}) \,.$$

11.4 Simplicial Constructions

We shall construct complexes and homotopies in additive categories by using the simplicial category Δ (see Exercise 1.21). For the reader's convenience, we recall its definition and some properties.

Definition 11.4.1. (a) *The simplicial category, denoted by* Δ, *is the category whose objects are the finite totally ordered sets and whose morphisms are the order-preserving maps.*
(b) *We denote by* Δ_{inj} *the subcategory of* Δ *such that* $\mathrm{Ob}(\Delta_{inj}) = \mathrm{Ob}(\Delta)$, *the morphisms being the injective order-preserving maps.*
(c) *We denote by* $\widetilde{\Delta}$ *the subcategory of* Δ *consisting of non-empty finite totally ordered sets, the morphisms being given by*

$$\mathrm{Hom}_{\widetilde{\Delta}}(\sigma, \tau) =$$
$$\left\{ u \in \mathrm{Hom}_{\Delta}(\sigma, \tau) \,;\, \begin{array}{l} u \text{ sends the smallest (resp. the largest)} \\ \text{element of } \sigma \text{ to the smallest (resp. the} \\ \text{largest) element of } \tau \end{array} \right\} \,.$$

For integers n, m, denote by $[n, m]$ the totally ordered set $\{k \in \mathbb{Z} \,;\, n \le k \le m\}$. The next results are obvious.

- the natural functor $\Delta \to \mathbf{Set}^f$ is faithful and half-full,

- the full subcategory of $\boldsymbol{\Delta}$ consisting of objects $\{[0, n]\}_{n \geq -1}$ is equivalent to $\boldsymbol{\Delta}$,
- $\boldsymbol{\Delta}$ admits an initial object, namely $\varnothing$, and a terminal object, namely $\{0\}$,
- $\widetilde{\boldsymbol{\Delta}}$ admits an initial object, namely $[0, 1]$, and a terminal object, namely $\{0\}$.

Let us recall that $\widetilde{\boldsymbol{\Delta}}$ is equivalent to $\boldsymbol{\Delta}^{\mathrm{op}}$ (see Exercise 1.21). We define the functor

$$\kappa : \boldsymbol{\Delta} \to \widetilde{\boldsymbol{\Delta}}$$

as follows: for $\tau \in \boldsymbol{\Delta}$, $\kappa(\tau) = \{0\} \sqcup \tau \sqcup \{\infty\}$ where 0 (resp. ∞) is the smallest (resp. largest) element in $\{0\} \sqcup \tau \sqcup \{\infty\}$. Note that the functor $\kappa : \boldsymbol{\Delta} \to \widetilde{\boldsymbol{\Delta}}$ sends $\varnothing$ to $[0, 1]$, sends $\{0\}$ to $[0, 2]$, etc.

Let us denote by

$$d_i^n : [0, n] \to [0, n + 1] \qquad (0 \leq i \leq n + 1)$$

the injective order-preserving map which does not take the value i. In other words

$$d_i^n(k) = \begin{cases} k & \text{for } k < i, \\ k + 1 & \text{for } k \geq i. \end{cases}$$

One checks immediately that

$$(11.4.1) \qquad d_j^{n+1} \circ d_i^n = d_i^{n+1} \circ d_{j-1}^n \text{ for } 0 \leq i < j \leq n + 2.$$

Indeed, each morphism is the unique injective order-preserving map which does not take the values i and j.

For $n > 0$, denote by

$$s_i^n : [0, n] \to [0, n - 1] \qquad (0 \leq i \leq n - 1)$$

the surjective order-preserving map which takes the same value at i and $i + 1$. In other words

$$s_i^n(k) = \begin{cases} k & \text{for } k \leq i, \\ k - 1 & \text{for } k > i. \end{cases}$$

One checks immediately that

$$(11.4.2) \qquad s_j^n \circ s_i^{n+1} = s_{i-1}^n \circ s_j^{n+1} \text{ for } 0 \leq j < i \leq n.$$

Moreover,

$$(11.4.3) \quad \begin{cases} s_j^{n+1} \circ d_i^n = d_i^{n-1} \circ s_{j-1}^n & \text{for } 0 \leq i < j \leq n, \\ s_j^{n+1} \circ d_i^n = \mathrm{id}_{[0,n]} & \text{for } 0 \leq i \leq n + 1, i = j, j + 1, \\ s_j^{n+1} \circ d_i^n = d_{i-1}^{n-1} \circ s_j^n & \text{for } 1 \leq j + 1 < i \leq n + 1. \end{cases}$$

Note that the maps d_i^n are morphisms in the category $\mathbf{\Delta}_{inj}$ and the maps s_i^n are morphisms in the category $\mathbf{\Delta}$.

The category $\mathbf{\Delta}_{inj}$ is visualized by the diagram below

$$(11.4.4) \qquad \emptyset \xrightarrow{d_0^{-1}} \{0\} \overset{d_0^0}{\underset{d_1^0}{\rightrightarrows}} \{0, 1\} \overset{d_0^1}{\underset{d_2^1}{\Rrightarrow}} \{0, 1, 2\} \cdots\Rrightarrow$$

Let $\mathcal{C}$ be an additive category and $F\colon \mathbf{\Delta}_{inj} \to \mathcal{C}$ a functor. We set

$$F^n = \begin{cases} F([0, n]) & \text{for } n \geq -1, \\ 0 & \text{otherwise,} \end{cases}$$

$$d_F^n \colon F^n \to F^{n+1} \quad \text{where } d_F^n = \sum_{i=0}^{n+1}(-1)^i F(d_i^n) \text{ for } n \geq -1.$$

Consider the differential object $F^\bullet$:

$$(11.4.5) \qquad F^\bullet := \cdots \to 0 \to F^{-1} \xrightarrow{d_F^{-1}} F^0 \xrightarrow{d_F^0} F^1 \to \cdots .$$

Proposition 11.4.2. *Let $F\colon \mathbf{\Delta}_{inj} \to \mathcal{C}$ be a functor.*

(i) *The differential object $F^\bullet$ is a complex.*

(ii) *Assume that there exist morphisms $s_F^n\colon F^n \to F^{n-1}$ satisfying:*

$$(11.4.6) \qquad \begin{cases} s_F^{n+1} \circ F(d_0^n) = \mathrm{id}_{F^n} & \text{for } n \geq -1, \\ s_F^{n+1} \circ F(d_{i+1}^n) = F(d_i^{n-1}) \circ s_F^n & \text{for } n \geq i \geq 0. \end{cases}$$

Then $F^\bullet$ is homotopic to zero.

Proof. (i) By (11.4.1), we have

$$d_F^{n+1} \circ d_F^n = \sum_{j=0}^{n+2}\sum_{i=0}^{n+1}(-1)^{i+j} F(d_j^{n+1} \circ d_i^n)$$

$$= \sum_{0 \leq j \leq i \leq n+1}(-1)^{i+j} F(d_j^{n+1} \circ d_i^n) + \sum_{0 \leq i < j \leq n+2}(-1)^{i+j} F(d_j^{n+1} \circ d_i^n)$$

$$= \sum_{0 \leq j \leq i \leq n+1}(-1)^{i+j} F(d_j^{n+1} \circ d_i^n) + \sum_{0 \leq i < j \leq n+2}(-1)^{i+j} F(d_i^{n+1} \circ d_{j-1}^n)$$

$$= 0 .$$

(ii) We have

$$s_F^{n+1} \circ d_F^n + d_F^{n-1} \circ s_F^n$$

$$= \sum_{i=0}^{n+1} (-1)^i s_F^{n+1} \circ F(d_i^n) + \sum_{i=0}^{n} (-1)^i F(d_i^{n-1}) \circ s_F^n$$

$$= s_F^{n+1} \circ F(d_0^n) + \sum_{i=0}^{n} (-1)^{i+1} s_F^{n+1} \circ F(d_{i+1}^n) + \sum_{i=0}^{n} (-1)^i F(d_i^{n-1}) \circ s_F^n$$

$$= \mathrm{id}_{F^n} + \sum_{i=0}^{n} (-1)^{i+1} F(d_i^{n-1}) \circ s_F^n + \sum_{i=0}^{n} (-1)^i F(d_i^{n-1}) \circ s_F^n$$

$$= \mathrm{id}_{F^n}.$$

q.e.d.

Corollary 11.4.3. *Let $F \colon \mathbf{\Delta}_{inj} \to \mathcal{C}$ be a functor. Assume that there exists a functor $\widetilde{F} \colon \widetilde{\mathbf{\Delta}} \to \mathcal{C}$ such that F is isomorphic to the composition $\mathbf{\Delta}_{inj} \to \mathbf{\Delta} \xrightarrow{\kappa} \widetilde{\mathbf{\Delta}} \xrightarrow{\widetilde{F}} \mathcal{C}$. Then the complex $F^\bullet$ is homotopic to zero.*

Proof. By identifying $\kappa([0,n])$ with $[0, n+2]$, we have $\kappa(d_i^n) = d_{i+1}^{n+2}$ and $F(d_i^n) = \widetilde{F}(d_{i+1}^{n+2})$. Set $s_F^n = \widetilde{F}(s_0^{n+2}) \colon F^n \to F^{n-1}$. Then (11.4.3) implies (11.4.6).

q.e.d.

11.5 Double Complexes

Let $\mathcal{C}$ be an additive category. A double complex X is the data of

$$\{X^{n,m}, d_X'^{n,m}, d_X''^{n,m}\}_{n,m \in \mathbb{Z}}$$

where $X^{n,m} \in \mathcal{C}$ and the pair of the "differentials" $d_X'^{n,m} \colon X^{n,m} \to X^{n+1,m}$, $d_X''^{n,m} \colon X^{n,m} \to X^{n,m+1}$ satisfy:

$$d_X'^{n+1,m} \circ d_X'^{n,m} = 0, \qquad d_X''^{n,m+1} \circ d_X''^{n,m} = 0,$$
$$d_X''^{n+1,m} \circ d_X'^{n,m} = d_X'^{n,m+1} \circ d_X''^{n,m}.$$

A double complex may be represented by a commutative diagram:

We shall sometimes write $X^{\bullet,\bullet}$ instead of X to emphasize the fact that we are dealing with a double complex.

There is a natural notion of a morphism of double complexes, and we obtain the additive category $C^2(\mathcal{C})$ of double complexes.

Notation 11.5.1. The functor

$$(11.5.1) \qquad F_I \text{ (resp. } F_{II}) : C^2(\mathcal{C}) \to C(C(\mathcal{C}))$$

is defined by associating to a double complex X the complex whose components are the rows (resp. the columns) of X.

For example $F_I(X)$ is the (simple) complex (X_I, d_I) in $C(\mathcal{C})$, where

$$\begin{cases} X_I^n \in C(\mathcal{C}) \text{ is given by } \{X^{n,\bullet}, d_X''^{n,\bullet}\} \text{ and} \\ d_I^n : X_I^n \to X_I^{n+1} \text{ is given by } d_X'^{n,\bullet}. \end{cases}$$

The two functors F_I and F_{II} are clearly equivalences of categories.

Notation 11.5.2. Denoting by T the shift functor in $C(\mathcal{C})$, we define the translation functors in $C^2(\mathcal{C})$:

$$T_a = F_a^{-1} \circ T \circ F_a \ (a = I, II) .$$

Hence,

$$(T_I X)^{n,m} = X^{n+1,m}, \ d_{T_I X}'^{n,m} = -d_X'^{n+1,m}, \ d_{T_I X}''^{n,m} = d_X''^{n+1,m} ,$$
$$(T_{II} X)^{n,m} = X^{n,m+1}, \ d_{T_{II} X}'^{n,m} = d_X'^{n,m+1}, \ d_{T_{II} X}''^{n,m} = -d_X''^{n,m+1} .$$

Assume that $\mathcal{C}$ admits countable direct sums. To a double complex $X \in C^2(\mathcal{C})$ we associate a differential object $\text{tot}_\oplus(X)$ by setting:

$$(11.5.2) \qquad \begin{aligned} \text{tot}_\oplus(X)^k &= \oplus_{m+n=k} X^{n,m} , \\ d_{\text{tot}_\oplus(X)}^k |_{X^{n,m}} &= d_X'^{n,m} \oplus (-1)^n d_X''^{n,m}. \end{aligned}$$

This is visualized by the diagram:

$$
\begin{array}{ccc}
X^{n,m} & \xrightarrow{(-1)^n d_X''} & X^{n,m+1} \\
\downarrow{\scriptstyle d_X'} & & \\
X^{n+1,m}. & &
\end{array}
$$

If there is no risk of confusion, we shall write $d_{\text{tot}(X)}$ instead of $d_{\text{tot}_\oplus(X)}$.

Proposition 11.5.3. *Assume that $\mathcal{C}$ admits countable direct sums. Then the differential object $\{\text{tot}_\oplus(X)^k, d_{\text{tot}_\oplus(X)}^k\}_{k \in \mathbb{Z}}$ is a complex (i.e., $d_{\text{tot}_\oplus(X)}^{k+1} \circ d_{\text{tot}_\oplus(X)}^k = 0$),*

Proof. Consider the restriction of $d^{k+1}_{\text{tot}(X)} \circ d^k_{\text{tot}(X)}$ to $X^{n,m}$:

$$d^{k+1}_{\text{tot}(X)} \circ d^k_{\text{tot}(X)} : X^{n,m} \to X^{n+2,m} \oplus X^{n+1,m+1} \oplus X^{n,m+2}$$

$$d^{k+1}_{\text{tot}(X)} \circ d^k_{\text{tot}(X)} = d' \circ d' \oplus \left(d' \circ (-1)^n d'' + (-1)^{n+1} d'' \circ d' \right) \oplus d'' \circ d''$$
$$= 0 .$$

q.e.d.

Assume that $\mathcal{C}$ admits countable products. To a double complex $X \in \mathrm{C}^2(\mathcal{C})$ we associate a differential object $\text{tot}_\pi(X)$ by setting:

$$\text{tot}_\pi(X)^k = \prod_{m+n=k} X^{n,m} ,$$

$$(d_{\text{tot}_\pi(X)})^{n+m-1} = d'^{\,n-1,m}_X + (-1)^n d''^{\,n,m-1}_X .$$

It means that the composition

$$\text{tot}_\pi(X)^{n+m-1} \xrightarrow{d^{n+m-1}_{\text{tot}_\pi(X)}} \text{tot}_\pi(X)^{n+m} \to X^{n,m}$$

is the sum of $\text{tot}_\pi(X)^{n+m-1} \to X^{n-1,m} \xrightarrow{d'^{\,n-1,m}_X} X^{n,m}$ and $\text{tot}_\pi(X)^{n+m-1} \to X^{n,m-1} \xrightarrow{(-1)^n d''^{\,n,m-1}_X} X^{n,m}$. This is visualized by the diagram:

$$\begin{array}{ccc} & & X^{n-1,m} \\ & & \downarrow{\scriptstyle d'_X} \\ X^{n,m-1} & \xrightarrow[(-1)^n d''_X]{} & X^{n,m} . \end{array}$$

Proposition 11.5.4. *Assume that $\mathcal{C}$ admits countable products. Then the differential object $\{\text{tot}_\pi(X)^k, d^k_{\text{tot}_\pi(X)}\}_{k\in\mathbb{Z}}$ is a complex (i.e., $d^{k+1}_{\text{tot}_\pi(X)} \circ d^k_{\text{tot}_\pi(X)} = 0$).*

The proof goes as for Proposition 11.5.3.

Assume that $\mathcal{C}$ admits countable direct sums and let $X \in \mathrm{C}^2(\mathcal{C})$. Define $v(X) \in \mathrm{C}^2(\mathcal{C})$ by setting

$$v(X^{n,m}) = X^{m,n}, \quad v(d'^{\,n,m}_X) = d'^{\,m,n}_X, \quad v(d''^{\,n,m}_X) = d''^{\,m,n}_X .$$

Now, for each $(n,m) \in \mathbb{Z} \times \mathbb{Z}$, define

(11.5.3) $r^{n,m} : X^{n,m} \to v(X)^{m,n}$ as $(-1)^{nm} \, \mathrm{id}_{X^{n,m}}$.

Proposition 11.5.5. *Assume that $\mathcal{C}$ admits countable direct sums. Let $X \in \mathrm{C}^2(\mathcal{C})$. The morphisms $r^{n,m}$ define an isomorphism in $\mathrm{C}(\mathcal{C})$:*

$$r : \text{tot}_\oplus(X) \xrightarrow{\sim} \text{tot}_\oplus(v(X)) .$$

If $\mathcal{C}$ admits countable products, the same isomorphism holds after replacing $\oplus$ by π.

Proof. It is enough to prove that the diagram below commutes, which is obvious:

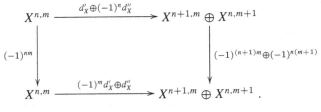

q.e.d.

Remark 11.5.6. One trick to treat signs is to introduce the formal notation $X[1] = \mathbb{Z}[1] \otimes X$ where $\mathbb{Z}[1]$ is $\mathbb{Z}$ viewed as a complex of $\mathbb{Z}$-modules concentrated in degree -1. For two complexes X, Y, let us define *formally* the differential of $x^p \otimes y^q \in X^p \otimes Y^q$ (of course, $x^p \otimes y^q$ has no meaning) by

$$d(x^p \otimes y^q) = dx^p \otimes y^q + (-1)^p (x^p \otimes dy^q) .$$

Then, Proposition 11.5.5 implies that the morphism $X \otimes Y \xrightarrow{\sim} Y \otimes X$ given by

$$X^p \otimes Y^q \ni x^p \otimes y^q \mapsto (-1)^{pq} y^q \otimes x^p \in Y^q \otimes X^p$$

commutes with the differential.
With this convention, the morphism

$$\mathbb{Z}[1] \otimes X \to X \otimes \mathbb{Z}[1] ,$$
$$1 \otimes x \mapsto x \otimes 1$$

does not commute with the differential, while the morphism defined by

$$\mathbb{Z}[1] \otimes X^n \to X^n \otimes \mathbb{Z}[1] ,$$
$$1 \otimes x \mapsto (-1)^n x \otimes 1$$

commutes.

Now consider the finiteness condition:

(11.5.4) $\{(n, m) \in \mathbb{Z} \times \mathbb{Z} ; n + m = k, X^{n,m} \neq 0\}$ is finite for all $k \in \mathbb{Z}$.

We denote by $C_f^2(\mathcal{C})$ the full subcategory of $C^2(\mathcal{C})$ consisting of objects X satisfying (11.5.4). Of course, if $X \in C_f^2(\mathcal{C})$, then $\mathrm{tot}_\oplus(X)$ and $\mathrm{tot}_\pi(X)$ are well defined and isomorphic. We simply denote this complex by $\mathrm{tot}(X)$.

Example 11.5.7. Let $f : X \to Y$ be a morphism in $C(\mathcal{C})$. Set:

$$Z^{-1,k} = X^k, \quad Z^{0,k} = Y^k$$

and consider the double complex Z:

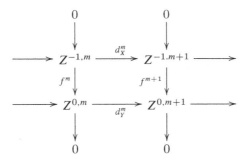

Then $\mathrm{tot}(Z)$ is $\mathrm{Mc}(f)$, the mapping cone of f. In other words, if $Z^{\bullet,\bullet}$ is a double complex such that $Z^{p,\bullet} = 0$ for $p \neq -1, 0$, then $\mathrm{tot}(Z^{\bullet,\bullet})$ is the mapping cone of $d'^{,\bullet}\colon Z^{-1,\bullet} \to Z^{0,\bullet}$.

Definition 11.5.8. *A morphism* $f\colon X \to Y$ *in* $\mathrm{C}^2(\mathcal{C})$ *is homotopic to zero if there exist morphisms* $t_1^{n,m}\colon X^{n,m} \to Y^{n-1,m}$ *and* $t_2^{n,m}\colon X^{n,m} \to Y^{n,m-1}$ *such that*

$$d''^{n-1,m}_Y \circ t_1^{n,m} = t_1^{n,m+1} \circ d''^{n,m}_X \,,$$
$$d''^{n,m-1}_Y \circ t_2^{n,m} = t_2^{n+1,m} \circ d'^{n,m}_X \,,$$
$$f^{n,m} = d'^{n-1,m}_Y \circ t_1^{n,m} + t_1^{n+1,m} \circ d'^{n,m}_X$$
$$+ d''^{n,m-1}_Y \circ t_2^{n,m} + t_2^{n,m+1} \circ d''^{n,m}_X \,.$$

It is easily checked that if f is homotopic to zero then $\mathrm{tot}_\oplus$ or tot_π is homotopic to zero whenever they exist.

11.6 Bifunctors

Let $F\colon \mathcal{C} \times \mathcal{C}' \to \mathcal{C}''$ be an additive bifunctor (i.e., $F(\cdot,\cdot)$ is additive with respect to each argument). It defines an additive bifunctor $\mathrm{C}^2(F)\colon \mathrm{C}(\mathcal{C}) \times \mathrm{C}(\mathcal{C}') \to \mathrm{C}^2(\mathcal{C}'')$. In other words, if $X \in \mathrm{C}(\mathcal{C})$ and $X' \in \mathrm{C}(\mathcal{C}')$ are complexes, then $\mathrm{C}^2(F)(X,X')$ is a double complex. If there is no risk of confusion, we often write F instead of $\mathrm{C}^2(F)$.

Assume that $\mathcal{C}''$ admits countable direct sums. We define the functor

$$F_\oplus^\bullet\ :\ \mathrm{C}(\mathcal{C}) \times \mathrm{C}(\mathcal{C}') \to \mathrm{C}(\mathcal{C}'')$$
$$F_\oplus^\bullet(X,Y) = \mathrm{tot}_\oplus(F(X,Y))\,.$$

Similarly, if $\mathcal{C}''$ admits countable products, we set

$$F_\pi^\bullet(X,Y) = \mathrm{tot}_\pi(F(X,Y))\,.$$

Let us denote by the same letter T the shift functors on $\mathrm{C}(\mathcal{C})$, $\mathrm{C}(\mathcal{C}')$ $\mathrm{C}(\mathcal{C}'')$.

Lemma 11.6.1. *The functor* $F_{\oplus}^{\bullet}$ *(resp.* $F_{\pi}^{\bullet}$*) induces an additive bifunctor from* $C(\mathcal{C}) \times C(\mathcal{C}')$ *to* $C(\mathcal{C}'')$.

Proof. This follows immediately from Proposition 11.5.3 (resp. 11.5.4). q.e.d.

The full subcategory $(C(\mathcal{C}) \times C(\mathcal{C}'))_f$ of $C(\mathcal{C}) \times C(\mathcal{C}')$ is defined similarly as the subcategory $C_f^2(\mathcal{C})$ of $C^2(\mathcal{C})$. Then the two functors $F_{\oplus}^{\bullet}$ and $F_{\pi}^{\bullet}$ are well defined on $(C(\mathcal{C}) \times C(\mathcal{C}'))_f$ and are isomorphic. We denote it by $F^{\bullet}$:

$$F^{\bullet}(X, Y) = \mathrm{tot}(F(X, Y)), \quad (X, Y) \in (C(\mathcal{C}) \times C(\mathcal{C}'))_f .$$

Hence, the functor F induces well defined bifunctors of additive categories, all denoted by $F^{\bullet}$:

$$F^{\bullet}: C^+(\mathcal{C}) \times C^+(\mathcal{C}') \to C^+(\mathcal{C}''), \quad F^{\bullet}: C^-(\mathcal{C}) \times C^-(\mathcal{C}') \to C^-(\mathcal{C}'') ,$$
$$F^{\bullet}: C^b(\mathcal{C}) \times C(\mathcal{C}') \to C(\mathcal{C}''), \quad F^{\bullet}: C(\mathcal{C}) \times C^b(\mathcal{C}') \to C(\mathcal{C}'') .$$

Examples 11.6.2. (i) Consider the bifunctor $\mathrm{Hom}_{\mathcal{C}}: \mathcal{C} \times \mathcal{C}^{\mathrm{op}} \to \mathrm{Mod}(\mathbb{Z})$, $(Y, X) \mapsto \mathrm{Hom}_{\mathcal{C}}(X, Y)$. We shall write $\mathrm{Hom}_{\mathcal{C}}^{\bullet,\bullet}$ instead of $C^2(\mathrm{Hom}_{\mathcal{C}})$. If X and Y are two objects of $C(\mathcal{C})$, we have

$$\mathrm{Hom}_{\mathcal{C}}^{\bullet,\bullet}(X, Y)^{n,m} = \mathrm{Hom}_{\mathcal{C}}(X^{-m}, Y^n) ,$$
$$d'^{n,m} = \mathrm{Hom}_{\mathcal{C}}(X^{-m}, d_Y^n) ,$$
$$d''^{n,m} = \mathrm{Hom}_{\mathcal{C}}((-1)^n d_X^{-n-1}, Y^m) .$$

Here, the calculation of d'' follows from Definition 11.3.12.

Note that $\mathrm{Hom}_{\mathcal{C}}^{\bullet,\bullet}(X, Y)$ is a double complex in the category $\mathrm{Mod}(\mathbb{Z})$, which should not be confused with the group $\mathrm{Hom}_{C(\mathcal{C})}(X, Y)$ (see Proposition 11.7.3 below).

(ii) Let R be a k-algebra. The functor $\cdot \otimes_R \cdot: \mathrm{Mod}(R^{\mathrm{op}}) \times \mathrm{Mod}(R) \to \mathrm{Mod}(k)$ defines an additive bifunctor

$$(\cdot \otimes_R \cdot)_{\oplus}: C(\mathrm{Mod}(R^{\mathrm{op}})) \times C(\mathrm{Mod}(R)) \to C(\mathrm{Mod}(k)) .$$

The functor $(\cdot \otimes_R \cdot)_{\oplus}$ is usually still denoted by $\cdot \otimes_R \cdot$.

The above result may be formulated as follows in terms of a bifunctor of additive categories with translation.

Assuming that $\mathcal{C}''$ admits countable direct sums, let us define the functor

$$F_{\oplus}: \mathrm{Gr}(\mathcal{C}) \times \mathrm{Gr}(\mathcal{C}') \to \mathrm{Gr}(\mathcal{C}'')$$

as above (i.e., $F_{\oplus}(X, Y)^n = \oplus_{i+j=n} F(X^i, Y^j)$). We define the functor

$$\theta_{X,Y}: F_{\oplus}(TX, Y) \xrightarrow{\sim} T F_{\oplus}(X, Y) \quad \text{and}$$
$$\theta'_{X,Y}: F_{\oplus}(X, TY) \xrightarrow{\sim} T F_{\oplus}(X, Y)$$

as follows. The composition

$$F((TX)^i, Y^j) \to F_\oplus(TX, Y)^{i+j} \xrightarrow{\theta_{X,Y}} \left(T F_\oplus(X, Y)\right)^{i+j}$$

is given by the canonical embedding $F(X^{i+1}, Y^j) \to \left(F_\oplus(X, Y)\right)^{i+j+1} = \left(T F_\oplus(X, Y)\right)^{i+j}$, and the composition

$$F(X^i, (TY)^j) \to F_\oplus(X, TY)^{i+j} \xrightarrow{\theta'_{X,Y}} \left(T F_\oplus(X, Y)\right)^{i+j}$$

is given by the canonical embedding $F(X^i, Y^{j+1}) \to \left(F_\oplus(X, Y)\right)^{i+j+1} = \left(T F_\oplus(X, Y)\right)^{i+j}$ multiplied by $(-1)^i$.

Lemma 11.6.3. *The functor $F_\oplus \colon \mathrm{Gr}(\mathcal{C}) \times \mathrm{Gr}(\mathcal{C}') \to \mathrm{Gr}(\mathcal{C}'')$ is a bifunctor of additive categories with translation.*

Proof. The diagram

$$
\begin{array}{ccc}
F(TX, T'Y) & \xrightarrow{\theta_{X,T'Y}} & T'' F(X, T'Y) \\
{\scriptstyle \theta'_{TX,Y}} \downarrow & & \downarrow {\scriptstyle T''\theta'_{X,Y}} \\
T'' F(TX, Y) & \xrightarrow[T''\theta_{X,Y}]{} & T''^2 F(X, Y)
\end{array}
$$

in Definition 10.1.1 (v) reduces to the following diagram when we restrict it to $F(X^{i+1}, Y^{j+1}) = F((TX)^i, (T'Y)^j)$:

$$
\begin{array}{ccc}
F(X^{i+1}, Y^{j+1}) & \xrightarrow{\mathrm{id}} & F(X^{i+1}, Y^{j+1}) \\
{\scriptstyle (-1)^i} \downarrow & & \downarrow {\scriptstyle (-1)^{i+1}} \\
F(X^{i+1}, Y^{j+1}) & \xrightarrow[\mathrm{id}]{} & F(X^{i+1}, Y^{j+1}),
\end{array}
$$

and this last diagram is anti-commutative.   q.e.d.

Note that the differential of $F_\oplus(X, Y)$ for $X \in C(\mathcal{C})$ and $Y \in C(\mathcal{C}')$ given by (11.2.2) coincides with the one given by (11.5.2).

Applying Proposition 11.2.11, we get

Proposition 11.6.4. *Let $F \colon \mathcal{C} \times \mathcal{C}' \to \mathcal{C}''$ be an additive bifunctor.*

(i) *The bifunctor F induces well defined triangulated bifunctors $\mathrm{K}^+(\mathcal{C}) \times \mathrm{K}^+(\mathcal{C}') \to \mathrm{K}^+(\mathcal{C}'')$, $\mathrm{K}^-(\mathcal{C}) \times \mathrm{K}^-(\mathcal{C}') \to \mathrm{K}^-(\mathcal{C}'')$, $\mathrm{K}^b(\mathcal{C}) \times \mathrm{K}(\mathcal{C}') \to \mathrm{K}(\mathcal{C}'')$ and $\mathrm{K}(\mathcal{C}) \times \mathrm{K}^b(\mathcal{C}') \to \mathrm{K}(\mathcal{C}'')$.*

(ii) *Assume that $\mathcal{C}''$ admits countable direct sums (resp. countable products). Then $F_\oplus^\bullet$ (resp. $F_\pi^\bullet$) induces a well defined triangulated bifunctor $\mathrm{K}(\mathcal{C}) \times \mathrm{K}(\mathcal{C}') \to \mathrm{K}(\mathcal{C}'')$.*

Denote by v the canonical isomorphism $\mathcal{C}' \times \mathcal{C} \simeq \mathcal{C} \times \mathcal{C}'$, $v(Y, X) = (X, Y)$ and let $G = F \circ v \colon \mathcal{C}' \times \mathcal{C} \to \mathcal{C}''$. In other words,

$$G(Y, X) = F(X, Y) .$$

Proposition 11.6.5. *Assume that $\mathcal{C}''$ admits countable direct sums. Let $X \in \mathrm{C}(\mathcal{C})$ and $Y \in \mathrm{C}(\mathcal{C}')$. For each $(n, m) \in \mathbb{Z} \times \mathbb{Z}$, define $r \colon F(X^n, Y^m) \to G(Y^m, X^n)$ as $(-1)^{nm}$. Then r defines an isomorphism of complexes in $\mathrm{C}(\mathcal{C}'')$:*

$$r \colon F_\oplus^\bullet(X, Y) \xrightarrow{\sim} G_\oplus^\bullet(Y, X) .$$

If $\mathcal{C}''$ admits countable products, the same isomorphism holds after replacing $\oplus$ by π.

Proof. This follows from Proposition 11.5.5.   q.e.d.

11.7 The Complex Hom$^\bullet$

We shall study the complex $(\mathrm{Hom}_\mathcal{C})^\bullet_\pi(X, Y)$, when X and Y are complexes in $\mathcal{C}$.

For short, we shall write $\mathrm{Hom}^\bullet_\mathcal{C}$ instead of $(\mathrm{Hom}_\mathcal{C})^\bullet_\pi$. Hence

$$\mathrm{Hom}^\bullet_\mathcal{C}(X, Y) = \mathrm{tot}_\pi(\mathrm{Hom}^{\bullet,\bullet}_\mathcal{C}(X, Y)) .$$

We shall also write for short $\mathrm{Hom}_\mathcal{C}(X, Y)^n$ instead of $(\mathrm{Hom}_\mathcal{C})^\bullet_\pi(X, Y)^n$ and d^n instead of $d^n_{\mathrm{Hom}^\bullet_\mathcal{C}(X,Y)}$.

Note that $\mathrm{Hom}^\bullet_\mathcal{C}$ defines functors

$$\mathrm{Hom}^\bullet_\mathcal{C} \colon \mathrm{C}(\mathcal{C}) \times \mathrm{C}(\mathcal{C})^{\mathrm{op}} \to \mathrm{C}(\mathrm{Mod}(\mathbb{Z})) ,$$
$$\mathrm{Hom}^\bullet_\mathcal{C} \colon \mathrm{K}(\mathcal{C}) \times \mathrm{K}(\mathcal{C})^{\mathrm{op}} \to \mathrm{K}(\mathrm{Mod}(\mathbb{Z})) .$$

Convention 11.7.1. When considering the bifunctor $\mathrm{Hom}^\bullet_\mathcal{C}$ (or its variants, such as $\mathcal{H}om$ or RHom, etc. in the subsequent chapters), we shall consider it as defined on $\mathrm{C}(\mathcal{C}) \times \mathrm{C}(\mathcal{C})^{\mathrm{op}}$ (or $\mathrm{K}(\mathcal{C}) \times \mathrm{K}(\mathcal{C})^{\mathrm{op}}$). Hence, to a pair $(X, Y) \in \mathrm{C}(\mathcal{C}) \times \mathrm{C}(\mathcal{C})^{\mathrm{op}}$, this functor associates $\mathrm{Hom}^\bullet_\mathcal{C}(Y, X)$. The reason of this convention is that, together with Definition 11.3.12, the differential whose components are given by (11.7.3) will satisfy the formula in Exercise 11.11. However, by Proposition 11.6.5, we may also regard $\mathrm{Hom}^\bullet_\mathcal{C}(\cdot, \cdot)$ as a bifunctor from $\mathrm{K}(\mathcal{C})^{\mathrm{op}} \times \mathrm{K}(\mathcal{C})$ to $\mathrm{K}(\mathrm{Mod}(\mathbb{Z}))$.

If X and Y are two objects of $\mathrm{C}(\mathcal{C})$, we get

(11.7.1) $$\mathrm{Hom}_\mathcal{C}(X, Y)^n = \prod_{k \in \mathbb{Z}} \mathrm{Hom}_\mathcal{C}(X^k, Y^{n+k})$$

and

(11.7.2) $d^n_{\mathrm{Hom}^{\bullet}_{\mathcal{C}}(X,Y)} : \mathrm{Hom}_{\mathcal{C}}(X,Y)^n \to \mathrm{Hom}_{\mathcal{C}}(X,Y)^{n+1}$

is given as follows. To $f = \{f^k\}_k \in \prod_{k \in \mathbb{Z}} \mathrm{Hom}_{\mathcal{C}}(X^k, Y^{n+k})$ we associate $d^n f = \{g^k\}_k \in \prod_{k \in \mathbb{Z}} \mathrm{Hom}_{\mathcal{C}}(X^k, Y^{n+k+1})$, with

$$g^k = d'^{n+k,-k} f^k + (-1)^{k+n+1} d''^{k+n+1,-k-1} f^{k+1} .$$

In other words, the component of $d^n f$ in $\mathrm{Hom}_{\mathcal{C}}(X,Y)^{n+1}$ will be

(11.7.3) $(d^n f)^k = d^{k+n}_Y \circ f^k + (-1)^{n+1} f^{k+1} \circ d^k_X \in \mathrm{Hom}_{\mathcal{C}}(X^k, Y^{n+k}) .$

Notation 11.7.2. Recall that we write d^n instead of $d^n_{\mathrm{Hom}^{\bullet}_{\mathcal{C}}}$. We set

$$Z^0(\mathrm{Hom}^{\bullet}_{\mathcal{C}}(X,Y)) = \mathrm{Ker}\, d^0 ,$$
$$B^0(\mathrm{Hom}^{\bullet}_{\mathcal{C}}(X,Y)) = \mathrm{Im}\, d^{-1} ,$$
$$H^0(\mathrm{Hom}^{\bullet}_{\mathcal{C}}(X,Y)) = (\mathrm{Ker}\, d^0)/(\mathrm{Im}\, d^{-1}) .$$

Proposition 11.7.3. *Let $\mathcal{C}$ be an additive category and let $X, Y \in C(\mathcal{C})$. There are isomorphisms:*

$$Z^0(\mathrm{Hom}^{\bullet}_{\mathcal{C}}(X,Y)) \simeq \mathrm{Hom}_{C(\mathcal{C})}(X,Y) ,$$
$$B^0(\mathrm{Hom}^{\bullet}_{\mathcal{C}}(X,Y)) \simeq \mathrm{Ht}(X,Y) ,$$
$$H^0(\mathrm{Hom}^{\bullet}_{\mathcal{C}}(X,Y)) \simeq \mathrm{Hom}_{K(\mathcal{C})}(X,Y) .$$

Proof. (i) Let us calculate $Z^0(\mathrm{Hom}^{\bullet}_{\mathcal{C}}(X,Y))$. By (11.7.3) the component of $d^0\{f^k\}_k$ in $\mathrm{Hom}_{\mathcal{C}}(X^k, Y^{k+1})$ will be zero if and only if $d^k_Y \circ f^k = f^{k+1} \circ d^k_X$, that is, if the family $\{f^k\}_k$ defines a morphism of complexes.
(ii) Let us calculate $B^0(\mathrm{Hom}^{\bullet}_{\mathcal{C}}(X,Y))$. An element $f^k \in \mathrm{Hom}^{\bullet}_{\mathcal{C}}(X^k, Y^k)$ will be in the image of d^{-1} if it can be written as $f^k = d^{k-1}_Y \circ s^k + s^{k+1} \circ d^k_X$ with $s^k \in \mathrm{Hom}_{\mathcal{C}}(X^k, Y^{k-1})$.
(iii) The last isomorphism follows from the others.   q.e.d.

Exercises

Exercise 11.1. Let $\mathcal{C}$ be a category and let $T: \mathcal{C} \to \mathcal{C}$ be a functor. Let $T^{-1}\mathcal{C}$ be the category defined as follows:

$$\mathrm{Ob}(T^{-1}\mathcal{C}) = \{(X, n)\, ; \, X \in \mathrm{Ob}(\mathcal{C}), \, n \in \mathbb{Z}\} ,$$
$$\mathrm{Hom}_{T^{-1}\mathcal{C}}\big((X,n),(Y,m)\big) = \varinjlim_{k \geq -n, -m} \mathrm{Hom}_{\mathcal{C}}(T^{n+k}X, T^{m+k}Y) .$$

(i) Prove that $T^{-1}\mathcal{C}$ is a well-defined category.

(ii) Prove that the functor $\widetilde{T}$ which sends (X, n) to $(X, n+1)$ is a well-defined translation functor.

(iii) Prove that the functor $\varphi \colon \mathcal{C} \to T^{-1}\mathcal{C}$ which sends X to $(X, 0)$ is well defined and $\widetilde{T} \circ \varphi \simeq \varphi \circ T$.

(iv) Prove that the category with translation $(T^{-1}\mathcal{C}, \widetilde{T})$ has the following universal property: for any category with translation $(\mathcal{A}, T')$ and any functor $\psi \colon \mathcal{C} \to \mathcal{A}$ such that $T' \circ \psi \simeq \psi \circ T$ there exist a functor $\psi' \colon (T^{-1}\mathcal{C}, \widetilde{T}) \to (\mathcal{A}, T')$ of categories with translation and an isomorphism $\psi' \circ \varphi \simeq \psi$. Moreover such a ψ' is unique up to an isomorphism.

Exercise 11.2. Let $(\mathcal{A}, T)$ be an additive category with translation and assume to be given a morphism of functors $\eta \colon \mathrm{id}_\mathcal{A} \to T^2$ such that $\eta \circ T = T \circ \eta$, that is, $\eta_{TX} = T(\eta_X)$ for any $X \in \mathcal{A}$. Let $\mathcal{A}_\eta$ be the full subcategory of $\mathcal{A}_d$ consisting of differential objects X such that $T(d_X) \circ d_X = \eta_X$.

(i) Let X and Y be objects of $\mathcal{A}_\eta$, and let $u \colon X \to T^{-1}Y$ be a morphism in $\mathcal{A}$. Prove that $T(u) \circ d_X + T^{-1}(d_Y) \circ u \colon X \to Y$ is a morphism in $\mathcal{A}_\eta$.

(ii) Prove that the mapping cone of any morphism in $\mathcal{A}_\eta$ belongs to $\mathcal{A}_\eta$.

(iii) Let $\mathrm{K}_\eta(\mathcal{A})$ be the full subcategory of $\mathrm{K}_d(\mathcal{A})$ given by $\mathrm{Ob}(\mathrm{K}_\eta(\mathcal{A})) = \mathrm{Ob}(\mathcal{A}_\eta)$. Prove that $\mathrm{K}_\eta(\mathcal{A})$ is a full triangulated subcategory of $\mathrm{K}_d(\mathcal{A})$.

Exercise 11.3. Let $(\mathcal{A}, T)$ be an additive category with translation. Let $\mathcal{B}$ be the category of pairs (X, e) of $X \in \mathcal{A}$ and $e \colon X \to T^2 X$.

(i) Define a translation functor T' on $\mathcal{B}$ such that $(\mathcal{B}, T')$ is an additive category with translation and that the functor $for \colon (\mathcal{B}, T') \to (\mathcal{A}, T)$, which forgets e, is a functor of additive categories with translation.

(ii) Let $\eta \colon \mathrm{id}_\mathcal{B} \to T'^2$ be the morphism of functors that associates to (X, e) the morphism e. Prove that η is well defined and satisfies $\eta \circ T = T \circ \eta$. Prove also that for induces an equivalence of triangulated categories $\mathrm{K}_\eta(\mathcal{B}) \xrightarrow{\sim} \mathrm{K}_d(\mathcal{A})$.

Exercise 11.4. Let $(\mathcal{A}, T)$ be an additive category with translation, and let $f, g \colon X \rightrightarrows Y$ be two morphisms in $\mathcal{A}_d$. Prove that f and g are homotopic if and only if there exists a commutative diagram in $\mathcal{A}_d$

$$
\begin{array}{ccccc}
Y & \xrightarrow{\;\alpha(f)\;} & \mathrm{Mc}(f) & \xrightarrow{\;\beta(f)\;} & X[1] \\
\| & & \downarrow{\scriptstyle u} & & \| \\
Y & \xrightarrow{\;\alpha(g)\;} & \mathrm{Mc}(g) & \xrightarrow{\;\beta(g)\;} & X[1] \, .
\end{array}
$$

In such a case, prove that u is an isomorphism in $\mathcal{A}_d$.

Exercise 11.5. Let $(\mathcal{A}, T)$ be an additive category with translation and $f \colon X \to Y$ a morphism in $\mathcal{A}_d$. By using Theorem 11.2.6, prove that f is an isomorphism in $\mathrm{K}_d(\mathcal{A})$ if and only if $\mathrm{Mc}(f)$, the mapping cone of f, is homotopic to zero.

Exercise 11.6. Let $(\mathcal{A}, T)$ be an additive category with translation and let $f \colon X \to Y$ be a morphism in $\mathcal{A}_d$.
(i) Prove that the following conditions are equivalent:

  (a) f is homotopic to zero,
  (b) f factors through $\alpha(\mathrm{id}_X) \colon X \to \mathrm{Mc}(\mathrm{id}_X)$,
  (c) f factors through $T^{-1}(\beta(\mathrm{id}_Y)) \colon T^{-1}\mathrm{Mc}(\mathrm{id}_Y) \to Y$,
  (d) f decomposes as $X \to Z \to Y$ with a differential object Z homotopic to zero.

(ii) Let $\mathcal{N}$ be the full subcategory of $\mathcal{A}_d$ consisting of differential objects homotopic to zero. Prove that the category $(\mathcal{A}_d)_{\mathcal{N}}$ defined in Exercise 8.6 is equivalent to $\mathrm{K}_d(\mathcal{A})$.

Exercise 11.7. Let $(\mathcal{A}, T)$ be an additive category with translation, and consider two morphisms in $\mathcal{A}_c$

$$X \underset{\psi}{\overset{\varphi}{\rightleftarrows}} Y \ .$$

Assume that $\psi \circ \varphi - \mathrm{id}_X$ is homotopic to zero. Prove that there exist an object Z in $\mathcal{A}_c$ and morphisms in $\mathcal{A}_c$

$$X \underset{\beta}{\overset{\alpha}{\rightleftarrows}} Y \oplus Z$$

such that $\beta \circ \alpha = \mathrm{id}_X$ in $\mathcal{A}_c$. (Hint: use Exercise 11.6.)

Exercise 11.8. Let $(\mathcal{A}, T)$ be an additive category with translation and let $0 \to X \xrightarrow{f} Y \xrightarrow{g} Z \to 0$ be a complex in $\mathcal{A}_c$.
(i) Prove that $u = (0, g) \colon \mathrm{Mc}(f) \to Z$ is a well-defined morphism in $\mathcal{A}_c$.
(ii) Assume that $0 \to X \xrightarrow{f} Y \xrightarrow{g} Z \to 0$ splits in $\mathcal{A}$ (see Exercise 8.34), i.e., there exist morphisms $k \colon Y \to X$ and $h \colon Z \to Y$ in $\mathcal{A}$ such that $\mathrm{id}_Y = f \circ k + h \circ g$, $g \circ h = \mathrm{id}_Z$ and $k \circ f = \mathrm{id}_X$. Prove that $u \colon \mathrm{Mc}(f) \to Z$ is an isomorphism in $\mathrm{K}_c(\mathcal{A})$. (Hint: $\left(\begin{smallmatrix} -T(k) \circ d_Y \circ h \\ h \end{smallmatrix} \right)$ defines a morphism $Z \to \mathrm{Mc}(f)$.)

Exercise 11.9. Let $(\mathcal{A}, T)$ be an additive category with translation and let $X \in \mathcal{A}_c$ be a complex. Assume that there exist morphisms $s, t \colon X \to T^{-1}X$ in $\mathcal{A}$ such that $\mathrm{id}_X = T(s) \circ d_X + T^{-1}(d_X) \circ t$. Prove that X is homotopic to zero. (Hint: consider $s \circ T^{-1}(d_X) \circ t$.)

Exercise 11.10. Let $\mathcal{C}$ be an additive category and let $X \in \mathrm{C}(\mathcal{C})$.
(i) Prove that there exists a morphism of functors $\xi \colon \mathrm{id}_{\mathrm{Gr}(\mathcal{C})} \to \mathrm{id}_{\mathrm{Gr}(\mathcal{C})}$ such that $T(\xi_X) - \xi_{TX} = \mathrm{id}_{TX}$ for any $X \in \mathrm{Gr}(\mathcal{C})$.
(ii) Prove that $d_X \colon X \to X[1]$ defines a morphism in $\mathrm{C}(\mathcal{C})$.
(iii) Prove that $d_X \colon X \to X[1]$ is homotopic to zero. (Hint: use (i).)

Exercise 11.11. Let $\mathcal{C} = \mathrm{Mod}(\mathbb{Z})$ and let $X \in C^b(\mathcal{C})$, $Y \in C(\mathcal{C})$. Prove that the family of morphisms

$$\mathrm{Hom}_{\mathcal{C}}(X^n, Y^m) \otimes X^n \to Y^m$$
$$f \otimes u \mapsto f(u) \,.$$

defines a morphism of complexes $\mathrm{Hom}^{\bullet}_{\mathcal{C}}(X, Y) \otimes X \to Y$. (Remark that the signs in Definition 11.3.12 are so chosen that the above map is a morphism of complexes.)

Exercise 11.12. Let $\mathcal{C}$ be an additive category, and let $X \in C(\mathcal{C})$, $a \in \mathbb{Z}$. The stupid truncation $\sigma^{\geq a} X$ has been defined in Definition 11.3.11.
(i) Show that $\sigma^{\geq a}$ does not induce a functor from $K(\mathcal{C})$ to itself in general.
(ii) Prove that for $X \in C(\mathcal{C})$ and $f \in \mathrm{Mor}(C(\mathcal{C}))$, there exist distinguished triangles in $K(\mathcal{C})$

$$\sigma^{>a} X \to X \to \sigma^{\leq a} X \to (\sigma^{>a} X)[1] \,,$$

$$\sigma^{>a} X \to \sigma^{\geq a} X \to X^a[-a] \to (\sigma^{>a} X)[1] \,,$$

$$\mathrm{Mc}(\sigma^{>a}(f)) \to \mathrm{Mc}(f) \to \mathrm{Mc}(\sigma^{\leq a}(f)) \to \mathrm{Mc}(\sigma^{>a}(f))[1] \,.$$

12

Complexes in Abelian Categories

In this chapter, we study complexes (and double complexes) in abelian categories and give tools to compute their cohomology. In particular, we prove the classical "Snake lemma" and we construct the long exact sequence associated with a short exact sequence of complexes.

As an application, we discuss Koszul complexes associated to functors defined on a category of finite subsets of a set S, with values in an abelian category C. The main result asserts that such a complex may be obtained as the mapping cone of a morphism acting on a simpler Koszul complex. We apply these results to the study of distributive families of subobjects of an object X in C.

We postpone the introduction of derived categories to the next chapter.

Note that we avoid the use of spectral sequences, using instead systematically the "truncation functors".

12.1 The Snake Lemma

Let C be an abelian category.

Lemma 12.1.1. [The snake Lemma] *Consider the commutative diagram in C with exact rows:*

$$
\begin{array}{ccccccc}
X' & \xrightarrow{f} & X & \xrightarrow{g} & X'' & \longrightarrow & 0 \\
\downarrow{\scriptstyle u} & & \downarrow{\scriptstyle v} & & \downarrow{\scriptstyle w} & & \\
0 & \longrightarrow & Y' & \xrightarrow{f'} & Y & \xrightarrow{g'} & Y'' .
\end{array}
$$

It gives rise to an exact sequence:

$$ \operatorname{Ker} u \xrightarrow{f_1} \operatorname{Ker} v \xrightarrow{g_1} \operatorname{Ker} w \xrightarrow{\varphi} \operatorname{Coker} u \xrightarrow{f_2} \operatorname{Coker} v \xrightarrow{g_2} \operatorname{Coker} w . $$

Proof. (a) First, we construct $\varphi\colon \operatorname{Ker} w \to \operatorname{Coker} u$. Set $W = X \times_{X''} \operatorname{Ker} w$, $Z = Y \oplus_{Y'} \operatorname{Coker} u$ and let $h\colon W \to \operatorname{Ker} w$ be the natural morphism. We get a commutative diagram

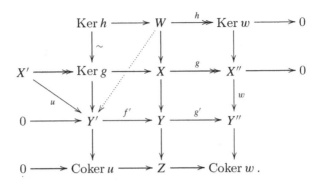

Then the composition $W \to X \to Y \to Z$ uniquely decomposes as

$$W \twoheadrightarrow \operatorname{Ker} w \xrightarrow{\varphi} \operatorname{Coker} u \rightarrowtail Z \ .$$

Indeed, since the composition $W \to Y \to Y''$ vanishes, the morphism $W \to Y$ factors uniquely through Y'. By Lemma 8.3.11, the morphism $\operatorname{Ker} h \to \operatorname{Ker} g$ is an isomorphism. Since $X' \to \operatorname{Ker} g$ is an epimorphism, $\operatorname{Ker} g \to Y' \to \operatorname{Coker} u$ vanishes. Hence the composition $\operatorname{Ker} h \to W \to \operatorname{Coker} u$ vanishes and $W \to \operatorname{Coker} u$ factors uniquely as $W \twoheadrightarrow \operatorname{Ker} w \xrightarrow{\varphi} \operatorname{Coker} u$. (Recall that $\operatorname{Ker} w \simeq \operatorname{Coker}(\operatorname{Ker} h \to W)$.)

(b) Let us show that the sequence $\operatorname{Ker} u \xrightarrow{f_1} \operatorname{Ker} v \xrightarrow{g_1} \operatorname{Ker} w \xrightarrow{\varphi} \operatorname{Coker} u \to \operatorname{Coker} v \to \operatorname{Coker} w$ is exact.

(i) The sequence $\operatorname{Ker} u \xrightarrow{f_1} \operatorname{Ker} v \xrightarrow{g_1} \operatorname{Ker} w$ is exact. Choose $S \in \mathcal{C}$ and a morphism $\psi\colon S \to \operatorname{Ker} v$ such that $g_1 \circ \psi = 0$. Consider the diagram

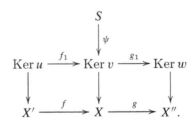

The composition $S \to \operatorname{Ker} v \to X \to X''$ is 0. Applying Lemma 8.3.12 we find an epimorphism $h\colon S' \to S$ and the commutative diagram below on the left:

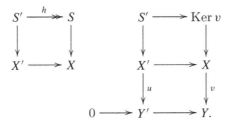

By considering the commutative diagram above on the right, we find that the composition $S' \to X' \xrightarrow{u} Y' \xrightarrow{f'} Y$ is 0, and therefore, the composition $S' \to X' \xrightarrow{u} Y'$ is 0. Hence, $S' \to X'$ factors through $\operatorname{Ker} u$ and it remains to apply Lemma 8.3.12.

(ii) The sequence $\operatorname{Ker} v \xrightarrow{g_1} \operatorname{Ker} w \xrightarrow{\varphi} \operatorname{Coker} u$ is exact. Let $\psi : S \to \operatorname{Ker} w$ be a morphism such that $\varphi \circ \psi = 0$. Since $W \to \operatorname{Ker} w$ is an epimorphism, we can find an epimorphism $S^1 \twoheadrightarrow S$ and a commutative diagram

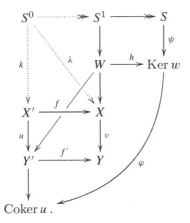

Since the composition $S^1 \to W \to Y' \to \operatorname{Coker} u$ vanishes, there exists an epimorphism $S^0 \twoheadrightarrow S^1$ such that the composition $S^0 \to S^1 \to W \to Y'$ decomposes into $S^0 \xrightarrow{k} X' \xrightarrow{u} Y'$. Denote by λ the composition $S^0 \to S^1 \to W \to X$. Then $v \circ \lambda = v \circ f \circ k$. Hence $\lambda - f \circ k : S^0 \to X$ factors through $\operatorname{Ker} v$. Therefore we obtain a commutative diagram

$$
\begin{array}{ccc}
S^0 & \longrightarrow\!\!\!\!\rightarrow & S \\
{\scriptstyle \lambda - f \circ k}\downarrow & & \downarrow \\
\operatorname{Ker} v & \longrightarrow & \operatorname{Ker} w \\
\downarrow & & \downarrow \\
X & \xrightarrow{\ \ g\ \ } & X''.
\end{array}
$$

It remains to apply Lemma 8.3.12.

(iii) The proof that $\mathrm{Ker}\, w \to \mathrm{Coker}\, u \to \mathrm{Coker}\, v \to \mathrm{Coker}\, w$ is exact follows by reversing the arrows.   q.e.d.

12.2 Abelian Categories with Translation

An *abelian category with translation* $(\mathcal{A}, T)$ is an additive category with translation $(\mathcal{A}, T)$ (see Definition 10.1.1) such that $\mathcal{A}$ is abelian. Hence T is an exact additive functor.

Proposition 12.2.1. *Let* $(\mathcal{A}, T)$ *be an abelian category with translation. Then the categories* $\mathcal{A}_d$ *and* $\mathcal{A}_c$ *(see* Definition 11.1.1*) are abelian categories with translation.*

The proof is straightforward.

Let $(\mathcal{A}, T)$ be an abelian category with translation and let $X \in \mathcal{A}_c$. We define (see Definition 8.3.8):

$$H(X) := H(T^{-1}X \to X \to TX)$$
$$\simeq \mathrm{Coker}(\mathrm{Im}\, T^{-1}d_X \to \mathrm{Ker}\, d_X)$$
$$(12.2.1) \qquad \simeq \mathrm{Coker}(T^{-1}X \to \mathrm{Ker}\, d_X) \simeq \mathrm{Coker}(\mathrm{Coker}\, T^{-2}d_X \to \mathrm{Ker}\, d_X)$$
$$\simeq \mathrm{Ker}(\mathrm{Coker}\, T^{-1}d_X \to \mathrm{Im}\, d_X)$$
$$\simeq \mathrm{Ker}(\mathrm{Coker}\, T^{-1}d_X \to TX) \simeq \mathrm{Ker}(\mathrm{Coker}\, T^{-1}d_X \to \mathrm{Ker}\, Td_X) \,.$$

The last isomorphism follows from the fact that $\mathrm{Ker}\, Td_X \to TX$ is a monomorphism, and similarly for the third isomorphism.

We shall also make use of the notations

$$Z(X) := \mathrm{Ker}\, d_X \,,$$
$$B(X) := \mathrm{Im}\, T^{-1}d_X \,.$$

Hence we have monomorphisms $B(X) \rightarrowtail Z(X) \rightarrowtail X$ and an exact sequence $0 \to B(X) \to Z(X) \to H(X) \to 0$.

We call $H(X)$ the *cohomology* of X. If $f: X \to Y$ is a morphism in $\mathcal{A}_c$, it induces morphisms $Z(f): \mathrm{Ker}\, d_X \to \mathrm{Ker}\, d_Y$ and $B(f): \mathrm{Im}\, T^{-1}d_X \to \mathrm{Im}\, T^{-1}d_Y$, thus a morphism $H(f): H(X) \to H(Y)$. We have obtained an additive functor:

$$H: \mathcal{A}_c \to \mathcal{A} \,.$$

The isomorphisms (12.2.1) give rise to the exact sequence:

$$(12.2.2) \qquad 0 \to H(X) \to \mathrm{Coker}(T^{-1}d_X) \xrightarrow{d_X} \mathrm{Ker}(Td_X) \to H(TX) \to 0 \,.$$

Lemma 12.2.2. *If* $f: X \to Y$ *is a morphism in* $\mathcal{A}_c$ *homotopic to zero, then* $H(f): H(X) \to H(Y)$ *is the zero morphism.*

Proof. Let $f = T(u) \circ d_X + T^{-1}(d_Y) \circ u$. The composition

$$\operatorname{Ker} d_X \to X \xrightarrow{T(u) \circ d_X} Y$$

is the zero morphism. Moreover, $T^{-1}(d_Y) \circ u$ factorizes through $\operatorname{Im} T^{-1}(d_Y)$.
q.e.d.

Hence the functor H defines a functor (denoted by the same symbol)

$$H \colon \mathrm{K}_c(\mathcal{A}) \to \mathcal{A} \ .$$

Definition 12.2.3. *A morphism* $f \colon X \to Y$ *in* $\mathcal{A}_c$ *or in* $\mathrm{K}_c(\mathcal{A})$ *is a quasi-isomorphism* (*a qis for short*), *if* $H(f)$ *is an isomorphism.*

An object X is qis to 0 if the natural morphism $X \to 0$ is a qis, or equivalently if $T^{-1}X \xrightarrow{T^{-1}d_X} X \xrightarrow{d_X} TX$ is exact.

Theorem 12.2.4. *Let* $0 \to X' \xrightarrow{f} X \xrightarrow{g} X'' \to 0$ *be an exact sequence in* $\mathcal{A}_c$.

(i) *The sequence* $H(X') \to H(X) \to H(X'')$ *is exact.*
(ii) *There exists* $\delta \colon H(X'') \to H(T(X'))$ *making the sequence:*

$$(12.2.3) \qquad H(X) \to H(X'') \xrightarrow{\delta} H(T(X')) \to H(T(X))$$

exact. Moreover, we can construct δ *functorial with respect to short exact sequences of* $\mathcal{A}_c$.

Proof. The exact sequence in $\mathcal{A}_c$ gives rise to a commutative diagram with exact rows:

$$
\begin{array}{ccccccc}
\operatorname{Coker} T^{-1}d_{X'} & \xrightarrow{f} & \operatorname{Coker} T^{-1}d_X & \xrightarrow{g} & \operatorname{Coker} T^{-1}d_{X''} & \longrightarrow & 0 \\
\downarrow{\scriptstyle d_{X'}} & & \downarrow{\scriptstyle d_X} & & \downarrow{\scriptstyle d_{X''}} & & \\
0 \longrightarrow \operatorname{Ker} T d_{X'} & \xrightarrow{f} & \operatorname{Ker} T d_X & \xrightarrow{g} & \operatorname{Ker} T d_{X''} \ . & &
\end{array}
$$

Applying the snake lemma (Lemma 12.1.1) with $u = d_{X'}$, $v = d_X$ and $w = d_{X''}$, the result follows from the exact sequence (12.2.2).   q.e.d.

Corollary 12.2.5. *Let* $(\mathcal{A}, T)$ *be an abelian category with translation. Then the functor*

$$H \colon \mathrm{K}_c(\mathcal{A}) \to \mathcal{A}$$

is cohomological.

It means that, if $X \to Y \to Z \to T(X)$ is a d.t. in $\mathrm{K}_c(\mathcal{A})$, then the functor H sends it to an exact sequence in $\mathcal{A}$:

$$\cdots \to H(X) \to H(Y) \to H(Z) \to H(T(X)) \to \cdots \ .$$

Proof. Let $X \to Y \to Z \to T(X)$ be a d.t. in $K_c(\mathcal{A})$. It is isomorphic to $V \xrightarrow{\alpha(u)} \mathrm{Mc}(u) \xrightarrow{\beta(u)} T(U) \to T(V)$ for a morphism $u \colon U \to V$ in $\mathcal{A}_c$. Since the sequence in $\mathcal{A}_c$:

$$0 \to V \to \mathrm{Mc}(u) \to T(U) \to 0$$

is exact, it follows from Theorem 12.2.4 that the sequence

$$H(V) \to H(\mathrm{Mc}(u)) \to H(T(U))$$

is exact. Therefore, $H(X) \to H(Y) \to H(Z)$ is exact.   q.e.d.

Corollary 12.2.6. *Let* $0 \to X \xrightarrow{f} Y \xrightarrow{g} Z \to 0$ *be an exact sequence in* $\mathcal{A}_c$ *and define* $\varphi \colon \mathrm{Mc}(f) \to Z$ *as* $\varphi = (0, g)$. *Then* φ *is a morphism in* $\mathcal{A}_c$ *and is a qis.*

Proof. The commutative diagram in $\mathcal{A}_c$ with exact rows

$$
\begin{array}{ccccccccc}
0 & \longrightarrow & X & \xrightarrow{\mathrm{id}_X} & X & \longrightarrow & 0 & \longrightarrow & 0 \\
& & \downarrow{\scriptstyle \mathrm{id}_X} & & \downarrow{\scriptstyle f} & & \downarrow & & \\
0 & \longrightarrow & X & \xrightarrow{f} & Y & \longrightarrow & Z & \longrightarrow & 0
\end{array}
$$

yields an exact sequence in $\mathcal{A}_c$:

$$0 \to \mathrm{Mc}(\mathrm{id}_X) \xrightarrow{\gamma} \mathrm{Mc}(f) \xrightarrow{\varphi} \mathrm{Mc}(0 \to Z) \to 0 .$$

Since $H(\mathrm{Mc}(\mathrm{id}_X)) \simeq 0$, φ is a qis by Theorem 12.2.4.   q.e.d.

12.3 Complexes in Abelian Categories

Let $\mathcal{C}$ be an abelian category. Recall (see Definition 11.3.1) that the category with translation $(\mathrm{Gr}(\mathcal{C}), T)$ is given by $\mathrm{Gr}(\mathcal{C}) = \mathcal{C}^{\mathbb{Z}_d}$, and that we set (see (11.3.2)):

$$\mathrm{C}(\mathcal{C}) := (\mathrm{Gr}(\mathcal{C}))_c .$$

The categories $\mathrm{C}^*(\mathcal{C})$ ($* = \mathrm{ub}, +, -, \mathrm{b}$) are obviously abelian categories with translation.

Let us translate the definitions and results of §12.2 in the case where $\mathcal{A} = \mathrm{Gr}(\mathcal{C})$ and hence, $\mathcal{A}_c = \mathrm{C}(\mathcal{C})$.

Applying the functors $\pi_n \colon \mathrm{Gr}(\mathcal{C}) \to \mathcal{C}$, which associates X^n to $X = \{X^l\}_{l \in \mathbb{Z}} \in \mathrm{Gr}(\mathcal{C})$, we find additive functors:

$$H^n \colon \mathrm{C}(\mathcal{C}) \to \mathcal{C}, \quad H^n(X) = H(X^{n-1} \to X^n \to X^{n+1}) ,$$
$$Z^n \colon \mathrm{C}(\mathcal{C}) \to \mathcal{C}, \quad Z^n(X) = \mathrm{Ker}\, d_X^n ,$$
$$B^n \colon \mathrm{C}(\mathcal{C}) \to \mathcal{C}, \quad B^n(X) = \mathrm{Im}\, d_X^{n-1} .$$

We call $H^n(X)$ the n-th cohomology object of X.

Notice that:

$$H^n(X) \simeq H^0(X[n])$$

by the commutative diagram

$$
\begin{array}{ccccc}
X^{n-1} & \xrightarrow{\;d_X^{n-1}\;} & X^n & \xrightarrow{\;d_X^n\;} & X^{n+1} \\
{\scriptstyle(-1)^n\,\mathrm{id}_{X^{n-1}}}\big\downarrow & & {\scriptstyle\mathrm{id}_{X^n}}\big\downarrow & & \big\downarrow{\scriptstyle(-1)^n\,\mathrm{id}_{X^{n+1}}} \\
X^{n-1} & \xrightarrow{\;(-1)^n d_X^{n-1}\;} & X^n & \xrightarrow{\;(-1)^n d_X^n\;} & X^{n+1}.
\end{array}
$$

Then the exact sequence (12.2.2) give rise to the exact sequence:

$$(12.3.1) \quad 0 \to H^n(X) \to \mathrm{Coker}(d_X^{n-1}) \xrightarrow{d_X^n} \mathrm{Ker}\, d_X^{n+1}(X) \to H^{n+1}(X) \to 0 \,.$$

Definition 12.3.1. *Let $\mathcal{C}$ be an abelian category and let $n \in \mathbb{Z}$. The* truncation *functors :*

$$\widetilde{\tau}^{\leq n}, \quad \tau^{\leq n} \;:\; \mathrm{C}(\mathcal{C}) \to \mathrm{C}^-(\mathcal{C})$$
$$\widetilde{\tau}^{\geq n}, \quad \tau^{\geq n} \;:\; \mathrm{C}(\mathcal{C}) \to \mathrm{C}^+(\mathcal{C})$$

are defined as follows. Let $X := \cdots \to X^{n-1} \to X^n \to X^{n+1} \to \cdots$. We set:

$$\tau^{\leq n} X := \cdots \to X^{n-2} \to X^{n-1} \to \mathrm{Ker}\, d_X^n \to 0 \to 0 \to \cdots$$
$$\widetilde{\tau}^{\leq n} X := \cdots \to X^{n-2} \to X^{n-1} \to X^n \to \mathrm{Im}\, d_X^n \to 0 \to \cdots$$
$$\widetilde{\tau}^{\geq n} X := \cdots \to 0 \to \mathrm{Im}\, d_X^{n-1} \to X^n \to X^{n+1} \to X^{n+2} \to \cdots$$
$$\tau^{\geq n} X := \cdots \to 0 \to 0 \to \mathrm{Coker}\, d_X^{n-1} \to X^{n+1} \to X^{n+2} \to \cdots \,.$$

There is a chain of morphisms in $\mathrm{C}(\mathcal{C})$:

$$\tau^{\leq n} X \to \widetilde{\tau}^{\leq n} X \to X \to \widetilde{\tau}^{\geq n} X \to \tau^{\geq n} X,$$

and there are exact sequences in $\mathrm{C}(\mathcal{C})$:

$$(12.3.2) \quad
\begin{cases}
0 \to \widetilde{\tau}^{\leq n-1} X \to \tau^{\leq n} X \to H^n(X)[-n] \to 0 \,, \\
0 \to H^n(X)[-n] \to \tau^{\geq n} X \to \widetilde{\tau}^{\geq n+1} X \to 0 \,, \\
0 \to \tau^{\leq n} X \to X \to \widetilde{\tau}^{\geq n+1} X \to 0 \,, \\
0 \to \widetilde{\tau}^{\leq n-1} X \to X \to \tau^{\geq n} X \to 0 \,, \\
0 \to \tau^{\leq n} X \to \widetilde{\tau}^{\leq n} X \to \mathrm{Mc}(\mathrm{id}_{\mathrm{Im}\, d_X^n}[-n-1]) \to 0 \,.
\end{cases}
$$

We have the isomorphisms

$$(12.3.3) \quad
\begin{aligned}
H^j(\tau^{\leq n} X) \xrightarrow{\;\sim\;} H^j(\widetilde{\tau}^{\leq n} X) &\simeq \begin{cases} H^j(X) & j \leq n \,, \\ 0 & j > n. \end{cases} \\
H^j(\widetilde{\tau}^{\geq n} X) \xrightarrow{\;\sim\;} H^j(\tau^{\geq n} X) &\simeq \begin{cases} H^j(X) & j \geq n \,, \\ 0 & j < n. \end{cases}
\end{aligned}
$$

The verification is straightforward.

Lemma 12.3.2. (i) *If* $f: X \to Y$ *is a morphism in* $\mathrm{C}(\mathcal{C})$ *homotopic to zero, then* $H^n(f): H^n(X) \to H^n(Y)$ *is the zero morphism.*

(ii) *If* $f: X \to Y$ *is a morphism in* $\mathrm{C}(\mathcal{C})$ *homotopic to zero, then* $\tau^{\leq n}(f)$, $\tau^{\geq n}(f)$, $\widetilde{\tau}^{\leq n}(f)$, $\widetilde{\tau}^{\geq n}(f)$ *are homotopic to zero.*

Proof. (i) is a particular case of Lemma 12.2.2.

(ii) The proof is straightforward.   q.e.d.

Hence the functor H^n defines a functor (denoted by the same symbol)

$$H^n: \mathrm{K}(\mathcal{C}) \to \mathcal{C} \ .$$

Similarly, the functors $\tau^{\leq n}$ and $\widetilde{\tau}^{\leq n}$ define functors, denoted by the same symbols, from $\mathrm{K}(\mathcal{C})$ to $\mathrm{K}^-(\mathcal{C})$, and the functors $\tau^{\geq n}$ and $\widetilde{\tau}^{\geq n}$ define functors, denoted by the same symbols, from $\mathrm{K}(\mathcal{C})$ to $\mathrm{K}^+(\mathcal{C})$.

Note that a morphism $f: X \to Y$ in $\mathrm{C}(\mathcal{C})$ or in $\mathrm{K}(\mathcal{C})$ is a qis if and only if $H^n(f)$ is an isomorphism for all $n \in \mathbb{Z}$ and a complex X is qis to 0 if and only if the complex X is exact.

There are qis in $\mathrm{C}(\mathcal{C})$:

(12.3.4)
$$\begin{cases} \tau^{\leq n} X \to \widetilde{\tau}^{\leq n} X \ , \\ \widetilde{\tau}^{\geq n} X \to \tau^{\geq n} X \ . \end{cases}$$

Theorem 12.2.4 and Corollaries 12.2.5 and 12.2.6 are translated as:

Theorem 12.3.3. *Let* $0 \to X' \xrightarrow{f} X \xrightarrow{g} X'' \to 0$ *be an exact sequence in* $\mathrm{C}(\mathcal{C})$.

(i) *For each* $n \in \mathbb{Z}$, *the sequence* $H^n(X') \to H^n(X) \to H^n(X'')$ *is exact.*

(ii) *For each* $n \in \mathbb{Z}$, *there exists* $\delta^n: H^n(X'') \to H^{n+1}(X')$ *making the sequence:*

(12.3.5) $H^n(X) \to H^n(X'') \xrightarrow{\delta^n} H^{n+1}(X') \to H^{n+1}(X)$

exact. Moreover, we can construct δ^n *functorial with respect to short exact sequences of* $\mathrm{C}(\mathcal{C})$.

Corollary 12.3.4. *Let* $\mathcal{C}$ *be an abelian category. Then the functor*

$$H^n: \mathrm{K}(\mathcal{C}) \to \mathcal{C}$$

is cohomological.

Corollary 12.3.5. *Let* $0 \to X \xrightarrow{f} Y \xrightarrow{g} Z \to 0$ *be an exact sequence in* $\mathrm{C}(\mathcal{C})$ *and define* $\varphi: \mathrm{Mc}(f) \to Z$ *as* $\varphi^n = (0, g^n)$. *Then* φ *is a morphism in* $\mathrm{C}(\mathcal{C})$ *and is a qis.*

Using Corollaries 12.3.4 and 12.3.5, we find a long exact sequence

(12.3.6) $\cdots \to H^n(Y) \to H^n(Z) \xrightarrow{\delta'^n} H^{n+1}(X) \to H^{n+1}(Y) \to \cdots$.

Here $H^n(Z) \xrightarrow{\delta'^n} H^{n+1}(X)$ is the composition $H^n(Z) \xleftarrow{\sim} H^n(\mathrm{Mc}(f)) \to H^n(X[1]) \simeq H^{n+1}(X)$.

Proposition 12.3.6. *The morphism δ'^n in (12.3.6) is related to the morphism δ^n constructed in* Theorem 12.3.3 *by the relation:* $\delta'^n = -\delta^n$.

Proof. The morphism $\delta^n \colon H^n(Z) \to H^{n+1}(X)$ is characterized as follows (see the proof of the "snake lemma" (12.1.1)). There exists a commutative diagram

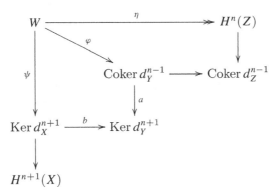

such that $\eta \colon W \to H^n(Z)$ is an epimorphism and the composition $W \xrightarrow{\eta} H^n(Z) \xrightarrow{\delta^n} H^{n+1}(X)$ is the same as the composition $W \to \mathrm{Ker}\, d_X^{n+1} \to H^{n+1}(X)$. On the other hand, δ'^n is given by

$$H^n(Z) \xleftarrow{\sim} H^n(\mathrm{Mc}(f)) \to H^n(X[1]) .$$

Now observe that the diagram below commutes.

$$
\begin{array}{ccc}
\mathrm{Ker}\, d_X^{n+1} \oplus \mathrm{Coker}\, d_Y^{n-1} & \longrightarrow & \mathrm{Coker}\, d_{\mathrm{Mc}(f)}^{n-1} \\
{\scriptstyle (b,a)} \downarrow & & \downarrow \\
\mathrm{Ker}\, d_Y^{n+1} & \longrightarrow & \mathrm{Ker}\, d_{\mathrm{Mc}(f)}^{n+1} .
\end{array}
$$

Let $\xi = (-\psi, \varphi) \colon W \to \mathrm{Ker}\, d_X^{n+1} \oplus \mathrm{Coker}\, d_Y^{n-1}$. Then the composition $W \xrightarrow{\xi} \mathrm{Ker}\, d_X^{n+1} \oplus \mathrm{Coker}\, d_Y^{n-1} \to \mathrm{Coker}\, d_{\mathrm{Mc}(f)}^{n-1} \to \mathrm{Ker}\, d_{\mathrm{Mc}(f)}^{n+1}$ vanishes. We get the diagram

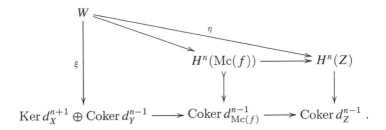

Note that the diagram below commutes

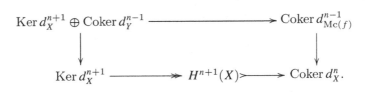

Hence, the composition $W \to H^n(\mathrm{Mc}(f)) \xrightarrow{\delta^n} H^{n+1}(X)$ is equal to the composition $W \xrightarrow{-\psi} \mathrm{Ker}\, d_X^{n+1} \to H^{n+1}(X)$. Therefore, we have the commutative diagram

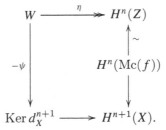

This completes the proof.   q.e.d.

12.4 Example: Koszul Complexes

We shall give some useful tools which permit us to construct and calculate the cohomology of some complexes. Such complexes appear in various contexts, such as Commutative Algebra (regular sequences of endomorphisms of a module over a ring) or Sheaf Theory (Čech cohomology of a sheaf associated with a closed or an open covering).

First of all, we recall that if L is a finite free $\mathbb{Z}$-module of rank n, the module $\bigwedge^n L$ is free of rank one, and is usually denoted by $\det L$. We understand $\det 0 = \mathbb{Z}$. Let I be a finite set and let $\{e_s\}_{s \in I}$ be the corresponding basis of the free module $\mathbb{Z}^I$. If π is a permutation of I, it induces an isomorphism of $\mathbb{Z}^I$, and the isomorphism $\pm \mathrm{id}$ on $\det \mathbb{Z}^I$, where $\pm$ is the signature of π. Note that when I is the empty set, $\mathbb{Z}^I = 0$.

If $I = \{s_1, \ldots, s_n\}$, then $e_{s_1} \wedge \cdots \wedge e_{s_n}$ is a basis of $\det \mathbb{Z}^I$. If $J = I \sqcup \{s\}$, $e_s \wedge$ denotes the linear isomorphism

$$\det \mathbb{Z}^I \to \det \mathbb{Z}^J \ ,$$

$$u \mapsto e_s \wedge u \ .$$

Recall that if L is a finitely generated $\mathbb{Z}$-module and X is an object of $\mathcal{C}$, the object $X \otimes L$ is well defined in $\mathcal{C}$. (See Remark 8.5.7.)

Let S be a set and let Σ be a family of *finite* subsets of S. We shall assume

(12.4.1)   If $I \subset J \subset K$ and I, K belong to Σ, then J belongs to Σ .

In the construction below, if Σ is not finite we have to assume

(12.4.2) $\mathcal{C}$ admits small projective limits .

The set Σ is ordered by inclusion, hence defines a category. Recall that

$$\operatorname{Hom}_\Sigma(I, J) = \begin{cases} \{\mathrm{pt}\} & \text{if } I \subset J \ , \\ \varnothing & \text{otherwise .} \end{cases}$$

Let $F \colon \Sigma \to \mathcal{C}$ be a functor. We shall write X_I instead of $F(I)$ and f_{JI} instead of $F(I \to J)$ (for $I \subset J$, $I, J \in \Sigma$). Hence $f_{II} = \operatorname{id}_{X_I}$ and $f_{KJ} \circ f_{JI} = f_{KI}$, for $I \subset J \subset K$ with $I, J, K \in \Sigma$.

To these data we associate a complex in $\mathcal{C}$ as follows. Let $|I|$ denote the cardinal of $I \in \Sigma$. We set

$$C^n(F) = \prod_{|I|=n, I \in \Sigma} X_I \otimes \det \mathbb{Z}^I \ .$$

If $J = I \sqcup \{s\}$, we have the morphism:

(12.4.3) $\varphi_{JI} := f_{JI} \otimes (e_s \wedge) \colon X_I \otimes \det \mathbb{Z}^I \to X_J \otimes \det \mathbb{Z}^J$.

Since for any J with $|J| = n + 1$ there are finitely many I with $I \subset J$, the family of morphisms (12.4.3) define a morphism

(12.4.4) $C^n(F) \to \prod_{|I|=n, I \subset J, I \in \Sigma} X_I \otimes \det \mathbb{Z}^I \xrightarrow{\varphi_{JI}} X_J \otimes \det \mathbb{Z}^J$,

from which we deduce the morphism

(12.4.5) $d_F^n \colon C^n(F) \to C^{n+1}(F)$.

Lemma 12.4.1. *We have* $d_F^{n+1} \circ d_F^n = 0$.

Proof. Let $I, J \in \Sigma$, with $J = I \sqcup \{s\} \sqcup \{t\}$, $s, t \in J \setminus I$, $s \neq t$, $|I| = n$. We shall show that the composition $\psi_{JI} \colon X_I \otimes \det \mathbb{Z}^I \to X_J \otimes \det \mathbb{Z}^J$ induced by $d_F^{n+1} \circ d_F^n$ is zero. Set for short $I_u = I \cup \{u\}$ $(u = s, t)$. We have

$$\psi_{JI} = f_{JI_s} \circ f_{I_s I} \otimes (e_t \wedge) \circ (e_s \wedge) + f_{JI_t} \circ f_{I_t I} \otimes (e_s \wedge) \circ (e_t \wedge)$$
$$= f_{JI} \otimes (e_t \wedge e_s \wedge + e_s \wedge e_t \wedge)$$
$$= 0 .$$

q.e.d.

We shall denote by $C^\bullet(F)$ the complex

$$(12.4.6) \qquad C^\bullet(F) \colon 0 \to C^0(F) \xrightarrow{d_F^0} \cdots \to C^n(F) \xrightarrow{d_F^n} C^{n+1}(F) \to \cdots .$$

Example 12.4.2. Let S be a finite set and let Σ be the family of all subsets of S. Let $X \in \mathcal{C}$ and let $\{f_s\}_{s \in S}$ be a family of endomorphisms of X satisfying $f_s \circ f_t = f_t \circ f_s$ for all $s, t \in S$. Define the functor $F \colon \Sigma \to \mathcal{C}$ as follows. For $I \in \Sigma$, set $X_I = X$. For $I \subset J$, define $f_{JI} \colon X_I \to X_J$ as $f_{JI} = \prod_{s \in J \setminus I} f_s$. In this situation, the complex $C^\bullet(F)$ is called the Koszul complex associated with the family $\{f_s\}_{s \in S}$. This complex is usually denoted by $\mathrm{K}^\bullet(X, \{f_s\}_{s \in S})$. If $H^n(\mathrm{K}^\bullet(X, \{f_s\}_{s \in S})) \simeq 0$ for $n \neq \mathrm{card}(S)$, we say that $\{f_s\}_{s \in S}$ is a quasi-regular family.

We shall now give a technique for computing these complexes. Let Σ be as above, and let Σ_0 be a subset of Σ satisfying

$$(12.4.7) \qquad\qquad \Sigma \ni I \subset J \in \Sigma_0 \Longrightarrow I \in \Sigma_0 .$$

Set $\Sigma_1 = \Sigma \setminus \Sigma_0$. We have $\Sigma_1 \ni I \subset J \in \Sigma \Longrightarrow J \in \Sigma_1$. Clearly, both Σ_0 and Σ_1 satisfy hypothesis (12.4.1).

If $F \colon \Sigma \to \mathcal{C}$ is a functor, we denote by F_i $(i = 0, 1)$ its restriction to Σ_i. Hence, we have the complexes $C^\bullet(F_i), i = 0, 1$ and $C^n(F) \simeq C^n(F_0) \oplus C^n(F_1)$. For $i = 0, 1$ there are natural morphisms $C^n(F_i) \to C^n(F)$ and $C^n(F) \to C^n(F_i)$. We define $\varphi^n \colon C^{n-1}(F_0) \to C^n(F_1)$ as the composition

$$\varphi^n \colon C^{n-1}(F_0) \to C^{n-1}(F) \xrightarrow{d_F^{n-1}} C^n(F) \to C^n(F_1) .$$

Theorem 12.4.3. (i) *The φ^n's define a morphism of complexes*

$$\varphi \colon C^\bullet(F_0)[-1] \to C^\bullet(F_1) .$$

(ii) *The complex $C^\bullet(F)$ is isomorphic to* $\mathrm{Mc}(\varphi)$, *the mapping cone of φ.*
(iii) *There is a d.t.*

$$C^\bullet(F_1) \to C^\bullet(F) \to C^\bullet(F_0) \xrightarrow{+1}$$

and a long exact sequence

$$\cdots \to H^n(C^\bullet(F_1)) \to H^n(C^\bullet(F)) \to H^n(C^\bullet(F_0)) \to \cdots .$$

Proof. (i) Applying Proposition 11.1.4, we obtain that φ is a morphism of complexes if and only if $\mathrm{Mc}(\varphi)$ is a complex. Hence (i) follows from (ii).

(ii) We have $\mathrm{Mc}(\varphi)^n = (C^\bullet(F_0)[-1])^{n+1} \oplus C^n(F_1) \simeq C^n(F_0) \oplus C^n(F_1) \simeq C^n(F)$. The differential $d^n_{\mathrm{Mc}(\varphi)} \colon \mathrm{Mc}(\varphi)^n \to \mathrm{Mc}(\varphi)^{n+1}$ is given by the matrix

$$d^n_{\mathrm{Mc}(\varphi)} = \begin{pmatrix} d^n_{F_0} & 0 \\ \varphi^{n+1} & d^n_{F_1} \end{pmatrix} .$$

The differential $d^n_F \colon C^n(F_0) \oplus C^n(F_1) \to C^{n+1}(F_0) \oplus C^{n+1}(F_1)$ is given by the same matrix since d^n_F induces the morphism 0 from $C^n(F_1)$ to $C^{n+1}(F_0)$.

(iii) follows from (ii). q.e.d.

Example 12.4.4. Let (S, Σ) be as in Example 12.4.2. Let $s_0 \in S$. Then $\Sigma_0 = \{I \in \Sigma; s_0 \notin I\}$ and $\Sigma_1 = \{J \in \Sigma; s_0 \in J\}$ satisfy the above conditions.

Let $I \in \Sigma_0$. Then $\varphi \colon C^n(F_0) \to C^{n+1}(F_1)$ induces the morphism

$$f_{(I \sqcup \{s_0\})I} \otimes e_{s_0} \wedge \colon X_I \otimes \det \mathbb{Z}^I \to X_{I \sqcup \{s_0\}} \otimes \det \mathbb{Z}^{I \sqcup \{s_0\}} .$$

Recall that $X_I = X$ and $f_{(I \sqcup \{s_0\})I} = f_{s_0}$. The morphisms

$$\mathrm{id}_X \otimes e_{s_0} \wedge \colon X_I \otimes \det \mathbb{Z}^I \to X_{I \sqcup \{s_0\}} \otimes \det \mathbb{Z}^{I \sqcup \{s_0\}}$$

induce an isomorphism $C^\bullet(F_0) \simeq C^\bullet(F_1)[1]$. Hence, we get a long exact sequence

$$\cdots \to H^n(C^\bullet(F_0)) \xrightarrow{f_{s_0}} H^n(C^\bullet(F_0)) \to H^{n+1}(C^\bullet(F)) \to \cdots .$$

Hence, if $\{f_s\}_{s \in S \setminus s_0}$ is a quasi-regular family and $f_{s_0} \colon X / \sum_{s \neq s_0} \mathrm{Im}\, f_s \to X / \sum_{s \neq s_0} \mathrm{Im}\, f_s$ is a monomorphism, then $\{f_s\}_{s \in S}$ is a quasi-regular family.

Remark 12.4.5. We may also encounter contravariant functors, that is, functors $G \colon \Sigma^{\mathrm{op}} \to \mathcal{C}$. In such a case if Σ is not finite, we have to assume:

(12.4.8) $\mathcal{C}$ admits small inductive limits .

For $I \in \Sigma$ and for $I \subset J$, we set $X_I = G(I)$, $g_{JI} = G(J \to I)$,

$$C_n(G) = \bigoplus_{|I|=n} X_I \otimes \det \mathbb{Z}^I .$$

If $J = I \sqcup \{s\}$ we denote by

$$e_s \lfloor \colon \det \mathbb{Z}^J \to \det \mathbb{Z}^I$$

the isomorphism inverse to $e_s \wedge$, and we get the morphism:

$$g_{JI} \otimes (e_s \lfloor) \colon X_J \otimes \det \mathbb{Z}^J \to X_I \otimes \det \mathbb{Z}^I$$

from which we deduce the morphism $d_n^G \colon C_n(G) \to C_{n-1}(G)$. We have $d_{n-1}^G \circ d_n^G = 0$ and denote by $C_\bullet(G)$ the complex

$$(12.4.9) \quad C_\bullet(G) \colon \cdots \to C_n(G) \xrightarrow{d_n^G} C_{n-1}(G) \to \cdots \xrightarrow{d_1^G} C_0(G) \to 0 .$$

We leave to the reader the translation of Theorem 12.4.3 in this framework.

Distributive Families of Subobjects

As an application of Theorem 12.4.3, we shall study distributive families of subobjects of an object. References are made to [33, 60].

Let $\mathcal{C}$ be an abelian category and let $X \in \mathcal{C}$. Let $\{X_s\}_{s \in S}$ be a family of subobjects (see Notation 8.3.10) of X indexed by a finite subset S. Let Σ be the set of subsets of S. For $I \in \Sigma$, set $X_I = \bigcap_{s \in I} X_s$ and $X_\emptyset = X$. Then $I \mapsto X_I$ gives a functor from Σ^{op} to $\mathcal{C}$. We get a complex

(12.4.10)
$$\mathfrak{X}_\bullet(X, \{X_s\}_{s \in S}) :=$$
$$\cdots \to \bigoplus_{|I|=n} X_I \otimes \det \mathbb{Z}^I \to \cdots \to \bigoplus_{s \in S} X_s \otimes \mathbb{Z}e_s \to X \to 0,$$

where X stands in degree 0.

Note that

(12.4.11)
$$H_0\big(\mathfrak{X}_\bullet(X, \{X_s\}_{s \in S})\big) \simeq X/\big(\sum_{s \in S} X_s\big) .$$

Definition 12.4.6. *A family* $\{X_s\}_{s \in S}$ *of subobjects of* X *is* distributive *if* $H_n\big(\mathfrak{X}_\bullet(X, \{X_s\}_{s \in S'})\big) \simeq 0$ *for any* $n \neq 0$ *and any subset* S' *of* S.

Remark that for a subobject Y of X and a finite family $\{X_s\}_{s \in S}$ of subobjects of Y, the family $\{X_s\}_{s \in S}$ is distributive as a family of subobjects of X if and only if it is so as a family of subobjects of Y.

Assuming $\mathrm{card}(S) \geq 1$, let us take $s_0 \in S$ and set $S_0 = S \setminus \{s_0\}$. Then for $I \subset S_0$,

$$X_{I \cup \{s_0\}} = \bigcap_{s \in I}(X_{s_0} \cap X_s) .$$

Applying Theorem 12.4.3, we obtain a d.t.

(12.4.12)
$$\mathfrak{X}_\bullet(X_{s_0}, \{X_{s_0} \cap X_s\}_{s \in S_0}) \to \mathfrak{X}_\bullet(X, \{X_s\}_{s \in S_0}) \to \mathfrak{X}_\bullet(X, \{X_s\}_{s \in S}) \xrightarrow{+1} .$$

Lemma 12.4.7. *Assume that* $H_n\big(\mathfrak{X}_\bullet(X, \{X_s\}_{s \in S_0})\big) \simeq 0$ *for* $n \neq 0$. *Then the following two conditions are equivalent:*

(a) $H_n\big(\mathfrak{X}_\bullet(X, \{X_s\}_{s \in S})\big) \simeq 0$ *for* $n \neq 0$,
(b) $H_n\big(\mathfrak{X}_\bullet(X_{s_0}, \{X_{s_0} \cap X_s\}_{s \in S_0})\big) \simeq 0$ *for* $n \neq 0$ *and* $X_{s_0} \cap \big(\sum_{s \in S_0} X_s\big) = \sum_{s \in S_0}(X_{s_0} \cap X_s)$.

Proof. By the d.t. (12.4.12), we have an exact sequence

$$0 \to H_1\big(\mathfrak{X}_\bullet(X, \{X_s\}_{s \in S})\big) \to H_0\big(\mathfrak{X}_\bullet(X_{s_0}, \{X_{s_0} \cap X_s\}_{s \in S_0})\big)$$
$$\to H_0\big(\mathfrak{X}_\bullet(X, \{X_s\}_{s \in S_0})\big)$$

and isomorphisms

$$H_n\big(\mathfrak{X}_\bullet(X, \{X_s\}_{s\in S})\big) \overset{\sim}{\longrightarrow} H_{n-1}\big(\mathfrak{X}_\bullet(X_{s_0}, \{X_{s_0} \cap X_s\}_{s\in S_0})\big)$$

for $n > 1$. Since

$$\mathrm{Ker}\Big(X_{s_0}/\sum_{s\in S_0}(X_{s_0} \cap X_s) \to X/\sum_{s\in S_0} X_s\Big)$$

$$= \big(X_{s_0} \cap \sum_{s\in S_0} X_s\big)/\big(\sum_{s\in S_0}(X_{s_0} \cap X_s)\big),$$

the result follows.   q.e.d.

The next result gives a tool to proceed by induction in order to prove that a finite family of subobjects of X is distributive.

Proposition 12.4.8. *Let $s_0 \in S$ and set $S_0 = S \setminus \{s_0\}$. Let $\{X_s\}_{s\in S}$ be a family of subobjects of X. Then the following conditions are equivalent.*

(a) *$\{X_s\}_{s\in S}$ is a distributive family of subobjects of X,*
(b) *the conditions (i)–(iii) below are satisfied:*
    (i) *$\{X_s\}_{s\in S_0}$ is a distributive family of subobjects of X,*
    (ii) *$\{X_{s_0} \cap X_s\}_{s\in S_0}$ is a distributive family of subobjects of X_{s_0},*
    (iii) *$X_{s_0} \cap (\sum_{s\in I} X_s) = \sum_{s\in I}(X_{s_0} \cap X_s)$ for any subset I of S_0.*

Proof. (a) $\Rightarrow$ (b). Condition (b) (i) is clearly satisfied. For any $I \subset S_0$, $H_n\big(\mathfrak{X}_\bullet(X, \{X_s\}_{s\in I})\big)$ and $H_n\big(\mathfrak{X}_\bullet(X, \{X_s\}_{s\in I\cup\{s_0\}})\big)$ vanish for $n \neq 0$. Hence (b) (ii) and (b) (iii) follow from Lemma 12.4.7.

(b) $\Rightarrow$ (a). Let $I \subset S$. Let us show that $H_n\big(\mathfrak{X}_\bullet(X, \{X_s\}_{s\in I})\big)$ vanishes for $n \neq 0$. If $s_0 \notin I$, it is obvious since $\{X_s\}_{s\in S_0}$ is distributive. Assume $s_0 \in I$. Then $H_n\big(\mathfrak{X}_\bullet(X, \{X_s\}_{s\in I\setminus\{s_0\}})\big) \simeq 0$ for $n \neq 0$ since $\{X_s\}_{s\in S_0}$ is distributive, and $H_n\big(\mathfrak{X}_\bullet(X_{s_0}, \{X_{s_0} \cap X_s\}_{s\in I\setminus\{s_0\}})\big) \simeq 0$ since $\{X_{s_0} \cap X_s\}_{s\in S_0}$ is distributive. Moreover,

$$X_{s_0} \cap \Big(\sum_{s\in I\setminus\{s_0\}} X_s\Big) = \sum_{s\in I\setminus\{s_0\}}(X_{s_0} \cap X_s).$$

Hence the result follows from Lemma 12.4.7.   q.e.d.

Example 12.4.9. (i) If $X = X_s$ for some s, then $H_n\big(\mathfrak{X}_\bullet(X, \{X_s\}_{s\in S})\big) \simeq 0$ for all n by (12.4.12).
(ii) $H_n\big(\mathfrak{X}_\bullet(X, \{X_s\}_{s\in S})\big) \simeq 0$ unless $0 \leq n \leq |S| - 2$.
(iii) If $\mathrm{card}(S) \leq 2$, the family $\{X_s\}_{s\in S}$ is distributive.
(iv) $\{X_1, X_2, X_3\}$ is a distributive family of subobjects of X if and only if $X_1 \cap (X_2 + X_3) \subset (X_1 \cap X_2) + (X_1 \cap X_3)$. Of course, the last condition is equivalent to $X_1 \cap (X_2 + X_3) = (X_1 \cap X_2) + (X_1 \cap X_3)$.
(v) $\{X_1, X_2, X_3, X_4\}$ is a distributive family of subobjects of X if and only

$$\begin{cases} X_i \cap (X_j + X_k) = (X_i \cap X_j) + (X_i \cap X_k) & \text{for } 1 \leq i < j < k \leq 4, \\ X_1 \cap (X_2 + X_3 + X_4) = (X_1 \cap X_2) + (X_1 \cap X_3) + (X_1 \cap X_4). \end{cases}$$

Let us give some properties of distributive families.

Proposition 12.4.10. *Let* $\{X_s\}_{s \in S}$ *be a distributive family of subobjects of* X *and let* $S' = S \sqcup \{0\}$.

  (i) *Setting* $X_0 = X$ *or* $X_0 = 0$, *the family* $\{X_s\}_{s \in S'}$ *is distributive.*
  (ii) *Let* $s_0 \in S$. *Setting* $X_0 = X_{s_0}$, *the family* $\{X_s\}_{s \in S'}$ *is distributive.*
  (iii) *Let* $s_1, s_2 \in S$. *Setting* $X_0 = X_{s_1} + X_{s_2}$, *the family* $\{X_s\}_{s \in S'}$ *is distributive.*
  (iv) *Let* $s_1, s_2 \in S$. *Setting* $X_0 = X_{s_1} \cap X_{s_2}$, *the family* $\{X_s\}_{s \in S'}$ *is distributive.*

Proof. (i) In both cases, $\{X_0 \cap X_s\}_{s \in S}$ is distributive and for any $I \subset S$, $X_0 \cap \left(\sum_{s \in I} X_s \right) = \sum_{s \in I} (X_0 \cap X_s)$.

(ii) $\{X_0 \cap X_s\}_{s \in S}$ is distributive by (i) and Proposition 12.4.8 (a)$\Rightarrow$(b)(i). Hence, it is enough to show that

$$(12.4.13) \qquad X_{s_0} \cap \left(\sum_{s \in I} X_s \right) = \sum_{s \in I} (X_{s_0} \cap X_s) \text{ for any } I \subset S .$$

If $s_0 \notin I$, it is a consequence of Proposition 12.4.8. If $s_0 \in I$, both terms of (12.4.13) are equal to X_{s_0}.

(iii) By (ii), we may assume that $s_1 \neq s_2$. We proceed by induction on $n = \operatorname{card}(S)$. If $n \leq 1$, the result is clear.
If $n = 2$, the result follows from $X_0 \cap (X_{s_1} + X_{s_2}) = (X_0 \cap X_{s_1}) + (X_0 \cap X_{s_2})$ (see Example 12.4.9 (iv)).
Assume $n \geq 3$. Take $s_0 \in S \setminus \{s_1, s_2\}$. Then $\{X_s\}_{s \in S \setminus \{s_0\}}$ is distributive, and hence $\{X_s\}_{s \in S' \setminus \{s_0\}}$ is distributive by the induction hypothesis. Since $\{X_0, X_{s_1}, X_{s_2}\}$ is distributive, $X_{s_0} \cap X_0 = (X_{s_0} \cap X_{s_1}) + (X_{s_0} \cap X_{s_2})$. Since $\{X_{s_0} \cap X_s\}_{s \in S \setminus \{s_0\}}$ is distributive, $\{X_{s_0} \cap X_s\}_{s \in S' \setminus \{s_0\}}$ is distributive by the induction hypothesis. In order to apply Proposition 12.4.8 and conclude, it remains to show that

$$(12.4.14) \qquad X_{s_0} \cap \left(\sum_{s \in I} X_s \right) = \sum_{s \in I} (X_{s_0} \cap X_s) \text{ for any } I \subset S' \setminus \{s_0\} .$$

If $0 \notin I$, this is obvious. If $0 \in I$, the left hand side of (12.4.14) is equal to

$$X_{s_0} \cap \left(X_{s_1} + X_{s_2} + \sum_{s \in I \setminus 0} X_s \right) = X_{s_0} \cap X_{s_1} + X_{s_0} \cap X_{s_2} + \sum_{s \in I \setminus 0} (X_{s_0} \cap X_s)$$

$$\subset \sum_{s \in I} (X_{s_0} \cap X_s) .$$

(iv) Since $\{X_{s_1} \cap X_s\}_{s \in S}$ is distributive, $\{X_{s_1} \cap X_{s_2} \cap X_s\}_{s \in S}$ is distributive. For any $I \subset S$,

$$X_0 \cap \left(\sum_{s \in I} X_s \right) = X_0 \cap \left(\sum_{s \in I} X_{s_1} \cap X_s \right)$$

$$= \sum_{s \in I} (X_0 \cap X_s) ,$$

where the last equality follows from the distributivity of $\{X_{s_1} \cap X_s\}_{s \in S}$. Hence, $\{X_s\}_{s \in S'}$ is distributive by Proposition 12.4.8.   q.e.d.

Corollary 12.4.11. *Let $\{X_s\}_{s \in S}$ be a finite family of subobjects of $X \in \mathcal{C}$. Let S be the smallest family of subobjects that contains the X_s's and closed by the operations $+$ and $\cap$. Then the following conditions are equivalent:*

(i) *$\{X_s\}_{s \in S}$ is distributive,*
(ii) *$U \cap (V + W) = (U \cap V) + (U \cap W)$ for any $U, V, W \in \mathcal{S}$.*

Proof. (i) $\Rightarrow$ (ii). By Proposition 12.4.10, for any finite subset J of $\mathcal{S}$, the family $\{X_s\}_{s \in J}$ is distributive.

(ii) $\Rightarrow$ (i). We argue by induction on card(S). Take $s_0 \in S$. Then $\{X_s\}_{s \in S \setminus \{s_0\}}$ and $\{X_{s_0} \cap X_s\}_{s \in S \setminus \{s_0\}}$ are distributive by the induction hypothesis and $X_{s_0} \cap (\sum_{s \in I} X_s) = \sum_{s \in I}(X_{s_0} \cap X_s)$ for any $I \subset S$. Hence, the result follows from Proposition 12.4.8.   q.e.d.

See also Exercises 12.5–12.7 for further properties of distributive families.

12.5 Double Complexes

Let $\mathcal{C}$ be an abelian category and consider a double complex in $\mathcal{C}$:

$$X := (X^{\bullet,\bullet}, d_X) = \{X^{n,m}, d_X'^{n,m}, d_X''^{n,m}; (n, m) \in \mathbb{Z} \times \mathbb{Z}\} .$$

We shall make use of the two functors F_I and F_{II} defined in Notation 11.5.1.
    The functors $\tau_I^{\leq n}, \widetilde{\tau}_I^{\leq n}, \tau_I^{\geq n}, \widetilde{\tau}_I^{\geq n}$ from $\mathrm{C}^2(\mathcal{C}) \to \mathrm{C}^2(\mathcal{C})$ and $H_I^n(\bullet)$ from $\mathrm{C}^2(\mathcal{C}) \to \mathrm{C}(\mathcal{C})$ are defined by using the functor F_I, and similarly $\tau_{II}^{\leq n}$, etc. by using F_{II}. For example, we set

$$\tau_I^{\leq n} = (F_I^{-1}) \circ \tau^{\leq n} \circ F_I .$$

Then $H_I^n(X)$ is the simple complex

$$\cdots \xrightarrow{d''^{\bullet,m-1}} H_I^n(X^{\bullet,m}) \xrightarrow{d''^{\bullet,m}} H_I^n(X^{\bullet,m+1}) \xrightarrow{d''^{\bullet,m+1}} \cdots ,$$

where $H_I^n(X^{\bullet,m})$ is the n-th cohomology object of the complex

$$\cdots \xrightarrow{d'^{p-1,m}} X^{p,m} \xrightarrow{d'^{p,m}} X^{p+1,m} \xrightarrow{d'^{p+1,m}} \cdots .$$

We denote by $H_I(X)$ the double complex whose rows are the $H_I^n(X)$'s and with zero vertical differentials $H_I^n(X) \to H_I^{n+1}(X)$. Iterating this operation, we find a complex $H_{II}H_I(X)$ with (vertical and horizontal) zero differentials.
    In order to prove Theorem 12.5.4 below, we prepare some lemmas.

Lemma 12.5.1. *The functor $\mathrm{tot}: \mathrm{C}_f^2(\mathcal{C}) \to \mathrm{C}(\mathcal{C})$ is exact.*

The proof is straightforward.

Lemma 12.5.2. *Let* $(X^{\bullet,\bullet}, d', d'') \in C_f^2(\mathcal{C})$ *be a double complex. Then the natural morphism* $\mathrm{tot}(\tau_I^{\leq q}(X)) \to \mathrm{tot}(\widetilde{\tau}_I^{\leq q}(X))$ *is a qis for all* q.

Proof. We have an exact sequence in $C(\mathrm{C}(\mathcal{C}))$

$$0 \to \tau_I^{\leq q}(X) \to \widetilde{\tau}_I^{\leq q}(X) \to \mathrm{Mc}(\mathrm{id}_{\mathrm{Im}\, d_X^{\prime q}[-q-1]}) \to 0 .$$

Applying Lemma 12.5.1, we get the exact sequence

$$0 \to \mathrm{tot}(\tau_I^{\leq q}(X)) \to \mathrm{tot}(\widetilde{\tau}_I^{\leq q}(X)) \to \mathrm{tot}(\mathrm{Mc}(\mathrm{id}_{\mathrm{Im}\, d_X^{\prime q}[-q-1]})) \to 0 .$$

Since $\mathrm{tot}(\mathrm{Mc}(\mathrm{id}_{\mathrm{Im}\, d_X^{\prime q}[-q-1]})) \simeq \mathrm{Mc}(\mathrm{id}_{\mathrm{Im}\, d_X^{\prime q}[-q-1]})$ and this complex is exact, we get the result by Theorem 12.3.3.   q.e.d.

Lemma 12.5.3. *Let* $X \in C_f^2(\mathcal{C})$ *be a double complex. Then for each* q*, there is an exact sequence in* $\mathrm{C}(\mathcal{C})$

$$0 \to \mathrm{tot}(\widetilde{\tau}_I^{\leq q-1}(X)) \to \mathrm{tot}(\tau_I^{\leq q}(X)) \to H_I^q(X)[-q] \to 0 .$$

Proof. Consider the functorial exact sequence in $\mathrm{C}(\mathrm{C}(\mathcal{C}))$:

$$0 \to \widetilde{\tau}^{\leq q-1}(F_I(X)) \to \tau^{\leq q}(F_I(X)) \to H^q(F_I(X))[-q] \to 0$$

and apply the exact functor $\mathrm{tot} \circ F_I^{-1}$. It is immediately checked that $(\mathrm{tot} \circ F_I^{-1})(H^q(F_I(X))[-q]) \simeq H_I^q(X)[-q]$.   q.e.d.

Theorem 12.5.4. *Let* $f : X \to Y$ *be a morphism in* $C_f^2(\mathcal{C})$ *and assume that* f *induces an isomorphism*

$$f : H_{II} H_I(X) \xrightarrow{\sim} H_{II} H_I(Y) .$$

Then $\mathrm{tot}(f) \colon \mathrm{tot}(X) \to \mathrm{tot}(Y)$ *is a qis.*

Proof. First note that the hypothesis is equivalent to saying that for each q the morphism of complexes $H_I^q(f) \colon H_I^q(X) \to H_I^q(Y)$ is a qis.

Since $H_I^n \tau_I^{\leq q}(X)$ is isomorphic to $H_I^n(X)$ or 0 depending whether $n \geq q$ or not, the hypothesis entails the isomorphisms

$$H_{II} H_I(\tau_I^{\geq q}(X)) \xrightarrow{\sim} H_{II} H_I(\tau_I^{\geq q}(Y))$$

for all $q \in \mathbb{Z}$. For a given k,

$$H^k(\mathrm{tot}(X)) \simeq H^k(\mathrm{tot}(\tau_I^{\geq q}(X)) \text{ for } q \ll 0 ,$$

and similarly with Y instead of X. Hence, replacing X and Y with $\tau_I^{\geq q}(X)$ and $\tau_I^{\geq q}(Y)$, we may assume from the beginning that $\tau_I^{\leq q}(X)$ and $\tau_I^{\leq q}(Y)$ are zero for $q \ll 0$.

Applying Lemma 12.5.3, we obtain a commutative diagram of exact sequences in $C(\mathcal{C})$:

$$(12.5.1)$$

$$
\begin{array}{ccccccccc}
0 & \longrightarrow & \mathrm{tot}(\widetilde{\tau}_I^{\leq q-1}(X)) & \longrightarrow & \mathrm{tot}(\tau_I^{\leq q}(X)) & \longrightarrow & H_I^q(X)[-q] & \longrightarrow & 0 \\
& & \Big\downarrow {\scriptstyle \mathrm{tot}(\widetilde{\tau}_I^{\leq q-1}(f))} & & \Big\downarrow {\scriptstyle \mathrm{tot}(\tau_I^{\leq q}(f))} & & \Big\downarrow {\scriptstyle H_I^q(f)[-q]} & & \\
0 & \longrightarrow & \mathrm{tot}(\widetilde{\tau}_I^{\leq q-1}(Y)) & \longrightarrow & \mathrm{tot}(\tau_I^{\leq q}(Y)) & \longrightarrow & H_I^q(Y)[-q] & \longrightarrow & 0.
\end{array}
$$

Let us denote by r_q, m_q, l_q the vertical arrow on the right, on the middle, and on the left in (12.5.1), respectively. By the hypothesis, the arrow r_q is a qis for all q. Assuming that the arrow l_q is a qis, we get that the arrow m_q is a qis. By Lemma 12.5.2, we deduce that the arrow l_{q+1} is a qis. Since l_q is the arrow $0 \to 0$ for $q \ll 0$, the induction proceeds and all l_q's are qis. Then the result follows from

$$H^k(\mathrm{tot}(\widetilde{\tau}_I^{\leq q}(f))) \simeq H^k(\mathrm{tot}(f)) \text{ for } q \gg 0 .$$

q.e.d.

Corollary 12.5.5. *Let X be a double complex in $\mathrm{C}_f^2(\mathcal{C})$.*

(i) *Assume that all rows of X are exact. Then $\mathrm{tot}(X)$ is qis to 0.*
(ii) *Assume that the rows $X^{j,\bullet}$ of X are are exact for all $j \neq p$. Then $\mathrm{tot}(X)$ is qis to $X^{p,\bullet}[-p]$.*
(iii) *Assume that all rows $X^{j,\bullet}$ and columns $X^{\bullet,j}$ of X are exact for $j \neq 0$. Then $H^p(X^{0,\bullet}) \simeq H^p(X^{\bullet,0})$ for all p.*

Proof. (i) is obvious.
(ii) We set $\sigma_I^{\geq p} = F_I^{-1} \circ \sigma^{\geq p} \circ F_I$, where $\sigma^{\geq p}$ is the stupid truncation functor given in Definition 11.3.11 and F_I is given in Notation 11.5.1. We define similarly $\sigma_I^{\leq p}$. Then the result follows by applying Theorem 12.5.4 to the morphism $\sigma_I^{\geq p}(X) \to X$, next to the morphism $\sigma_I^{\geq p}(X) \to \sigma_I^{\leq p}\sigma_I^{\geq p}(X) \simeq X^{p,\bullet}[-p]$.
(iii) Both $X^{0,\bullet}$ and $X^{\bullet,0}$ are qis to $\mathrm{tot}(X)$ by (ii).   q.e.d.

When $\mathcal{C}$ is the category $\mathrm{Mod}(R)$ of modules over a ring R and all rows $X^{j,\bullet}$ and columns $X^{\bullet,j}$ of X are 0 for $j < 0$, the isomorphism in Corollary 12.5.5 (iii) may be described by the so-called "Weil procedure".

Let $x^{p,0} \in X^{p,0}$, with $d'x^{p,0} = 0$ which represents $y \in H^p(X^{\bullet,0})$. Define $x^{p,1} = d''x^{p,0}$. Then $d'x^{p,1} = 0$, and the first column being exact, there exists $x^{p-1,1} \in X^{p-1,1}$ with $d'x^{p-1,1} = x^{p,1}$. This procedure can be iterated until getting $x^{0,p} \in X^{0,p}$. Since $d'd''x^{0,p} = 0$, and d' is injective on $X^{0,p}$ for $p > 0$ by the hypothesis, we get $d''x^{0,p} = 0$. The class of $x^{0,p}$ in $H^p(X^{0,\bullet})$ will be the image of y by the Weil procedure. Of course, it remains to check that this image does not depend of the various choices we have made, and that it induces an isomorphism.

This procedure can be visualized by the diagram:

$$X^{0,p} \xrightarrow{d''} 0$$
$$d' \downarrow$$
$$X^{1,p-1} \xrightarrow{d''} X^{1,p}$$
$$\vdots$$

$$X^{p-1,1} \dashrightarrow$$
$$d' \downarrow$$
$$X^{p,0} \xrightarrow{d''} X^{p,1}$$
$$d' \downarrow$$
$$0.$$

Exercises

Exercise 12.1. Let $\mathcal{C}$ be an abelian category and let X be a double complex with $X^{i,j} = 0$ for $i < 0$ or $j < 0$. Assume that all rows and all columns of X are exact, and denote by Y the double complex obtained by replacing $X^{0,j}$ and $X^{i,0}$ by 0 for all j and all i. Prove that there is a qis $X^{0,0} \to \text{tot}(Y)$.

Exercise 12.2. Let $\mathcal{C}, \mathcal{C}'$ and $\mathcal{C}''$ be abelian categories, $F : \mathcal{C} \times \mathcal{C}' \to \mathcal{C}''$ an exact bifunctor. Let $X \to I$ and $Y \to J$ be two qis in $C^+(\mathcal{C})$ and $C^+(\mathcal{C}')$ respectively. Prove that $F^{\bullet}(X, Y) \to F^{\bullet}(I, J)$ is a qis.

Exercise 12.3. Let k be a field and let $X, Y \in C^-(\text{Mod}(k))$.
(i) Prove the isomorphism $H^n(\text{tot}(X \otimes Y)) \simeq \bigoplus_{i+j=n} H^i(X) \otimes H^j(Y)$.
(ii) Denote by v the isomorphism $M \otimes N \to N \otimes M$ in $\text{Mod}(k)$. With the notations and the help of Proposition 11.6.5, prove that the diagram below commutes:

$$
\begin{array}{ccc}
H^n(\text{tot}(X \otimes Y)) & \xrightarrow{\sim} & \bigoplus_{i+j=n} H^i(X) \otimes H^j(Y) \\
r \downarrow & & (-)^{ij} v \downarrow \\
H^n(\text{tot}(Y \otimes X)) & \xrightarrow{\sim} & \bigoplus_{i+j=n} H^j(Y) \otimes H^i(X).
\end{array}
$$

Exercise 12.4. Let $\mathcal{C}$ be an abelian category, $X \in \mathcal{C}$, and let $\{f_i\}_{i=1,\dots,n}$ be a sequence of commuting endomorphisms of X. The sequence $\{f_1, \dots, f_n\}$ is *regular* if for each i $(1 \leq i \leq n)$, f_i induces a monomorphism on $X / \sum_{j<i} \text{Im } f_j$. (Hence, f_1 is a monomorphism, f_2 induces a monomorphism on $X / \text{Im } f_1$, etc.)

Prove that if $(f_1, \dots, f_n)$ is a regular sequence, then it is quasi-regular (see Example 12.4.2).

Exercise 12.5. Let $\mathcal{C}$ be an abelian category, and let $\mathcal{C}^{\text{op}}$ be its opposite category. Let $\{X_s\}_{s \in S}$ be a finite family of subobjects of an object X of $\mathcal{C}$. Let Y_s be the subobject of $X^{\text{op}} \in \mathcal{C}^{\text{op}}$ defined as the kernel of $X^{\text{op}} \to X_s^{\text{op}}$. Prove that $\{Y_s\}_{s \in S}$ is distributive if and only if $\{X_s\}_{s \in S}$ is distributive.

Exercise 12.6. Let $\mathcal{C}$ be an abelian category. Let $\{S_i\}_{i \in I}$ be a finite family of non-empty finite sets and $S = \bigsqcup_{i \in I} S_i$. Let $\{X_s\}_{s \in S}$ be a family of subobjects of an object X in $\mathcal{C}$. Assume that for any $i \in I$ and any $s, s' \in S_i$, we have either $X_s \subset X_{s'}$ or $X_{s'} \subset X_s$. Prove that the following two conditions are equivalent:

(a) the family $\{X_s\}_{s \in S}$ is distributive,
(b) for any $s_i \in S_i$ ($i \in I$), $\{X_{s_i}\}_{i \in I}$ is distributive.

Exercise 12.7. Let $\mathcal{C}$ be an abelian category. An object $X \in \mathcal{C}$ is *semisimple* if for any subobject $Y \hookrightarrow X$, there exist Z and an isomorphism $Y \oplus Z \xrightarrow{\sim} X$. Equivalently, any monomorphism $Y \rightarrowtail X$ admits a cosection, or any epimorphism $X \twoheadrightarrow W$ admits a section.

Assume that X is semisimple and let $\{X_s\}_{s \in S}$ be a finite family of subobjects of X. Prove that the two conditions below are equivalent:

(a) the family $\{X_s\}_{s \in S}$ is distributive,
(b) there exists a finite direct sum decomposition $X \simeq \bigoplus_{a \in A} Y_a$ such that each X_s is a direct sum of some of the Y_a's ($a \in A$).

Exercise 12.8. We regard the ordered set $\mathbb{N}$ as a category. Let A be a ring and let β be a projective system in $\mathrm{Mod}(A)$ indexed by $\mathbb{N}$, that is, $\beta \in \mathrm{Fct}(\mathbb{N}^{\mathrm{op}}, \mathrm{Mod}(A))$. Set $\beta(n) = M_n$ and denote by $v_{np} \colon M_p \to M_n$ the linear map associated with $n \leq p$. The projective system β satisfies the *Mittag-Leffler condition* (or *M-L condition*, for short) if for any $n \in \mathbb{N}$, the sequence $\{v_{np}(M_p)\}_{p \geq n}$ of submodules of M_n is stationary. Consider an exact sequence in $\mathrm{Fct}(\mathbb{N}^{\mathrm{op}}, \mathrm{Mod}(A))$:

$$0 \to \beta' \to \beta \to \beta'' \to 0 .$$

(i) Prove that if β' and β'' satisfy the M-L condition, then so does β.
(ii) Prove that if β satisfies the M-L condition, then so does β''.
(iii) Prove that if β' satisfies the M-L condition, then the sequence $0 \to \varprojlim \beta' \to \varprojlim \beta \to \varprojlim \beta'' \to 0$ is exact.
(iv) Prove that if β satisfies the M-L condition then $R^n \pi (\text{``}\varprojlim\text{''} \beta) \simeq 0$ for $n \neq 0$ (see page 335).
(Hint: see [29] or [38, Proposition 1.12.2].)

Exercise 12.9. We regard the ordered set $\mathbb{R}$ as a category. Let A be a ring. Consider a functor $\beta \colon \mathbb{R} \to \mathrm{Mod}(A)$ and set $X_s := \beta(s)$ for $s \in \mathbb{R}$. Consider the maps

$$\lambda_s \colon \varinjlim_{t < s} X_t \to X_s \, ,$$

$$\mu_s \colon X_s \to \varprojlim_{t > s} X_t \, .$$

Prove that the maps $\beta(t \to s) \colon X_t \to X_s$ are injective (resp. surjective) for all $t \leq s$ if λ_s and μ_s are injective (resp. surjective) for all $s \in \mathbb{R}$. (Hint: see [38, Proposition 1.12.6].)

Exercise 12.10. Let A be a ring. Consider a functor $\beta^\bullet : \mathbb{R} \to C(\mathrm{Mod}(A))$. Set $\beta_t^\bullet := \beta^\bullet(t) \in C(\mathrm{Mod}(A))$ for $t \in \mathbb{R}$. Denote by

$$a_s^n : \varinjlim_{t<s} H^n(\beta_t^\bullet) \to H^n(\beta_s^\bullet) ,$$

$$b_s^n : H^n(\beta_s^\bullet) \to \varprojlim_{t>s} H^n(\beta_t^\bullet) .$$

the natural maps. We assume

(a) $\beta_s^n \to \beta_t^n$ is surjective for all $n \in \mathbb{Z}$ and all $s \le t \in \mathbb{R}$,

(b) $\beta_s^n \xrightarrow{\sim} \varprojlim_{t>s} \beta_t^n$ for all $n \in \mathbb{Z}$ and all $s \in \mathbb{R}$,

(c) a_s^n is an isomorphism for all $n \in \mathbb{Z}$, all $s \in \mathbb{R}$.

(i) Prove that b_s^n is an epimorphism for all $n \in \mathbb{Z}$, all $\in \mathbb{R}$.

(ii) Let $n_0 \in \mathbb{Z}$, and assume that $b_s^{n_0-1}$ is an isomorphism for all $s \in \mathbb{R}$. Prove that $b_s^{n_0}$ is an isomorphism for all $s \in \mathbb{R}$.

(iii) Assume that there exists $n_0 \in \mathbb{Z}$ such that b_s^n is an isomorphism for all $s \in \mathbb{R}$ and all $n \le n_0$. Prove that $H^n(\beta_t^\bullet) \to H^n(\beta_s^\bullet)$ is an isomorphism for all $n \in \mathbb{Z}$ and all $t \le s$.

(Hint: use Exercises 12.8, 12.9 or see [38, Proposition 2.7.2].)

13

Derived Categories

In this chapter we study derived categories and derived functors. Most of the results concerning derived categories bounded form below (or from above) are well-known, besides perhaps Theorem 13.3.7 which is useful when deriving abelian categories which do not admit enough injectives, such as abelian categories of ind-objects.

13.1 Derived Categories

Let $(\mathcal{A}, T)$ be an abelian category with translation. Recall (Corollary 12.2.5) that the cohomology functor $H: \mathcal{A}_c \to \mathcal{A}$ induces a cohomological functor

$$(13.1.1) \qquad\qquad H: K_c(\mathcal{A}) \to \mathcal{A}.$$

Let $\mathcal{N}$ be the full subcategory of $K_c(\mathcal{A})$ consisting of objects X such that $H(X) \simeq 0$, that is, X is qis to 0. Since H is cohomological, the category $\mathcal{N}$ is a triangulated subcategory of $K_c(\mathcal{A})$. We shall localize $K_c(\mathcal{A})$ with respect to $\mathcal{N}$ (see Sects. 7.1 and 10.2).

Definition 13.1.1. *We denote by* $D_c(\mathcal{A})$ *the category* $K_c(\mathcal{A})/\mathcal{N}$ *and call it the* derived category *of* $(\mathcal{A}, T)$.

Note that $D_c(\mathcal{A})$ is triangulated by Theorem 10.2.3. By the properties of the localization, a quasi-isomorphism in $K_c(\mathcal{A})$ (or in $\mathcal{A}_c$) becomes an isomorphism in $D_c(\mathcal{A})$. One shall be aware that the category $D_c(\mathcal{A})$ may be a big category.

From now on, we shall restrict our study to the case where $\mathcal{A}_c$ is the category of complexes of an abelian category.

Let $\mathcal{C}$ be an abelian category. Recall that the categories $C^*(\mathcal{C})$ ($* = $ ub, b, $+$, $-$, $[a, b]$, $\geq a$, $\leq b$) are defined in Notations 11.3.4, and we define similarly the full subcategories $K^*(\mathcal{C})$ of $K(\mathcal{C})$ by $Ob(K^*(\mathcal{C})) = Ob(C^*(\mathcal{C}))$.

Therefore, $C^*(\mathcal{C}) \to K^*(\mathcal{C})$ is essentially surjective. Note that $K^*(\mathcal{C})$ is a triangulated category for $* = $ ub, $+$, $-$, b. For $* = $ ub, $+$, $-$, b, define

$$N^*(\mathcal{C}) = \left\{ X \in K^*(\mathcal{C}) \, ; \, H^k(X) \simeq 0 \text{ for all } k \right\} \, .$$

Clearly, $N^*(\mathcal{C})$ is a null system in $K^*(\mathcal{C})$.

Definition 13.1.2. *The triangulated categories* $D^*(\mathcal{C})$ ($* = $ ub, $+$, $-$, b) *are defined as* $K^*(\mathcal{C})/N^*(\mathcal{C})$ *and are called the* derived categories *of* $\mathcal{C}$.

Notation 13.1.3. (i) We denote by Q the localization functor $K^*(\mathcal{C}) \to D^*(\mathcal{C})$. If there is no risk of confusion, we still denote by Q the composition $C^*(\mathcal{C}) \to K^*(\mathcal{C}) \to D^*(\mathcal{C})$.
 (ii) Recall that when dealing with complexes in additive categories, we denote by [1] the translation functor and we write $X[k]$ instead of $T^k(X)$ for $k \in \mathbb{Z}$. We shall also write $X \to Y \to Z \overset{+1}{\longrightarrow}$ instead of $X \to Y \to Z \to X[1]$ to denote a triangle in the homotopy category associated with an additive category.

Recall that to a null system $\mathcal{N}$ we have associated in (10.2.1) a multiplicative system denoted by $\mathcal{N}Q$. It will be more intuitive to use here another notation for $\mathcal{N}Q$ when $\mathcal{N} = N^{\text{ub}}(\mathcal{C})$:

(13.1.2) $\text{Qis} := \left\{ f \in \text{Mor}(K(\mathcal{C})) \, ; \, f \text{ is a quasi-isomorphism} \right\} \, .$

Hence

$$\text{Hom}_{D(\mathcal{C})}(X, Y) \simeq \varinjlim_{(X' \to X) \in \text{Qis}} \text{Hom}_{K(\mathcal{C})}(X', Y) \, ,$$

$$\simeq \varinjlim_{(Y \to Y') \in \text{Qis}} \text{Hom}_{K(\mathcal{C})}(X, Y') \, ,$$

$$\simeq \varinjlim_{(X' \to X) \in \text{Qis}, (Y \to Y') \in \text{Qis}} \text{Hom}_{K(\mathcal{C})}(X', Y') \, .$$

Remark 13.1.4. (i) Let $X \in K(\mathcal{C})$, and let $Q(X)$ denote its image in $D(\mathcal{C})$. It follows from the result of Exercise 10.11 that:

$$Q(X) \simeq 0 \iff H^n(X) \simeq 0 \text{ for all } n \, .$$

(ii) Let $f \colon X \to Y$ be a morphism in $C(\mathcal{C})$. Then $f = 0$ in $D(\mathcal{C})$ if and only if there exist X' and a qis $g \colon X' \to X$ such that $f \circ g$ is homotopic to 0, or else, if and only if there exist Y' and a qis $h \colon Y \to Y'$ such that $h \circ f$ is homotopic to 0.

Proposition 13.1.5. (i) *For* $n \in \mathbb{Z}$, *the functor* $H^n \colon D(\mathcal{C}) \to \mathcal{C}$ *is well defined and is a cohomological functor.*
 (ii) *A morphism* $f \colon X \to Y$ *in* $D(\mathcal{C})$ *is an isomorphism if and only if* $H^n(f) \colon H^n(X) \to H^n(Y)$ *is an isomorphism for all* n.

(iii) *For $n \in \mathbb{Z}$, the functors $\tilde{\tau}^{\le n}, \tau^{\le n} \colon D(\mathcal{C}) \to D^-(\mathcal{C})$, as well as the functors $\tilde{\tau}^{\ge n}, \tau^{\ge n} \colon D(\mathcal{C}) \to D^+(\mathcal{C})$, are well defined and isomorphic.*
(iv) *For $n \in \mathbb{Z}$, the functor $\tau^{\le n}$ induces a functor $D^+(\mathcal{C}) \to D^b(\mathcal{C})$ and $\tau^{\ge n}$ induces a functor $D^-(\mathcal{C}) \to D^b(\mathcal{C})$.*

Proof. (i)–(ii) Since $H^n(X) \simeq 0$ for $X \in N(\mathcal{C})$, the first assertion holds. The second one follows from the result of Exercise 10.11.
(iii) If $f \colon X \to Y$ is a qis in $K(\mathcal{C})$, then $\tau^{\le n}(f)$ and $\tau^{\ge n}(f)$ are qis. Moreover, for $X \in K(\mathcal{C})$, the morphisms $\tau^{\le n}(X) \to \tilde{\tau}^{\le n}(X)$ and $\tilde{\tau}^{\ge n}(X) \to \tilde{\tau}^{\ge n}(X)$ are qis. (See (12.3.3).)
(iv) is obvious.   q.e.d.

Notation 13.1.6. We shall sometimes write $\tau^{>a}$ and $\tau^{<a}$ instead of $\tau^{\ge a+1}$ and $\tau^{\le a-1}$, respectively.

To a d.t. $X \xrightarrow{f} Y \xrightarrow{g} Z \xrightarrow{+1}$ in $D(\mathcal{C})$, the cohomological functor H^0 associates a long exact sequence in $\mathcal{C}$:

$$(13.1.3) \qquad \cdots \to H^k(X) \to H^k(Y) \to H^k(Z) \to H^{k+1}(X) \to \cdots .$$

For $X \in K(\mathcal{C})$, recall that the categories Qis_X and Qis^X are defined in Definition 7.1.9. They are filtrant full subcategories of $K(\mathcal{C})_X$ and $K(\mathcal{C})^X$, respectively. If $\mathcal{J}$ is a full subcategory of $K(\mathcal{C})_X$, we denote by $\mathrm{Qis}_X \cap \mathcal{J}$ the full subcategory of Qis_X consisting of objects which belong to $\mathcal{J}$. We use similar notations when replacing Qis_X and $K(\mathcal{C})_X$ with Qis^X and $K(\mathcal{C})^X$.

Lemma 13.1.7. (i) *For $X \in K^{\le a}(\mathcal{C})$, the categories $\mathrm{Qis}_X \cap K^{\le a}(\mathcal{C})_X$ and $\mathrm{Qis}_X \cap K^-(\mathcal{C})_X$ are co-cofinal to Qis_X.*
(ii) *For $X \in K^{\ge a}(\mathcal{C})$, the categories $\mathrm{Qis}^X \cap K^{\ge a}(\mathcal{C})^X$ and $\mathrm{Qis}^X \cap K^+(\mathcal{C})^X$ are cofinal to Qis^X.*

Proof. The two statements are equivalent by reversing the arrows. Let us prove (ii).
    The category $\mathrm{Qis}^X \cap K^{\ge a}(\mathcal{C})^X$ is a full subcategory of the filtrant category Qis^X, and for any object $(X \to Y)$ in Qis^X there exists a morphism $(X \to Y) \to (X \to \tau^{\ge a} Y)$. Hence, the result follows from Proposition 3.2.4.   q.e.d.

Proposition 13.1.8. *Let $a \in \mathbb{Z}$, $X \in K^{\le a}(\mathcal{C})$ and $Y \in K^{\ge a}(\mathcal{C})$. Then*

$$\mathrm{Hom}_{D(\mathcal{C})}(X, Y) \simeq \mathrm{Hom}_{\mathcal{C}}(H^a(X), H^a(Y)) .$$

Proof. Let $X \in C^{\le a}(\mathcal{C})$ and $Y \in C^{\ge a}(\mathcal{C})$. The map $\mathrm{Hom}_{C(\mathcal{C})}(X, Y) \to \mathrm{Hom}_{K(\mathcal{C})}(X, Y)$ is an isomorphism and

$$\mathrm{Hom}_{C(\mathcal{C})}(X, Y) \simeq \{u \in \mathrm{Hom}_{\mathcal{C}}(X^a, Y^a) ; u \circ d_X^{a-1} = 0, d_Y^a \circ u = 0\}$$
$$\simeq \mathrm{Hom}_{\mathcal{C}}(\mathrm{Coker}\, d_X^{a-1}, \mathrm{Ker}\, d_Y^a)$$
$$\simeq \mathrm{Hom}_{\mathcal{C}}(H^a(X), H^a(Y)) .$$

Hence, $\mathrm{Hom}_{\mathrm{K}(\mathcal{C})}(X, Y) \simeq \mathrm{Hom}_{\mathcal{C}}(H^a(X), H^a(Y))$.

On the other hand, $\mathrm{Hom}_{\mathrm{D}(\mathcal{C})}(X, Y) \simeq \varinjlim_{(Y \to Y') \in \mathrm{Qis}} \mathrm{Hom}_{\mathrm{K}(\mathcal{C})}(X, Y')$. Since $\mathrm{Qis}^Y \cap \mathrm{K}^{\geq a}(\mathcal{C})^Y$ is cofinal to Qis^Y by Lemma 13.1.7, we have

$$\mathrm{Hom}_{\mathrm{D}(\mathcal{C})}(X, Y) \simeq \varinjlim_{(Y \to Y') \in \mathrm{Qis} \cap \mathrm{K}^{\geq a}(\mathcal{C})} \mathrm{Hom}_{\mathrm{K}(\mathcal{C})}(X, Y')$$
$$\simeq \mathrm{Hom}_{\mathcal{C}}(H^a(X), H^a(Y)) \ .$$

q.e.d.

Notation 13.1.9. Let X, Y be objects of $\mathcal{C}$. We set

$$\mathrm{Ext}_{\mathcal{C}}^k(X, Y) = \mathrm{Hom}_{\mathrm{D}(\mathcal{C})}(X, Y[k]) \ .$$

Remark that the set $\mathrm{Ext}_{\mathcal{C}}^k(X, Y)$ is not necessarily $\mathcal{U}$-small.

Proposition 13.1.10. *Let X and Y be objects of $\mathcal{C}$. Then*

(i) $\mathrm{Ext}_{\mathcal{C}}^k(X, Y) \simeq 0$ *for $k < 0$,*
(ii) $\mathrm{Ext}_{\mathcal{C}}^0(X, Y) \simeq \mathrm{Hom}_{\mathcal{C}}(X, Y)$. *In other words, the natural functor $\mathcal{C} \to \mathrm{D}(\mathcal{C})$ is fully faithful.*

Proof. (i) and (ii) follow immediately from Proposition 13.1.8.   q.e.d.

Notation 13.1.11. For $-\infty \leq a \leq b \leq \infty$, we denote by $\mathrm{D}^{[a,b]}(\mathcal{C})$ the full additive subcategory of $\mathrm{D}(\mathcal{C})$ consisting of objects X satisfying $H^j(X) \simeq 0$ for $j \notin [a, b]$. We set $\mathrm{D}^{\leq a}(\mathcal{C}) := \mathrm{D}^{[-\infty, a]}(\mathcal{C})$ and $\mathrm{D}^{\geq a}(\mathcal{C}) := \mathrm{D}^{[a, \infty]}(\mathcal{C})$.

Proposition 13.1.12.   (i) *For $* = +, -, \mathrm{b}$, the triangulated category $\mathrm{D}^*(\mathcal{C})$ defined in Definition 13.1.2 is equivalent to the full triangulated subcategory of $\mathrm{D}(\mathcal{C})$ consisting of objects X satisfying $H^j(X) \simeq 0$ for $j \ll 0$ in case $* = +$, $H^j(X) \simeq 0$ for $j \gg 0$ in case $* = -$, and $H^j(X) \simeq 0$ for $|j| \gg 0$ in case $* = \mathrm{b}$.*
(ii) *For $-\infty \leq a \leq b \leq \infty$, the functor $Q \colon \mathrm{K}^{[a,b]}(\mathcal{C}) \to \mathrm{D}^{[a,b]}(\mathcal{C})$ is essentially surjective.*
(iii) *The category $\mathcal{C}$ is equivalent to the full subcategory of $\mathrm{D}(\mathcal{C})$ consisting of objects X satisfying $H^j(X) \simeq 0$ for $j \neq 0$.*
(iv) *For $a \in \mathbb{Z}$ and $X, Y \in \mathrm{D}(\mathcal{C})$, we have*

$$\mathrm{Hom}_{\mathrm{D}(\mathcal{C})}(\tau^{\leq a}X, \tau^{\geq a}Y) \simeq \mathrm{Hom}_{\mathcal{C}}(H^a(X), H^a(Y)) \ .$$

In particular, $\mathrm{Hom}_{\mathrm{D}(\mathcal{C})}(\tau^{\leq a}X, \tau^{\geq a+1}Y) \simeq 0$.

Proof. (i) (a) Let us treat the case $* = +$, the other cases being similar. For $Y \in \mathrm{K}^{\geq a}(\mathcal{C})$ and $Z \in N(\mathcal{C})$, any morphism $Z \to Y$ in $\mathrm{K}(\mathcal{C})$ factors through $\tau^{\geq a}Z \in N(\mathcal{C}) \cap \mathrm{K}^{\geq a}(\mathcal{C})$. Applying Proposition 10.2.6, we find that the natural functor $\mathrm{D}^+(\mathcal{C}) \to \mathrm{D}(\mathcal{C})$ is fully faithful.

(b) Clearly, if $Y \in D(\mathcal{C})$ belongs to the image of the functor $D^+(\mathcal{C}) \to D(\mathcal{C})$, then $H^j(X) = 0$ for $j \ll 0$.

(c) Conversely, let $X \in K(\mathcal{C})$ with $H^j(X) \simeq 0$ for $j < a$. Then $\tau^{\geq a} X \in K^+(\mathcal{C})$ and the morphism $X \to \tau^{\geq a} X$ in $K(\mathcal{C})$ is a qis, hence an isomorphism in $D(\mathcal{C})$.

(ii) The proof goes as in (i) (c).

(iii) By Proposition 13.1.10, the functor $\mathcal{C} \to D(\mathcal{C})$ is fully faithful and by the result in (ii) this functor is essentially surjective.

(iv) follows from (ii) and Proposition 13.1.8.   q.e.d.

By Proposition 13.1.12 (iii), we often regard $\mathcal{C}$ as a full subcategory of $D(\mathcal{C})$.

Proposition 13.1.13. *Let* $0 \to X \xrightarrow{f} Y \xrightarrow{g} Z \to 0$ *be an exact sequence in* $C(\mathcal{C})$. *Then there exists a d.t.* $X \xrightarrow{f} Y \xrightarrow{g} Z \xrightarrow{+1}$ *in* $D(\mathcal{C})$, *and* Z *is isomorphic to* $\mathrm{Mc}(f)$ *in* $D(\mathcal{C})$.

Proof. Define $\varphi \colon \mathrm{Mc}(f) \to Z$ in $C(\mathcal{C})$ by $\varphi^n = (0, g^n)$. By Corollary 12.3.5, φ is a qis, hence an isomorphism in $D(\mathcal{C})$.   q.e.d.

Remark 13.1.14. Let $0 \to X \to Y \to Z \to 0$ be an exact sequence in $\mathcal{C}$. We get a morphism $\gamma \colon Z \to X[1]$ in $D(\mathcal{C})$. The morphism $H^k(\gamma) \colon H^k(Z) \to H^{k+1}(X)$ is 0 for all $k \in \mathbb{Z}$ although γ is *not* the zero morphism in $D(\mathcal{C})$ in general ($\gamma = 0$ happens only if the short exact sequence splits). The morphism γ may be described in $K(\mathcal{C})$ by the morphisms with φ a qis:

$$X[1] \xleftarrow{\beta(f)} \mathrm{Mc}(f) \xrightarrow{\varphi} Z \quad \text{i.e. } \gamma = Q(\beta(f)) \circ Q(\varphi)^{-1}.$$

This is visualized by

$$
\begin{array}{ccccccccc}
Z := & & 0 & \longrightarrow & 0 & \longrightarrow & Z & \longrightarrow & 0 \\
 & \varphi \uparrow \, qis & & & \uparrow & & \uparrow & & \\
\mathrm{Mc}(f) := & & 0 & \longrightarrow & X & \xrightarrow{f} & Y & \longrightarrow & 0 \\
 & \beta(f) \downarrow & & & \mathrm{id} \downarrow & & \downarrow & & \\
X[1] := & & 0 & \longrightarrow & X & \longrightarrow & 0 & \longrightarrow & 0.
\end{array}
$$

Proposition 13.1.15. *Let* $X \in D(\mathcal{C})$.

(i) *There are d.t.'s in* $D(\mathcal{C})$:

(13.1.4) $\tau^{\leq n} X \to X \to \tau^{\geq n+1} X \xrightarrow{+1}$,

(13.1.5) $\tau^{\leq n-1} X \to \tau^{\leq n} X \to H^n(X)[-n] \xrightarrow{+1}$,

(13.1.6) $H^n(X)[-n] \to \tau^{\geq n} X \to \tau^{\geq n+1} X \xrightarrow{+1}$.

(ii) *Moreover,* $H^n(X)[-n] \simeq \tau^{\leq n} \tau^{\geq n} X \simeq \tau^{\geq n} \tau^{\leq n} X$.

Proof. This follows from (12.3.4) and (12.3.2).   q.e.d.

Proposition 13.1.16. *The functor* $\tau^{\leq n} \colon D(\mathcal{C}) \to D^{\leq n}(\mathcal{C})$ *is a right adjoint to the natural functor* $D^{\leq n}(\mathcal{C}) \to D(\mathcal{C})$ *and* $\tau^{\geq n} \colon D(\mathcal{C}) \to D^{\geq n}(\mathcal{C})$ *is a left adjoint to the natural functor* $D^{\geq n}(\mathcal{C}) \to D(\mathcal{C})$. *In other words, there are functorial isomorphisms*

$$\mathrm{Hom}_{D(\mathcal{C})}(X, Y) \simeq \mathrm{Hom}_{D(\mathcal{C})}(X, \tau^{\leq n} Y) \quad \text{for } X \in D^{\leq n}(\mathcal{C}) \text{ and } Y \in D(\mathcal{C}) \,,$$
$$\mathrm{Hom}_{D(\mathcal{C})}(X, Y) \simeq \mathrm{Hom}_{D(\mathcal{C})}(\tau^{\geq n} X, Y) \quad \text{for } X \in D(\mathcal{C}) \text{ and } Y \in D^{\geq n}(\mathcal{C}) \,.$$

Proof. By the d.t. (13.1.4) for Y, we have an exact sequence

$$(13.1.7) \quad \mathrm{Hom}_{D(\mathcal{C})}(X, \tau^{> n} Y[-1]) \to \mathrm{Hom}_{D(\mathcal{C})}(X, \tau^{\leq n} Y)$$
$$\to \mathrm{Hom}_{D(\mathcal{C})}(X, Y) \to \mathrm{Hom}_{D(\mathcal{C})}(X, \tau^{> n} Y) \,.$$

Since $\tau^{> n} Y[-1]$ and $\tau^{> n} Y$ belong to $D^{> n}(\mathcal{C})$, the first and fourth terms in (13.1.7) vanish by Proposition 13.1.12 (iv).
The second isomorphism follows by reversing the arrows.   q.e.d.

Lemma 13.1.17. *An abelian category* $\mathcal{C}$ *is semisimple* (*see* Definition 8.3.16) *if and only if* $\mathrm{Ext}_{\mathcal{C}}^{k}(X, Y) = 0$ *for any* $k \neq 0$ *and any* $X, Y \in \mathcal{C}$.

Proof. The condition is sufficient by the result of Exercise 13.5 and necessary by that of Exercise 13.15.   q.e.d.

Definition 13.1.18. *An abelian category* $\mathcal{C}$ *is hereditary if* $\mathrm{Ext}_{\mathcal{C}}^{k}(X, Y) = 0$ *for* $k \geq 2$ *and* $X, Y \in \mathcal{C}$.

Example 13.1.19. If a ring R is a principal ideal domain (such as a field, or $\mathbb{Z}$, or $k[x]$ for a field k), then the category $\mathrm{Mod}(R)$ is hereditary.

Corollary 13.1.20. *Let* $\mathcal{C}$ *be an abelian category and assume that* $\mathcal{C}$ *is hereditary. Let* $X \in D^{b}(\mathcal{C})$. *Then there exists a* (*non canonical*) *isomorphism*

$$X \simeq \oplus_{j} H^{j}(X) [-j] \,.$$

Proof. Arguing by induction on n, we shall prove the existence of an isomorphism $\tau^{\leq n} X \simeq \oplus_{j \leq n} H^{j}(X) [-j]$. Consider the d.t. (13.1.5):

$$\tau^{\leq n-1} X \to \tau^{\leq n} X \to H^{n}(X) [-n] \xrightarrow{+1}$$

and assume $\tau^{\leq n-1} X \simeq \oplus_{j < n} H^{j}(X) [-j]$. By the result of Exercise 10.4, it is enough to show that

$$\mathrm{Hom}_{D^{b}(\mathcal{C})}(H^{n}(X)[-n], H^{j}(X) [-j + 1]) = \mathrm{Ext}_{\mathcal{C}}^{n-j+1}(H^{n}(X), H^{j}(X)) = 0$$

for $j < n$, which follows from the assumption and $n - j + 1 \geq 2$.   q.e.d.

13.2 Resolutions

Lemma 13.2.1. *Let $\mathcal{J}$ be a full additive subcategory of $\mathcal{C}$ and let $X^\bullet \in \mathrm{C}^{\geq a}(\mathcal{C})$ for some $a \in \mathbb{Z}$. Assume the condition* (a) *or* (b) *below:*

(a) *$\mathcal{J}$ is cogenerating in $\mathcal{C}$ (i.e., for any $Y \in \mathcal{C}$ there exists a monomorphism $Y \rightarrowtail I$ with $I \in \mathcal{J}$),*

(b)   (i) *$\mathcal{J}$ is closed by extensions and by cokernels of monomorphisms,*

    (ii) *for any monomorphism $I' \rightarrowtail Y$ in $\mathcal{C}$ with $I' \in \mathcal{J}$, there exists a morphism $Y \to I$ with $I \in \mathcal{J}$ such that the composition $I' \to I$ is a monomorphism,*

    (iii) *$H^j(X^\bullet) \in \mathcal{J}$ for all $j \in \mathbb{Z}$.*

Then there exist $Y^\bullet \in \mathrm{C}^{\geq a}(\mathcal{J})$ and a qis $X^\bullet \to Y^\bullet$.

Proof. Let $X^\bullet \in \mathrm{C}^{\geq a}(\mathcal{C})$. We shall construct by induction on p a complex $Y^\bullet_{\leq p}$ in $\mathcal{J}$ and a morphism $X^\bullet \to Y^\bullet_{\leq p}$:

$$
\begin{array}{ccccccccc}
X^\bullet: = & \cdots \longrightarrow & X^{p-1} & \xrightarrow{d_X^{p-1}} & X^p & \xrightarrow{d_X^p} & X^{p+1} & \xrightarrow{d_X^{p+1}} & \cdots \\
& & \big\downarrow {\scriptstyle f^{p-1}} & & \big\downarrow {\scriptstyle f^p} & & & & \\
Y^\bullet_{\leq p}: = & \cdots \xrightarrow{d_Y^{p-2}} & Y^{p-1} & \xrightarrow{d_Y^{p-1}} & Y^p & & & &
\end{array}
$$

such that $H^k(X^\bullet) \to H^k(Y^\bullet_{\leq p})$ is an isomorphism for $k < p$ and is a monomorphism for $k = p$. We assume further that $H^p(Y^\bullet_{\leq p}) = \operatorname{Coker} d_Y^{p-1}$ belongs to $\mathcal{J}$ in case (b).

For $p < a$ it is enough to take $Y^\bullet_{\leq p} = 0$. Assuming that $Y^\bullet_{\leq p}$ have been constructed, we shall construct $Y^\bullet_{\leq p+1}$. Set:

$$
Z^p = \operatorname{Coker} d_Y^{p-1} \oplus_{\operatorname{Coker} d_X^{p-1}} \operatorname{Ker} d_X^{p+1}, \quad W^p = \operatorname{Coker} d_Y^{p-1} \oplus_{\operatorname{Coker} d_X^{p-1}} X^{p+1} .
$$

There is a monomorphism $Z^p \rightarrowtail W^p$. Consider the commutative diagram:

$$
\begin{array}{ccccccccc}
0 \longrightarrow & H^p(X^\bullet) & \longrightarrow & \operatorname{Coker} d_X^{p-1} & \longrightarrow & \operatorname{Ker} d_X^{p+1} & \longrightarrow & H^{p+1}(X^\bullet) & \longrightarrow 0 \\
& \big\downarrow {\scriptstyle \mathrm{id}} & & \big\downarrow & & \big\downarrow & & \big\downarrow {\scriptstyle \mathrm{id}} & \\
0 \longrightarrow & H^p(X^\bullet) & \longrightarrow & \operatorname{Coker} d_Y^{p-1} & \longrightarrow & Z^p & \longrightarrow & H^{p+1}(X^\bullet) & \longrightarrow 0 .
\end{array}
$$

By the result of Exercise 8.21 and the hypothesis that $H^p(X^\bullet) \to \operatorname{Coker} d_Y^{p-1}$ is a monomorphism, the rows are exact. Assuming hypothesis (b), we have $Z^p \in \mathcal{J}$ since $H^p(X^\bullet)$, $\operatorname{Coker} d_Y^{p-1}$ and $H^{p+1}(X^\bullet)$ belong to $\mathcal{J}$. Hence, assuming either (a) or (b), we may find a morphism $W^p \to Y^{p+1}$ with $Y^{p+1} \in \mathcal{J}$ such that the composition $Z^p \to W^p \to Y^{p+1}$ is a monomorphism. The above construction defines naturally $f^{p+1}: X^{p+1} \to Y^{p+1}$ and $d_Y^p: Y^p \to Y^{p+1}$. Let $Y^\bullet_{\leq p+1}$ be the complex so constructed. Then

$$H^p(Y^\bullet_{\leq p+1}) \simeq \mathrm{Ker}(\mathrm{Coker}\, d_Y^{p-1} \to Y^{p+1})$$
$$\simeq \mathrm{Ker}(\mathrm{Coker}\, d_Y^{p-1} \to Z^p) \simeq H^p(X^\bullet)\,.$$

Moreover,

$$H^{p+1}(X^\bullet) \simeq \mathrm{Coker}(\mathrm{Coker}\, d_Y^{p-1} \to Z^p)$$
$$\rightarrowtail \mathrm{Coker}(\mathrm{Coker}\, d_Y^{p-1} \to Y^{p+1}) \simeq H^{p+1}(Y^\bullet_{\leq p+1})\,.$$

In case (b),

$$\mathrm{Im}\, d_Y^p \simeq \mathrm{Im}(\mathrm{Coker}(d_Y^{p-1}) \to Z^p) \simeq \mathrm{Coker}(H^p(X^\bullet) \to \mathrm{Coker}(d_Y^{p-1}))$$

belongs to $\mathcal{J}$, and the exact sequence $0 \to \mathrm{Im}\, d_Y^p \to Y^{p+1} \to H^{p+1}(Y^\bullet_{\leq p+1}) \to 0$ implies that $H^{p+1}(Y^\bullet_{\leq p+1})$ belongs to $\mathcal{J}$.   q.e.d.

We shall also consider the extra condition:

(13.2.1) $\begin{cases} \text{there exists a non-negative integer } d \text{ such that, for any exact} \\ \text{sequence } Y_d \to \cdots \to Y_1 \to Y \to 0 \text{ with } Y_j \in \mathcal{J} \text{ for } 1 \leq j \leq d, \\ \text{we have } Y \in \mathcal{J}\,. \end{cases}$

It is clear that $N^+(\mathcal{J}) := N(\mathcal{C}) \cap \mathrm{K}^+(\mathcal{J})$ and $N^b(\mathcal{J}) := N(\mathcal{C}) \cap \mathrm{K}^b(\mathcal{J})$ are null systems in $\mathrm{K}^+(\mathcal{J})$ and $\mathrm{K}^b(\mathcal{J})$, respectively.

Proposition 13.2.2.   (i) *Assume that $\mathcal{J}$ is cogenerating in $\mathcal{C}$. Then the natural functor $\theta^+ \colon \mathrm{K}^+(\mathcal{J})/N^+(\mathcal{J}) \to \mathrm{D}^+(\mathcal{C})$ is an equivalence of categories.*
(ii) *If moreover $\mathcal{J}$ satisfies (13.2.1), then $\theta^b \colon \mathrm{K}^b(\mathcal{J})/N^b(\mathcal{J}) \to \mathrm{D}^b(\mathcal{C})$ is an equivalence of categories.*

Proof. Let $X \in \mathrm{K}^+(\mathcal{C})$. By Lemma 13.2.1, there exist $Y \in \mathrm{K}^+(\mathcal{J})$ and a qis $X \to Y$.
(i) follows by Proposition 10.2.7 (i).
(ii) Let $k \in \mathbb{Z}$ and assume that $X^j \simeq 0$ for $j \geq k$. Then $\tau^{\leq j} Y \to Y$ is a qis for $j \geq k$ and the hypothesis implies that $\tau^{\leq j} Y$ belongs to $\mathrm{K}^b(\mathcal{J})$ for $j > k + d$. This proves (ii) again by Proposition 10.2.7 (i).   q.e.d.

Let us apply the preceding proposition to the full subcategory of injective objects:

$$\mathcal{I}_\mathcal{C} = \{X \in \mathcal{C}; X \text{ is injective}\}\,.$$

Proposition 13.2.3. *Assume that $\mathcal{C}$ admits enough injectives. Then the functor $\mathrm{K}^+(\mathcal{I}_\mathcal{C}) \to \mathrm{D}^+(\mathcal{C})$ is an equivalence of categories.*
If moreover the category $\mathcal{I}_\mathcal{C}$ satisfies condition (13.2.1), then $\mathrm{K}^b(\mathcal{I}_\mathcal{C}) \to \mathrm{D}^b(\mathcal{C})$ is an equivalence of categories.

Proof. By Proposition 13.2.2, it is enough to prove that if $X^\bullet \in \mathrm{C}^+(\mathcal{I}_\mathcal{C})$ is qis to 0, then $X^\bullet$ is homotopic to 0. This is a particular case of the lemma below (choose $f = \mathrm{id}_{X^\bullet}$ in the lemma).   q.e.d.

Lemma 13.2.4. *Let* $f^\bullet\colon X^\bullet \to I^\bullet$ *be a morphism in* $\mathrm{C}(\mathcal{C})$. *Assume that* $I^\bullet$ *belongs to* $\mathcal{C}^+(\mathcal{I}_\mathcal{C})$ *and* $X^\bullet$ *is exact. Then* $f^\bullet$ *is homotopic to* 0.

Proof. Consider the diagram:

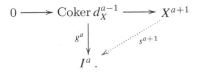

We shall construct by induction morphisms $s^k\colon X^k \to I^{k-1}$ satisfying:

$$(13.2.2) \qquad\qquad f^k = s^{k+1} \circ d_X^k + d_I^{k-1} \circ s^k .$$

For $k \ll 0$, we set $s^k = 0$. Assume that we have constructed the s^k $(k \le a)$ such that (13.2.2) is satisfied for $k < a$. We have $f^a \circ d_X^{a-1} = d_I^{a-1} \circ f^{a-1} = d_I^{a-1} \circ (s^a \circ d_X^{a-1} + d_I^{a-2} \circ s^{a-1}) = d_I^{a-1} \circ s^a \circ d_X^{a-1}$. Define $g^a = f^a - d_I^{a-1} \circ s^a$. Then $g^a \circ d_X^{a-1} = 0$. Hence, g^a factorizes through $\operatorname{Coker} d_X^{a-1}$, and since the complex $X^\bullet$ is exact, the sequence $0 \to \operatorname{Coker} d_X^{a-1} \to X^{a+1}$ is exact. Consider

$$0 \longrightarrow \operatorname{Coker} d_X^{a-1} \longrightarrow X^{a+1}$$

$$g^a \downarrow \quad\nearrow{\scriptstyle s^{a+1}}$$

$$I^a .$$

The dotted arrow may be completed since I^a is injective. Then (13.2.2) holds for $k = a$. \hfill q.e.d.

Corollary 13.2.5. *Let* $\mathcal{C}$ *be an abelian* $\mathcal{U}$*-category with enough injectives. Then* $\mathrm{D}^+(\mathcal{C})$ *is a* $\mathcal{U}$*-category.*

The next result will be useful when dealing with unbounded derived categories in Sect. 14.3.

Proposition 13.2.6. *Let* $\mathcal{J}$ *be a full additive subcategory of* $\mathcal{C}$ *and assume:*

(i) $\mathcal{J}$ *is cogenerating,*
(ii) (13.2.1) *holds.*

Then for any $X \in \mathrm{C}(\mathcal{C})$, *there exist* $Y \in \mathrm{C}(\mathcal{J})$ *and a qis* $X \to Y$. *In particular, there is an equivalence of triangulated categories* $\mathrm{K}(\mathcal{J})/(\mathrm{K}(\mathcal{J}) \cap \mathcal{N}) \xrightarrow{\sim} \mathrm{D}(\mathcal{C})$.

Proof. The second statement follows from the first one by Proposition 10.2.7. The proof of the first statement decomposes into several steps.

(a) For any $X \in \mathrm{C}(\mathcal{C})$ and any $n \in \mathbb{Z}$, there exists a qis $X \to Z$ such that $Z^i \in \mathcal{J}$ for $i \ge n$. Indeed, let $\sigma^{\ge n} X$ denote the stupid truncated complex given

328   13 Derived Categories

in Definition 11.3.11. By Lemma 13.2.1, there exists a qis $\sigma^{\geq n}X \to Y$ with $Y \in C^{\geq n}(\mathcal{J})$. Define Z as the complex

$$Z := \cdots \to X^{n-2} \to X^{n-1} \to Y^n \to Y^{n+1} \to \cdots$$

where the morphism $X^{n-1} \to Y^n$ is the composition $X^{n-1} \to X^n \to Y^n$. Then $X \to Z$ is a qis.

(b) Let $n' < n$ be integers. Let $X \in C(\mathcal{C})$ and assume that $X^i \in \mathcal{J}$ for $i \geq n$. Then there exists a qis $X \to Z$ such that $Z^i \in \mathcal{J}$ for $i \geq n'$ and $X^i \to Z^i$ is an isomorphism for $i \geq n+d$.

Indeed, by (a) there exists a qis $f : X \to Y$ such that $Y^i \in \mathcal{J}$ for $i \geq n'$. Let $M := \mathrm{Mc}(f)$, the mapping cone of f. Then M is an exact complex and $M^i = X^{i+1} \oplus Y^i$ belongs to $\mathcal{J}$ for $i \geq n-1$. Hence, $\mathrm{Ker}\, d_M^i$ belongs to $\mathcal{J}$ for $i \geq n-1+d$ since there exists an exact sequence $M^{i-d} \to \cdots \to M^{i-1} \to \mathrm{Ker}\, d_M^i \to 0$, and $M^{i-d}, \ldots, M^{i-1}$ belong to $\mathcal{J}$. We have

$$\mathrm{Coker}\, d_M^{i-2} \simeq X^i \oplus_{X^{i-1}} \mathrm{Coker}\, d_Y^{i-2} \,,$$

$$\mathrm{Ker}\, d_M^i \simeq \mathrm{Ker}\, d_X^{i+1} \times_{Y^{i+1}} Y^i \,,$$

and the natural isomorphism $\mathrm{Coker}\, d_M^{i-2} \to \mathrm{Ker}\, d_M^i$ is an isomorphism. Set $a = n+d$ and construct a complex Z as follows:

$$Z^i = \begin{cases} X^i & \text{for } i > a \,, \\ \mathrm{Ker}\, d_M^i & \text{for } i = a \,, \\ Y^i & \text{for } i < a, \end{cases}$$

the differentials d_Z^i being defined in an obvious way as seen in the diagram below.

$$
\begin{array}{ccccc}
X^{a-1} & \longrightarrow & X^a & \longrightarrow & X^{a+1} \\
\downarrow & & \downarrow & & \\
Y^{a-1} & \longrightarrow & \mathrm{Coker}\, d_M^{a-2} & & \mathrm{id} \Big\downarrow \\
 & & \sim \downarrow & & \\
\mathrm{id} \Big\downarrow & & \mathrm{Ker}\, d_M^a & \longrightarrow & X^{a+1} \\
\downarrow & & \downarrow & & \downarrow \\
Y^{a-1} & \longrightarrow & Y^a & \longrightarrow & Y^{a+1}.
\end{array}
$$

Then we get morphisms of complexes $X \to Z \to Y$, and $Z^i \in \mathcal{J}$ for $i \geq n'$. Let us show that $H^i(X) \to H^i(Z)$ is an isomorphism for all $i \in \mathbb{Z}$. This is clear for $i \neq a-1, a, a+1$. Since $Z^a \simeq \mathrm{Coker}\, d_M^{a-2} \simeq X^a \oplus_{X^{a-1}} \mathrm{Coker}\, d_Y^{a-2}$, we have $\mathrm{Im}\, d_Z^a \simeq \mathrm{Im}\, d_X^a$ and hence $H^{a+1}(Z) \simeq \mathrm{Ker}\, d_X^{a+1}/\mathrm{Im}\, d_X^a \simeq H^{a+1}(X)$. Since $\mathrm{Ker}\, d_Z^a \simeq \mathrm{Ker}(\mathrm{Ker}\, d_X^{a+1} \times_{Y^{a+1}} Y^a \to X^{a+1}) \simeq \mathrm{Ker}\, d_X^a$, we have $H^a(Z) \simeq H^a(Y) \simeq H^a(X)$. Finally, $\mathrm{Ker}\, d_Z^{a-1} \simeq \mathrm{Ker}\, d_Y^{a-1}$ implies $H^{a-1}(Z) \simeq H^{a-1}(Y) \simeq H^{a-1}(X)$.

(c) We can now complete the proof. Let us take an infinite sequence $n_0 > n_1 > \cdots$. By (a), there exists a qis $X \to Y_0$ such that $Y_0^i \in \mathcal{J}$ for $i \geq n_0$. By (b), we may construct inductively a chain of quasi-isomorphisms $Y_0 \to Y_1 \to \cdots$ such that $Y_k^i \in \mathcal{J}$ for all $i \geq n_k$ and $Y_k^i \xrightarrow{\sim} Y_{k+1}^i$ for $i > n_k + d$. Then $Y := \varinjlim_k Y_k$ exists in $C(\mathcal{C})$ and $Y^i \simeq Y_k^i$ for $i > n_k + d$. Hence, all Y^i's belong to $\mathcal{J}$ and $X \to Y$ is a qis.   q.e.d.

Derived Category of a Subcategory

Let $\mathcal{C}'$ be a thick full abelian subcategory of $\mathcal{C}$.

Definition 13.2.7. *For* $* = \mathrm{ub}, +, -, \mathrm{b}$, $\mathrm{D}^*_{\mathcal{C}'}(\mathcal{C})$ *denotes the full additive subcategory of* $\mathrm{D}^*(\mathcal{C})$ *consisting of objects* X *such that* $H^j(X) \in \mathcal{C}'$ *for all* j.

This is clearly a triangulated subcategory of $\mathrm{D}(\mathcal{C})$, and there is a natural functor

$$(13.2.3) \qquad \delta^* \colon \mathrm{D}^*(\mathcal{C}') \to \mathrm{D}^*_{\mathcal{C}'}(\mathcal{C}) \qquad \text{for } * = \mathrm{ub}, +, -, \mathrm{b} \, .$$

Theorem 13.2.8. *Let* $\mathcal{C}'$ *be a thick abelian subcategory of* $\mathcal{C}$ *and assume that for any monomorphism* $Y' \rightarrowtail X$, *with* $Y' \in \mathcal{C}'$, *there exists a morphism* $X \to Y$ *with* $Y \in \mathcal{C}'$ *such that the composition* $Y' \to Y$ *is a monomorphism.*
Then the functors δ^+ *and* δ^{b} *in* (13.2.3) *are equivalences of categories.*

Proof. The result for δ^+ is an immediate consequence of Proposition 7.2.1 and Lemma 13.2.1. The case of δ^{b} follows since $\mathrm{D}^{\mathrm{b}}(\mathcal{C}')$ is equivalent to the full subcategory of $\mathrm{D}^+(\mathcal{C}')$ of objects with bounded cohomology, and similarly for $\mathrm{D}^{\mathrm{b}}_{\mathcal{C}'}(\mathcal{C})$.   q.e.d.

Note that, by reversing the arrows in Theorem 13.2.8, the functors δ^- and δ^{b} in (13.2.3) are equivalences of categories if for any epimorphism $X \twoheadrightarrow Y$ with $Y \in \mathcal{C}'$, there exists a morphism $Y' \to X$ with $Y' \in \mathcal{C}'$ such that the composition $Y' \to Y$ is an epimorphism.

13.3 Derived Functors

In this section, $\mathcal{C}$, $\mathcal{C}'$ and $\mathcal{C}''$ denote abelian categories. Let $F \colon \mathcal{C} \to \mathcal{C}'$ be an additive functor. It defines naturally a triangulated functor

$$\mathrm{K}^*(F) \colon \mathrm{K}^*(\mathcal{C}) \to \mathrm{K}^*(\mathcal{C}') \, .$$

For short, we often write F instead of $\mathrm{K}^*(F)$. We shall denote by $Q \colon \mathrm{K}^*(\mathcal{C}) \to \mathrm{D}^*(\mathcal{C})$ the localization functor, and similarly with Q', Q'', when replacing $\mathcal{C}$ with $\mathcal{C}'$, $\mathcal{C}''$.

Definition 13.3.1. *Let* $* = $ ub, $+$, b. *The functor* F *is* right derivable (*or* F *admits a* right derived functor) *on* $\mathrm{K}^*(\mathcal{C})$ *if the triangulated functor* $\mathrm{K}^*(F)\colon \mathrm{K}^*(\mathcal{C}) \to \mathrm{K}^*(\mathcal{C}')$ *is universally right localizable with respect to* $N^*(\mathcal{C})$ *and* $N^*(\mathcal{C}')$.

In such a case the localization of F *is denoted by* R^*F *and* $H^k \circ R^*F$ *is denoted by* R^kF. *The functor* $R^*F\colon \mathrm{D}^*(\mathcal{C}) \to \mathrm{D}^*(\mathcal{C}')$ *is called the* right derived functor *of* F *and* R^kF *the* k-th derived functor *of* F.

We shall also say for short that "R^*F exists" instead of "F is right derivable on $\mathrm{K}^*(\mathcal{C})$*".*

By the definition, the functor F admits a right derived functor on $\mathrm{K}^*(\mathcal{C})$ if $\underset{(X \to X') \in \mathrm{Qis}, X' \in \mathrm{K}^*(\mathcal{C})}{\text{"}\varinjlim\text{"}} Q' \circ \mathrm{K}(F)(X')$ exists in $\mathrm{D}^*(\mathcal{C}')$ for all $X \in \mathrm{K}^*(\mathcal{C})$. In such a case, this object is isomorphic to $R^*F(X)$.

Note that R^*F is a triangulated functor from $\mathrm{D}^*(\mathcal{C})$ to $\mathrm{D}^*(\mathcal{C}')$ if it exists, and R^kF is a cohomological functor from $\mathrm{D}^*(\mathcal{C})$ to $\mathcal{C}'$.

Notation 13.3.2. In the sequel, we shall often write "$\varinjlim$" $F(X')$ instead of "$\varinjlim$" $Q' \circ \mathrm{K}(F)(X')$ in the above formula.

Corollary 13.3.3. *If* RF *exists, then* R^+F *exists and* R^+F *is the restriction of* RF *to* $\mathrm{D}^+(\mathcal{C})$.

Proof. For $X \in \mathrm{K}^+(\mathcal{C})$, the category $\mathrm{Qis}^X \cap \mathrm{K}^+(\mathcal{C})^X$ is cofinal to the category Qis^X by Lemma 13.1.7.   q.e.d.

Definition 13.3.4. *Let* $\mathcal{J}$ *be a full additive subcategory of* $\mathcal{C}$. *We say for short that* $\mathcal{J}$ *is* F-injective *if the subcategory* $\mathrm{K}^+(\mathcal{J})$ *of* $\mathrm{K}^+(\mathcal{C})$ *is* $\mathrm{K}^+(F)$-*injective with respect to* $N^+(\mathcal{C})$ *and* $N^+(\mathcal{C}')$ (*see* Definition 10.3.2). *We shall also say that* $\mathcal{J}$ *is* injective with respect to F.

We define similarly the notion of an F-projective *subcategory.*

By the definition, $\mathcal{J}$ is F-injective if and only if for any $X \in \mathrm{K}^+(\mathcal{C})$, there exists a qis $X \to Y$ with $Y \in \mathrm{K}^+(\mathcal{J})$ and $F(Y)$ is exact for any exact complex $Y \in \mathrm{K}^+(\mathcal{J})$.

Proposition 13.3.5. *Let* $F\colon \mathcal{C} \to \mathcal{C}'$ *be an additive functor of abelian categories and let* $\mathcal{J}$ *be a full additive subcategory of* $\mathcal{C}$.

(i) *If* $\mathcal{J}$ *is* F-injective, then $R^+F\colon \mathrm{D}^+(\mathcal{C}) \to \mathrm{D}^+(\mathcal{C}')$ *exists and*

$$(13.3.1) \quad R^+F(X) \simeq F(X') \text{ for any qis } X \to X' \text{ with } X' \in \mathrm{K}^+(\mathcal{J}).$$

(ii) *If* F *is left exact, the following two conditions are equivalent.*
   (a) $\mathcal{J}$ *is* F-injective,
   (b) *The following two conditions hold:*
      (1) *the category* $\mathcal{J}$ *is cogenerating in* $\mathcal{C}$,

(2) *for any exact sequence* $0 \to X' \to X \to X'' \to 0$, *the sequence* $0 \to F(X') \to F(X) \to F(X'') \to 0$ *is exact as soon as* $X \in \mathcal{J}$ *and there exists an exact sequence* $0 \to Y^0 \to \cdots \to Y^n \to X' \to 0$ *with the* Y^j's *in* $\mathcal{J}$.

Proof. (i) follows from Proposition 10.3.3.

(ii) (a) $\Rightarrow$ (b) (1). For $X \in \mathcal{C}$, there exists a qis $X \to Y$ with $Y \in \mathrm{K}^+(\mathcal{J})$. Then, the composition $X \to \mathrm{Ker}(d_Y^0) \to H^0(Y)$ is an isomorphism and hence $X \to Y^0$ is a monomorphism.
(a) $\Rightarrow$ (b) (2). By (1) and Lemma 13.2.1, there exists an exact sequence $0 \to X'' \to Z^0 \to Z^1 \to \cdots$ with $Z^j \in \mathcal{J}$ for all j. Then the sequence

$$0 \to Y^0 \to \cdots \to Y^n \to X \to Z^0 \to Z^1 \to \cdots$$

is exact and belongs to $\mathrm{K}^+(\mathcal{J})$. Hence $F(X) \to F(Z^0) \to F(Z^1)$ is exact. Since F is left exact, $F(X'') \simeq \mathrm{Ker}\big(F(Z^0) \to F(Z^1)\big)$ and this implies that $F(X) \to F(X'')$ is an epimorphism.

(ii) (b) $\Rightarrow$ (a). By (1) and Lemma 13.2.1, for any $X \in \mathrm{K}^+(\mathcal{C})$ there exists a qis $X \to Y$ with $Y \in \mathrm{K}^+(\mathcal{J})$. Hence, it is enough to show that $F(X)$ is exact if $X \in \mathrm{K}^+(\mathcal{J})$ is exact. For each $n \in \mathbb{Z}$, the sequences $\cdots \to X^{n-2} \to X^{n-1} \to \mathrm{Ker}(d_X^n) \to 0$ and $0 \to \mathrm{Ker}(d_X^n) \to X^n \to \mathrm{Ker}(d_X^{n+1}) \to 0$ are exact. By (2), the sequence $0 \to F(\mathrm{Ker}(d_X^n)) \to F(X^n) \to F(\mathrm{Ker}(d_X^{n+1})) \to 0$ is exact. q.e.d.

Remark 13.3.6. (i) Note that for $X \in \mathcal{C}$, $R^k F(X) \simeq 0$ for $k < 0$ and assuming that F is left exact, $R^0 F(X) \simeq F(X)$. Indeed for $X \in \mathcal{C}$ and for any qis $X \to Y$, the composition $X \to Y \to \tau^{\geq 0} Y$ is a qis.
(ii) If F is right (resp. left) derivable, an object X of $\mathcal{C}$ such that $R^k F(X) \simeq 0$ (resp. $L^k F(X) \simeq 0$) for all $k \neq 0$ is called *right* F-*acyclic* (resp. *left* F-*acyclic*). If $\mathcal{J}$ is an F-injective subcategory, then any object of $\mathcal{J}$ is right F-acyclic.
(iii) If $\mathcal{C}$ has enough injectives, then the full subcategory $\mathcal{I}_\mathcal{C}$ of injective objects of $\mathcal{C}$ is F-injective for any additive functor $F \colon \mathcal{C} \to \mathcal{C}'$. Indeed, any exact complex in $\mathrm{C}^+(\mathcal{I})$ is homotopic to zero by Lemma 13.2.4. In particular, $R^+ F \colon \mathrm{D}^+(\mathcal{C}) \to \mathrm{D}^+(\mathcal{C}')$ exists in this case.

We shall give sufficient conditions in order that $\mathcal{J}$ is F-injective.

Theorem 13.3.7. *Let* $\mathcal{J}$ *be a full additive subcategory of* $\mathcal{C}$ *and let* $F \colon \mathcal{C} \to \mathcal{C}'$ *be a left exact functor. Assume:*

(i) *the category* $\mathcal{J}$ *is cogenerating in* $\mathcal{C}$,
(ii) *for any monomorphism* $Y' \rightarrowtail X$ *with* $Y' \in \mathcal{J}$ *there exists an exact sequence* $0 \to Y' \to Y \to Y'' \to 0$ *with* Y, Y'' *in* $\mathcal{J}$ *such that* $Y' \to Y$ *factors through* $Y' \rightarrowtail X$ *and the sequence* $0 \to F(Y') \to F(Y) \to F(Y'') \to 0$ *is exact.*

Then $\mathcal{J}$ *is* F-*injective.*

Condition (ii) may be visualized as

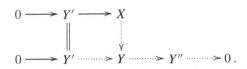

Condition (ii) is rather intricate. The following particular case is sufficient for many applications. It is also an immediate consequence of Proposition 13.3.5.

Corollary 13.3.8. *Let $\mathcal{J}$ be a full additive subcategory of $\mathcal{C}$ and let $F \colon \mathcal{C} \to \mathcal{C}'$ be a left exact functor. Assume:*

(i) *the category $\mathcal{J}$ is cogenerating in $\mathcal{C}$,*
(ii) *for any exact sequence $0 \to X' \to X \to X'' \to 0$ in $\mathcal{C}$ with $X', X \in \mathcal{J}$, we have $X'' \in \mathcal{J}$,*
(iii) *for any exact sequence $0 \to X' \to X \to X'' \to 0$ in $\mathcal{C}$ with $X', X \in \mathcal{J}$, the sequence $0 \to F(X') \to F(X) \to F(X'') \to 0$ is exact.*

Then $\mathcal{J}$ is F-injective.

Example 13.3.9. Let R be a ring and let N be a right R-module. The full additive subcategory of $\mathrm{Mod}(R)$ consisting of flat R-modules is $(N \otimes_R \cdot)$-projective.

The proof of Theorem 13.3.7 is decomposed into several lemmas.

Lemma 13.3.10. *Let $0 \to Y' \to X \to X'' \to 0$ be an exact sequence in $\mathcal{C}$ with $Y' \in \mathcal{J}$. Then the sequence $0 \to F(Y') \to F(X) \to F(X'') \to 0$ is exact.*

Proof. Choose an exact sequence $0 \to Y' \to Y \to Y'' \to 0$ as in Theorem 13.3.7. We get the commutative exact diagram:

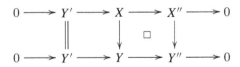

where the square labeled by $\square$ is Cartesian. Since F is left exact, it transforms this square to a Cartesian square and the bottom row to an exact row. Hence, the result follows from Lemma 8.3.11.   q.e.d.

Lemma 13.3.11. *Let $X^\bullet \in \mathrm{C}^+(\mathcal{C})$ be an exact complex, and assume $X^n = 0$ for $n < a$ and $X^a \in \mathcal{J}$. There exist an exact complex $Y^\bullet \in \mathrm{C}^+(\mathcal{J})$ and a morphism $f \colon X^\bullet \to Y^\bullet$ such that $Y^n = 0$ for $n < a$, $f^a \colon X^a \to Y^a$ is an isomorphism, and $\mathrm{Ker}\, d_Y^n \in \mathcal{J}$ for all n.*

Note that the complex $F(Y^\bullet)$ will be exact by Lemma 13.3.10.

Proof. We argue by induction. By the hypothesis, there exists a commutative exact diagram:

$$
\begin{array}{ccccccccc}
0 & \longrightarrow & X^a & \longrightarrow & X^{a+1} & & & & \\
& & \| & & \downarrow & & & & \\
0 & \longrightarrow & Y^a & \longrightarrow & Y^{a+1} & \longrightarrow & Z^{a+2} & \longrightarrow & 0
\end{array}
$$

with Y^{a+1}, Z^{a+2} in $\mathcal{J}$. Assume that we have constructed

$$
\begin{array}{ccccccccc}
0 & \longrightarrow & X^a & \longrightarrow & \cdots & \longrightarrow & X^n & & \\
& & \| & & & & \downarrow & & \\
0 & \longrightarrow & Y^a & \xrightarrow{d_Y^a} & \cdots & \xrightarrow{d_Y^{n-1}} & Y^n & \longrightarrow & Z^{n+1} & \longrightarrow & 0
\end{array}
$$

where the row in the bottom is exact and belongs to $C^+(\mathcal{J})$, and $\operatorname{Im} d_Y^j$ belongs to $\mathcal{J}$ for $a \le j \le n-1$.

Define $W^{n+1} = X^{n+1} \oplus_{\operatorname{Coker} d_X^{n-1}} Z^{n+1}$. In other words, we have a co-Cartesian exact diagram:

$$
\begin{array}{ccccc}
0 & \longrightarrow & \operatorname{Coker} d_X^{n-1} & \longrightarrow & X^{n+1} \\
& & \downarrow & & \downarrow \\
0 & \longrightarrow & Z^{n+1} & \longrightarrow & W^{n+1}.
\end{array}
$$

By the hypotheses, there exists an exact commutative diagram

$$
\begin{array}{ccccccccc}
0 & \longrightarrow & Z^{n+1} & \longrightarrow & W^{n+1} & & & & \\
& & \| & & \downarrow & & & & \\
0 & \longrightarrow & Z^{n+1} & \longrightarrow & Y^{n+1} & \longrightarrow & Z^{n+2} & \longrightarrow & 0
\end{array}
$$

with Y^{n+1} and Z^{n+2} in $\mathcal{J}$. Define d_Y^n as the composition $Y^n \to Z^{n+1} \to Y^{n+1}$. Then $\operatorname{Im}(d_Y^n : Y^n \to Y^{n+1}) \simeq Z^{n+1} \in \mathcal{J}$. Hence, the induction proceeds. q.e.d.

Now we can prove Theorem 13.3.7, using Proposition 13.3.5.

End of the proof of Theorem 13.3.7. Let $X^\bullet \in C^+(\mathcal{J})$ be an exact complex. We have to prove that $F(X^\bullet)$ is exact. Let us show by induction on $b - a$ that $H^b(F(X^\bullet)) \simeq 0$ if $X \in C^{\ge a}(\mathcal{J})$. If $b < a$, this is clear. Hence, we assume $b \ge a$.

By Lemma 13.3.11, there exists a morphism of complexes $f : X^\bullet \to Y^\bullet$ in $C^+(\mathcal{J})$ such that $Y^\bullet \in C^{\ge a}(\mathcal{J})$, $X^a \xrightarrow{\sim} Y^a$ and $F(Y^\bullet)$ is exact. Let $\sigma^{>a} X^\bullet$ and $\sigma^{>a} Y^\bullet$ denote the stupid truncated complexes given in Definition 11.3.11.

Let W denote the mapping cone of $\sigma^{>a}(f) : \sigma^{>a} X^\bullet \to \sigma^{>a} Y^\bullet$. Then $W^n = (\sigma^{>a} X^\bullet)^{n+1} \oplus (\sigma^{>a} Y^\bullet)^n \simeq 0$ for $n < a$. Let us consider the distinguished triangle in $K(\mathcal{J})$ (see Exercise 11.12)

$$W \to \mathrm{Mc}(f) \to \mathrm{Mc}(X^a[-a] \to Y^a[-a]) \xrightarrow{+1} .$$

Since $X^a \to Y^a$ is an isomorphism, $W \to \mathrm{Mc}(f)$ is an isomorphism in $\mathrm{K}(\mathcal{C})$. Applying the functor F, we obtain the isomorphism $F(W) \xrightarrow{\sim} F(\mathrm{Mc}(f))$ in $\mathrm{K}(\mathcal{C}')$. Therefore, $H^j(F(W)) \simeq H^j(F(\mathrm{Mc}(f)))$ for all j. On the other hand, there is a d.t. in $\mathrm{K}^+(\mathcal{C}')$

$$F(X) \to F(Y) \to F(\mathrm{Mc}(f)) \xrightarrow{+1} ,$$

and $H^j(F(Y)) \simeq 0$ for all j. Hence, $H^b(F(X)) \simeq H^{b-1}(F(\mathrm{Mc}(f))) \simeq H^{b-1}(F(W))$. Since W is an exact complex and belongs to $\mathrm{C}^{\geq a}(\mathcal{J})$, we have $H^{b-1}(F(W)) \simeq 0$ by the induction hypothesis.   q.e.d.

Lemma 13.3.12. *Let $F : \mathcal{C} \to \mathcal{C}'$ be a left exact functor of abelian categories and let $\mathcal{J}$ be an F-injective full subcategory of $\mathcal{C}$. Denote by $\mathcal{J}_F$ the full subcategory of $\mathcal{C}$ consisting of right F-acyclic objects. Then $\mathcal{J}_F$ contains $\mathcal{J}$ and $\mathcal{J}_F$ satisfies the conditions* (i)–(iii) *of Corollary 13.3.8. In particular, $\mathcal{J}_F$ is F-injective.*

Proof. Let us check the conditions (i)–(iii) of Corollary 13.3.8.
(i) Since $\mathcal{J}_F$ contains $\mathcal{J}$, $\mathcal{J}_F$ is cogenerating.
(ii)–(iii) Consider an exact sequence $0 \to X' \to X \to X'' \to 0$ in $\mathcal{C}$ with $X', X \in \mathcal{J}_F$. The exact sequences $R^j F(X) \to R^j F(X'') \to R^{j+1} F(X')$ for $j \geq 0$ imply that $R^k F(X'') \simeq 0$ for $k > 0$. Moreover, there is an exact sequence $0 \to F(X') \to F(X) \to F(X'') \to R^1 F(X')$ and $R^1 F(X') \simeq 0$.   q.e.d.

Hence a full additive subcategory $\mathcal{J}$ of $\mathcal{C}$ is F-injective if and only if it is cogenerating and any object of $\mathcal{J}$ is F-acyclic (assuming the right derivability of F). Note that even if F is right derivable, there may not exist an F-injective subcategory.

Derived Functor of a Composition

Let $F : \mathcal{C} \to \mathcal{C}'$ and $F' : \mathcal{C}' \to \mathcal{C}''$ be additive functors of abelian categories.

Proposition 13.3.13.   (i) *Let $* = \mathrm{ub}, +, \mathrm{b}$. Assume that the right derived functors $R^* F$, $R^* F'$ and $R^*(F' \circ F)$ exist. Then there is a canonical morphism in $\mathrm{Fct}(\mathrm{D}^*(\mathcal{C}), \mathrm{D}^*(\mathcal{C}''))$:*

$$(13.3.2) \qquad\qquad R^*(F' \circ F) \to R^* F' \circ R^* F .$$

(ii) *Assume that there exist full additive subcategories $\mathcal{J} \subset \mathcal{C}$ and $\mathcal{J}' \subset \mathcal{C}'$ such that $\mathcal{J}$ is F-injective, $\mathcal{J}'$ is F' injective and $F(\mathcal{J}) \subset \mathcal{J}'$. Then $\mathcal{J}$ is $F' \circ F$-injective and (13.3.2) induces an isomorphism*

$$(13.3.3) \qquad\qquad R^+(F' \circ F) \xrightarrow{\sim} R^+ F' \circ R^+ F .$$

Proof. Apply Proposition 10.3.5 to the functors $K^*(F)\colon K^*(\mathcal{C}) \to K^*(\mathcal{C}')$ and $K^*(F')\colon K^*(\mathcal{C}') \to K^*(\mathcal{C}'')$.   q.e.d.

Note that in many cases (even if F is exact), F may not send injective objects of $\mathcal{C}$ to injective objects of $\mathcal{C}'$. This is a reason why the notion of an "F-injective" category is useful.

Remark 13.3.14. The notion of the left derived functor L^*G ($* = \mathrm{ub}, -, \mathrm{b}$) of an additive functor G is defined similarly. Moreover, there is a similar result to Proposition 13.3.13 for the composition of $L^-G\colon \mathrm{D}^-(\mathcal{C}) \to \mathrm{D}^-(\mathcal{C}')$ and $L^-G'\colon \mathrm{D}^-(\mathcal{C}') \to \mathrm{D}^-(\mathcal{C}'')$. Note that

$$L^-G(X) \simeq \varprojlim_{(X'\to X)\in\mathrm{Qis}} G(X')\,.$$

Derived Functor of the Projective Limit

As an application of Theorem 13.3.7 we shall discuss the existence of the derived functor of projective limits.

Let $\mathcal{C}$ be an abelian $\mathcal{U}$-category. Recall that $\mathrm{Pro}(\mathcal{C})$ is an abelian category admitting small projective limits, and small filtrant projective limits as well as small products are exact (see § 8.6). Assume that $\mathcal{C}$ admits small projective limits. Then the natural exact functor $\mathcal{C} \to \mathrm{Pro}(\mathcal{C})$ admits a right adjoint

$$(13.3.4) \qquad\qquad \pi_{\mathcal{C}}\colon \mathrm{Pro}(\mathcal{C}) \to \mathcal{C}\,.$$

If $\beta\colon I^{\mathrm{op}} \to \mathcal{C}$ is a functor with I small and filtrant, then $\pi_{\mathcal{C}}(\text{``}\varprojlim\text{''}\,\beta) \simeq \varprojlim \beta$. The functor $\pi_{\mathcal{C}}$ is left exact and we shall give a condition in order that it is right derivable.

For a full additive subcategory $\mathcal{J}$ of $\mathcal{C}$, the full additive subcategory $\mathcal{J}_{pro}$ of $\mathrm{Pro}(\mathcal{C})$ is defined by

$$\mathcal{J}_{pro} := \{X \in \mathrm{Pro}(\mathcal{C}); X \simeq \text{``}\prod_{i\in I}\text{''}\, X_i \text{ for a small set } I \text{ and } X_i \in \mathcal{J}\}\,.$$

Here, "$\prod$" denotes the product in $\mathrm{Pro}(\mathcal{C})$. Hence for X_i, $Y \in \mathcal{C}$,

$$\mathrm{Hom}_{\mathrm{Pro}(\mathcal{C})}(\text{``}\prod_{i\in I}\text{''}\, X_i, Y) \simeq \bigoplus_{i\in I} \mathrm{Hom}_{\mathcal{C}}(X_i, Y)\,.$$

Proposition 13.3.15. *Let $\mathcal{C}$ be an abelian category admitting small projective limits and let $\mathcal{J}$ be a full additive subcategory of $\mathcal{C}$ satisfying:*

(i) *$\mathcal{J}$ is cogenerating in $\mathcal{C}$,*
(ii) *if $0 \to Y' \to Y \to Y'' \to 0$ is an exact sequence and $Y', Y \in \mathcal{J}$, then $Y'' \in \mathcal{J}$,*
(iii) *if $0 \to Y_i' \to Y_i \to Y_i'' \to 0$ is a family indexed by a small set I of sequences in $\mathcal{J}$ which are exact in $\mathcal{C}$, then the sequence $0 \to \prod_{i\in I} Y_i' \to \prod_{i\in I} Y_i \to \prod_{i\in I} Y_i'' \to 0$ is exact in $\mathcal{C}$.*

Then

(a) *the category $\mathcal{J}_{pro}$ is $\pi_{\mathcal{C}}$-injective,*

(b) *the left exact functor $\pi_{\mathcal{C}}$ admits a right derived functor*

$$R^{+}\pi_{\mathcal{C}} \colon D^{+}(\mathrm{Pro}(\mathcal{C})) \to D^{+}(\mathcal{C}) ,$$

and $R^{k}\pi_{\mathcal{C}}(\text{``}\prod_{i}\text{''}X_{i}) \simeq 0$ for $k > 0$ and $X_{i} \in \mathcal{J}$,

(c) *the composition $D^{+}(\mathcal{C}) \to D^{+}(\mathrm{Pro}(\mathcal{C})) \xrightarrow{R^{+}\pi_{\mathcal{C}}} D^{+}(\mathcal{C})$ is isomorphic to the identity.*

Proof. (a) We shall verify the hypotheses of Theorem 13.3.7.

(i) The category $\mathcal{J}_{pro}$ is cogenerating in $\mathrm{Pro}(\mathcal{C})$. Indeed, for $A = \text{``}\varprojlim_{i}\text{''} \alpha(i) \in \mathrm{Pro}(\mathcal{C})$, we obtain a monomorphism $A \rightarrowtail \text{``}\prod_{i\in\mathrm{Ob}(I)}\text{''} X_{i}$ by choosing a monomorphism $\alpha(i) \rightarrowtail X_{i}$ with $X_{i} \in \mathcal{J}$ for each $i \in I$.

(ii) Consider an exact sequence $0 \to Y \to A$ in $\mathrm{Pro}(\mathcal{C})$ with $A \in \mathrm{Pro}(\mathcal{C})$ and $Y = \text{``}\prod_{i}\text{''} Y_{i}$, $Y_{i} \in \mathcal{J}$. Applying Proposition 8.6.9 (with the arrows reversed), for each i, we find $X_{i} \in \mathcal{C}$ and a commutative exact diagram

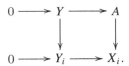

By hypothesis (i) on $\mathcal{J}$, we may assume $X_{i} \in \mathcal{J}$. Let $Z_{i} = \mathrm{Coker}(Y_{i} \to X_{i})$. Then $Z_{i} \in \mathcal{J}$. The functor $\text{``}\prod\text{''}$ being exact, we get the exact commutative diagram

$$
\begin{array}{ccccccccc}
0 & \longrightarrow & Y & \longrightarrow & A & & & & \\
& & \downarrow{\scriptstyle \mathrm{id}} & & \downarrow & & & & \\
0 & \longrightarrow & \text{``}\prod_{i}\text{''}Y_{i} & \longrightarrow & \text{``}\prod_{i}\text{''}X_{i} & \longrightarrow & \text{``}\prod_{i}\text{''}Z_{i} & \longrightarrow & 0.
\end{array}
$$

Applying $\pi_{\mathcal{C}}$ to the second row, we find the sequence $0 \to \prod_{i} Y_{i} \to \prod_{i} X_{i} \to \prod_{i} Z_{i} \to 0$ and this sequence is exact by hypothesis (iii).

(b) follows from (a).

(c) By the assumption, $\mathcal{J}$ is injective with respect to the exact functor $\mathcal{C} \to \mathrm{Pro}(\mathcal{C})$. Since the functor $\mathcal{C} \to \mathrm{Pro}(\mathcal{C})$ sends $\mathcal{J}$ to $\mathcal{J}_{pro}$, the result follows from Proposition 13.3.13 (ii).   q.e.d.

Corollary 13.3.16. *Let $\mathcal{J}$ be a full additive subcategory of $\mathcal{C}$ satisfying the hypotheses of* Proposition 13.3.15. *Let $\{X_n\}_{n\in\mathbb{N}}$ be a projective system in $\mathcal{J}$ indexed by $\mathbb{N}$. Then $R^k\pi_{\mathcal{C}}(\text{“}\varprojlim_n\text{”}\,X_n) \simeq 0$ for $k > 1$, and*

$$R^1\pi_{\mathcal{C}}(\text{“}\varprojlim_n\text{”}\,X_n) \simeq \operatorname{Coker}\left(\prod_{n\geq 0} X_n \xrightarrow{\ \text{id}-\text{sh}\ } \prod_{n\geq 0} X_n\right),$$

where sh *is the morphism associated with the family of morphisms $X_{n+1} \to X_n$.*

Proof. By Exercise 8.37, there exists an exact sequence in $\operatorname{Pro}(\mathcal{C})$

$$0 \to \text{“}\varprojlim_n\text{”}\,X_n \to \text{“}\prod_n\text{”}\,X_n \xrightarrow{\ \text{id}-\text{sh}\ } \text{“}\prod_n\text{”}\,X_n \to 0 \ .$$

Applying the functor $R^+\pi_{\mathcal{C}}$ we get a long exact sequence and the results follows since $R^k\pi_{\mathcal{C}}(\text{“}\prod_n\text{”}\,X_n) \simeq 0$ for $k \neq 0$.   q.e.d.

Example 13.3.17. (i) If the category $\mathcal{C}$ admits enough injectives, then the category $\mathcal{I}_{\mathcal{C}}$ of injective objects satisfies the conditions of Proposition 13.3.15 (see Exercise 13.6).
(ii) If R is a ring and $\mathcal{C} = \operatorname{Mod}(R)$, we may choose $\mathcal{J} = \mathcal{C}$. Hence $\pi_{\mathcal{C}}$ is right derivable in this case. (See also Exercise 12.8.)

13.4 Bifunctors

Let us begin the study of derived bifunctors with the functor Hom. Recall Convention 11.7.1.

Theorem 13.4.1. *Let $\mathcal{C}$ be an abelian category, let X, $Y \in D(\mathcal{C})$. Assume that the functor $\operatorname{Hom}_{\mathcal{C}}^{\bullet} : K(\mathcal{C}) \times K(\mathcal{C})^{\mathrm{op}} \to K(\operatorname{Mod}(\mathbb{Z}))$ given by $(Y', X') \mapsto \operatorname{tot}_{\pi}\operatorname{Hom}_{\mathcal{C}}^{\bullet,\bullet}(X', Y')$ (see § 11.7) is right localizable at (Y, X). Then*

$$(13.4.1) \qquad H^0\operatorname{RHom}_{\mathcal{C}}(X, Y) \simeq \operatorname{Hom}_{D(\mathcal{C})}(X, Y) \ .$$

Proof. By the hypothesis,

$$\operatorname{RHom}_{\mathcal{C}}(X, Y) \simeq \underset{(X'\to X)\in\mathrm{Qis},(Y\to Y')\in\mathrm{Qis}}{\text{“}\varinjlim\text{”}} \operatorname{tot}_{\pi}(\operatorname{Hom}_{\mathcal{C}}^{\bullet,\bullet}(X', Y')) \ .$$

Applying the functor H^0 and recalling that "$\varinjlim$" commutes with any functor, we get using Proposition 11.7.3:

$$H^0\operatorname{RHom}_{\mathcal{C}}(X, Y) \simeq \underset{(X'\to X)\in\mathrm{Qis},(Y\to Y')\in\mathrm{Qis}}{\text{“}\varinjlim\text{”}} H^0(\operatorname{tot}_{\pi}(\operatorname{Hom}_{\mathcal{C}}^{\bullet,\bullet}(X', Y')))$$

$$\simeq \underset{(X'\to X)\in\mathrm{Qis},(Y\to Y')\in\mathrm{Qis}}{\text{“}\varinjlim\text{”}} \operatorname{Hom}_{K(\mathcal{C})}(X', Y')$$

$$\simeq \operatorname{Hom}_{D(\mathcal{C})}(X, Y) \ .$$

q.e.d.

Notice that in the situation of Theorem 13.4.1, if $X, Y \in \mathcal{C}$, then we have

$$(13.4.2) \qquad\qquad \operatorname{Ext}^k_{\mathcal{C}}(X, Y) \simeq H^k(\operatorname{RHom}_{\mathcal{C}}(X, Y)) \, .$$

Consider now three abelian categories $\mathcal{C}, \mathcal{C}', \mathcal{C}''$ and an additive bifunctor

$$F \colon \mathcal{C} \times \mathcal{C}' \to \mathcal{C}'' \, .$$

By Proposition 11.6.3, the triangulated functor:

$$\mathrm{K}^+ F \colon \mathrm{K}^+(\mathcal{C}) \times \mathrm{K}^+(\mathcal{C}') \to \mathrm{K}^+(\mathcal{C}'')$$

is naturally defined by setting:

$$\mathrm{K}^+ F(X, X') = \operatorname{tot}(F(X, X')).$$

Similarly to the case of functors, if the triangulated bifunctor $\mathrm{K}^+ F \colon \mathrm{K}^+(\mathcal{C}) \times \mathrm{K}^+(\mathcal{C}') \to \mathrm{D}^+(\mathcal{C}'')$ is universally right localizable with respect to $(N^+(\mathcal{C}) \times N^+(\mathcal{C}'), N^+(\mathcal{C}''))$, F is said to be right derivable and its localization is denoted by $R^+ F$. We set $R^k F = H^k \circ R^+ F$ and call it the k-th derived bifunctor of F.

Definition 13.4.2. *Let $\mathcal{J}$ and $\mathcal{J}'$ be full additive subcategories of $\mathcal{C}$ and $\mathcal{C}'$ respectively. We say for short that $(\mathcal{J}, \mathcal{J}')$ is F-injective if $(\mathrm{K}^+(\mathcal{J}), \mathrm{K}^+(\mathcal{J}'))$ is $\mathrm{K}^+ F$-injective (see Definition 10.3.9).*

Proposition 13.4.3. *Let $\mathcal{J}$ and $\mathcal{J}'$ be full additive subcategories of $\mathcal{C}$ and $\mathcal{C}'$ respectively. Assume that $(\mathcal{J}, \mathcal{J}')$ is F-injective. Then F is right derivable and for $(X, X') \in \mathrm{D}^+(\mathcal{C}) \times \mathrm{D}^+(\mathcal{C}')$ we have:*

$$R^+ F(X, X') \simeq Q'' \circ \mathrm{K}^+ F(Y, Y')$$

for $(X \to Y) \in \mathrm{Qis}$ and $(X' \to Y') \in \mathrm{Qis}$ with $Y \in \mathrm{K}^+(\mathcal{J})$, $Y' \in \mathrm{K}^+(\mathcal{J}')$.

Proof. Apply Proposition 10.3.10 to the functor $Q'' \circ \mathrm{K}^+ F \colon \mathrm{K}^+(\mathcal{C}) \times \mathrm{K}^+(\mathcal{C}') \to \mathrm{D}^+(\mathcal{C}'')$.   q.e.d.

Proposition 13.4.4. *Let $\mathcal{J}$ and $\mathcal{J}'$ be full additive subcategories of $\mathcal{C}$ and $\mathcal{C}'$ respectively. Assume:*

(i) *for any $Y \in \mathcal{J}$, $\mathcal{J}'$ is $F(Y, \cdot)$-injective,*
(ii) *for any $Y' \in \mathcal{J}'$, $\mathcal{J}$ is $F(\cdot, Y')$-injective.*

Then $(\mathcal{J}, \mathcal{J}')$ is F-injective.

Proof. Let $(Y, Y') \in \mathrm{K}^+(\mathcal{J}) \times \mathrm{K}^+(\mathcal{J}')$. If either Y or Y' is qis to zero, then $\operatorname{tot}(F(Y, Y'))$ is qis to zero by Corollary 12.5.5.   q.e.d.

Choosing $\mathcal{J}' = \mathcal{C}'$, we get:

Corollary 13.4.5. *Let $\mathcal{J}$ be a full additive cogenerating subcategory of $\mathcal{C}$ and assume:*

(i) *for any $X \in \mathcal{J}$, $F(X, \cdot) \colon \mathcal{C}' \to \mathcal{C}''$ is exact,*
(ii) *for any $X' \in \mathcal{C}'$, $\mathcal{J}$ is $F(\cdot, X')$-injective.*

Then F is right derivable and for $X \in \mathrm{K}^+(\mathcal{C})$, $X' \in \mathrm{K}^+(\mathcal{C}')$

$$R^+ F(X, X') \simeq Q'' \circ \mathrm{K}^+ F(Y, X') \text{ for any } (X \to Y) \in \mathrm{Qis} \text{ with } Y \in \mathrm{K}^+(\mathcal{J}) \;.$$

In particular, for $X \in \mathcal{C}$ and $X' \in \mathcal{C}'$, $R^+ F(X, X')$ is the derived functor of $F(\cdot, X')$ calculated at X, that is, $R^+ F(X, X') = (R^+ F(\cdot, X'))(X)$.

Corollary 13.4.6. *Let $\mathcal{C}$ be an abelian category and assume that there are subcategories $\mathcal{P}$ in $\mathcal{C}$ and $\mathcal{J}$ in $\mathcal{C}$ such that $(\mathcal{J}, \mathcal{P}^{\mathrm{op}})$ is injective with respect to the functor $\mathrm{Hom}_{\mathcal{C}}$. Then the functor $\mathrm{Hom}_{\mathcal{C}}$ admits a right derived functor $\mathrm{R}^+\mathrm{Hom}_{\mathcal{C}} \colon \mathrm{D}^+(\mathcal{C}) \times \mathrm{D}^-(\mathcal{C})^{\mathrm{op}} \to \mathrm{D}^+(\mathbb{Z})$. In particular, $\mathrm{D}^{\mathrm{b}}(\mathcal{C})$ is a $\mathcal{U}$-category.*

Notation 13.4.7. Let R be a ring. We shall often write for short $\mathrm{D}^*(R)$ instead of $\mathrm{D}^*(\mathrm{Mod}(R))$, for $* = \mathrm{ub}, \mathrm{b}, +, -$.

Remark 13.4.8. Assume that $\mathcal{C}$ has enough injectives. Then

$$\mathrm{R}^+\mathrm{Hom}_{\mathcal{C}} \colon \mathrm{D}^+(\mathcal{C}) \times \mathrm{D}^-(\mathcal{C})^{\mathrm{op}} \to \mathrm{D}^+(\mathbb{Z})$$

exists and may be calculated as follows. Let $X \in \mathrm{D}^-(\mathcal{C})$, $Y \in \mathrm{D}^+(\mathcal{C})$. There exists a qis $Y \to I$ in $\mathrm{K}^+(\mathcal{C})$, the I^j's being injective. Then:

$$(13.4.3) \qquad \mathrm{R}^+\mathrm{Hom}_{\mathcal{C}}(X, Y) \simeq \mathrm{tot}(\mathrm{Hom}_{\mathcal{C}}^{\bullet,\bullet}(X, I)) \;.$$

If $\mathcal{C}$ has enough projectives, then $\mathrm{R}^+\mathrm{Hom}_{\mathcal{C}}$ exists. For a qis $P \to X$ in $\mathrm{K}^-(\mathcal{C})$ with the P^j's projective, we have:

$$(13.4.4) \qquad \mathrm{R}^+\mathrm{Hom}_{\mathcal{C}}(X, Y) \simeq \mathrm{tot}(\mathrm{Hom}_{\mathcal{C}}^{\bullet,\bullet}(P, Y)) \;.$$

These isomorphisms hold in $\mathrm{D}^+(\mathbb{Z})$, which means that $\mathrm{R}^+\mathrm{Hom}_{\mathcal{C}}(X, Y) \in \mathrm{D}^+(\mathbb{Z})$ is represented by the simple complex associated with the double complex $\mathrm{Hom}_{\mathcal{C}}^{\bullet,\bullet}(X, I)$, or $\mathrm{Hom}_{\mathcal{C}}^{\bullet,\bullet}(P, Y)$.

Example 13.4.9. Let R be a k-algebra. Since the category $\mathrm{Mod}(R)$ has enough projectives, the left derived functor of the functor $\cdot \otimes_R \cdot$ is well defined. It is denoted by $\cdot \overset{\mathrm{L}}{\otimes}_R \cdot$. Hence:

$$\cdot \overset{\mathrm{L}}{\otimes}_R \cdot \colon \mathrm{D}^-(R^{\mathrm{op}}) \times \mathrm{D}^-(R) \to \mathrm{D}^-(k)$$

may be calculated as follows:

$$\begin{aligned} N \overset{\mathrm{L}}{\otimes}_R M &\simeq \mathrm{tot}(N \otimes_R P) \\ &\simeq \mathrm{tot}(Q \otimes_R M) \\ &\simeq \mathrm{tot}(Q \otimes_R P), \end{aligned}$$

where P is a complex of projective R-modules quasi-isomorphic to M and Q is a complex of projective R^{op}-modules quasi-isomorphic to N.

A classical notation is:

$$(13.4.5) \qquad \mathrm{Tor}_n^R(N, M) := H^{-n}(N \overset{\mathrm{L}}{\otimes}_R M) \;.$$

Exercises

Exercise 13.1. Let $F: \mathcal{C} \to \mathcal{C}'$ be a left exact functor of abelian categories. Let $\mathcal{J}$ be an F-injective subcategory of $\mathcal{C}$ and let $Y^\bullet$ be an object of $\mathrm{C}^+(\mathcal{J})$. Assume that $H^k(Y^\bullet) = 0$ for all $k \neq p$ for some $p \in \mathbb{Z}$, and let $X = H^p(Y^\bullet)$. Prove that $R^k F(X) \simeq H^{k+p}(F(Y^\bullet))$.

Exercise 13.2. We consider the situation of Proposition 13.3.13 (ii).
(i) Let $X \in \mathcal{C}$ and assume that there is $q \in \mathbb{N}$ with $R^k F(X) = 0$ for $k \neq q$. Prove that $R^j(F' \circ F)(X) \simeq R^{j-q} F'(R^q F(X))$.
(ii) Assume now that $R^j F(X) = 0$ for $j \neq a, b$ for $a < b$. Prove that there is a long exact sequence: $\cdots \to R^{k-a} F'(R^a F(X)) \to R^k(F' \circ F)(X) \to R^{k-b} F'(R^b F(X)) \to R^{k-a+1} F'(R^a F(X)) \to \cdots$. (Hint: use $\tau^{\le a} R F(X) \to R F(X) \to \tau^{>a} R F(X) \xrightarrow{+1}$.)

Exercise 13.3. Let $F: \mathcal{C} \to \mathcal{C}'$ be a left exact and right derivable functor of abelian categories. Let $X \in \mathrm{D}^+(\mathcal{C})$ such that $H^k(X) = 0$ for $k < p$ for some $p \in \mathbb{Z}$. Prove that $R^n F(X) = 0$ for $n < p$ and $R^p F(X) \simeq F(H^p(X))$.

Exercise 13.4. In the situation of Proposition 13.3.13 (i), let $X \in \mathcal{C}$ and assume that $R^j F(X) \simeq 0$ for $j < n$. Prove that $R^n(F' \circ F)(X) \simeq F'(R^n F(X))$.

Exercise 13.5. Let $\mathcal{C}$ be an abelian category and let $0 \to X' \to X \to X'' \to 0$ be an exact sequence in $\mathcal{C}$. Assuming that $\mathrm{Ext}^1(X'', X') \simeq 0$, prove that the sequence splits.

Exercise 13.6. Let $\mathcal{C}$ be an abelian category.
(i) Prove that if $\{X_i\}_{i \in I}$ is a small family of injective objects in $\mathcal{C}$, then "$\prod\limits_i$" X_i is an injective object of $\mathrm{Pro}(\mathcal{C})$.
(ii) Prove that if $\mathcal{C}$ has enough injectives, then so does $\mathrm{Pro}(\mathcal{C})$.
(iii) Deduce that if $\mathcal{C}$ has enough injectives and admits small projective limits, then the functor $\pi_\mathcal{C}: \mathrm{Pro}(\mathcal{C}) \to \mathcal{C}$ (see (13.3.4)) admits a right derived functor $R^+\pi_\mathcal{C}: \mathrm{D}^+(\mathrm{Pro}(\mathcal{C})) \to \mathrm{D}^+(\mathcal{C})$.

Exercise 13.7. Let $\mathcal{C}$ be an abelian category. Prove that the following conditions on $X \in \mathcal{C}$ are equivalent.
(i) X is injective,
(ii) $\mathrm{Ext}^1_\mathcal{C}(Y, X) \simeq 0$ for all $Y \in \mathcal{C}$,
(iii) $\mathrm{Ext}^n_\mathcal{C}(Y, X) \simeq 0$ for all $Y \in \mathcal{C}$ and all $n \neq 0$.

Exercise 13.8. Let $\mathcal{C}$ be an abelian category and consider the following condition on an integer n:

$$(13.4.6) \qquad \text{for all } X \text{ and } Y \text{ in } \mathcal{C}, \mathrm{Ext}^j_\mathcal{C}(X, Y) \simeq 0 \text{ for all } j > n.$$

If such an integer n exists, we say that $\mathcal{C}$ has finite homological dimension, and the smallest $n \geq -1$ such that (13.4.6) is satisfied is called the homological dimension of $\mathcal{C}$ and is denoted by $\mathrm{hd}(\mathcal{C})$. (Note that $\mathrm{hd}(\mathcal{C}) = -1$ if and only if $\mathcal{C} \simeq \mathbf{Pt}$.)

(i) Prove that $\mathrm{hd}(\mathcal{C}) \leq n$ if and only if $\mathrm{Ext}_{\mathcal{C}}^{n+1}(X, Y) = 0$ for all $X, Y \in \mathcal{C}$. (Hint: use Exercise 13.16.)

(ii) Assume that $\mathcal{C}$ has enough injectives and let n be a non-negative integer. Prove that the conditions (a)–(c) below are equivalent:

(a) $\mathrm{hd}(\mathcal{C}) \leq n$,
(b) for all X in $\mathcal{C}$, there exists an exact sequence $0 \to X \to X^0 \to \cdots \to X^n \to 0$ with the X^j's injective,
(c) if $X^0 \to \cdots \to X^n \to 0$ is an exact sequence in $\mathcal{C}$ and X^j is injective for $j < n$, then X^n is injective.

(iii) Assume that $\mathcal{C}$ has enough projectives. Prove that $\mathrm{hd}(\mathcal{C}) \leq n$ if and only if, for all X in $\mathcal{C}$, there exists an exact sequence $0 \to X_n \to \cdots \to X_0 \to X \to 0$ with the X_j's projective.

Exercise 13.9. Let k be a field of characteristic 0 and let $W = W_n(k)$ be the Weyl algebra in n variables over k (see Exercise 8.39). Denote by $\mathcal{O}$ the left W-module W/I, where I is the left ideal generated by $(\partial_1, \ldots, \partial_n)$ (hence $\mathcal{O} \simeq k[x_1, \ldots, x_n]$) and denote by Ω the right W-module W/J, where J is the right ideal generated by $(\partial_1, \ldots, \partial_n)$.

(i) Let $\cdot\partial_i$ denote the multiplication on the right by ∂_i on W. Prove that $\varphi = (\cdot\partial_1, \ldots, \cdot\partial_n)$ is a regular sequence (see Exercise 12.4) and $H^n(\mathrm{K}^\bullet(W, \varphi)) \simeq \mathcal{O}$.

(ii) Calculate the k-vector spaces $\mathrm{Tor}_j^W(\Omega, \mathcal{O})$.

(iii) Calculate the k-vector spaces $\mathrm{Ext}_W^j(\mathcal{O}, \mathcal{O})$.

Exercise 13.10. Let k be a field, let $A = k[x_1, \ldots, x_n]$ and set $\mathcal{C} = \mathrm{Mod}^f(A)$. It is well known that $\mathcal{C}$ has enough projectives and finite homological dimension. Define the functor $* : \mathrm{D}^b(\mathcal{C})^{\mathrm{op}} \to \mathrm{D}^b(\mathcal{C})$ by $M^* = \mathrm{RHom}_A(M, A)$. Prove that the functor $* : \mathrm{D}^b(\mathcal{C})^{\mathrm{op}} \to \mathrm{D}^b(\mathcal{C})$ is well defined and satisfies $* \circ * \simeq \mathrm{id}_{\mathrm{D}^b(\mathcal{C})}$. In particular, it is an equivalence of categories.

Exercise 13.11. Let $\mathcal{C}$ be an abelian category with enough injectives and such that $\mathrm{hd}(\mathcal{C}) \leq 1$. Let $F \colon \mathcal{C} \to \mathcal{C}'$ be a left exact functor and let $X \in \mathrm{D}^+(\mathcal{C})$.

(i) Prove that $H^k(RF(X)) \simeq F(H^k(X)) \oplus R^1 F(H^{k-1}(X))$.

(ii) Recall that $\mathrm{hd}(\mathrm{Mod}(\mathbb{Z})) = 1$. Let $X \in \mathrm{D}^-(\mathrm{Mod}(\mathbb{Z}))$, and let $M \in \mathrm{Mod}(\mathbb{Z})$. Prove that $H^k(X \overset{L}{\otimes} M) \simeq (H^k(X) \otimes M) \oplus \mathrm{Tor}_1(H^{k+1}(X), M)$.

Exercise 13.12. Let $\mathcal{A}$ be an abelian category, $\mathcal{C}$ a thick abelian subcategory of $\mathcal{A}$. Assume that there is a category $\mathcal{J}' \times \mathcal{P}'^{\mathrm{op}}$ (resp. $\mathcal{J} \times \mathcal{P}^{\mathrm{op}}$) in $\mathcal{A} \times \mathcal{A}^{\mathrm{op}}$ (resp. $\mathcal{C} \times \mathcal{C}^{\mathrm{op}}$) injective with respect to the functor $\mathrm{Hom}_{\mathcal{A}}$ (resp. $\mathrm{Hom}_{\mathcal{C}}$), and satisfying $\mathcal{J} \subset \mathcal{J}'$ and $\mathcal{P} \subset \mathcal{P}'$. Prove that the natural functor $\mathrm{D}^b(\mathcal{C}) \to \mathrm{D}^b_{\mathcal{C}}(\mathcal{A})$ is an equivalence.

Exercise 13.13. (See Deligne [18]). Let $\mathcal{C}$ be an abelian category. Let $X \in$ $\mathrm{D}^b(\mathcal{C})$ and let $c\colon X \to X[2]$ be a morphism such that $c^p\colon X \to X[2p]$ induces an isomorphism

$$(13.4.7) \qquad\qquad H^{-p}(X) \xrightarrow{\sim} H^p(X) \text{ for all } p > 0.$$

(i) Assume that $H^j(X) = 0$ for $|j| > d$. By considering the morphisms

$$\tau^{\leq 0}(X[-d]) \to X[-d] \xrightarrow{c^d} X[d] \to \tau^{\geq 0}(X[d]),$$

construct morphisms $X \underset{\varphi}{\overset{\psi}{\rightleftarrows}} H^{-d}(X)[d] \oplus H^d(X)[-d]$ such that $\psi \circ \varphi$ is an

isomorphism and deduce that there exists a decomposition $X \simeq H^{-d}(X)[d] \oplus$ $Y \oplus H^d(X)[-d]$ such that $H^j(Y) = 0$ for $|j| \geq d$ and that c induces a morphism $Y \to Y[2]$ satisfying (13.4.7).
(ii) Prove that $X \simeq \bigoplus_k H^k(X)[-k]$.

Exercise 13.14. Let k be a field and let $\mathcal{D} = \mathrm{D}^b(\mathrm{Mod}^f(k))$. To $X \in \mathcal{D}$, associate its index $\chi(X) := \sum_i (-1)^i \dim H^i(X)$. Denote by $\mathcal{N}$ the full additive subcategory of $\mathcal{D}$ consisting of objects X such that $\chi(X)$ is even.
(i) Prove that $\mathcal{N}$ is a null system in $\mathcal{D}$.
(ii) Prove that $\mathcal{D}/\mathcal{N} \simeq 0$. (Hint: use the result of Exercise 10.11.)

Exercise 13.15. Let $\mathcal{C}$ be an abelian category. To an exact sequence $E\colon 0 \to$ $Y \to Z \to X \to 0$ in $\mathcal{C}$, associate $\theta(E) \in \mathrm{Ext}^1_{\mathcal{C}}(X, Y)$, the image of id_X by the morphism $\mathrm{Hom}_{\mathcal{C}}(X, X) \to \mathrm{Ext}^1_{\mathcal{C}}(X, Y)$.
(i) Prove that $\theta(E)\colon X \to Y[1]$ is described by the morphism of complexes (the complexes are the horizontal arrows) where the first morphism of complexes (given by the two vertical arrows on the top) is a qis:

(ii) Prove that $\theta(E) = 0$ if and only if the exact sequence E splits.
(iii) Prove that for any $u \in \mathrm{Ext}^1_{\mathcal{C}}(X, Y)$ there exists an exact sequence E such that $u = \theta(E)$.
(iv) Let us consider two exact sequences $E\colon 0 \to Y \to Z \to X \to 0$ and $E'\colon 0 \to Y \to Z' \to X \to 0$. Prove that $\theta(E) = \theta(E')$ if and only if there exists a commutative diagram

(v) Let E and E' be two exact sequences in $\mathcal{C}$. Construct an exact sequence E'' such that $\theta(E'') = \theta(E) + \theta(E')$.

Exercise 13.16. Let $\mathcal{C}$ be an abelian category. Let $X, Y \in \mathcal{C}$, let $n > 0$ and let $u \in \mathrm{Ext}^n_{\mathcal{C}}(X, Y) = \mathrm{Hom}_{D(\mathcal{C})}(X, Y[n])$.
(i) Prove that there exists an exact sequence $0 \to Y \to Z_1 \to \cdots \to Z_n \to X \to 0$ such that u is given by the composition

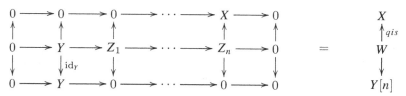

where the vertical arrows on the top define a qis.
(ii) Assume that $u \in \mathrm{Ext}^n_{\mathcal{C}}(X, Y)$ is defined as in (i). Prove that $u = 0$ if and only if there exists a commutative diagram with exact rows:

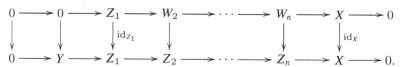

(iii) Prove that any morphism $u \colon X \to Y[n]$ in $D(\mathcal{C})$ decomposes into $X \to Y'[n-1] \to Y[n]$ for some $Y' \in \mathcal{C}$.

Exercise 13.17. Let $\mathcal{C}$ be an abelian category, $\mathcal{C}'$ a thick abelian subcategory.
(i) Prove that $\mathrm{Ext}^n_{\mathcal{C}'}(X, Y) \to \mathrm{Ext}^n_{\mathcal{C}}(X, Y)$ is an isomorphism if $X, Y \in \mathcal{C}'$ and $n \leq 1$.
(ii) Prove that the conditions below are equivalent.

(a) $D^b(\mathcal{C}') \to D^b_{\mathcal{C}'}(\mathcal{C})$ is an equivalence of categories,
(b) $D^b(\mathcal{C}') \to D^b_{\mathcal{C}'}(\mathcal{C})$ is fully faithful,
(c) for any X, Y in $\mathcal{C}'$ and any $n > 0$, $\mathrm{Ext}^n_{\mathcal{C}'}(X, Y) \to \mathrm{Ext}^n_{\mathcal{C}}(X, Y)$ is an isomorphism,
(d) for any X, Y in $\mathcal{C}'$, $n > 0$ and $u \in \mathrm{Ext}^n_{\mathcal{C}}(X, Y)$ there exist a monomorphism $Y \rightarrowtail Y'$ and an epimorphism $X' \twoheadrightarrow X$ in $\mathcal{C}'$ such that the image of u by the morphism $\mathrm{Ext}^n_{\mathcal{C}}(X, Y) \to \mathrm{Ext}^n_{\mathcal{C}}(X', Y')$ vanishes.

(Hint: for (a) $\Rightarrow$ (d) use Exercise 13.16, and for (d) $\Rightarrow$ (c) argue by induction on n. See [4].)

Exercise 13.18. Let $\mathcal{C}$ be an abelian category with enough injectives and let $\mathcal{C}'$ be an abelian category. Let F and G be additive functors from $\mathcal{C}$ to $\mathcal{C}'$ and let $\lambda \colon F \to G$ be a morphism of functors. Construct a functor $H \colon D^+(\mathcal{C}) \to D^+(\mathcal{C}')$ and morphisms of functors $R^+G \xrightarrow{\varphi} H \xrightarrow{\psi} [1] \circ R^+F$ such that the triangle $R^+F(X) \to R^+G(X) \xrightarrow{\varphi(X)} H(X) \xrightarrow{\psi(X)} R^+F(X)[1]$ is a d.t. for any $X \in D^+(\mathcal{C})$.

Exercise 13.19. Let C be an abelian category with enough injectives and let C' be an abelian category. Let $F' \xrightarrow{\varphi} F \xrightarrow{\psi} F''$ be a sequence of additive functors from C to C' such that the sequence $0 \to F'(X) \to F(X) \to F''(X) \to 0$ is exact for any injective object $X \in C$. Construct a morphism of functors $\xi \colon R^+F'' \to [1] \circ R^+F'$ such that $R^+F'(X) \to R^+F(X) \to R^+F''(X) \xrightarrow{\xi_X} R^+F'(X)[1]$ is a d.t. in $D^+(C')$ for all $X \in D^+(C)$.

Exercise 13.20. Let C be an abelian category, let $a < b \in \mathbb{Z}$ and let $X \in D^b(C)$.
(i) Assume that $H^j(X) \simeq 0$ for $j \neq a, b$. (We say that X is *concentrated in degrees* a, b.) By using the d.t. (13.1.4), construct a canonical element $c_X \in \operatorname{Ext}_C^{b-a+1}(H^b(X), H^a(X))$ and prove that $c_X = 0$ if and only if $X \simeq H^a(X)[-a] \oplus H^b(X)[-b]$.
(ii) Let $X, Y \in D(C)$ be concentrated in degrees a, b and let $u \colon X \to Y$ be a morphism. Prove that $c_Y \circ H^b(u) = (H^a(u)[b-a+1]) \circ c_X$.
(iii) Let $X, Y \in D(C)$ be concentrated in degrees a, b and assume that there are morphisms $\varphi \colon H^a(X) \to H^a(Y)$ and $\psi \colon H^b(X) \to H^b(Y)$ satisfying $c_Y \circ \psi = (\varphi[b-a+1]) \circ c_X$. Prove that there exists a morphism $f \colon X \to Y$ such that $\varphi = H^a(f)$ and $\psi = H^b(f)$.

Exercise 13.21. Let k be a commutative ring and set $A = k[x, y]$, $C = \operatorname{Mod}(A)$. Let $L_0 = A$, $L = A \oplus A$ and let $\mathfrak{a}$ be the ideal $Ax + Ay$.
(i) Construct the exact sequence

$$0 \to L_0 \xrightarrow{\varphi} L \xrightarrow{\psi} L_0 \to k \to 0 ,$$

where $\varphi(a) = (xa, ya)$ and k is identified with $A/\mathfrak{a}$, and deduce that $\operatorname{Ext}_C^2(k, k) \simeq k$.
(ii) Let E be the set of isomorphism classes of objects M of $D^b(C)$ with $H^i(M) = 0$ for $i \neq 0, 1$, $H^0(M) \simeq H^1(M) \simeq k$. Prove that E is in bijection with the quotient $k/k^\times$, where $k^\times$ is the group of invertible elements of k and acts on k by multiplication. In particular $k/k^\times$ is $\mathbb{N}$ for $k = \mathbb{Z}$, and it consists of two elements if k is a field. (Hint: use Exercise 13.20 (iii).)

Exercise 13.22. Let C be an abelian category and let $X \in D^b(C)$. Assume that $\operatorname{Ext}_C^{j-i+1}(H^j(X), H^i(X)) = 0$ for all $i, j \in \mathbb{Z}$ with $i < j$. Prove that $X \simeq \oplus_j H^j(X)[-j]$. (Hint: adapt the proof of Corollary 13.1.20.)

Exercise 13.23. Let C be an abelian category.
(i) Prove that $\operatorname{Mor}(C)$ is an abelian category.
(ii) Prove that $\operatorname{Ker} \colon \operatorname{Mor}(C) \to C$ is an additive left exact functor.
(iii) Let $\mathcal{J}$ be the full subcategory of $\operatorname{Mor}(C)$ consisting of epimorphisms. Prove that $\mathcal{J}$ is Ker-injective.
(iv) Prove that $R^+\operatorname{Ker} \colon D^+(\operatorname{Mor}(C)) \to D^+(C)$ exists and that for $f \in \operatorname{Mor}(C)$, we have

$$H^k(R^+ \operatorname{Ker})(f) \simeq \begin{cases} \operatorname{Ker} f & \text{for } k = 0 \,, \\ \operatorname{Coker} f & \text{for } k = 1 \,, \\ 0 & \text{otherwise} \,. \end{cases}$$

Exercise 13.24. Let A be a principal ideal domain (that is, a commutative ring without zero-divisors and such that any ideal admits a generator). For an A-module M, denote by M_{tor} its torsion part, that is,

$$M_{tor} = \{u \in M; au = 0 \text{ for some } a \in A \setminus \{0\}\} \,.$$

(i) Prove that if I is an injective A-module, then $I \otimes_A M \to I \otimes_A (M/M_{tor})$ is an isomorphism.
(ii) Let $\varphi \colon \operatorname{Mod}(A) \to \operatorname{Mod}(A)$ be the functor given by $\varphi(X) = X \otimes_A (M/M_{tor})$. Prove that φ is exact.
In the sequel, we shall keep the notation φ to denote the functor induced by φ on $D^+(A)$.
(iii) Let $\psi \colon \operatorname{Mod}(A) \to \operatorname{Mod}(A)$ be the functor given by $\psi(X) = X \otimes_A M$. Prove that $R^+\psi \colon D^+(A) \to D^+(A)$ exists and that $R^+\psi \simeq \varphi$.

Exercise 13.25. Let $\mathcal{C}$ be an abelian category with enough injectives, and let $\mathcal{I}$ be the full subcategory of injective objects. Let $\mathcal{C}_{\mathcal{I}}$ be the additive category defined in Exercise 8.6.
(i) For $X \in \mathcal{C}$, let us take an exact sequence $0 \to X \to I \to S \to 0$ with $I \in \mathcal{I}$. Prove that the functor which associates S to X is a well defined functor $T \colon \mathcal{C}_{\mathcal{I}} \to \mathcal{C}_{\mathcal{I}}$.
(ii) Let $\mathcal{T}$ be the category with translation $T^{-1}(\mathcal{C}_{\mathcal{I}})$, whose translation functor is still denoted by the same letter T (see Exercise 11.1). We say that a triangle in $\mathcal{T}$ is a d.t. if it is isomorphic to

$$T^n X \xrightarrow{(-1)^n T^n f} T^n Y \xrightarrow{(-1)^n T^n g} T^n Z \xrightarrow{(-1)^n T^n h} T^n S \simeq T(T^n X)$$

for an integer n and a commutative diagram in $\mathcal{C}$

$$
\begin{array}{ccccccccc}
0 & \longrightarrow & X & \xrightarrow{\ f\ } & Y & \xrightarrow{\ g\ } & Z & \longrightarrow & 0 \\
& & \downarrow{\scriptstyle \operatorname{id}_X} & & \downarrow & & \downarrow{\scriptstyle h} & & \\
0 & \longrightarrow & X & \longrightarrow & I & \longrightarrow & S & \longrightarrow & 0
\end{array}
$$

with exact rows and $I \in \mathcal{I}$. Prove that $\mathcal{T}$ is a triangulated category.
(iii) Prove that $K^b(\mathcal{I}) \to D^b(\mathcal{C})$ is fully faithful and that $\mathcal{T}$ is equivalent to $D^b(\mathcal{C})/K^b(\mathcal{I})$ as a triangulated category. (Hint: embed $D^b(\mathcal{C})$ in $D^+(\mathcal{C})$ and consider injective resolutions.)

14

Unbounded Derived Categories

In this chapter we study the unbounded derived categories of Grothendieck categories, using the results of Chap. 9. We prove the existence of enough homotopically injective objects in order to define unbounded right derived functors, and we prove that these triangulated categories satisfy the hypotheses of the Brown representability theorem. We also study unbounded derived functors in particular for pairs of adjoint functors. We start this study in the framework of abelian categories with translation, then we apply it to the case of the categories of unbounded complexes in abelian categories.

Many of the results in this Chapter are not new and many authors have contributed to the results presented here, in particular, Spaltenstein [65] who first considered unbounded complexes and unbounded derived functors. Other contributions are due to [2, 6, 21, 41, 44], [53]. Note that many of the ideas encountered here come from Topology, and the names of Adams, Bousfield, Kan, Thomason, and certainly many others, should be mentioned.

14.1 Derived Categories of Abelian Categories with Translation

Let $(\mathcal{A}, T)$ be an abelian category with translation. Recall (Definition 13.1.1) that, denoting by $\mathcal{N}$ the triangulated subcategory of the homotopy category $\mathrm{K}_c(\mathcal{A})$ consisting of objects X qis to 0, the derived category $\mathrm{D}_c(\mathcal{A})$ of $(\mathcal{A}, T)$ is the localization $\mathrm{K}_c(\mathcal{A})/\mathcal{N}$. Recall that X is qis to 0 if and only if $T^{-1}X \xrightarrow{T^{-1}d_X} X \xrightarrow{d_X} TX$ is exact.

For $X \in \mathcal{A}_c$, the differential $d_X \colon X \to TX$ is a morphism in $\mathcal{A}_c$. Hence its cohomology $H(X)$ is regarded as an object of $\mathcal{A}_c$ and similarly for $\operatorname{Ker} d_X$ and $\operatorname{Im} d_X$. Note that their differentials vanish.

Proposition 14.1.1. *Assume that $\mathcal{A}$ admits direct sums indexed by a set I and that such direct sums are exact. Then $\mathcal{A}_c$, $\mathrm{K}_c(\mathcal{A})$ and $\mathrm{D}_c(\mathcal{A})$ admit such* *direct sums and the two functors $\mathcal{A}_c \to \mathrm{K}_c(\mathcal{A})$ and $\mathrm{K}_c(\mathcal{A}) \to \mathrm{D}_c(\mathcal{A})$ commute with such direct sums.*

Proof. The result concerning $\mathcal{A}_c$ and $\mathrm{K}_c(\mathcal{A})$ is obvious, and that concerning $\mathrm{D}_c(\mathcal{A})$ follows from Proposition 10.2.8.   q.e.d.

For an object X of $\mathcal{A}$, we denote by $M(X)$ the mapping cone of $\mathrm{id}_{T^{-1}X}$, regarding $T^{-1}X$ as an object of $\mathcal{A}_c$ with the zero differential. Hence $M(X)$ is the object $X \oplus T^{-1}X$ of $\mathcal{A}_c$ with the differential

$$d_{M(X)} = \begin{pmatrix} 0 & 0 \\ \mathrm{id}_X & 0 \end{pmatrix} : X \oplus T^{-1}X \to TX \oplus X \ .$$

Therefore $M : \mathcal{A} \to \mathcal{A}_c$ is an exact functor. Moreover M is a left adjoint functor to the forgetful functor $\mathcal{A}_c \to \mathcal{A}$ as seen by the following lemma.

Lemma 14.1.2. *For $Z \in \mathcal{A}$ and $X \in \mathcal{A}_c$, we have the isomorphism*

(14.1.1) $\mathrm{Hom}_{\mathcal{A}_c}(M(Z), X) \xrightarrow{\sim} \mathrm{Hom}_{\mathcal{A}}(Z, X)$.

Proof. The morphism $(u, v) \colon M(Z) \to X$ in $\mathcal{A}_c$ satisfies $d_X \circ (u, v) = T(u, v) \circ d_{M(X)}$ which reads as $d_X \circ u = Tv$ and $d_X \circ v = 0$. Hence it is determined by $u \colon Z \to X$.   q.e.d.

Proposition 14.1.3. *Let $\mathcal{A}$ be a Grothendieck category. Then $\mathcal{A}_c$ is again a Grothendieck category.*

Proof. The category $\mathcal{A}_c$ is abelian and admits small inductive limits, and small filtrant inductive limits in $\mathcal{A}_c$ are clearly exact. Moreover, if G is a generator in $\mathcal{A}$, then $M(G)$ is a generator in $\mathcal{A}_c$ by Lemma 14.1.2.   q.e.d.

Definition 14.1.4. (i) *An object $I \in \mathrm{K}_c(\mathcal{A})$ is homotopically injective if* $\mathrm{Hom}_{\mathrm{K}_c(\mathcal{A})}(X, I) \simeq 0$ *for all $X \in \mathrm{K}_c(\mathcal{A})$ that is qis to 0.*
  (ii) *An object $P \in \mathrm{K}_c(\mathcal{A})$ is homotopically projective if P is homotopically injective in $\mathrm{K}_c(\mathcal{A}^{\mathrm{op}})$, that is, if $\mathrm{Hom}_{\mathrm{K}_c(\mathcal{A})}(P, X) \simeq 0$ for all $X \in \mathrm{K}_c(\mathcal{A})$ that is qis to 0.*

We shall denote by $\mathrm{K}_{c,\mathrm{hi}}(\mathcal{A})$ the full subcategory of $\mathrm{K}_c(\mathcal{A})$ consisting of homotopically injective objects and by $\iota \colon \mathrm{K}_{c,\mathrm{hi}}(\mathcal{A}) \to \mathrm{K}_c(\mathcal{A})$ the embedding functor. We denote by $\mathrm{K}_{c,\mathrm{hp}}(\mathcal{A})$ the full subcategory of $\mathrm{K}_c(\mathcal{A})$ consisting of homotopically projective objects.

Note that $\mathrm{K}_{c,\mathrm{hi}}(\mathcal{A})$ is obviously a full triangulated subcategory of $\mathrm{K}_c(\mathcal{A})$.

Lemma 14.1.5. *Let $(\mathcal{A}, T)$ be an abelian category with translation. If $I \in \mathrm{K}_c(\mathcal{A})$ is homotopically injective, then*

$$\mathrm{Hom}_{\mathrm{K}_c(\mathcal{A})}(X, I) \xrightarrow{\sim} \mathrm{Hom}_{\mathrm{D}_c(\mathcal{A})}(X, I)$$

for all $X \in \mathrm{K}_c(\mathcal{A})$.

Proof. Let $X \in K_c(\mathcal{A})$ and let $X' \to X$ be a qis. Then for $I \in K_{c,\mathrm{hi}}(\mathcal{A})$, the morphism $\mathrm{Hom}_{K_c(\mathcal{A})}(X, I) \to \mathrm{Hom}_{K_c(\mathcal{A})}(X', I)$ is an isomorphism, since there exists a d.t. $X' \to X \to N \to TX$ with N qis to 0 and $\mathrm{Hom}_{K_c(\mathcal{A})}(N, I) \simeq \mathrm{Hom}_{K_c(\mathcal{A})}(T^{-1}N, I) \simeq 0$. Therefore, for any $X \in K_c(\mathcal{A})$ and $I \in K_{c,\mathrm{hi}}(\mathcal{A})$, we have

$$\mathrm{Hom}_{D_c(\mathcal{A})}(X, I) \simeq \varinjlim_{(X' \to X) \in \mathrm{Qis}} \mathrm{Hom}_{K_c(\mathcal{A})}(X', I) \simeq \mathrm{Hom}_{K_c(\mathcal{A})}(X, I) \,.$$

q.e.d.

Let us introduce the notation

(14.1.2) $\mathrm{QM} = \{f \in \mathrm{Mor}(\mathcal{A}_c) \,; \, f \text{ is both a qis and a monomorphism}\}$.

Recall (see Definition 9.5.1) that an object $I \in \mathcal{A}_c$ is QM-injective if, for any morphism $f : X \to Y$ in QM, $\mathrm{Hom}_{\mathcal{A}_c}(Y, I) \xrightarrow{\circ f} \mathrm{Hom}_{\mathcal{A}_c}(X, I)$ is surjective.

Proposition 14.1.6. *Let $I \in \mathcal{A}_c$. Then I is QM-injective if and only if it satisfies the following two conditions:*

(a) *I is homotopically injective,*
(b) *I is injective as an object of $\mathcal{A}$.*

Proof. (i) Assume that I is QM-injective.
(a) Recall that for a morphism $f : X \to Y$ in $\mathcal{A}_c$, we have constructed a natural monomorphism $\alpha(f) : Y \to \mathrm{Mc}(f)$ in $\mathcal{A}_c$. Let $X \in \mathcal{A}_c$ be qis to 0. Then $u := \alpha(\mathrm{id}_X)$ is a monomorphism and it is also a qis since both X and $\mathrm{Mc}(\mathrm{id}_X)$ are qis to 0. Hence $u \in \mathrm{QM}$, and it follows that any morphism $f : X \to I$ factorizes through $\mathrm{Mc}(\mathrm{id}_X)$. Since $\mathrm{Mc}(\mathrm{id}_X) \simeq 0$ in $K_c(\mathcal{A})$, the morphism f vanishes in $K_c(\mathcal{A})$.
(b) Consider a monomorphism $v : U \to V$ in $\mathcal{A}$. The morphism v defines the morphism $M(v) : M(U) \to M(V)$ in $\mathcal{A}_c$ and $M(v)$ belongs to QM. Consider the commutative diagram

$$\begin{array}{ccc}
\mathrm{Hom}_{\mathcal{A}_c}(M(V), I) & \longrightarrow & \mathrm{Hom}_{\mathcal{A}_c}(M(U), I) \\
\downarrow{\scriptstyle \sim} & & \downarrow{\scriptstyle \sim} \\
\mathrm{Hom}_{\mathcal{A}}(V, I) & \longrightarrow & \mathrm{Hom}_{\mathcal{A}}(U, I).
\end{array}$$

Since $M(v)$ belongs to QM and I is QM-injective, the horizontal arrow on the top is surjective. Hence, the horizontal arrow in the bottom is also surjective, and we conclude that I is injective.

(ii) Assume that I satisfies conditions (a) and (b). Let $f : X \to Y$ be a morphism in $\mathcal{A}_c$ belonging to QM and let $\varphi : X \to I$ be a morphism in $\mathcal{A}_c$. Since I is injective as an object of $\mathcal{A}$, there exists a morphism $\psi : Y \to I$ in $\mathcal{A}$ such that $\varphi = \psi \circ f$. Let $h : T^{-1}Y \to I$ be the morphism in $\mathcal{A}$ given by

$$h = T^{-1}d_I \circ T^{-1}\psi - \psi \circ T^{-1}d_Y$$
$$= T^{-1}d_I \circ T^{-1}\psi + \psi \circ d_{T^{-1}Y}.$$

Then $h\colon T^{-1}Y \to I$ is a morphism in $\mathcal{A}_c$ and $h \circ T^{-1}f = 0$.

Let us consider an exact sequence $0 \to X \xrightarrow{f} Y \xrightarrow{g} Z \to 0$ in $\mathcal{A}_c$. Then, Z is qis to 0. Since $h \circ T^{-1}f = 0$, there exists a morphism $\tilde{h}\colon T^{-1}Z \to I$ in $\mathcal{A}_c$ such that $h = \tilde{h} \circ T^{-1}g$. Since Z is exact and I is homotopically injective, $\tilde{h}$ is homotopic to zero, i.e., there exists a morphism $\xi\colon Z \to I$ in $\mathcal{A}$ such that

$$\tilde{h} = T^{-1}d_I \circ T^{-1}\xi + \xi \circ d_{T^{-1}Z}.$$

Then the morphism $\tilde{\psi} = \psi - \xi \circ g$ gives a morphism $\tilde{\psi}\colon Y \to I$ in $\mathcal{A}_c$ which satisfies $\tilde{\psi} \circ f = \varphi$.   q.e.d.

Now we shall prove the following theorem.

Theorem 14.1.7. *Let $(\mathcal{A}, T)$ be an abelian category with translation and assume that $\mathcal{A}$ is a Grothendieck category. Then, for any $X \in \mathcal{A}_c$, there exists $u\colon X \to I$ such that $u \in \mathrm{QM}$ and I is QM-injective.*

Applying Proposition 14.1.6, we get:

Corollary 14.1.8. *Let $(\mathcal{A}, T)$ be an abelian category with translation and assume that $\mathcal{A}$ is a Grothendieck category. Then for any $X \in \mathcal{A}_c$, there exists a qis $X \to I$ such that I is homotopically injective.*

The proof of Theorem 14.1.7 decomposes into several steps.

Define a subcategory $\mathcal{A}_{c,0}$ of $\mathcal{A}_c$ as follows:

$$\mathrm{Ob}(\mathcal{A}_{c,0}) = \mathrm{Ob}(\mathcal{A}_c), \quad \mathrm{Mor}(\mathcal{A}_{c,0}) = \mathrm{QM}.$$

We shall apply Theorems 9.5.4 and 9.5.5 to the categories $\mathcal{A}_c$ and $\mathcal{A}_{c,0}$ (denoted by $\mathcal{C}$ and $\mathcal{C}_0$ in these theorems).

Let us check that hypothesis (9.5.2) is satisfied. Hypothesis (9.5.2) (i) is satisfied since small filtrant inductive limits are exact and hence $H\colon \mathcal{A}_c \to \mathcal{A}$ commutes with such limits. Hypothesis (9.5.2) (ii) follows from

(14.1.3) $\begin{cases} \text{if } u\colon X \to Y \text{ belongs to QM and } X \to X' \text{ is a morphism in } \mathcal{A}_c, \\ \text{then } u'\colon X' \to X' \oplus_X Y \text{ belongs to QM.} \end{cases}$

Set $Y' = X' \oplus_X Y$. Then $u'\colon X' \to Y'$ is a monomorphism. Note that u (resp. u') is a qis if and only if $\mathrm{Coker}(u)$ (resp. $\mathrm{Coker}(u')$) is qis to zero. Hence (14.1.3) follows from $\mathrm{Coker}(u) \simeq \mathrm{Coker}(u')$ (Lemma 8.3.11 (b)).

Since $\mathcal{A}_c$ is a Grothendieck category by Proposition 14.1.3, Theorem 9.6.1 implies that there exists an essentially small full subcategory $\mathcal{S}$ of $\mathcal{A}_c$ such that

(14.1.4) $\begin{cases}\end{cases}$
    (i) $\mathcal{S}$ contains a generator of $\mathcal{A}_c$,

    (ii) $\mathcal{S}$ is closed by subobjects and quotients in $\mathcal{A}_c$,

    (iii) for any solid diagram $Y' \overset{g}{\dashrightarrow} Y$ in which $f \colon X' \to X$ is

$$\begin{array}{ccc} Y' & \overset{g}{\dashrightarrow} & Y \\ \downarrow & & \downarrow \\ X' & \overset{f}{\longrightarrow} & X \end{array}$$

    an epimorphism in $\mathcal{A}_c$ and $Y \in \mathcal{S}$, the dotted arrow may be completed to a commutative diagram with $Y' \in \mathcal{S}$ and g an epimorphism,

    (iv) $\mathcal{S}$ is closed by countable direct sums.

In particular, $\mathcal{S}$ is a fully abelian subcategory of $\mathcal{A}_c$ closed by countable inductive limits.

Define the set

$$\mathcal{F}' = \{u \colon X \to Y; u \in \mathrm{QM}, \ X, Y \in \mathcal{S}\},$$

and take $\mathcal{F} \subset \mathcal{F}'$ by collecting a representative of each isomorphism class in $\mathcal{F}'$ (i.e., for the relation of being isomorphic in $\mathrm{Mor}(\mathcal{A}_c)$). Since $\mathcal{S}$ is essentially small, $\mathcal{F}$ is a small subset of $\mathcal{F}'$ such that any $u \in \mathcal{F}'$ is isomorphic to an element of $\mathcal{F}$.

By Theorem 9.6.1, there exists an infinite cardinal π such that if $u \colon X \to Y$ belongs to $\mathcal{F}$, then $X \in (\mathcal{A}_c)_\pi$. Applying Theorem 9.5.4, we find that for any $X \in \mathcal{A}_c$ there exists a morphism $u \colon X \to I$ such that $u \in \mathrm{QM}$ and I is $\mathcal{F}$-injective. In order to prove that I is QM-injective, we shall apply Theorem 9.5.5.

For $X \in \mathcal{A}_{c,0}$, an object of $(\mathcal{A}_{c,0})_X$ is given by a monomorphism $Y \rightarrowtail X$. Therefore $(\mathcal{A}_{c,0})_X$ is essentially small by Corollary 8.3.26, and hence hypothesis (9.5.6) is satisfied.

Let us check (9.5.7). We have an exact sequence $0 \to X' \to X \oplus Y' \overset{w}{\to} Y$. Then $\mathrm{Im}\, w \simeq X \oplus_{X'} Y'$ and $h \colon \mathrm{Im}\, w \to Y$ is a monomorphism. Hence (9.5.7) follows from (14.1.3).

Hypothesis (9.5.8) will be checked in Lemmas 14.1.9–14.1.11 below.

Lemma 14.1.9. *Let $X \in \mathcal{A}_c$ and let $j \colon V \rightarrowtail X$ be a monomorphism with $V \in \mathcal{S}$. Then there exist $V' \in \mathcal{S}$ and a monomorphism $V' \rightarrowtail X$ such that j decomposes as $V \rightarrowtail V' \rightarrowtail X$ and $\mathrm{Ker}\big(H(V) \to H(X)\big) \to H(V')$ vanishes.*

Proof. Since $V \cap \mathrm{Im}(T^{-1}d_X)$ belongs to $\mathcal{S}$, there exists $W \subset T^{-1}X$ such that $W \in \mathcal{S}$ and $(T^{-1}d_X)(W) = V \cap \mathrm{Im}(T^{-1}d_X)$. Set $V' = V + TW$. Then V' is a subobject of X, it belongs to $\mathcal{S}$ and satisfies the desired condition.   q.e.d.

Lemma 14.1.10. *Let $X \in \mathcal{A}_c$ and let $j \colon V \rightarrowtail X$ be a monomorphism with $V \in \mathcal{S}$. Then there exist $V' \in \mathcal{S}$ and a monomorphism $V' \rightarrowtail X$ such that j decomposes as $V \rightarrowtail V' \rightarrowtail X$ and $H(V') \to H(X)$ is a monomorphism.*

Proof. Set $V_0 = V$. Using Lemma 14.1.9, we construct by induction $V_n \in \mathcal{S}$ such that $V_{n-1} \subset V_n \subset X$ and the morphism

$$\operatorname{Ker}\big(H(V_{n-1}) \to H(X)\big) \to \operatorname{Ker}\big(H(V_n) \to H(X)\big)$$

vanishes.

Take $V' = \varinjlim_n V_n \subset X$. Then $V' \in \mathcal{S}$ and

$$\operatorname{Ker}\big(H(V') \to H(X)\big) \simeq \varinjlim_n \operatorname{Ker}\big(H(V_n) \to H(X)\big) \simeq 0 .$$

q.e.d.

Lemma 14.1.11. *Let $f \colon X \to Y$ be in* QM. *If f satisfies (9.5.5), then f is an isomorphism.*

Proof. Let $Z = \operatorname{Coker} f$. We get an exact sequence in $\mathcal{A}_c$

$$0 \to X \xrightarrow{f} Y \xrightarrow{g} Z \to 0$$

and Z is qis to 0.

Since $\mathcal{S}$ contains a generator of $\mathcal{A}_c$, it is enough to show that $\operatorname{Hom}_{\mathcal{A}_c}(W, Z) \simeq 0$ for any $W \in \mathcal{S}$. Moreover, replacing W with its image in Z, it is enough to check that any $W \subset Z$ with $W \in \mathcal{S}$ vanishes.

For $W \subset Z$ with $W \in \mathcal{S}$, there exists $W' \in \mathcal{S}$ such that $W \subset W' \subset Z$ and $H(W') \simeq 0$ by Lemma 14.1.10. Let us take $V \subset Y$ with $V \in \mathcal{S}$ and $g(V) = W'$. Set $U = f^{-1}(V)$. Thus we obtain a Cartesian square $\begin{array}{ccc} U & \xrightarrow{s} & V \\ \downarrow & & \downarrow \\ X & \xrightarrow{f} & Y \end{array}$. We have an

exact sequence $0 \to U \xrightarrow{s} V \to W' \to 0$. Since W' is qis to zero, $U \xrightarrow{s} V$ belongs to $\mathcal{F}$. Since f satisfies (9.5.5), $V \to Y$ factors through $X \to Y$ and hence $W' = g(V) \simeq 0$. This shows that $W \simeq 0$.   q.e.d.

Thus we have proved hypothesis (9.5.8), and the proof of Theorem 14.1.7 is now complete.

Corollary 14.1.12. *Let $(\mathcal{A}, T)$ be an abelian category with translation and assume that $\mathcal{A}$ is a Grothendieck category. Then:*

(i) *the localization functor $Q \colon \mathrm{K}_c(\mathcal{A}) \to \mathrm{D}_c(\mathcal{A})$ induces an equivalence $\mathrm{K}_{c,\mathrm{hi}}(\mathcal{A}) \xrightarrow{\sim} \mathrm{D}_c(\mathcal{A})$,*

(ii) *the category $\mathrm{D}_c(\mathcal{A})$ is a $\mathcal{U}$-category,*

(iii) *the functor $Q \colon \mathrm{K}_c(\mathcal{A}) \to \mathrm{D}_c(\mathcal{A})$ admits a right adjoint $R_q \colon \mathrm{D}_c(\mathcal{A}) \to \mathrm{K}_c(\mathcal{A})$, $Q \circ R_q \simeq \mathrm{id}$, and R_q is the composition of $\iota \colon \mathrm{K}_{c,\mathrm{hi}}(\mathcal{A}) \to \mathrm{K}_c(\mathcal{A})$ and a quasi-inverse of $Q \circ \iota$,*

(iv) *for any triangulated category $\mathcal{D}$, any triangulated functor $F \colon \mathrm{K}_c(\mathcal{A}) \to \mathcal{D}$ admits a right localization $RF \colon \mathrm{D}_c(\mathcal{A}) \to \mathcal{D}$, and $RF \simeq F \circ R_q$.*

Proof. (i) The functor $Q: \mathrm{K}_{c,\mathrm{hi}}(\mathcal{A}) \to \mathrm{D}_c(\mathcal{A})$ is fully faithful by Lemma 14.1.5 and essentially surjective by Corollary 14.1.8.
(ii)–(iii) follow immediately.
(iv) follows from Proposition 7.3.2.   q.e.d.

14.2 The Brown Representability Theorem

We shall show that the hypotheses of the Brown representability theorem (Theorem 10.5.2) are satisfied for $\mathrm{D}_c(\mathcal{A})$ when $\mathcal{A}$ is a Grothendieck abelian category with translation. Note that $\mathrm{D}_c(\mathcal{A})$ admits small direct sums and the localization functor $Q: \mathrm{K}_c(\mathcal{A}) \to \mathrm{D}_c(\mathcal{A})$ commutes with such direct sums by Proposition 14.1.1.

Theorem 14.2.1. *Let $(\mathcal{A}, T)$ be an abelian category with translation and assume that $\mathcal{A}$ is a Grothendieck category. Then the triangulated category $\mathrm{D}_c(\mathcal{A})$ admits small direct sums and a system of t-generators.*

Applying Theorem 10.5.2, we obtain

Corollary 14.2.2. *Let $(\mathcal{A}, T)$ be an abelian category with translation and assume that $\mathcal{A}$ is a Grothendieck category. Let $G: (\mathrm{D}_c(\mathcal{A}))^{\mathrm{op}} \to \mathrm{Mod}(\mathbb{Z})$ be a cohomological functor which commutes with small products (i.e., $G(\bigoplus_i X_i) \simeq \prod_i G(X_i)$ for any small family $\{X_i\}_i$ in $\mathrm{D}_c(\mathcal{A})$). Then G is representable.*

Applying Corollary 10.5.3, we obtain:

Corollary 14.2.3. *Let $(\mathcal{A}, T)$ be an abelian category with translation and assume that $\mathcal{A}$ is a Grothendieck category. Let $\mathcal{D}$ be a triangulated category and let $F: \mathrm{D}_c(\mathcal{A}) \to \mathcal{D}$ be a triangulated functor. Assume that F commutes with small direct sums. Then F admits a right adjoint.*

We shall prove a slightly more general statement than Theorem 14.2.1. Let $\mathcal{I}$ be a full subcategory of $\mathcal{A}$ closed by subobjects, quotients and extensions in $\mathcal{A}$, and also by small direct sums. Similarly to Definition 13.2.7, let us denote by $\mathrm{D}_{c,\mathcal{I}}(\mathcal{A})$ the full subcategory of $\mathrm{D}_c(\mathcal{A})$ consisting of objects $X \in \mathrm{D}_c(\mathcal{A})$ such that $H(X) \in \mathcal{I}$. Then $\mathrm{D}_{c,\mathcal{I}}(\mathcal{A})$ is a full triangulated subcategory of $\mathrm{D}_c(\mathcal{A})$ closed by small direct sums.

Proposition 14.2.4. *The triangulated category $\mathrm{D}_{c,\mathcal{I}}(\mathcal{A})$ admits a system of t-generators.*

In proving Proposition 14.2.4, we need preliminary lemmas. Recall that there exists an essentially small fully abelian subcategory $\mathcal{S}$ of $\mathcal{A}_c$ satisfying (14.1.4).

Lemma 14.2.5. *Assume that $X \in \mathcal{A}_c$ satisfies $H(X) \in \mathcal{S}$. Then there exists a morphism $j: Y \to X$ with $Y \in \mathcal{S}$ and $j \in \mathrm{QM}$.*

Proof. There exists $S \in \mathcal{S}$ such that $S \subset \operatorname{Ker} d_X$ and that the composition $S \to \operatorname{Ker} d_X \to H(X)$ is an epimorphism. Since the differential of S vanishes, $H(S)$ is isomorphic to S and $H(S) \to H(X)$ is an epimorphism. By Lemma 14.1.10, there exists $Y \in \mathcal{S}$ such that $S \subset Y \subset X$ and $H(Y) \to H(X)$ is a monomorphism. Hence, $H(Y) \to H(X)$ is an isomorphism.   q.e.d.

Lemma 14.2.6. *Let* $X \in \mathcal{A}_c$ *with* $H(X) \in \mathcal{I}$. *If* $\operatorname{Hom}_{\mathrm{D}_{c,\mathcal{I}}(\mathcal{A})}(Y, X) \simeq 0$ *for all* $Y \in \mathcal{S}$ *such that* $H(Y) \in \mathcal{I}$, *then* X *is qis to zero.*

Proof. It is enough to show that $\operatorname{Hom}_{\mathcal{A}_c}(S, H(X)) \simeq 0$ for all $S \in \mathcal{S}$. Let us show that any $u \colon S \to H(X)$ vanishes. Replacing S with the image of u, we may assume that u is a monomorphism. Since $\operatorname{Ker} d_X \to H(X)$ is an epimorphism, there exists $S' \in \mathcal{S}$ such that $S' \subset \operatorname{Ker} d_X$ and that the image of the composition $S' \to \operatorname{Ker} d_X \to H(X)$ is equal to S. By Lemma 14.1.10, there exists $V \in \mathcal{S}$ such that $S' \subset V \subset X$ and $H(V) \to H(X)$ is a monomorphism. Hence $H(V)$ belongs to $\mathcal{I}$. Since $\operatorname{Hom}_{\mathrm{D}_{c,\mathcal{I}}(\mathcal{A})}(V, X) \simeq 0$ by the assumption, the morphism $V \to X$ vanishes in $\mathrm{D}_c(\mathcal{A})$. Taking the cohomology, we find that $H(V) \to H(X)$ vanishes. Since the differentials of S' and S vanish, we have $H(S') \simeq S'$ and $H(S) \simeq S$. Since the composition $H(S') \to H(V) \to H(X)$ vanishes, the composition $S' \twoheadrightarrow S \xrightarrow{u} H(X)$ vanishes. Hence $u = 0$.   q.e.d.

Proof of Proposition 14.2.4. Denote by $\mathcal{T}$ the subset of $\mathrm{D}_{c,\mathcal{I}}(\mathcal{A})$ consisting of the image of objects $Y \in \mathcal{S}$ such that $H(Y) \in \mathcal{I}$. We shall show that $\mathcal{T}$ is a system of t-generators in $\mathrm{D}_{c,\mathcal{I}}(\mathcal{A})$.
(i) $\mathcal{T}$ is a system of generators. Indeed, $\operatorname{Hom}_{\mathrm{D}_{c,\mathcal{I}}(\mathcal{A})}(Y, X) \simeq 0$ for all $Y \in \mathcal{T}$ implies that $X \simeq 0$ by Lemma 14.2.6.
(ii) We shall check condition (iii)' in Remark 10.5.4. Consider a small set I and a morphism $C \to \bigoplus_{i \in I} X_i$ in $\mathrm{D}_{c,\mathcal{I}}(\mathcal{A})$, with $C \in \mathcal{T}$. This morphism is represented by morphisms in $\mathcal{A}_c$:

$$C \xleftarrow{u} Y \to \bigoplus_{i \in I} X_i$$

where $Y \in \mathcal{A}_c$ and u is a qis. By Lemma 14.2.5, there exists a qis $C' \to Y$ with $C' \in \mathcal{S}$. Replacing C with C', we may assume from the beginning that we have a morphism $C \to \bigoplus_{i \in I} X_i$ in $\mathcal{A}_c$. Set $Y_i = \operatorname{Im}(C \to X_i)$. Then Y_i belongs to $\mathcal{S}$. By Lemma 14.1.10, there exists $C_i \in \mathcal{S}$ such that $Y_i \subset C_i \subset X_i$ and that $H(C_i) \to H(X_i)$ is a monomorphism. Then $H(C_i)$ belongs to $\mathcal{I}$. By the result of Exercise 8.35, the morphism $C \to \bigoplus_i X_i$ factorizes through $\bigoplus_i Y_i \to \bigoplus_i X_i$, and hence through $\bigoplus_i C_i \to \bigoplus_i X_i$.   q.e.d.

14.3 Unbounded Derived Category

From now on and until the end of this chapter, $\mathcal{C}, \mathcal{C}'$, etc. are abelian categories.

We shall apply the results in the preceding Sects. 14.1 and 14.2 to the abelian category with translation $\mathcal{A} := \mathrm{Gr}(\mathcal{C})$. Then we have $\mathcal{A}_c \simeq \mathrm{C}(\mathcal{C})$, $\mathrm{K}_c(\mathcal{A}) \simeq \mathrm{K}(\mathcal{C})$ and $\mathrm{D}_c(\mathcal{A}) \simeq \mathrm{D}(\mathcal{C})$. Assume that $\mathcal{C}$ admits direct sums indexed by a set I and that such direct sums are exact. Then, clearly, $\mathrm{Gr}(\mathcal{C})$ has the same properties. It then follows from Proposition 14.1.1 that $\mathrm{C}(\mathcal{C})$, $\mathrm{K}(\mathcal{C})$ and $\mathrm{D}(\mathcal{C})$ also admit such direct sums and the two functors $\mathrm{C}(\mathcal{C}) \to \mathrm{K}(\mathcal{C})$ and $\mathrm{K}(\mathcal{C}) \to \mathrm{D}(\mathcal{C})$ commute with such direct sums.

We shall write $\mathrm{K}_{\mathrm{hi}}(\mathcal{C})$ for $\mathrm{K}_{c,\mathrm{hi}}(\mathcal{A})$. Hence $\mathrm{K}_{\mathrm{hi}}(\mathcal{C})$ is the full subcategory of $\mathrm{K}(\mathcal{C})$ consisting of homotopically injective objects. Let us denote by $\iota \colon \mathrm{K}_{\mathrm{hi}}(\mathcal{C}) \to \mathrm{K}(\mathcal{C})$ the embedding functor. Similarly we denote by $\mathrm{K}_{\mathrm{hp}}(\mathcal{C})$ the full subcategory of $\mathrm{K}(\mathcal{C})$ consisting of homotopically projective objects. Recall that $I \in \mathrm{K}(\mathcal{C})$ is homotopically injective if and only if $\mathrm{Hom}_{\mathrm{K}(\mathcal{C})}(X, I) \simeq 0$ for all $X \in \mathrm{K}(\mathcal{C})$ that is qis to 0.

Note that an object $I \in \mathrm{K}^+(\mathcal{C})$ whose components are all injective is homotopically injective in view of Lemma 13.2.4.

Let $\mathcal{C}$ be a Grothendieck abelian category. Then $\mathcal{A} := \mathrm{Gr}(\mathcal{C})$ is also a Grothendieck category. Applying Corollary 14.1.8 and Theorem 14.2.1, we get the following theorem.

Theorem 14.3.1. *Let $\mathcal{C}$ be a Grothendieck category.*

(i) *if $I \in \mathrm{K}(\mathcal{C})$ is homotopically injective, then we have an isomorphism*

$$\mathrm{Hom}_{\mathrm{K}(\mathcal{C})}(X, I) \xrightarrow{\sim} \mathrm{Hom}_{\mathrm{D}(\mathcal{C})}(X, I) \quad \textit{for any } X \in \mathrm{K}(\mathcal{C}) \,,$$

(ii) *for any $X \in \mathrm{C}(\mathcal{C})$, there exists a qis $X \to I$ such that I is homotopically injective,*

(iii) *the localization functor $Q \colon \mathrm{K}(\mathcal{C}) \to \mathrm{D}(\mathcal{C})$ induces an equivalence*

$$\mathrm{K}_{\mathrm{hi}}(\mathcal{C}) \xrightarrow{\sim} \mathrm{D}(\mathcal{C}) \,,$$

(iv) *the category $\mathrm{D}(\mathcal{C})$ is a $\mathcal{U}$-category,*

(v) *the functor $Q \colon \mathrm{K}(\mathcal{C}) \to \mathrm{D}(\mathcal{C})$ admits a right adjoint $R_q \colon \mathrm{D}(\mathcal{C}) \to \mathrm{K}(\mathcal{C})$, $Q \circ R_q \simeq \mathrm{id}$, and R_q is the composition of $\iota \colon \mathrm{K}_{\mathrm{hi}}(\mathcal{C}) \to \mathrm{K}(\mathcal{C})$ and a quasi-inverse of $Q \circ \iota$,*

(vi) *for any triangulated category $\mathcal{D}$, any triangulated functor $F \colon \mathrm{K}(\mathcal{C}) \to \mathcal{D}$ admits a right localization $RF \colon \mathrm{D}(\mathcal{C}) \to \mathcal{D}$ and $RF \simeq F \circ R_q$,*

(vii) *the triangulated category $\mathrm{D}(\mathcal{C})$ admits small direct sums and a system of t-generators,*

(viii) *any cohomological functor $G \colon (\mathrm{D}(\mathcal{C}))^{\mathrm{op}} \to \mathrm{Mod}(\mathbb{Z})$ is representable as soon as G commutes with small products (i.e., $G(\bigoplus_i X_i) \simeq \prod_i G(X_i)$ for any small family $\{X_i\}_i$ in $\mathrm{D}(\mathcal{C})$),*

(ix) *for any triangulated category $\mathcal{D}$, any triangulated functor $F \colon \mathrm{D}(\mathcal{C}) \to \mathcal{D}$ admits a right adjoint as soon as F commutes with small direct sums.*

Corollary 14.3.2. *Let k be a commutative ring and let C be a Grothendieck k-abelian category. Then $\left(\mathrm{K}_{\mathrm{hi}}(\mathcal{C}), \mathrm{K}(\mathcal{C})^{\mathrm{op}}\right)$ is $\mathrm{Hom}_{\mathcal{C}}$-injective, and the functor $\mathrm{Hom}_{\mathcal{C}}$ admits a right derived functor $\mathrm{RHom}_{\mathcal{C}} \colon \mathrm{D}(\mathcal{C}) \times \mathrm{D}(\mathcal{C})^{\mathrm{op}} \to \mathrm{D}(k)$.*
  Moreover, $H^0(\mathrm{RHom}_{\mathcal{C}}(X, Y)) \simeq \mathrm{Hom}_{\mathrm{D}(\mathcal{C})}(X, Y)$ for $X, Y \in \mathrm{D}(\mathcal{C})$.

Proof. (i) The functor $\mathrm{Hom}_{\mathcal{C}}$ defines a functor $\mathrm{Hom}_{\mathcal{C}}^{\bullet} \colon \mathrm{K}(\mathcal{C}) \times \mathrm{K}(\mathcal{C})^{\mathrm{op}} \to \mathrm{K}(k)$ and $H^0(\mathrm{Hom}_{\mathcal{C}}^{\bullet}) \simeq \mathrm{Hom}_{\mathrm{K}(\mathcal{C})}$ by Proposition 11.7.3. Let $I \in \mathrm{K}_{\mathrm{hi}}(\mathcal{C})$. If $X \in \mathrm{K}(\mathcal{C})$ is qis to 0, we find $\mathrm{Hom}_{\mathrm{K}(\mathcal{C})}(X, I) \simeq 0$. Moreover, if $I \in \mathrm{K}_{\mathrm{hi}}(\mathcal{C})$ is qis to 0, then I is isomorphic to 0. Therefore $\left(\mathrm{K}_{\mathrm{hi}}(\mathcal{C}), \mathrm{K}(\mathcal{C})^{\mathrm{op}}\right)$ is $\mathrm{Hom}_{\mathcal{C}}$-injective, and we can apply Corollary 10.3.11 to the functor $\mathrm{Hom}_{\mathcal{C}}^{\bullet} \colon \mathrm{K}(\mathcal{C}) \times \mathrm{K}(\mathcal{C})^{\mathrm{op}} \to \mathrm{K}(k)$ and conclude.

(ii) The last assertion follows from Theorem 13.4.1.   q.e.d.

Remark 14.3.3. Let $\mathcal{I}$ be a full subcategory of a Grothendieck category $\mathcal{C}$ and assume that $\mathcal{I}$ is closed by subobjects, quotients and extensions in $\mathcal{C}$, and also by small direct sums. Then by Proposition 14.2.4, the triangulated category $\mathrm{D}_{\mathcal{I}}(\mathcal{C})$ admits small direct sums and a system of t-generators. Hence $\mathrm{D}_{\mathcal{I}}(\mathcal{C}) \to \mathrm{D}(\mathcal{C})$ has a right adjoint.

We shall now give another criterion for the existence of derived functors in the unbounded case, when the functor has finite cohomological dimension.

Proposition 14.3.4. *Let C and C' be abelian categories and $F \colon C \to C'$ a left exact functor. Let $\mathcal{J}$ be an F-injective full additive subcategory of C satisfying the finiteness condition (13.2.1). Then*

(i) *$\mathrm{K}(\mathcal{J})$ is $\mathrm{K}(F)$-injective. In particular, the functor F admits a right derived functor $RF \colon \mathrm{D}(C) \to \mathrm{D}(C')$ and*

$$RF(X) \simeq \mathrm{K}(F)(Y) \quad for\ (X \to Y) \in \mathrm{Qis}\ with\ Y \in \mathrm{K}(\mathcal{J})\ .$$

(ii) *Assume that C and C' admit direct sums indexed by a set I and such direct sums are exact. (Hence, $\mathrm{D}(C)$ and $\mathrm{D}(C')$ admit such direct sums by Proposition 10.2.8.) If F commutes with direct sums indexed by I and $\mathcal{J}$ is closed by such direct sums, then $RF \colon \mathrm{D}(C) \to \mathrm{D}(C')$ commutes with such direct sums.*

Note that by Proposition 13.3.5, the conditions on the full additive subcategory $\mathcal{J}$ are equivalent to the conditions (a)–(c) below:

(14.3.1)
$$\begin{cases}
\text{(a)} & \mathcal{J} \text{ is cogenerating in } C, \\
\text{(b)} & \text{there exists a non-negative integer } d \text{ such that if } Y^0 \to \\
& Y^1 \to \cdots \to Y^d \to 0 \text{ is an exact sequence and } Y^j \in \mathcal{J} \\
& \text{for } j < d, \text{ then } Y^d \in \mathcal{J}, \\
\text{(c)} & \text{for any exact sequence } 0 \to X' \to X \to X'' \to 0 \text{ in } C \\
& \text{with } X', X \in \mathcal{J}, \text{ the sequence } 0 \to F(X') \to F(X) \to \\
& F(X'') \to 0 \text{ is exact.}
\end{cases}$$

Proof. (i) By Proposition 13.2.6, it remains to prove that if $X \in C(\mathcal{J})$ is exact, then $F(X)$ is exact. Consider the truncated complex

$$X^{i-d} \to \cdots \to X^{i-1} \to \operatorname{Coker} d_X^{i-1} \to 0 \ .$$

By the assumption, $\operatorname{Coker} d_X^{i-1}$ belongs to $\mathcal{J}$. Hence,

$$\tau^{\geq i} X := 0 \to \operatorname{Coker} d_X^{i-1} \to X^{i+1} \to \cdots$$

belongs to $\mathrm{K}^+(\mathcal{J})$ and is an exact complex. Therefore,

$$0 \to F(\operatorname{Coker} d_X^{i-1}) \to F(X^{i+1}) \to \cdots$$

is exact.
(ii) Let $\{X_i\}_{i \in I}$ be a family of objects in $C(\mathcal{C})$. For each $i \in I$, choose a qis $X_i \to Y_i$ with $Y_i \in C(\mathcal{J})$. Since direct sums indexed by I are exact in $C(\mathcal{C})$, $\bigoplus_i X_i \to \bigoplus_i Y_i$ is a qis, and by the hypothesis, $\bigoplus_i Y_i$ belongs to $C(\mathcal{J})$. Then $Q(\bigoplus_i X_i) \simeq \bigoplus_i Q(X_i)$ by Proposition 14.1.1 and

$$RF(\bigoplus_i X_i) \simeq F(\bigoplus_i Y_i) \simeq \bigoplus_i F(Y_i) \simeq \bigoplus_i RF(X_i)$$

in $D(\mathcal{C}')$.   q.e.d.

Corollary 14.3.5. *Let $\mathcal{C}$ and $\mathcal{C}'$ be abelian categories and let $F: \mathcal{C} \to \mathcal{C}'$ and $F': \mathcal{C}' \to \mathcal{C}''$ be left exact functors of abelian categories. Let $\mathcal{J}$ and $\mathcal{J}'$ be full additive subcategories of $\mathcal{C}$ and $\mathcal{C}'$ respectively, and assume that $\mathcal{J}$ satisfies the conditions* (a)–(c) *of* (14.3.1) *and similarly for $\mathcal{J}'$ with respect to $\mathcal{C}', \mathcal{C}'', F'$. Assume moreover that $F(\mathcal{J}) \subset \mathcal{J}'$. Then $R(F' \circ F) \simeq RF' \circ RF$.*

Remark 14.3.6. Applying Proposition 14.3.4 and Corollary 14.3.5 with $\mathcal{C}, \mathcal{C}'$ and $\mathcal{C}''$ replaced with the opposite categories, we obtain similar results for left derived functors of right exact functors.

By Proposition 14.3.4 together with Theorem 14.3.1, we obtain the following corollary.

Corollary 14.3.7. *Let $\mathcal{C}$ be a Grothendieck category and let $F: \mathcal{C} \to \mathcal{C}'$ be a left exact functor of abelian categories which commutes with small direct sums. Let $\mathcal{J}$ be a full additive subcategory of $\mathcal{C}$ satisfying the conditions* (a)– (c) *of* (14.3.1). *Assume moreover that $\mathcal{J}$ is closed by small direct sums. Then $RF: D(\mathcal{C}) \to D(\mathcal{C}')$ admits a right adjoint.*

14.4 Left Derived Functors

In this section, we shall give a criterion for the existence of the left derived functor $LG: D(\mathcal{C}) \to D(\mathcal{C}')$ of an additive functor $G: \mathcal{C} \to \mathcal{C}'$ of abelian categories, assuming that G admits a right adjoint.

Let $\mathcal{C}$ be an abelian category. We shall assume

(14.4.1) $\mathcal{C}$ admits small direct sums and small direct sums are exact in $\mathcal{C}$.

Hence, by Proposition 14.1.1, $C(\mathcal{C})$, $K(\mathcal{C})$ and $D(\mathcal{C})$ admit small direct sums. Note that Grothendieck categories satisfy (14.4.1).

Lemma 14.4.1. *Assume* (14.4.1) *and let $\mathcal{P}$ be a full additive generating subcategory of $\mathcal{C}$. For any $X \in C(\mathcal{C})$, there exists a quasi-isomorphism $X' \to X$ such that X' is the mapping cone of a morphism $Q \to P$, where P and Q are countable direct sums of objects of $C^-(\mathcal{P})$.*

Proof. By Lemma 13.2.1 (with the arrows reversed), for each $n \in \mathbb{Z}$, there exists a quasi-isomorphism $p_n \colon P_n \to \tau^{\leq n} X$ with $P_n \in C^-(\mathcal{P})$. Then there exists a quasi-isomorphism

$$Q_n \to \mathrm{Mc}\big(P_n \oplus P_{n+1} \xrightarrow{(p_n, -p_{n+1})} \tau^{\leq n+1} X\big)[-1]$$

with $Q_n \in C^-(\mathcal{P})$. Hence, we have a commutative diagram in $K^-(\mathcal{C})$:

$$
\begin{array}{ccccc}
Q_n & \longrightarrow & P_n & \longrightarrow & \tau^{\leq n} X \\
& \searrow & & & \downarrow \\
& & P_{n+1} & \longrightarrow & \tau^{\leq n+1} X \ .
\end{array}
$$

By the octahedral axiom of triangulated categories, there exists a d.t. in $K(\mathcal{C})$

$$\mathrm{Mc}(P_{n+1} \to P_n \oplus P_{n+1}) \to \mathrm{Mc}(P_{n+1} \to \tau^{\leq n+1} X)$$
$$\to \mathrm{Mc}(P_n \oplus P_{n+1} \to \tau^{\leq n+1} X) \xrightarrow{+1} \ .$$

Since $P_{n+1} \to \tau^{\leq n+1} X$ is a qis, the morphism

$$\mathrm{Mc}(P_n \oplus P_{n+1} \to \tau^{\leq n+1} X)[-1] \to \mathrm{Mc}(P_{n+1} \to P_n \oplus P_{n+1})$$

is an isomorphism in $D(\mathcal{C})$. Hence, $Q_n \to P_n$ is a qis.
    Set $Q = \bigoplus_{n \in \mathbb{Z}} Q_n$ and $P = \bigoplus_{n \in \mathbb{Z}} P_n$. Then $Q_n \to P_n$ and $Q_n \to P_{n+1}$ define morphisms $u_0, u_1 \colon Q \to P$. Set

$$R := \mathrm{Mc}(Q \xrightarrow{u_0 - u_1} P) \ .$$

There is a d.t. $Q \to P \to R \to Q[1]$. Since the composition $Q \xrightarrow{u_0 - u_1} P \to X$ is zero in $K(\mathcal{C})$, $P \to X$ factors as $P \to R \to X$ in $K(\mathcal{C})$. Let us show that $R \to X$ is a qis. For $i \in \mathbb{Z}$, set $\varphi_i := H^i(u_0 - u_1)$. We have an exact sequence

$$H^i(Q) \xrightarrow{\varphi_i} H^i(P) \to H^i(R) \to H^{i+1}(Q) \xrightarrow{\varphi_{i+1}} H^{i+1}(P) \ .$$

The hypothesis (14.4.1) implies

$$H^i(Q) \simeq \bigoplus_{n \in \mathbb{Z}} H^i(Q_n) \simeq \bigoplus_{i \leq n} H^i(X) ,$$

$$H^i(P) \simeq \bigoplus_{n \in \mathbb{Z}} H^i(P_n) \simeq \bigoplus_{i \leq n} H^i(X) .$$

Hence, φ_{i+1} is a monomorphism by Exercise 8.37. Note that $\mathrm{id} - \sigma$ in Exercise 8.37 corresponds to φ_i and $X_0 \to X_1 \to \cdots$ corresponds to $H^i(X) \overset{\mathrm{id}}{\to} H^i(X) \to \cdots$. Therefore, $\mathrm{Coker}\, \varphi_i \simeq \varinjlim_n H^i(P_n) \to H^i(X)$ is an isomorphism.

Hence, $H^i(R) \to H^i(X)$ is an isomorphism.   q.e.d.

Lemma 14.4.2. *Assume* (14.4.1). *Let $\mathcal{P}$ be the full subcategory of $\mathcal{C}$ consisting of projective objects and let $\widetilde{\mathcal{P}}$ be the smallest full triangulated subcategory of* $\mathrm{K}(\mathcal{C})$ *closed by small direct sums and containing* $\mathrm{K}^-(\mathcal{P})$. *Then any object of $\widetilde{\mathcal{P}}$ is homotopically projective.*

Proof. The full subcategory $\mathrm{K}_{\mathrm{hp}}(\mathcal{C})$ of $\mathrm{K}(\mathcal{C})$ consisting of homotopically projective objects is closed by small direct sums and contains $\mathrm{K}^-(\mathcal{P})$. Hence, it contains $\widetilde{\mathcal{P}}$.   q.e.d.

Theorem 14.4.3. *Let $\mathcal{C}$ be an abelian category satisfying* (14.4.1) *and admitting enough projectives. Then,*

(i) *for any $X \in \mathrm{K}(\mathcal{C})$, there exist $P \in \mathrm{K}_{\mathrm{hp}}(\mathcal{C})$ and a qis $P \to X$,*
(ii) *for any additive functor $G \colon \mathcal{C} \to \mathcal{C}'$, the left derived functor $LG \colon \mathrm{D}(\mathcal{C}) \to \mathrm{D}(\mathcal{C}')$ exists, and $LG(X) \simeq G(X)$ if X is homotopically projective.*

Proof. Apply Lemmas 14.4.1 and 14.4.2.   q.e.d.

By reversing the arrows in Theorem 14.4.3, we obtain

Theorem 14.4.4. *Let $\mathcal{C}$ be an abelian category. Assume that $\mathcal{C}$ admits enough injectives, small products exist in $\mathcal{C}$ and such products are exact in $\mathcal{C}$. Then*

(i) *for any $X \in \mathrm{K}(\mathcal{C})$, there exist $I \in \mathrm{K}_{\mathrm{hi}}(\mathcal{C})$ and a qis $X \to I$,*
(ii) *for any additive functor $F \colon \mathcal{C} \to \mathcal{C}'$, the right derived functor $RF \colon \mathrm{D}(\mathcal{C}) \to \mathrm{D}(\mathcal{C}')$ exists, and $RF(X) \simeq F(X)$ if X is homotopically injective.*

Note that Grothendieck categories always admit small products, but small products may not be exact.

Theorem 14.4.5. *Let k be a commutative ring and let $G \colon \mathcal{C} \to \mathcal{C}'$ and $F \colon \mathcal{C}' \to \mathcal{C}$ be k-additive functors of k-abelian categories such that (G, F) is a pair of adjoint functors. Assume that $\mathcal{C}'$ is a Grothendieck category and $\mathcal{C}$ satisfies* (14.4.1). *Let $\mathcal{P}$ be a G-projective full subcategory of $\mathcal{C}$.*

(a) *Let $\widetilde{\mathcal{P}}$ be the smallest full triangulated subcategory of $\mathrm{K}(\mathcal{C})$ closed by small direct sums and containing $\mathrm{K}^-(\mathcal{P})$. Then $\widetilde{\mathcal{P}}$ is $\mathrm{K}(G)$-projective.*
(b) *The left derived functor $LG \colon \mathrm{D}(\mathcal{C}) \to \mathrm{D}(\mathcal{C}')$ exists and (LG, RF) is a pair of adjoint functors.*

(c) *We have an isomorphism in* $D(k)$, *functorial with respect to* $X \in D(\mathcal{C})$ *and* $Y \in D(\mathcal{C}')$:

$$\mathrm{RHom}_{\mathcal{C}}(X, RF(Y)) \simeq \mathrm{RHom}_{\mathcal{C}'}(LG(X), Y) \,.$$

Proof. (i) Let us denote by $\widetilde{\mathcal{P}}'$ the full subcategory of $K(\mathcal{C})$ consisting of objects X such that

(14.4.2) $$\mathrm{Hom}_{K(\mathcal{C})}(X, F(I)) \to \mathrm{Hom}_{D(\mathcal{C})}(X, F(I))$$

is bijective for any homotopically injective object $I \in C(\mathcal{C}')$. Then $\widetilde{\mathcal{P}}'$ is a triangulated subcategory of $K(\mathcal{C})$ closed by small direct sums.
(ii) Let us show that $\widetilde{\mathcal{P}}'$ contains $K^-(\mathcal{P})$. If $X \in K^-(\mathcal{P})$, then $\mathrm{Qis}_X \cap K^-(\mathcal{P})_X$ is co-cofinal to Qis_X, and hence we have

$$\mathrm{Hom}_{D(\mathcal{C})}(X, F(I)) \simeq \varinjlim_{(X' \to X) \in \mathrm{Qis}, X' \in K^-(\mathcal{P})} \mathrm{Hom}_{K(\mathcal{C})}(X', F(I)) \,.$$

Let $X' \to X$ be a qis with $X' \in K^-(\mathcal{P})$. Let X'' be the mapping cone of $X' \to X$. Then X'' is an exact complex in $K^-(\mathcal{P})$. Hence

$$\mathrm{Hom}_{K(\mathcal{C})}(X'', F(I)) \simeq \mathrm{Hom}_{K(\mathcal{C}')}(G(X''), I) \simeq 0 \,,$$

where the second isomorphism follows from the fact that $\mathcal{P}$ being G-projective, $G(X'')$ is an exact complex. Hence, for $X, X' \in K^-(\mathcal{P})$ and for a qis $X' \to X$, the map $\mathrm{Hom}_{K(\mathcal{C})}(X, F(I)) \to \mathrm{Hom}_{K(\mathcal{C})}(X', F(I))$ is bijective. It follows that the map in (14.4.2) is bijective.
(iii) By (ii), $\widetilde{\mathcal{P}}'$ contains $\widetilde{\mathcal{P}}$.
(iv) We shall prove that if $X \in \widetilde{\mathcal{P}}'$ is exact, then $G(X) \simeq 0$ in $D(\mathcal{C}')$. Indeed, for any homotopically injective object I in $C(\mathcal{C}')$, we have

$$\mathrm{Hom}_{D(\mathcal{C}')}(G(X), I) \simeq \mathrm{Hom}_{K(\mathcal{C}')}(G(X), I) \simeq \mathrm{Hom}_{K(\mathcal{C})}(X, F(I))$$
$$\simeq \mathrm{Hom}_{D(\mathcal{C})}(X, F(I)) \simeq 0 \,.$$

(v) By Lemma 14.4.1, for every $X \in C(\mathcal{C})$, there exists a quasi-isomorphism $P \to X$ with $P \in \widetilde{\mathcal{P}}$. Hence $\widetilde{\mathcal{P}}$ is $K(G)$-projective and LG exists. Moreover, we have $LG(X) \simeq G(X)$ for any $X \in \widetilde{\mathcal{P}}$. For a homotopically injective object $I \in C(\mathcal{C}')$ and $X \in \widetilde{\mathcal{P}}$, we have

$$\mathrm{RHom}_{\mathcal{C}}(X, RF(I)) \simeq \mathrm{Hom}_{\mathcal{C}}^{\bullet}(X, F(I))$$
$$\simeq \mathrm{Hom}_{\mathcal{C}'}^{\bullet}(G(X), I) \simeq \mathrm{RHom}_{\mathcal{C}'}(LG(X), I) \,.$$

Hence we obtain (c). By taking the cohomologies, we obtain (b).   q.e.d.

Corollary 14.4.6. *Let $\mathcal{C}$ and $\mathcal{C}'$ be Grothendieck categories and let $G : \mathcal{C} \to \mathcal{C}'$ be an additive functor commuting with small inductive limits. Assume that there exists a G-projective subcategory $\mathcal{P}$ of $\mathcal{C}$. Then*

(i) $LG \colon \mathrm{D}(\mathcal{C}) \to \mathrm{D}(\mathcal{C}')$ *exists and commutes with small direct sums,*
(ii) *for any small filtrant inductive system* $\alpha \colon I \to \mathcal{C}$, $\varinjlim H^n(LG(\alpha)) \to$
   $H^n(LG(\varinjlim \alpha))$ *is an isomorphism for all* $n \in \mathbb{Z}$.

Proof. (i) By Theorem 8.3.27, G admits a right adjoint functor and we may apply Theorem 14.4.5.
(ii) Let $\mathcal{P}'$ be the full subcategory of $\mathcal{C}$ consisting of left G-acyclic objects (see Remark 13.3.6). Then $\mathcal{P}'$ is also G-projective by Lemma 13.3.12 and closed by small direct sums by (i). For each $i \in I$, let us take an epimorphism $P_i \twoheadrightarrow \alpha(i)$ with $P_i \in \mathcal{P}'$. Set $p_0(i) = \bigoplus_{i' \to i} P_{i'}$. Then $p_0 \colon I \to \mathcal{C}$ is a functor and $p_0 \to \alpha$ is an epimorphism in $\mathrm{Fct}(I, \mathcal{C})$. It is easily checked that $\varinjlim p_0 \simeq \bigoplus_{i \in I} P_i$ (see Exercise 2.21). By this procedure, we construct an exact sequence in $\mathrm{Fct}(I, \mathcal{C})$

$$(14.4.3) \qquad \cdots \to p_{n+1} \to p_n \to \cdots \to p_0 \to \alpha \to 0$$

such that any $p_k(i)$ as well as $\varinjlim_i p_k(i)$ belongs to $\mathcal{P}'$.

Define the complex in $\mathrm{Fct}(I, \mathcal{C})$

$$p_\bullet := \cdots \to p_{n+1} \to p_n \to \cdots \to p_0 \to 0 \,.$$

Hence we have

$$H^n(LG(\varinjlim \alpha)) \simeq H^n(G(\varinjlim p_\bullet)) \simeq \varinjlim H^n(G(p_\bullet)) \simeq \varinjlim H^n(LG(\alpha)) \,.$$

q.e.d.

Proposition 14.4.7. *Let* $\mathcal{C}$, $\mathcal{C}'$, $\mathcal{C}''$ *be Grothendieck categories and let* $F \colon \mathcal{C} \to \mathcal{C}'$, $F' \colon \mathcal{C}' \to \mathcal{C}''$, $G \colon \mathcal{C}' \to \mathcal{C}$, $G' \colon \mathcal{C}'' \to \mathcal{C}'$ *be additive functors such that* (G, F) *and* (G', F') *are pairs of adjoint functors. Assume that there exist a* G-*projective subcategory* $\mathcal{P}'$ *of* $\mathcal{C}'$ *and a* G'-*projective subcategory* $\mathcal{P}''$ *of* $\mathcal{C}''$ *such that* $G'(\mathcal{P}'') \subset \mathcal{P}'$. *Then* $R(F' \circ F) \to RF' \circ RF$ *and* $LG \circ LG' \to L(G \circ G')$ *are isomorphisms of functors.*

Proof. Since $R(F' \circ F)$, RF', RF are left adjoint functors to $L(G \circ G')$, LG', LG, it is enough to prove the isomorphism $LG \circ LG' \xrightarrow{\sim} L(G \circ G')$. Let $\widetilde{\mathcal{P}}''$ (resp. $\widetilde{\mathcal{P}}'$) denote the smallest full triangulated subcategory of $\mathrm{K}(\mathcal{C}'')$ (resp. $\mathrm{K}(\mathcal{C}')$) closed by small direct sums and containing $\mathrm{K}^-(\mathcal{P}'')$ (resp. $\mathrm{K}^-(\mathcal{P}')$). Then $\widetilde{\mathcal{P}}''$ (resp. $\widetilde{\mathcal{P}}'$) is projective with respect to the functor $\mathrm{K}(G')$ (resp. $\mathrm{K}(G)$). Moreover, $\mathrm{K}(G')(\widetilde{\mathcal{P}}'') \subset \widetilde{\mathcal{P}}'$. Hence $LG \circ LG' \to L(G \circ G')$ is an isomorphism by Proposition 10.3.5. q.e.d.

Theorem 14.4.8. *Let* k *be a commutative ring and let* $\mathcal{C}_1$, $\mathcal{C}_2$ *and* $\mathcal{C}_3$ *be* k-*abelian categories. We assume that* $\mathcal{C}_3$ *is a Grothendieck category and that* $\mathcal{C}_1$ *and* $\mathcal{C}_2$ *satisfy* (14.4.1). *Let* $G \colon \mathcal{C}_1 \times \mathcal{C}_2 \to \mathcal{C}_3$, $F_1 \colon \mathcal{C}_2^{\mathrm{op}} \times \mathcal{C}_3 \to \mathcal{C}_1$ *and*

$F_2\colon \mathcal{C}_1^{\mathrm{op}} \times \mathcal{C}_3 \to \mathcal{C}_2$ *be k-additive functors. Assume that there are isomorphisms, functorial with respect to* $X_i \in \mathcal{C}_i$ $(i = 1, 2, 3)$:

(14.4.4)
$$\mathrm{Hom}_{\mathcal{C}_3}(G(X_1, X_2), X_3) \simeq \mathrm{Hom}_{\mathcal{C}_1}(X_1, F_1(X_2, X_3))$$
$$\simeq \mathrm{Hom}_{\mathcal{C}_2}(X_2, F_2(X_1, X_3)) \,.$$

Let $\mathrm{K}(G)\colon \mathrm{K}(\mathcal{C}_1) \times \mathrm{K}(\mathcal{C}_2) \to \mathrm{K}(\mathcal{C}_3)$ *be the triangulated functor associated with* $\mathrm{tot}_\oplus G(X_1, X_2)$, *and let* $\mathrm{K}(F_1)\colon \mathrm{K}(\mathcal{C}_2)^{\mathrm{op}} \times \mathrm{K}(\mathcal{C}_3) \to \mathrm{K}(\mathcal{C}_1)$ *be the triangulated functor associated with* $\mathrm{tot}_\pi F_1(X_2, X_3)$ *and similarly for* $\mathrm{K}(F_2)$.
Let $\mathcal{P}_i \subset \mathcal{C}_i$ $(i = 1, 2)$ *be a full subcategory such that* $(\mathcal{P}_1, \mathcal{P}_2)$ *is* $\mathrm{K}(G)$-*projective. Denote by* $\widetilde{\mathcal{P}}_i$ *the smallest full triangulated subcategory of* $\mathrm{K}(\mathcal{C}_i)$ *that contains* $\mathrm{K}^-(\mathcal{P}_i)$ *and is closed by small direct sums* $(i = 1, 2)$.
    Then:

(i) $(\widetilde{\mathcal{P}}_1, \widetilde{\mathcal{P}}_2)$ *is* $\mathrm{K}(G)$-*projective. In particular* $LG\colon \mathrm{D}(\mathcal{C}_1) \times \mathrm{D}(\mathcal{C}_2) \to \mathrm{D}(\mathcal{C}_3)$ *exists and* $LG(X_1, X_2) \simeq \mathrm{K}(G)(X_1, X_2)$ *for* $X_1 \in \widetilde{\mathcal{P}}_1$ *and* $X_2 \in \widetilde{\mathcal{P}}_2$.
(ii) $(\widetilde{\mathcal{P}}_2^{\mathrm{op}}, \mathrm{K}_{\mathrm{hi}}(\mathcal{C}_3))$ *is* $\mathrm{K}(F_1)$-*injective. In particular,* $RF_1\colon \mathrm{D}(\mathcal{C}_2)^{\mathrm{op}} \times \mathrm{D}(\mathcal{C}_3) \to \mathrm{D}(\mathcal{C}_1)$ *exists and* $RF_1(X_2, X_3) \simeq \mathrm{K}(F_1)(X_2, X_3)$ *for* $X_2 \in \widetilde{\mathcal{P}}_2$ *and* $X_3 \in \mathrm{K}_{\mathrm{hi}}(\mathcal{C}_3)$. *Similar statements hold for* F_2.
(iii) *There are isomorphisms, functorial with respect to* $X_i \in \mathrm{D}(\mathcal{C}_i)$ $(i = 1, 2, 3)$

(14.4.5) $\mathrm{Hom}_{\mathrm{D}(\mathcal{C}_3)}(LG(X_1, X_2), X_3) \simeq \mathrm{Hom}_{\mathrm{D}(\mathcal{C}_1)}(X_1, RF_1(X_2, X_3))$
$$\simeq \mathrm{Hom}_{\mathrm{D}(\mathcal{C}_2)}(X_2, RF_2(X_1, X_3)),$$

and

(14.4.6) $\mathrm{RHom}_{\mathcal{C}_3}(LG(X_1, X_2), X_3) \simeq \mathrm{RHom}_{\mathcal{C}_1}(X_1, RF_1(X_2, X_3))$
$$\simeq \mathrm{RHom}_{\mathcal{C}_2}(X_2, RF_2(X_1, X_3)) \,.$$

(iv) *Moreover, if* $\mathcal{P}_i = \mathcal{C}_i$ *for* $i = 1$ *or* $i = 2$, *we can take* $\widetilde{\mathcal{P}}_i = \mathrm{K}(\mathcal{C}_i)$ *in* (i) *and* (ii).

Proof. In the sequel, we shall write for short G and F_i instead of $\mathrm{K}(G)$ and $\mathrm{K}(F_i)$, respectively. The isomorphism (14.4.4) gives rise to an isomorphism

(14.4.7) $\mathrm{Hom}_{\mathrm{K}(\mathcal{C}_3)}(G(X_1, X_2), X_3) \simeq \mathrm{Hom}_{\mathrm{K}(\mathcal{C}_1)}(X_1, F_1(X_2, X_3))$

functorial with respect to $X_i \in \mathrm{K}(\mathcal{C}_i)$ $(i = 1, 2, 3)$.
    Note also that for any $X_2 \in \mathcal{C}_2$, the functor $X_1 \mapsto G(X_1, X_2)$ commutes with small direct sums. Indeed this functor has a right adjoint $X_3 \mapsto F_1(X_2, X_3)$.

(a) Let us first prove the following statement:

(14.4.8)   if $X_1 \in \mathrm{K}^-(\mathcal{P}_1)$ is an exact complex and $X_2 \in \widetilde{\mathcal{P}}_2$,
            then $G(X_1, X_2)$ is exact.

Indeed, for such an X_1, the category

$$\widetilde{\mathcal{P}}_2' = \left\{Y \in K(\mathcal{C}_2)\,;\, G(X_1, Y) \text{ is exact}\right\}$$

is a triangulated subcategory of $K(\mathcal{C}_2)$ which contains $K^-(\mathcal{P}_2)$ and is closed by small direct sums. Hence, $\widetilde{\mathcal{P}}_2'$ contains $\widetilde{\mathcal{P}}_2$.

(b) Set

$$\widetilde{\mathcal{P}}_1' = \{X_1 \in K(\mathcal{C}_1); \operatorname{Hom}_{K(\mathcal{C}_1)}(X_1, F_1(X_2, X_3)) \to \operatorname{Hom}_{D(\mathcal{C}_1)}(X_1, F_1(X_2, X_3))$$
$$\text{is an isomorphism for all } X_2 \in \widetilde{\mathcal{P}}_2,\, X_3 \in K_{\mathrm{hi}}(\mathcal{C}_3)\}\,.$$

Let us show that $\widetilde{\mathcal{P}}_1 \subset \widetilde{\mathcal{P}}_1'$.

Since the category $\widetilde{\mathcal{P}}_1'$ is a full triangulated subcategory of $K(\mathcal{C}_1)$ closed by small direct sums, it is enough to show that $K^-(\mathcal{P}_1) \subset \widetilde{\mathcal{P}}_1'$. If $Y_1 \in K^-(\mathcal{P}_1)$ is exact, then

$$(14.4.9) \quad \operatorname{Hom}_{K(\mathcal{C}_1)}(Y_1, F_1(X_2, X_3)) \simeq \operatorname{Hom}_{K(\mathcal{C}_3)}(G(Y_1, X_2), X_3) \simeq 0\,,$$

where the last isomorphism follows from (14.4.8) and $X_3 \in K_{\mathrm{hi}}(\mathcal{C}_3)$. Hence, if $X_1' \to X_1$ is a qis in $K^-(\mathcal{P}_1)$, then

$$\operatorname{Hom}_{K(\mathcal{C}_1)}(X_1, F_1(X_2, X_3)) \xrightarrow{\sim} \operatorname{Hom}_{K(\mathcal{C}_1)}(X_1', F_1(X_2, X_3))\,.$$

Hence we obtain for any $X_1 \in K^-(\mathcal{P}_1)$

$$\operatorname{Hom}_{D(\mathcal{C}_1)}(X_1, F_1(X_2, X_3))$$
$$(14.4.10) \qquad \simeq \varinjlim_{(X_1' \to X_1) \in \operatorname{Qis} \cap K^-(\mathcal{P}_1)} \operatorname{Hom}_{K(\mathcal{C}_1)}(X_1', F_1(X_2, X_3))$$
$$\simeq \operatorname{Hom}_{K(\mathcal{C}_1)}(X_1, F_1(X_2, X_3))\,.$$

Thus $K^-(\mathcal{P}_1) \subset \widetilde{\mathcal{P}}_1'$ and hence $\widetilde{\mathcal{P}}_1 \subset \widetilde{\mathcal{P}}_1'$.

(c) Next let us show

$$(14.4.11) \quad \begin{array}{l} \text{for } X_i \in \widetilde{\mathcal{P}}_i \ (i = 1, 2) \text{ and } X_3 \in K_{\mathrm{hi}}(\mathcal{C}_3), \text{ we have} \\ \operatorname{Hom}_{D(\mathcal{C}_3)}(G(X_1, X_2), X_3) \simeq \operatorname{Hom}_{D(\mathcal{C}_1)}(X_1, F_1(X_2, X_3))\,. \end{array}$$

There are isomorphisms

$$\operatorname{Hom}_{D(\mathcal{C}_3)}(G(X_1, X_2), X_3) \simeq \operatorname{Hom}_{K(\mathcal{C}_3)}(G(X_1, X_2), X_3)$$
$$\simeq \operatorname{Hom}_{K(\mathcal{C}_1)}(X_1, F_1(X_2, X_3))$$
$$\simeq \operatorname{Hom}_{D(\mathcal{C}_1)}(X_1, F_1(X_2, X_3))\,.$$

Here the first isomorphism follows from $X_3 \in K_{\mathrm{hi}}(\mathcal{C}_3)$ and the last isomorphism follows from $\widetilde{\mathcal{P}}_1 \subset \widetilde{\mathcal{P}}_1'$.

(d) Let us prove (i). It is enough to show that for $X_i \in \widetilde{\mathcal{P}}_i$ $(i = 1, 2)$, $G(X_1, X_2)$ is exact as soon as X_1 or X_2 is exact. Assume that X_1 is exact. Then, for any $X_3 \in \mathrm{K}_{\mathrm{hi}}(\mathcal{C}_3)$, we have by (14.4.11)

$$\mathrm{Hom}_{\mathrm{D}(\mathcal{C}_3)}(G(X_1, X_2), X_3) \simeq \mathrm{Hom}_{\mathrm{D}(\mathcal{C}_1)}(X_1, F_1(X_2, X_3)) \simeq 0 \,.$$

This implies that $G(X_1, X_2)$ is exact. The proof in the case where X_2 is exact is similar.

(e) Let us prove (ii). It is enough to show that for $X_2 \in \widetilde{\mathcal{P}}_2$ and $X_3 \in \mathrm{K}_{\mathrm{hi}}(\mathcal{C}_3)$, $F_1(X_2, X_3)$ is exact as soon as X_2 or X_3 is exact.
(e1) Assume that X_2 is exact. For any $X_1 \in \widetilde{\mathcal{P}}_1$, $G(X_1, X_2)$ is exact by (i), and hence $\mathrm{Hom}_{\mathrm{D}(\mathcal{C}_1)}(X_1, F_1(X_2, X_3)) \simeq \mathrm{Hom}_{\mathrm{D}(\mathcal{C}_3)}(G(X_1, X_2), X_3)$ vanishes. This implies that $F_1(X_2, X_3)$ is exact.
(e2) Assume that $X_3 \in \mathrm{K}_{\mathrm{hi}}(\mathcal{C}_3)$ is exact. Then $X_3 \simeq 0$ in $\mathrm{K}(\mathcal{C}_3)$ and $F_1(X_2, X_3)$ is exact.

(f) Let us show (iii). The isomorphisms (14.4.5) immediately follow from (14.4.11). The adjunction morphism $X_1 \to RF_1(X_2, LG(X_1, X_2))$ induces the morphisms

$$\mathrm{RHom}_{\mathcal{C}_3}(LG(X_1, X_2), X_3) \to \mathrm{RHom}_{\mathcal{C}_1}\big(RF_1(X_2, LG(X_1, X_2)), RF_1(X_2, X_3)\big)$$
$$\to \mathrm{RHom}_{\mathcal{C}_1}(X_1, RF_1(X_2, X_3)) \,.$$

By taking the cohomologies, it induces isomorphisms by (14.4.5) and Theorem 13.4.1.

(g) Let us prove (iv). Assume $\mathcal{P}_1 = \mathrm{K}^-(\mathcal{C}_1)$.
(g1) Let us show that $(\mathrm{K}(\mathcal{C}_1), \widetilde{\mathcal{P}}_2)$ is $\mathrm{K}(G)$-projective. For that purpose it is enough to show that $G(X_1, X_2)$ is exact for $X_1 \in \mathrm{K}(\mathcal{C}_1)$ and $X_2 \in \widetilde{\mathcal{P}}_2$ as soon as X_1 or X_2 is exact. Since $\tau^{\leq n} X_1$ or X_2 is exact and $(\widetilde{\mathcal{P}}_1, \widetilde{\mathcal{P}}_2)$ is $\mathrm{K}(G)$-projective, $G(\tau^{\leq n} X_1, X_2)$ is exact. Hence $G(X_1, X_2) \simeq \varinjlim_n G(\tau^{\leq n} X_1, X_2)$ is exact.

(g2) Let us show that $(\mathrm{K}(\mathcal{C}_1)^{\mathrm{op}}, \mathrm{K}_{\mathrm{hi}}(\mathcal{C}_3))$ is $\mathrm{K}(F_2)$-injective. Let $X_1 \in \mathrm{K}(\mathcal{C}_1)$ and $X_3 \in \mathrm{K}_{\mathrm{hi}}(\mathcal{C}_3)$. If X_3 is exact, then $X_3 \simeq 0$, and hence $F_2(X_1, X_3)$ is exact. If $X_1 \in \mathrm{K}(\mathcal{C}_1)$ is exact, then for any $X_2 \in \widetilde{\mathcal{P}}_2$ we have

$$\mathrm{Hom}_{\mathrm{K}(\mathcal{C}_2)}(X_2, F_2(X_1, X_3)) \simeq \mathrm{Hom}_{\mathrm{K}(\mathcal{C}_3)}(G(X_1, X_2), X_3) \simeq 0 \,,$$

where the last isomorphism follows from the fact that $G(X_1, X_2)$ is exact by (g1). Hence $F_2(X_1, X_3)$ is exact.   q.e.d.

Corollary 14.4.9. *Let $\mathcal{C}_1$, $\mathcal{C}_2$ and $\mathcal{C}_3$ be Grothendieck categories. Let $G\colon \mathcal{C}_1 \times \mathcal{C}_2 \to \mathcal{C}_3$ be an additive functor which commutes with small inductive limits with respect to each variable. Let $\mathcal{P}_i \subset \mathcal{C}_i$ $(i = 1, 2)$ be a full subcategory such that $(\mathcal{P}_1, \mathcal{P}_2)$ is G-projective. Denote by $\widetilde{\mathcal{P}}_i$ the smallest full triangulated subcategory of $\mathrm{K}(\mathcal{C}_i)$ that contains $\mathrm{K}^-(\mathcal{P}_i)$ and is closed by small direct sums $(i = 1, 2)$. Let $\mathrm{K}(G)\colon \mathrm{K}(\mathcal{C}_1) \times \mathrm{K}(\mathcal{C}_2) \to \mathrm{K}(\mathcal{C}_3)$ be the functor associated with $\mathrm{tot}_\oplus G(X_1, X_2)$. Then*

(i) $(\widetilde{\mathcal{P}}_1, \widetilde{\mathcal{P}}_2)$ *is* $K(G)$-*projective. In particular* $LG \colon D(\mathcal{C}_1) \times D(\mathcal{C}_2) \to D(\mathcal{C}_3)$
   exists and $LG(X_1, X_2) \simeq G(X_1, X_2)$ *for* $X_1 \in \widetilde{\mathcal{P}}_1$ *and* $X_2 \in \widetilde{\mathcal{P}}_2$.
(ii) LG *commutes with small direct sums.*
(iii) *Moreover, if* $\mathcal{P}_i = \mathcal{C}_i$ *for* $i = 1$ *or* $i = 2$*, we can take* $\widetilde{\mathcal{P}}_i = K(\mathcal{C}_i)$.

Proof. By Theorem 8.3.27, the two functors $X_1 \mapsto G(X_1, X_2)$ and $X_2 \mapsto$
$G(X_1, X_2)$ have right adjoints.   q.e.d.

Example 14.4.10. Let R denote a k-algebra. The functor $\cdot \otimes_R \cdot \colon \mathrm{Mod}(R^{\mathrm{op}}) \times$
$\mathrm{Mod}(R) \to \mathrm{Mod}(k)$ defines a functor

$$(14.4.12) \quad \cdot \otimes_R \cdot \colon K(\mathrm{Mod}(R^{\mathrm{op}})) \times K(\mathrm{Mod}(R)) \to K(\mathrm{Mod}(k)) \,,$$
$$(X^\bullet, Y^\bullet) \mapsto \mathrm{tot}_\oplus(X^\bullet \otimes_R Y^\bullet) \,.$$

Then

$$\mathrm{Hom}_k(N \otimes_R M, L) \simeq \mathrm{Hom}_{R^{\mathrm{op}}}(N, \mathrm{Hom}_k(M, L))$$
$$\simeq \mathrm{Hom}_R(M, \mathrm{Hom}_k(N, L))$$

for any $N \in \mathrm{Mod}(R^{\mathrm{op}})$, $M \in \mathrm{Mod}(R)$ and $L \in \mathrm{Mod}(k)$.
   Let $\mathcal{P}_{proj}$ denote the full additive subcategory of $\mathrm{Mod}(R)$ consisting of
projective modules and $\widetilde{\mathcal{P}}_{proj}$ the smallest full triangulated subcategory of
$K(\mathrm{Mod}(R))$ closed by small direct sums and containing $K^-(\mathcal{P}_{proj})$. We may
apply Theorem 14.4.8 with $\mathcal{C}_1 = \mathrm{Mod}(R^{\mathrm{op}})$, $\mathcal{C}_2 = \mathrm{Mod}(R)$ and $\mathcal{C}_3 = \mathrm{Mod}(k)$.
Then $(K(\mathrm{Mod}(R^{\mathrm{op}})), \widetilde{\mathcal{P}}_{proj})$ is $(\cdot \otimes_R \cdot)$-projective and the functor in (14.4.12)
admits a left derived functor

$$\cdot \overset{\mathrm{L}}{\otimes}_R \cdot \colon D(R^{\mathrm{op}}) \times D(R) \to D(k) \,,$$

and

$$N \overset{\mathrm{L}}{\otimes}_R M \simeq \mathrm{tot}_\oplus(N \otimes_R P) \quad \text{for} \quad P \in \widetilde{\mathcal{P}}_{proj}, \quad (P \to M) \in \mathrm{Qis} \,.$$

Moreover, the functor

$$\mathrm{Hom}_k(\cdot, \cdot) \colon K(\mathrm{Mod}(R))^{\mathrm{op}} \times K(\mathrm{Mod}(k)) \to K(\mathrm{Mod}(R^{\mathrm{op}}))$$

admits a right adjoint functor and we have

$$\mathrm{RHom}_k(N \overset{\mathrm{L}}{\otimes}_R M, L) \simeq \mathrm{RHom}_{R^{\mathrm{op}}}(N, \mathrm{RHom}_k(M, L))$$
$$\simeq \mathrm{RHom}_R(M, \mathrm{RHom}_k(N, L))$$

for any $N \in D(R^{\mathrm{op}})$, $M \in D(R)$ and $L \in D(k)$.

Exercises

Exercise 14.1. Let $\mathcal{C}$ be an abelian category and let $a \leq b$ be integers.
(i) Prove that for $X \in D^{\geq b}(\mathcal{C})$ and $Y \in D^{\leq a}(\mathcal{C})$, any morphism $f \colon X \to Y$ in $D(\mathcal{C})$ decomposes as $X \to U[-b] \to V[-a] \to Y$ for some $U, V \in \mathcal{C}$. (Hint: to prove the existence of V, represent X by an object of $C^{\geq b}(\mathcal{C})$ and use $\sigma^{\geq a}$.)
(ii) Assume that $\mathrm{hd}(\mathcal{C}) < b - a$. Prove that $\mathrm{Hom}_{D(\mathcal{C})}(X, Y) \simeq 0$ for $X \in D^{\geq b}(\mathcal{C})$ and $Y \in D^{\leq a}(\mathcal{C})$.

Exercise 14.2. Let $\mathcal{C}$ be an abelian category with enough projectives and which satisfies (14.4.1). Let $\mathcal{P}$ denote the full subcategory of $\mathcal{C}$ consisting of projective objects. Denote by $\widetilde{\mathcal{P}}$ the smallest full triangulated category of $K(\mathcal{C})$ that contains $K^-(\mathcal{P})$ and is closed by small direct sums. Prove that the derived functor $\mathrm{RHom}_{\mathcal{C}} \colon D(\mathcal{C}) \times D(\mathcal{C})^{\mathrm{op}} \to D(\mathbb{Z})$ exists and prove that if $P \to X$ is a qis in $K(\mathcal{C})$ with $P \in \widetilde{\mathcal{P}}$, then $\mathrm{RHom}_{\mathcal{C}}(X, Y) \simeq \mathrm{tot}_{\pi}(\mathrm{Hom}_{\mathcal{C}}^{\bullet, \bullet}(P, Y))$.

Exercise 14.3. Let $\mathcal{C}$ be an abelian category which admits countable direct sums and assume that such direct sums are exact. Let $X \in D(\mathcal{C})$.
(i) Prove that there is a d.t. in $D(\mathcal{C})$:

$$(14.4.13) \qquad \bigoplus_{n \geq 0} \tau^{\leq n} X \xrightarrow{\mathrm{id} - \sigma} \bigoplus_{n \geq 0} \tau^{\leq n} X \xrightarrow{w} X \xrightarrow{v} \bigoplus_{n \geq 0} \tau^{\leq n} X[1] \,,$$

where σ is defined in Notation 10.5.10 and w is induced by the canonical morphisms $\tau^{\leq n} X \to X$.
(ii) Assume further that the cohomological dimension of $\mathcal{C}$ is less than or equal to 1. Prove that any $X \in D(\mathcal{C})$ is isomorphic to $\bigoplus_{n \in \mathbb{Z}} H^n(X)[-n]$. (Hint: applying Exercise 14.1 to $\tau^{<n} X \to X \to \tau^{\geq n} X \xrightarrow{+1}$, construct $H^n(X)[-n] \to X$.)

Exercise 14.4. Let k be a field, $A = k[x, y]$, $\mathcal{C} = \mathrm{Mod}(A)$ and denote by $D_{\mathrm{coh}}^b(\mathcal{C})$ the full triangulated subcategory of $D(\mathcal{C})$ consisting of objects X such that $H^j(X)$ is finitely generated over A for any $j \in \mathbb{Z}$. Let $L_0 = A$, $L = A \oplus A$, and consider the exact sequence $0 \to L_0 \xrightarrow{\varphi} L \xrightarrow{\psi} L_0 \to k \to 0$ in Exercise 13.21. Let $p := \varphi \circ \psi \colon L \to L$ and denote by X the object of $K(\mathcal{C})$:

$$X := 0 \to L_0 \xrightarrow{\varphi} L \xrightarrow{p} L \xrightarrow{p} L \to \cdots$$

where L_0 stands in degree -2.
(i) For $Z \in D_{\mathrm{coh}}^b(\mathcal{C})$ and $Y_n \in D(\mathcal{C})$ $(n \in \mathbb{Z})$, prove the isomorphism

$$\bigoplus_n \mathrm{Hom}_{D(\mathcal{C})}(Z, Y_n) \xrightarrow{\sim} \mathrm{Hom}_{D(\mathcal{C})}(Z, \bigoplus_n Y_n) \,.$$

(ii) Prove that
(a) $H^i(X) \simeq k$ for $i \geq 0$ and $H^i(X) \simeq 0$ for $i < 0$,

(b) for $i \geq 0$ and the d.t. in $D(\mathcal{C})$

$$H^i(X)[-i] \to \tau^{\leq i+1}\tau^{\geq i}(X) \to H^{i+1}(X)[-i-1] \xrightarrow{u_i} H^i(X)[-i+1] ,$$

the morphism u_i does not vanish in $D(\mathcal{C})$.

(iii) Prove that the object $\tau^{\leq n}X$ of $K(\mathcal{C})$ is isomorphic to the complex

$$0 \to L_0 \xrightarrow{\varphi} L \xrightarrow{p} L \xrightarrow{p} L \to \cdots \xrightarrow{p} L \to L_0 \to 0$$

where L_0 on the right stands in degree n and L_0 on the left in degree -2.
(iv) Prove the isomorphism $\tau^{\geq n}X \simeq X[-n]$ in $D(\mathcal{C})$ for $n \geq 0$.
(v) Prove that for any $n > 0$ and any morphism $f: X \to X[n]$ in $D(\mathcal{C})$, $H^i(f)$ vanishes for all $i \in \mathbb{Z}$. (Hint: use the commutative diagram

$$\begin{array}{ccc} H^k(X) & \longrightarrow & H^{k-1}(X)[2] \\ \downarrow{\scriptstyle H^k(f)} & & \downarrow{\scriptstyle H^{k-1}(f)} \\ H^k(X[n]) & \longrightarrow & H^{k-1}(X[n])[2] \end{array}$$

deduced from (ii) (b).)
(vi) Prove that the morphism v in (14.4.13) does not vanish in $D(\mathcal{C})$ using the following steps.

(a) If $v = 0$, then there exists $s: X \to \bigoplus_n \tau^{\leq n}X$ such that $w \circ s = \mathrm{id}_X$.

(b) For any $a > 0$, there exists b such that the composition $\tau^{<a}X \to X \xrightarrow{s} \bigoplus_n \tau^{\leq n}X$ factors through $\bigoplus_{n<b} \tau^{\leq n}X \to \bigoplus_n \tau^{\leq n}X$. (Hint: use (i).)
(c) For any $a > 0$, there exist $b > 0$ and morphisms $\tau^{\geq a}X \to X$ and $X \to \tau^{\leq b}X$ such that the composition $X \to \tau^{\geq a}X \oplus \tau^{\leq b}X \to X$ is id_X. (Hint: s is the sum of two morphisms $X \to \bigoplus_{n<b} \tau^{\leq n}X$ and $X \to \bigoplus_{n\geq b} \tau^{\leq n}X$.)
(d) For any $a > 0$, there exists a morphism $\tau^{\geq a}X \to X$ such that the composition $\tau^{\geq a}X \to X \to \tau^{\geq b}X$ is the canonical morphism for some $b > a$.
(e) Using (v) and (iv), conclude.

(vii) Prove that $\tau^{\leq n}v = 0$ in $D(\mathcal{C})$ for all $n \in \mathbb{Z}$.
(viii) Prove that the natural functor $D^+(\mathcal{C}) \to \mathrm{Ind}(D^b(\mathcal{C}))$, given by $X \mapsto \underrightarrow{\text{"lim"}}_n \tau^{\leq n}X$, is not faithful.

Exercise 14.5. Let $\mathcal{C}$ be an abelian category and let $\mathrm{Gr}(\mathcal{C})$ be the associated graded category (see Definition 11.3.1). Consider the functor

$$\Theta: \mathrm{Gr}(\mathcal{C}) \to D(\mathcal{C})$$
$$\{X^n\}_{n\in\mathbb{Z}} \mapsto \bigoplus_n X^n[-n] .$$

(i) Prove that Θ is an equivalence if and only if $\mathcal{C}$ is semisimple.
(ii) Prove that Θ is essentially surjective if and only if $\mathcal{C}$ is hereditary.

Exercise 14.6. Let C be an abelian category which has enough injectives and denote by $\mathcal{I}_C$ the full additive subcategory of injective objects of C. Assume moreover that C has finite homological dimension (see Exercise 13.8). Prove that any $X \in \mathrm{K}(\mathcal{I}_C)$ is homotopically injective.

Exercise 14.7. Let k be a commutative ring and $C = \mathrm{Mod}(k)$. Let $x \in k$ be a non-zero-divisor. Consider the additive functor $F : C \to C$ given by $M \mapsto x \cdot M$ (see Example 8.3.19). Prove that $RF \simeq \mathrm{id}_{\mathrm{D}(C)}$, $LF \simeq \mathrm{id}_{\mathrm{D}(C)}$ and the canonical morphism $LF \to RF$ (see (7.3.3)) is given by the multiplication by x.

Exercise 14.8. Let C be a Grothendieck category. Prove that an object I of $\mathrm{C}(C)$ is an injective object if and only if I is homotopic to zero and all I^n are injective objects of C. (Hint: consider $I \to \mathrm{Mc}(\mathrm{id}_I)$.)

Indization and Derivation
of Abelian Categories

In this chapter we study the derived category $D^b(\mathrm{Ind}(\mathcal{C}))$ of the category of ind-objects of the abelian category $\mathcal{C}$. The main difficulty comes from the fact that, as we shall see, the category $\mathrm{Ind}(\mathcal{C})$ does not have enough injectives in general. This difficulty is partly overcome by introducing the weaker notion of "quasi-injective objects", and these objects are sufficient to derive functors on $\mathrm{Ind}(\mathcal{C})$ which are indization of functors on $\mathcal{C}$.

As a byproduct, we shall give a sufficient condition which ensures that the right derived functor of a left exact functor commutes with small filtrant inductive limits.

Finally, we study the relations between $D^b(\mathrm{Ind}(\mathcal{C}))$ and the category $\mathrm{Ind}(D^b(\mathcal{C}))$ of ind-objects of $D^b(\mathcal{C})$.

15.1 Injective Objects in $\mathrm{Ind}(\mathcal{C})$

In this chapter, $\mathcal{C}$ is an abelian category and recall that by the hypothesis, $\mathcal{C}$ is a $\mathcal{U}$-category (see Convention 1.4.1). It follows that $\mathrm{Ind}(\mathcal{C})$ is again an abelian $\mathcal{U}$-category.

Recall that we denote by "$\bigoplus$" the coproduct in $\mathrm{Ind}(\mathcal{C})$ (see Notation 8.6.1).

As in Chap. 6, we denote by $\iota_\mathcal{C}\colon \mathcal{C} \to \mathrm{Ind}(\mathcal{C})$ the natural functor. This functor is fully faithful and exact. By Proposition 6.3.1, if $\mathcal{C}$ admits small inductive limits, the functor $\iota_\mathcal{C}$ admits a left adjoint, denoted by $\sigma_\mathcal{C}$. It follows from Proposition 8.6.6 that if the small filtrant inductive limits are exact in $\mathcal{C}$, then the functor $\sigma_\mathcal{C}$ is exact.

Proposition 15.1.1. *Assume that $\mathcal{C}$ admits small inductive limits and that small filtrant inductive limits are exact. Let $X \in \mathcal{C}$. Then*

(i) *X is injective in $\mathcal{C}$ if and only if $\iota_\mathcal{C}(X)$ is injective in $\mathrm{Ind}(\mathcal{C})$,*
(ii) *X is projective in $\mathcal{C}$ if and only if $\iota_\mathcal{C}(X)$ is projective in $\mathrm{Ind}(\mathcal{C})$.*

Proof. (i) Let X be an injective object of $\mathcal{C}$. For $A \in \mathrm{Ind}(\mathcal{C})$, we have $\mathrm{Hom}_{\mathrm{Ind}(\mathcal{C})}(A, \iota_{\mathcal{C}}(X)) \simeq \mathrm{Hom}_{\mathcal{C}}(\sigma_{\mathcal{C}}(A), X)$ and the result follows since $\sigma_{\mathcal{C}}$ is exact. The converse statement is obvious.

(ii) Let X be a projective object of $\mathcal{C}$, let $f \colon A \to B$ be an epimorphism in $\mathrm{Ind}(\mathcal{C})$ and $u \colon X \to B$ a morphism. Let us show that u factors through f. By Proposition 8.6.9, there exist an epimorphism $f' \colon Y \to X$ in $\mathcal{C}$ and a morphism $v \colon Y \to A$ such that $u \circ f' = f \circ v$. Since X is projective, there exists a section $s \colon X \to Y$ of f'. Therefore, $f \circ (v \circ s) = u \circ f' \circ s = u$. This is visualized by the diagram

$$
\begin{array}{ccc}
Y & \underset{s}{\overset{f'}{\longleftarrow\!\!\!\!-\!\!\!\!\longrightarrow}} & X \\
{\scriptstyle v}\downarrow & & \downarrow{\scriptstyle u} \\
A & \underset{f}{\longrightarrow} & B \, .
\end{array}
$$

The converse statement is obvious.   q.e.d.

In the simple case where $\mathcal{C} = \mathrm{Mod}(k)$ with a field k, we shall show that the category $\mathrm{Ind}(\mathcal{C})$ does not have enough injectives. In the sequel, we shall write $\mathrm{Ind}(k)$ instead of $\mathrm{Ind}(\mathrm{Mod}(k))$, for short.

Proposition 15.1.2. *Assume that k is a field. Let $Z \in \mathrm{Ind}(k)$. Then Z is injective if and only if Z belongs to $\mathrm{Mod}(k)$.*

Proof. Assume that $Z \in \mathrm{Ind}(k)$ is injective. Any object in $\mathrm{Ind}(k)$ is a quotient of "$\bigoplus_i$" M_i with $M_i \in \mathrm{Mod}(k)$, and the natural morphism "$\bigoplus_i$" $M_i \to \bigoplus_i M_i$ is a monomorphism. Since Z is injective, "$\bigoplus_i$" $M_i \to Z$ factorizes through $\bigoplus_i M_i$. Hence we can assume from the beginning that

$$ Z = X/Y \text{ with } X \in \mathrm{Mod}(k), \ Y \in \mathrm{Ind}(k). $$

Since $Y \to X$ is a monomorphism, $\sigma_{\mathcal{C}}(Y)$ is a sub-object of X. Hence, there exits a decomposition $X = X' \oplus \sigma_{\mathcal{C}}(Y)$ in $\mathrm{Mod}(k)$. Then $Z = X' \oplus (\sigma_{\mathcal{C}}(Y)/Y)$ and $\sigma_{\mathcal{C}}(Y)/Y$ is injective. Thus we may assume from the beginning that

$$ Z = X/Y \text{ with } X \in \mathrm{Mod}(k), \ Y \subset X \text{ and } \sigma_{\mathcal{C}}(Y) = X. $$

Let $\kappa_{\mathcal{C}} \colon \mathrm{Mod}(k) \to \mathrm{Ind}(k)$ be the functor introduced in Sect. 6.3, $V \mapsto$ "$\varinjlim$" W, where W ranges over the family of finite-dimensional vector subspaces of V. Then we have $\kappa_{\mathcal{C}}(V) \subset Y$ for any $V \in \mathrm{Mod}(k)$ with $V \subset X$.

Assuming $Y \neq X$, we shall derive a contradiction. Set

$$
\begin{aligned}
\mathcal{K} &= \{ V \, ; \, V \in \mathrm{Mod}(k), \, V \subset Y \}, \\
N &= k^{\oplus \mathcal{K}} = \bigoplus_{V \in \mathcal{K}} k e_V, \\
\Phi &= \mathrm{Hom}_k(N, X) \, .
\end{aligned}
$$

For $\varphi \in \Phi$, let N_φ be a copy of N and let $a_\varphi : N \xrightarrow{\sim} N_\varphi$ be the isomorphism. We denote by $c_\varphi : N \to \bigoplus_{\varphi' \in \Phi} N_{\varphi'}$ the composition $N \xrightarrow{\sim}_{a_\varphi} N_\varphi \to \bigoplus_{\varphi' \in \Phi} N_{\varphi'}$. Set

$$T = \text{``}\bigoplus_{V \in \mathcal{K}}\text{''}\Big(\bigoplus_{\varphi \in \Phi} kc_\varphi(e_V)\Big) \subset \bigoplus_{\varphi \in \Phi} N_\varphi .$$

Then, for any finite subset A of Φ, we have

$$T \cap \Big(\bigoplus_{\varphi \in A} N_\varphi\Big) = \text{``}\bigoplus_{V \in \mathcal{K}}\text{''}\Big(\bigoplus_{\varphi \in A} kc_\varphi(e_V)\Big)$$
$$= \bigoplus_{\varphi \in A}\Big(\text{``}\bigoplus_{V \in \mathcal{K}}\text{''} kc_\varphi(e_V)\Big) = \bigoplus_{\varphi \in A} \kappa_\mathcal{C}(N_\varphi) .$$

Hence, we have a monomorphism

$$\text{``}\bigoplus_{\varphi \in \Phi}\text{''}\big(N_\varphi/\kappa_\mathcal{C}(N_\varphi)\big) \hookrightarrow \Big(\bigoplus_{\varphi \in \Phi} N_\varphi\Big)/T .$$

Let $f : \bigoplus_{\varphi \in \Phi} N_\varphi \to X$ be the morphism defined by $f \circ c_\varphi(u) = \varphi(u)$ for $u \in N$. It induces a morphism

$$\tilde{f} : \text{``}\bigoplus_{\varphi \in \Phi}\text{''}\big(N_\varphi/\kappa_\mathcal{C}(N_\varphi)\big) \to Z .$$

Since Z is injective, the morphism $\tilde{f}$ factors through $\big(\bigoplus_{\varphi \in \Phi} N_\varphi\big)/T$. Note that any object in $\mathrm{Mod}(k)$ is a projective object in $\mathrm{Ind}(k)$ by Proposition 15.1.1. Hence $\bigoplus_{\varphi \in \Phi} N_\varphi$ is a projective object of $\mathrm{Ind}(k)$, and the composition $\bigoplus_{\varphi \in \Phi} N_\varphi \to \big(\bigoplus_{\varphi \in \Phi} N_\varphi\big)/T \to Z$ factors through X. Thus we obtain the commutative diagram

(15.1.1)

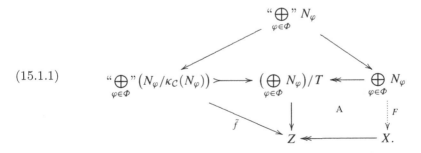

The morphism $F : \bigoplus_{\varphi \in \Phi} N_\varphi \to X$ has the following properties:

- $F_\varphi := F \circ c_\varphi : N \to X$ satisfies the condition: for any $V \in \mathcal{K}$, there exists $K(V) \in \mathcal{K}$ such that $F_\varphi(e_V) \in K(V)$ for any $\varphi \in \Phi$,
- $G_\varphi := (F_\varphi - \varphi)(N) \subset X$ belongs to $\mathcal{K}$ for any $\varphi \in \Phi$.

Indeed, the first property follows from the fact that the composition

$$\bigoplus_{\varphi \in \Phi} kc_\varphi(e_V) \to \bigoplus_{\varphi \in \Phi} N_\varphi \xrightarrow{F} X \to Z = X/Y$$

vanishes by the commutativity of the square labeled by A in (15.1.1), and the second follows from the fact that the two compositions $N \underset{F_\varphi}{\overset{\varphi}{\rightrightarrows}} X \longrightarrow Z$ coincide.

Hence we have

(15.1.2) $\varphi(e_V) \in K(V) + G_\varphi$ for any $V \in \mathcal{K}$ and $\varphi \in \Phi$.

Since $Y \neq X$, we have $K(V) + V \neq X$ for any $V \in \mathcal{K}$. Hence there exists $x(V) \in X$ such that $x(V) \notin K(V) + V$. Define $\varphi_0 \in \Phi$ by $\varphi_0(e_V) = x(V)$. Then for $V = G_{\varphi_0}$, we have

$$\varphi_0(e_V) = x(V) \notin K(V) + V = K(V) + G_{\varphi_0} .$$

This contradicts (15.1.2).   q.e.d.

Corollary 15.1.3. *The category* $\mathrm{Ind}(k)$ *does not have enough injectives.*

Proof. Let us take $V \in \mathrm{Mod}(k)$ with $\dim V = \infty$ and let $U = \kappa_\mathcal{C}(V)$. Define $W \in \mathrm{Ind}(k)$ by the exact sequence

$$0 \to U \to V \to W \to 0 .$$

Then, we have $\sigma_\mathcal{C}(W) \simeq 0$, but W does not vanish. Assume that there exists a monomorphism $W \rightarrowtail Z$ with an injective object $Z \in \mathrm{Ind}(k)$. Then Z belongs to $\mathrm{Mod}(k)$ by Proposition 15.1.2. The morphism of functors $\mathrm{id} \to \sigma_\mathcal{C}$ (we do not write $\iota_\mathcal{C}$) induces the commutative diagram in $\mathrm{Ind}(k)$

$$
\begin{array}{ccc}
W & \rightarrowtail & Z \\
\downarrow & & \downarrow \\
\sigma_\mathcal{C}(W) & \longrightarrow & \sigma_\mathcal{C}(Z).
\end{array}
$$

Since $Z \to \sigma_\mathcal{C}(Z)$ is an isomorphism, we get $W \simeq 0$, which is a contradiction.
                                                                         q.e.d.

15.2 Quasi-injective Objects

Let $\mathcal{C}$ be an abelian category. We have seen in Sect. 15.1 that the abelian category $\mathrm{Ind}(\mathcal{C})$ does not have enough injectives in general. However, quasi-injective objects, which we introduce below, are sufficient for many purposes.

Definition 15.2.1. *Let $A \in \mathrm{Ind}(\mathcal{C})$. We say that A is* quasi-injective *if the functor*

$$\mathcal{C} \to \mathrm{Mod}(\mathbb{Z}),$$
$$X \mapsto A(X) = \mathrm{Hom}_{\mathrm{Ind}(\mathcal{C})}(X, A),$$

is exact.

Clearly, a small filtrant inductive limit of quasi-injective objects is quasi-injective.

Lemma 15.2.2. *Let $0 \to A' \xrightarrow{f} A \xrightarrow{g} A'' \to 0$ be an exact sequence in $\mathrm{Ind}(\mathcal{C})$ and assume that A' is quasi-injective. Then*

(i) *the sequence $0 \to A'(X) \to A(X) \to A''(X) \to 0$ is exact for any $X \in \mathcal{C}$,*
(ii) *A is quasi-injective if and only if A'' is quasi-injective.*

Proof. (i) It is enough to prove the surjectivity of $A(X) \to A''(X)$. Let $u \in A''(X)$. Using Proposition 8.6.9, we get a commutative solid diagram with exact rows and with $Z, Y \in \mathcal{C}$

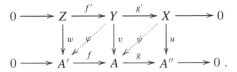

Since A' is quasi-injective, there exists a morphism $\varphi \colon Y \to A'$ such that $w = \varphi \circ f'$. Therefore, $(v - f \circ \varphi) \circ f' = v \circ f' - f \circ w = 0$, and the morphism $v - f \circ \varphi$ factors through $\mathrm{Coker}\, f' \simeq X$. Hence, there exists $\psi \colon X \to A$ such that $v - f \circ \varphi = \psi \circ g'$. Then $g \circ \psi \circ g' = g \circ (v - f \circ \varphi) = u \circ g'$, and this implies $u = g \circ \psi$.

(ii) The proof is left as an easy exercise.   q.e.d.

Proposition 15.2.3. *Assume that $\mathcal{C}$ has enough injectives and let $A \in \mathrm{Ind}(\mathcal{C})$. Then the conditions below are equivalent.*

(i) *A is quasi-injective,*
(ii) *there exist a small and filtrant category J and a functor $\alpha \colon J \to \mathcal{C}$ such that $A \simeq \text{``}\varinjlim\text{''}\, \alpha$ and $\alpha(j)$ is injective in $\mathcal{C}$ for all $j \in J$,*
(iii) *any morphism $a \colon X \to A$ with $X \in \mathcal{C}$ factorizes through an injective object Y of $\mathcal{C}$, (i.e., $a = b \circ f$ with $X \xrightarrow{f} Y \xrightarrow{b} A$).*

Proof. Let $\mathcal{I}$ denote the full subcategory of $\mathcal{C}$ consisting of injective objects.
(i) $\Rightarrow$ (iii). By the hypothesis, there exists a monomorphism $X \rightarrowtail Y$ with $Y \in \mathcal{I}$. Since A is quasi-injective, $X \to A$ factorizes through Y.
(iii) $\Rightarrow$ (ii) follows from Exercise 6.11.
(ii) $\Rightarrow$ (i). Let $X \in \mathcal{C}$. We have $A(X) \simeq \varinjlim_{j \in J} \mathrm{Hom}_{\mathcal{C}}(X, \alpha(j))$. Since $\alpha(j)$ is injective and the functor $\varinjlim$ is exact, A is exact.   q.e.d.

Definition 15.2.4. *We say that* $\mathrm{Ind}(\mathcal{C})$ *has enough quasi-injectives if the full subcategory of quasi-injective objects is cogenerating in* $\mathrm{Ind}(\mathcal{C})$.

Theorem 15.2.5. *Let* $\mathcal{J}$ *be a cogenerating full subcategory of* $\mathcal{C}$. *Then* $\mathrm{Ind}(\mathcal{J})$ *is cogenerating in* $\mathrm{Ind}(\mathcal{C})$.

In order to prove this result, we need a lemma.

Lemma 15.2.6. *For any small subset* $\mathcal{S}$ *of* $\mathrm{Ob}(\mathcal{C})$, *there exists a small fully abelian subcategory* $\mathcal{C}_0$ *of* $\mathcal{C}$ *such that*

(i) $\mathcal{S} \subset \mathrm{Ob}(\mathcal{C}_0)$,
(ii) $\mathcal{C}_0 \cap \mathcal{J}$ *is cogenerating in* $\mathcal{C}_0$.

Proof. We shall define an increasing sequence $\{\mathcal{S}_n\}_{n\geq 0}$ of full subcategories of $\mathcal{C}$ by induction on n. For any $X \in \mathcal{S}$, let us take $I_X \in \mathcal{J}$ and a monomorphism $X \rightarrowtail I_X$. We define $\mathcal{S}_0$ as the full subcategory of $\mathcal{C}$ such that $\mathrm{Ob}(\mathcal{S}_0) = \mathcal{S}$. For $n > 0$, let $\mathcal{S}_n$ be the full subcategory of $\mathcal{C}$ such that

$$\mathrm{Ob}(\mathcal{S}_n) = \mathrm{Ob}(\mathcal{S}_{n-1}) \cup \{I_X ; X \in \mathcal{S}_{n-1}\} \cup \{X \oplus Y ; X, Y \in \mathcal{S}_{n-1}\}$$
$$\cup \{\mathrm{Ker}\, u ; u \in \mathrm{Mor}(\mathcal{S}_{n-1})\} \cup \{\mathrm{Coker}\, u ; u \in \mathrm{Mor}(\mathcal{S}_{n-1})\} \ .$$

Then $\mathcal{C}_0 = \bigcup_n \mathcal{S}_n$ satisfies the desired conditions.   q.e.d.

Proof of Theorem 15.2.5. Let $A \in \mathrm{Ind}(\mathcal{C})$. There exist a small filtrant category I and a functor $\alpha : I \to \mathcal{C}$ such that $A \simeq \text{``}\underrightarrow{\lim}\text{''}\, \alpha$. By Lemma 15.2.6, there exists a small fully abelian subcategory $\mathcal{C}_0$ of $\mathcal{C}$ such that $\alpha(i) \in \mathcal{C}_0$ for all $i \in I$ and $\mathcal{J} \cap \mathcal{C}_0$ is cogenerating in $\mathcal{C}_0$. Then $A \in \mathrm{Ind}(\mathcal{C}_0)$, and $\mathrm{Ind}(\mathcal{C}_0)$ admits enough injectives by Corollary 9.6.5. Hence, there exist an injective object B of $\mathrm{Ind}(\mathcal{C}_0)$ and a monomorphism $A \rightarrowtail B$. In order to prove that $B \in \mathrm{Ind}(\mathcal{C}_0 \cap \mathcal{J})$, it is enough to check that any morphism $Z \to B$ with $Z \in \mathcal{C}_0$, factorizes through an object of $\mathcal{C}_0 \cap \mathcal{J}$ (see Exercise 6.11). Take a monomorphism $Z \to Y$ with $Y \in \mathcal{C}_0 \cap \mathcal{J}$. Since B is injective, $Z \to B$ factors through $Z \to Y$.   q.e.d.

Corollary 15.2.7. *Let* $\mathcal{C}$ *be an abelian category which admits enough injectives. Then* $\mathrm{Ind}(\mathcal{C})$ *admits enough quasi-injectives.*

15.3 Derivation of Ind-categories

As above, $\mathcal{C}$ denotes an abelian category.

Theorem 15.3.1.   (i) *The natural functor* $\mathrm{D}^*(\mathcal{C}) \to \mathrm{D}^*_{\mathcal{C}}(\mathrm{Ind}(\mathcal{C}))$ *is an equivalence for* $* = \mathrm{b}, -$.
   (ii) *Assume that* $\mathcal{C}$ *admits small inductive limits and small filtrant inductive limits are exact. Then* $\mathrm{D}^+(\mathcal{C}) \to \mathrm{D}^+_{\mathcal{C}}(\mathrm{Ind}(\mathcal{C}))$ *is an equivalence.*

Proof. (i) By Theorem 13.2.8 (with the arrows reversed), it is enough to show that for any epimorphism $A \twoheadrightarrow Y$ in $\mathrm{Ind}(\mathcal{C})$ with $Y \in \mathcal{C}$, there exist $X \in \mathcal{C}$ and a morphism $X \to A$ such that the composition $X \to Y$ is an epimorphism. This follows from Proposition 8.6.9.

(ii) Let us apply Theorem 13.2.8 and consider a monomorphism $X \rightarrowtail A$ with $X \in \mathcal{C}$ and $A \in \mathrm{Ind}(\mathcal{C})$. Then $X \simeq \sigma_{\mathcal{C}}(X) \to \sigma_{\mathcal{C}}(A)$ is a monomorphism and factors through $X \to A$. (Recall that the functor $\sigma_{\mathcal{C}}$ is defined in Proposition 6.3.1.)   q.e.d.

Let $F: \mathcal{C} \to \mathcal{C}'$ be a left exact functor, and let $IF: \mathrm{Ind}(\mathcal{C}) \to \mathrm{Ind}(\mathcal{C}')$ be the associated left exact functor. We shall consider the following hypothesis:

(15.3.1)   there exists an F-injective subcategory $\mathcal{J}$ of $\mathcal{C}$.

Hypothesis (15.3.1) implies that the right derived functor $R^+F: \mathrm{D}^+(\mathcal{C}) \to \mathrm{D}^+(\mathcal{C}')$ exists and $R^k F: \mathcal{C} \to \mathcal{C}'$ induces a functor $I(R^k F): \mathrm{Ind}(\mathcal{C}) \to \mathrm{Ind}(\mathcal{C}')$.

Proposition 15.3.2. *Let $F: \mathcal{C} \to \mathcal{C}'$ be a left exact functor of abelian categories and let $\mathcal{J}$ be an F-injective subcategory of $\mathcal{C}$. Then*

(a) *$\mathrm{Ind}(\mathcal{J})$ is IF-injective,*
(b) *the functor IF admits a right derived functor $R^+(IF): \mathrm{D}^+(\mathrm{Ind}(\mathcal{C})) \to \mathrm{D}^+(\mathrm{Ind}(\mathcal{C}'))$,*
(c) *the diagram below commutes*

$$
\begin{array}{ccc}
\mathrm{D}^+(\mathcal{C}) & \xrightarrow{\ \ R^+F\ \ } & \mathrm{D}^+(\mathcal{C}') \\
\downarrow & & \downarrow \\
\mathrm{D}^+(\mathrm{Ind}(\mathcal{C})) & \xrightarrow{\ \ R^+(IF)\ \ } & \mathrm{D}^+(\mathrm{Ind}(\mathcal{C}')),
\end{array}
$$

(d) *there is an isomorphism $I(R^k F) \simeq R^k(IF)$ for all $k \in \mathbb{Z}$. In particular, $R^k(IF)$ commutes with small filtrant inductive limits.*

Proof. (a) First, note that $\mathrm{Ind}(\mathcal{J})$ is cogenerating by Theorem 15.2.5. Set

$$\widetilde{\mathcal{J}} := \{A \in \mathrm{Ind}(\mathcal{C}); I(R^k F)(A) \simeq 0 \text{ for all } k > 0\} \ .$$

Since $\widetilde{\mathcal{J}}$ contains $\mathrm{Ind}(\mathcal{J})$, it is cogenerating. Let us check that $\widetilde{\mathcal{J}}$ satisfies the conditions (ii) and (iii) in Corollary 13.3.8. Consider an exact sequence $0 \to A \to B \to C \to 0$ in $\mathrm{Ind}(\mathcal{C})$. By Proposition 8.6.6 (a), there exist a small filtrant category I and an exact sequence of functors from I to $\mathcal{C}$

(15.3.2) $0 \to \alpha \to \beta \to \gamma \to 0$

such that the exact sequence in $\mathrm{Ind}(\mathcal{C})$ is obtained by applying the functor "$\varinjlim$" to (15.3.2). Consider the long exact sequence for $i \in I$

$$0 \to R^0 F(\alpha(i)) \to R^0 F(\beta(i)) \to R^0 F(\gamma(i)) \to R^1 F(\alpha(i)) \to \cdots \ .$$

Applying the functor "$\varinjlim$", we obtain the long exact sequence

(15.3.3)
$$0 \to I(R^0 F)(A) \to I(R^0 F)(B) \to I(R^0 F)(C) \to I(R^1 F)(A) \to \cdots .$$

Assuming $A, B \in \widetilde{\mathcal{J}}$, we deduce $C \in \widetilde{\mathcal{J}}$. Assuming $A \in \widetilde{\mathcal{J}}$, we deduce the exact sequence $0 \to IF(A) \to IF(B) \to IF(C) \to 0$. Therefore, $\widetilde{\mathcal{J}}$ is IF-injective and it follows from Proposition 13.3.5 (ii) that $\text{Ind}(\mathcal{J})$ is itself IF-injective.

(b) follows from Proposition 13.3.5 (i).

(c) follows from Proposition 13.3.13. Indeed, $R^+(\iota_{\mathcal{C}'} \circ F) \simeq \iota_{\mathcal{C}'} \circ R^+ F$ since $\iota_{\mathcal{C}'}$ is exact and $R^+(IF \circ \iota_{\mathcal{C}}) \simeq R^+(IF) \circ \iota_{\mathcal{C}}$ since $\text{Ind}(\mathcal{J})$ contains $\mathcal{J}$ and is IF-injective.

(d) We construct a morphism $I(R^k F) \to R^k(IF)$ as follows. For $A \in \text{Ind}(\mathcal{C})$,
$$I(R^k F)(A) \simeq \underset{(X \to A) \in \mathcal{C}_A}{\text{"}\varinjlim\text{"}} R^k F(X) \simeq \underset{(X \to A) \in \mathcal{C}_A}{\text{"}\varinjlim\text{"}} R^k(IF)(X) \to R^k(IF)(A) .$$

The isomorphism in (d) obviously holds for $k = 0$. We shall prove that it holds for $k = 1$, then for all k.

Consider an exact sequence $0 \to A \to B \to C \to 0$ with $B \in \widetilde{\mathcal{J}}$. Then $I(R^k F)(B) \simeq 0$ for all $k > 0$ by definition and $R^k(IF)(B) \simeq 0$ for all $k > 0$ since $\widetilde{\mathcal{J}}$ is IF-injective. There exists an exact sequence

(15.3.4)
$$0 \to R^0(IF)(A) \to R^0(IF)(B) \to R^0(IF)(C) \to R^1(IF)(A) \to \cdots .$$

By comparing the exact sequences (15.3.3) and (15.3.4), we get the result for $k = 1$.

We have the isomorphisms $I(R^k F)(A) \simeq I(R^{k-1} F)(C)$ and $R^k(IF)(A) \simeq R^{k-1}(IF)(C)$ for $k \geq 2$. By induction on k, we may assume $I(R^{k-1} F)(C) \simeq R^{k-1}(IF)(C)$. Therefore, $I(R^k F)(A) \simeq R^k(IF)(A)$.   q.e.d.

Proposition 15.3.3. *Let $\mathcal{C}$ and $\mathcal{C}'$ be abelian categories admitting small inductive limits and assume that small filtrant inductive limits are exact in $\mathcal{C}$ and $\mathcal{C}'$. Let $F : \mathcal{C} \to \mathcal{C}'$ be a left exact functor commuting with small filtrant inductive limits and let $\mathcal{J}$ be an F-injective additive subcategory of $\mathcal{C}$ closed by small filtrant inductive limits. Then $R^k F : \mathcal{C} \to \mathcal{C}'$ commutes with small filtrant inductive limits for all $k \in \mathbb{Z}$.*

Proof. The functor $\sigma_{\mathcal{C}} : \text{Ind}(\mathcal{C}) \to \mathcal{C}$ is exact and induces a triangulated functor $D^+(\text{Ind}(\mathcal{C})) \to D^+(\mathcal{C})$ that we still denote by $\sigma_{\mathcal{C}}$, and similarly with $\mathcal{C}$ replaced with $\mathcal{C}'$. Consider the diagram

(15.3.5)
$$
\begin{array}{ccc}
D^+(\text{Ind}(\mathcal{C})) & \xrightarrow{\;R^+(IF)\;} & D^+(\text{Ind}(\mathcal{C}')) \\
\sigma_{\mathcal{C}} \downarrow & & \downarrow \sigma_{\mathcal{C}'} \\
D^+(\mathcal{C}) & \xrightarrow{\;R^+ F\;} & D^+(\mathcal{C}').
\end{array}
$$

We shall show that this diagram commutes. Note that $\sigma_{C'} \circ IF \simeq F \circ \sigma_C$ by the assumption, and $\sigma_{C'} \circ R^+(IF) \simeq R^+(\sigma_{C'} \circ IF)$. Hence, it is enough to show that

$$(15.3.6) \qquad (R^+F) \circ \sigma_C \simeq R^+(F \circ \sigma_C),$$

and this follows from Proposition 13.3.13 since σ_C sends $\mathrm{Ind}(\mathcal{J})$ to $\mathcal{J}$.

To conclude, consider a small filtrant inductive system $\{X_i\}_{i \in I}$ in $\mathcal{C}$. We have the chain of isomorphisms

$$\varinjlim_i R^k F(X_i) \simeq \sigma_{C'}(\text{``}\varinjlim_i\text{''} \, R^k F(X_i))$$

$$\simeq \sigma_{C'} R^k(IF)(\text{``}\varinjlim_i\text{''} \, X_i)$$

$$\simeq (R^k F)\sigma_C(\text{``}\varinjlim_i\text{''} \, X_i) \simeq R^k F(\varinjlim_i X_i).$$

Here, the second isomorphism follows from Proposition 15.3.2 (d) and the third one from the commutativity of (15.3.5).   q.e.d.

Notation 15.3.4. We shall denote by $\mathcal{I}_{qinj}$ the full subcategory of $\mathrm{Ind}(\mathcal{C})$ consisting of quasi-injective objects.

Consider the hypothesis

$$(15.3.7) \qquad \text{the category } \mathcal{I}_{qinj} \text{ is cogenerating in } \mathrm{Ind}(\mathcal{C}).$$

This condition is a consequence of one of the following hypotheses

$$(15.3.8) \qquad\qquad\qquad \mathcal{C} \text{ has enough injectives,}$$
$$(15.3.9) \qquad\qquad\qquad \mathcal{C} \text{ is small.}$$

Indeed, (15.3.8) implies (15.3.7) by Corollary 15.2.7, and (15.3.9) implies (15.3.7) by Theorem 9.6.2.

Proposition 15.3.5. *Assume* (15.3.7) *and let $F: \mathcal{C} \to \mathcal{C}'$ be a left exact functor. Then the category $\mathcal{I}_{qinj}$ of quasi-injective objects is IF-injective. In particular, $R^+(IF): \mathrm{D}^+(\mathrm{Ind}(\mathcal{C})) \to \mathrm{D}^+(\mathrm{Ind}(\mathcal{C}'))$ exists.*

Proof. (i) We shall verify the hypotheses (i)–(iii) of Corollary 13.3.8. The first one is nothing but (15.3.7).
(ii) follows from Lemma 15.2.2 (ii).
(iii) Consider an exact sequence $0 \to A \to B \to C \to 0$ in $\mathrm{Ind}(\mathcal{C})$ and assume that $A \in \mathcal{I}_{qinj}$. For any $X \in \mathcal{C}$ and any morphism $u: X \to C$, Lemma 15.2.2 implies that u factors through $X \xrightarrow{w} B \to C$.

This defines a morphism $F(w): F(X) \to IF(B)$ such that the composition $F(X) \to IF(B) \to IF(C)$ is the canonical morphism. Therefore, we get the exact sequence

$$IF(B) \times_{IF(C)} F(X) \to F(X) \to 0$$

Applying the functor "$\varinjlim_{(X \to C) \in \mathcal{C}_C}$", we find that $IF(B) \to IF(C)$ is an epimorphism by Lemma 3.3.9.   q.e.d.

Corollary 15.3.6. *Assume* (15.3.7). *Then for any* $A \in \mathrm{Ind}(\mathcal{C})$ *there is a natural isomorphism*

$$\text{``}\varinjlim_{(X \to A) \in \mathcal{C}_A}\text{''} \ R^k(IF)(X) \xrightarrow{\sim} R^k(IF)(A) .$$

In particular, $R^k(IF)$ *commutes with small filtrant inductive limits.*

Proof. Consider the functor $IF : \mathrm{Ind}(\mathcal{C}) \to \mathrm{Ind}(\mathcal{C}')$. The subcategory $\mathcal{I}_{qinj}$ of $\mathrm{Ind}(\mathcal{C})$ is closed by small filtrant inductive limits and is IF-injective. Hence, the result follows from Proposition 15.3.3.   q.e.d.

We consider now a right exact functor $G : \mathcal{C} \to \mathcal{C}'$ of abelian categories.

Proposition 15.3.7. *Let* $G : \mathcal{C} \to \mathcal{C}'$ *be a right exact functor of abelian categories and let* $\mathcal{K}$ *be a* G-*projective additive subcategory of* $\mathcal{C}$. *Then*

(a) *the category* $\mathrm{Ind}(\mathcal{K})$ *is* IG-*projective,*
(b) *the functor* IG *admits a left derived functor* $L^-(IG) : \mathrm{D}^-(\mathrm{Ind}(\mathcal{C})) \to \mathrm{D}^-(\mathrm{Ind}(\mathcal{C}'))$,
(c) *the diagram below commutes*

$$
\begin{array}{ccc}
\mathrm{D}^-(\mathcal{C}) & \xrightarrow{\ L^- G\ } & \mathrm{D}^-(\mathcal{C}') \\
\downarrow & & \downarrow \\
\mathrm{D}^-(\mathrm{Ind}(\mathcal{C})) & \xrightarrow{\ L^-(IG)\ } & \mathrm{D}^-(\mathrm{Ind}(\mathcal{C}')),
\end{array}
$$

(d) *there is a natural isomorphism* $I(L^k G) \simeq L^k(IG)$ *for all* $k \in \mathbb{Z}$. *In particular,* $L^k(IG)$ *commutes with small filtrant inductive limits.*

Proof. The proof is very similar to that of Proposition 15.3.2, but we partly repeat it for the reader's convenience.

(a) Set

$$\widetilde{\mathcal{K}} = \{A \in \mathrm{Ind}(\mathcal{C}); I(L^k G)(A) \simeq 0 \text{ for all } k < 0\} .$$

Then $\widetilde{\mathcal{K}}$ contains $\mathrm{Ob}(\mathrm{Ind}(\mathcal{K}))$. Let us show that $\widetilde{\mathcal{K}}$ satisfies the conditions (i)–(iii) (with the arrows reversed) of Corollary 13.3.8.
(i) The category $\widetilde{\mathcal{K}}$ is generating. Indeed, if $A \in \mathrm{Ind}(\mathcal{C})$, there exists an epimorphism "$\bigoplus_{i \in I}$" $X_i \twoheadrightarrow A$ with a small set I and $X_i \in \mathcal{C}$. For each i choose an epimorphism $Y_i \twoheadrightarrow X_i$ with $Y_i \in \mathcal{K}$. Then

$$I(L^kG)(``\bigoplus_i" Y_i) \simeq ``\bigoplus_i" L^kG(Y_i) \simeq 0$$

for all $k < 0$, hence $``\bigoplus_i" Y_i \in \widetilde{\mathcal{K}}$.

(ii)–(iii) Consider an exact sequence $0 \to A \to B \to C \to 0$ in $\mathrm{Ind}(\mathcal{C})$. We may assume that this sequence is obtained by applying the functor "$\varinjlim$" to (15.3.2). Consider the long exact sequences for $i \in I$

$$\cdots \to L^{-1}G(\gamma(i)) \to L^0G(\alpha(i)) \to L^0G(\beta(i)) \to L^0G(\gamma(i)) \to 0 .$$

Applying the functor "$\varinjlim$", we obtain the long exact sequence

(15.3.10)
$$\cdots \to I(L^{-1}G)(C) \to I(L^0G)(A) \to I(L^0G)(B) \to I(L^0G)(C) \to 0 .$$

Assuming $B, C \in \widetilde{\mathcal{J}}$, we deduce $A \in \widetilde{\mathcal{J}}$. Assuming $C \in \widetilde{\mathcal{J}}$, we deduce the exact sequence $0 \to IG(A) \to IG(B) \to IG(C) \to 0$.

(b)–(c) go as in Proposition 15.3.2.

(d) The isomorphism in (d) clearly holds for $k = 0$. We shall prove that it holds for $k = 1$, then for all k.

Consider an exact sequence $0 \to A \to B \to C \to 0$ with $B \in \widetilde{\mathcal{K}}$. Then $I(L^kG)(B) \simeq 0$ for all $k < 0$ by definition and $L^k(IG)(B) \simeq 0$ for all $k < 0$ since $\widetilde{\mathcal{K}}$ is IG-projective. There exists an exact sequence

(15.3.11)
$$\cdots \to L^{-1}(IG)(C) \to L^0(IG)(A) \to L^0(IG)(B) \to L^0(IG)(C) \to 0 .$$

By comparing the exact sequences (15.3.11) and (15.3.10), we get the result for $k = 1$. Then the proof goes as in Proposition 15.3.2.   q.e.d.

Theorem 15.3.8. *Assume* (15.3.7).

(i) *The bifunctor* $\mathrm{Hom}_{\mathrm{Ind}(\mathcal{C})}$ *admits a right derived functor*

$$\mathrm{R}^+\mathrm{Hom}_{\mathrm{Ind}(\mathcal{C})} \colon \mathrm{D}^+(\mathrm{Ind}(\mathcal{C})) \times \mathrm{D}^-(\mathrm{Ind}(\mathcal{C}))^{\mathrm{op}} \to \mathrm{D}^+(\mathrm{Mod}(\mathbb{Z})) .$$

(ii) *Moreover, for* $X \in \mathrm{D}^-(\mathrm{Ind}(\mathcal{C}))$ *and* $Y \in \mathrm{D}^+(\mathrm{Ind}(\mathcal{C}))$,

$$H^0\mathrm{R}^+\mathrm{Hom}_{\mathrm{Ind}(\mathcal{C})}(X, Y) \simeq \mathrm{Hom}_{\mathrm{D}(\mathrm{Ind}(\mathcal{C}))}(X, Y) .$$

(iii) $\mathrm{D}^b(\mathcal{C})$ *and* $\mathrm{D}^b(\mathrm{Ind}(\mathcal{C}))$ *are* $\mathcal{U}$-*categories*.

Proof. Let $\mathcal{P}$ denote the full additive subcategory of $\mathrm{Ind}(\mathcal{C})$ defined by:

$$\mathcal{P} = \{A \in \mathrm{Ind}(\mathcal{C}); A \simeq ``\bigoplus_{i \in I}" X_i, \ I \text{ small}, \ X_i \in \mathcal{C}\} .$$

Clearly, the category $\mathcal{P}$ is generating in $\mathrm{Ind}(\mathcal{C})$.

We shall apply Proposition 13.4.4 and Theorem 13.4.1 to the subcategory $\mathcal{I}_{qinj} \times \mathcal{P}^{\mathrm{op}}$ of $\mathrm{Ind}(\mathcal{C}) \times \mathrm{Ind}(\mathcal{C})^{\mathrm{op}}$.

(A) For $B \in \mathcal{P}$, the functor $\mathrm{Hom}_{\mathrm{Ind}(\mathcal{C})}(B, \cdot)$ is exact on $\mathcal{I}_{qinj}$. Indeed, we have

$$\mathrm{Hom}_{\mathrm{Ind}(\mathcal{C})}\big(\text{``}\bigoplus_i\text{''}\, X_i, A\big) \simeq \prod_i \mathrm{Hom}_{\mathrm{Ind}(\mathcal{C})}(X_i, A) \,,$$

the functor $\prod_i$ is exact on $\mathrm{Mod}(\mathbb{Z})$ and the functor $\mathrm{Hom}_{\mathrm{Ind}(\mathcal{C})}(X_i, \cdot)$ is exact on the category $\mathcal{I}_{qinj}$.

(B) Let A be quasi-injective. In order to see that $\mathcal{P}^{\mathrm{op}}$ is injective with respect to the functor $\mathrm{Hom}_{\mathrm{Ind}(\mathcal{C})}(\cdot, A)$, we shall apply Theorem 13.3.7.

Consider an epimorphism $B \twoheadrightarrow P''$ with $P'' \in \mathcal{P}$. We shall show that there exists an exact sequence $0 \to P' \to P \to P'' \to 0$ in $\mathcal{P}$ such that $P \to P''$ factorizes through $B \to P''$. Let $P'' = \text{``}\bigoplus_i\text{''}\, X_i''$. By Proposition 8.6.9, there exist an epimorphism $X_i \twoheadrightarrow X_i''$ and a morphism $X_i \to B$ making the diagram below commutative

$$\begin{array}{ccc} X_i & \longrightarrow & X_i'' \\ \downarrow & & \downarrow \\ B & \longrightarrow & \text{``}\bigoplus_i\text{''}\, X_i''. \end{array}$$

Define X_i' as the kernel of $X_i \to X_i''$, and define $P' = \text{``}\bigoplus_i\text{''}\, X_i'$, $P = \text{``}\bigoplus_i\text{''}\, X_i$. Then the sequence $0 \to P' \to P \to P'' \to 0$ is exact.

Let us apply the functor $\mathrm{Hom}_{\mathrm{Ind}(\mathcal{C})}(\cdot, A)$ to this sequence. The formula

$$\mathrm{Hom}\big(\text{``}\bigoplus_i\text{''}\, X_i, A\big) \simeq \prod_i \mathrm{Hom}\,(X_i, A)$$

and the fact that the functor $\prod$ is exact on $\mathrm{Mod}(\mathbb{Z})$ show that the sequence $0 \to \mathrm{Hom}_{\mathrm{Ind}(\mathcal{C})}(P'', A) \to \mathrm{Hom}_{\mathrm{Ind}(\mathcal{C})}(P, A) \to \mathrm{Hom}_{\mathrm{Ind}(\mathcal{C})}(P', A) \to 0$ remains exact.

Hence we have proved (i). The other statements easily follow from (i). q.e.d.

Corollary 15.3.9. *Assume* (15.3.7). *For any* $X \in \mathcal{C}$ *and* $A \in \mathrm{Ind}(\mathcal{C})$, *there is an isomorphism*

$$\varinjlim_{(Y \to A) \in \mathcal{C}_A} \mathrm{Ext}^k_{\mathcal{C}}(X, Y) \xrightarrow{\sim} \mathrm{Ext}^k_{\mathrm{Ind}(\mathcal{C})}(X, A) \,.$$

Proof. For $X \in \mathcal{C}$, let $F \colon \mathrm{Ind}(\mathcal{C}) \to \mathrm{Mod}(\mathbb{Z})$ be the functor $\mathrm{Hom}_{\mathrm{Ind}(\mathcal{C})}(X, \cdot)$. Then $R^+ F \colon \mathrm{D}^+(\mathrm{Ind}(\mathcal{C})) \to \mathrm{D}^+(\mathbb{Z})$ exists and $\mathrm{Ext}^k_{\mathrm{Ind}(\mathcal{C})}(X, \cdot) \simeq R^k F$. On the other hand, $\mathcal{I}_{qinj}$ being F-injective and closed by small filtrant inductive limits, Proposition 15.3.3 implies the isomorphism $\varinjlim_{(Y \to A) \in \mathcal{C}_A} R^k F(Y) \xrightarrow{\sim} R^k F(A)$. Hence, we obtain

$$\mathrm{Ext}^k_{\mathrm{Ind}(\mathcal{C})}(X, A) \simeq \varinjlim_{(Y \to A) \in \mathcal{C}_A} R^k F(Y)$$

$$\simeq \varinjlim_{(Y \to A) \in \mathcal{C}_A} \mathrm{Ext}^k_{\mathrm{Ind}(\mathcal{C})}(X, Y) \,.$$

Finally, Theorem 15.3.1 (i) implies $\mathrm{Ext}^k_{\mathrm{Ind}(\mathcal{C})}(X, Y) \simeq \mathrm{Ext}^k_{\mathcal{C}}(X, Y)$.   q.e.d.

15.4 Indization and Derivation

In this section we shall study some links between the derived category $\mathrm{D}^b(\mathrm{Ind}(\mathcal{C}))$ and the category $\mathrm{Ind}(\mathrm{D}^b(\mathcal{C}))$ associated with an abelian category $\mathcal{C}$. Notice that we do not know whether $\mathrm{Ind}(\mathrm{D}^b(\mathcal{C}))$ is a triangulated category.

Throughout this section we assume that $\mathcal{C}$ satisfies condition (15.3.7). Then $\mathrm{D}^b(\mathrm{Ind}(\mathcal{C}))$ and $\mathrm{D}^b(\mathcal{C})$ are $\mathcal{U}$-categories by Theorem 15.3.8.

The shift automorphism $[n]\colon \mathrm{D}^b(\mathcal{C}) \to \mathrm{D}^b(\mathcal{C})$ gives an automorphism of $\mathrm{Ind}(\mathrm{D}^b(\mathcal{C}))$ that we denote by the same symbol $[n]$.

Let $\tau^{\leq a}$ and $\tau^{\geq b}$ denote the truncation functors from $\mathrm{D}^b(\mathcal{C})$ to itself. They define functors $I\tau^{\leq a}$ and $I\tau^{\geq b}$ from $\mathrm{Ind}(\mathrm{D}^b(\mathcal{C}))$ to itself. If $A \simeq \text{``}\varinjlim\text{''} X_i$ with $X_i \in \mathrm{D}^b(\mathcal{C})$, then $I\tau^{\leq a} A \simeq \text{``}\varinjlim\text{''} \tau^{\leq a} X_i$ and similarly for $\tau^{\geq b}$.

Let $Y \in \mathrm{D}^b(\mathcal{C})$ and let $A \simeq \text{``}\varinjlim_i\text{''} X_i \in \mathrm{Ind}(\mathrm{D}^b(\mathcal{C}))$. The distinguished triangles in $\mathrm{D}^b(\mathcal{C})$

$$\tau^{<a} X_i \to X_i \to \tau^{\geq a} X_i \to (\tau^{<a} X_i)[1]$$

give rise to morphisms

$$\tau^{<a} A \to A \to \tau^{\geq a} A \to (\tau^{<a} A)[1]$$

and to a long exact sequence

$$(15.4.1) \quad \cdots \to \mathrm{Hom}_{\mathrm{Ind}(\mathrm{D}^b(\mathcal{C}))}(Y, I\tau^{<a} A) \to \mathrm{Hom}_{\mathrm{Ind}(\mathrm{D}^b(\mathcal{C}))}(Y, A) \to$$
$$\mathrm{Hom}_{\mathrm{Ind}(\mathrm{D}^b(\mathcal{C}))}(Y, I\tau^{\geq a} A) \to \mathrm{Hom}_{\mathrm{Ind}(\mathrm{D}^b(\mathcal{C}))}(Y, I\tau^{<a} A[1]) \to \cdots \,.$$

There are similar long exact sequences corresponding to the other distinguished triangles in Proposition 13.1.15.

Lemma 15.4.1. *Let $\mathcal{A}$ be an additive category and let $n_0, n_1 \in \mathbb{Z}$ with $n_0 \leq n_1$. There is a natural equivalence*

$$\mathrm{Ind}(\mathrm{C}^{[n_0, n_1]}(\mathcal{A})) \xrightarrow{\sim} \mathrm{C}^{[n_0, n_1]}(\mathrm{Ind}(\mathcal{A})) \,.$$

Proof. Let K be the category associated with the ordered set

$$\{n \in \mathbb{Z}\,; n_0 \leq n \leq n_1\} \,.$$

The natural functors $C^{[n_0,n_1]}(\mathcal{A}) \to \text{Fct}(K, \mathcal{A})$ and $C^{[n_0,n_1]}(\text{Ind}(\mathcal{A})) \to \text{Fct}(K, \text{Ind}(\mathcal{A}))$ are fully faithful, and it follows from Proposition 6.4.1 that $\text{Ind}(C^{[n_0,n_1]}(\mathcal{A})) \to C^{[n_0,n_1]}(\text{Ind}(\mathcal{A}))$ is fully faithful.

Let us show that this last functor is essentially surjective. Theorem 6.4.3 implies that $\text{Ind}(\text{Fct}(K, \mathcal{A})) \to \text{Fct}(K, \text{Ind}(\mathcal{A}))$ is an equivalence of categories, and we obtain the quasi-commutative diagram:

$$
\begin{array}{ccc}
\text{Ind}(C^{[n_0,n_1]}(\mathcal{A})) & \xhookrightarrow{\text{ f.f. }} & \text{Ind}(\text{Fct}(K, \mathcal{A})) \\
{\scriptstyle\text{f.f.}}\downarrow & & \downarrow{\scriptstyle\sim} \\
C^{[n_0,n_1]}(\text{Ind}(\mathcal{A})) & \xhookrightarrow{\text{ f.f. }} & \text{Fct}(K, \text{Ind}(\mathcal{A}))
\end{array}
$$

where the arrows labeled by f.f. are fully faithful functors.

Let $A \in C^{[n_0,n_1]}(\text{Ind}(\mathcal{A}))$, and regard it as an object of $\text{Ind}(\text{Fct}(K, \mathcal{A}))$. By Exercise 6.11, it is enough to show that for $X \in \text{Fct}(K, \mathcal{A})$, any morphism $u \colon X \to A$ factors through an object of $C^{[n_0,n_1]}(\mathcal{A})$.

We shall construct by induction on i an object $Y = Y^{n_0} \to \cdots \to Y^i$ in $C^{[n_0,i]}(\mathcal{A})$ and a diagram $\sigma^{\leq i} X \xrightarrow{w} Y \xrightarrow{v} \sigma^{\leq i} A$ whose composition is equal to $\sigma^{\leq i}(u)$. Assume that we have constructed the diagram of solid arrows

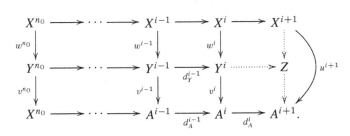

Since the category $\mathcal{A}_{A^{i+1}}$ is filtrant, the dotted arrows may be completed making the diagram commutative. Since the composition $d_A^i \circ d_A^{i-1}$ is zero, the composition $Y^{i-1} \to Y^i \to Z \to A^{i+1}$ is zero. This implies that the morphism $Z \to A$ factorizes through a morphism $Z \to Y^{i+1}$ such that the composition $Y^{i-1} \to Y^i \to Y^{i+1}$ is zero.   q.e.d.

Recall that $Q \colon C^b(\mathcal{C}) \to D^b(\mathcal{C})$ denotes the localization functor. We shall denote by the same letter Q the localization functor $C^b(\text{Ind}(\mathcal{C})) \to D^b(\text{Ind}(\mathcal{C}))$.

Proposition 15.4.2. *Assume* (15.3.7). *Consider integers* $n_0, n_1 \in \mathbb{Z}$ *with* $n_0 \leq n_1$ *and a small and filtrant inductive system* $\{X_i\}_{i \in I}$ *in* $C^{[n_0,n_1]}(\mathcal{C})$. *Let* $Y \in D^b(\mathcal{C})$. *Then:*

$$(15.4.2) \quad \varinjlim_i \text{Hom}_{D^b(\mathcal{C})}(Y, Q(X_i)) \xrightarrow{\sim} \text{Hom}_{D^b(\text{Ind}(\mathcal{C}))}(Y, Q(\text{``}\varinjlim_i\text{''} X_i)).$$

Proof. By dévissage, we may assume $Y \in \mathcal{C}$. By using the truncation functors we are reduced to prove the isomorphisms below for $Y, X_i \in \mathcal{C}$:

$$(15.4.3) \qquad \mathrm{Ext}^k_{\mathrm{Ind}(\mathcal{C})}(Y, \text{``}\varinjlim_i\text{''} X_i) \simeq \varinjlim_i \mathrm{Ext}^k_{\mathcal{C}}(Y, X_i) .$$

These isomorphisms follow from Corollary 15.3.9.   q.e.d.

We define the functor $J: \mathrm{D}^b(\mathrm{Ind}(\mathcal{C})) \to (\mathrm{D}^b(\mathcal{C}))^\wedge$ by setting for $A \in \mathrm{D}^b(\mathrm{Ind}(\mathcal{C}))$ and $Y \in \mathrm{D}^b(\mathcal{C})$

$$(15.4.4) \qquad J(A)(Y) = \mathrm{Hom}_{\mathrm{D}^b(\mathrm{Ind}(\mathcal{C}))}(Y, A) .$$

Hence,

$$J(A) \simeq \text{``}\varinjlim_{(Y \to A) \in \mathrm{D}^b(\mathcal{C})_A}\text{''} \quad Y .$$

Theorem 15.4.3. *Assume* (15.3.7).

(i) *Consider integers $n_0, n_1 \in \mathbb{Z}$ with $n_0 \le n_1$ and a small and filtrant inductive system $\{X_i\}_{i \in I}$ in $\mathrm{C}^{[n_0,n_1]}(\mathcal{C})$. Setting $A := Q(\text{``}\varinjlim_i\text{''} X_i) \in \mathrm{D}^b(\mathrm{Ind}(\mathcal{C}))$, we have $J(A) \simeq \text{``}\varinjlim_i\text{''} Q(X_i)$.*

(ii) *The functor J takes its values in $\mathrm{Ind}(\mathrm{D}^b(\mathcal{C}))$. In particular, for any $A \in \mathrm{D}^b(\mathrm{Ind}(\mathcal{C}))$, the category $\mathrm{D}^b(\mathcal{C})_A$ is cofinally small and filtrant.*

(iii) *For each $k \in \mathbb{Z}$, the diagram below commutes*

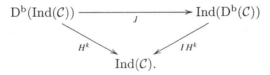

Proof. (i) By Proposition 15.4.2, we have for $Y \in \mathrm{D}^b(\mathcal{C})$

$$\mathrm{Hom}_{\mathrm{Ind}(\mathrm{D}^b(\mathcal{C}))}(Y, J(A)) = \mathrm{Hom}_{\mathrm{D}^b(\mathrm{Ind}(\mathcal{C}))}(Y, A)$$
$$\simeq \varinjlim_i \mathrm{Hom}_{\mathrm{D}^b(\mathcal{C})}(Y, Q(X_i))$$
$$\simeq \mathrm{Hom}_{\mathrm{Ind}(\mathrm{D}^b(\mathcal{C}))}(Y, \text{``}\varinjlim_i\text{''} Q(X_i)) .$$

Therefore, $J(A) \simeq \text{``}\varinjlim_i\text{''} Q(X_i)$.

(ii) Let $A \in \mathrm{D}^b(\mathrm{Ind}(\mathcal{C}))$. There exists A' in $\mathrm{C}^{[n_0,n_1]}(\mathrm{Ind}(\mathcal{C}))$ with $A \simeq Q(A')$. Using Lemma 15.4.1 we may write $A' = \text{``}\varinjlim_{i \in I}\text{''} X_i$ with a small filtrant inductive system $\{X_i\}_{i \in I}$ in $\mathrm{C}^{[n_0,n_1]}(\mathcal{C})$. Then $J(A) \simeq \text{``}\varinjlim_i\text{''} Q(X_i)$ by (i). This object belongs to $\mathrm{Ind}(\mathrm{D}^b(\mathcal{C}))$.

(iii) The morphism $IH^k \circ J \to H^k$ is constructed by the sequence of morphisms

$$IH^k \circ J(A) \simeq IH^k(\underset{(Y \to A) \in D^b(\mathcal{C})_A}{\text{"}\varinjlim\text{"}} Y)$$

$$\simeq \underset{(Y \to A) \in D^b(\mathcal{C})_A}{\text{"}\varinjlim\text{"}} H^k(Y) \to H^k(A) \, .$$

In order to see that it is an isomorphism, let us take an inductive system $\{X_i\}_{i \in I}$ as above. By (i) we have $J(A) \simeq \text{"}\varinjlim_i\text{"} Q(X_i)$. Hence, $IH^k(J(A)) \simeq \text{"}\varinjlim_i\text{"} H^k(Q(X_i)) \simeq \text{"}\varinjlim_i\text{"} H^k(X_i)$. On the other hand, we have $H^k(A) \simeq H^k(Q(\text{"}\varinjlim_i\text{"} X_i)) \simeq H^k(\text{"}\varinjlim_i\text{"} X_i) \simeq \text{"}\varinjlim_i\text{"} H^k(X_i)$.   q.e.d.

Corollary 15.4.4. *Assume* (15.3.7). *Then the functor* $J \colon D^b(\text{Ind}(\mathcal{C})) \to \text{Ind}(D^b(\mathcal{C}))$ *is conservative.*

Remark 15.4.5. The functor $J \colon D^b(\text{Ind}(\mathcal{C})) \to \text{Ind}(D^b(\mathcal{C}))$ is not faithful in general (see Exercise 15.2).

Lemma 15.4.6. *Assume* (15.3.7). *Let* A, $B \in \text{Ind}(D^b(\mathcal{C}))$ *and let* $\varphi \colon A \to B$ *be a morphism in* $\text{Ind}(D^b(\mathcal{C}))$ *such that* $IH^k(\varphi) \colon IH^k(A) \to IH^k(B)$ *is an isomorphism for all* $k \in \mathbb{Z}$. *Assume one of the conditions* (a) *and* (b) *below:*

(a) $A \simeq I\tau^{\geq a} A$ *and* $B \simeq I\tau^{\geq a} B$ *for some* $a \in \mathbb{Z}$,
(b) *the homological dimension of* $\mathcal{C}$ *is finite.*

Then φ *is an isomorphism in* $\text{Ind}(D^b(\mathcal{C}))$.

Proof. Let $Y \in D^b(\mathcal{C})$. It is enough to prove that φ induces an isomorphism $\text{Hom}_{\text{Ind}(D^b(\mathcal{C}))}(Y, A) \xrightarrow{\sim} \text{Hom}_{\text{Ind}(D^b(\mathcal{C}))}(Y, B)$.
(i) Assume (a). By the hypothesis, it is enough to prove the isomorphisms

$$(15.4.5) \quad \text{Hom}_{\text{Ind}(D^b(\mathcal{C}))}(Y, I\tau^{\geq k} A) \xrightarrow{\sim} \text{Hom}_{\text{Ind}(D^b(\mathcal{C}))}(Y, I\tau^{\geq k} B)$$

for all $k \in \mathbb{Z}$, all $m \in \mathbb{Z}$ and all $Y \in D^{\leq m}(\mathcal{C})$. Fixing m, let us prove this result by descending induction on k. If $k > m$, then both sides vanish. Assume that $\text{Hom}_{\text{Ind}(D^b(\mathcal{C}))}(Y, I\tau^{\geq k} A) \to \text{Hom}_{\text{Ind}(D^b(\mathcal{C}))}(Y, I\tau^{\geq k} B)$ is an isomorphism for all $k > n$ and all $Y \in D^{\leq m}(\mathcal{C})$. Applying the long exact sequence (15.4.1) we find a commutative diagram (we shall write Hom instead of $\text{Hom}_{\text{Ind}(D^b(\mathcal{C}))}$ for short)

$$\begin{array}{ccccc} \text{Hom}\,(Y[1], I\tau^{>n} A) & \twoheadrightarrow & \text{Hom}\,(Y, IH^n(A)[-n]) & \longrightarrow & \text{Hom}\,(Y, I\tau^{\geq n} A) \\ \downarrow & & \downarrow & & \downarrow \\ \text{Hom}\,(Y[1], I\tau^{>n} B) & \twoheadrightarrow & \text{Hom}\,(Y, IH^n(B)[-n]) & \longrightarrow & \text{Hom}\,(Y, I\tau^{\geq n} B) \end{array}$$

$$\begin{array}{ccc} \longrightarrow \text{Hom}\,(Y, I\tau^{>n} A) & \longrightarrow & \text{Hom}\,(Y, IH^n(A)[1-n]) \\ \downarrow & & \downarrow \\ \longrightarrow \text{Hom}\,(Y, I\tau^{>n} B) & \longrightarrow & \text{Hom}\,(Y, IH^n(B)[1-n]) \, . \end{array}$$

Since $Y[1]$ and Y belong to $D^{\le m}(\mathcal{C})$, the first and the fourth vertical arrows are isomorphisms by the induction hypothesis. The second and the fifth vertical arrows are isomorphisms by the hypothesis. Hence, the third vertical arrow is an isomorphism, and the induction proceeds.

(ii) Assume (b) and let d denote the homological dimension of $\mathcal{C}$. If $Y \in D^{\ge n_0}(\mathcal{C})$ then $\operatorname{Hom}_{\operatorname{Ind}(D^b(\mathcal{C}))}(Y, \tau^{\le n}A) \simeq 0$ for $n < n_0 - d$. We get the isomorphism $\operatorname{Hom}_{\operatorname{Ind}(D^b(\mathcal{C}))}(Y, A) \simeq \operatorname{Hom}_{\operatorname{Ind}(D^b(\mathcal{C}))}(Y, \tau^{\ge n}A)$, and similarly with A replaced with B. Then the result follows from the case (i).   q.e.d.

Proposition 15.4.7. *Assume that $\mathcal{C}$ and $\mathcal{C}'$ satisfy (15.3.7). Consider a triangulated functor $\psi\colon D^b(\operatorname{Ind}(\mathcal{C})) \to D^b(\operatorname{Ind}(\mathcal{C}'))$ which satisfies:*

(15.4.6) $H^k\psi\colon \operatorname{Ind}(\mathcal{C})) \to \operatorname{Ind}(\mathcal{C}')$ *commutes with small filtrant inductive limits,*

(15.4.7) ψ *sends $D^{\ge 0}(\operatorname{Ind}(\mathcal{C})) \cap D^b(\operatorname{Ind}(\mathcal{C}))$ to $D^{\ge n}(\operatorname{Ind}(\mathcal{C}))$ for some n .*

Then there exists a unique functor $\lambda\colon \operatorname{Ind}(D^b(\mathcal{C})) \to \operatorname{Ind}(D^b(\mathcal{C}'))$ which commutes with small filtrant "$\varinjlim$" and such that the diagram below commutes:

$$
\begin{array}{ccc}
D^b(\operatorname{Ind}(\mathcal{C})) & \xrightarrow{\ \psi\ } & D^b(\operatorname{Ind}(\mathcal{C}')) \\
{\scriptstyle J}\downarrow & & {\scriptstyle J}\downarrow \\
\operatorname{Ind}(D^b(\mathcal{C})) & \xrightarrow{\ \lambda\ } & \operatorname{Ind}(D^b(\mathcal{C}')).
\end{array}
$$

Proof. First, notice that (15.4.6) implies that, for $n_0, n_1 \in \mathbb{Z}$ with $n_0 \le n_1$ and for any small filtrant inductive system $\{X_i\}_{i \in I}$ in $C^{[n_0, n_1]}(\mathcal{C})$, there is an isomorphism

$$
\varinjlim_i H^k(\psi \circ Q(X_i)) \simeq H^k(\psi \circ Q(\varinjlim_i X_i)) \ .
$$

Denote by $\varphi\colon D^b(\mathcal{C}) \to \operatorname{Ind}(D^b(\mathcal{C}'))$ the restriction of $J \circ \psi$ to $D^b(\mathcal{C})$. The functor φ naturally extends to a functor $\lambda\colon \operatorname{Ind}(D^b(\mathcal{C})) \to \operatorname{Ind}(D^b(\mathcal{C}'))$ such that λ commutes with small filtrant inductive limits. We construct a natural morphism of functors

$$
u\colon \lambda \circ J \to J \circ \psi
$$

as follows. For $A \in D^b(\operatorname{Ind}(\mathcal{C}))$,

$$
\lambda \circ J(A) \simeq \lambda(\varinjlim_{(Y \to A) \in D^b(\mathcal{C})_A} Y) \simeq \varinjlim_{(Y \to A) \in D^b(\mathcal{C})_A} J \circ \psi(Y)
$$
$$
\to J \circ \psi(A) \ .
$$

Let us show that u is an isomorphism. Consider a small filtrant inductive system $\{X_i\}_{i \in I}$ in $C^{[n_0, n_1]}(\mathcal{C})$ such that $A \simeq Q(\varinjlim_i X_i) \in D^b(\operatorname{Ind}(\mathcal{C}))$. We have the chain of isomorphisms

$$IH^k(\lambda \circ J(A)) \simeq \text{``}\varinjlim_i\text{''} H^k(\psi(Q(X_i))) \simeq H^k(\psi(A))$$

$$\simeq IH^k(J \circ \psi(A)) .$$

Since $\lambda \circ J(A) \simeq I\tau^{\geq a}(\lambda \circ J(A))$ and $J \circ \psi(A) \simeq I\tau^{\geq a}(J \circ \psi(A))$ for $a \ll 0$, the result follows by Lemma 15.4.6. q.e.d.

Let $\mathcal{T}$ be a full triangulated subcategory of $D^b(\mathcal{C})$. We identify $\mathrm{Ind}(\mathcal{T})$ with a full subcategory of $\mathrm{Ind}(D^b(\mathcal{C}))$. For $A \in D^b(\mathrm{Ind}(\mathcal{C}))$, we denote as usual by $\mathcal{T}_A$ the category of arrows $Y \to A$ with $Y \in \mathcal{T}$. We know by Proposition 10.1.18 that $\mathcal{T}_A$ is filtrant.

Notation 15.4.8. Let $\mathcal{T}$ be a full triangulated subcategory of $D^b(\mathcal{C})$. We denote by $J^{-1}\mathrm{Ind}(\mathcal{T})$ the full subcategory of $D^b(\mathrm{Ind}(\mathcal{C}))$ consisting of objects $A \in D^b(\mathrm{Ind}(\mathcal{C}))$ such that $J(A)$ is isomorphic to an object of $\mathrm{Ind}(\mathcal{T})$.

Note that $A \in D^b(\mathrm{Ind}(\mathcal{C}))$ belongs to $J^{-1}\mathrm{Ind}(\mathcal{T})$ if and only if any morphism $X \to A$ with $X \in D^b(\mathcal{C})$ factors through an object of $\mathcal{T}$ by Exercise 6.11.

Proposition 15.4.9. *Assume* (15.3.7). *The category* $J^{-1}\mathrm{Ind}(\mathcal{T})$ *is a triangulated subcategory of* $D^b(\mathrm{Ind}(\mathcal{C}))$.

Proof. Let $A \xrightarrow{f} B \xrightarrow{g} C \to A[1]$ be a d.t. in $D^b(\mathrm{Ind}(\mathcal{C}))$ with B, C in $J^{-1}\mathrm{Ind}(\mathcal{T})$. Let us show that $A \in J^{-1}\mathrm{Ind}(\mathcal{T})$. Let $u \colon X \to A$ be a morphism with $X \in D^b(\mathcal{C})$. Since $B \in J^{-1}\mathrm{Ind}(\mathcal{T})$, the composition $X \to A \to B$ factors through $Y \in \mathcal{T}$. We have thus a commutative diagram

$$
\begin{array}{ccccccc}
X & \longrightarrow & Y & \longrightarrow & Z & \longrightarrow & X[1] \\
\downarrow{\scriptstyle u} & & \downarrow & & \downarrow & & \downarrow \\
A & \xrightarrow{f} & B & \xrightarrow{g} & C & \longrightarrow & A[1]
\end{array}
$$

in which the rows are d.t.'s and $X, Z \in D^b(\mathcal{C})$, $Y \in \mathcal{T}$. Since $C \in J^{-1}\mathrm{Ind}(\mathcal{T})$, the arrow $Z \to C$ factors through $Z' \in \mathcal{T}$. Let us embed the composition $Y \to Z \to Z'$ into a d.t. $X' \to Y \to Z' \to X'[1]$ in $\mathcal{T}$. We thus have a commutative diagram whose rows are d.t.'s

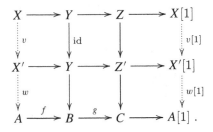

Since $x := u - w \circ v$ satisfies $x \circ f = 0$, it factors through $C[-1] \to A$. Since $C[-1] \in J^{-1}\mathrm{Ind}(\mathcal{T})$, the morphism $X \to C[-1]$ factors through $X'' \in \mathcal{T}$. Thus $x \colon X \to A$ factors through X''. It follows that $u = x + w \circ v$ factors through $X' \oplus X'' \in \mathcal{T}$. q.e.d.

Exercises

Exercise 15.1. Let $\mathcal{C}$ be an abelian category and assume that $\mathrm{D}^b(\mathcal{C})$ is a $\mathcal{U}$-category. Let $A \in \mathrm{Ind}(\mathrm{D}^b(\mathcal{C}))$ which satisfies the two conditions
(a) there exist $a, b \in \mathbb{Z}$ such that $I\tau^{\leq b}A \xrightarrow{\sim} A \xrightarrow{\sim} I\tau^{\geq a}A$,
(b) $IH^n(A) \in \mathcal{C}$ for any $n \in \mathbb{Z}$.
Prove that $A \in \mathrm{D}^b(\mathcal{C})$. (Hint: argue by induction on $b - a$ and use Exercise 10.14.)

Exercise 15.2. In this exercise, we shall give an example for Remark 15.4.5. Let k be a field and set $\mathcal{C} = \mathrm{Mod}(k)$. Let $J \colon \mathrm{D}^b(\mathrm{Ind}(\mathcal{C})) \to \mathrm{Ind}(\mathrm{D}^b(\mathcal{C}))$ be the canonical functor.
(i) Prove that, for any $X, Y \in \mathrm{Ind}(\mathcal{C})$, $\mathrm{Hom}_{\mathrm{Ind}(\mathrm{D}^b(\mathcal{C}))}(J(X), J(Y[n])) \simeq 0$ for any $n \neq 0$. (Hint: any object of $\mathrm{D}^b(\mathcal{C})$ is a finite direct sum of $Z[m]$'s where $Z \in \mathcal{C}$.)
(ii) Let $Z \in \mathcal{C}$. Prove that the short exact sequence $0 \to \kappa_{\mathcal{C}}(Z) \to Z \to Z/(\kappa_{\mathcal{C}}(Z)) \to 0$ splits in $\mathrm{Ind}(\mathcal{C})$ if and only if Z is a finite-dimensional vector space.
(iii) Deduce that J is not faithful.

16

Grothendieck Topologies

As already mentioned, sheaves on topological spaces were invented by Leray and this notion was extended to sheaves on categories by Grothendieck who noticed that the notion of sheaves on a topological space X essentially relies on the category Op_X of open subsets of X and on the notion of open coverings, nothing else. Hence to define sheaves on a category $\mathcal{C}$, it is enough to axiomatize the notion of a covering which defines a so-called Grothendieck topology on $\mathcal{C}$.

Notice that, even in the topological case, if $\{U_i\}_{i \in I}$ is a covering of an open subset U, there is no natural object describing it in the category Op_X, but it is possible to consider the coproduct of the U_i's in the category $(\mathrm{Op}_X)^\wedge$. Hence, to define the notion of a covering on $\mathcal{C}$, we work in $\mathcal{C}^\wedge$, the category of presheaves of sets on $\mathcal{C}$.

Here, we first give the axioms of Grothendieck topologies using sieves and then introduce the notions of local epimorphisms and local isomorphisms. We give several examples and study in some details the properties of the family of local isomorphisms, showing in particular that this family is stable by inductive limits.

Important related topics, such as Topos Theory, will not be approached in this book.

References are made to [64].

16.1 Sieves and Local Epimorphisms

Let $\mathcal{C}$ be a category.

Definition 16.1.1. *Let $U \in \mathrm{Ob}(\mathcal{C})$. A sieve[1] S over U is a subset of $\mathrm{Ob}(\mathcal{C}_U)$ such that the composition $W \to V \to U$ belongs to S as soon as $V \to U$ belongs to S.*

---

[1] " Un crible" in French

To a sieve S over U, we associate a subobject A_S of U in $\mathcal{C}^\wedge$ by taking

$$(16.1.1) \quad A_S(V) = \left\{ s \in \mathrm{Hom}_{\mathcal{C}}(V, U) \, ; \, (V \xrightarrow{s} U) \in S \right\} \quad \text{for any } V \in \mathcal{C} .$$

If $\mathcal{C}$ is small, we have $A_S = \mathrm{Im}\left(\underset{(V \to U) \in S}{\text{"}\bigsqcup\text{"}} V \to U \right)$.

Conversely, to an object $A \to U$ of $(\mathcal{C}^\wedge)_U$ we associate a sieve S_A by taking

$$(16.1.2) \quad (V \to U) \in S_A \text{ if and only if } V \to U \text{ decomposes as } V \to A \to U .$$

Note that $S_A = S_{\mathrm{Im}(A \to U)}$. Hence, there is a one-to-one correspondence between the family of sieves over U and the family of subobjects of U in $\mathcal{C}^\wedge$.

Definition 16.1.2. *A* Grothendieck topology (*or simply a topology*) *on a category $\mathcal{C}$ is the data of a family $\{\mathcal{S}\mathrm{Cov}_U\}_{U \in \mathrm{Ob}(\mathcal{C})}$, where $\mathcal{S}\mathrm{Cov}_U$ is a family of sieves over U, these data satisfying the axioms* GT1–GT4 *below.*

GT1 $\mathrm{Ob}(\mathcal{C}_U)$ *belongs to $\mathcal{S}\mathrm{Cov}_U$.*

GT2 *If $S_1 \subset S_2 \subset \mathrm{Ob}(\mathcal{C}_U)$ are sieves and if S_1 belongs to $\mathcal{S}\mathrm{Cov}_U$, then S_2 belongs to $\mathcal{S}\mathrm{Cov}_U$.*

GT3 *Let $U \to V$ be a morphism in $\mathcal{C}$. If S belongs to $\mathcal{S}\mathrm{Cov}_V$, then $S \times_V U$ belongs to $\mathcal{S}\mathrm{Cov}_U$. Here,*

$$S \times_V U := \{W \to U \, ; \, \text{the composition } W \to U \to V \text{ belongs to } \mathcal{S}\mathrm{Cov}_V\} .$$

GT4 *Let S and S' be sieves over U. Assume that $S' \in \mathcal{S}\mathrm{Cov}_U$ and that $S \times_U V \in \mathcal{S}\mathrm{Cov}_V$ for any $(V \to U) \in S'$. Then $S \in \mathcal{S}\mathrm{Cov}_U$.*

A sieve S over U is called a covering sieve *if $S \in \mathcal{S}\mathrm{Cov}_U$.*

Definition 16.1.3. *Let $\mathcal{C}$ be a category endowed with a Grothendieck topology.*

(i) *A morphism $A \to U$ in $\mathcal{C}^\wedge$ with $U \in \mathcal{C}$ is called a* local epimorphism *if the sieve S_A given by (16.1.2) is a covering sieve over U.*

(ii) *A morphism $A \to B$ in $\mathcal{C}^\wedge$ is called a* local epimorphism *if for any $V \in \mathcal{C}$ and any morphism $V \to B$, $A \times_B V \to V$ is a local epimorphism.*

Consider a local epimorphism $A \to U$ as in Definition 16.1.3 (i) and let $V \to U$ be a morphism in $\mathcal{C}$. The sieve $S_{A \times_U V} = S_A \times_U V$ is a covering sieve over V by GT3 and it follows that $A \times_U V \to V$ is a local epimorphism. Therefore, if we take $B = U \in \mathcal{C}$ in Definition 16.1.3 (ii), we recover Definition 16.1.3 (i).

The family of local epimorphisms associated with a Grothendieck topology will satisfy the following properties (the verification is left to the reader):

LE1 For any $U \in \mathcal{C}$, $\mathrm{id}_U : U \to U$ is a local epimorphism.

LE2 Let $A_1 \xrightarrow{u} A_2 \xrightarrow{v} A_3$ be morphisms in $\mathcal{C}^\wedge$. If u and v are local epimorphisms, then $v \circ u$ is a local epimorphism.

LE3 Let $A_1 \xrightarrow{u} A_2 \xrightarrow{v} A_3$ be morphisms in $\mathcal{C}^\wedge$. If $v \circ u$ is a local epimorphism, then v is a local epimorphism.

LE4 A morphism $u \colon A \to B$ in $\mathcal{C}^\wedge$ is a local epimorphism if and only if for any $U \in \mathcal{C}$ and any morphism $U \to B$, the morphism $A \times_B U \to U$ is a local epimorphism.

Conversely, consider a family of morphisms in $\mathcal{C}^\wedge$ satisfying LE1–LE4. Let us say that a sieve S over U is a covering sieve if $A_S \to U$ is a local epimorphism, where A_S is given by (16.1.1). Then it is easily checked that the axioms GT1–GT4 will be satisfied. In other words, a Grothendieck topology can alternatively be defined by starting from a family of morphisms in $\mathcal{C}^\wedge$ satisfying LE1–LE4.

Note that a family of morphisms in $\mathcal{C}^\wedge$ satisfies LE1–LE4 if and only if it satisfies LE2–LE4 and LE1' below:

LE1' If $u \colon A \to B$ is an epimorphism in $\mathcal{C}^\wedge$, then u is a local epimorphism.

Indeed, LE1' implies LE1. Conversely, assume that $u \colon A \to B$ is an epimorphism in $\mathcal{C}^\wedge$. If $w \colon U \to B$ is a morphism with $U \in \mathcal{C}$, there exists $v \colon U \to A$ such that $w = u \circ v$. Hence, $\mathrm{id}_U \colon U \to U$ factors as $U \to A \times_B U \to U$. Therefore $A \times_B U \to U$ is a local epimorphism by LE1 and LE3, and this implies that $A \to B$ is a local epimorphism by LE4. This is visualized by:

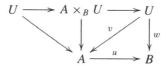

Definition 16.1.4. *Let $\mathcal{C}$ be a small category and $U \in \mathcal{C}$. Consider two small families of objects of $\mathcal{C}_U$, $\mathcal{S}_1 = \{U_i\}_{i \in I}$ and $\mathcal{S}_2 = \{V_j\}_{j \in J}$. The family $\mathcal{S}_1$ is a refinement of $\mathcal{S}_2$ if for any $i \in I$ there exist $j \in J$ and a morphism $U_i \to V_j$ in $\mathcal{C}_U$. In such a case, we write $\mathcal{S}_1 \preceq \mathcal{S}_2$.*

Note that $\mathcal{S}_1 = \{U_i\}_{i \in I}$ is a refinement of $\mathcal{S}_2 = \{V_j\}_{j \in J}$ if and only if

$$(16.1.3) \qquad \mathrm{Hom}_{\mathcal{C}_U^\wedge}\left(\text{``}\coprod_i\text{''}\, U_i, \text{``}\coprod_j\text{''}\, V_j\right) \neq \emptyset .$$

Definition 16.1.5. *Let $\mathcal{C}$ be a small category which admits fiber products. Assume that $\mathcal{C}$ is endowed with a Grothendieck topology and let $U \in \mathcal{C}$. A small family $\mathcal{S} = \{U_i\}_{i \in I}$ of objects of $\mathcal{C}_U$ is a covering of U if the morphism $\text{``}\coprod_i\text{''}\, U_i \to U$ is a local epimorphism.*

Denote by Cov_U the family of coverings of U. The family of coverings will satisfy the axioms COV1–COV4 below.

COV1 $\{U\}$ belongs to Cov_U.

COV2 If $\mathcal{S}_1 \in \mathrm{Cov}_U$ is a refinement of a family $\mathcal{S}_2 \subset \mathrm{Ob}(\mathcal{C}_U)$, then $\mathcal{S}_2 \in \mathrm{Cov}_U$.

COV3 If $\mathcal{S} = \{U_i\}_{i \in I}$ belongs to Cov_U, then $\mathcal{S} \times_U V := \{U_i \times_U V\}_{i \in I}$ belongs to Cov_V for any morphism $V \to U$ in $\mathcal{C}$.

COV4 If $\mathcal{S}_1 = \{U_i\}_{i \in I}$ belongs to Cov_U, $\mathcal{S}_2 = \{V_j\}_{j \in J}$ is a small family of objects of $\mathcal{C}_U$, and $\mathcal{S}_2 \times_U U_i$ belongs to Cov_{U_i} for any $i \in I$, then $\mathcal{S}_2$ belongs to Cov_U.

Conversely, to a covering $\mathcal{S} = \{U_i\}_{i \in I}$ of U, we associate a sieve S over U by setting

$$S = \{\varphi \in \mathrm{Hom}_{\mathcal{C}}(V, U); \ \varphi \text{ factors through } U_i \to U \text{ for some } i \in I\} \, .$$

If the family of coverings satisfies COV1–COV4, it is easily checked that the associated family of sieves $S\mathrm{Cov}_U$ will satisfy the axioms GT1–GT4.

In this book, we shall mainly use the notion of local epimorphisms. However, we started by introducing sieves, because this notion does not depend on the choice of a universe.

In the sequel, $\mathcal{C}$ is a category endowed with a Grothendieck topology.

Lemma 16.1.6. *Let* $u \colon A \to B$ *be a morphism in* $\mathcal{C}^\wedge$. *The conditions below are equivalent.*

(i) *u is a local epimorphism,*
(ii) *for any* $t \colon U \to B$ *with* $U \in \mathcal{C}$, *there exist a local epimorphism* $u \colon C \to U$ *and a morphism* $s \colon C \to A$ *such that* $u \circ s = t \circ u$,
(iii) *$\mathrm{Im}\, u \to B$ is a local epimorphism.*

Proof. (i) $\Rightarrow$ (ii) is obvious.
(ii) $\Rightarrow$ (i). Let $C \to U$ be a local epimorphism. It factorizes through $A \times_B U \to U$ by the hypothesis. Therefore $A \times_B U \to U$ is a local epimorphism by LE3 and the result follows from LE4.
(i) $\Rightarrow$ (iii) follows from LE3.
(iii) $\Rightarrow$ (i) follows from LE1' and LE2.   q.e.d.

Example 16.1.7. Let X be a topological space, $\mathcal{C}_X := \mathrm{Op}_X$ the category of its open subsets. Note that $\mathcal{C}_X$ admits a terminal object, namely X, and the products of two objects U, $V \in \mathcal{C}_X$ is $U \cap V$. Also note that if U is an open subset of X, then $(\mathcal{C}_X)_U \simeq \mathrm{Op}_U$. We define a Grothendieck topology by deciding that a small family $\mathcal{S} = \{U_i\}_{i \in I}$ of objects of Op_U belongs to Cov_U if $\bigcup_i U_i = U$.

We may also define a Grothendieck topology as follows. A morphism $u \colon A \to B$ in $(\mathcal{C}_X)^\wedge$ is a local epimorphism if for any $U \in \mathrm{Op}_X$ and any $t \in B(U)$, there exist a covering $U = \bigcup_i U_i$ and for each i an $s_i \in A(U_i)$ with $u(s_i) = t|_{U_i}$. (Here, $t|_{U_i}$ is the image of t by $B(U) \to B(U_i)$.) Hence, a morphism $A \to U$ in $(\mathcal{C}_X)^\wedge$ ($U \in \mathcal{C}$) is a local epimorphism if there exists an open covering $U = \bigcup_{i \in I} U_i$ such that $U_i \to U$ factorizes through A for every $i \in I$.

These two definitions give the same topology. We shall call this Grothendieck topology the "associated Grothendieck topology" on X.

Example 16.1.8. For a real analytic manifold X, denote by $\mathcal{C}_{X_{\mathrm{sa}}}$ the full subcategory of $\mathcal{C}_X = \mathrm{Op}_X$ consisting of open subanalytic subsets (see [38] for an exposition). We define a Grothendieck topology on the category $\mathcal{C}_{X_{\mathrm{sa}}}$ by deciding that a small family $\mathcal{S} = \{U_i\}_{i \in I}$ of subobjects of $U \in \mathcal{C}_{\mathrm{sa}}$ belongs to Cov_U if for any compact subset K of X, there is a finite subset $J \subset I$ such that $\bigcup_{j \in J} U_j \cap K = U \cap K$. We call this Grothendieck topology the subanalytic topology on X. This topology naturally arises in Analysis, for example when studying temperate holomorphic functions. References are made to [39].

Examples 16.1.9. Let $\mathcal{C}$ be a category.
(i) We may endow $\mathcal{C}$ with a Grothendieck topology by deciding that the local epimorphisms in $\mathcal{C}^{\wedge}$ are the epimorphisms. This topology is called the final topology.
(ii) We may endow $\mathcal{C}$ with a Grothendieck topology by deciding that all morphisms are local epimorphisms. This topology is called the initial topology.
(iii) Recall that $\mathbf{Pt}$ denotes the category with one object c and one morphism. We endow this category with the final topology. Note that this topology is different from the initial one. Indeed, the morphism $\emptyset_{\mathbf{Pt}^{\wedge}} \to c$ in $\mathbf{Pt}^{\wedge}$ is a local epimorphism for the initial topology, not for the final one. In other words, the empty covering of pt is a covering for the initial topology, not for the final one.

Examples 16.1.10. The following examples are extracted from [51].
Let G be a finite group and denote by $G\text{-}\mathbf{Top}$ the category of small G-topological spaces. An object is a small topological space X endowed with a continuous action of G, and a morphism $f \colon X \to Y$ is a continuous map which commutes with the action of G. Such an f is said to be G-equivariant.

The category $\mathcal{E}t_G$ is defined as follows. Its objects are those of $G\text{-}\mathbf{Top}$ and its morphisms $f \colon V \to U$ are the G-equivariant maps such that f is a local homeomorphism. Note that $f(V)$ is open in U.

The category $\mathcal{E}t_G$ admits fiber products. If $U \in G\text{-}\mathbf{Top}$, then the category $\mathcal{E}t_G(U) := (\mathcal{E}t_G)_U$ admits finite projective limits.

(i) The étale topology on $\mathcal{E}t_G$ is defined as follows. A sieve S over $U \in \mathcal{E}t_G$ is a covering sieve if for any x, there exists a morphism $f \colon V \to U$ in S such that $x \in f(V)$.
(ii) The Nisnevich topology on $\mathcal{E}t_G$ is defined as follows. A sieve S over $U \in \mathcal{E}t_G$ is a covering sieve if for any $x \in U$ there exist a morphism $V \to U$ in S and $y \in V$ such that $f(y) = x$ and y has the same isotropy group as x. (The isotropy group G_y of y is the subgroup of G consisting of $g \in G$ satisfying $g \cdot y = y$.)
(iii) The Zariski topology on $\mathcal{E}t_G$ is defined as follows. A sieve S over $U \in \mathcal{E}t_G$ is a covering sieve if for any $x \in U$, there exists an open embedding $f \colon V \to U$ in S such that $x \in f(V)$.

It is easily checked that the axioms of Grothendieck topologies are satisfied in these three cases.

Proposition 16.1.11. (i) *Let* $u: A \to B$ *be a local epimorphism and let* $v: C \to B$ *be a morphism. Then* $A \times_B C \to C$ *is a local epimorphism.*
(ii) *If* $u: A \to B$ *is a morphism in* $C^\wedge$, $v: C \to B$ *is a local epimorphism and* $w: A \times_B C \to C$ *is a local epimorphism, then* u *is a local epimorphism.*

Property (i) is translated by saying that "local epimorphisms are stable by base change" and property (ii) by saying that for $u: A \to B$ to be a local epimorphism is a local property on B.

Proof. (i) For any $U \to C$ with $U \in C$, $(A \times_B C) \times_C U \simeq A \times_B U \to U$ is a local epimorphism.
(ii) It follows from the hypothesis that $v \circ w$ is a local epimorphism. Denote by $s: A \times_B C \to A$ the natural morphism. Then $v \circ w = u \circ s$, and u is a local epimorphism by LE3.   q.e.d.

Proposition 16.1.12. *Let* I *be a small category and let* $\alpha: I \to \mathrm{Mor}(C^\wedge)$ *be a functor. Assume that for each* $i \in I$, $\alpha(i): A_i \to B_i$ *is a local epimorphism. Let* $u: A \to B$ *denote the inductive limit in* $\mathrm{Mor}(C^\wedge)$ *of* α. *Then* u *is a local epimorphism.*

Proof. Consider a morphism $v: V \to B$ with $V \in C$. There exists $i \in I$ such that v factorizes as $V \to B_i \to B$. By the hypothesis, $A_i \times_{B_i} V \to V$ is a local epimorphism. Since this morphism factorizes through $A \times_B V \to V$, this last morphism is a local epimorphism by LE3.   q.e.d.

16.2 Local Isomorphisms

Consider a morphism $u: A \to B$ in $C^\wedge$. Recall (see Exercise 2.4) that the associated diagonal morphism $A \to A \times_B A$ is a monomorphism. It is an epimorphism if and only if u is a monomorphism. This naturally leads to the following:

Definition 16.2.1. (i) *We say that a morphism* $u: A \to B$ *in* $C^\wedge$ *is a* local monomorphism *if* $A \to A \times_B A$ *is a local epimorphism.*
(ii) *We say that a morphism* $u: A \to B$ *in* $C^\wedge$ *is a* local isomorphism *if it is both a local epimorphism and a local monomorphism.*

Example 16.2.2. Let X be a topological space and let $C = \mathrm{Op}_X$ with the associated Grothendieck topology (see Example 16.1.7). Let $A = \text{"}\coprod_{i \in I}\text{"} U_i$ and $B = \text{"}\coprod_{j \in J}\text{"} V_j$, where the U_i's and V_j's are open in X. Any morphism $u: A \to B$ is induced by a map $\varphi: I \to J$ such that $U_i \subset V_{\varphi(i)}$ for all $i \in I$. Notice that

(i) u is a local epimorphism if and only if, for any $j \in J$, $V_j = \bigcup_{i \in \varphi^{-1}(j)} U_i$,

(ii) let U be an open subset, $\{U_i\}_{i \in I}$ an open covering of U, and for each $i, i' \in I$ let $\{W_j\}_{j \in J(i,i')}$ be an open covering of $U_i \cap U_{i'}$. Set

$$C := \mathrm{Coker}\Big(\underset{i,i' \in I, j \in J(i,i')}{\text{"}\coprod\text{"}} W_j \rightrightarrows \underset{i \in I}{\text{"}\coprod\text{"}} U_i \Big).$$

Then $C \to U$ is a local isomorphism (see Exercise 16.6). Conversely, for any local isomorphism $A \to U$, we can find families $\{U_i\}_{i \in I}$ and $\{W_j\}_{j \in J(i,i')}$ as above such that $C \to U$ factors as $C \to A \to U$. It is a classical result (see [27], Lemma 3.8.1) that if U is normal and paracompact, we can take $W_j = U_i \times_U U_{i'}$, i.e., $C = \mathrm{Im}(\underset{i \in I}{\text{"}\coprod\text{"}} U_i \to U)$.

Lemma 16.2.3. (i) *If $u \colon A \to B$ is a monomorphism, then it is a local monomorphism. In particular, a monomorphism which is a local epimorphism is a local isomorphism.*

(ii) *If $u \colon A \to B$ is a local epimorphism, then $\mathrm{Im}(A \to B) \to B$ is a local isomorphism.*

(iii) *For a morphism $u \colon A \to B$, the conditions below are equivalent.*

(a) *$u \colon A \to B$ is a local monomorphism,*

(b) *for any diagram $U \rightrightarrows A \to B$ with $U \in C$ such that the two compositions coincide, there exists a local epimorphism $S \to U$ such that the two compositions $S \to U \rightrightarrows A$ coincide,*

(c) *for any diagram $Z \rightrightarrows A \to B$ with $Z \in C^\wedge$ such that the two compositions coincide, there exists a local epimorphism $S \to Z$ such that the two compositions $S \to Z \rightrightarrows A$ coincide.*

Proof. (i)–(ii) are obvious.

(iii) Notice first that a morphism $U \to A \times_B A$ is nothing but a diagram $U \rightrightarrows A \to B$ such that the two compositions coincide, and then any diagram $S \to U \rightrightarrows A$ such that the two compositions coincide factorizes as $S \to A \underset{A \times_B A}{\times} U \to U$.

(b) $\Rightarrow$ (a). Let $U \to A \times_B A$ be a morphism. Let $S \to U$ be a local epimorphism such that the two compositions $S \to U \rightrightarrows A$ coincide. Then $S \to U$ factorizes through $A \underset{A \times_B A}{\times} U \to U$ and this morphism will be a local epimorphism. By LE4, this implies (a).

(a) $\Rightarrow$ (c). Given $Z \to A \times_B A$, take $A \underset{A \times_B A}{\times} Z \to Z$ as $S \to Z$.   q.e.d.

Lemma 16.2.4. (i) *Let $u \colon A \to B$ be a local monomorphism (resp. local isomorphism) and let $v \colon C \to B$ be a morphism. Then $A \times_B C \to C$ is a local monomorphism (resp. local isomorphism).*

(ii) *Conversely, let $u \colon A \to B$ be a morphism and let $v \colon C \to B$ be a local epimorphism. If $A \times_B C \to C$ is a local monomorphism (resp. local isomorphism), then u is a local monomorphism (resp. local isomorphism).*

(iii) *Let $A_1 \overset{u}{\to} A_2 \overset{v}{\to} A_3$ be morphisms in $C^\wedge$. If u and v are local monomorphisms, then $v \circ u$ is a local monomorphism.*

(iv) *Let $A_1 \xrightarrow{u} A_2 \xrightarrow{v} A_3$ be morphisms in $\mathcal{C}^\wedge$. If $v \circ u$ is a local epimorphism and v is a local monomorphism, then u is a local epimorphism.*

(v) *Let $A_1 \xrightarrow{u} A_2 \xrightarrow{v} A_3$ be morphisms in $\mathcal{C}^\wedge$. If $v \circ u$ is a local monomorphism, then u is a a local monomorphism.*

(vi) *Let $A_1 \xrightarrow{u} A_2 \xrightarrow{v} A_3$ be morphisms in $\mathcal{C}^\wedge$. If $v \circ u$ is a local monomorphism and u is a local epimorphism, then v is a local monomorphism.*

(vii) *Let $A_1 \xrightarrow{u} A_2 \xrightarrow{v} A_3$ be morphisms in $\mathcal{C}^\wedge$. If two of the three morphisms u, v, $v \circ u$ are local isomorphisms, then all are local isomorphisms.*

Proof. (i) (a) Assume that u is a local monomorphism. Let $D = A \times_B C$. Consider the commutative diagram

(16.2.1)
$$
\begin{array}{ccccc}
D & \xrightarrow{w'} & D \times_C D & \longrightarrow & C \\
\downarrow & & \downarrow{\scriptstyle h} & & \downarrow{\scriptstyle v} \\
A & \xrightarrow{w} & A \times_B A & \longrightarrow & B.
\end{array}
$$

Since both squares (A, B, C, D) and $(A \times_B A, B, C, D \times_C D)$ are Cartesian, the square $(A, A \times_B A, D \times_C D, D)$ is Cartesian. Since $A \to A \times_B A$ is a local epimorphism, $D \to D \times_C D$ is also a local epimorphism.

(i) (b) Since both local epimorphisms and local monomorphisms are stable by base change, the same result holds for local isomorphisms.

(ii) It is enough to treat the case where $A \times_B C \to C$ is a local monomorphism. In the diagram (16.2.1), h is a local epimorphism. Since w' is a local epimorphism, so is w by Proposition 16.1.11 (ii).

(iii) Consider the diagram

$$
\begin{array}{ccccc}
A_1 & \xrightarrow{u'} & A_1 \times_{A_2} A_1 & \xrightarrow{w'} & A_1 \times_{A_3} A_1 \\
 & & \downarrow & & \downarrow \\
 & & A_2 & \xrightarrow{v'} & A_2 \times_{A_3} A_2.
\end{array}
$$

Since the square is Cartesian and v' is a local epimorphism, w' is also a local epimorphism. Therefore $w' \circ u'$ is again a local epimorphism.

(iv) Consider the Cartesian squares

$$
\begin{array}{ccc}
A_1 & \xrightarrow{w_1} & A_1 \times_{A_3} A_2 \\
\downarrow & & \downarrow \\
A_2 & \xrightarrow{v'} & A_2 \times_{A_3} A_2,
\end{array}
\qquad
\begin{array}{ccc}
A_1 \times_{A_3} A_2 & \xrightarrow{w_2} & A_2 \\
\downarrow & & \downarrow{\scriptstyle v} \\
A_1 & \xrightarrow{v \circ u} & A_3.
\end{array}
$$

Since v' and $v \circ u$ are local epimorphisms, w_1 and w_2 as well as $w_2 \circ w_1 = u$ are local epimorphisms.

(v) Consider the Cartesian square

$$
\begin{array}{ccc}
A_1 & \xrightarrow{\; w_2 \;} & A_1 \times_{A_2} A_1 \\
\downarrow & & \downarrow \\
A_1 & \xrightarrow{\; w_3 \;} & A_1 \times_{A_3} A_1 .
\end{array}
$$

Since w_3 is a local epimorphism so is w_2.

(vi) The composition of the local epimorphisms $A_1 \times_{A_3} A_1 \to A_2 \times_{A_3} A_1 \to A_2 \times_{A_3} A_2$ is a local epimorphism. Consider the commutative diagram

$$
\begin{array}{ccc}
A_1 & \xrightarrow{\; w_2 \;} & A_1 \times_{A_3} A_1 \\
{\scriptstyle u}\downarrow & & \downarrow{\scriptstyle w_3} \\
A_2 & \xrightarrow{\; v' \;} & A_2 \times_{A_3} A_2 .
\end{array}
$$

Hence, $w_3 \circ w_2 = v' \circ u$ is a local epimorphism and this implies that v' is a local epimorphism.

(vii) (a) Assume that u and v are local isomorphisms. Then $v \circ u$ is a local epimorphism by LE2, and a local monomorphism by (iii).

(vii) (b) Assume that v and $v \circ u$ are local isomorphisms. We know by (iv) that u is a local epimorphism. It is a local monomorphism by (v).

(vii) (c) Assume that u and $v \circ u$ are local isomorphisms. We already know that v is a local epimorphism. It is a local monomorphism by (vi).   q.e.d.

Notations 16.2.5. (i) We denote by $\mathcal{LI}$ the set of local isomorphisms.

(ii) Following Definition 7.1.9, for $A \in \mathcal{C}^\wedge$, we denote by $\mathcal{LI}_A$ the category given by

$$
\mathrm{Ob}(\mathcal{LI}_A) = \{\text{the local isomorphisms } B \to A\} ,
$$

$$
\mathrm{Hom}_{\mathcal{LI}_A}((B \xrightarrow{u} A), (C \xrightarrow{v} A)) = \{w \colon B \to C; u = v \circ w\} .
$$

Note that such a w is necessarily a local isomorphism.

(iii) The category $\mathcal{LI}^A$ is defined similarly.

Lemma 16.2.6. *The family $\mathcal{LI}$ of local isomorphisms in $\mathcal{C}^\wedge$ is a left saturated multiplicative system.*

Proof. Let us check the axioms S1–S5 of Definitions 7.1.5 and 7.1.19. Axiom S1 is obviously satisfied, S2 follows from Lemma 16.2.4 (iii), and S3 (with the arrows reversed, as in Remark 7.1.7) follows from Lemma 16.2.4 (i).

S4 Consider a pair of parallel morphisms $f, g \colon A \rightrightarrows B$ and a local isomorphism $t \colon B \to C$ such that $t \circ f = t \circ g$. Consider the Cartesian square

$$
\begin{array}{ccc}
\mathrm{Ker}(f, g) & \xrightarrow{\hspace{2cm}} & B \\
{\scriptstyle s}\downarrow & & \downarrow{\scriptstyle u} \\
A & \xrightarrow{\hspace{2cm}} & B \times_C B .
\end{array}
$$

By the hypothesis, u is a local epimorphism, and it is a monomorphism. Hence u is a local isomorphism. Since local isomorphisms are stable by base change (Lemma 16.2.4 (i)), s is a local isomorphism.

S5 Consider morphisms

$$A \xrightarrow{f} B \xrightarrow{g} C \xrightarrow{h} D$$

and assume that $g \circ f$ as well as $h \circ g$ are local isomorphisms. It follows that g is both a local epimorphism and a local monomorphism. Then both $g \circ h$ and g are local isomorphisms, and this implies that h is a local isomorphism.
q.e.d.

Lemma 16.2.7. *The category $\mathcal{LI}_A$ admits finite projective limits. In particular, $\mathcal{LI}_A$ is cofiltrant.*

Proof. (i) The category $\mathcal{LI}_A$ admits a terminal object, namely $A \xrightarrow{\mathrm{id}} A$.
(ii) The category $\mathcal{LI}_A$ admits fiber products. Indeed, if $C \to B$, $D \to B$ and $B \to A$ are local isomorphisms, then $C \times_B D \to A$ is a local isomorphism by Lemma 16.2.4.
q.e.d.

Lemma 16.2.8. *Assume that $\mathcal{C}$ is small. Then, for any $A \in \mathcal{C}^\wedge$, the category $(\mathcal{LI}_A)^{\mathrm{op}}$ is cofinally small.*

Proof. Set $I = \{(U, s); U \in \mathcal{C}, s \in A(U)\}$. For $i = (U, s) \in I$, set $U_i = U$. Note that I is a small set and there exists a canonical epimorphism

$$\text{``}\coprod_{i \in I}\text{''}\, U_i \twoheadrightarrow A.$$

For a subset $J \subset I$, we set

$$C_J = \text{``}\coprod_{j \in J}\text{''}\, U_j .$$

Let us consider the set $\mathcal{S}$ of (J, S, v, w) where J is a subset of I, $v \colon C_J \to S$ is an epimorphism and $w \colon S \to A$ is a local isomorphism:

$$C_J \xrightarrow{v} S \xrightarrow{w} A .$$

By Proposition 5.2.9 and the result of Exercise 5.1, the set of quotients of any object of $\mathcal{C}^\wedge$ is small, and hence $\mathcal{S}$ is a small set. On the other hand we have a map

$$\varphi \colon \mathcal{S} \to \mathrm{Ob}(\mathcal{LI}_A) ,$$
$$(C_J \xrightarrow{v} S \xrightarrow{w} A) \mapsto (S \xrightarrow{w} A) .$$

Let us show that $\varphi(\mathcal{S})$ satisfies the condition in Proposition 3.2.6. Let $B \to A$ be a local isomorphism. Set $B_1 = \mathrm{Im}(B \to A)$. Then we have $B_1(U) \subset A(U)$ for any $U \in \mathcal{C}$. Set $J = \{(U, s) \in I \,;\, s \in B_1(U)\}$. Then $C_J \to B_1$ is an epimorphism, and it decomposes into $C_J \to B \to B_1$ since $B(U) \to B_1(U)$ is surjective for any $U \in \mathcal{C}$. Thus we obtain the following commutative diagram:

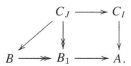

Set $S = \mathrm{Im}(C_J \to B)$. Since $B_1 \to A$ and $B \to A$ are local isomorphisms, $B \to B_1$ is a local isomorphism. Since $C_J \to B_1$ is an epimorphism, $C_J \to B$ is a local epimorphism. Therefore $S \to B$ is a local epimorphism, hence a local isomorphism as well as $S \to A$. This shows that $C_J \twoheadrightarrow S \to A$ belongs to $\mathcal{S}$. q.e.d.

16.3 Localization by Local Isomorphisms

In this section, $\mathcal{C}$ is assumed to be a *small* category endowed with a Grothendieck topology. Recall that $\mathcal{LI}$ denotes the set of local isomorphisms. We shall construct a functor

$$(\cdot)^a : \mathcal{C}^\wedge \to \mathcal{C}^\wedge .$$

Since $\mathcal{LI}$ is a left multiplicative system and $(\mathcal{LI}_A)^{\mathrm{op}}$ is cofinally small for any $A \in \mathcal{C}^\wedge$, the left localization $(\mathcal{C}^\wedge)_{\mathcal{LI}}$ is a well-defined $\mathcal{U}$-category. We denote as usual by

$$Q : \mathcal{C}^\wedge \to (\mathcal{C}^\wedge)_{\mathcal{LI}}$$

the localization functor. For $A \in \mathcal{C}^\wedge$, we define $A^a \in \mathcal{C}^\wedge$ by

$$A^a : \mathcal{C} \ni U \mapsto \mathrm{Hom}_{(\mathcal{C}^\wedge)_{\mathcal{LI}}}(Q(U), Q(A)) .$$

By the definition of $(\mathcal{C}^\wedge)_{\mathcal{LI}}$, we get

$$(16.3.1) \qquad A^a(U) \simeq \varinjlim_{(B \to U) \in \mathcal{LI}_U} \mathrm{Hom}_{\mathcal{C}^\wedge}(B, A) .$$

For a morphism $U \to U'$ in $\mathcal{C}$, the map $A^a(U') \to A^a(U)$ is given as follows:

$$
\begin{aligned}
A^a(U') &\simeq \varinjlim_{(B' \to U') \in \mathcal{LI}_{U'}} \mathrm{Hom}_{\mathcal{C}^\wedge}(B', A) \\
&\to \varinjlim_{(B' \to U') \in \mathcal{LI}_{U'}} \mathrm{Hom}_{\mathcal{C}^\wedge}(B' \times_{U'} U, A) \\
&\to \varinjlim_{(B \to U) \in \mathcal{LI}_U} \mathrm{Hom}_{\mathcal{C}^\wedge}(B, A) \simeq A^a(U) ,
\end{aligned}
$$

where the first morphism is associated with $B' \times_{U'} U \to B'$ and the second one is the natural morphism induced by $\mathcal{LI}_{U'} \ni (B' \to U') \mapsto (B' \times_{U'} U \to U) \in \mathcal{LI}_U$.

The identity morphism $U \xrightarrow{\mathrm{id}} U \in \mathcal{LI}_U$ defines $A(U) \to A^a(U)$. We thus obtain a morphism of functors

$$(16.3.2) \qquad\qquad \varepsilon\colon \mathrm{id}_{\mathcal{C}^\wedge} \to (\bullet)^a \, .$$

In this section, we shall study the properties of the functor $(\bullet)^a$. Since we shall treat this functor in a more general framework in Chap. 17, we restrict ourselves to the study of the properties that we need later.

Lemma 16.3.1. *Let* $u\colon B \to A$ *be a morphism in* $\mathcal{C}^\wedge$ *and let* $s\colon B \to U$ *be a local isomorphism with* $U \in \mathcal{C}$. *Denote by* $v \in A^a(U)$ *the corresponding element (using* (16.3.1)*). Then the diagram*

$$(16.3.3)$$

$$
\begin{array}{ccc}
B & \xrightarrow{\;\; s \;\;} & U \\
{\scriptstyle u}\downarrow & & \downarrow{\scriptstyle v} \\
A & \xrightarrow[\varepsilon(A)]{} & A^a
\end{array}
$$

commutes.

Proof. It is enough to show that, for any $t\colon V \to B$ with $V \in \mathcal{C}$, we have $\varepsilon(A) \circ u \circ t = v \circ s \circ t$. The element $v \circ s \circ t \in A^a(V)$ is given by the pair $(B \times_U V \to V, B \times_U V \to A)$ of the local isomorphism $B \times_U V \to V$ and the morphism $B \times_U V \to B \xrightarrow{u} A$.

Let $w\colon V \to B \times_U V$ be the morphism such that the composition $V \to B \times_U V \to V$ is id_V and $V \to B \times_U V \to B$ is t. Then w gives a morphism $(V \xrightarrow{\mathrm{id}_V} V) \to (B \times_U V \to V)$ in $\mathcal{LI}_V$. Hence, $v \circ s \circ t$ is given by the pair $(V \xrightarrow{\mathrm{id}_V} V, V \xrightarrow{t} B \xrightarrow{u} A)$, which is equal to $\varepsilon(A) \circ u \circ t$.   q.e.d.

Lemma 16.3.2. *For any* $A \in \mathcal{C}^\wedge$, *the natural morphism* $\varepsilon(A)\colon A \to A^a$ *is a local isomorphism.*

Proof. (i) Consider a morphism $U \to A^a$. By the definition of A^a, there exist a local isomorphism $B \to U$ and a commutative diagram (16.3.3). Therefore, $A \to A^a$ is a local epimorphism by Lemma 16.1.6.

(ii) Consider a diagram $U \rightrightarrows A \to A^a$ such that the two compositions coincide. The two morphisms $U \rightrightarrows A$ define $s_1, s_2 \in A(U)$ with the same image in $A^a(U)$. Since $\mathcal{LI}_U$ is cofiltrant, there exist a local isomorphism $B \to U$ and a diagram $B \to U \rightrightarrows A$ such that the two compositions coincide. Therefore $A \to A^a$ is a local monomorphism by Lemma 16.2.3 (iii).   q.e.d.

Proposition 16.3.3. *Let* $w\colon A_1 \to A_2$ *be a local isomorphism. Then* $w^a\colon A_1^a \to A_2^a$ *is an isomorphism.*

Proof. It is enough to show that $A_1^a(U) \to A_2^a(U)$ is bijective for any $U \in \mathcal{C}$.

(i) Injectivity. Let $v_1, v_2 \in A_1^a(U)$ and assume they have the same image in $A_2^a(U)$. Since $\mathcal{L}\mathcal{I}_U$ is cofiltrant, there exist a local isomorphism $s \colon B \to U$ and $u_i \colon B \to A_1$ $(i = 1, 2)$ such that (u_i, s) gives $v_i \in A_1^a(U)$. Since $w^a(v_1) = w^a(v_2) \in A_2^a(U)$, there exists a local isomorphism $t \colon B' \to B$ such that the two compositions $B' \xrightarrow{\ } B \underset{u_2}{\overset{u_1}{\rightrightarrows}} A_1 \longrightarrow A_2$ coincide. Since $A_1 \to A_2$ is a local monomorphism, there exists a local isomorphism $B'' \to B$ such that the two compositions $B'' \longrightarrow B' \longrightarrow B \underset{u_2}{\overset{u_1}{\rightrightarrows}} A_1$ coincide. Hence, $v_1 = v_2$.

(ii) Surjectivity. Let $v \in A_2^a$. Then v is represented by a local isomorphism $s \colon B \to U$ and a morphism $u \colon B \to A_2$. In the following commutative diagram

$$
\begin{array}{ccccc}
A_1 \times_{A_2} B & \xrightarrow{\ w'\ } & B & \xrightarrow{\ s\ } & U \\
{\scriptstyle u'}\downarrow & & \downarrow{\scriptstyle u} & & \\
A_1 & \xrightarrow{\ w\ } & A_2 & &
\end{array}
$$

w' is a local isomorphism and $(u', s \circ w')$ defines an element of $A_1^a(U)$ whose image in $A_2^a(U)$ coincides with v.   q.e.d.

Proposition 16.3.4. *Let I be a small category and let $\alpha \colon I \to \mathrm{Mor}(\mathcal{C}^\wedge)$ be a functor. Assume that for each $i \in I$, $\alpha(i) \colon A_i \to B_i$ is a local isomorphism. Let $u \colon A \to B$ denote the inductive limit in $\mathrm{Mor}(\mathcal{C}^\wedge)$ of $\alpha(i) \colon A_i \to B_i$. Then u is a local isomorphism.*

In other words, $\mathcal{L}\mathcal{I}$, considered as a full subcategory of $\mathrm{Mor}(\mathcal{C}^\wedge)$, is closed by small inductive limits in $\mathrm{Mor}(\mathcal{C}^\wedge)$.

Proof. Since $A_i^a \to B_i^a$ is an isomorphism by Proposition 16.3.3, we get the following commutative diagram on the left:

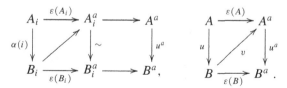

Taking the inductive limit with respect to i, we obtain the commutative diagram on the right. Since $\varepsilon(A) = v \circ u$ is a local isomorphism, v is a local epimorphism. Since $u^a \circ v = \varepsilon(B)$ is a local monomorphism, v is a local monomorphism. Hence v as well as u is a local isomorphism.   q.e.d.

Exercises

Exercise 16.1. Prove that the axioms LE1–LE4 are equivalent to the axioms GT1–GT4, and also prove that they are equivalent to the axioms COV1–COV4 when $\mathcal{C}$ is small and admits fiber products.

Exercise 16.2. Prove that the axioms LE1', LE2 and LE4 imply LE3.

Exercise 16.3. Let C be a category and C_0 a subcategory of C. Let us say that a morphism $u \colon A \to B$ in $C^\wedge$ is a local epimorphism if for any $U \in C_0$ and any morphism $U \to B$ in $C^\wedge$, there exist a morphism $s \colon V \to U$ in C_0 and a commutative diagram

$$
\begin{array}{ccc}
V & \xrightarrow{\;s\;} & U \\
\downarrow & & \downarrow \\
A & \longrightarrow & B
\end{array}
\quad \text{in } C^\wedge.
$$

Prove that the family of local epimorphisms defined above satisfies the axioms LE1–LE4.

Exercise 16.4. Let C be a category. Let us say that a morphism $f \colon B \to A$ in $C^\wedge$ is a local epimorphism if for any morphism $U \to A$ with $U \in C$, there exist $V \in C$, an epimorphism $g \colon V \to U$ in C and a morphism $V \to B$ such that the diagram below commutes:

$$
\begin{array}{ccc}
V & \xrightarrow{\;g\;} & U \\
\downarrow & & \downarrow \\
B & \xrightarrow{\;f\;} & A.
\end{array}
$$

(i) Check that the axioms LE1–LE4 are satisfied. We call this topology the epitopology on C.
(ii) Assume that C admits finite coproducts. Show that it is also possible to define a topology, replacing V above by "$\bigsqcup_{i \in I}$" V_i with I finite, under the condition that $\bigsqcup_{i \in I} V_i \to U$ is an epimorphism in C.

Exercise 16.5. Let C be a category. Let $\mathcal{LI}$ be a subset of $\mathrm{Ob}(\mathrm{Mor}(C^\wedge))$ satisfying:

LI 1 every isomorphism belongs to $\mathcal{LI}$,
LI 2 let $A \xrightarrow{u} B \xrightarrow{v} C$ be morphisms in $C^\wedge$. If two of the morphisms u, v and $v \circ u$ belong to $\mathcal{LI}$, then all belong to $\mathcal{LI}$,
LI 3 a morphism $u \colon A \to B$ in $C^\wedge$ belongs to $\mathcal{LI}$ if and only if for any $U \in C$ and any morphism $U \to B$, the morphism $A \times_B U \to U$ belongs to $\mathcal{LI}$.

Let us say that a morphism $u \colon A \to B$ in $C^\wedge$ is a local epimorphism if the morphism $\mathrm{Im}\, u \to B$ belongs to $\mathcal{LI}$.
  Prove that the family of local epimorphisms so defined satisfies LE1–LE4 and $\mathcal{LI}$ coincides with the set of local isomorphisms for this Grothendieck topology.
  Hence, we have an alternative definition of Grothendieck topologies, using LI1–LI3.

Exercise 16.6. Let C be a category endowed with a Grothendieck topology. Let $B \to A$ and $C \to B \times_A B$ be local epimorphisms. Prove that the induced morphism $\mathrm{Coker}(C \rightrightarrows B) \to A$ is a local isomorphism.

Exercise 16.7. Let $\mathcal{C}$ be a small category endowed with a Grothendieck topology and let $A \in \mathcal{C}^\wedge$. Recall the morphism of functors ε of (16.3.2).

(i) Prove that (a, ε) is a projector on $\mathcal{C}^\wedge$ (see Definition 4.1.1).

(ii) Prove that a morphism $A_1 \to A_2$ is a local isomorphism if and only if $A_1^a \to A_2^a$ is an isomorphism.

(iii) Prove that, for any local isomorphism $B_1 \to B_2$, the induced map $\mathrm{Hom}_{\mathcal{C}^\wedge}(B_2, A^a) \to \mathrm{Hom}_{\mathcal{C}^\wedge}(B_1, A^a)$ is bijective.

(iv) Prove that A^a is a terminal object in $\mathcal{L}\mathcal{I}^A$.

17

Sheaves on Grothendieck Topologies

Historically, a presheaf was defined as a contravariant functor on the category of open subsets of a topological space with values in the category **Set**. By extension, "a presheaf" is any contravariant functor defined on a category $\mathcal{C}$ with values in another category $\mathcal{A}$. Therefore, presheaves are nothing but functors, although the two notions play different roles.

A site X is a small category $\mathcal{C}_X$ endowed with a Grothendieck topology. The aim of this chapter is to construct presheaves and sheaves on a site with values in a category $\mathcal{A}$ satisfying suitable properties.

A presheaf F on X is a sheaf if $F(U) \to F(A)$ is an isomorphism for any local isomorphism $A \to U$. This definition is shown to be equivalent to the classical (and more intuitive) one by using coverings. We construct the sheaf F^a associated with a presheaf F with values in $\mathcal{A}$, we show that the functor $(\,\cdot\,)^a$ is left adjoint to the forgetful functor which associates the underlying presheaf to a sheaf on X, and we show that the functor $(\,\cdot\,)^a$ is exact.

We also study the direct and inverse images functors for sheaves, particularly for a morphism of sites $X \to A$ associated with $A \in (\mathcal{C}_X)^\wedge$ and the internal $\mathcal{H}om$ functor. We shall glue sheaves in Chap. 19.

Let us mention that when the sites admit finite projective limits, the theory of sheaves resembles the one on topological spaces, and a concise exposition in this case may be found in [67].

17.1 Presites and Presheaves

Definition 17.1.1. (i) *A* presite *X is nothing but a category which we denote by $\mathcal{C}_X$.*
  (ii) *A morphism of presites $f \colon X \to Y$ is a functor $f^t \colon \mathcal{C}_Y \to \mathcal{C}_X$.*
  (iii) *A presite X is small if $\mathcal{C}_X$ is small. More generally, we say that a presite has a property "P" if the category $\mathcal{C}_X$ has the property "P".*

Notation 17.1.2. (i) Let X be a small presite. We denote by $\widehat{X}$ the presite associated with the category $(\mathcal{C}_X)^\wedge$, and we set $\mathrm{h}_X^t := \mathrm{h}_{\mathcal{C}_X} \colon \mathcal{C}_X \to (\mathcal{C}_X)^\wedge$. Hence we denote by

$$\mathrm{h}_X \colon \widehat{X} \to X$$

the associated morphism of presites. If there is no risk of confusion, we write $\mathcal{C}_X^\wedge$ instead of $(\mathcal{C}_X)^\wedge$.

(ii) Let $f \colon X \to Y$ be a morphism of small presites. We denote by $\widehat{f} \colon \widehat{X} \to \widehat{Y}$ the associated morphism of presites given by Proposition 2.7.1 (using Notation 2.7.2). Hence we have

$$(\widehat{f}^t A)(U) \simeq \varinjlim_{(V \to A) \in (\mathcal{C}_Y)_A^\wedge} \mathrm{Hom}_{(\mathcal{C}_Y)^\wedge}(f^t(V), U)$$

$$\simeq \varinjlim_{(U \to f^t(V)) \in (\mathcal{C}_Y)^U} A(V)$$

for any $A \in (\mathcal{C}_Y)^\wedge$ and $U \in \mathcal{C}_X$. Note that $\widehat{f}^t \colon \mathcal{C}_Y^\wedge \to \mathcal{C}_X^\wedge$ commutes with small inductive limits.

(iii) For a presite X, we denote by pt_X the terminal object of $\mathcal{C}_X^\wedge$.

In all this section, $\mathcal{A}$ denotes a category.

Definition 17.1.3. *Let X be a presite. We set* $\mathrm{PSh}(X, \mathcal{A}) = \mathrm{Fct}(\mathcal{C}_X^{\mathrm{op}}, \mathcal{A})$ *and call an object of this category a presheaf on X with values in $\mathcal{A}$.*

We set $\mathrm{PSh}(X) := \mathrm{PSh}(X, \mathcal{U}\text{-}\mathbf{Set}) = \mathcal{C}_X^\wedge$.

Note that when X is $\mathcal{U}$-small and $\mathcal{A}$ is a $\mathcal{U}$-category, $\mathrm{PSh}(X, \mathcal{A})$ is a $\mathcal{U}$-category.

As already noticed, if $\mathcal{A}$ admits small inductive (resp. projective) limits, then the category $\mathrm{PSh}(X, \mathcal{A})$ admits small inductive (resp. projective) limits. If $i \mapsto F_i$ (resp. $i \mapsto G_i$) is a small inductive (resp. projective) system of presheaves, then we have for $U \in \mathcal{C}_X$

$$(\varinjlim_i F_i)(U) \simeq \varinjlim_i (F_i(U)) \, ,$$

$$(\varprojlim_i G_i)(U) \simeq \varprojlim_i (G_i(U)) \, .$$

Here, $\varinjlim$ and $\varprojlim$ on the right-hand side are taken in the category $\mathcal{A}$. If moreover small filtrant inductive limits are exact in $\mathcal{A}$, then these limits are exact in $\mathrm{PSh}(X, \mathcal{A})$.

Example 17.1.4. To a small topological space X we associate the small category Op_X of its open subsets, the morphisms in Op_X being the inclusion morphisms. Let $\mathcal{A} = \mathrm{Mod}(k)$, where k is a commutative ring. A presheaf F on X with values in $\mathrm{Mod}(k)$ thus associates to each open subset U of X a k-module $F(U)$, and to each open inclusion $V \subset U$ a k-linear map $F(U) \to F(V)$ called the restriction map. An element $s \in F(U)$ is called a section of F on U. Its image by the restriction map is often denoted by $s|_V$ and called the restriction of s to V. Let us give elementary explicit examples.

(i) Let $\mathcal{C}^0(U)$ denote the $\mathbb{C}$-vector space of $\mathbb{C}$-valued continuous functions on $U \in \mathrm{Op}_X$. Then $U \mapsto \mathcal{C}^0(U)$, with the usual restriction morphisms, is a presheaf on X with values in the category $\mathrm{Mod}(\mathbb{C})$ of $\mathbb{C}$-vector spaces.

(ii) The constant presheaf with values $\mathbb{Z}$ associates $\mathbb{Z}$ to each open subset U. This is the presheaf of constant $\mathbb{Z}$-valued functions on X. A more interesting presheaf, denoted by $\mathbb{Z}_X$, is that of locally constant $\mathbb{Z}$-valued functions on X.

Let us paraphrase Definition 2.3.1 in the framework of presheaves. For simplicity, we shall assume that all presites are small and also that

(17.1.1) $\mathcal{A}$ admits small inductive and small projective limits.

Let $f : X \to Y$ be a morphism of small presites, that is, a functor $f^t : \mathcal{C}_Y \to \mathcal{C}_X$. One has the functors

$$f^t_* \ : \ \mathrm{PSh}(X, \mathcal{A}) \to \mathrm{PSh}(Y, \mathcal{A}) \,,$$
$$f^{t\dagger} \ : \ \mathrm{PSh}(Y, \mathcal{A}) \to \mathrm{PSh}(X, \mathcal{A}) \,,$$
$$f^{t\ddagger} \ : \ \mathrm{PSh}(Y, \mathcal{A}) \to \mathrm{PSh}(X, \mathcal{A}) \,.$$

Notation 17.1.5. Let $f : X \to Y$ be a morphism of small presites. We shall write for short:

$$f_* := f^t_*, \quad f^\dagger := f^{t\dagger}, \quad f^\ddagger := f^{t\ddagger} \,.$$

Recall that these functors are defined as follows. For $F \in \mathrm{PSh}(X, \mathcal{A})$, $G \in \mathrm{PSh}(Y, \mathcal{A})$, $U \in \mathcal{C}_X$ and $V \in \mathcal{C}_Y$:

$$(17.1.2) \qquad\qquad f_* F(V) = F(f^t(V)) \,,$$
$$(17.1.3) \qquad\qquad f^\dagger G(U) = \varinjlim_{(U \to f^t(V)) \in (\mathcal{C}_Y)^U} G(V) \,,$$
$$(17.1.4) \qquad\qquad f^\ddagger G(U) = \varprojlim_{(f^t(V) \to U) \in (\mathcal{C}_Y)_U} G(V) \,.$$

Thus, we get functors

$$\mathrm{PSh}(X, \mathcal{A}) \ \underset{\substack{\longleftarrow \\ f^\dagger}}{\overset{\substack{f^\ddagger \\ \longleftarrow \\ \xrightarrow{\quad f_* \quad}}}{}} \ \mathrm{PSh}(Y, \mathcal{A}) \,.$$

Applying Theorem 2.3.3 we find that $(f^\dagger, f_*)$ and $(f_*, f^\ddagger)$ are two pairs of adjoint functors.

For two morphisms of presites $f : X \to Y$ and $g : Y \to Z$, we have

(17.1.5) $(g \circ f)_* \simeq g_* \circ f_*,$ $(g \circ f)^\dagger \simeq f^\dagger \circ g^\dagger,$ $(g \circ f)^\ddagger \simeq f^\ddagger \circ g^\ddagger .$

We extend presheaves on X to presheaves on $\widehat{X}$, as in § 2.7, using the functor $h_X^\ddagger$ associated with the Yoneda functor $h_X' = h_{\mathcal{C}_X}$. Hence, for $F \in \mathrm{PSh}(X, \mathcal{A})$ and $A \in \mathcal{C}_X^\wedge$, we have:

$$ h_X^\ddagger F(A) \simeq \varprojlim_{(U \to A) \in (\mathcal{C}_X)_A} F(U) . $$

By Corollary 2.7.4, the functor $h_X^\ddagger$ induces an equivalence of categories between the category $\mathrm{PSh}(X, \mathcal{A})$ and the full subcategory of $\mathrm{PSh}(\widehat{X}, \mathcal{A})$ consisting of presheaves which commute with small projective limits.

Convention 17.1.6. In the sequel, we shall identify a presheaf $F \in \mathrm{PSh}(X, \mathcal{A})$ and its image by $h_X^\ddagger$ in $\mathrm{PSh}(\widehat{X}, \mathcal{A})$. In other words, we shall write F instead of $h_X^\ddagger F$.

With this convention, we have

(17.1.6) $(f_* F)(B) \simeq F(\widehat{f^t}(B))$ for $F \in \mathrm{PSh}(X, \mathcal{A})$ and $B \in \mathcal{C}_Y^\wedge$.

Indeed, we have

$$ (f_* F)(B) \simeq \varprojlim_{V \in (\mathcal{C}_Y)_B} f_* F(V) \simeq \varprojlim_{V \in (\mathcal{C}_Y)_B} F(f^t(V)) $$
$$ \simeq F\big(\varinjlim_{V \in (\mathcal{C}_Y)_B} f^t(V) \big) \simeq F(\widehat{f^t}(B)) . $$

Let $A \in (\mathcal{C}_X)^\wedge$. Recall that $j_A : (\mathcal{C}_X)_A \to \mathcal{C}_X$ is the forgetful functor.

Notations 17.1.7. We denote by A the presite associated with the category $(\mathcal{C}_X)_A$ and by

(17.1.7) $j_{A \to X} : X \to A$

the morphism of presites associated with the functor j_A. Hence

(17.1.8) $\mathcal{C}_A := (\mathcal{C}_X)_A,$ $j_{A \to X}^t := j_A : (\mathcal{C}_X)_A \to \mathcal{C}_X .$

To avoid confusing it with the functor $j_A : \mathcal{C}_A \to \mathcal{C}_X$, we denote by

(17.1.9) $j_{\widehat{A}} : (\mathcal{C}_X^\wedge)_A \to \mathcal{C}_X^\wedge$

the forgetful functor. By Proposition 2.7.1, the functor $j_A : \mathcal{C}_A \to \mathcal{C}_X$ extends to a functor $\widehat{j}_A : (\mathcal{C}_A)^\wedge \to \mathcal{C}_X^\wedge$. We shall compare these functors.

Lemma 17.1.8. *There is a quasi-commutative diagram of categories*

$$
\begin{array}{ccccc}
\mathcal{C}_A & \xrightarrow{\ (h'_X)_A\ } & (\mathcal{C}_X^\wedge)_A & \xrightarrow{\ j_{\widehat{A}}\ } & \mathcal{C}_X^\wedge \ .\\
& \searrow{\scriptstyle h'_A} & \ \downarrow{\scriptstyle \sim}\,{\scriptstyle \lambda} & \nearrow{\scriptstyle \widehat{j}_A} & \\
& & (\mathcal{C}_A)^\wedge & &
\end{array}
$$

Proof. By Lemma 1.4.12, we already know that $\lambda \circ (h'_X)_A \simeq h_A$. Hence, it remains to prove that $\widehat{j}_A \circ \lambda \simeq j_{\widehat{A}}$. Since $\widehat{j}_A \circ \lambda \circ (h'_X)_A \simeq j_{\widehat{A}} \circ (h'_X)_A$, by Proposition 2.7.1 it is enough to check that $j_{\widehat{A}} \circ \lambda^{-1}$ commutes with small inductive limits. This follows from Lemma 2.1.13.   q.e.d.

The morphism of presites $j_{A \to X}$ gives rise to the functors

$$
\begin{aligned}
j_{A \to X*} &: \ \mathrm{PSh}(X, \mathcal{A}) \to \mathrm{PSh}(A, \mathcal{A}) \ , \\
j_{A \to X}^\dagger &: \ \mathrm{PSh}(A, \mathcal{A}) \to \mathrm{PSh}(X, \mathcal{A}) \ , \\
j_{A \to X}^\ddagger &: \ \mathrm{PSh}(A, \mathcal{A}) \to \mathrm{PSh}(X, \mathcal{A}) \ .
\end{aligned}
$$

Proposition 17.1.9. *Let $G \in \mathrm{PSh}(A, \mathcal{A})$ and $F \in \mathrm{PSh}(X, \mathcal{A})$. We have the isomorphisms*

(17.1.10) $j_{A \to X*}(F)(B \to A) \simeq F(B) \quad \text{for } (B \to A) \in \mathcal{C}_A^\wedge,$

(17.1.11) $j_{A \to X}^\dagger(G)(U) \simeq \coprod_{s \in A(U)} G(U \xrightarrow{s} A) \quad \text{for } U \in \mathcal{C}_X,$

(17.1.12) $j_{A \to X}^\ddagger(G)(B) \simeq G(B \times A \to A) \quad \text{for } B \in \mathcal{C}_X^\wedge \ .$

Proof. (i) Isomorphism (17.1.10) is obvious when $B \in \mathcal{C}_A$. If $B \in \mathcal{C}_A^\wedge$, we have

$$
\begin{aligned}
j_{A \to X*}(F)(B \to A) &\simeq \varprojlim_{(U \to B) \in \mathcal{C}_B} j_{A \to X*}(F)(U \to A) \\
&\simeq \varprojlim_{(U \to B) \in \mathcal{C}_B} F(U) \simeq F(B) \ .
\end{aligned}
$$

(ii) Let us check (17.1.11). By (17.1.3), we have

$$
\begin{aligned}
j_{A \to X}^\dagger(G)(U) &\simeq \varinjlim_{(U \to j_A(V \to A)) \in (\mathcal{C}_A)^U} G(V \to A) \simeq \varinjlim_{U \to V \to A} G(V \to A) \\
&\simeq \varinjlim_{U \to A} G(U \to A) \simeq \coprod_{s \in A(U)} G(U \xrightarrow{s} A) \ .
\end{aligned}
$$

Here, we use the fact that the discrete category $\mathrm{Hom}_{\mathcal{C}_X^\wedge}(U, A)$ is cofinal in $((\mathcal{C}_A)^U)^{\mathrm{op}}$.

(iii) Let us check (17.1.12). If $B \in \mathcal{C}_X$, we have

$$(j^{\ddagger}_{A \to X} G)(B) \simeq \varprojlim_{j_A(V \xrightarrow{} A) \to B} G(V \to A)$$

$$\simeq \varprojlim_{A \leftarrow V \to B} G(V \to A) \simeq G(B \times A \to A) \, .$$

In the general case, we have:

$$(j^{\ddagger}_{A \to X} G)(B) \simeq \varprojlim_{(U \to B) \in \mathcal{C}_B} (j^{\ddagger}_{A \to X} G)(U)$$

$$\simeq \varprojlim_{(U \to B) \in \mathcal{C}_B} G(U \times A \to A)$$

$$\simeq G(\ \underset{(U \to B) \in \mathcal{C}_B}{\text{``}\varinjlim\text{''}} U \times A \to A) \simeq G(B \times A \to A) \, .$$

q.e.d.

More generally, for a morphism $u \colon A \to B$ in $\mathcal{C}^{\wedge}_X$, we have a functor $(j_{A \xrightarrow{u} B})^t \colon \mathcal{C}^{\wedge}_A \to \mathcal{C}^{\wedge}_B$ given by $(C \to A) \mapsto (C \to A \to B)$, which induces a morphism or presites

$$j_{A \xrightarrow{u} B} \colon B \to A \, .$$

For morphisms $A \to B \to C$ in $\mathcal{C}^{\wedge}_X$, we have

$$j_{A \to C} \simeq j_{A \to B} \circ j_{B \to C} \, ,$$

and hence (17.1.5) implies the isomorphisms:

$$\begin{aligned}(17.1.13) \qquad &(j_{A \to C})_* \simeq (j_{A \to B})_* \circ (j_{B \to C})_*, \\ &(j_{A \to C})^{\dagger} \simeq (j_{B \to C})^{\dagger} \circ (j_{A \to B})^{\dagger}, \\ &(j_{A \to C})^{\ddagger} \simeq (j_{B \to C})^{\ddagger} \circ (j_{A \to B})^{\ddagger}.\end{aligned}$$

Internal $\mathcal{H}om$

Let X be a presite. For $F, G \in \mathrm{PSh}(X, \mathcal{A})$ and $U \in \mathcal{C}_X$, we set

$$(17.1.14) \quad \mathcal{H}om_{\mathrm{PSh}(X,\mathcal{A})}(F, G)(U) = \mathrm{Hom}_{\mathrm{PSh}(U,\mathcal{A})}(j_{U \to X*}F, j_{U \to X*}G) \, .$$

Definition 17.1.10. *The presheaf* $\mathcal{H}om_{\mathrm{PSh}(X,\mathcal{A})}(F, G)$ *given by* (17.1.14) *is called the* internal hom *of* (F, G).

Note the isomorphism

$$(17.1.15) \qquad \mathrm{Hom}_{\mathrm{PSh}(X,\mathcal{A})}(F, G) \simeq \varprojlim_{U \in \mathcal{C}_X} \mathcal{H}om_{\mathrm{PSh}(U,\mathcal{A})}(F, G)(U) \, .$$

Lemma 17.1.11. *For* $F, G \in \mathrm{PSh}(X, \mathcal{A})$ *and* $A \in \mathcal{C}^{\wedge}_X$, *there are isomorphisms*

(i) $j_{A \to X*}\mathcal{H}om_{\mathrm{PSh}(X,\mathcal{A})}(F, G) \simeq \mathcal{H}om_{\mathrm{PSh}(A,\mathcal{A})}(j_{A \to X*}F, j_{A \to X*}G)$,

(ii) $\mathcal{H}om_{\mathrm{PSh}(X,\mathcal{A})}(F, G)(A) \simeq \mathrm{Hom}_{\mathrm{PSh}(A,\mathcal{A})}(j_{A \to X*}F, j_{A \to X*}G)$.

Proof. (i) Let $(U \to A) \in \mathcal{C}_A$. There is a chain of isomorphisms

$$j_{A \to X*} \mathcal{H}om_{\mathrm{PSh}(X,\mathcal{A})}(F, G)(U \to A)$$
$$\simeq \mathcal{H}om_{\mathrm{PSh}(X,\mathcal{A})}(F, G)(U)$$
$$\simeq \mathrm{Hom}_{\mathrm{PSh}(U,\mathcal{A})}(j_{U \to X*}F, j_{U \to X*}G)$$
$$\simeq \mathrm{Hom}_{\mathrm{PSh}(U,\mathcal{A})}(j_{U \to A*}j_{A \to X*}F, j_{U \to A*}j_{A \to X*}G)$$
$$\simeq \mathcal{H}om_{\mathrm{PSh}(A,\mathcal{A})}(j_{A \to X*}F, j_{A \to X*}G)(U \to A) .$$

(ii) There is a chain of isomorphisms

$$\mathcal{H}om_{\mathrm{PSh}(X,\mathcal{A})}(F, G)(A) \simeq \varprojlim_{U \to A} \mathcal{H}om_{\mathrm{PSh}(X,\mathcal{A})}(F, G)(U)$$
$$\simeq \varprojlim_{U \to A} j_{A \to X*} \mathcal{H}om_{\mathrm{PSh}(X,\mathcal{A})}(F, G)(U \to A)$$
$$\simeq \varprojlim_{U \to A} \mathcal{H}om_{\mathrm{PSh}(A,\mathcal{A})}(j_{A \to X*}F, j_{A \to X*}G)(U \to A)$$
$$\simeq \mathrm{Hom}_{\mathrm{PSh}(A,\mathcal{A})}(j_{A \to X*}F, j_{A \to X*}G) ,$$

where the third isomorphism follows from (i) and the last one from (17.1.15).

q.e.d.

17.2 Sites

Definition 17.2.1. (i) *A site X is a small presite endowed with a Grothendieck topology.*

(ii) *A morphism of sites $f : X \to Y$ is a functor $f^t : \mathcal{C}_Y \to \mathcal{C}_X$ such that for any local isomorphism $B \to A$ in $\mathcal{C}_Y^\wedge$, $\widehat{f^t}(B) \to \widehat{f^t}(A)$ is a local isomorphism in $\mathcal{C}_X^\wedge$.*

Clearly the family of sites and morphisms of sites defines a category.

The above definition of morphisms of sites depends on the choice of a universe in appearance. However it does not as we shall show by the following lemmas.

Lemma 17.2.2. *Let X and Y be sites, and let $f : X \to Y$ be a morphism of presites.*

(i) *If f is a morphism of sites, then $\widehat{f^t}$ sends the local epimorphisms in $\mathcal{C}_Y^\wedge$ to local epimorphisms in $\mathcal{C}_X^\wedge$.*

(ii) *The following conditions are equivalent.*
  (a) *f is a morphism of sites,*
  (b) *for any $V \in \mathcal{C}_Y$ and any morphism $B \to V$ in $\mathcal{C}_Y^\wedge$ which is both a monomorphism and a local isomorphism, $\widehat{f^t}(B) \to f^t(V)$ is a local isomorphism.*

Proof. The proof of (i) being given in the course of the proof of (ii), we shall prove (ii). The implication (a)$\Rightarrow$(b) is obvious. Let us prove the converse. Assume (b).

(1) First, let us show that for any $V \in \mathcal{C}_Y$, $\widehat{f}^t$ sends any local epimorphism $u: A \to V$ in $\mathcal{C}_Y^\wedge$ to a local epimorphism in $\mathcal{C}_X^\wedge$. The morphism u decomposes as $A \to \operatorname{Im} u \to V$, where $A \to \operatorname{Im} u$ is an epimorphism and $\operatorname{Im} u \to V$ is both a monomorphism and a local epimorphism. The assumption (b) implies that $\widehat{f}^t(\operatorname{Im} u) \to f^t(V)$ is a local isomorphism. Since $\widehat{f}^t(A) \to \widehat{f}^t(\operatorname{Im} u)$ is an epimorphism (see Exercise 3.4), $\widehat{f}^t(A) \to f^t(V)$ is a local epimorphism.

(2) Let us show that $\widehat{f}^t$ sends any local epimorphism $u: A \to B$ in $\mathcal{C}_Y^\wedge$ to a local epimorphism in $\mathcal{C}_X^\wedge$. Let $V \to B$ be a morphism in $\mathcal{C}_Y^\wedge$ with $V \in \mathcal{C}_Y$ and set $u_V: A \times_B V \to V$. Then $\widehat{f}^t(u_V)$ is a local epimorphism by (1). Taking the inductive limit with respect to $(V \to B) \in (\mathcal{C}_Y)_B$, the morphism $\widehat{f}^t(A) \to f^t(B)$ is a local epimorphism by Proposition 16.1.12. Note that "$\varinjlim_{V \in (\mathcal{C}_Y)_B}$" $A \times_B V \simeq A$ by Exercise 3.2.

(3) Next, let us show that if a local isomorphism $u: A \to B$ in $\mathcal{C}_Y^\wedge$ is either a monomorphism or an epimorphism, then $\widehat{f}^t(u)$ is a local isomorphism in $\mathcal{C}_X^\wedge$. As in (2), let $V \to B$ be a morphism in $\mathcal{C}_Y^\wedge$ with $V \in \mathcal{C}_Y$. Then $u_V: A \times_B V \to V$ is either a monomorphism or an epimorphism. Let us show that $\widehat{f}^t(u_V)$ is a local isomorphism. If u_V is a monomorphism, it follows from (b). Assume that u_V is an epimorphism. Then u_V has a section $s: V \to A \times_B V$. Since u_V is a local isomorphism, s is a local isomorphism. Since $\widehat{f}^t(u_V) \circ \widehat{f}^t(s) \simeq \operatorname{id}_{f^t(V)}$ is a local monomorphism and $\widehat{f}^t(s)$ is a local epimorphism by (2), Lemma 16.2.4 (vi) implies that $\widehat{f}^t(u_V)$ is a local monomorphism. Since $\widehat{f}^t(u_V)$ is an epimorphism by (2), $\widehat{f}^t(u_V)$ is a local isomorphism. Thus in the both cases, $\widehat{f}^t(u_V)$ is a local isomorphism. Taking the inductive limit with respect to $V \in (\mathcal{C}_Y)_B$, $\widehat{f}^t(u)$ is a local isomorphism by Proposition 16.3.4.

(4) Finally let us show that if $u: A \to B$ in $\mathcal{C}_Y^\wedge$ is a local isomorphism, then $\widehat{f}^t(u)$ is a local isomorphism in $\mathcal{C}_X^\wedge$. Since $A \to \operatorname{Im} u$ is an epimorphism and $\operatorname{Im} u \to B$ is a monomorphism and they are local isomorphisms, their images by $\widehat{f}^t$ are local isomorphisms by (3). Therefore their composition $\widehat{f}^t(u)$ is also a local isomorphism.   q.e.d.

Lemma 17.2.3. *Let $\mathcal{U}$ and $\mathcal{V}$ be universes such that $\mathcal{U} \subset \mathcal{V}$. Let $f: X \to Y$ be a morphism of $\mathcal{U}$-small presites. Then $\widehat{f}^t$ sends the local isomorphisms in $(\mathcal{C}_Y)_\mathcal{U}^\wedge$ (see Definition 1.4.2) to local isomorphisms in $(\mathcal{C}_X)_\mathcal{U}^\wedge$ if and only if $\widehat{f}^t$ sends the local isomorphisms in $(\mathcal{C}_Y)_\mathcal{V}^\wedge$ to local isomorphisms in $(\mathcal{C}_X)_\mathcal{V}^\wedge$.*

Proof. The assertion follows from the preceding lemma, since any subobject of $V \in \mathcal{C}_Y$ in $(\mathcal{C}_Y)_\mathcal{V}^\wedge$ is isomorphic to an object of $(\mathcal{C}_Y)_\mathcal{U}^\wedge$. Indeed, if $A \in (\mathcal{C}_Y)_\mathcal{V}^\wedge$ is a subobject of V, then $A(U) \subset V(U)$ for any $U \in \mathcal{C}$, and since $V(U)$ belongs to $\mathcal{U}$, we obtain that $A(U)$ belongs to $\mathcal{U}$.   q.e.d.

We shall also encounter not necessarily small presites endowed with a Grothendieck topology. A *big site* is a presite endowed with a Grothendieck topology.

Definition 17.2.4. *Let $f\colon X \to Y$ be a morphism of presites.*

(i) *We say that f is* left exact *if the functor $f^t\colon \mathcal{C}_Y \to \mathcal{C}_X$ is left exact.*

(ii) *We say that f is* weakly left exact *if the functor $\mathcal{C}_Y \to (\mathcal{C}_X)_{\widehat{f^t}(\mathrm{pt}_Y)}$ induced by f^t is left exact.*

Lemma 17.2.5. *Let $f\colon X \to Y$ be a morphism of presites.*

(i) *If $f\colon X \to Y$ is left exact, then f is weakly left exact.*

(ii) *If $f\colon X \to Y$ is weakly left exact, then $\widehat{f^t}\colon \mathcal{C}_Y^\wedge \to \mathcal{C}_X^\wedge$ commutes with fiber products and sends the monomorphisms to monomorphisms.*

(iii) *For any $A \in \mathcal{C}_X^\wedge$, the morphism $\mathrm{j}_{A \to X}\colon X \to A$ is weakly left exact.*

Proof. (i) Since f is left exact, $\widehat{f^t}(\mathrm{pt}_Y) \simeq \mathrm{pt}_X$.

(ii) By Corollary 3.3.19, the functor $(\mathcal{C}_Y)^\wedge \to ((\mathcal{C}_X)_{\widehat{f^t}(\mathrm{pt}_Y)})^\wedge$ is left exact and $((\mathcal{C}_X)_{\widehat{f^t}(\mathrm{pt}_Y)})^\wedge \simeq (\mathcal{C}_X^\wedge)_{\widehat{f^t}(\mathrm{pt}_Y)}$ by Lemma 1.4.12. Since the functor $(\mathcal{C}_Y)^\wedge \to ((\mathcal{C}_X)_{\widehat{f^t}(\mathrm{pt}_Y)})^\wedge$ as well as $((\mathcal{C}_X)_{\widehat{f^t}(\mathrm{pt}_Y)})^\wedge \to (\mathcal{C}_X)^\wedge$ commutes with fiber products and sends the monomorphisms to monomorphisms, so does their composition $\widehat{f^t}$. (See Exercise 17.14 for a converse statement.)

(iii) is translated by saying that the identity functor $(\mathcal{C}_X)_A \to (\mathcal{C}_X)_A$ is left exact.   q.e.d.

In practice, it is easier to manipulate local epimorphisms than local isomorphisms.

Proposition 17.2.6. *Let X and Y be two sites and let $f\colon X \to Y$ be a morphism of presites. Assume that*

(i) *f is weakly left exact,*

(ii) *$\widehat{f^t}$ sends local epimorphisms $B \to V$ with $V \in \mathcal{C}_Y$ to local epimorphisms.*

Then f is a morphism of sites.

Proof. Since $\widehat{f^t}$ sends the monomorphisms to monomorphisms, the condition (ii) (b) in Lemma 17.2.2 is satisfied.   q.e.d.

Examples 17.2.7. (i) Let $f\colon X \to Y$ be a continuous map of topological spaces, identified with the functor of presites $f^t\colon \mathrm{Op}_Y \to \mathrm{Op}_X$ given by $\mathrm{Op}_Y \ni V \mapsto f^{-1}(V) \in \mathrm{Op}_X$ (see Example 16.1.7). Then f is a left exact morphism of sites by Proposition 17.2.6.

(ii) Consider the topologies defined in Example 16.1.10. We denote by $\mathcal{E}t_G{}^{et}$, $\mathcal{E}t_G{}^{nis}$, and $\mathcal{E}t_G{}^{zar}$ the category $\mathcal{E}t_G$ endowed with the étale topology, the Nisnevich topology and the Zariski topology, respectively. We obtain the big sites called the étale site, the Nisnevich site, and the Zariski site, respectively. There are natural morphisms of big sites

$$\mathcal{E}t_G{}^{et} \to \mathcal{E}t_G{}^{nis} \to \mathcal{E}t_G{}^{zar} \ .$$

(iii) Let $\mathcal{C}_X$ be a small category. The site obtained by endowing $\mathcal{C}_X$ with the initial (resp. final) topology is denoted by X_{ini} (resp. X_{fin}) (see Example 16.1.9). There are natural morphisms of sites

$$X_{ini} \to X \to X_{fin} \ .$$

Definition 17.2.8. *Let T and T' be two topologies on a presite X and denote by X_T and $X_{T'}$ the associated sites. The topology T is stronger than T', or the topology T' is weaker than T, if the identity functor on $\mathcal{C}_X$ induces a morphism of sites $X_T \to X_{T'}$, that is, if the local isomorphisms with respect to T' are local isomorphisms with respect to T.*

By Lemma 16.2.3, T is stronger than T' if and only if the local epimorphisms with respect to T' are local epimorphisms with respect to T.

Let $\{T_i\}_{i \in I}$ be a family of topologies on a presite X. The intersection topology $\cap_i T_i$ is defined as follows: a morphism is a local epimorphism with respect to $\cap_i T_i$ if and only if it is a local epimorphism with respect to all the T_i's.

The topology $\cup_i T_i$ on X is the weakest topology among the topologies stronger than all T_i's, i.e., $\cup_i T_i = \cap T$ where T ranges over the family of topologies which are stronger than all T_i's.

There are morphisms of sites

(17.2.1) $$X_{ini} \to X_{\cup T_i} \to X_{T_i} \to X_{\cap T_i} \to X_{fin} \ .$$

17.3 Sheaves

Let X be a site and let $\mathcal{A}$ be a category admitting small projective limits. Using Corollary 2.7.4, we shall identify the presheaves on X with the presheaves on $\widehat{X}$ commuting with projective limits. In the sequel, for a presheaf F on X with values in $\mathcal{A}$, we write F instead of $h_X^{\ddagger} F$. Hence we have

$$F(A) = \varprojlim_{(U \to A) \in \mathcal{C}_A} F(U) \quad \text{for } A \in \mathcal{C}_X^{\wedge} \ .$$

We have thus obtained a functor $F \colon (\mathcal{C}_X^{\wedge})^{\mathrm{op}} \to \mathcal{A}$ which commutes with small projective limits. In particular, if $A \to B$ is an epimorphism in $\mathcal{C}_X^{\wedge}$, then $F(B) \to F(A)$ is a monomorphism in $\mathcal{A}$ (see Exercise 3.4).

Definition 17.3.1.   (i) *A presheaf $F \in \mathrm{PSh}(X, \mathcal{A})$ is separated if for any local isomorphism $A \to U$ with $U \in \mathcal{C}_X$ and $A \in \mathcal{C}_X^{\wedge}$, the morphism $F(U) \to F(A)$ is a monomorphism.*

(ii) *A presheaf $F \in \mathrm{PSh}(X, \mathcal{A})$ is a sheaf if for any local isomorphism $A \to U$ with $U \in \mathcal{C}_X$ and $A \in \mathcal{C}_X^{\wedge}$, the morphism $F(U) \to F(A)$ is an isomorphism.*

(iii) *We denote by* $\mathrm{Sh}(X, \mathcal{A})$ *the full subcategory of* $\mathrm{PSh}(X, \mathcal{A})$ *consisting of sheaves on* X. *We shall simply write* $\mathrm{Sh}(X)$ *instead of* $\mathrm{Sh}(X, \mathbf{Set})$. *We denote by* $\iota \colon \mathrm{Sh}(X, \mathcal{A}) \to \mathrm{PSh}(X, \mathcal{A})$ *the forgetful functor.*

Recall that for $F \in \mathrm{PSh}(X, \mathcal{A})$ and $M \in \mathcal{A}$, we have introduced the **Set**-valued presheaf $\mathcal{H}om_{\mathcal{A}}(M, F) \in \mathcal{C}_X^{\wedge}$ which satisfies:

$$(17.3.1) \quad \mathrm{Hom}_{\mathcal{C}_X^{\wedge}}(A, \mathcal{H}om_{\mathcal{A}}(M, F)) \simeq \mathrm{Hom}_{\mathcal{A}}(M, F(A)) \quad \text{for } A \in \mathcal{C}_X \ .$$

Proposition 17.3.2. *An $\mathcal{A}$-valued presheaf F is separated (resp. is a sheaf) if and only if the presheaf of sets $\mathcal{H}om_{\mathcal{A}}(M, F)$ is separated (resp. is a sheaf) for any $M \in \mathcal{A}$.*

Proof. This follows from (17.3.1).   q.e.d.

Proposition 17.3.3. *Let F be a separated presheaf (resp. a sheaf) on X. Then for any local epimorphism (resp. local isomorphism) $A \to B$, the morphism $F(B) \to F(A)$ is a monomorphism (resp. an isomorphism).*

Proof. (i) Let F be a separated presheaf and let $A \to B$ be a local epimorphism. Assume first that $B = U \in \mathcal{C}_X$. Set $S := \mathrm{Im}(A \to U)$. Then $S \to U$ is a local isomorphism. Therefore, $F(U) \to F(S)$ is a monomorphism, and F being left exact, $F(S) \to F(A)$ is a monomorphism.

(ii) In the general case, $B \simeq$ "$\varinjlim$" U (here and in the sequel, $U \in \mathcal{C}_X$), and
$$A \simeq A \times_B B \simeq \text{``}\varinjlim\text{''} A \times_B U.$$
If F is a presheaf, we get $F(B) \simeq \varprojlim_{U \to B} F(U)$ and $F(A) \simeq \varprojlim_{U \to B} F(A \times_B U)$. If $A \to B$ is a local epimorphism (resp. isomorphism), then so is $A \times_B U \to U$. If F is a separated presheaf (resp. a sheaf), then $F(U) \to F(A \times_B U)$ is a monomorphism (resp. an isomorphism). Taking the projective limit with respect to $U \to B$, the result follows.   q.e.d.

Proposition 17.3.4. (i) *Let F be a sheaf and let $B \to A$ be a local epimorphism in $\mathcal{C}_X^{\wedge}$. Then the sequence $F(A) \to F(B) \rightrightarrows F(B \times_A B)$ is exact.*

(ii) *Conversely, let F be a presheaf and assume that for any local isomorphism $B \to U$ with $U \in \mathcal{C}_X$, the sequence $F(U) \to F(B) \rightrightarrows F(B \times_U B)$ is exact. Then F is a sheaf.*

Proof. (i) Set $Z := \mathrm{Im}(B \to A) \simeq \mathrm{Coker}(B \times_A B \rightrightarrows A)$. Since F is left exact on $\mathcal{C}_X^{\wedge}$, the sequence below is exact

$$F(Z) \to F(B) \rightrightarrows F(B \times_A B) \ .$$

Since $Z \to A$ is a local isomorphism, $F(A) \to F(Z)$ is an isomorphism.

(ii) For any local isomorphism $B \to U$, $F(U) \to F(B)$ is a monomorphism. This implies that F is separated. Let $B \to U$ be a local isomorphism with

$U \in \mathcal{C}_X$. Then $B \to B \times_U B$ is a local epimorphism. Since F is separated, the morphism $q \colon F(B \times_U B) \to F(B)$ is a monomorphism. The two morphisms $(p_1, p_2) \colon F(B) \rightrightarrows F(B \times_U B)$ coincide since $q \circ p_1 = q \circ p_2 = \mathrm{id}_{F(B)}$. This implies $F(B) \simeq \mathrm{Ker}\big(F(B) \rightrightarrows F(B \times_U B)\big)$. Hence $F(B) \simeq F(U)$.   q.e.d.

Using Definition 16.1.5, we shall give a more intuitive criterion to recognize sheaves.

Proposition 17.3.5. *Assume that $\mathcal{C}_X$ admits fiber products. A presheaf F is a sheaf if and only if, for any covering $\{U_i\}_{i \in I}$ of U, the sequence below is exact:*

$$(17.3.2) \qquad F(U) \to \prod_{i \in I} F(U_i) \rightrightarrows \prod_{j,k \in I} F(U_j \times_U U_k) \,.$$

Proof. (i) Set $S' = \text{``}\coprod\text{''}_{i \in I} U_i$. Then $S' \times_U S' \simeq \text{``}\coprod\text{''}_{j,k \in I} U_j \times_U U_k$, $F(S') \simeq \prod_{i \in I} F(U_i)$ and $F(S' \times_U S') \simeq \prod_{j,k \in I} F(U_i \times_U U_j)$. Hence (17.3.2) is exact if F is a sheaf.

(ii) Conversely, for any local epimorphism $S \to U$, there exists an epimorphism $S' \to S$ with S' as above. We get the monomorphism

$$\mathrm{Ker}\big(F(S) \rightrightarrows F(S \times_U S)\big) \rightarrowtail \mathrm{Ker}\big(F(S') \rightrightarrows F(S' \times_U S')\big) \,.$$

Therefore the isomorphism $F(U) \xrightarrow{\sim} \mathrm{Ker}\big(F(S') \rightrightarrows F(S' \times_U S')\big)$ entails the isomorphism $F(U) \xrightarrow{\sim} \mathrm{Ker}\big(F(S) \rightrightarrows F(S \times_U S)\big)$ (see Exercise 1.7).   q.e.d.

Example 17.3.6. Let X_{fin} be a small category $\mathcal{C}_X$ endowed with the final topology (see Example 16.1.9). Then any presheaf on X_{fin} is a sheaf and the natural functor $\mathrm{Sh}(X_{fin}, \mathcal{A}) \to \mathrm{PSh}(X, \mathcal{A})$ is an equivalence. Indeed, any local isomorphism in $\mathcal{C}^\wedge_{X_{fin}}$ is an isomorphism.

Example 17.3.7. Let X be a topological space. By choosing $\mathcal{C}_X = \mathrm{Op}_X$ endowed with the Grothendieck topology given in Example 16.1.7 (i), we regard X as a site. By Proposition 17.3.5, a presheaf F with values in **Set** is separated if and only if it satisfies the property S1 below, and F is a sheaf if and only if it satisfies the properties S1 and S2 below.

S1 For any open subset $U \subset X$, any open covering $U = \bigcup_i U_i$, any $s, t \in F(U)$ satisfying $s|_{U_i} = t|_{U_i}$ for all i, we have $s = t$.

S2 For any open subset $U \subset X$, any open covering $U = \bigcup_i U_i$, any family $\{s_i \in F(U_i)\}_{i \in I}$ satisfying $s_i|_{U_{ij}} = s_j|_{U_{ij}}$ for all i, j, there exists $s \in F(U)$ such that $s|_{U_i} = s_i$ for all i. Here, $U_{ij} = U_i \cap U_j$.

Roughly speaking, S1 is translated by saying that uniqueness is a local property, and S2 by saying that natural patching conditions give existence.

Let us give some explicit examples.

(a) The presheaf C_X^0 of $\mathbb{C}$-valued continuous functions on a topological space X is a sheaf with values in the category $\mathrm{Mod}(\mathbb{C})$ of $\mathbb{C}$-vector spaces. The presheaf $U \mapsto C_X^{0,b}(U)$ of continuous bounded functions is not a sheaf in general, to be bounded being not a local property and axiom S2 is not satisfied.

(b) Let $M \in \mathrm{Mod}(k)$. The presheaf of locally constant functions on a topological space X with values in M is a sheaf, called the constant sheaf with stalk M and denoted M_X. Note that the constant presheaf with stalk M is not a sheaf in general.

(c) On a real manifold X of class C^∞, the presheaf C_X^∞ of complex valued functions of class C^∞ is a sheaf with values in $\mathrm{Mod}(\mathbb{C})$. On a complex manifold X, the presheaf $\mathcal{O}_X$ of holomorphic functions is a sheaf with values in $\mathrm{Mod}(\mathbb{C})$.

(d) Let $X = \mathbb{C}$, the complex line, denote by z a holomorphic coordinate and by $\frac{\partial}{\partial z}$ the holomorphic derivation. Consider the presheaf F given by $U \mapsto \mathcal{O}(U)/\frac{\partial}{\partial z}\mathcal{O}(U)$. For U any open disc, $F(U) = 0$ since the equation $\frac{\partial}{\partial z}f = g$ is always solvable. However $F(U) \neq 0$ for a punctured disk $U = \{z \in X; 0 < |z - a| < c\}$ for $a \in X$ and $c > 0$. Hence the presheaf F does not satisfy axiom S1.

17.4 Sheaf Associated with a Presheaf

From now on and *until the end of this chapter*, we shall assume that the category $\mathcal{A}$ satisfies:

(17.4.1) $\begin{cases} \mathcal{A} \text{ admits small projective and small inductive limits,} \\ \text{small filtrant inductive limits are exact,} \\ \mathcal{A} \text{ satisfies the IPC-property (see Definition 3.1.10).} \end{cases}$

For example, the category **Set**, the category **Group** of groups, the category k-**Alg** of algebras over a commutative ring k, or the category $\mathrm{Mod}(R)$ of modules over a ring R satisfies these conditions.

Let X be a site, let $A, A' \in \mathcal{C}^\wedge$, and let $u \colon A' \to A$ be a morphism. For a local isomorphism $B \to A$, $B \times_A A' \to A'$ is a local isomorphism. We thus obtain a functor

(17.4.2) $\lambda_u \colon \mathcal{LI}_A \to \mathcal{LI}_{A'}, \quad (B \to A) \mapsto (B \times_A A' \to A') .$

If moreover u is a local isomorphism, we define by

(17.4.3) $\mu_u \colon \mathcal{LI}_{A'} \to \mathcal{LI}_A, \quad (B \to A') \mapsto (B \to A' \overset{u}{\to} A)$

the functor associated with u. In such a case, (μ_u, λ_u) is clearly a pair of adjoint functors.

Recall that we identify $\mathrm{PSh}(X, \mathcal{A})$ with the full subcategory of $\mathrm{PSh}(\widehat{X}, \mathcal{A})$ consisting of presheaves commuting with small projective limits.

Let $F \in \mathrm{PSh}(\widehat{X}, \mathcal{A})$ and let $A \in \mathcal{C}_X^\wedge$. We set

$$(17.4.4) \qquad F^b(A) = \varinjlim_{(B \to A) \in \mathcal{LI}_A} F(B) \, .$$

(Recall that $(\mathcal{LI}_A)^{\mathrm{op}}$ is cofinally small by Lemma 16.2.8.) Equivalently, the presheaf F defines a functor $\alpha \colon (\mathcal{LI}_A)^{\mathrm{op}} \to \mathcal{A}$ and

$$F^b(A) \simeq \varinjlim \alpha \, .$$

For a morphism $u \colon A' \to A$, we define the morphism $F^b(A) \to F^b(A')$ by the chain of morphisms

$$
\begin{aligned}
F^b(A) = &\varinjlim_{(B \to A) \in \mathcal{LI}_A} F(B) \to \varinjlim_{(B \to A) \in \mathcal{LI}_A} F(B \times_A A') \\
\to &\varinjlim_{(B' \to A') \in \mathcal{LI}_{A'}} F(B') = F^b(A') \, .
\end{aligned}
$$

The second arrow is given by λ_u. Hence, $F^b \in \mathrm{PSh}(\widehat{X}, \mathcal{A})$.

Definition 17.4.1. *We denote by* $(\,\cdot\,)^b \colon \mathrm{PSh}(\widehat{X}, \mathcal{A}) \to \mathrm{PSh}(\widehat{X}, \mathcal{A})$ *the functor given by* (17.4.4).

Note that there is a natural morphism of functors

$$(17.4.5) \qquad \varepsilon_b \colon \mathrm{id} \to (\,\cdot\,)^b \, .$$

Lemma 17.4.2. *Let* $F \in \mathrm{PSh}(\widehat{X}, \mathcal{A})$ *and let* $u \colon A' \to A$ *be a local isomorphism. Then* $F^b(A) \to F^b(A')$ *is an isomorphism.*

Proof. The morphism $F^b(A) \to F^b(A')$ is obtained as the composition

$$\varinjlim \alpha \to \varinjlim \alpha \circ \mu_u^{\mathrm{op}} \circ \lambda_u^{\mathrm{op}} \to \varinjlim \alpha \circ \mu_u^{\mathrm{op}} \, .$$

Since $(\lambda_u^{\mathrm{op}}, \mu_u^{\mathrm{op}})$ is a pair of adjoint functors, the composition is an isomorphism by the result of Exercise 2.15.   q.e.d.

Lemma 17.4.3. *The pair* (b, ε_b) *is a projector on* $\mathrm{PSh}(\widehat{X}, \mathcal{A})$ (*see Definition 4.1.1*). *Namely, for any* $F \in \mathrm{PSh}(\widehat{X}, \mathcal{A})$, $\varepsilon_b(F^b) = (\varepsilon_b(F))^b$ *in* $\mathrm{Hom}_{\mathrm{PSh}(\widehat{X}, \mathcal{A})}(F^b, F^{bb})$ *and this morphism is an isomorphism.*

Proof. (i) The morphism $(\varepsilon_b(F))^b$ is obtained by

$$\varinjlim \varepsilon_b(F)(B) \colon \varinjlim_{(B \to A) \in \mathcal{LI}_A} F(B) \to \varinjlim_{(B \to A) \in \mathcal{LI}_A} F^b(B) \, .$$

On the other hand, we have the isomorphism

$$(17.4.6) \qquad \varinjlim_{(B \to A) \in \mathcal{LI}_A} F^b(B) \xrightarrow{\sim} \varinjlim_{(B \to A) \in \mathcal{LI}_A} \varinjlim_{(B' \to B) \in \mathcal{LI}_B} F(B') \ .$$

Hence, applying Corollary 2.3.4 to $\theta = \mathrm{id}_{\mathcal{LI}_A} \colon \mathcal{LI}_A \to \mathcal{LI}_A$, the right hand side of (17.4.6) is isomorphic to $\varinjlim_{(B' \to A) \in \mathcal{LI}_A} F(B')$. This shows that $(\varepsilon_b(F))^b$ is an isomorphism.

(ii) The morphism $\varepsilon_b(F^b)(A) \colon F^b(A) \to F^{bb}(A)$ is obtained as the composition

$$F^b(A) \xrightarrow{\sim} \varinjlim_{(B \to A) \in \mathcal{LI}_A} F^b(A) \to \varinjlim_{(B \to A) \in \mathcal{LI}_A} F^b(B) \ .$$

This morphism is an isomorphism by Lemma 17.4.2.

Hence, (b, ε_b) is a projector and $\varepsilon_b(F^b) = (\varepsilon_b(F))^b$ by Lemma 4.1.2. q.e.d.

In the proof of the following proposition, we need the assumption that $\mathcal{A}$ satisfies the IPC property.

Proposition 17.4.4. *Let* $F \in \mathrm{PSh}(\widehat{X}, \mathcal{A})$. *If* F *commutes with small projective limits, then so does* $F^b \in \mathrm{PSh}(\widehat{X}, \mathcal{A})$.

Proof. It is enough to check that F^b commutes with small products and with fiber products.

(i) F^b commutes with small products. Let $\{A_i\}_{i \in I}$ be a small family of objects in $(\mathcal{C}_X)^\wedge$ and set $A := \text{``}\coprod_{i \in I}\text{''} A_i$. We shall show the isomorphism

$$F^b(A) \simeq \prod_{i \in I} F^b(A_i) \ .$$

Set $K = \prod_{i \in I} \mathcal{LI}_{A_i}$. Then K^{op} is cofinally small and filtrant. Let $\xi \colon K \to \mathcal{LI}_A$ be the functor

$$(17.4.7) \qquad K \ni \{(B_i \to A_i)\}_{i \in I} \mapsto \left(\text{``}\coprod_{i \in I}\text{''} B_i \to A\right) \in \mathcal{LI}_A \ .$$

Since an inductive limit of local isomorphisms is a local isomorphism by Proposition 16.3.4, the functor ξ is well-defined.

The functor ξ has a right adjoint, namely the functor

$$\mathcal{LI}_A \ni C \mapsto \{C \times_A A_i\}_{i \in I} \in K \ .$$

By Lemma 3.3.10, ξ is co-cofinal. We get the isomorphisms (using the fact that $\mathcal{A}$ satisfies the IPC property):

$$\prod_i F^b(A_i) \simeq \prod_i \varinjlim_{(B_i \to A_i) \in \mathcal{LI}_{A_i}} F(B_i) \simeq \varinjlim_{\{(B_i \to A_i)\}_{i \in I} \in K} \prod_i F(B_i)$$

$$\simeq \varinjlim_{\{(B_i \to A_i)\}_{i \in I} \in K} F(\text{``}\coprod_i\text{''} B_i) \simeq \varinjlim_{(B \to A) \in \mathcal{LI}_A} F(B) \simeq F^b(A) \ .$$

(ii) F^b commutes with fiber products. Let $A \xrightarrow{u} B$ and $A \xrightarrow{v} C$ be morphisms in $(\mathcal{C}_X)^\wedge$. We shall show the isomorphism

$$F^b(B \text{ "}\bigsqcup_A\text{" } C) \xrightarrow{\sim} F^b(B) \times_{F^b(A)} F^b(C) \,.$$

Consider the category $\mathcal{E}$ whose objects are the commutative diagrams

(17.4.8)
$$\begin{array}{ccccc} B' & \xleftarrow{\ u'\ } & A' & \xrightarrow{\ v'\ } & C' \\ \downarrow{\scriptstyle \beta} & & \downarrow{\scriptstyle \alpha} & & \downarrow{\scriptstyle \gamma} \\ B & \xleftarrow{\ u\ } & A & \xrightarrow{\ v\ } & C \end{array}$$

with α, β, γ local isomorphisms. The morphisms in $\mathcal{E}$ are the natural ones. Set $D = B \text{ "}\bigsqcup_A\text{" } C$. For $e = A, B, C, D$, we have functors $p_e \colon \mathcal{E} \to \mathcal{LI}_e$. Here, p_D associates to the object (17.4.8) the morphism $B' \text{ "}\bigsqcup_{A'}\text{" } C' \to B \text{ "}\bigsqcup_A\text{" } C$, which is a local isomorphism by Proposition 16.3.4.

Set $I = (\mathcal{LI}_B)^{\mathrm{op}}$, $J = (\mathcal{LI}_C)^{\mathrm{op}}$ and $K = (\mathcal{LI}_A)^{\mathrm{op}}$. Then I, J and K are cofinally small and filtrant. The morphisms u and v induce functors $I \xrightarrow{\varphi} K \xleftarrow{\psi} J$, $\varphi(B' \to B) = B' \times_B A$ and $\psi(C' \to C) = C' \times_C A$. Then $\mathcal{E}^{\mathrm{op}}$ is equivalent to the category of $\{(i, j, k, \xi, \eta)\}$ with $i \in I$, $j \in J$, $k \in K$, $\xi \colon \varphi(i) \to k$, $\eta \colon \psi(j) \to k$. Hence $\mathcal{E}^{\mathrm{op}}$ is equivalent to the category

$$M[J \to K \leftarrow M[I \xrightarrow{\varphi} K \xleftarrow{\mathrm{id}} K]] \,.$$

Applying Proposition 3.4.5, the category $\mathcal{E}^{\mathrm{op}}$ is cofinally small and filtrant, and the three functors from $\mathcal{E}^{\mathrm{op}}$ to I, J and K are cofinal. The functor $p_D \colon \mathcal{E} \to \mathcal{LI}_D$ admits a right adjoint

$$\mathcal{LI}_D \ni (D' \to D) \mapsto (A \times_D D', B \times_D D', C \times_D D') \in \mathcal{E} \,.$$

Hence, $p_D^{\mathrm{op}} \colon \mathcal{E}^{\mathrm{op}} \to \mathcal{LI}_D^{\mathrm{op}}$ is cofinal by Lemma 3.3.10. Therefore,

$$\begin{aligned} F^b(B \text{ "}\bigsqcup_A\text{" } C) &\simeq \varinjlim_{x \in \mathcal{E}} F(p_D(x)) \simeq \varinjlim_{x \in \mathcal{E}} F\big(p_B(x) \text{ "}\bigsqcup_{p_A(x)}\text{" } p_C(x)\big) \\ &\simeq \varinjlim_{x \in \mathcal{E}} F(p_B(x)) \times_{F(p_A(x))} F(p_C(x)) \\ &\simeq \Big(\varinjlim_{x \in \mathcal{E}} F(p_B(x))\Big) \times_{\varinjlim_{x \in \mathcal{E}} F(p_A(x))} \Big(\varinjlim_{x \in \mathcal{E}} F(p_C(x))\Big) \\ &\simeq F^b(B) \times_{F^b(A)} F^b(C) \,. \end{aligned}$$

Here, the third isomorphism follows from the fact that the functor F is left exact and the fourth isomorphism follows from the fact that filtrant inductive limits are exact in $\mathcal{A}$.   q.e.d.

Definition 17.4.5. *We define the functor*

$$(\cdot)^a \colon \mathrm{PSh}(X, \mathcal{A}) \ni F \mapsto \mathrm{h}_{X*}\big((\mathrm{h}_X^\dagger F)^b\big) \in \mathrm{PSh}(X, \mathcal{A}) \,.$$

Hence, for $F \in \mathrm{PSh}(X, \mathcal{A})$ and $U \in \mathcal{C}_X$, we have

$$F^a(U) \simeq \varinjlim_{A \in \mathcal{LI}_U} F(A) .$$

Note that this definition agrees with (16.3.1) when $\mathcal{A} = \mathbf{Set}$. Proposition 17.4.4 together with Proposition 2.7.1 implies

(17.4.9) $(\mathrm{h}_X^\ddagger F)^b \simeq \mathrm{h}_X^\ddagger(F^a) .$

Hence, we have

(17.4.10) $F^a(A) \simeq \varinjlim_{(B \to A) \in \mathcal{LI}_A} F(B)$ for any $A \in \mathcal{C}_X^\wedge$.

The morphism of functors (17.4.5) gives rise to the morphism of functors

(17.4.11) $\varepsilon\colon \mathrm{id}_{\mathrm{PSh}(X,\mathcal{A})} \to (\,\cdot\,)^a .$

Lemma 17.4.6. *Let $F \in \mathrm{PSh}(X, \mathcal{A})$.*

 (i) *If F is separated, then $F \to F^a$ is a monomorphism.*
 (ii) *If F is a sheaf, then $F \to F^a$ is an isomorphism.*

Proof. (i) For any local isomorphism $A \to U$ with $U \in \mathcal{C}_X$, $F(U) \to F(A)$ is a monomorphism by the definition. Since $\mathcal{LI}_U$ is cofiltrant, $F(U) \to F^a(U)$ is a monomorphism.
(ii) For any local isomorphism $A \to U$, $F(U) \to F(A)$ is an isomorphism. Hence $F(U) \to F^a(U)$ is an isomorphism.   q.e.d.

Theorem 17.4.7. *Let $F \in \mathrm{PSh}(X, \mathcal{A})$.*

 (i) *We have $(\mathrm{h}_X^\ddagger F)^b \simeq \mathrm{h}_X^\ddagger(F^a)$, and $F^a \in \mathrm{Sh}(X, \mathcal{A})$, that is, F^a is a sheaf.*
 (ii) *(a, ε) is a projector, namely $\varepsilon(F^a) = \varepsilon(F)^a$ in $\mathrm{Hom}_{\mathrm{PSh}(X,\mathcal{A})}(F^a, F^{aa})$ and this morphism is an isomorphism.*
 (iii) *The functor $(\,\cdot\,)^a\colon \mathrm{PSh}(X, \mathcal{A}) \to \mathrm{Sh}(X, \mathcal{A})$ is left adjoint to the functor $\iota\colon \mathrm{Sh}(X, \mathcal{A}) \to \mathrm{PSh}(X, \mathcal{A})$. In other words, if $F \in \mathrm{PSh}(X, \mathcal{A})$ and $G \in \mathrm{Sh}(X, \mathcal{A})$, the morphism $F \to F^a$ induces the isomorphism:*

(17.4.12) $\mathrm{Hom}_{\mathrm{Sh}(X,\mathcal{A})}(F^a, G) \xrightarrow{\sim} \mathrm{Hom}_{\mathrm{PSh}(X,\mathcal{A})}(F, G) .$

 (iv) *The functor $(\,\cdot\,)^a\colon \mathrm{PSh}(X, \mathcal{A}) \to \mathrm{Sh}(X, \mathcal{A})$ is exact.*

Proof. (i) The first statement has already been obtained in (17.4.9). The second statement follows from Lemma 17.4.2.

(ii) follows from Lemma 17.4.3.

(iii) By (i) and Lemma 17.4.6, $F \in \mathrm{PSh}(X, \mathcal{A})$ is a sheaf if and only if $F \to F^a$ is an isomorphism. Hence, (iii) follows from Proposition 4.1.3.

(iv) The functor $\mathrm{PSh}(X, \mathcal{A}) \ni F \mapsto F(B) \in \mathcal{A}$ is left exact for $B \in (\mathcal{C}_X)^\wedge$. Since filtrant inductive limits are exact in $\mathcal{A}$, $F \mapsto F^a(U) \simeq \varinjlim_{A \in \mathcal{L}\mathcal{I}_U} F(A)$ is left exact for any $U \in \mathcal{C}_X$. This implies that the functor $(\cdot)^a \colon \mathrm{PSh}(X, \mathcal{A}) \to \mathrm{PSh}(X, \mathcal{A})$ is left exact. Since this functor admits a right adjoint, it is right exact. q.e.d.

We shall give a converse statement to Lemma 17.4.6.

Corollary 17.4.8. *Let $F \in \mathrm{PSh}(X, \mathcal{A})$. Then*

(i) *F is separated if and only if $F \to F^a$ is a monomorphism,*
(ii) *F is a sheaf if and only if $F \to F^a$ is an isomorphism.*

Proof. (i) Assume that $F \to F^a$ is a monomorphism. Let $A \to U$ be a local isomorphism. Since the composition $F(U) \to F^a(U) \to F^a(A)$ is a monomorphism and is equal to the composition $F(U) \to F(A) \to F^a(A)$, we conclude that $F(U) \to F(A)$ is a monomorphism. The converse statement follows from Lemma 17.4.6 (i).
(ii) Assume that $F \to F^a$ is an isomorphism. Then F^a is a sheaf by Theorem 17.4.7 (i). The converse statement follows from Lemma 17.4.6 (ii). q.e.d.

Theorem 17.4.9. (i) *The category $\mathrm{Sh}(X, \mathcal{A})$ admits small projective limits and the functor $\iota \colon \mathrm{Sh}(X, \mathcal{A}) \to \mathrm{PSh}(X, \mathcal{A})$ commutes with such limits.*
(ii) *The category $\mathrm{Sh}(X, \mathcal{A})$ admits small inductive limits and the functor $^a \colon \mathrm{PSh}(X, \mathcal{A}) \to \mathrm{Sh}(X, \mathcal{A})$ commutes with such limits. Moreover, the inductive limit of an inductive system $\{F_i\}_{i \in I}$ in $\mathrm{Sh}(X, \mathcal{A})$ is the sheaf associated with the inductive limit in $\mathrm{PSh}(X, \mathcal{A})$.*
(iii) *Filtrant inductive limits in $\mathrm{Sh}(X, \mathcal{A})$ are exact.*
(iv) *Assume that $\mathcal{A}$ has the following property:*

$$(17.4.13) \quad \begin{cases} \text{any morphism } u \text{ is strict, i.e., the natural morphism} \\ \mathrm{Coim}\, u \to \mathrm{Im}\, u \text{ is an isomorphism } (\textit{see } \text{Definition } 5.1.4). \end{cases}$$

Then $\mathrm{Sh}(X, \mathcal{A})$ satisfies (17.4.13). In particular if $\mathcal{A}$ is abelian, then $\mathrm{Sh}(X, \mathcal{A})$ is abelian.

Proof. (i) It is enough to show that, for a small projective system $\{F_i\}_{i \in I}$ in $\mathrm{Sh}(X, \mathcal{A})$, its projective limit $\varprojlim_i F_i$ in $\mathrm{PSh}(X, \mathcal{A})$ is a sheaf. For $A \in \mathcal{C}_X^\wedge$, we have $(\varprojlim_i F_i)(A) \simeq \varprojlim_i (F_i(A))$. Let $B \to A$ be a local isomorphism. The family of isomorphisms

$$F_i(A) \xrightarrow{\sim} F_i(B)$$

defines a similar isomorphism with F_i replaced with $\varprojlim_i F_i$.

(ii) Since the functor $^a \colon \mathrm{PSh}(X, \mathcal{A}) \to \mathrm{Sh}(X, \mathcal{A})$ admits a right adjoint, it commutes with small inductive limits. The other statements follow.

(iii) Filtrant inductive limits are exact in $\mathrm{PSh}(X, \mathcal{A})$ and the functor $(\,\cdot\,)^a$ is exact.

(iv) Let $u\colon F \to G$ be a morphism in $\mathrm{Sh}(X, \mathcal{A})$. Let us prove that its image and coimage are isomorphic. Let L and K be the presheaves defined by $L(U) = \mathrm{Im}(F(U) \to G(U)) = \mathrm{Ker}\big(G(U) \rightrightarrows G(U) \sqcup_{F(U)} G(U)\big)$ and $K(U) = \mathrm{Coim}(F(U) \to G(U)) = \mathrm{Coker}\big(F(U) \times_{G(U)} F(U) \rightrightarrows F(U)\big)$. Then L and K are the image and coimage of u in $\mathrm{PSh}(X, \mathcal{A})$. Since $K(U) \to L(U)$ is an isomorphism for any $U \in \mathcal{C}_X$ by the hypothesis, $K \to L$ is an isomorphism of presheaves. Therefore, it is enough to remark that

$$L^a \simeq \mathrm{Im}(F \to G) = \mathrm{Ker}\big(G \rightrightarrows G \sqcup_F G\big) \text{ in the category } \mathrm{Sh}(X, \mathcal{A}),$$

$$K^a \simeq \mathrm{Coim}(F \to G) = \mathrm{Coker}(F \times_G F \rightrightarrows F) \text{ in the category } \mathrm{Sh}(X, \mathcal{A}),$$

which follow from (i), (ii) and the exactness of the functor ^{a}.   q.e.d.

Remark that the functor $\iota\colon \mathrm{Sh}(X, \mathcal{A}) \to \mathrm{PSh}(X, \mathcal{A})$ does not commute with inductive limits in general.

17.5 Direct and Inverse Images

Let $f\colon X \to Y$ be a morphism of sites. Recall that the direct image functor for presheaves $f_*\colon \mathrm{PSh}(X, \mathcal{A}) \to \mathrm{PSh}(Y, \mathcal{A})$ satisfies

$$(f_* F)(A) = F(\widehat{f}^{\,t}(A)) \quad \text{for } A \in \mathcal{C}_Y^{\wedge} \text{ and } F \in \mathrm{PSh}(X, \mathcal{A})$$

by (17.1.6).

Proposition 17.5.1. *Let* $f\colon X \to Y$ *be a morphism of sites and let* $F \in \mathrm{Sh}(X, \mathcal{A})$. *Then* $f_* F \in \mathrm{Sh}(Y, \mathcal{A})$.

The functor $f_*\colon \mathrm{Sh}(X, \mathcal{A}) \to \mathrm{Sh}(Y, \mathcal{A})$ is called the *direct image* functor for sheaves.

Proof. Let $A \to B$ be a local isomorphism in $\mathcal{C}_Y^{\wedge}$. Since f is a morphism of sites, $\widehat{f}^{\,t}(A) \to \widehat{f}^{\,t}(B)$ is a local isomorphism and F being a sheaf, $F(\widehat{f}^{\,t}(A)) \simeq F(\widehat{f}^{\,t}(B))$. We get the chain of isomorphisms

$$(f_* F)(B) \simeq F(\widehat{f}^{\,t}(B)) \xrightarrow{\sim} F(\widehat{f}^{\,t}(A)) \simeq (f_* F)(A) \, .$$

q.e.d.

The functor $f^{\dagger}\colon \mathrm{PSh}(Y, \mathcal{A}) \to \mathrm{PSh}(X, \mathcal{A})$ is defined in Sect. 17.1. Recall that

$$(f^{\dagger} G)(U) \simeq \varinjlim_{(U \to f^t(V)) \in (\mathcal{C}_Y)^V} G(V) \quad \text{for } G \in \mathrm{PSh}(Y, \mathcal{A}), \, U \in \mathcal{C}_X.$$

The *inverse image* functor for sheaves $f^{-1}\colon \mathrm{Sh}(Y, \mathcal{A}) \to \mathrm{Sh}(X, \mathcal{A})$ is defined by setting for $G \in \mathrm{Sh}(Y, \mathcal{A})$

$$f^{-1} G = (f^{\dagger} G)^a \, .$$

Theorem 17.5.2. *Let $f\colon X \to Y$ be a morphism of sites.*

(i) *The functor $f^{-1}\colon \mathrm{Sh}(Y, \mathcal{A}) \to \mathrm{Sh}(X, \mathcal{A})$ is left adjoint to the functor f_*. In other words, we have an isomorphism, functorial with respect to $F \in \mathrm{Sh}(X, \mathcal{A})$ and $G \in \mathrm{Sh}(Y, \mathcal{A})$:*

$$\mathrm{Hom}_{\mathrm{Sh}(X,\mathcal{A})}(f^{-1}G, F) \simeq \mathrm{Hom}_{\mathrm{Sh}(Y,\mathcal{A})}(G, f_*F) \,.$$

(ii) *The functor f_* is left exact and commutes with small projective limits.*
(iii) *The functor f^{-1} is right exact and commutes with small inductive limits.*
(iv) *Assume that $f\colon X \to Y$ is left exact. Then the functor f^{-1} is exact.*

Proof. (i) The functor $f^\dagger\colon \mathrm{PSh}(Y, \mathcal{A}) \to \mathrm{PSh}(X, \mathcal{A})$ is left adjoint to f_* by Proposition 2.3.3. Hence we have the chain of isomorphisms

$$\begin{aligned}
\mathrm{Hom}_{\mathrm{Sh}(Y,\mathcal{A})}(G, f_*F) &\simeq \mathrm{Hom}_{\mathrm{PSh}(Y,\mathcal{A})}(G, f_*F) \\
&\simeq \mathrm{Hom}_{\mathrm{PSh}(X,\mathcal{A})}(f^\dagger G, F) \\
&\simeq \mathrm{Hom}_{\mathrm{Sh}(X,\mathcal{A})}((f^\dagger G)^a, F) \\
&= \mathrm{Hom}_{\mathrm{Sh}(X,\mathcal{A})}(f^{-1}G, F) \,.
\end{aligned}$$

(ii)–(iii) are obvious by the adjunction property.
(iv) The functor $f^\dagger$ is left exact by Theorem 3.3.18. Since the functor $(\bullet)^a$ is exact, the result follows.   q.e.d.

Consider morphisms of sites $X \xrightarrow{f} Y \xrightarrow{g} Z$.

Proposition 17.5.3. *There are natural isomorphisms of functors*

$$\begin{cases} g_* \circ f_* \simeq (g \circ f)_* \,, \\ f^{-1} \circ g^{-1} \simeq (g \circ f)^{-1}. \end{cases}$$

Proof. The first isomorphism follows from Proposition 17.5.1. The second one is deduced by adjunction.   q.e.d.

17.6 Restriction and Extension of Sheaves

Let $A \in \mathcal{C}_X^\wedge$. We follow the notations and the results in Proposition 17.1.9. In particular, A is regarded as a presite and the forgetful functor $(j_{A \to X})^t\colon \mathcal{C}_A := (\mathcal{C}_X)_A \to \mathcal{C}_X$ gives a morphism of presites $j_{A \to X}\colon X \to A$.

Definition 17.6.1. *We endow the presite A with the following topology: a morphism $C \to B$ in $\mathcal{C}_A^\wedge$ is a local epimorphism if and only if $\widehat{j}_{A \to X}{}^t(C) \to \widehat{j}_{A \to X}{}^t(B)$ is a local epimorphism in $\mathcal{C}_X^\wedge$.*

It is easily checked that we obtain a Grothendieck topology on $\mathcal{C}_A$. The morphism of presites $j_{A \to X}\colon X \to A$ is weakly left exact by Lemma 17.2.5. It is a morphism of sites by Proposition 17.2.6.

Proposition 17.6.2. *Let* $G \in \mathrm{Sh}(A, \mathcal{A})$. *Then* $\mathrm{j}^{\ddagger}_{A \to X} G \in \mathrm{Sh}(X, \mathcal{A})$. *In other words, the presheaf* $\mathrm{j}^{\ddagger}_{A \to X} G$ *is a sheaf. Moreover,* $\mathrm{j}^{\ddagger}_{A \to X} \colon \mathrm{Sh}(A, \mathcal{A}) \to \mathrm{Sh}(X, \mathcal{A})$ *is a right adjoint to* $\mathrm{j}_{A \to X *}$.

Proof. Let $C \to B$ be a local isomorphism in $\mathcal{C}_X^{\wedge}$. There is a chain of isomorphisms

$$\mathrm{j}^{\ddagger}_{A \to X} G(B) \simeq G(B \times A) \xrightarrow{\sim} G(C \times A) \simeq \mathrm{j}^{\ddagger}_{A \to X} G(C) \ .$$

The last assertion follows from its counterpart for presheaves.   q.e.d.

Proposition 17.6.3. *Assume that* $\mathcal{A}$ *is an additive category which satisfies* (17.4.1). *Let* $f \colon X \to Y$ *be a weakly left exact morphism of sites. Then* $f^{-1} \colon \mathrm{Sh}(Y, \mathcal{A}) \to \mathrm{Sh}(X, \mathcal{A})$ *is exact.*
    In particular, for $A \in (\mathcal{C}_X)^{\wedge}$, *the functor* $\mathrm{j}^{-1}_{A \to X} \colon \mathrm{Sh}(A, \mathcal{A}) \to \mathrm{Sh}(X, \mathcal{A})$ *is exact.*

Proof. (i) First, we treat the case of $\mathrm{j}_{A \to X}$. Since small coproducts are exact in $\mathcal{A}$ by the assumption, the functor $\mathrm{j}^{\ddagger}_{A \to X}$ is exact by (17.1.11). Since $(\,\cdot\,)^a$ is exact, $\mathrm{j}^{-1}_{A \to X}$ is exact.

(ii) Set $A = \widehat{f}^t(\mathrm{pt}_Y) \in \mathcal{C}_X^{\wedge}$. Then f factors as $X \xrightarrow{\mathrm{j}_{A \to X}} A \xrightarrow{g} Y$. Since g is left exact, the functor g^{-1} is exact by Theorem 17.5.2 (iv) and $\mathrm{j}^{-1}_{A \to X}$ is exact by (i). Hence, $f^{-1} \simeq \mathrm{j}^{-1}_{A \to X} \circ g^{-1}$ is exact.   q.e.d.

Remark 17.6.4. In Proposition 17.6.3, we have assumed that $\mathcal{A}$ is additive since we need the condition that small coproducts are exact.

Example 17.6.5. Let X be a topological space identified with the site associated with Op_X and let $i_U \colon U \hookrightarrow X$ be an open embedding. The map i_U defines a morphism of sites, and $i_U^t \colon \mathrm{Op}_X \to \mathrm{Op}_U$ is given by $i_U^t(V) = V \cap U$. On the other hand, we have the morphism of sites $\mathrm{j}_{U \to X} \colon X \to U$. There are isomorphisms of functors

$$\mathrm{j}_{U \to X *} \simeq i_U^{-1} \quad \text{and} \quad \mathrm{j}^{\ddagger}_{U \to X} \simeq i_{U *} \ .$$

The functor $\mathrm{j}^{-1}_{U \to X}$ is isomorphic to a functor usually denoted by $i_{U !}$ in the literature.

Hence, we have two pairs of adjoint functors $(\mathrm{j}^{-1}_{A \to X}, \mathrm{j}_{A \to X *})$, $(\mathrm{j}_{A \to X *}, \mathrm{j}^{\ddagger}_{A \to X})$:

$$(17.6.1) \qquad \mathrm{Sh}(A, \mathcal{A}) \underset{\underset{\mathrm{j}^{\ddagger}_{A \to X}}{\xleftarrow{\mathrm{j}_{A \to X *}}}}{\overset{\mathrm{j}^{-1}_{A \to X}}{\xrightarrow{\hspace{2cm}}}} \mathrm{Sh}(X, \mathcal{A}) \ .$$

Proposition 17.6.6. *The functor* $\mathrm{j}_{A \to X *} \colon \mathrm{Sh}(X, \mathcal{A}) \to \mathrm{Sh}(A, \mathcal{A})$ *is exact. Moreover it commutes with small inductive limits and small projective limits.*

Proof. The functor $j_{A \to X *}$ has both a right and a left adjoint.   q.e.d.

Now consider a morphism in $\mathcal{C}_X^\wedge$

$$u : A_1 \to A_2 .$$

The functor $u : \mathcal{C}_{A_1} \to \mathcal{C}_{A_2}$ given by $(U \to A_1) \mapsto (U \to A_1 \to A_2)$ defines a morphism of sites by

$$j_u : A_2 \to A_1 .$$

Next consider morphisms $u : A_1 \to A_2$ and $v : A_2 \to A_3$ in $\mathcal{C}_X^\wedge$. Then the composition $A_3 \xrightarrow{j_v} A_2 \xrightarrow{j_u} A_1$ is a morphism of sites, and we have

$$j_{v \circ u} \simeq j_u \circ j_v .$$

We have the isomorphisms:

(i) $j_{v \circ u *} \simeq j_{u *} \circ j_{v *} : \mathrm{Sh}(A_3, \mathcal{A}) \to \mathrm{Sh}(A_1, \mathcal{A})$,
(ii) $j_{v \circ u}^{-1} \simeq j_v^{-1} \circ j_u^{-1} : \mathrm{Sh}(A_1, \mathcal{A}) \to \mathrm{Sh}(A_3, \mathcal{A})$,
(iii) $j_{v \circ u}^\ddagger \simeq j_v^\ddagger \circ j_u^\ddagger : \mathrm{Sh}(A_1, \mathcal{A}) \to \mathrm{Sh}(A_3, \mathcal{A})$.

Let $f : X \to Y$ be a morphism of sites and let $B \in \mathcal{C}_Y^\wedge$. Set $A := \widehat{f}^t(B)$. We get a diagram of sites:

(17.6.2)

$$
\begin{array}{ccc}
X & \xrightarrow{\ j_{A \to X}\ } & A \\
{\scriptstyle f} \downarrow & & \downarrow {\scriptstyle f_B} \\
Y & \xrightarrow{\ j_{B \to Y}\ } & B .
\end{array}
$$

This diagram of sites clearly commutes. We deduce the isomorphisms of functors:

(i) $f^{-1} \circ j_{B \to Y}^{-1} \simeq j_{A \to X}^{-1} \circ f_B^{-1}$,
(ii) $j_{B \to Y *} \circ f_* \simeq f_{B *} \circ j_{A \to X *}$.

Proposition 17.6.7. *Let $f : X \to Y$ be a left exact morphism of sites. Then, using the notations in diagram (17.6.2), we have the isomorphisms of functors*

(i) $j_{B \to Y}^\ddagger \circ f_{B *} \simeq f_* \circ j_{A \to X}^\ddagger$,
(ii) $j_{A \to X *} \circ f^{-1} \simeq f_B^{-1} \circ j_{B \to Y *}$.

Proof. (i) Let $G \in \mathrm{Sh}(A, \mathcal{A})$. Then for any $V \in \mathcal{C}_Y$, the hypothesis implies the isomorphism $\widehat{f}^t(V \times B) \simeq f^t(V) \times \widehat{f}^t(B)$. Hence,

$$
\begin{aligned}
(j_{B \to Y}^\ddagger \circ f_{B *} G)(V) &\simeq f_{B *}(G)(V \times B \to B) \\
&\simeq G\big(\widehat{f}^t(V \times B) \to \widehat{f}^t(B)\big) \\
&\simeq G\big(f^t(V) \times A \to A\big) ,
\end{aligned}
$$

and

$$(f_* \circ j^{\ddagger}_{A \to X} G)(V) \simeq (j^{\ddagger}_{A \to X} G)(f^t(V))$$
$$\simeq G(f^t(V) \times A \to A) .$$

(ii) follows by adjunction.   q.e.d.

Recall that pt_X denotes the terminal object of $\mathcal{C}_X^{\wedge}$.

Proposition 17.6.8. *Let* $A \in \mathcal{C}_X^{\wedge}$ *and assume that* $A \to \mathrm{pt}_X$ *is a local iso-morphism. Then the functor* $j_{A \to X *} \colon \mathrm{Sh}(X, \mathcal{A}) \to \mathrm{Sh}(A, \mathcal{A})$ *is an equivalence of categories.*

Proof. (i) Let us first show that $G \to j^{\ddagger}_{A \to X} \circ j_{A \to X *} G$ is an isomorphism for any $G \in \mathrm{Sh}(X, \mathcal{A})$. For any $U \in \mathcal{C}_X$, we have

$$j^{\ddagger}_{A \to X} \circ j_{A \to X *} G(U) \simeq j_{A \to X *} G(U \times A \to A)$$
$$\simeq G(U \times A) .$$

Since $p \colon U \times A \to U$ is a local isomorphism by the hypothesis, $G(U) \xrightarrow{\sim} G(U \times A)$.

(ii) Let us show that $j_{A \to X *} \circ j^{\ddagger}_{A \to X} F \to F$ is an isomorphism for any $F \in \mathrm{Sh}(A, \mathcal{A})$. For any $u \colon V \to A$ in $\mathcal{C}_A$, we have

$$(j_{A \to X *} \circ j^{\ddagger}_{A \to X} F)(V \to A) \simeq (j^{\ddagger}_{A \to X} F)(V) \simeq F(V \times A \to A) .$$

The morphism $s = (\mathrm{id}_V, u) \colon (V \to A) \to (V \times A \to A)$ is a local isomorphism by the hypothesis. Hence, $F(V \times A \to A) \to F(V \to A)$ is an isomorphism. q.e.d.

Corollary 17.6.9. *Let* $A \in \mathcal{C}_X^{\wedge}$ *and assume that* $A \to \mathrm{pt}_X$ *is a local epimor-phism. Then the functor* $j_{A \to X *} \colon \mathrm{Sh}(X, \mathcal{A}) \to \mathrm{Sh}(A, \mathcal{A})$ *is conservative and faithful.*

Proof. By Proposition 2.2.3, it is enough to prove that $j_{A \to X *}$ is conservative.

We decompose the morphism $A \to \mathrm{pt}_X$ as $A \xrightarrow{u} B \xrightarrow{v} \mathrm{pt}_X$, with $B = \mathrm{Im}(A \to \mathrm{pt}_X)$. Then v is a local isomorphism, and by Proposition 17.6.8 it remains to show that the epimorphism u induces a conservative functor $j_{u *} \colon \mathrm{Sh}(B, \mathcal{A}) \to \mathrm{Sh}(A, \mathcal{A})$.

Let $\varphi \colon F \to G$ be a morphism in $\mathrm{Sh}(B, \mathcal{A})$, and assume that $j_{u *}\varphi \colon j_{u *} F \to j_{u *} G$ is an isomorphism. Let $(U \to B) \in \mathcal{C}_B$. Since $u \colon A \to B$ is an epi-morphism and $U \in \mathcal{C}_X$, the morphism $U \to B$ factorizes as $U \to A \xrightarrow{u} B$. Therefore,

$$F(U \to B) \simeq (j_{u *} F)(U \to A) \xrightarrow{\sim} (j_{u *} G)(U \to A) \simeq G(U \to B) .$$

                                                                             q.e.d.

Definition 17.6.10. *Let X be a site, let $A \in \mathcal{C}_X^\wedge$ and let $F \in \mathrm{Sh}(X, \mathcal{A})$. We set*

(i) $F_A = j_{A \to X}^{-1} j_{A \to X *}(F)$,

(ii) $\Gamma_A(F) = j_{A \to X}^{\ddagger} j_{A \to X *}(F)$.

These two functors $(\bullet)_A$ and $\Gamma_A(\bullet)$ are functors from $\mathrm{Sh}(X, \mathcal{A})$ to itself, and $((\bullet)_A, \Gamma_A(\bullet))$ is a pair of adjoint functors. When $\mathcal{A}$ is additive the functor $(\bullet)_A$ is exact by Propositions 17.6.3 and 17.6.6.

Note that the adjunction morphisms $j_{A \to X}^{-1} \circ j_{A \to X *} \to \mathrm{id}_{\mathrm{Sh}(X, \mathcal{A})}$ and $\mathrm{id}_{\mathrm{Sh}(X, \mathcal{A})} \to j_{A \to X}^{\ddagger} \circ j_{A \to X *}$ give morphisms

$$(17.6.3) \qquad\qquad F_A \to F \to \Gamma_A(F).$$

Moreover, if $A \xrightarrow{u} B$ is a morphism in $\mathcal{C}_X^\wedge$, the adjunction morphisms $j_u^{-1} j_{u *} \to \mathrm{id}_{\mathrm{Sh}(B, \mathcal{A})}$ and $\mathrm{id}_{\mathrm{Sh}(B, \mathcal{A})} \to j_u^{\ddagger} j_{u *}$ give the natural morphisms

$$(17.6.4) \qquad\qquad F_A \to F_B, \ \ \Gamma_B(F) \to \Gamma_A(F).$$

Applying Proposition 17.1.9 and Proposition 17.6.2, we obtain:

Lemma 17.6.11.   (i) *The sheaf F_A is the sheaf associated with the presheaf*
$$U \mapsto F(U)^{\coprod A(U)}.$$
(ii) *We have $\Gamma_A(F)(B) \simeq F(A \times B)$ for $B \in \mathcal{C}_X^\wedge$.*

Remark 17.6.12. Recall that **Pt** is endowed with the final topology (see Example 16.1.9). Let us denote by c the unique object of **Pt**. There is a natural equivalence

$$\mathrm{Sh}(\mathbf{Pt}, \mathcal{A}) \xrightarrow{\sim} \mathcal{A}, \ \ F \mapsto F(c).$$

In the sequel, we shall identify these two categories.

Notations 17.6.13. (i) Let X be a site and let $A \in \mathcal{C}_X^\wedge$. For $M \in \mathcal{A}$, let us denote by M_A the sheaf associated with the constant presheaf $\mathcal{C}_X \ni U \mapsto M$. We define the object M_{XA} of $\mathrm{Sh}(X, \mathcal{A})$ by $j_{A \to X}^{-1}(M_A)$. (See Exercise 17.11.)

(ii) For $A \in \mathcal{C}_X^\wedge$, we set

$$(17.6.5) \qquad\qquad \bullet|_A := j_{A \to X *}.$$

In other words, we shall often write $F|_A$ instead of $j_{A \to X *} F$ for $F \in \mathrm{Sh}(X, \mathcal{A})$.

(iii) We introduce the functor

$$\Gamma(X; \bullet) \colon \mathrm{Sh}(X, \mathcal{A}) \to \mathcal{A}, \ \ F \mapsto \Gamma(X; F) := F(\mathrm{pt}_X) = \varprojlim_{U \in \mathcal{C}_X} F(U).$$

For $A \in \mathcal{C}_X^\wedge$ and $F \in \mathrm{Sh}(X, \mathcal{A})$, we set

$$\Gamma(A; F) = \Gamma(A; F|_A) .$$

Hence,

$$\Gamma(A; F) \simeq \varprojlim_{(U \to A) \in \mathcal{C}_A} F(U) \simeq F(A) .$$

With these notations, we get for $F \in \mathrm{Sh}(X, \mathcal{A})$:

$$\Gamma(X; F) = F(\mathrm{pt}_X) = \Gamma(\mathrm{pt}_X; F) = \varprojlim_{U \in \mathcal{C}_X} F(U) .$$

Proposition 17.6.14. *Let X be a site, let $A, B \in \mathcal{C}_X^\wedge$ and let $F \in \mathrm{Sh}(X, \mathcal{A})$. There are natural isomorphisms*

(17.6.6) $\qquad\qquad (F_A)_B \simeq F_{A \times B} ,$

(17.6.7) $\qquad\qquad \Gamma_B(\Gamma_A(F)) \simeq \Gamma_{A \times B}(F) ,$

(17.6.8) $\qquad\qquad \Gamma(X; \Gamma_A(F)) \simeq \Gamma(A; F) .$

Proof. (i) Let us first prove (17.6.7). For $U \in \mathcal{C}_X$, we have

$$\Gamma_B(\Gamma_A(F))(U) \simeq \Gamma_A(F)(B \times U) \simeq F(A \times B \times U)$$
$$\simeq \Gamma_{A \times B}(F)(U) .$$

(ii) (17.6.6) follows from (17.6.7) by adjunction.

(iii) We have the isomorphisms

$$\Gamma(X; \Gamma_A(F)) \simeq F(A \times \mathrm{pt}_X) \simeq F(A) .$$

$\qquad\qquad\qquad\qquad\qquad\qquad\qquad\qquad\qquad\qquad\qquad\qquad$ q.e.d.

17.7 Internal $\mathcal{H}om$

Recall that, for $F, G \in \mathrm{PSh}(X, \mathcal{A})$, we have defined $\mathcal{H}om_{\mathrm{PSh}(X,\mathcal{A})}(F, G) \in \mathrm{PSh}(X)$ which satisfies (see Lemma 17.1.11)

$$\mathcal{H}om_{\mathrm{PSh}(X,\mathcal{A})}(F, G)(A) \simeq \mathrm{Hom}_{\mathrm{PSh}(A,\mathcal{A})}(\mathrm{j}_{A \to X *} F, \mathrm{j}_{A \to X *} G) \quad \text{for } A \in \mathcal{C}_X^\wedge.$$

Proposition 17.7.1. (i) *Let F, G be objects of $\mathrm{Sh}(X, \mathcal{A})$. Then the presheaf $\mathcal{H}om_{\mathrm{PSh}(X,\mathcal{A})}(F, G)$ is a sheaf on X.*

(ii) *Let $F \in \mathrm{PSh}(X, \mathcal{A})$ and let $G \in \mathrm{Sh}(X, \mathcal{A})$. The morphism $F \to F^a$ gives the isomorphism $\mathcal{H}om_{\mathrm{PSh}(X,\mathcal{A})}(F^a, G) \xrightarrow{\sim} \mathcal{H}om_{\mathrm{PSh}(X,\mathcal{A})}(F, G)$. In particular, this last presheaf is a sheaf.*

Proof. (i) For a local isomorphism $A \to U$ in $\mathcal{C}_X^\wedge$ with $U \in \mathcal{C}_X$, $\mathrm{Sh}(U, \mathcal{A}) \to \mathrm{Sh}(A, \mathcal{A})$ is an equivalence by Proposition 17.6.8. Therefore, we have

$$
\begin{aligned}
\mathcal{H}om_{\mathrm{PSh}(X,\mathcal{A})}(F, G)(U) &= \mathrm{Hom}_{\mathrm{Sh}(U,\mathcal{A})}(j_{U \to X*}F, j_{U \to X*}G) \\
&\simeq \mathrm{Hom}_{\mathrm{Sh}(A,\mathcal{A})}(j_{A \to X*}F, j_{A \to X*}G) \\
&\simeq \mathcal{H}om_{\mathrm{PSh}(X,\mathcal{A})}(F, G)(A) .
\end{aligned}
$$

(ii) Let $U \in \mathcal{C}_X$. Then

$$
\begin{aligned}
\mathcal{H}om_{\mathrm{PSh}(X,\mathcal{A})}(F, G)(U) &= \mathrm{Hom}_{\mathrm{PSh}(X,\mathcal{A})}(j_{U \to X*}F, j_{U \to X*}G) \\
&\simeq \mathrm{Hom}_{\mathrm{PSh}(X,\mathcal{A})}(F, j^{\ddagger}_{U \to X}j_{U \to X*}G) .
\end{aligned}
$$

Since $j^{\ddagger}_{U \to X}j_{U \to X*}G$ is a sheaf by Proposition 17.6.2, we also have

$$
\begin{aligned}
\mathrm{Hom}_{\mathrm{PSh}(X,\mathcal{A})}(F, j^{\ddagger}_{U \to X}j_{U \to X*}G) &\simeq \mathrm{Hom}_{\mathrm{Sh}(X,\mathcal{A})}(F^a, j^{\ddagger}_{U \to X}j_{U \to X*}G) \\
&\simeq \mathrm{Hom}_{\mathrm{Sh}(A,\mathcal{A})}(j_{U \to X*}F^a, j_{U \to X*}G) \\
&\simeq \mathcal{H}om_{\mathrm{PSh}(X,\mathcal{A})}(F^a, G)(U) .
\end{aligned}
$$

q.e.d.

Notation 17.7.2. For F, G in $\mathrm{Sh}(X, \mathcal{A})$, we shall write $\mathcal{H}om_{\mathrm{Sh}(X,\mathcal{A})}(F, G)$ instead of $\mathcal{H}om_{\mathrm{PSh}(X,\mathcal{A})}(F, G)$.

Proposition 17.7.3. *Let $f : X \to Y$ be a left exact morphism of sites, let $F \in \mathrm{Sh}(X, \mathcal{A})$ and let $G \in \mathrm{Sh}(Y, \mathcal{A})$. There is a natural isomorphism in $\mathrm{Sh}(Y)$:*

$$(17.7.1) \qquad f_* \mathcal{H}om_{\mathrm{Sh}(X,\mathcal{A})}(f^{-1}G, F) \simeq \mathcal{H}om_{\mathrm{Sh}(Y,\mathcal{A})}(G, f_*F) .$$

Proof. Let $V \in \mathcal{C}_Y$ and set $U = f^t(V)$. Denote by $f_V^t : \mathcal{C}_U \to \mathcal{C}_V$ the functor induced by f^t. We have the chain of isomorphisms:

$$
\begin{aligned}
(f_* \mathcal{H}om_{\mathrm{Sh}(X,\mathcal{A})}(f^{-1}G, F))(V) &\simeq \mathcal{H}om_{\mathrm{Sh}(X,\mathcal{A})}(f^{-1}G, F)(U) \\
&\simeq \mathrm{Hom}_{\mathrm{Sh}(U,\mathcal{A})}(j_{U \to X*}f^{-1}G, j_{U \to X*}F) \\
&\simeq \mathrm{Hom}_{\mathrm{Sh}(U,\mathcal{A})}(f_V^{-1}j_{V \to Y*}G, j_{U \to X*}F) \\
&\simeq \mathrm{Hom}_{\mathrm{Sh}(V,\mathcal{A})}(j_{V \to Y*}G, f_{V*}j_{U \to X*}F) \\
&\simeq \mathrm{Hom}_{\mathrm{Sh}(V,\mathcal{A})}(j_{V \to Y*}G, j_{V \to Y*}f_*F) \\
&\simeq \mathcal{H}om_{\mathrm{Sh}(Y,\mathcal{A})}(G, f_*F)(V) .
\end{aligned}
$$

The third isomorphism follows from Proposition 17.6.7. These isomorphisms being functorial with respect to V, the isomorphism (17.7.1) follows.   q.e.d.

Exercises

In these exercises, $\mathcal{A}$ is a category satisfying (17.4.1).

Exercise 17.1. Let X be a small presite and let F, G, $H \in \mathrm{PSh}(X) = \mathcal{C}_X^\wedge$ and $U \in \mathcal{C}_X$. Prove the isomorphisms

$$\mathcal{H}om\,(F, G)(U) \simeq \mathrm{Hom}_{\mathcal{C}^\wedge}(F \times U, G)\,,$$
$$\mathcal{H}om\,(F \times H, G) \simeq \mathcal{H}om\,(F, \mathcal{H}om\,(H, G))\,,$$
$$\mathrm{Hom}_{\mathcal{C}^\wedge}(F \times H, G) \simeq \mathrm{Hom}_{\mathcal{C}^\wedge}(F, \mathcal{H}om\,(H, G))\,.$$

Exercise 17.2. Let X be a site and $F \in \mathrm{PSh}(X, \mathcal{A})$. Assume that $F|_A$ is a sheaf on A for some local epimorphism $A \to \mathrm{pt}_X$. Prove that F is a sheaf.

Exercise 17.3. Let X be a site and let $A \in \mathcal{C}_X^\wedge$. Assume that $U \times A$ is representable in $\mathcal{C}_X$ for any $U \in \mathcal{C}_X$. Denote by $i_A \colon A \to X$ the morphism of presites given by $\mathcal{C}_X \ni U \mapsto U \times A \in \mathcal{C}_A$.
(i) Prove that i_A is a morphism of sites.
(ii) Prove the isomorphisms of functors $\mathrm{j}_{A \to X}^\ddagger \simeq i_{A*}$ and $\mathrm{j}_{A \to X*} \simeq i_A^{-1}$.

Exercise 17.4. Let X be a site and let X_{fin} be the site $\mathcal{C}_X$ endowed with the final topology. Recall that $\mathrm{Sh}(X_{fin}, \mathcal{A}) \xrightarrow{\sim} \mathrm{PSh}(X, \mathcal{A})$ (Example 17.3.6). Denote by $f \colon X \to X_{fin}$ the natural morphism of sites. Prove that the following diagrams quasi-commute:

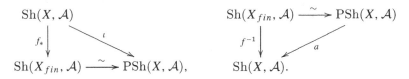

Exercise 17.5. (i) Let $f \colon X \to Y$ be a morphism of sites and let $G \in \mathrm{PSh}(Y, \mathcal{A})$. Prove that $f^{-1}(G^a) \simeq (f^\ddagger G)^a$.
(ii) Let X be a site, $A \in \mathcal{C}_X^\wedge$ and let $G \in \mathrm{PSh}(X, \mathcal{A})$. Prove that $(\mathrm{j}_{A \to X*} G)^a \simeq \mathrm{j}_{A \to X*}(G^a)$.

Exercise 17.6. Let $\mathcal{C}$ be a small category and let $F \in \mathrm{PSh}(\mathcal{C})$. Let us say that a morphism $u \colon A \to B$ in $\mathcal{C}^\wedge$ is an F-epimorphism if for any morphism $U \to B$ with $U \in \mathcal{C}$, we have $F(U) \xrightarrow{\sim} F\big(\mathrm{Im}(A \times_B U \to U)\big)$.
(i) Prove that the family of F-epimorphisms defines a topology on $\mathcal{C}$. Let us call it the F-topology.
(ii) Prove that a morphism $u \colon A \to B$ in $\mathcal{C}^\wedge$ is a local isomorphism with respect to the F-topology if and only if it satisfies:

(a) $F(U) \xrightarrow{\sim} F(A \times_B U)$ for any $U \to B$ with $U \in \mathcal{C}$,

(b) for any pair of arrows $U \rightrightarrows A$ in $\mathcal{C}^{\wedge}$ such that $U \in \mathcal{C}$ and the two compositions $U \rightrightarrows A \rightarrow B$ coincide, there is a natural isomorphism $F(U) \xrightarrow{\sim} F(\mathrm{Ker}(U \rightrightarrows A))$.

(iii) Prove that F is a sheaf for the F-topology.
(iv) Prove that the F-topology is the strongest topology with respect to which F is a sheaf.
(v) Let $\mathcal{F}$ be a family of presheaves on $\mathcal{C}$. Prove that there exists a strongest topology for which all presheaves in $\mathcal{F}$ are sheaves.
(vi) Let $\mathcal{T}$ be a Grothendieck topology on $\mathcal{C}$. Prove that $\mathcal{T}$ is the strongest among the topologies $\mathcal{T}'$ such that all sheaves with respect to $\mathcal{T}$ are sheaves with respect to $\mathcal{T}'$.

When $\mathcal{F}$ is the set of all representable functors, the topology in (v) is called the canonical Grothendieck topology.

Exercise 17.7. Let X be a site. Prove that $\mathrm{Sh}(X, \mathcal{A})$ is equivalent to the full subcategory of $\mathrm{Fct}(\mathrm{Sh}(X)^{\mathrm{op}}, \mathcal{A})$ consisting of objects which commute with small projective limits.

Exercise 17.8. Let $f \colon X \rightarrow Y$ be a morphism of sites and assume that $f_* \colon \mathrm{Sh}(X) \rightarrow \mathrm{Sh}(Y)$ is an equivalence of categories. Prove that the functors $f_* \colon \mathrm{Sh}(X, \mathcal{A}) \rightarrow \mathrm{Sh}(Y, \mathcal{A})$ and $f^{-1} \colon \mathrm{Sh}(Y, \mathcal{A}) \rightarrow \mathrm{Sh}(X, \mathcal{A})$ are equivalences of categories.

Exercise 17.9. Let X be a site and let $A \in \mathcal{C}_X^{\wedge}$. We consider a local epimorphism $u \colon B \rightarrow A$ and $F \in \mathrm{Sh}(A, \mathcal{A})$. Let p_1 and p_2 be the projections $B \times_A B \rightrightarrows B$ and let $v := u \circ p_1 = u \circ p_2$. For $i = 1, 2$, consider the morphisms

$$j_u^{\dagger} j_{u*} F \rightarrow j_u^{\dagger} j_{p_i}^{\dagger} j_{p_i*} j_{u*} F \simeq j_v^{\dagger} j_{v*} F .$$

Prove that the sequence of sheaves below is exact:

$$F \rightarrow j_u^{\dagger} j_{u*} F \rightrightarrows j_v^{\dagger} j_{v*} F.$$

Exercise 17.10. Consider two sites X and Y and a morphism of presites $f \colon X \rightarrow Y$. Assume that for any $F \in \mathrm{Sh}(X)$, the presheaf $f_* F$ on Y is a sheaf. Prove that f is a morphism of sites.

Exercise 17.11. We follow the Notations 17.6.13. Let $A \in \mathcal{C}_X^{\wedge}$, $M \in \mathcal{A}$ and $F \in \mathrm{Sh}(X, \mathcal{A})$. Prove the isomorphism

(17.7.2) $\mathrm{Hom}_{\mathrm{Sh}(X,\mathcal{A})}(M_{XA}, F) \simeq \mathrm{Hom}_{\mathcal{A}}(M, F(A))$.

Exercise 17.12. Let X be a site.
(i) Prove that the terminal object $\mathrm{pt}_{\widehat{X}}$ of $\mathrm{PSh}(X)$ is a sheaf.
(ii) Let $\emptyset_{\widehat{X}}$ be the initial object of $\mathcal{C}_X^{\wedge}$. Prove that the associated sheaf $(\emptyset_{\widehat{X}})^a$ is an initial object in $\mathrm{Sh}(X)$ and that $(\emptyset_{\widehat{X}})^a(U) \simeq \{\mathrm{pt}\}$ if $\emptyset_{\widehat{X}} \rightarrow U$ is a local epimorphism and $(\emptyset_{\widehat{X}})^a(U) \simeq \emptyset$ otherwise.

Exercise 17.13. Denote by **Top** the big category of topological spaces and continuous maps. Let X be a topological space. Prove that the category $\mathrm{Sh}(X)$ is equivalent to the full subcategory of $\mathbf{Top}_X$ consisting of pairs (Y, p) such that $p \colon Y \to X$ is a local homeomorphism.

Exercise 17.14. Let $\varphi \colon \mathcal{C} \to \mathcal{C}'$ be a functor of small categories. Prove that φ is weakly left exact if and only if $\widehat{\varphi} \colon \mathcal{C}^\wedge \to \mathcal{C}'^\wedge$ commutes with fiber products. (Hint: use Exercise 3.5.)

Exercise 17.15. Let $\mathcal{A}$ and $\mathcal{B}$ be categories satisfying (17.4.1), and $\theta \colon \mathcal{A} \to \mathcal{B}$ a functor which commutes with small projective limits and small inductive limits.
(i) Let X be a site. Prove that θ induces a functor $\theta_X \colon \mathrm{Sh}(X, \mathcal{A}) \to \mathrm{Sh}(X, \mathcal{B})$ and that θ_X commutes with small projective limits and small inductive limits.
(ii) Let $f \colon X \to Y$ be a morphism of sites. Prove that the following diagrams quasi-commute:

$$
\begin{array}{ccc}
\mathrm{Sh}(X, \mathcal{A}) & \xrightarrow{\ \theta_X\ } & \mathrm{Sh}(X, \mathcal{B}) \\
{\scriptstyle f_*}\big\downarrow & & \big\downarrow {\scriptstyle f_*} \\
\mathrm{Sh}(Y, \mathcal{A}) & \xrightarrow{\ \theta_Y\ } & \mathrm{Sh}(Y, \mathcal{B}),
\end{array}
\qquad
\begin{array}{ccc}
\mathrm{Sh}(Y, \mathcal{A}) & \xrightarrow{\ \theta_Y\ } & \mathrm{Sh}(Y, \mathcal{B}) \\
{\scriptstyle f^{-1}}\big\downarrow & & \big\downarrow {\scriptstyle f^{-1}} \\
\mathrm{Sh}(X, \mathcal{A}) & \xrightarrow{\ \theta_X\ } & \mathrm{Sh}(X, \mathcal{B}) \ .
\end{array}
$$

18

Abelian Sheaves

In this chapter we introduce sheaves of $\mathcal{R}$-modules, where $\mathcal{R}$ is a sheaf of rings on X. We prove that the category $\mathrm{Mod}(\mathcal{R})$ of $\mathcal{R}$-modules is a Grothendieck category and we construct in this framework the functors of internal hom, tensor product, inverse image and direct image.

Then we prove that $\mathrm{Mod}(\mathcal{R})$ has enough flat objects, and we derive the previous functors in the unbounded derived categories by applying the tools obtained in Chap. 14. In particular, we prove adjunction formulas for the derived functors of the internal hom and the tensor product as well as for the direct and inverse image functors.

For the sake of simplicity, when treating the derived categories we mainly consider the case where $\mathcal{R}$ is commutative, although many results extend to the non commutative case.

We end this chapter by constructing complexes associated to a local epimorphism (these complexes are classically known as "Čech complexes") and by proving in this framework the classical "Leray's acyclic covering theorem".

Such results are (almost) classical for bounded derived categories. The unbounded case was first considered by Spaltenstein [65].

We follow the notations introduced in Chap. 17.

18.1 $\mathcal{R}$-modules

As in the previous chapters, X denotes a site and $\mathcal{C}_X$ the underlying category. As usual k denotes a commutative unital ring.

A *sheaf of k-algebras* on X is an object $\mathcal{R} \in \mathrm{Sh}(X)$ such that for each $U \in \mathcal{C}_X$, $\mathcal{R}(U)$ is a k-algebra and for any morphism $U \to V$ in $\mathcal{C}_X$, the map $\mathcal{R}(V) \to \mathcal{R}(U)$ is a k-algebra morphism. The notion of a morphism of sheaves of k-algebras is naturally defined. Hence, a sheaf of k-algebras is nothing but a sheaf with values in the category k-**Alg** of k-algebras. A sheaf of $\mathbb{Z}$-algebras is simply called a sheaf of rings. A sheaf of k-algebras is also called a k_X-algebra.

Example 18.1.1. The constant sheaf k_X on a site X is a sheaf of k-algebras.

If $\mathcal{R}$ is a k_X-algebra, the opposite k_X-algebra $\mathcal{R}^{\mathrm{op}}$ is defined by setting for $U \in \mathcal{C}_X$

$$\mathcal{R}^{\mathrm{op}}(U) := \mathcal{R}(U)^{\mathrm{op}} .$$

Let $\mathcal{R}$ be a k_X-algebra. A presheaf F of $\mathcal{R}$-modules is a presheaf F such that for each $U \in \mathcal{C}_X$, $F(U)$ has a structure of a left $\mathcal{R}(U)$-module and for any morphism $U \to V$ in $\mathcal{C}_X$, the morphism $F(V) \to F(U)$ commutes with the action of $\mathcal{R}$. A morphism $\varphi \colon F \to G$ of presheaves of $\mathcal{R}$-modules is a morphism of presheaves such that for each $U \in \mathcal{C}_X$, $\varphi(U) \colon F(U) \to G(U)$ is $\mathcal{R}(U)$-linear.

A presheaf of $\mathcal{R}$-modules which is a sheaf is called a sheaf of $\mathcal{R}$-modules, or simply, an $\mathcal{R}$-*module*. A right $\mathcal{R}$-module is a left $\mathcal{R}^{\mathrm{op}}$-module.

If $\mathcal{O}$ is a sheaf of commutative rings on X, an $\mathcal{O}$-*algebra* is a sheaf of rings $\mathcal{R}$ with a morphism of sheaves of rings $\mathcal{O} \to \mathcal{R}$ such that the image of $\mathcal{O}(U) \to \mathcal{R}(U)$ is contained in the center of $\mathcal{R}(U)$ for any $U \in \mathcal{C}_X$, i.e., $\mathcal{R}(U)$ is an $\mathcal{O}(U)$-algebra.

Example 18.1.2. On a complex manifold X, the sheaf $\mathcal{O}_X$ of holomorphic functions is a $\mathbb{C}_X$-algebra and the sheaf $\mathcal{D}_X$ of holomorphic differential operators is a $\mathbb{C}_X$-algebra. The sheaf $\mathcal{O}_X$ is a left $\mathcal{D}_X$-module and the sheaf Ω_X of holomorphic forms of maximal degree is an $\mathcal{O}_X$-module and also a right $\mathcal{D}_X$-module. (See [37].)

Notations 18.1.3. (i) Let $\mathcal{R}$ be a sheaf of rings on X. We denote by $\mathrm{PSh}(\mathcal{R})$ the category of presheaves of $\mathcal{R}$-modules and by $\mathrm{Mod}(\mathcal{R})$ the category of sheaves of $\mathcal{R}$-modules.
(ii) We write $\mathrm{Hom}_{\mathcal{R}}$ instead of $\mathrm{Hom}_{\mathrm{Mod}(\mathcal{R})}$.

In particular, we have

$$\mathrm{Mod}(k_X) = \mathrm{Sh}(X, \mathrm{Mod}(k)) .$$

Note that if $\mathcal{R}$ is a k_X-algebra, the forgetful functor

(18.1.1) $for \colon \mathrm{Mod}(\mathcal{R}) \to \mathrm{Mod}(k_X)$

is faithful and conservative but not fully faithful in general.

Form now on, we denote by $\mathcal{R}$ a sheaf of k-algebras on X.

Lemma 18.1.4. *The functor* $^a \colon \mathrm{PSh}(k_X) \to \mathrm{Mod}(k_X)$ *in Definition 17.4.5 induces a functor (we keep the same notation)*

$$(\cdot)^a \colon \mathrm{PSh}(\mathcal{R}) \to \mathrm{Mod}(\mathcal{R}) ,$$

and this functor is left adjoint to the canonical inclusion functor $\mathrm{Mod}(\mathcal{R}) \to \mathrm{PSh}(\mathcal{R})$.

The proof follows easily from Theorem 17.4.7. Details are left to the reader.

Recall that we have set in Notations 17.6.13:

$$\bullet|_A := j_{A \to X *} \, .$$

Clearly, $\mathcal{R}|_A$ is a sheaf of k-algebras on A.

Lemma 18.1.5. *The functors* $j_{A \to X *} \colon \mathrm{Mod}(k_X) \to \mathrm{Mod}(k_A)$ *and* $j_{A \to X}^{-1}$, $j_{A \to X}^{\ddagger} \colon \mathrm{Mod}(k_A) \to \mathrm{Mod}(k_X)$ *induce well-defined functors (we keep the same notations)*

(18.1.2) $j_{A \to X *} \colon \mathrm{Mod}(\mathcal{R}) \to \mathrm{Mod}(\mathcal{R}|_A) \, ,$

(18.1.3) $j_{A \to X}^{-1} \colon \mathrm{Mod}(\mathcal{R}|_A) \to \mathrm{Mod}(\mathcal{R}) \, ,$

(18.1.4) $j_{A \to X}^{\ddagger} \colon \mathrm{Mod}(\mathcal{R}|_A) \to \mathrm{Mod}(\mathcal{R}) \, .$

Proof. (i) The assertion concerning $j_{A \to X *}$ is obvious.

(ii) Let us treat $j_{A \to X}^{-1}$. It is enough to check that $j_{A \to X}^{\ddagger}$ induces a well-defined functor from $\mathrm{PSh}(\mathcal{R}|_A)$ to $\mathrm{PSh}(\mathcal{R})$. By (17.1.11), one has $j_{A \to X}^{\ddagger}(G)(U) \simeq \coprod_{s \in A(U)} G(U \overset{s}{\to} A)$ for $U \in \mathcal{C}_X$ and $G \in \mathrm{PSh}(\mathcal{R}|_A)$. Since $\mathcal{R}|_A(U \overset{s}{\to} A) \simeq \mathcal{R}(U)$, $G(U \overset{s}{\to} A)$, as well as $j_{A \to X}^{\ddagger}(G)(U)$, is an $\mathcal{R}(U)$-module.

(iii) The case of the functor $j_{A \to X}^{\ddagger}$ is similar to that of $j_{A \to X}^{-1}$.   q.e.d.

Theorem 18.1.6. *Let $\mathcal{R}$ be a k_X-algebra.*

(i) *The category $\mathrm{Mod}(\mathcal{R})$ is an abelian category and the forgetful functor* $\mathit{for} \colon \mathrm{Mod}(\mathcal{R}) \to \mathrm{Mod}(k_X)$ *is exact.*

(ii) *The functor* $^a \colon \mathrm{PSh}(\mathcal{R}) \to \mathrm{Mod}(\mathcal{R})$ *is exact.*

(iii) *The category $\mathrm{Mod}(\mathcal{R})$ admits small projective limits and the functor* $\iota \colon \mathrm{Mod}(\mathcal{R}) \to \mathrm{PSh}(\mathcal{R})$ *commutes with such limits.*

(iv) *The category $\mathrm{Mod}(\mathcal{R})$ admits small inductive limits, and the functor* $^a \colon \mathrm{PSh}(\mathcal{R}) \to \mathrm{Mod}(\mathcal{R})$ *commutes with such limits. Moreover, filtrant inductive limits in $\mathrm{Mod}(\mathcal{R})$ are exact.*

(v) *The category $\mathrm{Mod}(\mathcal{R})$ is a Grothendieck category.*

Proof. (i)–(iv) follow easily from Theorems 17.4.7 and 17.4.9. Details are left to the reader.

(v) For $U \in \mathcal{C}_X$, recall that $\mathcal{R}_U = j_{U \to X}^{-1} j_{U \to X *} \mathcal{R} = j_{U \to X}^{-1}(\mathcal{R}|_U)$ (see Definition 17.6.10). For $F \in \mathrm{Mod}(\mathcal{R})$, we have

$$\mathrm{Hom}_{\mathcal{R}}(\mathcal{R}_U, F) \simeq F(U) \, .$$

Hence, the family $\{\mathcal{R}_U\}_{U \in \mathcal{C}_X}$ is a small system of generators in $\mathrm{Mod}(\mathcal{R})$. q.e.d.

Lemma 18.1.7. *Let $A \in \mathcal{C}_X^{\wedge}$.*

(i) $(j_{A \to X}^{-1}, j_{A \to X *})$ *and* $(j_{A \to X *}, j_{A \to X}^{\ddagger})$ *are pairs of adjoint functors. In particular, $j_{A \to X *}$ is exact.*

(ii) *Let $F \in \mathrm{Mod}(\mathcal{R}|_A)$. Then the morphism $F \to j_{A \to X *} j_{A \to X}^{-1} F$ is a monomorphism.*

(iii) *The functor $j_{A \to X}^{-1} \colon \mathrm{Mod}(\mathcal{R}|_A) \to \mathrm{Mod}(\mathcal{R})$ is exact and faithful.*

Proof. (i) By Proposition 17.5.1, there are natural maps, inverse to each others

$$(18.1.5) \qquad \mathrm{Hom}_{k_X}(j_{A \to X}^{-1} G, F) \overset{\Phi}{\underset{\Psi}{\rightleftarrows}} \mathrm{Hom}_{k_A}(G, F|_A) \,.$$

An element of $\mathrm{Hom}_{k_X}(j_{A \to X}^{-1} G, F)$ is given by a family of k_X-linear maps $\{\coprod_{U \to A} G(U \to A) \to F(U)\}_{U \in \mathcal{C}_X}$ compatible with the restriction morphisms. To such a family, Φ associates a family of k-linear maps $\{G(U \to A) \to F(U)\}_{(U \to A) \in \mathcal{C}_A}$ compatible with the restriction morphisms. Clearly, if all maps $\coprod_{U \to A} G(U \to A) \to F(U)$ are $\mathcal{R}(U)$-linear, then all maps $G(U \to A) \to F(U)$ will be $\mathcal{R}(U)$-linear. Hence, Φ sends $\mathrm{Hom}_{\mathcal{R}}(j_{A \to X}^{-1} G, F)$ to $\mathrm{Hom}_{\mathcal{R}|_A}(G, F|_A)$. One checks similarly that Ψ sends $\mathrm{Hom}_{\mathcal{R}|_A}(G, F|_A)$ to $\mathrm{Hom}_{\mathcal{R}}(j_{A \to X}^{-1} G, F)$.

(ii) By the result of Exercise 17.5, we have an isomorphism

$$j_{A \to X *} j_{A \to X}^{-1} F \simeq (j_{A \to X *} j_{A \to X}^{\dagger} F)^a \,.$$

Since the functor ^{a} is exact, it is enough to prove that the morphism $F \to j_{A \to X *} j_{A \to X}^{\dagger} F$ is a monomorphism. For $(U \overset{t}{\to} A) \in (\mathcal{C}_X)_A$, we have

$$j_{A \to X *} j_{A \to X}^{\dagger} F(U \overset{t}{\to} A) \simeq \bigoplus_{s \in A(U)} F(U \overset{s}{\to} A) \,.$$

(iii) The functor $j_{A \to X}^{-1} \colon \mathrm{Mod}(k_A) \to \mathrm{Mod}(k_X)$ is exact by Proposition 17.6.3. It follows that the functor $j_{A \to X}^{-1} \colon \mathrm{Mod}(\mathcal{R}|_A) \to \mathrm{Mod}(\mathcal{R})$ is exact. Then it is faithful by (ii).   q.e.d.

In Definition 17.6.10 and Notations (17.6.13) we have introduced the functors $F \mapsto F_A$, $F \mapsto \Gamma_A F$ and $F \mapsto \Gamma(A; F)$. They induce functors

$$(\bullet)_A \colon \mathrm{Mod}(\mathcal{R}) \to \mathrm{Mod}(\mathcal{R}) \,,$$
$$\Gamma_A \colon \mathrm{Mod}(\mathcal{R}) \to \mathrm{Mod}(\mathcal{R}) \,,$$
$$\Gamma(A; \bullet) \colon \mathrm{Mod}(\mathcal{R}) \to \mathrm{Mod}(\mathcal{R}(A)) \,.$$

Note that $(\bullet)_A$ is exact by Lemma 18.1.7 and Γ_A, $\Gamma(A; \bullet)$ are left exact. Proposition 17.6.14 remains true in the category $\mathrm{Mod}(\mathcal{R})$.

18.2 Tensor Product and Internal $\mathcal{H}om$

For $F, G \in \mathrm{PSh}(\mathcal{R})$, we define the presheaf of k_X-modules $\mathcal{H}om_{\mathcal{R}}(F, G)$ similarly as in Definition 17.1.10. More precisely, we set

$$(18.2.1) \qquad \mathcal{H}om_{\mathcal{R}}(F, G)(U) = \mathrm{Hom}_{\mathrm{PSh}(\mathcal{R}|_U)}(F|_U, G|_U) \quad \text{for } U \in \mathcal{C}_X.$$

If $\mathcal{R}$ is a sheaf of commutative rings, then $\mathcal{H}om_{\mathcal{R}}(F, G) \in \mathrm{PSh}(\mathcal{R})$.

Lemma 18.2.1. *Let $F \in \mathrm{PSh}(\mathcal{R})$ and let $G \in \mathrm{Mod}(\mathcal{R})$. Then*

(i) *the presheaf $\mathcal{H}om_{\mathcal{R}}(F, G)$ is a sheaf,*

(ii) *the natural morphism $\mathcal{H}om_{\mathcal{R}}(F^a, G) \to \mathcal{H}om_{\mathcal{R}}(F, G)$ is an isomorphism.*

The proof goes as for Proposition 17.7.1.

Note that

(18.2.2) $\mathrm{Hom}_{\mathcal{R}}(F, G) \simeq \Gamma(X; \mathcal{H}om_{\mathcal{R}}(F, G))$ for $F, G \in \mathrm{Mod}(\mathcal{R})$.

Also note that for $F \in \mathrm{Mod}(\mathcal{R})$, $G \in \mathrm{Mod}(\mathcal{R}^{\mathrm{op}})$ and $K \in \mathrm{Mod}(k_X)$, we have $\mathcal{H}om_{k_X}(K, F) \in \mathrm{Mod}(\mathcal{R})$ and $\mathcal{H}om_{k_X}(G, K) \in \mathrm{Mod}(\mathcal{R})$.

If $\mathcal{R}$ is a sheaf of commutative rings, then $\mathcal{H}om_{\mathcal{R}}(F, G) \in \mathrm{Mod}(\mathcal{R})$.

Let F' be a presheaf of $\mathcal{R}^{\mathrm{op}}$-modules and F a presheaf of $\mathcal{R}$-modules. The presheaf $F' \overset{\mathrm{psh}}{\otimes}_{\mathcal{R}} F$ of k_X-modules is defined by the formula

$$(F' \overset{\mathrm{psh}}{\otimes}_{\mathcal{R}} F)(U) := F'(U) \otimes_{\mathcal{R}(U)} F(U) \text{ for } U \in \mathcal{C}_X.$$

If $\mathcal{R}$ is commutative, $F' \overset{\mathrm{psh}}{\otimes}_{\mathcal{R}} F$ is a presheaf of $\mathcal{R}$-modules.

Definition 18.2.2. *For $F' \in \mathrm{Mod}(\mathcal{R}^{\mathrm{op}})$ and $F \in \mathrm{Mod}(\mathcal{R})$, we denote by $F' \otimes_{\mathcal{R}} F$ the sheaf associated with the presheaf $F' \overset{\mathrm{psh}}{\otimes}_{\mathcal{R}} F$ and call this sheaf the tensor product of F' and F.*

Hence, we have constructed a bifunctor

$$\cdot \otimes_{\mathcal{R}} \cdot : \mathrm{Mod}(\mathcal{R}^{\mathrm{op}}) \times \mathrm{Mod}(\mathcal{R}) \to \mathrm{Mod}(k_X).$$

If $\mathcal{R}$ is commutative, we get a bifunctor

$$\cdot \otimes_{\mathcal{R}} \cdot : \mathrm{Mod}(\mathcal{R}) \times \mathrm{Mod}(\mathcal{R}) \to \mathrm{Mod}(\mathcal{R}).$$

Note that for $F \in \mathrm{Mod}(\mathcal{R})$ and $K \in \mathrm{Mod}(k_X)$, we have $K \otimes_{k_X} F \in \mathrm{Mod}(\mathcal{R})$.

Proposition 18.2.3.   (i) *There are isomorphisms, functorial with respect to $F \in \mathrm{PSh}(\mathcal{R}^{\mathrm{op}})$, $G \in \mathrm{PSh}(\mathcal{R})$ and $H \in \mathrm{PSh}(k_X)$:*

$$\mathrm{Hom}_{\mathrm{PSh}(k_X)}(F \overset{\mathrm{psh}}{\otimes}_{\mathcal{R}} G, H) \simeq \mathrm{Hom}_{\mathrm{PSh}(\mathcal{R})}(G, \mathcal{H}om_{k_X}(F, H)),$$

$$\mathcal{H}om_{k_X}(F \overset{\mathrm{psh}}{\otimes}_{\mathcal{R}} G, H) \simeq \mathcal{H}om_{\mathcal{R}}(G, \mathcal{H}om_{k_X}(F, H)).$$

(ii) *There are isomorphisms, functorial with respect to $F \in \mathrm{Mod}(\mathcal{R}^{\mathrm{op}})$, $G \in \mathrm{Mod}(\mathcal{R})$ and $H \in \mathrm{Mod}(k_X)$:*

(18.2.3) $\mathrm{Hom}_{k_X}(F \otimes_{\mathcal{R}} G, H) \simeq \mathrm{Hom}_{\mathcal{R}}(G, \mathcal{H}om_{k_X}(F, H)),$

(18.2.4) $\mathcal{H}om_{k_X}(F \otimes_{\mathcal{R}} G, H) \simeq \mathcal{H}om_{\mathcal{R}}(G, \mathcal{H}om_{k_X}(F, H)).$

(iii) *Let F' be a presheaf of $\mathcal{R}^{\mathrm{op}}$-modules and F a presheaf of $\mathcal{R}$-modules. Then the natural morphism $(F \overset{\mathrm{psh}}{\otimes}_{\mathcal{R}} F')^a \to F'^a \otimes_{\mathcal{R}} F^a$ is an isomorphism.*
(iv) *The functor $(\cdot \otimes_{\mathcal{R}} \cdot) : \mathrm{Mod}(\mathcal{R})^{\mathrm{op}} \times \mathrm{Mod}(\mathcal{R}) \to \mathrm{Mod}(k_X)$ is right exact.*

Proof. (i) Since the second isomorphism follows from the first one, we prove only the first isomorphism. Let us define a map

$$\lambda : \mathrm{Hom}_{\mathrm{PSh}(\mathcal{R})}(G, \mathcal{H}om_{k_X}(F, H)) \to \mathrm{Hom}_{\mathrm{PSh}(k_X)}(F \overset{\mathrm{psh}}{\otimes}_{\mathcal{R}} G, H) .$$

For $U \in \mathcal{C}_X$, we have the chain of morphisms

$$\mathrm{Hom}_{\mathrm{PSh}(\mathcal{R})}(G, \mathcal{H}om_{k_X}(F, H))$$
$$\to \mathrm{Hom}_{\mathcal{R}(U)}(G(U), \mathcal{H}om_{k_X}(F, H)(U))$$
$$\to \mathrm{Hom}_{\mathcal{R}(U)}(G(U), \mathrm{Hom}_k(F(U), H(U)))$$
$$\simeq \mathrm{Hom}_{\mathcal{R}(U)}(G(U) \otimes_k F(U), H(U)) .$$

Since these morphisms are functorial with respect to U, they define λ.
Let us define a map

$$\mu : \mathrm{Hom}_{\mathrm{PSh}(k_X)}(F \overset{\mathrm{psh}}{\otimes}_{\mathcal{R}} G, H) \to \mathrm{Hom}_{\mathrm{PSh}(\mathcal{R})}(G, \mathcal{H}om_{k_X}(F, H)) .$$

For $V \to U$ in $\mathcal{C}_X$, we have the chain of morphisms

$$\mathrm{Hom}_{\mathrm{PSh}(k_X)}(F \overset{\mathrm{psh}}{\otimes}_{\mathcal{R}} G, H) \to \mathrm{Hom}_k(F(V) \overset{\mathrm{psh}}{\otimes}_{\mathcal{R}(V)} G(V), H(V))$$
$$\simeq \mathrm{Hom}_{\mathcal{R}(V)}(G(V), \mathrm{Hom}_k(F(V), H(V)))$$
$$\to \mathrm{Hom}_{\mathcal{R}(U)}(G(U), \mathrm{Hom}_k(F(V), H(V))) .$$

Since these morphism are functorial with respect to $(V \to U) \in \mathcal{C}_U$, they define μ.
It is easily checked that λ and μ are inverse to each other.
(ii) follows from (i) since $\mathcal{H}om_{k_X}(F \overset{\mathrm{psh}}{\otimes}_{\mathcal{R}} G, H) \simeq \mathcal{H}om_{k_X}(F \otimes_{\mathcal{R}} G, H)$ by Lemma 18.2.1 (ii).
(iii) Let $G \in \mathrm{Mod}(k_X)$. Using Lemma 18.2.1, we obtain the chain of isomorphisms

$$\mathrm{Hom}_{k_X}((F \overset{\mathrm{psh}}{\otimes}_{\mathcal{R}} F')^a, G) \simeq \mathrm{Hom}_{k_X}(F \overset{\mathrm{psh}}{\otimes}_{\mathcal{R}} F', G)$$
$$\simeq \mathrm{Hom}_{k_X}(F, \mathcal{H}om_{\mathcal{R}}(F', G))$$
$$\simeq \mathrm{Hom}_{k_X}(F^a, \mathcal{H}om_{\mathcal{R}}(F'^a, G))$$
$$\simeq \mathrm{Hom}_{k_X}(F^a \otimes_{\mathcal{R}} F'^a, G) .$$

(iv) The functor $\cdot \overset{\mathrm{psh}}{\otimes}_{\mathcal{R}} \cdot : \mathrm{Mod}(\mathcal{R})^{\mathrm{op}} \times \mathrm{Mod}(\mathcal{R}) \to \mathrm{PSh}(X, \mathrm{Mod}(k))$ is clearly right exact. Its composition with the exact functor ^{a} remains right exact.
q.e.d.

Lemma 18.2.4. *Let $F' \in \mathrm{Mod}(\mathcal{R}^{\mathrm{op}})$, $F \in \mathrm{Mod}(\mathcal{R})$ and let $A \in \mathcal{C}_X^{\wedge}$. Then*

$$(18.2.5) \qquad (F' \otimes_{\mathcal{R}} F)|_A \simeq F'|_A \otimes_{\mathcal{R}|_A} F|_A \ .$$

Proof. The isomorphism

$$j_{A \to X *}(F' \overset{\mathrm{psh}}{\otimes}_{\mathcal{R}} F) \simeq F'|_A \overset{\mathrm{psh}}{\otimes}_{\mathcal{R}|_A} F|_A$$

is clear. The result follows by applying the functor ^{a} which commutes with $j_{A \to X *}$ (Exercise 17.5). q.e.d.

Proposition 18.2.5. *Let G be an $(\mathcal{R}^{\mathrm{op}}|_A)$-module and F an $\mathcal{R}$-module. There is a natural isomorphism*

$$(18.2.6) \qquad j_{A \to X}^{-1}(G \otimes_{\mathcal{R}|_A} (F|_A)) \simeq (j_{A \to X}^{-1} G) \otimes_{\mathcal{R}} F \quad \text{in } \mathrm{Mod}(k_X) \ .$$

Proof. The right hand side of (18.2.6) is the sheaf associated with the presheaf

$$\mathcal{C}_X \ni U \mapsto \big(\bigoplus_{s \in A(U)} G(U \overset{s}{\to} A) \big) \otimes_{\mathcal{R}(U)} F(U)$$

and the left hand side is the sheaf associated with the presheaf

$$\mathcal{C}_X \ni U \mapsto \bigoplus_{s \in A(U)} \big(G(U \overset{s}{\to} A) \otimes_{\mathcal{R}(U)} F(U) \big) \ .$$

q.e.d.

Remark 18.2.6. There are general formulas using various sheaves of rings on X. Here, we state the main results, leaving the proofs to the readers. Consider a commutative sheaf of rings $\mathcal{O}_X$, four $\mathcal{O}_X$-algebras $\mathcal{R}_\nu$ ($\nu = 1, \ldots, 4$), and for $i, j \in \{1, \ldots, 4\}$, denote by ${}_i M_j$ an object of $\mathrm{Mod}(\mathcal{R}_i \otimes_{\mathcal{O}_X} \mathcal{R}_j^{\mathrm{op}})$. Then the functors below are well defined:

$$(18.2.7) \quad \cdot \otimes_{\mathcal{R}_2} \cdot : \mathrm{Mod}(\mathcal{R}_1 \otimes_{\mathcal{O}_X} \mathcal{R}_2^{\mathrm{op}}) \times \mathrm{Mod}(\mathcal{R}_2 \otimes_{\mathcal{O}_X} \mathcal{R}_3^{\mathrm{op}})$$
$$\to \mathrm{Mod}(\mathcal{R}_1 \otimes_{\mathcal{O}_X} \mathcal{R}_3^{\mathrm{op}}),$$

$$(18.2.8) \quad \mathcal{H}om_{\mathcal{R}_1} : \mathrm{Mod}(\mathcal{R}_1 \otimes_{\mathcal{O}_X} \mathcal{R}_2^{\mathrm{op}})^{\mathrm{op}} \times \mathrm{Mod}(\mathcal{R}_1 \otimes_{\mathcal{O}_X} \mathcal{R}_3^{\mathrm{op}})$$
$$\to \mathrm{Mod}(\mathcal{R}_2 \otimes_{\mathcal{O}_X} \mathcal{R}_3^{\mathrm{op}}),$$

$$(18.2.9) \quad \mathrm{Hom}_{\mathcal{R}_1} : \mathrm{Mod}(\mathcal{R}_1 \otimes_{\mathcal{O}_X} \mathcal{R}_2^{\mathrm{op}})^{\mathrm{op}} \times \mathrm{Mod}(\mathcal{R}_1 \otimes_{\mathcal{O}_X} \mathcal{R}_3^{\mathrm{op}})$$
$$\to \mathrm{Mod}\big((\mathcal{R}_2 \otimes_{\mathcal{O}_X} \mathcal{R}_3^{\mathrm{op}})(X)\big) \ .$$

Moreover, there are natural isomorphisms in $\mathrm{Mod}(\mathcal{R}_1 \otimes_{\mathcal{O}_X} \mathcal{R}_4^{\mathrm{op}})$:

$$(18.2.10)$$
$$({}_1 M_2 \otimes_{\mathcal{R}_2} {}_2 M_3) \otimes_{\mathcal{R}_3} {}_3 M_4 \simeq {}_1 M_2 \otimes_{\mathcal{R}_2} ({}_2 M_3 \otimes_{\mathcal{R}_3} {}_3 M_4),$$
$$(18.2.11)$$
$$\mathcal{H}om_{\mathcal{R}_2}({}_2 M_1, \mathcal{H}om_{\mathcal{R}_3}({}_3 M_2, {}_3 M_4)) \simeq \mathcal{H}om_{\mathcal{R}_3}({}_3 M_2 \otimes_{\mathcal{R}_2} {}_2 M_1, {}_3 M_4) \ .$$

Proposition 18.2.7. *Let $F \in \mathrm{Mod}(\mathcal{R})$ and let $A \in \mathcal{C}_X^\wedge$. There are natural isomorphisms in $\mathrm{Mod}(\mathcal{R})$:*

$$(18.2.12) \qquad F_A \simeq \mathcal{R}_A \otimes_{\mathcal{R}} F \simeq k_{XA} \otimes_{k_X} F\ ,$$

$$(18.2.13) \qquad \Gamma_A(F) \simeq \mathcal{H}om_{\mathcal{R}}(\mathcal{R}_A, F) \simeq \mathcal{H}om_{k_X}(k_{XA}, F)\ .$$

Proof. (i) By Proposition 18.2.5, we have

$$F_A \simeq \mathrm{j}^{-1}_{A \to X}(F|_A) \simeq \mathrm{j}^{-1}_{A \to X}(\mathcal{R}|_A \otimes_{\mathcal{R}|_A} F|_A)$$
$$\simeq \left(\mathrm{j}^{-1}_{A \to X}(\mathcal{R}|_A)\right) \otimes_{\mathcal{R}} F \simeq \mathcal{R}_A \otimes_{\mathcal{R}} F\ .$$

The second isomorphism in (18.2.12) is similarly proved.

(ii) By (18.2.11), we have the isomorphisms $\mathcal{H}om_{\mathcal{R}}(\mathcal{R}_A, F) \simeq \mathcal{H}om_{\mathcal{R}}(\mathcal{R} \otimes k_{XA}, F) \simeq \mathcal{H}om_{k_X}(k_{XA}, \mathcal{H}om_{\mathcal{R}}(\mathcal{R}, F)) \simeq \mathcal{H}om_{k_X}(k_{XA}, F)$. On the other hand, using (18.2.12), we get the chain of isomorphisms, functorial with respect to $G \in \mathrm{Mod}(\mathcal{R})$:

$$\mathrm{Hom}_{\mathcal{R}}(G, \mathcal{H}om_{k_X}(k_{XA}, F)) \simeq \mathrm{Hom}_{\mathcal{R}}(G \otimes_{k_X} k_{XA}, F)$$
$$\simeq \mathrm{Hom}_{\mathcal{R}}(\mathrm{j}^{-1}_{A \to X}\, \mathrm{j}_{A \to X*}G, F)$$
$$\simeq \mathrm{Hom}_{\mathcal{R}}(G, \mathrm{j}_{A \to X!}\, \mathrm{j}^{-1}_{A \to X} F)$$
$$= \mathrm{Hom}_{\mathcal{R}}(G, \Gamma_A(F))\ .$$

Then the result follows from the Yoneda lemma.   q.e.d.

18.3 Direct and Inverse Images

Lemma 18.3.1. *Let $\mathcal{R}$ be a k_X-algebra.*

(i) *Let $f\colon X \to Y$ be a morphism of sites. Then $f_*\mathcal{R}$ is a k_Y-algebra.*

(ii) *Let $g\colon Z \to X$ be a left exact morphism of sites. Then*

    (a) *$g^{-1}\mathcal{R}$ is a sheaf of rings on Z,*

    (b) *for any $\mathcal{R}$-module F, $g^{-1}F$ is a $g^{-1}\mathcal{R}$-module,*

    (c) *for an $\mathcal{R}^{\mathrm{op}}$-module F' and an $\mathcal{R}$-module F,*

$$g^{-1}(F' \otimes_{\mathcal{R}} F) \simeq g^{-1}F' \otimes_{g^{-1}\mathcal{R}} g^{-1}F\ ,$$

    (d) *the functor $g^{-1}\colon \mathrm{Mod}(\mathcal{R}) \to \mathrm{Mod}(g^{-1}\mathcal{R})$ is exact.*

Proof. (i) is obvious.

(ii) (a)–(b) Since $g^{-1}\colon \mathrm{Sh}(X) \to \mathrm{Sh}(Z)$ is exact by Theorem 17.5.2, the action of $\mathcal{R}$ on F induces a morphism $g^{-1}\mathcal{R} \times g^{-1}F \simeq g^{-1}(\mathcal{R} \times F) \to g^{-1}F$ in $\mathrm{Sh}(Z)$. Taking $\mathcal{R}$ as F, it induces a structure of a k_Z-algebra on $g^{-1}\mathcal{R}$. Moreover, $g^{-1}F$ has a structure of a $g^{-1}\mathcal{R}$-module.

(ii) (c) Let K be a k_Z-module. We have the chain of isomorphisms

$$\mathrm{Hom}_{k_Z}(g^{-1}(F' \otimes_{\mathcal{R}} F), K) \simeq \mathrm{Hom}_{k_X}(F' \otimes_{\mathcal{R}} F, g_* K)$$
$$\simeq \mathrm{Hom}_{\mathcal{R}}(F, \mathcal{H}om_{k_X}(F', g_* K))$$
$$\simeq \mathrm{Hom}_{\mathcal{R}}(F, g_* \mathcal{H}om_{k_X}(g^{-1} F', K))$$
$$\simeq \mathrm{Hom}_{g^{-1}\mathcal{R}}(g^{-1}F, \mathcal{H}om_{k_X}(g^{-1}F', K))$$
$$\simeq \mathrm{Hom}_{k_Z}(g^{-1}F' \otimes_{g^{-1}\mathcal{R}} g^{-1}F, K),$$

where the third isomorphism follows from Proposition 17.7.3. Then the result follows from the Yoneda lemma.

(ii) (d) By Proposition 17.6.3, the functor $g^{-1}\colon \mathrm{Mod}(k_X) \to \mathrm{Mod}(k_Z)$ is exact and the result follows.   q.e.d.

Remark that if $g\colon Z \to X$ is not left exact (e.g., $j_{A \to X}$), $g^{-1}\mathcal{R}$ is not necessarily a ring.

Lemma 18.3.2. *Let $f\colon X \to Y$ be a left exact morphism of sites and let $\mathcal{R}_Y$ be a k_Y-algebra. There are isomorphisms, functorial with respect to $G \in \mathrm{Mod}(\mathcal{R}_Y)$ and $F \in \mathrm{Mod}(f^{-1}\mathcal{R}_Y)$:*

(18.3.1) $\mathrm{Hom}_{f^{-1}\mathcal{R}_Y}(f^{-1}G, F) \simeq \mathrm{Hom}_{\mathcal{R}_Y}(G, f_* F),$

(18.3.2) $f_* \mathcal{H}om_{f^{-1}\mathcal{R}_Y}(f^{-1}G, F) \simeq \mathcal{H}om_{\mathcal{R}_Y}(G, f_* F).$

In particular, the functor $f^{-1}\colon \mathrm{Mod}(\mathcal{R}_Y) \to \mathrm{Mod}(f^{-1}\mathcal{R}_Y)$ is a left adjoint to the functor $f_*\colon \mathrm{Mod}(f^{-1}\mathcal{R}_Y) \to \mathrm{Mod}(\mathcal{R}_Y)$.

Proof. (i) By Theorem 17.5.2, we have an isomorphism

$$\mathrm{Hom}_{k_X}(f^{-1}G, F) \simeq \mathrm{Hom}_{k_Y}(G, f_* F).$$

One checks easily that for $G \in \mathrm{Mod}(\mathcal{R}_Y)$ and $F \in \mathrm{Mod}(f^{-1}\mathcal{R}_Y)$, this isomorphism induces (18.3.1).

(ii) follows from (i) similarly as Proposition 17.7.3 is deduced from Theorem 17.5.2.   q.e.d.

If a site X (resp. Y) is endowed with a k_X-algebra $\mathcal{R}_X$ (resp. a k_Y-algebra $\mathcal{R}_Y$) and one is given an $(\mathcal{R}_X \otimes_{k_X} f^{-1}(\mathcal{R}_Y^{\mathrm{op}}))$-module $K_{X \to Y}$, we can pass from $\mathcal{R}_Y$-modules to $\mathcal{R}_X$-modules, and conversely.

Proposition 18.3.3. *Let $f\colon X \to Y$ be a left exact morphism of sites. Let $\mathcal{R}_Y$ be a k_Y-algebra, $\mathcal{R}_X$ a k_X-algebra and $K_{X \to Y}$ an $(\mathcal{R}_X \otimes_{k_X} f^{-1}(\mathcal{R}_Y^{\mathrm{op}}))$-module. Let $F \in \mathrm{Mod}(\mathcal{R}_X)$ and $G \in \mathrm{Mod}(\mathcal{R}_Y)$. There are natural isomorphisms*

(18.3.3) $\mathrm{Hom}_{\mathcal{R}_X}(K_{X \to Y} \otimes_{f^{-1}\mathcal{R}_Y} f^{-1}G, F)$
$$\simeq \mathrm{Hom}_{\mathcal{R}_Y}(G, f_* \mathcal{H}om_{\mathcal{R}_X}(K_{X \to Y}, F)),$$

(18.3.4) $f_* \mathcal{H}om_{\mathcal{R}_X}(K_{X \to Y} \otimes_{f^{-1}\mathcal{R}_Y} f^{-1}G, F)$
$$\simeq \mathcal{H}om_{\mathcal{R}_Y}(G, f_* \mathcal{H}om_{\mathcal{R}_X}(K_{X \to Y}, F)).$$

In particular, the functor

$$K_{X \to Y} \otimes_{f^{-1}\mathcal{R}_Y} f^{-1}(\cdot) \colon \mathrm{Mod}(\mathcal{R}_Y) \to \mathrm{Mod}(\mathcal{R}_X)$$

is left adjoint to the functor

$$f_* \mathcal{H}om_{\mathcal{R}_X}(K_{X \to Y}, \cdot) \colon \mathrm{Mod}(\mathcal{R}_X) \to \mathrm{Mod}(\mathcal{R}_Y) \, .$$

Proof. By (18.2.11) and (18.3.1), we have the isomorphisms:

$$\begin{aligned}
\mathrm{Hom}_{\mathcal{R}_X}&(K_{X \to Y} \otimes_{f^{-1}\mathcal{R}_Y} f^{-1}G, F) \\
&\simeq \mathrm{Hom}_{f^{-1}\mathcal{R}_Y}(f^{-1}G, \mathcal{H}om_{\mathcal{R}_X}(K_{X \to Y}, F)) \\
&\simeq \mathrm{Hom}_{\mathcal{R}_Y}(G, f_* \mathcal{H}om_{\mathcal{R}_X}(K_{X \to Y}, F)) \, .
\end{aligned}$$

The proof of (18.3.4) is similar. The last statement follows from (18.3.3). q.e.d.

18.4 Derived Functors for Hom and $\mathcal{H}om$

Notation 18.4.1. Let $\mathcal{R}$ be a sheaf of k-algebras. We shall often write for short $\mathrm{D}^*(\mathcal{R})$ instead of $\mathrm{D}^*(\mathrm{Mod}(\mathcal{R}))$ for $* = \mathrm{ub}, \mathrm{b}, +, -$. In particular, we set $\mathrm{D}(k) = \mathrm{D}(\mathrm{Mod}(k))$. We also write $\mathrm{K}^*(\mathcal{R})$, $\mathrm{K}^*_{\mathrm{hi}}(\mathcal{R})$, etc. for short.

By Theorem 18.1.6, we may apply the results of Chap. 14. In particular, we may consider the derived category $\mathrm{D}(\mathcal{R})$ to $\mathrm{Mod}(\mathcal{R})$ and construct the right derived functor to any additive functor defined on $\mathrm{Mod}(\mathcal{R})$.

We shall follow the notations of § 11.7. For $F_1, F_2 \in \mathrm{K}(\mathcal{R})$, we set

$$\mathrm{Hom}^\bullet_{\mathcal{R}}(F_1, F_2) := \mathrm{tot}_\pi \mathrm{Hom}^{\bullet\bullet}_{\mathcal{R}}(F_1, F_2),$$

an object of $\mathrm{K}(k)$.

Proposition 18.4.2. *The pair* $(\mathrm{K}_{\mathrm{hi}}(\mathcal{R}), \mathrm{K}(\mathcal{R})^{\mathrm{op}})$ *is* $\mathrm{Hom}^\bullet_{\mathcal{R}}$*-injective and the functor* $\mathrm{Hom}_{\mathcal{R}}$ *admits a right derived functor*

$$\mathrm{RHom}_{\mathcal{R}}(\cdot, \cdot) \colon \mathrm{D}(\mathcal{R}) \times \mathrm{D}(\mathcal{R})^{\mathrm{op}} \to \mathrm{D}(k) \, .$$

Note that for $F_1 \in \mathrm{K}(\mathcal{R})$ and $F_2 \in \mathrm{K}_{\mathrm{hi}}(\mathcal{R})$, $\mathrm{RHom}_{\mathcal{R}}(F_1, F_2) \simeq \mathrm{Hom}^\bullet_{\mathcal{R}}(F_1, F_2)$.

Lemma 18.4.3. *Let* $A \in \mathcal{C}^\wedge_X$. *Then* $F|_A \in \mathrm{K}_{\mathrm{hi}}(\mathcal{R}|_A)$ *for any* $F \in \mathrm{K}_{\mathrm{hi}}(\mathcal{R})$.

Proof. By the hypothesis, $\mathrm{Hom}_{\mathrm{K}(\mathcal{R})}(F', F) \simeq 0$ for all $F' \in \mathrm{K}(\mathcal{R})$ qis to zero. Let $G \in \mathrm{K}(\mathcal{R}|_A)$ be qis to zero. We have by Lemma 18.1.7 (i)

$$\mathrm{Hom}_{\mathrm{K}(\mathcal{R}|_A)}(G, F|_A) \simeq \mathrm{Hom}_{\mathrm{K}(\mathcal{R})}(\mathrm{j}^{-1}_{A \to X} G, F) \, ,$$

and the right hand side vanishes since $\mathrm{j}^{-1}_{A \to X}$ is exact.   q.e.d.

Similarly as for the functor $\text{Hom}^\bullet_{\mathcal{R}}$, for $F_1, F_2 \in \text{K}(\mathcal{R})$ we set

$$\mathcal{H}om^\bullet_{\mathcal{R}}(F_1, F_2) := \text{tot}_\pi \mathcal{H}om^{\bullet\bullet}_{\mathcal{R}}(F_1, F_2) ,$$

an object of $\text{K}(k_X)$.

Lemma 18.4.4. *Let $F_1, F_2 \in \text{K}(\mathcal{R})$ with F_2 homotopically injective and F_1 qis to 0. Then $\mathcal{H}om^\bullet_{\mathcal{R}}(F_1, F_2) \in \text{K}(k_X)$ is qis to 0.*

Proof. Let $U \in \mathcal{C}_X$. Applying Lemma 17.1.11 (ii), we obtain

$$\mathcal{H}om^\bullet_{\mathcal{R}}(F_1, F_2)(U) \simeq \text{Hom}^\bullet_{\mathcal{R}}(F_1|_U, F_2|_U) .$$

Since $F_1|_U$ is qis to 0 by Lemma 18.1.7 and $F_2|_U$ is homotopically injective by Lemma 18.4.3, the result follows.   q.e.d.

Hence we have the following proposition.

Proposition 18.4.5. *The pair $(\text{K}_{\text{hi}}(\mathcal{R}), \text{K}(\mathcal{R})^{\text{op}})$ is $\mathcal{H}om^\bullet_{\mathcal{R}}$-injective and the functor $\mathcal{H}om_{\mathcal{R}}$ admits a right derived functor*

$$R\mathcal{H}om_{\mathcal{R}}(\cdot, \cdot) : \text{D}(\mathcal{R}) \times \text{D}(\mathcal{R})^{\text{op}} \to \text{D}(k_X) .$$

Moreover, if $\mathcal{R}$ is commutative, this functor takes its values in $\text{D}(\mathcal{R})$.

Note that for $F_1 \in \text{K}(\mathcal{R})$ and $F_2 \in \text{K}_{\text{hi}}(\mathcal{R})$, $R\mathcal{H}om_{\mathcal{R}}(F_1, F_2) \simeq \mathcal{H}om^\bullet_{\mathcal{R}}(F_1, F_2)$.

Proposition 18.4.6. *For $F_1, F_2 \in \text{D}(\mathcal{R})$ and $A \in \mathcal{C}_X^\wedge$, we have the isomorphism*

$$R\mathcal{H}om_{\mathcal{R}}(F_1, F_2)|_A \simeq R\mathcal{H}om_{\mathcal{R}|_A}(F_1|_A, F_2|_A) .$$

Proof. By Proposition 18.4.5, we may assume that $F_1 \in \text{K}(\mathcal{R})$ and $F_2 \in \text{K}_{\text{hi}}(\mathcal{R})$. Then $R\mathcal{H}om_{\mathcal{R}}(F_1, F_2) \simeq \mathcal{H}om^\bullet_{\mathcal{R}}(F_1, F_2)$. We then have

$$\begin{aligned}
R\mathcal{H}om_{\mathcal{R}}(F_1, F_2)|_A &\simeq \mathcal{H}om^\bullet_{\mathcal{R}}(F_1, F_2)|_A \\
&\simeq \mathcal{H}om^\bullet_{\mathcal{R}|_A}(F_1|_A, F_2|_A) \\
&\simeq R\mathcal{H}om_{\mathcal{R}|_A}(F_1|_A, F_2|_A) ,
\end{aligned}$$

where the last isomorphism follows from the fact that $F_2|_A \in \text{K}_{\text{hi}}(\mathcal{R}|_A)$ by Lemma 18.4.3.   q.e.d.

18.5 Flatness

In this section, $\mathcal{R}$ denotes a sheaf of rings.

Definition 18.5.1. (i) *An $\mathcal{R}$-module F is locally free (resp. locally free of finite rank) if there exists a local epimorphism $A \to \mathrm{pt}_X$ such that for any $U \to A$ with $U \in \mathcal{C}_X$, $F|_U$ is isomorphic to a direct sum (resp. a finite direct sum) of copies of $\mathcal{R}|_U$.*

(ii) *An $\mathcal{R}$-module F is locally of finite presentation if there exists a local epimorphism $A \to \mathrm{pt}_X$ such that for any $U \in \mathcal{C}_A$, there exists an exact sequence*

$$(\mathcal{R}|_U)^{\oplus m} \to (\mathcal{R}|_U)^{\oplus n} \to F|_U \to 0 \ .$$

(iii) *An $\mathcal{R}$-module F is flat if the functor $\mathrm{Mod}(\mathcal{R}^{\mathrm{op}}) \ni F' \mapsto F' \otimes_{\mathcal{R}} F \in \mathrm{Mod}(\mathbb{Z}_X)$ is exact.*

We shall study the properties of flat modules.

Proposition 18.5.2. (i) *If P is a flat $\mathcal{R}$-module, then $P|_A$ is a flat $(\mathcal{R}|_A)$-module for any $A \in \mathcal{C}_X^\wedge$.*

(ii) *Let P be an $\mathcal{R}$-module and let $A \in \mathcal{C}_X^\wedge$. Assume that $A \to \mathrm{pt}_X$ is a local epimorphism and $P|_U$ is a flat $(\mathcal{R}|_U)$-module for any $U \to A$ with $U \in \mathcal{C}_X$. Then P is a flat $\mathcal{R}$-module.*

(iii) *If Q is a flat $\mathcal{R}|_A$-module, then $j_{A \to X}^{-1} Q$ is a flat $\mathcal{R}$-module.*

(iv) *Small direct sums of flat $\mathcal{R}$-modules as well as small filtrant inductive limits of flat $\mathcal{R}$-modules are flat.*

Proof. (i) Let $0 \to F' \to F \to F''$ be an exact sequence in $\mathrm{Mod}(\mathcal{R}^{\mathrm{op}}|_A)$. Let us show that the sequence

$$(18.5.1) \qquad 0 \to F' \otimes_{\mathcal{R}|_A} (P|_A) \to F \otimes_{\mathcal{R}|_A} (P|_A) \to F'' \otimes_{\mathcal{R}|_A} (P|_A)$$

is exact. Since the functor $j_{A \to X}^{-1}$ is faithful and exact, it is enough to check that the image of (18.5.1) by $j_{A \to X}^{-1}$ is exact. This follows from Proposition 18.2.5 and the fact that the sequence below is exact

$$0 \to (j_{A \to X}^{-1} F') \otimes_{\mathcal{R}} P \to (j_{A \to X}^{-1} F) \otimes_{\mathcal{R}} P \to (j_{A \to X}^{-1} F'') \otimes_{\mathcal{R}} P \ .$$

(ii) Let $0 \to F' \to F \to F''$ be an exact sequence in $\mathrm{Mod}(\mathcal{R}^{\mathrm{op}})$. Let us show that the sequence

$$(18.5.2) \qquad 0 \to F' \otimes_{\mathcal{R}} P \to F \otimes_{\mathcal{R}} P \to F'' \otimes_{\mathcal{R}} P$$

is exact. The functor $\cdot|_A$ is exact and it is faithful by Corollary 17.6.9. Hence, it is enough check that the image of (18.5.2) by $\cdot|_A$ is exact. This follows from the hypothesis and Lemma 18.2.4.

(iii) follows from the fact that the functor $\cdot \otimes_{\mathcal{R}} j_{A \to X}^{-1} Q$ is isomorphic to the exact functor $j_{A \to X}^{-1}((\cdot)|_A \otimes_{\mathcal{R}|_A} Q)$ by (18.2.6).

(iv) Small direct sums and small filtrant inductive limits are exact, and the functor $\otimes_{\mathcal{R}}$ commutes with these limits.   q.e.d.

Properties (i) and (ii) above may be translated by saying that flatness is a local property.

Lemma 18.5.3. *Let P be an $\mathcal{R}$-module. Assume that for any $U \in \mathcal{C}_X$ and any morphism $u\colon (\mathcal{R}|_U)^{\oplus m} \to (\mathcal{R}|_U)^{\oplus n}$, the sequence*

$$0 \to \mathrm{Ker}(u) \otimes_{\mathcal{R}|_U} (P|_U) \to (\mathcal{R}|_U)^{\oplus m} \otimes_{\mathcal{R}|_U} (P|_U) \to (\mathcal{R}|_U)^{\oplus n} \otimes_{\mathcal{R}|_U} (P|_U)$$

is exact. Then P is a flat $\mathcal{R}$-module.

Proof. Let $0 \to M' \to M \to M'' \to 0$ be an exact sequence in $\mathrm{Mod}(\mathcal{R}^{\mathrm{op}})$. Let us show that $M' \otimes_{\mathcal{R}} P \to M \otimes_{\mathcal{R}} P$ is a monomorphism. For $U \in \mathcal{C}_X$, set

$$K(U) := \mathrm{Ker}\big(M'(U) \otimes_{\mathcal{R}(U)} P(U) \to M(U) \otimes_{\mathcal{R}(U)} P(U)\big)\,.$$

Then K is a presheaf and $K^a \simeq \mathrm{Ker}(M' \otimes_{\mathcal{R}} P \to M \otimes_{\mathcal{R}} P)$. Hence, it is enough to check that the morphism $K(U) \to (M' \otimes_{\mathcal{R}} P)(U)$ vanishes for any $U \in \mathcal{C}_X$. Let $s \in K(U) \subset M'(U) \otimes_{\mathcal{R}(U)} P(U)$. Then, there exist a morphism $\mathcal{R}^{\mathrm{op}}(U)^{\oplus n} \xrightarrow{f} M'(U)$ and $s_1 \in \mathcal{R}^{\mathrm{op}}(U)^{\oplus n} \otimes_{\mathcal{R}(U)} P(U)$ whose image by $f \otimes P(U)$ is s. Since its image in $M(U) \otimes_{\mathcal{R}(U)} P(U)$ vanishes, there exists a commutative diagram whose right column is a complex (i.e., $q \circ g = 0$)

(18.5.3)

$$
\begin{array}{ccc}
 & & \mathcal{R}^{\mathrm{op}}(U)^{\oplus m} \\
 & & \downarrow{\scriptstyle g} \\
\mathcal{R}^{\mathrm{op}}(U)^{\oplus n} & \xrightarrow{\;h\;} & \mathcal{R}^{\mathrm{op}}(U)^{\oplus l} \\
\downarrow{\scriptstyle f} & & \downarrow{\scriptstyle q} \\
M'|_U & \longrightarrow & M|_U \\
\uparrow & & \\
0 & &
\end{array}
$$

and there exists $t_2 \in \mathcal{R}^{\mathrm{op}}(U)^{\oplus m} \otimes_{\mathcal{R}(U)} P(U)$ whose image by $g \otimes P(U)$ coincides with the image of s_1 by $h \otimes P(U)$ in $\mathcal{R}^{\mathrm{op}}(U)^{\oplus l} \otimes_{\mathcal{R}(U)} P(U)$. Consider the diagram below, in which the square labeled $\square$ is Cartesian. .

(18.5.4)

$$
\begin{array}{ccc}
N & \longrightarrow & (\mathcal{R}^{\mathrm{op}}|_U)^{\oplus m} \\
\downarrow & \square & \downarrow{\scriptstyle g} \\
(\mathcal{R}^{\mathrm{op}}|_U)^{\oplus n} & \xrightarrow{\;h\;} & (\mathcal{R}^{\mathrm{op}}|_U)^{\oplus l} \\
\downarrow{\scriptstyle f} & & \downarrow{\scriptstyle q} \\
M'|_U & \longrightarrow & M|_U\,. \\
\uparrow & & \\
0 & &
\end{array}
$$

The composition $N \to (\mathcal{R}^{\mathrm{op}}|_U)^{\oplus n} \to M'|_U$ vanishes and the sequence in $\mathrm{Mod}(\mathcal{R}^{\mathrm{op}}|_U)$

$$0 \to N \to (\mathcal{R}^{\mathrm{op}}|_U^{\oplus m}) \oplus (\mathcal{R}^{\mathrm{op}}|_U^{\oplus n}) \to (\mathcal{R}^{\mathrm{op}}|_U^{\oplus l})$$

is exact. By the hypothesis, this sequence remains exact after applying the functor $\cdot \otimes_{\mathcal{R}|_U} P|_U$. Since the functor $\Gamma(U; \cdot)$ is left exact, we obtain an exact sequence:

$$0 \to (N \otimes_{\mathcal{R}|_U} P|_U)(U)$$
$$\longrightarrow \left((\mathcal{R}^{\mathrm{op}}|_U)^{\oplus m} \otimes_{\mathcal{R}|_U} P|_U\right)(U) \oplus \left((\mathcal{R}^{\mathrm{op}}|_U)^{\oplus n} \otimes_{\mathcal{R}|_U} P|_U\right)(U)$$
$$\longrightarrow \left((\mathcal{R}^{\mathrm{op}}|_U)^{\oplus l} \otimes_{\mathcal{R}|_U} P|_U\right)(U) \, .$$

Hence $s_1 \in ((\mathcal{R}^{\mathrm{op}})^{\oplus m} \otimes_{\mathcal{R}} P)(U)$ lifts to an element of $(N \otimes_{\mathcal{R}|_U} P|_U)(U)$. Since the composition

$$N \otimes_{\mathcal{R}|_U} (P|_U) \to (\mathcal{R}^{\mathrm{op}}|_U)^{\oplus n} \otimes_{\mathcal{R}|_U} (P|_U) \to (M'|_U) \otimes_{\mathcal{R}|_U} (P|_U)$$
$$\simeq (M' \otimes_{\mathcal{R}} P)|_U$$

vanishes, the image of $s \in K(U)$ in $(M' \otimes_{\mathcal{R}} P)(U)$ vanishes.   q.e.d.

Let $\mathcal{P}$ be the full subcategory of $\mathrm{Mod}(\mathcal{R})$ consisting of flat $\mathcal{R}$-modules. Clearly, $\mathcal{P}$ is closed by small direct sums.

Proposition 18.5.4. (i) *For any* $N \in \mathrm{Mod}(\mathcal{R}^{\mathrm{op}})$, *the category* $\mathcal{P}$ *is* $(N \otimes_{\mathcal{R}} \cdot)$-*projective. More precisely, the category* $\mathcal{P}$ *satisfies properties* (i)–(iii) *of* Corollary 13.3.8.
(ii) *Let* $0 \to M' \to M \to M'' \to 0$ *be an exact sequence in* $\mathrm{Mod}(\mathcal{R})$ *and assume that* M'' *is flat. Then, for any* $N \in \mathrm{Mod}(\mathcal{R}^{\mathrm{op}})$, *the sequence* $0 \to N \otimes_{\mathcal{R}} M' \to N \otimes_{\mathcal{R}} M \to N \otimes_{\mathcal{R}} M'' \to 0$ *is exact* .

Proof. (a) The object $\mathcal{G} := \bigoplus_{U \in \mathcal{C}_X} \mathcal{R}_U$ is a generator in $\mathrm{Mod}(\mathcal{R})$ and a flat module by Proposition 18.5.2. Hence, for any $M \in \mathrm{Mod}(\mathcal{R})$, there exist a small set I and an epimorphism $G^{\oplus I} \twoheadrightarrow M$. Therefore the category $\mathcal{P}$ is generating in $\mathrm{Mod}(\mathcal{R})$.
(b) Let us prove (ii). Applying (a) to $\mathcal{R}^{\mathrm{op}}$, there exists an exact sequence $0 \to K \to Q \to N \to 0$ with a flat $\mathcal{R}^{\mathrm{op}}$-module Q. Consider the commutative exact diagram in $\mathrm{Mod}(\mathbb{Z}_X)$:

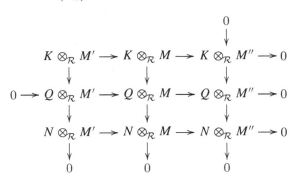

It follows from the snake lemma (Lemma 12.1.1) that $N \otimes_{\mathcal{R}} M' \to N \otimes_{\mathcal{R}} M$ is a monomorphism.

(c) Let us complete the proof of (i). It remains to prove that if $0 \to M' \to M \to M'' \to 0$ is an exact sequence in $\text{Mod}(\mathcal{R})$ with M and M'' flat, then M' is flat. Consider an exact sequence $0 \to N' \to N \to N'' \to 0$ in $\text{Mod}(\mathcal{R}^{\text{op}})$. We get the commutative diagram

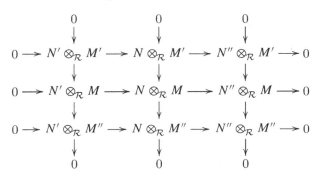

The middle and the bottom rows are exact, and so are all the columns by (ii). It follows that the top row is exact.   q.e.d.

18.6 Ringed Sites

Definition 18.6.1. (i) *A ringed site* $(X, \mathcal{O}_X)$ *is a site* X *endowed with a sheaf of commutative rings* $\mathcal{O}_X$ *on* X.

 (ii) *Let* $(X, \mathcal{O}_X)$ *and* $(Y, \mathcal{O}_Y)$ *be two ringed sites. A morphism of ringed sites is a* left exact *morphism of sites* $f : X \to Y$ *together with a morphism of sheaves of rings* $f^{-1}\mathcal{O}_Y \to \mathcal{O}_X$ (*or equivalently, a morphism of sheaves of rings* $\mathcal{O}_Y \to f_*\mathcal{O}_X$).

Note that $f^{-1}\mathcal{O}_Y$ is a sheaf of rings by Lemma 18.3.1.
    For a ringed site $(X, \mathcal{O}_X)$, we have functors

$$\text{Hom}_{\mathcal{O}_X} : \text{Mod}(\mathcal{O}_X) \times \text{Mod}(\mathcal{O}_X)^{\text{op}} \to \text{Mod}(\mathcal{O}_X(X)) \,,$$
$$\mathcal{H}om_{\mathcal{O}_X} : \text{Mod}(\mathcal{O}_X) \times \text{Mod}(\mathcal{O}_X)^{\text{op}} \to \text{Mod}(\mathcal{O}_X) \,.$$

For a ringed site $(X, \mathcal{O}_X)$ and $A \in \mathcal{C}_X^{\wedge}$, we regard A as a ringed site $(A, \mathcal{O}_A) :=$ $(A, \mathcal{O}_X|_A)$. The functor $\bullet|_A$ gives an exact functor $\text{Mod}(\mathcal{O}_X) \to \text{Mod}(\mathcal{O}_A)$.
    Note that for $A \in \mathcal{C}_X^{\wedge}$, the morphism $j_{A \to X} : X \to A$ is *not* a morphism of ringed sites in general (even for $A \in \mathcal{C}_X$).

Proposition 18.6.2. *There are isomorphisms, functorial with respect to* F, G, $H \in \text{Mod}(\mathcal{O}_X)$:

(18.6.1) $\text{Hom}_{\mathcal{O}_X}(F \otimes_{\mathcal{O}_X} G, H) \simeq \text{Hom}_{\mathcal{O}_X}(F, \mathcal{H}om_{\mathcal{O}_X}(G, H)) \,,$

(18.6.2) $\mathcal{H}om_{\mathcal{O}_X}(F \otimes_{\mathcal{O}_X} G, H) \simeq \mathcal{H}om_{\mathcal{O}_X}(F, \mathcal{H}om_{\mathcal{O}_X}(G, H)) \,.$

In particular, the functors $\cdot \otimes_{\mathcal{O}_X} G$ *and* $\mathcal{H}om_{\mathcal{O}_X}(G, \cdot)$ *are adjoint.*

This follows immediately from Proposition 18.2.3 (see Remark 18.2.6).

The direct image functor f_* induces a functor (we keep the same notation)

$$(18.6.3) \qquad f_* \colon \mathrm{Mod}(\mathcal{O}_X) \to \mathrm{Mod}(\mathcal{O}_Y) \,.$$

The *inverse image functor*

$$(18.6.4) \qquad f^* \colon \mathrm{Mod}(\mathcal{O}_Y) \to \mathrm{Mod}(\mathcal{O}_X)$$

is given by

$$G \mapsto \mathcal{O}_X \otimes_{f^{-1}\mathcal{O}_Y} f^{-1}G \,.$$

As a particular case of Proposition 18.3.3, we obtain:

Proposition 18.6.3. *There are isomorphisms, functorial with respect to* $F \in \mathrm{Mod}(\mathcal{O}_X)$ *and* $G \in \mathrm{Mod}(\mathcal{O}_Y)$:

$$(18.6.5) \qquad \mathrm{Hom}_{\mathcal{O}_X}(f^*G, F) \simeq \mathrm{Hom}_{\mathcal{O}_Y}(G, f_*F) \,,$$
$$(18.6.6) \qquad f_*\mathcal{H}om_{\mathcal{O}_X}(f^*G, F) \simeq \mathcal{H}om_{\mathcal{O}_Y}(G, f_*F) \,.$$

In particular, the functors f^* *and* f_* *are adjoint.*

It follows that the functors $\mathrm{Hom}_{\mathcal{O}_X}$ and f_* are left exact and the functors $\otimes_{\mathcal{O}_X}$ and f^* are right exact.

Derived Functors for $\otimes$

Let $\mathcal{P}_X$ be the full subcategory of $\mathrm{Mod}(\mathcal{O}_X)$ consisting of flat $\mathcal{O}_X$-modules and let $\widetilde{\mathcal{P}}_X$ be the smallest full triangulated subcategory of $\mathrm{K}(\mathcal{O}_X)$ stable by small direct sums and containing $\mathrm{K}^-(\mathcal{P}_X)$.

For $F_1, F_2 \in \mathrm{K}(\mathcal{O}_X)$, we shall write for short:

$$F_1 \otimes_{\mathcal{O}_X} F_2 := \mathrm{tot}_\oplus(F_1 \otimes_{\mathcal{O}_X} F_2) \,.$$

The hypotheses of Theorem 14.4.8 with $\mathcal{C}_i = \mathrm{Mod}(\mathcal{O}_X)$ $(i = 1, 2, 3)$, $G = \otimes_{\mathcal{O}_X}$ and $F_1 = F_2 = \mathcal{H}om_{\mathcal{O}_X}$, are satisfied with $\mathcal{P}_2 = \mathcal{P}_X$, $\mathcal{P}_1 = \mathrm{Mod}(\mathcal{O}_X)$ by Proposition 18.5.4.

Theorem 18.6.4. (i) $(\widetilde{\mathcal{P}}_X, \mathrm{K}(\mathcal{O}_X))$ *is* $(\cdot \otimes_{\mathcal{O}_X} \cdot)$-*projective,*

(ii) *the derived functor* $\cdot \overset{L}{\otimes}_{\mathcal{O}_X} \cdot \colon \mathrm{D}(\mathcal{O}_X) \times \mathrm{D}(\mathcal{O}_X) \to \mathrm{D}(\mathcal{O}_X)$ *exists and for* F_1 *or* F_2 *in* $\widetilde{\mathcal{P}}_X$ *we have* $F_1 \overset{L}{\otimes}_{\mathcal{O}_X} F_2 \simeq F_1 \otimes_{\mathcal{O}_X} F_2$,

(iii) $(\mathrm{K}_{\mathrm{hi}}(\mathcal{O}_X), \mathrm{K}(\mathcal{O}_X)^{\mathrm{op}})$ *is* $\mathcal{H}om_{\mathcal{O}_X}$-*injective,*

(iv) *the derived functor* $R\mathcal{H}om_{\mathcal{O}_X}\colon \mathrm{D}(\mathcal{O}_X) \times \mathrm{D}(\mathcal{O}_X)^{\mathrm{op}} \to \mathrm{D}(\mathcal{O}_X)$ *exists and for* $F_2 \in \mathrm{K}(\mathcal{O}_X)$ *and* $F_3 \in \mathrm{K}_{\mathrm{hi}}(\mathcal{O}_X)$, *we have* $R\mathcal{H}om_{\mathcal{O}_X}(F_2, F_3) \simeq \mathcal{H}om^\bullet_{\mathcal{O}_X}(F_2, F_3)$,

(v) *for* $F_2 \in \widetilde{\mathcal{P}}_X$ *and* $F_3 \in \mathrm{K}_{\mathrm{hi}}(\mathcal{O}_X)$, $\mathcal{H}om^\bullet_{\mathcal{O}_X}(F_2, F_3) \in \mathrm{K}_{\mathrm{hi}}(\mathcal{O}_X)$,

(vi) *for* $F_1, F_2 \in \widetilde{\mathcal{P}}_X$, $F_1 \otimes_{\mathcal{O}_X} F_2 \in \widetilde{\mathcal{P}}_X$,

(vii) *for* $F_1, F_2, F_3 \in \mathrm{D}(\mathcal{O}_X)$, *we have the isomorphisms*

$$(18.6.7)\quad \mathrm{Hom}_{\mathrm{D}(\mathcal{O}_X)}(F_1 \overset{\mathrm{L}}{\otimes}_{\mathcal{O}_X} F_2, F_3) \simeq \mathrm{Hom}_{\mathrm{D}(\mathcal{O}_X)}(F_1, R\mathcal{H}om_{\mathcal{O}_X}(F_2, F_3)) \,,$$

$$(18.6.8)\quad \mathrm{RHom}_{\mathcal{O}_X}(F_1 \overset{\mathrm{L}}{\otimes}_{\mathcal{O}_X} F_2, F_3) \simeq \mathrm{RHom}_{\mathcal{O}_X}(F_1, R\mathcal{H}om_{\mathcal{O}_X}(F_2, F_3)) \,,$$

$$(18.6.9)\quad R\mathcal{H}om_{\mathcal{O}_X}(F_1 \overset{\mathrm{L}}{\otimes}_{\mathcal{O}_X} F_2, F_3) \simeq R\mathcal{H}om_{\mathcal{O}_X}(F_1, R\mathcal{H}om_{\mathcal{O}_X}(F_2, F_3)) \,.$$

Proof. (i)–(iv) as well as (18.6.7) and (18.6.8) follow from Theorem 14.4.8 and Proposition 18.6.2.

(v) Assume that $F_1 \in \mathrm{K}(\mathcal{O}_X)$ is qis to 0. Then $F_1 \otimes_{\mathcal{O}_X} F_2$ is qis to 0 by (i). Hence, we have

$$\mathrm{Hom}_{\mathrm{K}(\mathcal{O}_X)}(F_1, \mathcal{H}om^\bullet_{\mathcal{O}_X}(F_2, F_3)) \simeq \mathrm{Hom}_{\mathrm{K}(\mathcal{O}_X)}(F_1 \otimes_{\mathcal{O}_X} F_2, F_3) \simeq 0 \,.$$

(vi) is obvious.

(vii) Let us prove the isomorphism (18.6.9). For any $K \in \mathrm{D}(\mathcal{O}_X)$, we have

$$\mathrm{Hom}_{\mathrm{D}(\mathcal{O}_X)}\big(K, R\mathcal{H}om_{\mathcal{O}_X}(F_1 \overset{\mathrm{L}}{\otimes}_{\mathcal{O}_X} F_2, F_3)\big)$$
$$\simeq \mathrm{Hom}_{\mathrm{D}(\mathcal{O}_X)}\big(K \overset{\mathrm{L}}{\otimes}_{\mathcal{O}_X} F_1 \overset{\mathrm{L}}{\otimes}_{\mathcal{O}_X} F_2, F_3\big)$$
$$\simeq \mathrm{Hom}_{\mathrm{D}(\mathcal{O}_X)}\big(K \overset{\mathrm{L}}{\otimes}_{\mathcal{O}_X} F_1, R\mathcal{H}om_{\mathcal{O}_X}(F_2, F_3)\big)$$
$$\simeq \mathrm{Hom}_{\mathrm{D}(\mathcal{O}_X)}\big(K, R\mathcal{H}om_{\mathcal{O}_X}(F_1, R\mathcal{H}om_{\mathcal{O}_X}(F_2, F_3))\big) \,,$$

from which (18.6.9) follows by the Yoneda lemma.   q.e.d.

Notation 18.6.5. The functor $\Gamma(A; \cdot)\colon \mathrm{Mod}(\mathcal{O}_X) \to \mathrm{Mod}(\mathcal{O}_X(X))$ defined in Notations 17.6.13 (with $A \in \mathcal{C}_X^\wedge$) is clearly left exact. The right derived functor of $\Gamma(A; \cdot)$ is denoted by $\mathrm{R}\Gamma(A; \cdot)$. Hence,

$$\mathrm{R}\Gamma(A; \cdot)\colon \mathrm{D}(\mathcal{O}_X) \to \mathrm{D}(\mathcal{O}_X(X)) \,.$$

Proposition 18.6.6. *There is an isomorphism* $\mathrm{R}\Gamma(X; R\mathcal{H}om_{\mathcal{O}_X}(F_1, F_2)) \simeq \mathrm{RHom}_{\mathcal{O}_X}(F_1, F_2)$ *in* $\mathrm{D}(\mathcal{O}_X(X))$ *for* $F_1, F_2 \in \mathrm{D}(\mathcal{O}_X)$.

Proof. We may assume that $F_1 \in \widetilde{\mathcal{P}}_X$ and $F_2 \in \mathrm{K}_{\mathrm{hi}}(\mathcal{O}_X)$. By Theorem 18.6.4 (v), $\mathcal{H}om^\bullet_{\mathcal{O}_X}(F_1, F_2)$ belongs to $\mathrm{K}_{\mathrm{hi}}(\mathcal{O}_X)$, and we obtain

$$\mathrm{R}\Gamma(X; R\mathcal{H}om_{\mathcal{O}_X}(F_1, F_2)) \simeq \Gamma(X; \mathcal{H}om^\bullet_{\mathcal{O}_X}(F_1, F_2))$$
$$\simeq \mathrm{Hom}^\bullet_{\mathcal{O}_X}(F_1, F_2) \simeq \mathrm{RHom}_{\mathcal{O}_X}(F_1, F_2) \,.$$

q.e.d.

Derived Functors for Direct and Inverse Images

Lemma 18.6.7. *Let $f \colon X \to Y$ be a morphism of ringed sites.*

(i) *If P is a flat $\mathcal{O}_Y$-module, then f^*P is a flat $\mathcal{O}_X$-module,*

(ii) *the functor f^* sends $\widetilde{\mathcal{P}}_Y$ to $\widetilde{\mathcal{P}}_X$.*

Proof. (i) (a) Assume first that $\mathcal{O}_X = f^{-1}\mathcal{O}_Y$. Then $f^*P = f^{-1}P$. By Lemma 18.5.3, it is enough to show that for any $U \in \mathcal{C}_X$ and any exact sequence

$$0 \to N \to \mathcal{O}_U^{\oplus n} \xrightarrow{u} \mathcal{O}_U^{\oplus m}$$

in $\mathrm{Mod}(\mathcal{O}_U)$, the sequence

(18.6.10)

$$0 \to N \otimes_{\mathcal{O}_U} (f^{-1}P)|_U \to \mathcal{O}_U^{\oplus n} \otimes_{\mathcal{O}_U} (f^{-1}P)|_U \to \mathcal{O}_U^{\oplus m} \otimes_{\mathcal{O}_U} (f^{-1}P)|_U$$

is exact. The morphism u is given by an element of

$$\mathcal{O}_X^{\oplus nm}(U) \simeq f^{-1}(\mathcal{O}_Y^{\oplus nm})(U) \simeq \varinjlim_{A \in \mathcal{L}\mathcal{I}_U} f^{\dagger}(\mathcal{O}_Y^{\oplus nm})(A) \ .$$

Hence, there exist $A \in \mathcal{L}\mathcal{I}_U$ and an element $s \in f^{\dagger}(\mathcal{O}_Y^{\oplus nm})(A)$ whose image is u. Let $W \in (\mathcal{C}_X)_A$. Then

$$f^{\dagger}(\mathcal{O}_Y^{\oplus nm})(W) \simeq \varinjlim_{V \in (\mathcal{C}_Y)^W} \mathcal{O}_Y^{\oplus nm}(V) \ .$$

Hence, there exists $V \in (\mathcal{C}_Y)^W$ and $s' \in \mathcal{O}_Y^{\oplus nm}(V)$ such that the image of s' coincides with s. Then s' gives a morphism $\mathcal{O}_V^{\oplus n} \xrightarrow{u_1} \mathcal{O}_V^{\oplus m}$. Let $f_W \colon W \to V$ be the morphism of sites induced by f. Then $f_W^{-1}(u_1) \colon \mathcal{O}_U^{\oplus n} \xrightarrow{u} \mathcal{O}_U^{\oplus m}$ is equal to $u|_W$. Let N_1 be the kernel of u_1. Since f_W^{-1} is exact, $N|_W \simeq f_W^{-1}N_1$. Then the sequence

$$0 \to N_1 \otimes_{\mathcal{O}_Y|_V} (P|_V) \to \mathcal{O}_V^{\oplus n} \otimes_{\mathcal{O}_Y|_V} (P|_V) \to \mathcal{O}_V^{\oplus m} \otimes_{\mathcal{O}_Y|_V} (P|_V)$$

is exact. Applying f_W^{-1} and recalling that this functor commutes with $\otimes$ (Lemma 18.3.1), we find that the sequence (18.6.10) in which U is replaced with W is exact. Since this property holds for any $W \in (\mathcal{C}_X)_A$ and $A \to U$ is a local isomorphism, the sequence (18.6.10) is exact.

(i) (b) We have seen that $f^{-1}P$ is a flat $(f^{-1}\mathcal{O}_Y)$-module. Hence, the functor

$$\mathrm{Mod}(\mathcal{O}_X) \ni M \mapsto M \otimes_{\mathcal{O}_X} f^*P \simeq M \otimes_{f^{-1}\mathcal{O}_Y} f^{-1}P$$

is exact.

(ii) obviously follows from (i).   q.e.d.

Lemma 18.6.8. *Let* $f\colon X \to Y$ *be a morphism of ringed sites. Then the category* $\mathcal{P}_Y$ *of flat* $\mathcal{O}_Y$*-modules is* f^**-projective.*

Proof. By Proposition 18.5.4 and Corollary 13.3.8, it is enough to check that if $0 \to G'' \to G \to G' \to 0$ is an exact sequence in $\mathrm{Mod}(\mathcal{O}_Y)$ and G' is $\mathcal{O}_Y$-flat, then the sequence remains exact after applying f^*.

Since f^{-1} is exact, the sequence $0 \to f^{-1}G'' \to f^{-1}G \to f^{-1}G' \to 0$ is exact in $\mathrm{Mod}(f^{-1}\mathcal{O}_Y)$. Since $f^{-1}G'$ is $f^{-1}\mathcal{O}_Y$-flat by Lemma 18.6.7, this sequence remains exact after applying the functor $\mathcal{O}_X \otimes_{f^{-1}\mathcal{O}_Y} \cdot$ by Proposition 18.5.4 (ii).   q.e.d.

By Theorem 18.1.6, the functor f_* admits a right derived functor

$$Rf_*\colon \mathrm{D}(\mathcal{O}_X) \to \mathrm{D}(\mathcal{O}_Y) .$$

Theorem 18.6.9. *Let* $f\colon X \to Y$ *be a morphism of ringed sites.*

(i) *The functor* f^* *has a left derived functor*

$$Lf^*\colon \mathrm{D}(\mathcal{O}_Y) \to \mathrm{D}(\mathcal{O}_X)$$

and for $G \in \widetilde{\mathcal{P}}_Y$, *we have* $Lf^*G \simeq f^*G$,
(ii) *for* $G, G' \in \mathrm{D}(\mathcal{O}_Y)$ *we have an isomorphism*

$$Lf^*(G\overset{L}{\otimes}_{\mathcal{O}_Y} G') \simeq Lf^*G\overset{L}{\otimes}_{\mathcal{O}_X} Lf^*G' ,$$

(iii) *there are isomorphisms, functorial with respect to* $F \in \mathrm{D}(\mathcal{O}_X)$ *and* $G \in \mathrm{D}(\mathcal{O}_Y)$

(18.6.11) $\mathrm{Hom}_{\mathrm{D}(\mathcal{O}_Y)}(G, Rf_*F) \simeq \mathrm{Hom}_{\mathrm{D}(\mathcal{O}_X)}(Lf^*G, F) ,$

(18.6.12) $\mathrm{RHom}_{\mathcal{O}_Y}(G, Rf_*F) \simeq \mathrm{RHom}_{\mathcal{O}_X}(Lf^*G, F) ,$

(18.6.13) $R\mathcal{H}om_{\mathcal{O}_Y}(G, Rf_*F) \simeq Rf_*R\mathcal{H}om_{\mathcal{O}_X}(Lf^*G, F) ,$

(18.6.14) $R\Gamma(Y; Rf_*F) \simeq R\Gamma(X; F) .$

In particular, (Lf^*, Rf_*) *is a pair of adjoint functors.*

Proof. (i) By Lemma 18.6.8, the category $\mathcal{P}_Y$ is f^*-projective. Hence, we may apply Theorem 14.4.5.
(ii) We may assume that $G, G' \in \widetilde{\mathcal{P}}_Y$. Then

$$Lf^*G\overset{L}{\otimes}_{\mathcal{O}_X} Lf^*G' \simeq f^*G \otimes_{\mathcal{O}_X} f^*G'$$

$$\simeq f^*(G \otimes_{\mathcal{O}_Y} G') \simeq Lf^*(G\overset{L}{\otimes}_{\mathcal{O}_Y} G') .$$

Here, the first isomorphism follows from Lemma 18.6.7 (ii) and Lemma 18.6.8, the second isomorphism follows from Lemma 18.3.1 (ii) (c) and the third one follows from Theorem 18.6.4 (vi).

(iii) The isomorphisms (18.6.11) and (18.6.12) follow from Theorem 14.4.5. Setting $G = \mathcal{O}_Y$ in (18.6.12), we obtain (18.6.14). Let us prove (18.6.13). For $K \in D(\mathcal{O}_Y)$, we have the chain of isomorphisms

$$\mathrm{Hom}_{D(\mathcal{O}_Y)}(K, Rf_* R\mathcal{H}om_{\mathcal{O}_X}(Lf^*G, F))$$
$$\simeq \mathrm{Hom}_{D(\mathcal{O}_X)}(Lf^*K, R\mathcal{H}om_{\mathcal{O}_X}(Lf^*G, F))$$
$$\simeq \mathrm{Hom}_{D(\mathcal{O}_X)}(Lf^*K \overset{L}{\otimes}_{\mathcal{O}_X} Lf^*G, F)$$
$$\simeq \mathrm{Hom}_{D(\mathcal{O}_X)}(Lf^*(K \overset{L}{\otimes}_{\mathcal{O}_Y} G), F)$$
$$\simeq \mathrm{Hom}_{D(\mathcal{O}_Y)}(K \overset{L}{\otimes}_{\mathcal{O}_Y} G, Rf_* F)$$
$$\simeq \mathrm{Hom}_{D(\mathcal{O}_Y)}(K, R\mathcal{H}om_{\mathcal{O}_Y}(G, Rf_* F)) .$$

Then the Yoneda lemma implies (18.6.13).   q.e.d.

Proposition 18.6.10. *Let* $f\colon (X, \mathcal{O}_X) \to (Y, \mathcal{O}_Y)$ *and* $g\colon (Y, \mathcal{O}_Y) \to (Z, \mathcal{O}_Z)$ *be two morphisms of ringed sites. Set* $h := g \circ f$. *Then*

(i) $h\colon (X, \mathcal{O}_X) \to (Z, \mathcal{O}_Z)$ *is a morphism of ringed sites,*
(ii) $Rh_* \simeq Rg_* \circ Rf_*$ *and* $Lh^* \simeq Lf^* \circ Lg^*$.

Proof. (i) is obvious.
(ii) Apply Proposition 14.4.7 and Lemma 18.6.7.   q.e.d.

For $F \in D(\mathcal{O}_X)$, $A \in \mathcal{C}_X^\wedge$ and $j \in \mathbb{Z}$, we set

$$(18.6.15) \qquad\qquad H^j(A; F) := H^j(R\Gamma(A; F)).$$

Remark 18.6.11. Many results of this section may be generalized to the case of sheaves of not necessarily commutative rings. Following the notations in Remark 18.2.6, we have the results below whose proofs are left to the readers. Consider four $\mathcal{O}_X$-algebras $\mathcal{R}_\nu$ ($\nu = 1, \ldots, 4$), and for $i, j \in \{1, \ldots, 4\}$, denote by ${}_iM_j$ an object of $D(\mathcal{R}_i \otimes_{\mathcal{O}_X} \mathcal{R}_j^{\mathrm{op}})$. We make the assumption:

$$(18.6.16) \qquad\qquad \mathcal{R}_\nu \text{ is a flat } \mathcal{O}_X\text{-module for all } \nu = 1, \ldots, 4 .$$

Then the functors below are well-defined

$$\overset{L}{\otimes}_{\mathcal{R}_2} : \ D(\mathcal{R}_1 \otimes_{\mathcal{O}_X} \mathcal{R}_2^{\mathrm{op}}) \times D(\mathcal{R}_2 \otimes_{\mathcal{O}_X} \mathcal{R}_3^{\mathrm{op}}) \to D(\mathcal{R}_1 \otimes_{\mathcal{O}_X} \mathcal{R}_3^{\mathrm{op}}) ,$$
$$R\mathcal{H}om_{\mathcal{R}_1} : \ D(\mathcal{R}_1 \otimes_{\mathcal{O}_X} \mathcal{R}_3^{\mathrm{op}}) \times D(\mathcal{R}_1 \otimes_{\mathcal{O}_X} \mathcal{R}_2^{\mathrm{op}})^{\mathrm{op}} \to D(\mathcal{R}_2 \otimes_{\mathcal{O}_X} \mathcal{R}_3^{\mathrm{op}}) ,$$
$$\mathrm{RHom}_{\mathcal{R}_1} : \ D(\mathcal{R}_1 \otimes_{\mathcal{O}_X} \mathcal{R}_3^{\mathrm{op}}) \times D(\mathcal{R}_1 \otimes_{\mathcal{O}_X} \mathcal{R}_2^{\mathrm{op}})^{\mathrm{op}} \to D((\mathcal{R}_2 \otimes_{\mathcal{O}_X} \mathcal{R}_3^{\mathrm{op}})(X)) ,$$

and there are natural isomorphisms in $D(\mathcal{R}_1 \otimes_{\mathcal{O}_X} \mathcal{R}_4^{\mathrm{op}})$:

$$({}_1M_2 \overset{L}{\otimes}_{\mathcal{R}_2} {}_2M_3) \overset{L}{\otimes}_{\mathcal{R}_3} {}_3M_4 \simeq {}_1M_2 \overset{L}{\otimes}_{\mathcal{R}_2} ({}_2M_3 \overset{L}{\otimes}_{\mathcal{R}_3} {}_3M_4),$$

$$R\mathcal{H}om_{\mathcal{R}_2}({}_2M_1, R\mathcal{H}om_{\mathcal{R}_3}({}_3M_2, {}_3M_4)) \simeq R\mathcal{H}om_{\mathcal{R}_3}({}_3M_2 \overset{L}{\otimes}_{\mathcal{R}_2} {}_2M_1, {}_3M_4) .$$

Note that the hypothesis (18.6.16) implies that a flat (resp. injective) module $_1M_2$ over $\mathcal{R}_1 \otimes_{\mathcal{O}_X} \mathcal{R}_2^{\mathrm{op}}$ is flat (resp. injective) over $\mathcal{R}_1$. Indeed, the functors $\cdot \underset{\mathcal{R}_1}{\otimes} {_1M_2} \simeq (\cdot \otimes_{\mathcal{O}_X} \mathcal{R}_2) \underset{\mathcal{R}_1 \otimes_{\mathcal{O}_X} \mathcal{R}_2^{\mathrm{op}}}{\otimes} {_1M_2}$ and $\mathrm{Hom}_{\mathcal{R}_1}(\cdot, {_1M_2}) \simeq \mathrm{Hom}_{\mathcal{R}_1 \otimes_{\mathcal{O}_X} \mathcal{R}_2^{\mathrm{op}}}$ $(\cdot \underset{\mathcal{O}_X}{\otimes} \mathcal{R}_2, {_1M_2})$ are exact on $\mathrm{Mod}(\mathcal{R}_1^{\mathrm{op}})$ and $\mathrm{Mod}(\mathcal{R}_1)$, respectively.

18.7 Čech Coverings

We end this chapter with a classical construction, known as Čech cohomology or Čech coverings. In order to calculate the cohomology of a sheaf on a site X, we shall replace X with a local epimorphism $A \to \mathrm{pt}_X$ with $A \in \mathcal{C}_X^\wedge$ (which corresponds in the classical theory to an open covering), the sheaf F having nice properties on A.

In this section, we consider again sheaves of k-modules.

Let $u \colon A \to B$ be a morphism in $\mathcal{C}_X^\wedge$. In the sequel, we shall often use the restriction of the functor $A^{\coprod_B} \colon (\mathbf{Set}^f)^{\mathrm{op}} \to \mathcal{C}_B^\wedge$ constructed in (2.2.15) to the simplicial category $\mathbf{\Delta}_{inj}$ constructed in §11.4.

Recall that the objects of the category $\mathbf{\Delta}$ are the finite totally ordered sets and the morphisms are the order-preserving maps.

The category $\mathbf{\Delta}_{inj}$ is the subcategory of $\mathbf{\Delta}$ whose objects are those of $\mathbf{\Delta}$, but the morphisms are the injective ones.

Notation 18.7.1. If $u \colon A \to B$ is a morphism in $\mathcal{C}_X^\wedge$, we denote by

(18.7.1)
$$\lambda_u \colon \mathbf{\Delta}_{inj}^{\mathrm{op}} \to \mathcal{C}_B^\wedge$$

the composition of the functor $A^{\coprod_B}$ in (2.2.15) with the embedding $\mathbf{\Delta}_{inj}^{\mathrm{op}} \to (\mathbf{Set}^f)^{\mathrm{op}}$.

The functor λ_u is visualized by the diagram

(18.7.2)
$$\begin{array}{ccccccc} & \cdots\cdots\to & & \xrightarrow{\quad p_{12} \quad} & & \xrightarrow{\quad p_1 \quad} & \\ \cdots\cdots\to & A \times_B A \times_B A & \xrightarrow{\quad p_{13} \quad} & A \times_B A & \underset{p_2}{\overset{}{\rightrightarrows}} & A & \xrightarrow{\; u \;} B \,. \\ \cdots\cdots\to & & \xrightarrow{\quad p_{23} \quad} & & & \end{array}$$

Here, p_i corresponds to the i-th projection and p_{ij} to the (i, j)-th projection.

Recall that $\widetilde{\mathbf{\Delta}}$ is the subcategory of $\mathbf{\Delta}$ consisting of non-empty sets, the morphisms $u \colon \sigma \to \tau$ of $\widetilde{\mathbf{\Delta}}$ being those of $\mathbf{\Delta}$ sending the smallest (resp. the largest) element of σ to the smallest (resp. the largest) element of τ.

In §11.4, we have constructed the functor

(18.7.3)
$$\kappa \colon \mathbf{\Delta} \to \widetilde{\mathbf{\Delta}}$$
$$\tau \mapsto \{0\} \sqcup \tau \sqcup \{\infty\}$$

where $\{0\}$ is the smallest element of $\kappa(\tau)$ and $\{\infty\}$ the largest. The functor κ is left adjoint to the embedding functor $\iota \colon \widetilde{\mathbf{\Delta}} \to \mathbf{\Delta}$. We shall still denote by κ its restriction to $\mathbf{\Delta}_{inj}$.

We shall also encounter the situation where $u\colon A \to B$ admits a section $v\colon B \to A$ (i.e., $v \circ u = \mathrm{id}_B$). In such a case, the functor λ_u extends to the category $\widetilde{\Delta}$.

Lemma 18.7.2. *Assume that the morphism* $u\colon A \to B$ *admits a section* $v\colon B \to A$. *Then there exists a functor* $\lambda_{u,v}\colon \widetilde{\Delta}^{\mathrm{op}} \to \mathcal{C}_B^{\wedge}$ *such that* $\lambda_{u,v} \circ \kappa = \lambda_u$.

Proof. Let $\tau \in \widetilde{\Delta}$. Let us denote by 0_τ the smallest element of τ and by ∞_τ the largest. Also set

$$\xi(\tau) = \tau \setminus \{0_\tau, \infty_\tau\} .$$

We define

$$\lambda_{u,v}(\tau) = A^{\prod_B \xi(\tau)} .$$

Let $f\colon \tau \to \sigma$ be a morphism in $\widetilde{\Delta}$. In order to define $\lambda_{u,v}(f)\colon A^{\prod_B \xi(\sigma)} \to A^{\prod_B \xi(\tau)}$, it is enough to define for all $i \in \xi(\tau)$ its composition with the i-th projection $p_i\colon A^{\prod_B \xi(\tau)} \to A$. We set

$$p_i \circ \lambda_{u,v}(f) = \begin{cases} \text{the composition } A^{\prod_B \xi(\tau)} \to B \xrightarrow{v} A & \text{if } f(i) \notin \xi(\sigma) , \\ \text{the projection } p_{f(i)} & \text{if } f(i) \in \xi(\sigma) . \end{cases}$$

It is easily checked that $\lambda_{u,v}$ is a functor, and this functor extends λ_u.   q.e.d.

Let $F \in \mathrm{Mod}(k_X)$ and let $u\colon A \to B$ be a morphism in $\mathcal{C}_X^{\wedge}$. We have a functor $\mathcal{C}_X^{\wedge} \to \mathrm{Mod}(k_X)$ given by $C \mapsto F_C$ (see (17.6.4)). Hence we obtain a functor

(18.7.4) $$\lambda_u^F\colon \Delta_{inj}^{\mathrm{op}} \to \mathrm{Mod}(k_X)$$

as the composition $\Delta_{inj}^{\mathrm{op}} \xrightarrow{\lambda_u} \mathcal{C}_B^{\wedge} \to \mathcal{C}_X^{\wedge} \to \mathrm{Mod}(k_X)$. As in §11.4, we can construct a complex $F_\bullet^u$ in $\mathrm{C}^-(\mathrm{Mod}(k_X))$. Recall its construction. Set

$$A^{\prod_B n} = \underbrace{A \times_B \cdots \times_B A}_{n}, \text{ in particular } A^{\prod_B 0} = B .$$

For $n \geq -1$, set

$$F_n^u := \lambda_u^F([0,n]) = F_{A^{\prod_B n+1}} = F_{\underbrace{A \times_B \cdots \times_B A}_{n+1}} .$$

Denote by $p_i\colon A^{\prod_B n+1} \to A^{\prod_B n}$ the projection which forgets the i-th components $(1 \leq i \leq n+1)$. This projection induces a morphism $\delta_n^i\colon F_n^u \to F_{n-1}^u$, which coincides with $\lambda_u^F(d_{i-1}^n)$. We define

$$d_n^u \colon F_n^u \to F_{n-1}^u, \qquad d_n^u = \sum_{i=1}^{n+1} (-1)^{i-1} \delta_n^i \ .$$

By Proposition 11.4.2 (i), we have

$$d_{n-1}^u \circ d_n^u = 0 \quad \text{for } n > 0.$$

(Note that the notations here slightly differ from those in Proposition 11.4.2.)
    Hence, we have constructed a complex

$$(18.7.5) \qquad F_\bullet^u := \cdots \to F_n^u \xrightarrow{d_n^u} \cdots \xrightarrow{d_1^u} F_0^u \xrightarrow{d_0^u} F_{-1}^u \to 0 \ .$$

By adjunction, we also have a complex

$$(18.7.6) \qquad F_u^\bullet := 0 \to F_u^{-1} \xrightarrow{d_u^{-1}} F_u^0 \xrightarrow{d_u^0} \cdots \to F_u^n \xrightarrow{d_u^n} \cdots \ ,$$

with

$$F_u^n = \Gamma_{A \prod_B {}^{n+1}}(F) = \Gamma_{\underbrace{A \times_B \cdots \times_B A}_{n+1}}(F) \ .$$

Note that

$$F_\bullet^u \in C^-(\mathrm{Mod}(k_X)), \quad F_u^\bullet \in C^+(\mathrm{Mod}(k_X)) \ .$$

More intuitively, we may write

$$F_\bullet^u := \cdots \to F_{A \times_B A} \to F_A \to F_B \to 0 \ ,$$
$$F_u^\bullet := 0 \to \Gamma_B F \to \Gamma_A F \to \Gamma_{A \times_B A} F \to \cdots \ .$$

Proposition 18.7.3. *Assume that the morphism* $u \colon A \to B$ *has a section* v
(i.e., $u \circ v = \mathrm{id}_B$*). Then the complexes* $F_\bullet^u$ *and* $F_u^\bullet$ *are homotopic to* 0.

Proof. This follows immediately from Lemma 18.7.2 and Corollary 11.4.3.
q.e.d.

Theorem 18.7.4. *Let* $F \in \mathrm{Mod}(k_X)$. *Assume that the morphism* $u \colon A \to \mathrm{pt}_X$
is a local epimorphism. Then

(i) *the complexes* $\mathrm{j}_{u*} F_\bullet^u$ *and* $\mathrm{j}_{u*} F_u^\bullet$ *are homotopic to* 0 *in* $C(\mathrm{Mod}(k_A))$,
(ii) *the complexes* $F_\bullet^u$ *and* $F_u^\bullet$ *are exact.*

Roughly speaking, (i) means that the complexes $F_\bullet^u$ and $F_u^\bullet$ are locally homotopic to 0, "locally" meaning "after a base change by a local epimorphism".

Proof. (i) Let us treat the complex $j_{u*} F_\bullet^u$. Denote by w the second projection $A \times A \to A$. Then $w = u \times A$, and Proposition 17.6.14 implies

$$j_u^{-1} j_{u*} F_\bullet^u \simeq (F_\bullet^u)_A \simeq (F|_A)_\bullet^w .$$

Since w admits a section (namely, the diagonal morphism $A \to A \times A$), the complex $(F|_A)_\bullet^w$ is homotopic to 0. Since the composition $j_{u*} \to j_{u*} j_u^{-1} j_{u*} \to j_{u*}$ is the identity, the identity of $j_{u*} F_\bullet^u$ factorizes through $j_{u*}(F|_A)_\bullet^w$. It follows that $j_{u*} F_\bullet^u$ is homotopic to 0.

The proof for $j_{u*} F_\bullet^\bullet$ is similar.

(ii) follows from (i) since j_{u*} is exact and faithful by Lemma 17.6.9.   q.e.d.

The stupid truncated complex $\sigma^{\geq 0} F_u^\bullet$ is the complex obtained by replacing F_u^{-1} with 0 in $F_u^\bullet$:

(18.7.7) $$\sigma^{\geq 0} F_u^\bullet = 0 \to F_u^0 \xrightarrow{d_u^0} F_u^1 \xrightarrow{d_u^1} \cdots \to F_u^n \xrightarrow{d_u^n} \cdots .$$

Then we have a d.t.

$$\Gamma_B F \to \sigma^{\geq 0} F_u^\bullet \to F_u^\bullet \xrightarrow{+1} .$$

Hence Theorem 18.7.4 asserts that $\Gamma_B F \to \sigma^{\geq 0} F_u^\bullet$ is a qis if u is a local epimorphism.

Corollary 18.7.5. (Leray's acyclic covering theorem) *Let $u: A \to \mathrm{pt}_X$ be a local epimorphism and let $F \in \mathrm{Mod}(k_X)$. Assume that $H^j(\mathrm{R}\Gamma(A^n; F)) \simeq 0$ for all $j > 0$ and all $n > 0$. Then there is a natural isomorphism*

$$\mathrm{R}\Gamma(X; F) \simeq \Gamma(X; \sigma^{\geq 0} F_u^\bullet) .$$

Proof. Let us take an injective resolution $F \to I^\bullet$ in $\mathrm{Mod}(k_X)$. Theorem 18.7.4 implies that $I^n \to \sigma^{\geq 0}(I^n)_u^\bullet$ is a qis, and hence $I^\bullet \to \mathrm{tot}(\sigma^{\geq 0}(I^\bullet)_u^\bullet)$ is a qis by Theorem 12.5.4. Thus $F \to I^\bullet \to \mathrm{tot}(\sigma^{\geq 0}(I^\bullet)_u^\bullet)$ are qis's, and $\Gamma_{A^n} I^m$ are injective by Exercise 18.2. Hence $\mathrm{R}\Gamma(X; F)$ is represented by the complex $\Gamma(X; \mathrm{tot}(\sigma^{\geq 0}(I^\bullet)_u^\bullet))$.

On the other hand, there is an isomorphism (see Exercise 18.2):

(18.7.8) $$\mathrm{R}\Gamma(X; \mathrm{R}\Gamma_{A^n}(F)) \simeq \mathrm{R}\Gamma(A^n; F) ,$$

and $\mathrm{R}\Gamma(A^n; F)$ is represented by the complex $\Gamma(A^n; I^\bullet)$ for any $n > 0$. Hence the assumption implies that $\Gamma(A^n; F) \to \Gamma(A^n; I^\bullet)$ is a qis. Applying Theorem 12.5.4 again, it follows that $\Gamma(X; \sigma^{\geq 0} F_u^\bullet) \to \Gamma(X; \mathrm{tot}(\sigma^{\geq 0}(I^\bullet)_u^\bullet))$ is a qis.
                                                                              q.e.d.

Exercises

Exercise 18.1. Prove that Lemma 18.5.3 remains true when assuming $n = 1$ in the hypothesis.

Exercise 18.2. Let X be a site, let $A \in \mathcal{C}_X^\wedge$ and let $\mathcal{R}$ be a k_X-algebra.
(i) Let $F \in K_{hi}(\mathcal{R})$. Prove that $\Gamma_A(F) \in K_{hi}(\mathcal{R})$.
(ii) Prove the isomorphism $R\Gamma(X; R\Gamma_A(F)) \simeq R\Gamma(A; F)$ for $F \in D(\mathcal{R})$.

Exercise 18.3. Let X be a topological space. For a subset Z of X, we endow Z with the induced topology and we denote by $i_Z \colon Z \to X$ the embedding. In the sequel, we work in the category $\mathrm{Mod}(k_X)$ of k_X-modules on X. For $x \in X$ and $F \in \mathrm{Mod}(k_X)$, let us write F_x instead of $i_{\{x\}}^{-1} F$. Recall (see Remark 17.6.12) that we have identified $\mathrm{Mod}(k_{\{x\}})$ with $\mathrm{Mod}(k)$. Hence, $F_x \in \mathrm{Mod}(k)$. The k-module F_x is called the germ of F at x.

(i)   Prove that $F_x \simeq \varinjlim_U F(U)$, where U ranges over the ordered set of open neighborhoods of x.

(ii)   Prove that a complex $F' \to F \to F''$ in $\mathrm{Mod}(k_X)$ is exact if and only if the sequence $F_x' \to F_x \to F_x''$ is exact in $\mathrm{Mod}(k)$ for any $x \in X$.

(iii)   Let Z be a subset of X. Prove that the functor $i_Z^t \colon \mathrm{Op}_X \to \mathrm{Op}_Z$, $V \mapsto V \cap Z$, defines a left exact morphism of sites $i_Z \colon Z \to X$.

(iv)   Let U be an open subset of X. Prove that the composition of morphisms of sites $U \xrightarrow{i_U} X \xrightarrow{j_{U \to X}} U$ is isomorphic to the identity and that $i_U^{-1} \simeq j_{U \to X *}$, $i_{U *} \simeq j_{U \to X}^\ddagger$. Set $i_{U!} := j_{U \to X}^{-1}$. Prove that $i_U^{-1} i_{U!} \simeq \mathrm{id}_{\mathrm{Mod}(k_U)}$ and $i_U^{-1} i_{U *} \simeq \mathrm{id}_{\mathrm{Mod}(k_U)}$.

(v)   Let S be a closed subset of X and let $U = X \setminus S$. Prove that the functors i_S^{-1} and $i_{S *}$ are exact. Define the functor $(\bullet)_S \colon \mathrm{Mod}(k_X) \to \mathrm{Mod}(k_X)$ by $(\bullet)_S = i_{S *} \circ i_S^{-1}$. Prove that there is an exact sequence $0 \to F_U \to F \to F_S \to 0$ for $F \in \mathrm{Mod}(k_X)$.

(vi)   Let S and U be as in (v). Prove that the functor $(\bullet)_S$ admits a left adjoint. Denoting it by Γ_S, prove that $\Gamma_S(\bullet) \simeq \mathcal{H}om_{k_X}((k_X)_S, \bullet)$ and prove that there is an exact sequence $0 \to \Gamma_S(F) \to F \to \Gamma_U(F)$ for $F \in \mathrm{Mod}(k_X)$. Prove moreover that $F \to \Gamma_U(F)$ is an epimorphism when F is injective.

(vii)   Let S and U be as in (v). Prove that the functor $\mathrm{Mod}(k_X) \to \mathrm{Mod}(k_U) \times \mathrm{Mod}(k_S)$ given by $F \mapsto (i_U^{-1} F, i_S^{-1} F)$ is exact, faithful and conservative.

(viii)   Let S and U be as in (v). Prove that the triangulated functor $D(k_X) \to D(k_U) \times D(k_S)$ given by $F \mapsto (i_U^{-1} F, i_S^{-1} F)$ is conservative but is not faithful in general.

(ix)   Let S and U be as in (v). Prove that $D(k_X)$ is equivalent to the category of triples (F, G, u) where $F \in D(k_S)$, $G \in D(k_U)$ and $u \colon Ri_{S *} F \to Ri_{U!} G$ is a morphism in $D(k_X)$. (Hint: use Exercise 10.15.)

Exercise 18.4. Let X be a Hausdorff compact space and let $\{F_i\}_{i \in I}$ be an inductive system in $\mathrm{Mod}(\mathbb{Z}_X)$ indexed by a small filtrant category I. Prove the isomorphism $\varinjlim_i \Gamma(X; F_i) \xrightarrow{\sim} \Gamma(X; \varinjlim_i F_i)$.

Exercise 18.5. Let X be the topological space $\mathbb{R}$.
(i) Prove that small products are not exact in $\mathrm{Mod}(\mathbb{Z}_X)$.
(ii) Prove that $\mathrm{Mod}(\mathbb{Z}_X)$ satisfies the IPC-property. (Hint: use Exercise 18.4.)

Stacks and Twisted Sheaves

Roughly speaking, a stack is a sheaf of categories. However, when replacing a set with a category, we have to replace the equalities with isomorphisms. This creates new difficulties, since these isomorphisms are not unique and it is necessary to control compatibility conditions among them.

Here, we define prestacks on a presite X and various associated notions. In particular, for a prestack $\mathfrak{S}$ and $A \in \mathcal{C}_X^{\wedge}$, we define $\mathfrak{S}(A)$ as a projective limit in the 2-category **Cat**. In the course of this study, we need a higher dimensional analogue of the notions of connected categories and cofinal functors, and we introduce 1-connected categories and 1-cofinal functors.

A prestack $\mathfrak{S}$ on a site X is a stack if for any local isomorphism $A \to U$ in $\mathcal{C}_X^{\wedge}$, $\mathfrak{S}(A) \to \mathfrak{S}(U)$ is an equivalence of categories. We give equivalent definitions and prove in particular that the prestack $\mathcal{C}_X \ni U \mapsto \mathrm{Sh}(U, \mathcal{A})$ of sheaves with values in a category $\mathcal{A}$ (satisfying hypothesis (17.4.1)) is a stack.

As an application of the general theory of stacks, we study twisted sheaves. We start by proving a "Morita equivalence" in the framework of stacks. It asserts that for two sheaves of algebras $\mathcal{R}_1$ and $\mathcal{R}_2$, any equivalence of stacks $\mathfrak{Mod}(\mathcal{R}_1) \simeq \mathfrak{Mod}(\mathcal{R}_2)$ is associated to a suitable module over $\mathcal{R}_1 \otimes \mathcal{R}_2^{\mathrm{op}}$. On a site X, a twisted sheaf is an object of a stack locally equivalent to the stack of sheaves on X.

The theory presented here is rather sketchy, and may be thought of as a first introduction to a vast subject which certainly should deserve further developments.

References are made to [26, 34, 63], [10, 12, 36].

19.1 Prestacks

A general definition of a prestack would be a contravariant 2-functor $\mathfrak{S}$ from a category to a 2-category. However, we shall not develop here such a general theory, and restrict ourselves to 2-functors with values in the 2-category **Cat**. As already noticed, the notion of isomorphism of categories is not natural and has to be weakened to that of equivalence of categories. In other words, the notion of equality of functors has to be replaced by that of isomorphism. For example, for two morphisms $U_1 \xrightarrow{u} U_2 \xrightarrow{v} U_3$ in $\mathcal{C}$, we have to replace the equality of functors $\mathfrak{S}(v) \circ \mathfrak{S}(u) = \mathfrak{S}(v \circ u)$ in **Cat** with the data of an isomorphism $\mathfrak{S}(v) \circ \mathfrak{S}(u) \xrightarrow{\sim} \mathfrak{S}(v \circ u)$. Moreover. we have to control the compatibility of such isomorphisms when considering three morphisms in $\mathcal{C}$. More precisely:

Definition 19.1.1. *Let X be a presite. A* prestack $\mathfrak{S}$ *on X consists of the following data:*

(a) *for any $U \in \mathcal{C}_X$, a category $\mathfrak{S}(U)$,*
(b) *for any morphism $u \colon U_1 \to U_2$ in $\mathcal{C}_X$, a functor $r_u \colon \mathfrak{S}(U_2) \to \mathfrak{S}(U_1)$, called the* restriction functor,
(c) *for any morphisms $u \colon U_1 \to U_2$ and $v \colon U_2 \to U_3$ in $\mathcal{C}_X$, an isomorphism of functors $c_{u,v} \colon r_u \circ r_v \xrightarrow{\sim} r_{v \circ u}$, called the* composition isomorphism,

these data satisfying:

(i) $r_{\mathrm{id}_U} = \mathrm{id}_{\mathfrak{S}(U)}$ *and* $c_{\mathrm{id}_U,\mathrm{id}_U} = \mathrm{id}_{\mathrm{id}_{\mathfrak{S}(U)}}$ *for any $U \in \mathcal{C}_X$,*
(ii) *for any $U_1 \xrightarrow{u} U_2 \xrightarrow{v} U_3 \xrightarrow{w} U_4$, the following diagram of functors commutes:*

$$(19.1.1) \qquad \begin{array}{ccc} r_u \circ r_v \circ r_w & \xrightarrow{\ c_{v,w}\ } & r_u \circ r_{w \circ v} \\ {\scriptstyle c_{u,v}}\downarrow & & \downarrow{\scriptstyle c_{u,v \circ w}} \\ r_{v \circ u} \circ r_w & \xrightarrow{\ c_{v \circ u,w}\ } & r_{w \circ v \circ u} . \end{array}$$

Note that $c_{u,\mathrm{id}_{U_2}} = \mathrm{id}_{r_u}$ by setting $U_3 = U_4 = U_2$ and $v = w = \mathrm{id}_{U_2}$ in (19.1.1). Similarly, $c_{\mathrm{id}_{U_1},u} = \mathrm{id}_{r_u}$.

An *additive prestack* is a prestack such that every $\mathfrak{S}(U)$ is an additive category and r_u is an additive functor. An additive prestack is called an *abelian prestack* if every $\mathfrak{S}(U)$ is an abelian category and every r_u is an *exact* functor. For a commutative ring k, we define in an obvious way the notions of k-additive and k-abelian prestacks.

If $\mathfrak{S}$ is a prestack, the *opposite prestack* $\mathfrak{S}^{\mathrm{op}}$ is given by $\mathfrak{S}^{\mathrm{op}}(U) = (\mathfrak{S}(U))^{\mathrm{op}}$ with the natural restriction functors and the natural composition isomorphisms of such functors.

Examples 19.1.2. (i) Let X be a small presite and let $\mathcal{A}$ be a category. Then $\mathcal{C}_X \ni U \mapsto \mathrm{PSh}(U, \mathcal{A})$ together with the natural restriction functors $\mathrm{PSh}(U_2, \mathcal{A}) \to \mathrm{PSh}(U_1, \mathcal{A})$ is a prestack. If X is endowed with a topology and $\mathcal{A}$ satisfies the hypothesis (17.4.1), then $\mathcal{C}_X \ni U \mapsto \mathrm{Sh}(U, \mathcal{A})$ is also a prestack. If $\mathcal{A}$ is abelian, these prestacks are abelian.
(ii) Let $\mathcal{G}$ be a sheaf of groups on a site X. To $\mathcal{G}$ we associate a prestack $\mathcal{S}$ on X by setting for $U \in \mathcal{C}_X$, $\mathrm{Ob}(\mathcal{S}(U)) = \{\mathrm{pt}\}$ and $\mathrm{Hom}_{\mathcal{S}(U)}(\{\mathrm{pt}\}, \{\mathrm{pt}\}) = \mathcal{G}(U)$.

Definition 19.1.3. *Let $\mathfrak{S}$ be a prestack on a presite X and let $f \colon X \to Y$ be a morphism of presites, that is, a functor $f^t \colon \mathcal{C}_Y \to \mathcal{C}_X$. We define $f_* \mathfrak{S}$, the* direct image *of $\mathfrak{S}$, as the prestack on Y given as follows. For $V \in \mathcal{C}_Y$, set $(f_* \mathfrak{S})(V) = \mathfrak{S}(f^t(V))$, for $u \colon V_1 \to V_2$, set $r_u = r_{f^t(u)} \colon \mathfrak{S}(f^t(V_2)) \to \mathfrak{S}(f^t(V_1))$ and for $u \colon V_1 \to V_2$, $v \colon V_2 \to V_3$, set $c_{u,v} = c_{f^t(u), f^t(v)} \colon r_{f^t(u)} \circ r_{f^t(v)} \xrightarrow{\sim} r_{f^t(v \circ u)}$.*

For $A \in \mathcal{C}_X^\wedge$, we write $\mathfrak{S}|_A$ instead of $\mathrm{j}_{A \to X *} \mathfrak{S}$.

Hence, $\mathfrak{S}|_A$ is the prestack which associates $\mathfrak{S}(U)$ with $(U \to A) \in \mathcal{C}_A$.

Definition 19.1.4. *Let $\mathfrak{S}_\nu$ ($\nu = 1, 2$) be prestacks on X with the restriction functors r_u^ν and the composition isomorphisms $c_{u,v}^\nu$. A* functor of prestacks *$\Phi \colon \mathfrak{S}_1 \to \mathfrak{S}_2$ is the data of:*

 (i) *for any $U \in \mathcal{C}_X$, a functor $\Phi(U) \colon \mathfrak{S}_1(U) \to \mathfrak{S}_2(U)$,*
 (ii) *for any morphism $u \colon U_1 \to U_2$, an isomorphism Φ_u of functors from $\mathfrak{S}_1(U_2)$ to $\mathfrak{S}_2(U_1)$*

$$\Phi_u \colon \Phi(U_1) \circ r_u^1 \xrightarrow{\sim} r_u^2 \circ \Phi(U_2) \,,$$

these data satisfying: for any morphisms $U_1 \xrightarrow{u} U_2 \xrightarrow{v} U_3$ the following diagram commutes in $\mathrm{Fct}(\mathfrak{S}_1(U_3), \mathfrak{S}_2(U_1))$

(19.1.2)
$$
\begin{array}{ccc}
\Phi(U_1) \circ r_u^1 \circ r_v^1 \xrightarrow{\Phi_u} r_u^2 \circ \Phi(U_2) \circ r_v^1 \xrightarrow{\Phi_v} r_u^2 \circ r_v^2 \circ \Phi(U_3) \\
{\scriptstyle c_{u,v}^1} \downarrow \qquad\qquad\qquad\qquad\qquad\qquad\qquad\qquad \downarrow {\scriptstyle c_{u,v}^2} \\
\Phi(U_1) \circ r_{v \circ u}^1 \xrightarrow{\qquad\qquad \Phi_{v \circ u} \qquad\qquad} r_{v \circ u}^2 \circ \Phi(U_3) \,.
\end{array}
$$

Note that for any $U \in \mathcal{C}_X$, $\Phi_{\mathrm{id}_U} = \mathrm{id}_{\Phi(U)}$ (set $U_1 = U_2 = U_3 = U$ and $u = v = \mathrm{id}_U$ in (19.1.2)).

Definition 19.1.5. *Let $\Phi_\nu \colon \mathfrak{S}_1 \to \mathfrak{S}_2$ ($\nu = 1, 2$) be two functors of prestacks on X. A* morphism of functors of prestacks *$\theta \colon \Phi_1 \to \Phi_2$ is the data for any $U \in \mathcal{C}_X$ of a morphism of functors $\theta(U) \colon \Phi_1(U) \to \Phi_2(U)$ such that for any morphism $u \colon U_1 \to U_2$ in $\mathcal{C}$, the following diagram commutes*

(19.1.3)
$$
\begin{array}{ccc}
\Phi_1(U_1) \circ r_u^1 & \xrightarrow{\theta(U_1)} & \Phi_2(U_1) \circ r_u^1 \\
{\scriptstyle \Phi_u^1} \downarrow & & \downarrow {\scriptstyle \Phi_u^2} \\
r_u^2 \circ \Phi_1(U_2) & \xrightarrow{\theta(U_2)} & r_u^2 \circ \Phi_2(U_2) \,.
\end{array}
$$

The set of functors and morphisms of functors forms a category $\mathrm{Fct}(\mathfrak{S}_1, \mathfrak{S}_2)$ on X, and $U \mapsto \mathrm{Fct}(\mathfrak{S}_1|_U, \mathfrak{S}_2|_U)$ is a prestack on X that we denote by $\mathcal{F}ct(\mathfrak{S}_1, \mathfrak{S}_2)$.

We denote by $\mathrm{End}\,(\mathrm{id}_{\mathfrak{S}})$ the set of endomorphisms of the identity functor $\mathrm{id}_{\mathfrak{S}} : \mathfrak{S} \to \mathfrak{S}$, that is,

$$\mathrm{End}\,(\mathrm{id}_{\mathfrak{S}}) = \mathrm{Hom}_{\,\mathrm{Fct}(\mathfrak{S},\mathfrak{S})}(\mathrm{id}_{\mathfrak{S}}, \mathrm{id}_{\mathfrak{S}})\,.$$

We denote by $\mathcal{E}nd(\mathrm{id}_{\mathfrak{S}})$ the presheaf on X given by $\mathcal{C}_X \ni U \mapsto \mathrm{End}\,(\mathrm{id}_{\mathfrak{S}\,|_U})$.

We denote by $\mathrm{Aut}\,(\mathrm{id}_{\mathfrak{S}})$ the subset of $\mathrm{End}\,(\mathrm{id}_{\mathfrak{S}})$ consisting of isomorphisms from $\mathrm{id}_{\mathfrak{S}}$ to $\mathrm{id}_{\mathfrak{S}}$ and by $\mathcal{A}ut(\mathrm{id}_{\mathfrak{S}})$ the presheaf on X given by $\mathcal{C}_X \ni U \mapsto \mathrm{Aut}\,(\mathrm{id}_{\mathfrak{S}\,|_U})$.

A functor of prestacks $\Phi \colon \mathfrak{S}_1 \to \mathfrak{S}_2$ is called an *equivalence of prestacks* if there exists a functor $\Psi \colon \mathfrak{S}_2 \to \mathfrak{S}_1$ such that $\Psi \circ \Phi \simeq \mathrm{id}_{\mathfrak{S}_1}$ and $\Phi \circ \Psi \simeq \mathrm{id}_{\mathfrak{S}_2}$.

It is easily checked that a functor of prestacks $\Phi \colon \mathfrak{S}_1 \to \mathfrak{S}_2$ is an equivalence if and only if $\Phi(U) \colon \mathfrak{S}_1(U) \to \mathfrak{S}_2(U)$ is an equivalence of categories for all $U \in \mathcal{C}_X$.

Definition 19.1.6. *Let $\mathfrak{S}$ be a prestack on X. We denote by $\varprojlim_{U \in \mathcal{C}_X} \mathfrak{S}(U)$ the category defined as follows.*

(a) *An object F of $\varprojlim_{U \in \mathcal{C}_X} \mathfrak{S}(U)$ is a family $\{(F_U, \varphi_u)\}_{U \in \mathcal{C}_X, u \in \mathrm{Mor}(\mathcal{C}_X)}$ where:*

   (i) *for any $U \in \mathcal{C}_X$, F_U is an object of $\mathfrak{S}(U)$,*

   (ii) *for any morphism $u \colon U_1 \to U_2$ in $\mathcal{C}_X$, $\varphi_u \colon r_u F_{U_2} \xrightarrow{\sim} F_{U_1}$ is an isomorphism such that for any sequence $U_1 \xrightarrow{u} U_2 \xrightarrow{v} U_3$ of morphisms in $\mathcal{C}_X$, the following diagram commutes (this is a so-called cocycle condition):*

(19.1.4)
$$
\begin{array}{ccc}
r_u r_v F_{U_3} & \xrightarrow{\ r_u(\varphi_v)\ } & r_u F_{U_2} \\
{\scriptstyle c_{u,v}}\big\downarrow & & \big\downarrow{\scriptstyle \varphi_u} \\
r_{v \circ u} F_{U_3} & \xrightarrow{\ \varphi_{v \circ u}\ } & F_{U_1}\,.
\end{array}
$$

(*Note that $\varphi_{\mathrm{id}_U} = \mathrm{id}_{F_U}$ for any $U \in \mathcal{C}_X$. Indeed, set $U_1 = U_2 = U_3 = U$ and $u = v = \mathrm{id}_U$ in (19.1.4).*)

(b) *For two objects $F = \{(F_U, \varphi_u)\}$ and $F' = \{(F'_U, \varphi'_u)\}$ in $\varprojlim_{U \in \mathcal{C}_X} \mathfrak{S}(U)$, $\mathrm{Hom}_{\varprojlim_{U \in \mathcal{C}_X} \mathfrak{S}(U)}(F, F')$ is the set of families $f = \{f_U\}_{U \in \mathcal{C}_X}$ such that $f_U \in \mathrm{Hom}_{\mathfrak{S}(U)}(F_U, F'_U)$ and the following diagram commutes for any $u \colon U_1 \to U_2$*

$$
\begin{array}{ccc}
r_u F_{U_2} & \xrightarrow{\ \varphi_u\ } & F_{U_1} \\
{\scriptstyle r_u(f_{U_2})}\big\downarrow & & \big\downarrow{\scriptstyle f_{U_1}} \\
r_u F'_{U_2} & \xrightarrow{\ \varphi'_u\ } & F'_{U_1}\,.
\end{array}
$$

Therefore,

$$\mathrm{Hom}_{\varprojlim_{U \in \mathcal{C}_X} \mathfrak{S}(U)}(F, F') \simeq \varprojlim_{U \in \mathcal{C}_X} \mathrm{Hom}_{\mathfrak{S}|_U}(F_U, F'_U) \,.$$

For any $A \in \mathcal{C}_X^\wedge$, we set

$$\mathfrak{S}(A) = \varprojlim_{(U \to A) \in \mathcal{C}_A} (\mathfrak{S}|_A)(U) = \varprojlim_{(U \to A) \in \mathcal{C}_A} \mathfrak{S}(U) \,.$$

Hence, $\varprojlim_{U \in \mathcal{C}_X} \mathfrak{S}(U) = \mathfrak{S}(\mathrm{pt}_X)$, where pt_X denotes as usual the terminal object of $\mathcal{C}_X^\wedge$. Similarly as in Notations 17.6.13 (iii), we set

$$(19.1.5) \qquad\qquad \mathfrak{S}(X) := \mathfrak{S}(\mathrm{pt}_X) = \varprojlim_{U \in \mathcal{C}_X} \mathfrak{S}(U).$$

A morphism $v \colon A \to A'$ in $\mathcal{C}_X^\wedge$ defines a functor

$$r_v \colon \mathfrak{S}(A') = \varprojlim_{U \to A'} \mathfrak{S}(U) \to \varprojlim_{U \to A} \mathfrak{S}(U) = \mathfrak{S}(A)$$

and it is easily checked that the conditions in Definition 19.1.1 are satisfied. Therefore

Proposition 19.1.7. *Let $\mathfrak{S}$ be a prestack on the small presite X. Then $\mathfrak{S}$ extends naturally to a prestack on $\widehat{X}$.*

Note that, for a small family of objects $\{A_i\}_{i \in I}$ of $\mathcal{C}_X^\wedge$, we have

$$(19.1.6) \qquad\qquad \mathfrak{S}\left(\text{``}\bigsqcup_i\text{''} A_i\right) \simeq \prod_i \mathfrak{S}(A_i) \,.$$

For $F \in \mathfrak{S}(X)$, we denote by $F|_U$ its image in $\mathfrak{S}(U)$ by the morphism associated with the unique morphism $U \to \mathrm{pt}_X$. For $u \colon A \to A'$, we sometimes use the notation

$$(19.1.7) \qquad\qquad u^* := r_u \colon \mathfrak{S}(A') \to \mathfrak{S}(A),$$

where r_u is the restriction functor of the stack $\mathfrak{S}$ on $\widehat{X}$.

Definition 19.1.8. *For $F_1, F_2 \in \mathfrak{S}(X)$, the presheaf of sets that associates $\mathrm{Hom}_{\mathfrak{S}(U)}(F_1|_U, F_2|_U)$ with $U \in \mathcal{C}_X$ is denoted by $\mathcal{H}om\,_{\mathfrak{S}}(F_1, F_2)$.*

Note that we have

$$(19.1.8) \qquad\qquad \mathcal{H}om\,_{\mathfrak{S}}(F_1, F_2)(A) \simeq \mathrm{Hom}_{\mathfrak{S}(A)}(F_1|_A, F_2|_A),$$

since $A \simeq \varinjlim_{(U \to A) \in \mathcal{C}_A} \text{``} U$ and the both-hand-sides of (19.1.8) are isomorphic to $\varprojlim_{(U \to A) \in \mathcal{C}_A} \mathrm{Hom}_{\mathfrak{S}(U)}(F_1|_U, F_2|_U)$.

19.2 Simply Connected Categories

The notions of connected category and cofinal functor were sufficient to treat inductive or projective limits in a category. However, when working with stacks, that is, essentially with 2-categories, we need a higher dimensional analogue of these notions.

Definition 19.2.1. *Let I be a small category. We say that I is simply connected if it satisfies:*

(i) *I is non empty,*

(ii) *for any category C and any functor $\alpha\colon I \to C$ such that $\alpha(u)$ is an isomorphism for any $u \in \mathrm{Mor}(I)$, $\varinjlim \alpha$ exists in C and $\alpha(i) \to \varinjlim \alpha$ is an isomorphism for any $i \in I$.*

We also say "1-connected" instead of "simply connected".

We remark the following facts whose proofs are similar to those of Proposition 2.5.2 and left to the reader:

- a small category I is 1-connected if and only if I^{op} is 1-connected,
- a 1-connected category I is connected by Corollary 2.4.5,
- condition (ii) in Definition 19.2.1 is equivalent to the similar condition with "any functor $\alpha\colon I \to C$" replaced by "any functor $\alpha\colon I \to \mathbf{Set}$" or else "any functor $\alpha\colon I \to \mathbf{Set}^{\mathrm{op}}$ ".

Similarly as in Definition 2.5.1, we set:

Definition 19.2.2.   (i) *We say that a functor $\varphi\colon J \to I$ is 1-cofinal if the category J^i is 1-connected for any $i \in I$.*

(ii) *We say that a functor $\varphi\colon J \to I$ is co-1-cofinal if the functor $\varphi^{\mathrm{op}}\colon J^{\mathrm{op}} \to I^{\mathrm{op}}$ is 1-cofinal, or equivalently, if the category J_i is 1-connected for any $i \in I$.*

We shall not develop here a systematic study of 1-connected categories and 1-cofinal functors, but only give the following result which will be used later.

Proposition 19.2.3. *Let $f\colon X \to Y$ be a 1-cofinal morphism of presites (i.e., $f^t\colon C_Y \to C_X$ is 1-cofinal). Then, for any prestack $\mathfrak{S}$ on X, the canonical functor $\Phi\colon \mathfrak{S}(X) \to (f_*\mathfrak{S})(Y)$ is an equivalence of categories.*

Proof. We shall construct a quasi-inverse to the functor Φ. Recall that an object of $(f_*\mathfrak{S})(Y)$ is a family $G = \{(G_V, \varphi_v)\}_{V\in C_Y, v\in\mathrm{Mor}(C_Y)}$ with $G_V \in (f_*\mathfrak{S})(V) = \mathfrak{S}(f^t(V))$ and for $v\colon V' \to V$, $\varphi_v\colon r_{f^t(v)}G_V \xrightarrow{\sim} G_{V'}$, such that the diagram (19.1.4) (with suitable modifications) commutes. Let us define $\Psi(G) = F := \{(F_U, \psi_u)\}_{U\in C_X, u\in\mathrm{Mor}(C_X)}$ as follows.

For $U \in C_X$, the category $(C_Y)^U$ is simply connected by the hypothesis. For $(V, u) = (U \xrightarrow{u} f^t(V)) \in (C_Y)^U$, we have the functor $r_u\colon \mathfrak{S}(f^t(V)) \to$

$\mathfrak{S}(U)$. Let $\beta \colon (\mathcal{C}_Y)^U \to \mathfrak{S}(U)$ be the contravariant functor which associates $r_u G_V \in \mathfrak{S}(U)$ to $(V, u) \in (\mathcal{C}_Y)^U$. Then, for any morphism $v \colon (V, u) \to (V', u')$ in $(\mathcal{C}_Y)^U$, $\beta(v) \colon \beta((V', u')) \to \beta((V, u))$ is an isomorphism. Indeed,

$$r_{u'} G_{V'} \simeq r_u \circ r_{f^t(v)} G_V \ .$$

Hence $F_U := \varprojlim \beta$ exists in $\mathfrak{S}(U)$. For a morphism $u \colon U \to U'$ in $\mathcal{C}_X$, we construct similarly a morphism $\psi_u \colon r_u(F_{U'}) \to F_U$ and $F := \{(F_U, \psi_u)\}$ defines an object $\Psi(G) \in \mathfrak{S}(X)$. It is easy to check that the functors Φ and Ψ are quasi-inverse to each other.   q.e.d.

19.3 Simplicial Constructions

We follow the notations introduced in §11.4 (see also § 18.7). For $0 \leq n \leq m$, we denote by $\boldsymbol{\Delta}_{inj}^{[n,m]}$ the full subcategory of $\boldsymbol{\Delta}_{inj}$ consisting of objects τ with $n \leq \mathrm{card}(\tau) \leq m$.

The category $\boldsymbol{\Delta}_{inj}^{[1,3]}$ is equivalent to the category with three objects $\{1\}$, $\{1, 2\}$, $\{1, 2, 3\}$ and morphisms other than identities visualized by the diagram

(19.3.1)
$$\{1\} \xrightarrow[\;\;p_2\;\;]{\;\;p_1\;\;} \{1,2\} \xrightarrow[\substack{p_{13} \\ p_{23}}]{p_{12}} \{1,2,3\}$$

where p_i is the map which sends 1 to i $(i = 1, 2)$ and p_{ij} is the map which sends $(1, 2)$ to (i, j) $(1 \leq i < j \leq 3)$. Hence, we have the relations:

(19.3.2)
$$\begin{cases} p_{12} \circ p_2 = p_{23} \circ p_1 \ , \\ p_{23} \circ p_2 = p_{13} \circ p_2 \ , \\ p_{12} \circ p_1 = p_{13} \circ p_1 \ . \end{cases}$$

We shall also make use of the category $\boldsymbol{\Delta}_{inj}^{[0,3]}$, visualized by

$$\emptyset \xrightarrow{\;p\;} \{1\} \xrightarrow[\;\;p_2\;\;]{\;\;p_1\;\;} \{1,2\} \xrightarrow[\substack{p_{13} \\ p_{23}}]{p_{12}} \{1,2,3\} \ .$$

with the same relations (19.3.2) together with the new relation:

(19.3.3)
$$p_1 \circ p = p_2 \circ p.$$

Convention 19.3.1. In the sequel, we shall employ the same notations $\boldsymbol{\Delta}$, $\boldsymbol{\Delta}_{inj}^{[n,m]}$ etc, for the associated presites.

A prestack $\mathfrak{S}$ on $(\boldsymbol{\Delta}_{inj}^{[1,3]})^{\mathrm{op}}$ is thus the data of:

- categories $\mathcal{C}_0, \mathcal{C}_1, \mathcal{C}_2$,
- functors $r_1, r_2 \colon \mathcal{C}_0 \to \mathcal{C}_1$, and $r_{12}, r_{13}, r_{23} \colon \mathcal{C}_1 \to \mathcal{C}_2$,

- isomorphisms of functors

$$(19.3.4) \qquad \begin{cases} r_{12} \circ r_2 \simeq r_{23} \circ r_1 \,, \\ r_{23} \circ r_2 \simeq r_{13} \circ r_2 \,, \\ r_{12} \circ r_1 \simeq r_{13} \circ r_1 \,. \end{cases}$$

It is sometimes visualized by a diagram of categories

$$(19.3.5) \qquad \mathcal{C}_0 \underset{r_2}{\overset{r_1}{\rightrightarrows}} \mathcal{C}_1 \underset{r_{23}}{\overset{r_{12}}{\underset{r_{13}}{\rightrightarrows}}} \mathcal{C}_2 \,.$$

A prestack $\mathfrak{S}$ on $(\mathbf{\Delta}_{inj}^{[0,3]})^{\mathrm{op}}$ is a prestack on $(\mathbf{\Delta}_{inj}^{[1,3]})^{\mathrm{op}}$ together with a category $\mathcal{C}$, a functor $r \colon \mathcal{C} \to \mathcal{C}_0$ and an isomorphism of functors

$$(19.3.6) \qquad u \colon r_1 \circ r \xrightarrow{\sim} r_2 \circ r$$

such that the diagram of functors below (corresponding to diagram 19.1.1) commutes:

$$(19.3.7) \qquad \begin{array}{ccc} r_{12}r_1r & \xrightarrow{\ r_{12}u\ } r_{12}r_2r \xrightarrow{\ \ \sim\ \ } r_{23}r_1r \\ \Big\downarrow{\scriptstyle\sim} & \Big\downarrow{\scriptstyle r_{23}u} \\ r_{13}r_1r & \xrightarrow{\ r_{13}u\ } r_{13}r_2r \xrightarrow{\ \ \sim\ \ } r_{23}r_2r. \end{array}$$

It is sometimes visualized by a diagram of categories

$$(19.3.8) \qquad \mathcal{C} \xrightarrow{\ r\ } \mathcal{C}_0 \underset{r_2}{\overset{r_1}{\rightrightarrows}} \mathcal{C}_1 \underset{r_{23}}{\overset{r_{12}}{\underset{r_{13}}{\rightrightarrows}}} \mathcal{C}_2 \,.$$

Notation 19.3.2. Consider a prestack $\mathfrak{S}$ on $(\mathbf{\Delta}_{inj}^{[1,3]})^{\mathrm{op}}$. With the notations of Diagram 19.3.5, we set

$$(19.3.9) \qquad \mathrm{Kern}(\mathcal{C}_0, \mathcal{C}_1, \mathcal{C}_2) := \mathfrak{S}((\mathbf{\Delta}_{inj}^{[1,3]})^{\mathrm{op}}) \,.$$

To be more precise, we may write

$$\mathrm{Kern}\left(\mathcal{C}_0 \underset{r_2}{\overset{r_1}{\rightrightarrows}} \mathcal{C}_1 \underset{r_{23}}{\overset{r_{12}}{\underset{r_{13}}{\rightrightarrows}}} \mathcal{C}_2 \right) \quad \text{instead of } \mathrm{Kern}(\mathcal{C}_0, \mathcal{C}_1, \mathcal{C}_2) \,.$$

By Definition 19.1.6, we get

(i) An object of $\mathrm{Kern}(\mathcal{C}_0, \mathcal{C}_1, \mathcal{C}_2)$ is a pair (F, u) of $F \in \mathcal{C}_0$ and an isomorphism $u \colon r_1 F \xrightarrow{\sim} r_2 F$ such that the diagram below commutes:

$$(19.3.10) \qquad \begin{array}{ccc} r_{12}r_1F & \xrightarrow{\ r_{12}u\ } r_{12}r_2F \xrightarrow{\ \ \sim\ \ } r_{23}r_1F \\ \Big\downarrow{\scriptstyle\sim} & \Big\downarrow{\scriptstyle r_{23}u} \\ r_{13}r_1F & \xrightarrow{\ r_{13}u\ } r_{13}r_2F \xrightarrow{\ \ \sim\ \ } r_{23}r_2F. \end{array}$$

(ii) A morphism $(F, u) \to (G, v)$ in $\mathrm{Kern}(\mathcal{C}_0, \mathcal{C}_1, \mathcal{C}_2)$ is a morphism $\varphi\colon F \to G$ such that the diagram below commutes:

$$
\begin{array}{ccc}
r_1 F & \xrightarrow{\;u\;} & r_2 F \\
{\scriptstyle r_1(\varphi)}\big\downarrow & & \big\downarrow{\scriptstyle r_2(\varphi)} \\
r_1 G & \xrightarrow{\;v\;} & r_2 G.
\end{array}
$$

By its construction, there exists a faithful functor $r\colon \mathrm{Kern}(\mathcal{C}_0, \mathcal{C}_1, \mathcal{C}_2) \to \mathcal{C}_0$ and an isomorphism of functors $r_1 \circ r \simeq r_2 \circ r$ such that (19.3.7) is satisfied. If $\mathfrak{S}$ is a prestack on $(\mathbf{\Delta}_{inj}^{[0,3]})^{\mathrm{op}}$ as in (19.3.8), we have a functor $\mathcal{C} \to \mathrm{Kern}(\mathcal{C}_0, \mathcal{C}_1, \mathcal{C}_2)$.

Definition 19.3.3. *Consider a prestack $\mathfrak{S}$ on $(\mathbf{\Delta}_{inj}^{[0,3]})^{\mathrm{op}}$. If the functor $\mathcal{C} \to \mathrm{Kern}(\mathcal{C}_0, \mathcal{C}_1, \mathcal{C}_2)$ is an equivalence, then we say that (19.3.8) is an exact sequence of categories.*

Now consider a morphism $u\colon A \to B$ in $\mathcal{C}_X^{\wedge}$. We follow Notation 18.7.1. Let $\mathfrak{S}$ be a prestack on $\mathcal{C}_X^{\wedge}$. We denote by $\mathfrak{S}_u$ the direct image of $\mathfrak{S}$ by the functor of presites $\lambda_u\colon (\mathbf{\Delta}_{inj})^{\mathrm{op}} \to \mathcal{C}_B^{\wedge}$. Hence,

$$
\mathfrak{S}_u := \lambda_{u*}(\mathfrak{S} \,|_B)
$$

is a prestack on $\mathbf{\Delta}_{inj}^{\mathrm{op}}$. We denote by $\mathfrak{S}_u^{[n,m]}$ the direct image of $\mathfrak{S}_u$ by the inclusion functor $(\mathbf{\Delta}_{inj}^{[n,m]})^{\mathrm{op}} \to (\mathbf{\Delta}_{inj})^{\mathrm{op}}$.

For example, $\mathfrak{S}_u^{[1,3]}$ is visualized by the diagram of categories:

$$
(19.3.11) \qquad \mathfrak{S}(A) \underset{p_2{}^*}{\overset{p_1{}^*}{\rightrightarrows}} \mathfrak{S}(A \times_B A) \underset{\overset{p_{23}{}^*}{\longrightarrow}}{\overset{\overset{p_{12}{}^*}{\longrightarrow}}{\underset{p_{13}{}^*}{\longrightarrow}}} \mathfrak{S}(A \times_B A \times_B A) .
$$

Here $p_i\colon A \times_B A \to A$ and $p_{ij}\colon A \times_B A \times_B A \to A \times_B A$ are the i-th projection and the (i, j)-the projection respectively, and we used the notation $p_i{}^*$ in (19.1.7).

Assume that $u\colon A \to B$ has a section $v\colon B \to A$ (that is, $v \circ u = \mathrm{id}_B$). Let $\mathfrak{S}$ be a prestack on $\mathcal{C}_X^{\wedge}$. We denote by $\mathfrak{S}_{u,v}$ the direct image of $\mathfrak{S}$ by the functor of presites $\lambda_{u,v}\colon \widetilde{\mathbf{\Delta}}^{\mathrm{op}} \to \mathcal{C}_B^{\wedge}$ (see Lemma 18.7.2). Hence,

$$
\mathfrak{S}_{u,v} := \lambda_{u,v*}(\mathfrak{S} \,|_B)
$$

is a prestack on $\widetilde{\mathbf{\Delta}}^{\mathrm{op}}$. We denote by $\mathfrak{S}_{u,v}^{[n,m]}$ the direct image of $\mathfrak{S}_{u,v}$ by the natural functor $(\mathbf{\Delta}_{inj}^{[n,m]})^{\mathrm{op}} \to \widetilde{\mathbf{\Delta}}^{\mathrm{op}}$.

Since $\{0, 1\}$ is a terminal object of $\widetilde{\mathbf{\Delta}}^{\mathrm{op}}$, we have:

$$
(19.3.12) \qquad \mathfrak{S}_{u,v}(\widetilde{\mathbf{\Delta}}^{\mathrm{op}}) \simeq \mathfrak{S}(\lambda_{u,v}(\{0, 1\})) \simeq \mathfrak{S}(B).
$$

The next statement is an analogue of an easy result on presheaves (see Exercise 2.24) in the framework of prestacks.

Proposition 19.3.4. *Let $\mathfrak{S}$ be a prestack on a presite X and let $u\colon A \to B$ be an epimorphism in $\mathcal{C}_X^{\wedge}$. Then the sequence of categories below is exact.*

(19.3.13) $\mathfrak{S}(B) \xrightarrow{u^*} \mathfrak{S}(A) \underset{\overrightarrow{p_2{}^*}}{\overset{\overrightarrow{p_1{}^*}}{\rightrightarrows}} \mathfrak{S}(A \times_B A) \underset{\overset{\overrightarrow{p_{13}{}^*}}{\overrightarrow{p_{23}{}^*}}}{\overset{\overrightarrow{p_{12}{}^*}}{\longrightarrow}} \mathfrak{S}(A \times_B A \times_B A) .$

In other words, $\mathfrak{S}(B) \simeq \mathfrak{S}_u^{[1,3]}((\Delta_{inj}^{[1,3]})^{\mathrm{op}})$.

Proof. By replacing $\mathcal{C}_X$ with $\mathcal{C}_B$, we may assume that B is the terminal object of $\mathcal{C}_X^{\wedge}$. Let $\mathfrak{K}$ be the prestack on X

$$\mathcal{C}_X \ni U \mapsto \mathrm{Kern}(\mathfrak{S}(U \times A), \mathfrak{S}(U \times A \times A), \mathfrak{S}(U \times A \times A \times A)).$$

Then the category $\mathrm{Kern}(\mathfrak{S}(A), \mathfrak{S}(A \times A), \mathfrak{S}(A \times A \times A))$ is equivalent to $\mathfrak{K}(B)$. It is thus enough to show that the functor of prestacks $\mathfrak{S} \to \mathfrak{K}$ is an equivalence. Hence it is enough to show that $\mathfrak{S}(U) \to \mathfrak{K}(U)$ is an equivalence of categories for any $U \in \mathfrak{S}$. Replacing $\mathcal{C}_X$ with $\mathcal{C}_U$, it is enough to prove the result when $B = U \in \mathcal{C}_X$. Then $A \to B$ has a section $v\colon B \to A$.

The functor $\kappa\colon \Delta_{inj}^{[1,3]} \to \widetilde{\Delta}$ is co-1-cofinal by the result of Exercise 19.6. Applying Proposition 19.2.3 to κ and the stack $\mathfrak{S}_{u,v}$, we obtain the equivalence $\mathfrak{S}_{u,v}(\widetilde{\Delta}^{\mathrm{op}}) \simeq \mathfrak{S}_u^{[1,3]}((\Delta_{inj}^{[1,3]})^{\mathrm{op}})$. Since $\mathrm{Kern}\big(\mathfrak{S}(A), \mathfrak{S}(A \times_B A), \mathfrak{S}(A \times_B A \times_B A)\big) = \mathfrak{S}_u^{[1,3]}((\Delta_{inj}^{[1,3]})^{\mathrm{op}})$, the result follows from the isomorphisms (19.3.12) q.e.d.

19.4 Stacks

Let X be a site and let $\mathcal{C}_X$ be the associated small category.

Definition 19.4.1. *A prestack $\mathfrak{S}$ on X is* separated *if for any $U \in \mathcal{C}_X$ and any $F_1, F_2 \in \mathfrak{S}(U)$, $\mathcal{H}om_{\mathfrak{S}|_U}(F_1, F_2)$ is a sheaf on U.*

Lemma 19.4.2. *Let $\mathfrak{S}$ be a separated prestack. For $A \in \mathcal{C}_X^{\wedge}$ and $F_1, F_2 \in \mathfrak{S}(A)$, the presheaf on A:*

$$\mathcal{H}om_{\mathfrak{S}|_A}(F_1, F_2) : (U \to A) \mapsto \mathrm{Hom}_{\mathfrak{S}(U)}(F_1|_U, F_2|_U),$$

is a sheaf on A.

Proof. For any $U \in \mathcal{C}_A$, $\mathcal{H}om_{\mathfrak{S}|_A}(F_1, F_2)|_U \simeq \mathcal{H}om_{\mathfrak{S}|_U}(F_1|_U, F_2|_U)$ is a sheaf. q.e.d.

Proposition 19.4.3. *A prestack $\mathfrak{S}$ is separated if and only if for any local isomorphism $A \to A'$ in $\mathcal{C}_X^{\wedge}$, $\mathfrak{S}(A') \to \mathfrak{S}(A)$ is fully faithful.*

Proof. (i) Assume that $\mathfrak{S}$ is separated. For $F_1, F_2 \in \mathfrak{S}(A')$, there are isomorphisms by (19.1.8)

$$\operatorname{Hom}_{\mathfrak{S}(A')}(F_1, F_2) \simeq \mathcal{H}om_{\mathfrak{S}|_{A'}}(F_1, F_2)(A')$$
$$\simeq \mathcal{H}om_{\mathfrak{S}|_{A'}}(F_1, F_2)(A)$$
$$\simeq \operatorname{Hom}_{\mathfrak{S}(A)}(F_1|_A, F_2|_A) \ .$$

(ii) Let $U \in \mathcal{C}_X$ and let $F_1, F_2 \in \mathfrak{S}(U)$. For any local isomorphism $A \to V$ in $\mathcal{C}_U^\wedge$, the map $\operatorname{Hom}_{\mathfrak{S}|_V}(F_1|_V, F_2|_V) \to \operatorname{Hom}_{\mathfrak{S}|_A}(F_1|_A, F_2|_A)$ is an isomorphism, and hence the presheaf $\mathcal{H}om_{\mathfrak{S}|_U}(F_1, F_2)$ is a sheaf.   q.e.d.

Definition 19.4.4. *A prestack is a stack if for any $U \in \mathcal{C}_X$ and any local isomorphism $A \to U$ in $\mathcal{C}_X^\wedge$, $\mathfrak{S}(U) \to \mathfrak{S}(A)$ is an equivalence of categories.*

Proposition 19.4.5.   (i) *A stack is a separated prestack.*
   (ii) *If $\mathfrak{S}$ is a stack on X, then for any $A \in \mathcal{C}_X^\wedge$, $\mathfrak{S}|_A$ is a stack on A.*

Proof. (i) follows from Proposition 19.4.3.
(ii) is obvious.   q.e.d.

   Clearly, if $\mathfrak{S}$ is a stack, then so is $\mathfrak{S}^{\mathrm{op}}$.

Proposition 19.4.6. *The conditions below are equivalent.*

   (i) $\mathfrak{S}$ *is a stack,*
   (ii) *for any local epimorphism $A \to U$ with $A \in \mathcal{C}_X^\wedge$ and $U \in \mathcal{C}_X$, the sequence below is exact:*

$$\mathfrak{S}(U) \longrightarrow \mathfrak{S}(A) \rightrightarrows \mathfrak{S}(A \times_U A) \underset{\longrightarrow}{\overset{\longrightarrow}{\longrightarrow}} \mathfrak{S}(A \times_U A \times_U A) \ ,$$

   (iii) *for any local epimorphism $A \to B$ in $\mathcal{C}_X^\wedge$, the sequence below is exact:*

$$\mathfrak{S}(B) \longrightarrow \mathfrak{S}(A) \rightrightarrows \mathfrak{S}(A \times_B A) \underset{\longrightarrow}{\overset{\longrightarrow}{\longrightarrow}} \mathfrak{S}(A \times_B A \times_B A) \ ,$$

   (iv) *for any local isomorphism $A \to B$ in $\mathcal{C}_X^\wedge$, $\mathfrak{S}(B) \to \mathfrak{S}(A)$ is an equivalence.*

Moreover, if $\mathcal{C}_X$ admits fiber products, these conditions are equivalent to

   (v) *for any covering $\{U_i\}_{i \in I}$ of $U \in \mathcal{C}_X$, setting $U_{ij} = U_i \times_U U_j$ and $U_{ijk} = U_i \times_U U_j \times_U U_k$, the sequence below is exact:*

$$\mathfrak{S}(U) \longrightarrow \prod_i \mathfrak{S}(U_i) \rightrightarrows \prod_{ij} \mathfrak{S}(U_{ij}) \underset{\longrightarrow}{\overset{\longrightarrow}{\longrightarrow}} \prod_{ijk} \mathfrak{S}(U_{ijk}) \ .$$

Proof. (iii) $\Rightarrow$ (ii) is obvious as well as (iv) $\Rightarrow$ (i).

(i) $\Rightarrow$ (iv). Since $U \times_A B \to U$ is a local isomorphism for any $(U \to A) \in \mathcal{C}_A$, we have

$$\mathfrak{S}(A) \simeq \varprojlim_{(U \to A) \in \mathcal{C}_A} \mathfrak{S}(U) \simeq \varprojlim_{(U \to A) \in \mathcal{C}_A} \mathfrak{S}(U \times_A B) .$$

Hence, $U \times_A B \to B$ defines the functor $\mathfrak{S}(B) \to \varprojlim_{(U \to A) \in \mathcal{C}_A} \mathfrak{S}(U \times_A B) \simeq \mathfrak{S}(A)$.

On the other hand, $B \to A$ defines $\mathfrak{S}(A) \to \mathfrak{S}(B)$. Clearly, these two functors are quasi-inverse to each other.

(iv) $\Rightarrow$ (iii). Set $S = \operatorname{Im}(A \to B)$. Since $S \to B$ is a local isomorphism, $\mathfrak{S}(B) \to \mathfrak{S}(S)$ is an equivalence. It remains to apply Proposition 19.3.4 (with B replaced by S).

(ii) $\Rightarrow$ (i). Since the proof is similar to the case (v) $\Rightarrow$ (i) (assuming that $\mathcal{C}_X$ admits fiber products), we shall prove this last implication.

(a) Given $F, F' \in \mathfrak{S}(V)$, $U \to V$ and a covering "$\bigsqcup_i$" $U_i \to U$, the sequence below is exact:

$$\mathcal{H}om\,(F, F')(U) \to \prod_i \mathcal{H}om\,(F, F')(U_i) \rightrightarrows \prod_{j,k} \mathcal{H}om\,(F, F')(U_{jk}) .$$

Therefore, $\mathcal{H}om\,(F, F')$ is a sheaf and $\mathfrak{S}$ is a separated prestack.

(b) Let $U \in \mathcal{C}_X$ and let $S \to U$ be a local isomorphism in $\mathcal{C}_X^\wedge$. There exists an epimorphism $A := "\bigsqcup_i" U_i \to S$ with $U_i \in \mathcal{C}_X$. Then $\{U_i\}_i$ is a covering of U. Note that $\mathfrak{S}(A) \simeq \prod_i \mathfrak{S}(U_i)$, $\mathfrak{S}(A \times_S A) \simeq \prod_{ij} \mathfrak{S}(U_i \times_S U_j)$, etc. Consider the following diagram:

$$
\begin{array}{ccccccc}
\mathfrak{S}(U) & \longrightarrow & \mathfrak{S}(A) & \rightrightarrows & \mathfrak{S}(A \times_U A) & \rightrightarrows & \mathfrak{S}(A \times_U A \times_U A) \\
\downarrow & & \downarrow{\scriptstyle\mathrm{id}} & & \downarrow & & \downarrow \\
\mathfrak{S}(S) & \longrightarrow & \mathfrak{S}(A) & \rightrightarrows & \mathfrak{S}(A \times_S A) & \rightrightarrows & \mathfrak{S}(A \times_S A \times_S A) .
\end{array}
$$

The row in the top is exact by the assumption and the row in the bottom is exact by Proposition 19.3.4. On the other hand, the third and fourth vertical arrows are fully faithful by (a), because $A \times_S A \to A \times_U A$ and $A \times_S A \times_S A \to A \times_U A \times_U A$ are local isomorphisms. Hence the first vertical arrow is an equivalence of categories.   q.e.d.

Proposition 19.4.7. (a) *Let $\mathcal{A}$ be a category satisfying (17.4.1) and denote by $\mathfrak{S}$ the prestack* $: U \mapsto \operatorname{Sh}(U, \mathcal{A})$. *Then*

   (i) *for any $A \in \mathcal{C}_X^\wedge$, $\mathfrak{S}(A)$ is equivalent to the category $\operatorname{Sh}(A, \mathcal{A})$,*

   (ii) *$\mathfrak{S}$ is a stack.*

(b) *Let $\mathcal{R}$ be a sheaf of rings on X. Then the prestack $U \mapsto \operatorname{Mod}(\mathcal{R}|_U)$ is a stack.*

Proof. (a) (i) We shall first construct a functor $\theta_1 \colon \mathfrak{S}(A) \to \mathrm{Sh}(A, \mathcal{A})$. Let $F = \{F_U, \varphi_u\}_{U \in \mathcal{C}_A, u \in \mathrm{Mor}(\mathcal{C}_A)} \in \mathrm{Ob}(\mathfrak{S}(A))$ (see Definition 19.1.6). Hence, $F_U \in \mathrm{Sh}(U, \mathcal{A})$ and $\varphi_u \colon r_u F_{U_2} \to F_{U_1}$ is an isomorphism, where $r_u = j_{u*}$ is the restriction morphism $\mathrm{Sh}(U_2, \mathcal{A}) \to \mathrm{Sh}(U_1, \mathcal{A})$. We define $F' = \theta_1(F) \in \mathrm{PSh}(A, \mathcal{A})$ as follows. For $U \in \mathcal{C}_A$, set $F'(U) = F_U(U)$. For $u \colon U_1 \to U_2$, define the morphism $F'(U_2) \to F'(U_1)$ by the sequence of morphisms:

$$F'(U_2) = F_{U_2}(U_2) \to r_u(F_{U_2})(U_1) \xrightarrow[\varphi_u]{\sim} F_{U_1}(U_1) = F'(U_1) \,.$$

Since $F'|_U \simeq F_U$ for any $U \in \mathcal{C}_A$, the presheaf F' is a sheaf.

Next we construct a functor $\theta_2 \colon \mathrm{Sh}(A, \mathcal{A}) \to \mathfrak{S}(A)$ by associating to $F \in \mathrm{Sh}(A, \mathcal{A})$ the family $\{F|_U\}$ with the obvious isomorphisms $\varphi_u \colon (F|_{U_2})|_{U_1} \simeq F|_{U_1}$.

It is easily checked that the functors θ_1 and θ_2 are quasi-inverse to each other.

(a) (ii) By Proposition 17.6.8, if $u \colon A \to A'$ is a local isomorphism, then $\mathrm{Sh}(A', \mathcal{A}) \to \mathrm{Sh}(A, \mathcal{A})$ is an equivalence. Hence, $\mathfrak{S}$ is a stack.

(b) The proof is similar.   q.e.d.

Notation 19.4.8. For a sheaf $\mathcal{R}$ of rings on X, we denote by $\mathfrak{Mod}(\mathcal{R})$ the stack $U \mapsto \mathrm{Mod}(\mathcal{R}|_U)$.

Let us denote by $\mathcal{S}h_X$ the stack on X: $U \mapsto \mathrm{Sh}(U)$. Let $\mathfrak{S}$ be a prestack on X. For any $F \in \mathfrak{S}(X)$, let us denote by $h_X(F) \colon \mathfrak{S}^{\mathrm{op}} \to \mathcal{S}h_X$ the functor of prestacks which associates $\mathcal{H}om_{\mathfrak{S}|_U}(F', F|_U) \in \mathcal{S}h_X(U)$ to $U \in \mathcal{C}_X$ and $F' \in \mathfrak{S}(U)$.

Similarly to Yoneda's lemma, we have

(19.4.1) $$\mathrm{Hom}_{\mathrm{Fct}(\mathfrak{S}^{\mathrm{op}}, \mathcal{S}h_X)}(h_X(F), h_X(F')) \simeq \mathrm{Hom}_{\mathfrak{S}(X)}(F, F')$$
$$\text{for any } F, F' \in \mathfrak{S}(X) \,.$$

Definition 19.4.9. *Let $\Phi \colon \mathfrak{S}^{\mathrm{op}} \to \mathcal{S}h_X$ be a functor of prestacks. If there exists an object $F \in \mathfrak{S}(X)$ such that Φ is isomorphic to $h_X(F)$, we say that Φ is* representable *and F* represents Φ. *If there exists a local epimorphism $A \to \mathrm{pt}_X$ such that $\Phi|_U \colon (\mathfrak{S}|_U)^{\mathrm{op}} \to \mathcal{S}h_U$ is representable for any $U \in \mathcal{C}_X$ and $U \to A$, then we say that Φ is* locally representable.

Proposition 19.4.10. *Let $\mathfrak{S}$ be a stack on X. If a functor $\Phi \colon \mathfrak{S}^{\mathrm{op}} \to \mathcal{S}h_X$ is locally representable, then Φ is representable.*

Proof. By replacing A with "$\bigsqcup_{U \in (\mathcal{C}_X)_A}$" U, we may assume from the beginning that $\Phi|_A$ is representable. Let $F_0 \in \mathfrak{S}(A)$ be its representative. Let $p \colon A \to \mathrm{pt}_X$ be a canonical morphism and let $p_i \colon A_1 := A \times A \to A$ be the i-th projection ($i = 1, 2$). Then we have $\Phi|_{A_1} \simeq p_i^* \Phi$, and isomorphism (19.4.1) induces an isomorphism $u\colon p_1^* F_0 \xrightarrow{\sim} p_2^* F_0$. By the same argument, u satisfies (19.3.10). Hence it gives an object of $\mathrm{Kern}(\mathfrak{S}(A), \mathfrak{S}(A \times A), \mathfrak{S}(A \times A \times A))$. By Proposition 19.4.5 (iii), there exists $F \in \mathfrak{S}(X)$ such that $p^* F \simeq F_0$. It is easily checked that F represents Φ.   q.e.d.

19.5 Morita Equivalence

Let $(X, \mathcal{O}_X)$ be a ringed site (see Definition 18.6.1) and let $\mathfrak{S}$ be an additive stack on X. We call $\mathfrak{S}$ an $\mathcal{O}_X$-*stack* if for any $U \in \mathcal{C}_X$, $\mathfrak{S}(U)$ has a structure of an $\mathcal{O}_X(U)$-category, i.e., $\mathfrak{S}(U)$ is endowed with a ring morphism $\mathcal{O}_X(U) \to \mathrm{End}(\mathrm{id}_{\mathfrak{S}(U)})$ and for any morphism $u\colon U_1 \to U_2$ and any $F \in \mathfrak{S}(U_2)$, the diagram below commutes

$$
\begin{array}{ccc}
\mathcal{O}_X(U_2) & \longrightarrow & \mathrm{End}_{\mathrm{id}_{\mathfrak{S}(U_2)}}(F) \\
\downarrow & & \downarrow{\scriptstyle r_u} \\
\mathcal{O}_X(U_1) & \longrightarrow & \mathrm{End}_{\mathrm{id}_{\mathfrak{S}(U_1)}}(r_u F).
\end{array}
$$

Using Lemma 1.3.8, we see that $\mathcal{E}nd(\mathrm{id}_{\mathfrak{S}})$ is a sheaf of commutative rings. Saying that $\mathfrak{S}$ is an $\mathcal{O}_X$-stack is equivalent to saying that one is given a morphism of sheaves of rings $\mathcal{O}_X \to \mathcal{E}nd(\mathrm{id}_{\mathfrak{S}})$.

For two $\mathcal{O}_X$-stacks $\mathfrak{S}_1$ and $\mathfrak{S}_2$, a functor of $\mathcal{O}_X$-stacks $\Phi\colon \mathfrak{S}_1 \to \mathfrak{S}_2$ is a functor of stacks such that, for any $U \in \mathcal{C}_X$ and $F \in \mathfrak{S}_1(U)$, the composition $\mathcal{O}_X|_U \to \mathcal{E}nd_{\mathfrak{S}_1}(F) \to \mathcal{E}nd_{\mathfrak{S}_2}(\Phi(U)(F))$ coincides with the one given by the $\mathcal{O}_X$-stack structure on $\mathfrak{S}_2$.

Let X be a site and let $\mathcal{R}$ be a sheaf of (not necessarily commutative) rings on X. For an $\mathcal{R}$-module F, we have introduced in Definition 18.5.1 the property of being locally free or of being locally of finite presentation. We define similarly other "local" properties such as of being a direct summand, or of having sections with a given property. The precise formulation is left to the reader.

Let us recall that an $\mathcal{R}$-module M is flat if the functor

$$
\cdot \otimes_{\mathcal{R}} M \colon \mathrm{Mod}(\mathcal{R}^{\mathrm{op}}) \to \mathrm{Mod}(\mathbb{Z}_X)
$$

is exact. If this functor is exact and faithful, we say that M is *faithfully flat*. It is a local property.

Lemma 19.5.1. *Let P be a flat $\mathcal{R}$-module locally of finite presentation. Then*

(i) *P is locally a direct summand of $\mathcal{R}^{\oplus n}$ for some n.*
(ii) *For any $\mathcal{R}$-module M,*

$$
\mathcal{H}om_{\mathcal{R}}(P, \mathcal{R}) \otimes_{\mathcal{R}} M \to \mathcal{H}om_{\mathcal{R}}(P, M)
$$

is an isomorphism.

Proof. (i) Locally there exists an exact sequence: $L_1 \to L_0 \to P \to 0$ where $L_\nu \simeq \mathcal{R}^{\oplus m_\nu}$. Then we have a commutative diagram with exact rows

$$0 \to \mathcal{H}om_{\mathcal{R}}(P, \mathcal{R}) \otimes_{\mathcal{R}} P \to \mathcal{H}om_{\mathcal{R}}(L_0, \mathcal{R}) \otimes_{\mathcal{R}} P \to \mathcal{H}om_{\mathcal{R}}(L_1, \mathcal{R}) \otimes_{\mathcal{R}} P$$

$$0 \longrightarrow \mathcal{H}om_{\mathcal{R}}(P, P) \longrightarrow \mathcal{H}om_{\mathcal{R}}(L_0, P) \longrightarrow \mathcal{H}om_{\mathcal{R}}(L_1, P).$$

Since the middle and the right arrows are isomorphisms, $\mathcal{H}om_{\mathcal{R}}(P, \mathcal{R}) \otimes_{\mathcal{R}} P \to \mathcal{H}om_{\mathcal{R}}(P, P)$ is an isomorphism. Hence there exists locally a section $\sum_{i=1}^n t_i \otimes s_i \in \mathcal{H}om_{\mathcal{R}}(P, \mathcal{R}) \otimes_{\mathcal{R}} P$ which corresponds to $\mathrm{id} \in \mathcal{H}om_{\mathcal{R}}(P, P)$. It means that $P \xrightarrow{\mathrm{id}} P$ decomposes into $P \xrightarrow{(t_i)} \mathcal{R}^{\oplus n} \xrightarrow{(s_i)} P$.
(ii) easily follows from (i).   q.e.d.

Let $(X, \mathcal{O}_X)$ be a ringed site and let now $\mathcal{R}_i$ ($i = 1, 2$) be a sheaf of $\mathcal{O}_X$-algebras on X.

Proposition 19.5.2. *Let P be an $(\mathcal{R}_1 \otimes_{\mathcal{O}_X} \mathcal{R}_2^{\mathrm{op}})$-module. Then the following conditions are equivalent.*

(i) *There is an $(\mathcal{R}_2 \otimes_{\mathcal{O}_X} \mathcal{R}_1^{\mathrm{op}})$-module Q such that $P \otimes_{\mathcal{R}_2} Q \simeq \mathcal{R}_1$ as an $\mathcal{R}_1 \otimes_{\mathcal{O}_X} \mathcal{R}_1^{\mathrm{op}}$-module and $Q \otimes_{\mathcal{R}_1} P \simeq \mathcal{R}_2$ as an $\mathcal{R}_2 \otimes_{\mathcal{O}_X} \mathcal{R}_2^{\mathrm{op}}$-module.*
(ii) *For $Q_0 := \mathcal{H}om_{\mathcal{R}_1}(P, \mathcal{R}_1) \in \mathrm{Mod}(\mathcal{R}_2 \otimes_{\mathcal{O}_X} \mathcal{R}_1^{\mathrm{op}})$, the canonical morphism $P \otimes_{\mathcal{R}_2} Q_0 \to \mathcal{R}_1$ is an isomorphism and $Q_0 \otimes_{\mathcal{R}_1} P \simeq \mathcal{R}_2$ as an $\mathcal{R}_2 \otimes_{\mathcal{O}_X} \mathcal{R}_2^{\mathrm{op}}$-module.*
(iii) *P is a faithfully flat $\mathcal{R}_1$-module of locally finite presentation and $\mathcal{R}_2^{\mathrm{op}} \xrightarrow{\sim} \mathcal{E}nd_{\mathcal{R}_1}(P)$.*
(iv) *P is a faithfully flat $\mathcal{R}_2^{\mathrm{op}}$-module of locally finite presentation and $\mathcal{R}_1 \xrightarrow{\sim} \mathcal{E}nd_{\mathcal{R}_2^{\mathrm{op}}}(P)$.*
(v) *$\bullet \otimes_{\mathcal{R}_1} P : \mathfrak{Mod}(\mathcal{R}_1^{\mathrm{op}}) \to \mathfrak{Mod}(\mathcal{R}_2^{\mathrm{op}})$ is an equivalence of $\mathcal{O}_X$-stacks.*
(vi) *$P \otimes_{\mathcal{R}_2} \bullet : \mathfrak{Mod}(\mathcal{R}_2) \to \mathfrak{Mod}(\mathcal{R}_1)$ is an equivalence of $\mathcal{O}_X$-stacks.*
(vii) *$\mathcal{H}om_{\mathcal{R}_1}(P, \bullet) : \mathfrak{Mod}(\mathcal{R}_1) \to \mathfrak{Mod}(\mathcal{R}_2)$ is an equivalence of $\mathcal{O}_X$-stacks.*
(viii) *$\mathcal{H}om_{\mathcal{R}_2^{\mathrm{op}}}(P, \bullet) : \mathfrak{Mod}(\mathcal{R}_2^{\mathrm{op}}) \to \mathfrak{Mod}(\mathcal{R}_1^{\mathrm{op}})$ is an equivalence of $\mathcal{O}_X$-stacks.*

Moreover, under the condition of (i), Q is isomorphic to $\mathcal{H}om_{\mathcal{R}_1}(P, \mathcal{R}_1)$ and to $\mathcal{H}om_{\mathcal{R}_2^{\mathrm{op}}}(P, \mathcal{R}_2)$ as an $(\mathcal{R}_2 \otimes_{\mathcal{O}_X} \mathcal{R}_1^{\mathrm{op}})$-module.

Proof. (i) $\Rightarrow$ (v) is obvious.

(v) $\Rightarrow$ (i). By the hypothesis, P is faithfully flat over $\mathcal{R}_1$. Take $Q \in \mathrm{Mod}(\mathcal{R}_1^{\mathrm{op}})$ such that $Q \otimes_{\mathcal{R}_1} P \simeq \mathcal{R}_2$ as an $\mathcal{R}_2^{\mathrm{op}}$-module. The isomorphisms of sheaves of rings

$$\mathcal{R}_2 \simeq \mathcal{E}nd_{\mathcal{R}_2^{\mathrm{op}}}(\mathcal{R}_2) \simeq \mathcal{E}nd_{\mathcal{R}_1^{\mathrm{op}}}(Q)$$

give a structure of $\mathcal{R}_2$-module over Q. Hence Q is an $(\mathcal{R}_2 \otimes_{\mathcal{O}_X} \mathcal{R}_1^{\mathrm{op}})$-module and we have an isomorphism $Q \otimes_{\mathcal{R}_1} P \simeq \mathcal{R}_2$ in $\mathrm{Mod}(\mathcal{R}_2 \otimes_{\mathcal{O}_X} \mathcal{R}_2^{\mathrm{op}})$. We have

(19.5.1) $P \otimes_{\mathcal{R}_2} Q \otimes_{\mathcal{R}_1} P \simeq P \otimes_{\mathcal{R}_2} \mathcal{R}_2 \simeq \mathcal{R}_1 \otimes_{\mathcal{R}_1} P$,

and hence there is an isomorphism $\psi \colon P \otimes_{\mathcal{R}_2} Q \xrightarrow{\sim} \mathcal{R}_1$ in $\mathrm{Mod}(\mathcal{R}_1^{\mathrm{op}})$. Since
(19.5.1) is $\mathcal{R}_1$-linear, ψ is also $\mathcal{R}_1$-linear. Hence ψ is an isomorphism of $\mathcal{R}_1 \otimes_{\mathcal{O}_X} \mathcal{R}_1^{\mathrm{op}}$-modules.

We have thus proved that (i) and (v) are equivalent. By replacing $(\mathcal{R}_1, \mathcal{R}_2)$ with $(\mathcal{R}_2^{\mathrm{op}}, \mathcal{R}_1^{\mathrm{op}})$, these properties are also equivalent to (vi).

(iii) $\Rightarrow$ (ii). Set $Q_0 = \mathcal{H}om_{\mathcal{R}_1}(P, \mathcal{R}_1)$. By Lemma 19.5.1, we have the isomorphisms

$$Q_0 \otimes_{\mathcal{R}_1} P \xrightarrow{\sim} \mathcal{E}nd_{\mathcal{R}_1}(P) \simeq \mathcal{R}_2^{\mathrm{op}} \ .$$

We get the isomorphisms

$$P \otimes_{\mathcal{R}_2} Q_0 \otimes_{\mathcal{R}_1} P \simeq P \otimes_{\mathcal{R}_2} \mathcal{R}_2 \simeq \mathcal{R}_1 \otimes_{\mathcal{R}_1} P \ .$$

Therefore $P \otimes_{\mathcal{R}_2} Q_0 \to \mathcal{R}_1$ is an isomorphism.

(i)+(v)+(vi) $\Rightarrow$ (iii). By the hypothesis (v), P is faithfully flat over $\mathcal{R}_1$. Let us show that P is of finite presentation. There exist locally finitely many sections $s_i \otimes t_i \in Q \otimes_{\mathcal{R}_1} P$ $(i = 1, \dots, m)$ which generate $Q \otimes_{\mathcal{R}_1} P$ as a left $\mathcal{R}_2$-module. Set $P' = \sum_i \mathcal{R}_1 t_i$. Then $Q \otimes_{\mathcal{R}_1} P' \to Q \otimes_{\mathcal{R}_1} P$ is an epimorphism. Applying the functor $P \otimes_{\mathcal{R}_2} \cdot$, we get that $P' \to P$ is an epimorphism. Hence P is locally finitely generated. Let $L := \mathcal{R}_1^m \to P$ be an epimorphism of $\mathcal{R}_1$-modules. Since the $\mathcal{R}_2$-linear morphism $Q \otimes_{\mathcal{R}_1} L \to Q \otimes_{\mathcal{R}_1} P \simeq \mathcal{R}_2$ is an epimorphism, it has locally a section. Hence, tensoring P from the left, the $\mathcal{R}_1$-linear morphism $L \simeq P \otimes_{\mathcal{R}_2} Q \otimes_{\mathcal{R}_1} L \to P \otimes_{\mathcal{R}_2} Q \otimes_{\mathcal{R}_1} P \simeq P$ has also locally a section. Hence P is locally of finite presentation.

(v)$\Leftrightarrow$(viii) follows from the fact that $\cdot \otimes_{\mathcal{R}_1} P \colon \mathfrak{Mod}(\mathcal{R}_1^{\mathrm{op}}) \to \mathfrak{Mod}(\mathcal{R}_2^{\mathrm{op}})$ is a left adjoint to $\mathcal{H}om_{\mathcal{R}_2^{\mathrm{op}}}(P, \cdot) \colon \mathfrak{Mod}(\mathcal{R}_2^{\mathrm{op}}) \to \mathfrak{Mod}(\mathcal{R}_1^{\mathrm{op}})$.
The other implications are now obvious.   q.e.d.

Definition 19.5.3. *If an $(\mathcal{R}_1 \otimes_{\mathcal{O}_X} \mathcal{R}_2^{\mathrm{op}})$-module P satisfies the equivalent conditions (i)–(viii) in Proposition 19.5.2, we say that P is* invertible. *An $\mathcal{O}_X$-module is called* invertible *if it is invertible as an $(\mathcal{O}_X \otimes_{\mathcal{O}_X} \mathcal{O}_X^{\mathrm{op}})$-module.*

Theorem 19.5.4. (Morita equivalence.) *Let $\Phi \colon \mathfrak{Mod}(\mathcal{R}_2) \to \mathfrak{Mod}(\mathcal{R}_1)$ be an equivalence of $\mathcal{O}_X$-stacks. Then there exists an invertible $(\mathcal{R}_1 \otimes_{\mathcal{O}_X} \mathcal{R}_2^{\mathrm{op}})$-module P such that $P \otimes_{\mathcal{R}_2} \cdot$ is isomorphic to Φ and $\mathcal{H}om_{\mathcal{R}_1}(P, \cdot)$ is isomorphic to Φ^{-1}.*

Proof. For any $U \in \mathcal{C}_X$, $\Phi(U)$ commutes with inductive limits and projective limits, and Φ commutes with $j_{U \to X *}$. Hence Φ commutes with $j_{U \to X}^{-1}$, a left adjoint of $j_{U \to X *}$. Set $P = \Phi(\mathcal{R}_2)$. Then the $\mathcal{R}_2^{\mathrm{op}}$-module structure of $\mathcal{R}_2$ induces $\mathcal{O}_X$-algebra morphisms $\mathcal{R}_2^{\mathrm{op}} \to \mathcal{E}nd_{\mathcal{R}_2}(\mathcal{R}_2) \to \mathcal{E}nd_{\mathcal{R}_1}(P)$. Hence P is an $(\mathcal{R}_1 \otimes_{\mathcal{O}_X} \mathcal{R}_2^{\mathrm{op}})$-module. Consider the functor

$$\Phi' := P \otimes_{\mathcal{R}_2} \cdot : \mathfrak{Mod}(\mathcal{R}_2) \to \mathfrak{Mod}(\mathcal{R}_1) \ .$$

Let $U \in \mathcal{C}_X$ and let $M \in \mathrm{Mod}(\mathcal{R}_2|_U)$. For any $V \in \mathcal{C}_U$ and $s \in M(V)$, we have a morphism $\mathcal{R}_2|_V \to M|_V$. Hence we have a morphism $P|_V = \Phi(\mathcal{R}_2)|_V \to \Phi(M)|_V$. We have thus defined a morphism $P(V) \otimes_{\mathcal{R}_1(V)} M(V) \to \Phi(M)(V)$, functorial with respect to $V \in \mathcal{C}_U$. This gives a morphism $P|_U \otimes_{\mathcal{R}_1|_U} M \to \Phi(M)$. Thus we obtain a morphism $\Phi' \to \Phi$.

Let us show that $\Phi'(M) \to \Phi(M)$ is an isomorphism for any $M \in \mathfrak{Mod}(\mathcal{R}_2)$. By the construction, this is true for $M = \mathcal{R}_2$. Since any M is isomorphic to the cokernel of morphisms of $\mathcal{R}_2$-modules of the form $\oplus_v (\mathcal{R}_2)_{U_v}$, we may assume that $M = \oplus_v (\mathcal{R}_2)_{U_v}$. Since Φ' and Φ commute with direct sum, we may assume that $M = (\mathcal{R}_2)_U$ for some $U \in \mathcal{C}_X$. Since Φ' and Φ commute with the functor $j_{U \to X}^{-1}$, $\Phi'((\mathcal{R}_2)_U) \to \Phi((\mathcal{R}_2)_U)$ is an isomorphism. The other assertions are obvious by the preceding proposition.   q.e.d.

19.6 Twisted Sheaves

Let $(X, \mathcal{O}_X)$ be a ringed site.

Definition 19.6.1. *An $\mathcal{O}_X$-stack $\mathfrak{S}$ on X is called an $\mathcal{O}_X$-stack of* twisted sheaves, *or else a stack of twisted $\mathcal{O}_X$-modules, if $\mathfrak{S}$ is locally equivalent to $\mathfrak{Mod}(\mathcal{O}_X)$, that is, if there is a local epimorphism $A \to \mathrm{pt}_X$ such that for any $U \to A$ with $U \in \mathcal{C}_X$, there is an equivalence $\Phi_U \colon \mathfrak{S}|_U \xrightarrow{\sim} \mathfrak{Mod}(\mathcal{O}_X)|_U$, or equivalently, there exists a local epimorphism $A \to \mathrm{pt}_X$ in $\mathcal{C}_X^\wedge$ such that $\mathfrak{S}|_A \xrightarrow{\sim} \mathfrak{Mod}(\mathcal{O}_X)|_A$.*

For a stack $\mathfrak{S}$ of twisted $\mathcal{O}_X$-modules, an object of $\mathfrak{S}(X)$ is called a twisted $\mathcal{O}_X$-module.

Let $\mathfrak{S}$ be an $\mathcal{O}_X$-stack of twisted sheaves. Then for any $M \in \mathrm{Mod}(\mathcal{O}_X)$ and $F \in \mathfrak{S}(X)$, the functor

$$\mathfrak{S}(X) \ni L \mapsto \mathrm{Hom}_{\mathcal{O}_X}(M, \mathcal{H}om_{\mathfrak{S}}(F, L))$$

is representable. Indeed, it is obvious that the functor from $\mathfrak{S}$ to $\mathcal{S}h_X$ given by $U \mapsto \mathcal{H}om_{\mathcal{O}_U}(M|_U, \mathcal{H}om_{\mathfrak{S}|_U}(F, \cdot))$ is locally representable, and hence it is representable by Proposition 19.4.10 (applied to $\mathfrak{S}^{\mathrm{op}}$). We shall denote a representative of this functor by $M \otimes_{\mathcal{O}_X} F$. Then it defines a bifunctor

$$\cdot \otimes_{\mathcal{O}_X} \cdot : \mathfrak{Mod}(\mathcal{O}_X) \times \mathfrak{S} \to \mathfrak{S} \ .$$

In the sequel, we write $\otimes$ instead of $\otimes_{\mathcal{O}_X}$ for simplicity.

Lemma 19.6.2. *Let $\mathfrak{S}$ be an $\mathcal{O}_X$-stack of twisted sheaves. Then, any equivalence of $\mathcal{O}_X$-stacks $\Phi : \mathfrak{S} \to \mathfrak{S}$ is isomorphic to $P \otimes \cdot$ for some invertible $\mathcal{O}_X$-module P.*

Proof. Set $P = \mathcal{H}om\,(\mathrm{id}_{\mathfrak{S}}, \Phi)$. This sheaf is an $\mathcal{O}_X$-module. Then P is an invertible $\mathcal{O}_X$-module and we have a morphism $P \otimes M \to \Phi(M)$. Moreover this morphism is an isomorphism. Indeed, this fact is a local property which holds when $\mathfrak{S} \simeq \mathfrak{Mod}(\mathcal{O}_X)$ by Theorem 19.5.4.   q.e.d.

Let $p\colon A \to \mathrm{pt}_X$ be a local epimorphism in $\mathcal{C}_X^{\wedge}$. Set $A_0 = A$, $A_1 = A \times_X A$, $A_2 = A \times_X A \times_X A$ and $A_3 = A \times_X A \times_X A \times_X A$. Denote by $p_i\colon A_1 \to A$ the i-th projection $(i = 1, 2)$, by $p_{ij}\colon A_2 \to A_1$ the (i, j)-th projection $(i, j = 1, 2, 3, 4)$ by $p_{ijk}\colon A_3 \to A_2$ the (i, j, k)-th projection $(i, j, k = 1, 2, 3)$. This is visualized by a diagram similar to (18.7.2)

$$(19.6.1) \qquad A_3 \begin{array}{c} \xrightarrow{\;\;p_{234}\;\;} \\[-2pt] \xrightarrow{\;\;p_{134}\;\;} \\[-2pt] \xrightarrow{\;\;p_{124}\;\;} \\[-2pt] \xrightarrow{\;\;p_{123}\;\;} \end{array} A_2 \begin{array}{c} \xrightarrow{\;p_{23}\;} \\[-2pt] \xrightarrow{\;p_{13}\;} \\[-2pt] \xrightarrow{\;p_{12}\;} \end{array} A_1 \begin{array}{c} \xrightarrow{\;p_2\;} \\[-2pt] \xrightarrow{\;p_1\;} \end{array} A \xrightarrow{\;p\;} X\,.$$

We also introduce $q_{ij}\colon A_3 \to A_1$ the (i, j)-th projection $(i, j = 1, 2, 3, 4)$, and $q_i\colon A_2 \to A$ the i-th projection $(i = 1, 2, 3)$. Let L be an invertible $\mathcal{O}_{A_1}$-module, and let

$$(19.6.2) \qquad\qquad \varphi\colon p_{12}{}^*L \otimes p_{23}{}^*L \xrightarrow{\;\sim\;} p_{13}{}^*L$$

be an isomorphism in $\mathfrak{Mod}(\mathcal{O}_{A_2})$ satisfying the chain condition given by the commutativity of the diagram:

$$
\begin{array}{ccc}
q_{12}{}^*L \otimes q_{23}{}^*L \otimes q_{34}{}^*L & \xrightarrow{\;\sim\;} & p_{123}{}^*(p_{12}{}^*L \otimes p_{23}{}^*L) \otimes q_{34}{}^*L \\
\downarrow & & \downarrow{\scriptstyle\varphi} \\
q_{12}{}^*L \otimes p_{234}{}^*(p_{12}{}^*L \otimes p_{23}{}^*L) & & p_{123}{}^*p_{13}{}^*L \otimes q_{34}{}^*L \\
\downarrow{\scriptstyle\varphi} & & \downarrow \\
q_{12}{}^*L \otimes p_{234}{}^*p_{13}{}^*L & & q_{13}{}^*L \otimes q_{34}{}^*L \\
\downarrow & & \downarrow \\
q_{12}{}^*L \otimes q_{24}{}^*L & & p_{134}{}^*(p_{12}{}^*L \otimes p_{23}{}^*L) \\
\downarrow & & \downarrow{\scriptstyle\varphi} \\
p_{124}{}^*(p_{12}{}^*L \otimes p_{23}{}^*L) & & p_{134}{}^*p_{13}{}^*L \\
\downarrow{\scriptstyle\varphi} & & \downarrow \\
p_{124}{}^*p_{13}{}^*L & \xrightarrow{\;\sim\;} & q_{14}{}^*L\,.
\end{array}
$$

These conditions are paraphrased as follows. For $U \in \mathcal{C}$ and $x, y \in A(U)$, let $L(x, y)$ be $L(U \xrightarrow{(x,y)} A_1)$. Then φ gives $\varphi(x_1, x_2, x_3)\colon L(x_1, x_2) \otimes L(x_2, x_3) \to L(x_1, x_3)$ for $x_1, x_2, x_3 \in A(U)$. The commutativity of the diagram above is equivalent to the commutativity of the following diagram

$$L(x_1, x_2) \otimes L(x_2, x_3) \otimes L(x_3, x_4) \xrightarrow{\varphi(x_1, x_2, x_3)} L(x_1, x_3) \otimes L(x_3, x_4)$$

$$\downarrow \varphi(x_2, x_3, x_4) \qquad\qquad\qquad\qquad \downarrow \varphi(x_1, x_3, x_4)$$

$$L(x_1, x_2) \otimes L(x_2, x_4) \xrightarrow{\varphi(x_1, x_2, x_4)} L(x_1, x_4)$$

for $x_1, x_2, x_3, x_4 \in A(U)$.

With these data, we define the stack $\mathfrak{S}$ as follows. To $U \in \mathcal{C}_X$ we associate

$$\mathfrak{S}(U) = \mathrm{Kern}\Big(\mathrm{Mod}(\mathcal{O}_X|_{A \times U}) \rightrightarrows \mathrm{Mod}(\mathcal{O}_X|_{A_1 \times U}) \substack{\longrightarrow \\ \longrightarrow \\ \longrightarrow} \mathrm{Mod}(\mathcal{O}_X|_{A_2 \times U}) \Big)$$

$$\simeq \left\{ (F, s) \, ; \, \begin{array}{l} F \in \mathrm{Mod}(\mathcal{O}_A|_{A \times U}) \text{ and } s \colon L \otimes p_2^* F \xrightarrow{\sim} p_1^* F \text{ is an iso-} \\ \text{morphism in } \mathrm{Mod}(\mathcal{O}_{X_1}|_{A_1 \times U}) \text{ satisfying the following} \\ \text{chain condition (19.6.3)} \end{array} \right\}$$

$$(19.6.3)$$

$$p_{12}^* L \otimes p_{23}^* L \otimes q_3^* F \longrightarrow p_{12}^* L \otimes p_{23}^*(L \otimes p_2^* F)$$

$$\downarrow \varphi \qquad\qquad\qquad\qquad \downarrow s$$

$$p_{13}^* L \otimes q_3^* F \qquad\qquad p_{12}^* L \otimes p_{23}^* p_1^* F$$

$$\downarrow \qquad\qquad\qquad\qquad \downarrow$$

$$p_{13}^*(L \otimes p_2^* F) \qquad\qquad p_{12}^* L \otimes q_2^* F \qquad\qquad \text{commutes.}$$

$$\downarrow s \qquad\qquad\qquad\qquad \downarrow$$

$$p_{13}^* p_1^* F \qquad\qquad p_{12}^*(L \otimes p_2^* F)$$

$$\downarrow \qquad\qquad\qquad\qquad \downarrow s$$

$$q_1^* F \longrightarrow p_{12}^* p_1^* F$$

Proposition 19.6.3. $\mathfrak{S}$ *is an $\mathcal{O}_X$-stack of twisted sheaves.*

Proof. It is obvious that $\mathfrak{S}$ is an $\mathcal{O}_X$-stack. We shall show that $\mathfrak{S}$ is locally equivalent to $\mathfrak{Mod}(\mathcal{O}_X)$. We may assume that $p \colon A \to \mathrm{pt}_X$ has a section s. Then s defines morphisms $s_0 \colon A \to A_1$ and $s_1 \colon A_1 \to A_2$ by $s_0(x) = (x, spx)$ and $s_1(x_1, x_2) = (x_1, x_2, spx_2)$. Then $p_{12} \circ s_1 = \mathrm{id}_{A_1}$, $s_0 \circ p_2 = p_{23} \circ s_1 \colon A_1 \to A$, $p_{13} \circ s_1 = s_0 \circ p_1$. Let $r_1 \colon A_1 \to X$ be the projection. For any $G \in \mathfrak{Mod}(\mathcal{O}_X)$, define $F = s_0^* L \otimes p^* G$. Then

$$L \otimes p_2^* F \simeq L \otimes p_2^*(s_0^* L \otimes p^* G)$$
$$\simeq s_1^* p_{12}^* L \otimes s_1^* p_{23}^* L \otimes r_1^* G$$
$$\simeq s_1^*(p_{12}^* L \otimes p_{23}^* L) \otimes r_1^* G$$
$$\simeq s_1^* p_{13}^* L \otimes p_1^* p^* G \simeq p_1^*(s_0^* L \otimes p^* G) \,.$$

We can easily see that the chain condition (19.6.3) is satisfied. Hence $G \mapsto F$ defines a functor $\mathfrak{S} \to \mathfrak{Mod}(\mathcal{O}_X)$. Conversely $F \mapsto s^* F$ defines a functor $\mathfrak{S} \to \mathfrak{Mod}(\mathcal{O}_X)$. We can easily check that they are quasi-inverse to each other.   q.e.d.

Moreover any $\mathcal{O}_X$-stack of twisted sheaves is obtained in this way.

Remark 19.6.4. Let X be a site and let $\mathcal{O}_X$ be a sheaf of commutative rings on X. Assume that any invertible $\mathcal{O}_X$-module is locally isomorphic to $\mathcal{O}_X$. (This assumption is satisfied when $\mathcal{O}_X$ is a local ring, see Exercise 19.2.) Denote by $\mathcal{O}_X^\times$ the abelian sheaf of invertible sections of $\mathcal{O}_X$.
(i) For a stack $\mathfrak{S}$ of twisted $\mathcal{O}_X$-modules, there is an isomorphism of abelian groups $\operatorname{Aut}(\operatorname{id}_{\mathfrak{S}}) \simeq \Gamma(X;\mathcal{O}_X^\times)$.
(ii) The set of equivalence classes of invertible $\mathcal{O}_X$-modules is isomorphic to $H^1(X;\mathcal{O}_X^\times)$.
(iii) The set of equivalence classes of stacks of twisted $\mathcal{O}_X$-modules is isomorphic to $H^2(X;\mathcal{O}_X^\times)$.
    We shall not give the proofs of these facts here and refer to Breen [10] (see also [34]). Note that (i) is clear and (ii) implies by Lemma 19.6.2 that for a stack $\mathfrak{S}$ of twisted $\mathcal{O}_X$-modules, the set of isomorphism classes of equivalences of stacks from $\mathfrak{S}$ to itself is isomorphic to $H^1(X;\mathcal{O}_X^\times)$.

Example 19.6.5. Let X be a complex manifold, and denote by Ω_X the sheaf of holomorphic forms of maximal degree. Take an open covering $X = \bigcup_{i \in I} U_i$ such that there are nowhere vanishing sections $\omega_i \in \Omega_{U_i}$. Let $t_{ij} \in \mathcal{O}_{U_{ij}}^\times$ be the transition functions given by $\omega_j|_{U_{ij}} = t_{ij}\omega_i|_{U_{ij}}$. Choose determinations $s_{ij} \in \mathcal{O}_{U_{ij}}^\times$ for the multivalued functions $t_{ij}^{1/2}$. Since $s_{ij}s_{jk}$ and s_{ik} are both determinations of $t_{ik}^{1/2}$, there exists $c_{ijk} \in \{-1, 1\}$ such that $s_{ij}s_{jk} = c_{ijk}s_{ik}$.
    By choosing $A = \text{``}\bigsqcup\text{''}\, U_i$ and $L = \mathbb{Z}_X|_{A \times_X A}$, $\varphi = (c_{ijk})$ in (19.6.2), we obtain a $\mathbb{Z}_X$-stack $\mathfrak{S}$ of twisted sheaves on X. The twisted sheaf of holomorphic half-forms is given by

$$\Omega_X^{1/2} = (\{\mathcal{O}_{U_i}\}_{i \in I}, \{s_{ij}\}_{i,j \in I}),$$

which is regarded as an object of $\mathfrak{S}(X)$. (See [34] for more explanation.)

Exercises

Exercise 19.1. Let $\mathfrak{S}_1$ be a prestack on a site X and let $\mathfrak{S}_2$ be a stack on X. Prove that $\mathcal{F}ct(\mathfrak{S}_1, \mathfrak{S}_2)$ is a stack on X.

Exercise 19.2. Let $\mathcal{O}_X$ be a sheaf of commutative rings on a site X. We say that $\mathcal{O}_X$ is a local ring if for any $U \in \mathcal{C}_X$ and any $a \in \mathcal{O}_X(U)$, there exists a covering sieve $S \in \mathcal{S}\mathrm{Cov}(U)$ such that for any $V \in S$, at least one of the sections $a|_V$ and $(1 - a)|_V$ is invertible in $\mathcal{O}_X(V)$. Prove that any invertible module over a local ring $\mathcal{O}_X$ is locally isomorphic to $\mathcal{O}_X$.
(Hint: prove that if $1 = \sum_{k=1}^n a_i$ with $a_i \in \mathcal{O}_X$, then one of the $\mathcal{O}_X a_i$'s is locally equal to $\mathcal{O}_X$.)

Exercise 19.3. Let X be a topological space and assume that X is locally arcwise connected. To X, we associate a category $\mathcal{C}_X$ as follows: $\mathrm{Ob}(\mathcal{C}_X) = X$, and for $x, y \in X$, a morphism $x \to y$ in $\mathcal{C}_X$ is a homotopy class of paths from x to y. Prove that the category $\mathcal{C}_X$ is simply connected (see Definition 19.2.1) if and only if the topological space X is simply connected in the classical sense.

Exercise 19.4. Prove that if a groupoid $\mathcal{C}$ (i.e., a category in which all morphisms are isomorphisms) is simply connected, then $\mathcal{C}$ is equivalent to **Pt**.

Exercise 19.5. Prove that any filtrant category is simply connected.

Exercise 19.6. We shall follow the notations in §19.3 (see also Exercise 1.21).
(i) Prove that the natural functor $\mathbf{\Delta}_{inj}^{[1,n]} \to \mathbf{\Delta}^{[1,\infty]}$ is co-1-cofinal for $n \geq 3$. Here $\mathbf{\Delta}^{[1,\infty]}$ is the full subcategory of $\mathbf{\Delta}$ consisting of non-empty finite totally ordered sets.
(ii) Prove that the functor $\kappa : \mathbf{\Delta}_{inj}^{[1,n]} \to \widetilde{\mathbf{\Delta}}$ (the composition $\mathbf{\Delta}_{inj}^{[1,n]} \to \mathbf{\Delta} \to \widetilde{\mathbf{\Delta}}$) is co-1-cofinal for $n \geq 3$.
(iii) Show that the natural functor $\mathbf{\Delta}_{inj}^{[1,2]} \to \mathbf{\Delta}$ is not co-1-cofinal.

Exercise 19.7. Let $\mathfrak{S}$ be a prestack on a site X. Define the prestack $\mathfrak{S}^a$ by setting:

$$\mathfrak{S}^a(U) = \{(F, A) \, ; \, A \to U \text{ is a local isomorphism and } F \in \mathfrak{S}(A)\} \, ,$$
$$\mathrm{Hom}_{\mathfrak{S}^a(U)}((F, A), (G, B)) = \Gamma(A \times_U B \, ; \, \mathcal{H}om_{\mathfrak{S}|_{A\times_U B}}(F|_{A\times_U B}, G|_{A\times_U B})^a) \, .$$

(i) Prove that $\mathfrak{S}^a$ is a well-defined stack and construct a natural functor of prestacks $\mathfrak{S} \to \mathfrak{S}^a$.
(ii) Prove that $\mathrm{Fct}(\mathfrak{S}^a, \mathfrak{S}') \to \mathrm{Fct}(\mathfrak{S}, \mathfrak{S}')$ is an equivalence of categories for any stack $\mathfrak{S}'$ on X.

Exercise 19.8. Let G be a group and denote by Γ the category with one object g and such that $\mathrm{Hom}_\Gamma(g, g) = G$. Let $\mathcal{C}$ be a category. Prove that giving an action of G on $\mathcal{C}$ (see Exercise 4.10) is equivalent to giving a prestack $\mathfrak{S}$ on Γ such that $\mathfrak{S}(g) = \mathcal{C}$.

References

1. J. Adámek and J. Rosický, *Locally presentable and accessible categories,* London Mathematical Society Lecture Note Series **189**, Cambridge University Press, Cambridge (1994).
2. L. Alonso, A Jeremias, and M-J. Souto, *Localization in categories of complexes and unbounded resolutions,* Canad. J. Math. **52** (2000) pp 225–247.
3. M. Artin and B. Mazur, *Etale homology,* Lecture Notes in Math. **100**, Springer-Verlag (1969).
4. A. Beilinson, J. Bernstein, and P. Deligne, *Faisceaux pervers,* in Analysis and topology on singular spaces, Luminy 1981, Astérisque **100**, Soc. Math. France (1982).
5. J. Bénabou, *Introduction to bicategories,* Lecture Notes in Math. **47**, Springer-Verlag (1969) pp 1–77.
6. M. Bökstedt and A. Neeman, *Homotopy limits in triangulated categories,* Compositio Mathematica **86** (1993) pp 209–234.
7. A. I. Bondal and M. Kapranov, *Enhanced triangulated categories,* Math USSR Sbornik **70** (1991) pp 93–107.
8. A. I. Bondal and M. Van den Bergh, *Generators and representability of functors in commutative and noncommutative geometry,* Moscow Math. Journ. **3** (2003) pp 1–36.
9. F. Borceux, *Handbook of categorical algebra I, II, III,* Encyclopedia of Mathematics and its Applications **51**, Cambridge University Press, Cambridge (1994).
10. L. Breen, *On the Classification of 2-gerbes and 2-stacks,* Astérisque **225**, Soc. Math. France (1994).
11. E. Brown, *Cohomologies theory,* Annals of Math. **75** (1962) pp 467–484.
12. J-L. Brylinski, *Loop spaces, characteristic classes and geometric quantization,* Progress in Math. **107**, Birkhauser (1993).
13. D. A. Buchsbaum, *Exact categories and duality,* Trans. Amer. Math. Soc. **80** (1955) pp 1–34.
14. H. Cartan and S. Eilenberg, *Homological algebra,* Princeton University Press (1956).
15. V. Chari and A. Pressley, *Quantum groups,* Cambridge Univ. Press (1994).
16. P. Colmez and J-P. Serre (Ed.), *Correspondance Grothendieck-Serre,* Documents Mathématiques, Soc. Math. France **2** (2001).
17. P. Deligne, *Cohomologie à supports propres,* in S-G-A 4.

484   References

18. P. Deligne, *Décompositions dans la catégorie dérivée*, Motives (Seattle, WA, 1991), Proc. Sympos. Pure Math., Part 1 **55**, Amer. Math. Soc., Providence, RI (1994) pp 115–128.
19. S. Eilenberg and S. Mac Lane, *Natural isomorphisms in group theory*, Proc. Nat. Acad. Sci. USA **28** (1942) pp 537–543.
20. S. Eilenberg and S. Mac Lane, *General theory of natural equivalences*, Trans. Amer. Math. Soc. **58** (1945) pp 231–294.
21. J. Franke, *On the Brown representability theorem for triangulated categories*, Topology **40** (2001) pp 667–680.
22. J-P. Freyd, *Abelian categories*, Harper & Row (1964).
23. P. Gabriel and F. Ulmer, *Lokal präsentibare categorien*, Lecture Notes in Math. **221**, Springer-Verlag (1971).
24. P. Gabriel and M. Zisman, *Calculus of fractions and homotopy theory*, Ergebnisse der Math. und ihrer Grenzgebiete **35**, Springer-Verlag (1967).
25. S.I. Gelfand and Yu.I. Manin, *Methods of homological algebra*, Springer-Verlag (1996).
26. J. Giraud, *Cohomologie non abélienne*, Grundlehren der Math. Wiss. **179**, Springer-Verlag (1971).
27. R. Godement, *Topologie algébrique et théorie des faisceaux*, Hermann (1958).
28. A. Grothendieck, *Sur quelques points d'algèbre homologique*, Tohoku Math. Journ. (2) **9** (1957) pp 119-221.
29. A. Grothendieck, *Elements de géomtrie algébrique III*, Inst. Hautes Etudes Sci. Publ. Math. **11** (1961), **17** (1963).
30. A. Grothendieck and J-L. Verdier, *Préfaisceaux*, in S-G-A 4.
31. R. Hartshorne, *Residues and duality*, Lecture Notes in Math. **20**, Springer-Verlag (1966).
32. M. Hovey, *Model Categories*, Math. Surveys and Monograph A.M.S. **63** (1999).
33. M. Kashiwara, *The asymptotic behavior of a variation of polarized Hodge structure*, Publ. Res. Inst. Math. Sci. Kyoto Univ. **21** (1985) pp.853–875.
34. M. Kashiwara, *Representation theory and D-modules on flag varieties* in Astérique **173-174**, Soc. Math. France (1989) pp 55-109.
35. M. Kashiwara, *D-modules and representation theory of Lie groups*, Ann. Inst. Fourier (Grenoble) **43** (1993) pp 1597–1618.
36. M. Kashiwara, *Quantization of contact manifolds*, Publ. RIMS, Kyoto Univ. **32** (1996) pp 1-5.
37. M. Kashiwara, *D-modules and microlocal calculus*, Translations of Mathematical Monograph **217**, A.M.S. (2000).
38. M. Kashiwara and P. Schapira, *Sheaves on manifolds*, Grundlehren der Math. Wiss. **292**, Springer-Verlag (1990).
39. M. Kashiwara and P. Schapira, *Ind-sheaves*, Astérisque **271**, Soc. Math. France (2001).
40. C. Kassel, *Quantum groups*, Graduate Texts in Math. **155**, Springer-Verlag 2nd ed. (1995).
41. B. Keller, *Introduction to A-infinity algebras and modules*, Homology, Homotopy and Applications **3** (2001) pp 1-35 (electronic).
42. B. Keller, *Deriving DG categories*, Ann. Sci. Ec. Norm. Sup. **27** (1994) pp 63–102.
43. M. Kontsevich, *Homological algebra of mirror symmetry*, Proceedings of the International Congress of Mathematicians (Zürich, 1994), Birkhäuser (1995) pp 120–139.

44. H. Krause, *A Brown representability theorem via coherent functors,* Topology **41** (2002) pp 853–861.

45. G. Laumon and L. Moret-Bailly, *Champs algébriques,* Ergebnisse der Mathematik und ihrer Grenzgebiete, 3 Folge **39**, Springer-Verlag, Berlin (2000).

46. J. Leray, *Oeuvres Scientiques, Vol I* , Springer-Verlag & Soc. Math. France (1997).

47. S. Mac Lane, *Categories for the working mathematician,* Graduate Texts in Math. **5**, Springer-Verlag 2nd ed. (1998).

48. S. Mac Lane and I. Moerdijk, *Sheaves in Geometry and Logic,* Universitext Springer-Verlag, New York (1994).

49. M. Makkai and R. Par, *Accessible categories: the foundations of categorical model theory,* Contemporary Mathematics, **104**, American Mathematical Society, Providence, RI, (1989).

50. B. Mitchell, *Theory of categories,* Academic Press (1965).

51. F. Morel and V. Voevodsky, A^1-*homotopy of schemes,* Inst. Hautes Etudes Sci. Publ. Math. **90** (1999) pp 1-143.

52. A. Neeman, *The connection between the K-theory localization theorem of Thomason, Trobaugh and Yao and the smashing subcategories of Bousfield and Ravenel,* Ann. Sci. École Norm. Sup. **25** (1992) pp 547–566.

53. A. Neeman, *Triangulated categories,* Annals of Mathematics Studies **148**, Princeton University Press, Princeton, NJ, (2001).

54. N. Popescu, *Abelian categories with applications to rings and modules,* Academic Press, New York (1973).

55. D. Puppe, *On the formal structure of stable homotopy theory,* Colloquium on Algebraic Topology, Aarhus Univ. (1962) pp 65–71.

56. D. Quillen, *Homotopical Algebras,* Lecture Notes in Math. **43**, Springer-Verlag (1967) pp 85–147.

57. R. Rouquier, *Catégories dérivées et géométrie birationnelle,* Sém. Bourbaki **947** (2004-2005).

58. M. Sato, T. Kawai and M. Kashiwara, *Hyperfunctions and pseudo-differential equations,* in: Hyperfunctions and Pseudo-Differential Equations, H. Komatsu, editor. Lecture Notes in Math. **287**, Springer-Verlag (1973) pp 265–529.

59. N. Saavedra Rivano, *Catégories tannakiennes,* Lecture Notes in Mathematics **265**, Springer-Verlag, Berlin-New York (1972).

60. M. Saito, *Modules de Hodge polarisables,* Publ. Res. Inst. Math. Sci. Kyoto Univ. **24** (1988)–(1989) pp 849–995.

61. J-P. Schneiders, *Quasi-abelian categories and sheaves,* Mémoires Soc. Math. France **76** (1999).

62. J-P. Serre, *Espaces fibrés algébriques,* in Sem. C. Chevalley, Exp. 1 (1958), Documents Mathématiques, Soc. Math. France **1** (2001).

63. S-G-A 1, Sém. Géom. Algébrique by A. Grothendieck, Exposé XIII *Revêtements étales et groupe fondamental* Lecture Notes in Math. **224**, Springer-Verlag (1971).

64. S-G-A 4, Sém. Géom. Algébrique (1963-64) by M. Artin, A. Grothendieck and J-L. Verdier, *Théorie des topos et cohomologie étale des schémas,* Lecture Notes in Math. **269, 270, 305**, Springer-Verlag (1972/73).

65. N. Spaltenstein, *Resolutions of unbounded complexes*, Compositio Mathematica **65** (1988) pp 121–154.

66. R. Stenstrom, *Rings of quotients,* Grundlehren der Math. Wiss. **217**, Springer-Verlag (1975).

67. G. Tamme, *Introduction to étale cohomology*, Universitext, Springer-Verlag (1994).
68. J-L. Verdier, *Catégories dérivées, état 0* in SGA $4\frac{1}{2}$ Lecture Notes in Math. **569**, Springer-Verlag (1977).
69. J-L. Verdier, *Des catégories dérivées des catégories abéliennes*, Astérisque **239**, Soc. Math. France (1996).
70. V. Voevodsky, A. Suslin and E.M. Friedlander, *Cycles, transfers, and motivic homology theories*, Annals of Mathematics Studies **143**, Princeton University Press, Princeton, NJ (2000).
71. C. Weibel, *An introduction to homological algebra*, Cambridge Studies in Advanced Math. **38** (1994).

List of Notations

Index

1-cofinal functor, 466
1-connected, 466
2-category, 20
F-injective, 255
F-projective, 256
$\mathcal{O}$-algebra, 436
$\mathcal{R}$-module, 436
k-bilinear, 203

abelian
    category, 177
    category with translation, 300
    prestack, 462
action
    of a group on a category, 110
    of a tensor category, 100
additive
    bifunctor, 289
    category, 171
    functor, 171
    prestack, 462
adjoint functor, 28
adjunction, 29
    isomorphism, 28
    morphisms, 29
algebra, 11
anti-commutative diagram, 174
anti-distinguished triangle, 245
arrow, 12
automorphism, 12

Barr-Beck Theorem, 105
base change
    functor, 47

stable by, 47
bifunctor, 17
    ... with translation, 242
    additive, 289
    localizable, 255
    triangulated, 255
big category, 23
bounded complex, 279
braiding, 101
Brown representability Theorem, 258

cardinal, 216
    regular, 217
Cartesian square, 48, 179
category, 11
    1-connected, 466
    2-, 20
    QM-injective, 349
    π-filtrant, 218
    k-abelian, 189
    k-additive, 189
    k-pre-additive, 188
    abelian, 177
    abelian with translation, 300
    additive, 171
    associated graded, 278
    big, 23
    co-cofinally small, 59
    cofiltrant, 72
    cofinally small, 59
    commutative tensor, 102
    connected, 13
    derived, 319, 320
    discrete, 13